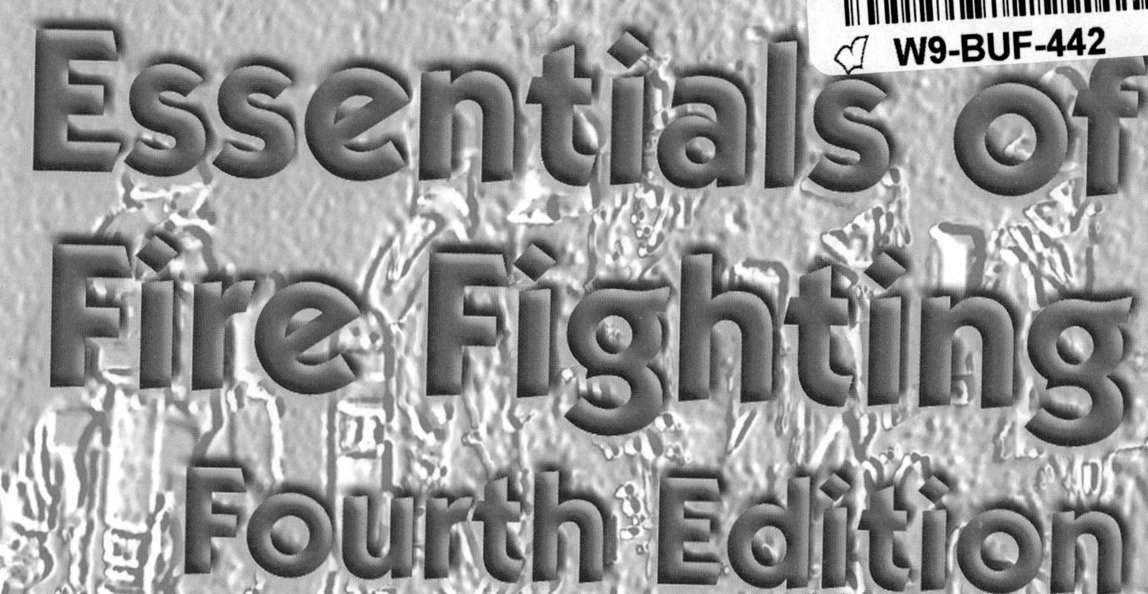

Essentials of Fire Fighting

Fourth Edition

Edited By
Richard Hall
and
Barbara Adams

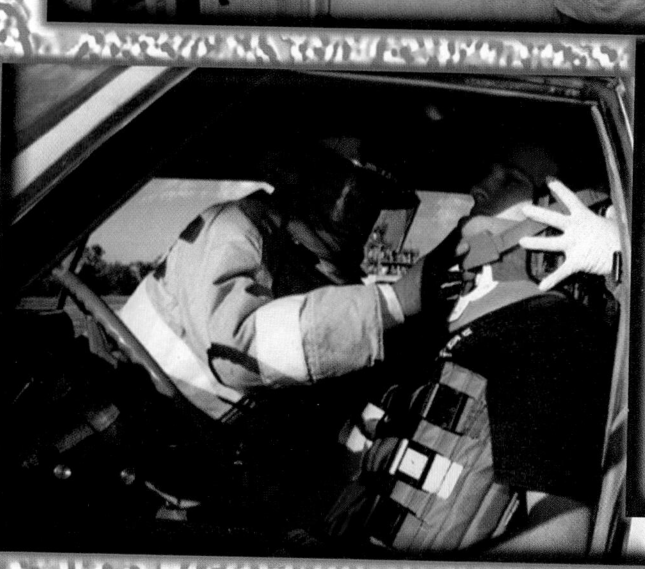

Validated by the International Fire Service Training Association
Published by Fire Protection Publications, Oklahoma State University

Special Dedication

The **Essentials of Fire Fighting, 4th** edition, is dedicated to the memory of Gerald E. Monigold, past Director of the University of Illinois, Fire Service Institute. Jerry was a graduate of the Fire Protection Technology and Trade and Industrial Education program at Oklahoma State University. Jerry attended and participated in IFSTA Validation Conferences for 30 years where he served as Chair and Vice-Chair on several validation committees and as a member of the Executive Board. Mr. Monigold was the recipient of the Everett E. Hudiburg Memorial Award in 1987. He was a founding member and Past Chair of the Association of State, Provincial, and Territorial Directors of Fire Training.

ISBN 0-87939-149-9
Library of Congress 98-70164

Fourth Edition

Printed in the United States of America

10 9 8

Dedication

This manual is dedicated to the members of that unselfish organization of men and women who hold devotion to duty above personal risk, who count on sincerity of service above personal comfort and convenience, who strive unceasingly to find better ways of protecting the lives, homes and property of their fellow citizens from the ravages of fire and other disasters . . . ***The Firefighters of All Nations.***

Dear Firefighter:

The International Fire Service Training Association (IFSTA) is an organization that exists for the purpose of serving firefighters' training needs. Fire Protection Publications is the publisher of IFSTA materials. Fire Protection Publications staff members participate in the National Fire Protection Association and the International Association of Fire Chiefs.

If you need additional information concerning our organization or assistance with manual orders, contact:

Customer Services
Fire Protection Publications
Oklahoma State University
930 N. Willis
Stillwater, OK 74078-8045
1 (800) 654-4055
Fax: (405) 744-8204

For assistance with training materials, recommended material for inclusion in a manual, or questions on manual content, contact:

Technical Services
Fire Protection Publications
Oklahoma State University
930 N. Willis
Stillwater, OK 74078-8045
(405) 744-5723
Fax: (405) 744-4112
email: editors@osufpp.org

THE INTERNATIONAL FIRE SERVICE TRAINING ASSOCIATION

The International Fire Service Training Association (IFSTA) was established as a "nonprofit educational association of fire fighting personnel who are dedicated to upgrading fire fighting techniques and safety through training." This training association was formed in November 1934, when the Western Actuarial Bureau sponsored a conference in Kansas City, Missouri. The meeting was held to determine how all the agencies interested in publishing fire service training material could coordinate their efforts. Four states were represented at this initial conference. Because the representatives from Oklahoma had done some pioneering in fire training manual development, it was decided that other interested states should join forces with them. This merger made it possible to develop training materials broader in scope than those published by individual agencies. This merger further made possible a reduction in publication costs, because it enabled each state or agency to benefit from the economy of relatively large printing orders. These savings would not be possible if each individual state or department developed and published its own training material.

To carry out the mission of IFSTA, Fire Protection Publications was established as an entity of Oklahoma State University. Fire Protection Publications' primary function is to publish and disseminate training texts as proposed and validated by IFSTA. As a secondary function, Fire Protection Publications researches, acquires, produces, and markets high-quality learning and teaching aids as consistent with IFSTA's mission. The IFSTA Executive Director is officed at Fire Protection Publications.

IFSTA's purpose is to validate training materials for publication, develop training materials for publication, check proposed rough drafts for errors, add new techniques and developments, and delete obsolete and outmoded methods. This work is carried out at the annual Validation Conference.

The IFSTA Validation Conference is held in July in the state of Oklahoma. Fire Protection Publications, the IFSTA publisher, establishes the revision schedule for manuals and introduces new manuscripts. Delegates are selected for technical input by the Delegate Selection Committee. The Delegate Selection Committee consists of three Board members and two conference delegates; the committee is chaired by the Vice-Chair of IFSTA. Applications are reviewed by the committee, and delegates are selected based upon technical expertise and demographics. Committees meet and work at the conference addressing the current standards of the National Fire Protection Association and other standard-making groups as applicable.

Most of the delegates are affiliated with other international fire protection organizations. The Validation Conference brings together individuals from several related and allied fields, such as:

- Key fire department executives and training officers
- Educators from colleges and universities
- Representatives from governmental agencies
- Delegates of firefighter associations and industrial organizations
- Engineers from the fire insurance industry

Delegates are not paid nor are they reimbursed for their expenses by IFSTA or Fire Protection Publications. They come because of commitment to the fire service and its future through training. Being on a committee is prestigious in the fire service community, and delegates are acknowledged leaders in their fields. This unique feature provides a close relationship between the International Fire Service Training Association and other fire protection agencies, which helps to correlate the efforts of all concerned.

IFSTA manuals are now the official teaching texts of most of the states and provinces of North America. Additionally, numerous U.S. and Canadian government agencies as well as other English-speaking countries have officially accepted the IFSTA manuals.

Table of Contents

Tables

NFPA 1001 References Within Essentials of Fire Fighting

NFPA 1001 Firefighter I and II Professional Qualifications Standards

1997 Edition	2002 Edition	Essentials Chapter Location
3-1.1	5.1.1	1, 4, 6
3-1.1.1	5.1.1.1	1, 6
3-1.1.2	5.1.1.2	4, 6
3-2.1	5.2.1	18
3-2.1(a)	5.2.1(A)	18
3-2.1(b)	5.2.1(B)	18
3-2.2	5.2.2	18
3-2.2 (a)	5.2.2(A)	18
3-2.2 (b)	5.2.2(B)	18
3-2.3	5.2.3	18
3-2.3 (a)	5.2.3(A)	18
3-2.3 (b)	5.2.3(B)	18
3-3.1	5.3.1	4
3-3.1 (a)	5.3.1(A)	4
3-3.1 (b)	5.3.1(B)	4
3-3.2	5.3.2	1, 4
3-3.2 (a)	5.3.2(A)	1, 4
3-3.2 (b)	5.3.2(B)	1, 4
None	5.3.3 (New)	1, 4, 7, 14
None	5.3.3(A) (New)	1, 4, 7, 14
None	5.3.3 (B) (New)	1, 4, 7, 14
3-3.3	5.3.4	8
3-3.3 (a)	5.3.4(A)	8
3-3.3 (b)	5.3.4(B)	8
3-3.4	5.3.5	1, 4, 7
3-3.4 (a)	5.3.5(A)	1, 4, 7
3-3.4 (b)	5.3.5(B)	4, 7
3-3.5	5.3.6	9
3-3.5 (a)	5.3.6(A)	9
3-3.5 (b)	5.3.6(B)	9
3-3.6	5.3.7	12, 13, 14
3-3.6 (a)	5.3.7(A)	14
3-3.6 (b)	5.3.7(B)	12, 13, 14
3-3.7	5.3.8	4, 7, 8, 12, 13, 14, 16, 17
3-3.7 (a)	5.3.8(A)	4, 7, 8, 12, 13, 14, 17
3-3.7 (b)	5.3.8(B)	8, 12, 14, 16, 17
3-3.8	5.3.9	4, 7, 9
3-3.8 (a)	5.3.9(A)	4, 7
3-3.8 (b)	5.3.9(B)	4, 7, 9
3-3.9	5.3.10	1, 2, 3, 4, 12, 13, 14
3-3.9 (a)	5.3.10(A)	1, 2, 3, 4, 12, 13, 14
3-3.9 (b)	5.3.10(B)	8, 12, 13, 14
3-3.10	5.3.11	1, 4, 8, 9, 10
3-3.10 (a)	5.3.11(A)	1, 2, 4, 10
3-3.10 (b)	5.3.11(B)	8, 9, 10
3-3.11	5.3.12	1, 3, 6, 8, 9, 10
3-3.11 (a)	5.3.12(A)	1, 2, 3, 10
3-3.11 (b)	5.3.12(B)	6, 8, 9, 10
3-3.12	5.3.13	8, 12, 16, 17
3-3.12 (a)	5.3.13(A)	8, 16, 17
3-3.12 (b)	5.3.13(B)	8, 12, 16, 17
3-3.13	5.3.14	15, 16, 17
3-3.13 (a)	5.3.14(A)	15, 16
3-3.13 (b)	5.3.14(B)	15, 16, 17
3-3.14	5.3.15	11, 12
3-3.14 (a)	5.3.15(A)	11, 12
3-3.14 (b)	5.3.15(B)	11, 12
3-3.15	5.3.16	1, 5
3-3.15 (a)	5.3.16(A)	1, 2, 5
3-3.15 (b)	5.3.16(B)	5
3-3.16	5.3.17	7
3-3.16 (a)	5.3.17(A)	7
3-3.16 (b)	5.3.17(B)	7
3-3.17	5.3.18	14
3-3.17 (a)	5.3.18(A)	14
3-3.17 (b)	5.3.18(B)	14
3-3.18	5.3.19	14
3-3.18 (a)	5.3.19(A)	14
3-3.18 (b)	5.3.19(B)	14
3-5.1	5.5.1	19
3-5.1 (a)	5.5.1(A)	19
3-5.1 (b)	5.5.1(B)	19
3-5.2	5.5.2	19
3-5.2 (a)	5.5.2(A)	19
3-5.2 (b)	5.5.2(B)	19
3-5.3	5.5.3	4, 6, 8, 9, 16
3-5.3 (a)	5.5.3(A)	4, 6, 8, 9, 16
3-5.3 (b)	5.5.3(B)	4, 6, 8, 9, 16
3-5.4	5.5.4	12
3-5.4 (a)	5.5.4(A)	12
3-5.4 (b)	5.5.4(B)	12

Firefighter II

1997 Edition	2002 Edition	Essentials Chapter Location
4-1.1	6.1.1	1
4-1.1.1	**6.1.1.1**	1
4-1.1.2	**6.1.1.2**	1
4-2.1	**6**.2.1	18
4-2.1(a)	6.2.1(A)	18
4-2.1(b)	6.2.1(B)	18
4-2.2	6.2.2	18
4-2.2 (a)	6.2.2(A)	18
4-2.2 (b)	6.2.2(B)	18
4-3.1	6.3.1	13
4-3.1 (a)	6.3.1(A)	13
4-3.1 (b)	6.3.1(B)	13
4-3.2	6.3.2	1, 3, 7, 8, 10, 12, 13, 14, 16
4-3.2 (a)	6.3.2(A)	3, 7, 8, 10, 12, 13, 14, 16
4-3.2 (b)	6.3.2(B)	1, 2, 3, 7, 8, 10, 14,
4-3.3	6.3.3	14
4-3.3 (a)	6.3.3(A)	14
4-3.3 (b)	6.3.3(B)	14
4.3.4	6.3.4	17
4-3.4 (a)	6.3.4(A)	17
4-3.4 (b)	6.3.4(B)	17
4-4.1	6.4.1	7, 8
4-4.1(a)	6.4.1(A)	7
4-4.1(b)	6.4.1(B)	7, 8
4-4.2	6.4.2	1, 7, 8
4-4.2(a)	6.4.2(A)	7, 8
4-4.2(b)	6.4.2(B)	1, 7, 8
4-5.1	6.5.1	11, 15, 19
4-5.1(a)	6.5.1(A)	11, 15, 19
4-5.1(b)	6.5.1(B)	15, 19
4-5.2	6.5.2	7
4-5.2(a)	6.5.2(A)	7
4-5.2(b)	6.5.2(B)	7
4-5.3	6.5.3	12
4-5.3(a)	6.5.3(A)	12
4-5.3(b)	6.5.3(B)	12
4-5.4	6.5.4	11
4-5.4(a)	6.5.4(A)	11
4-5.4(b)	6.5.4(B)	11

Preface

This fourth edition of **Essentials of Fire Fighting** is intended to serve as a primary text for the firefighter candidate or as a reference text for fire fighting personnel who are already on the job. This book addresses most of the fire fighting objectives found in NFPA 1001, *Standard for Fire Fighter Professional Qualifications*, 1997 Edition. The scope of the objectives concerning competencies for the First Responder at the Awareness and Operations Levels made it impossible for us to cover them in one manual. In order to meet all the objectives of NFPA 1001, the student will require IFSTA's **Hazardous Materials for First Responders**.

Acknowledgment and special thanks are extended to the members of the IFSTA validating committee who contributed their time, wisdom, and knowledge to this manual.

Chairman
Robert H. Noll
Yukon Fire Department
Yukon, OK

Vice-Chair
Frederick S. Richards
Office of Fire Prevention and Control
Albany, NY

Secretary
Russell Strickland
Maryland Fire and Rescue Institute
College Park, MD

Committee Members

Stan Amos
Scarborough Fire Department
Scarborough, Ontario, Canada

Stephen M. Ashbrock
Reading Fire Department
Reading, OH

Robert C. Barr
FireScope, Inc.
Hingham, MA

Deward Beeler
Michigan Firefighters Training Council
Saginaw, MI

Paul H. Boecker III
Montgomery Fire District
Montgomery, Illinois

Bradd Clark
Sand Springs Fire Department
Sand Springs, OK

Brian Ellis
DOD Fire Academy
Goodfellow AFB, TX

Richard A. Fritz
University of Illinois at Urbana-Champaign
Fire Service Institute
Champaign, IL

David Horton
Claremore Fire Department
Claremore, OK

George O. Lyon
Arlington Fire Department
Arlington, VA

Robert J. Madden
Bend Fire Department
Bend, OR

Jeff Morrissette
Commission on Fire Prevention and Control
Windsor Locks, CT

Joseph L. Murabito
Delaware State Fire School
Dover, DE

Dan Murphy
Justice Institute of British Columbia Fire and Safety Division
Maple Ridge, British Columbia, Canada

Andy O'Donnell
Chicago Fire Department
Chicago, Illinois

Tom Ruane
Peoria Fire Department
Peoria, AZ

William St. George
Nassau County Fire Service Academy
Old Bethpage, NY

Special thanks go to Gordon Earhart and Mike Wieder for shooting many of the new photographs in this edition. The following individuals and organizations have also contributed information, photographs, or other assistance that made the completion of this manual possible:

ALACO Ladder Company, Chino, CA
Ansul Inc., Marinette, WI
Tony Bacon, Novato (CA) Fire Protection District
Robert J. Bennett, Bridgeville, DE
Ron Bogardus, Albany, NY
Calgary (Alta) Fire Department
Champaign (IL) Fire Department
Conoco, Inc., Ponca City, OK
Scott Copeland, Livermore, CO
Cutters Edge, Division of Edge Industries, Inc.
Scott L. Davidson, York (PA) Fire Department
Des Plaines (IL) Fire Department
Detector Electronics Corporation, Minneapolis, MN
East Brady (PA) Volunteer Fire Department
East Greenville (PA) Fire Department
Harvey Eisner, Tenafly, NJ
Elk Grove Village (IL) Fire Department
Elkhart Brass Manufacturing Company, Elkhart, IN
Emergency One Inc., Ocala, FL
Bob Esposito, Warrior Run, PA
Keith Flood, Santa Rosa (CA) Fire Department
Gainesville (FL) Fire-Rescue, George Braun, Public Information Officer
Steve George, Oklahoma State University, Fire Service Training
Hale Fire Pump Co. Inc., Hurst Rescue Tool Division
Jim Hanson, Stillwater, OK
Harrington, Inc., Erie, PA
Ron Jeffers, New Jersey Metro Fire Photographers Association
Jon Jones and Associates, Lunenburg, MA.
KK Productions, a division of Task Force Tips, Inc.
Lambert Construction Co., Stillwater, OK
Lukas Of America, Inc.
Joseph Marino, New Britain, CT
Laura Mauri, Spokane County (WA) Fire District #9
Fred Myers, Stillwater, OK
Midwest City (OK) Fire Department
Monterey (CA) Fire Department
Monterey County (CA) Fire Training Officers Association
Mount Shasta (CA) Fire Protection District
Mustang (OK) Fire Department
National Interagency Fire Center, Boise, ID
Bob Norman, Elkton (MD) Volunteer Fire Department
Oklahoma State University, Fire Service Training
Peoria (AZ) Fire Department
Phoenix (AZ) Fire Department
Plano (TX) Fire Department
Ed Prendergast, Chicago (IL) Fire Department
Greg Russell, DOD Fire Academy, TX
Safety Corporation of America, Pittsburgh, PA
Santa Rosa (CA) Fire Department
SKEDCO, Inc., Portland, OR
Stillwater (OK) Fire Department
Stillwater Central Communications Center
Larry E. Stohler, Lebanon, PA
Superior Flamefighter Corp.
Bill Tompkins, Bergenfield, NJ
Tulsa (OK) Fire Department
University of Illinois at Urbana-Champaign, Fire Service Institute
University of Illinois Fire Department
Vespra (ONT) Fire Department
Warrington Group, LTD.
Washington (MO) Volunteer Fire Department
Wellington Leisure Products, Inc.
Joel Woods, Maryland Fire and Rescue Institute
Ziamatic Corporation

Last, but not least, gratitude also is extended to the following members of the Fire Protection Publications staff whose contributions made the final publication of this manual possible:

Susan S. Walker, Instructional Development Coordinator
Carol Smith, Senior Publications Editor
Don Davis, Coordinator, Publications Production
Ann Moffat, Graphic Design Analyst
Desa Porter, Senior Graphic Designer
Connie Cook, Senior Graphic Designer
Dean Clark, Senior Graphic Designer
Susan F. Walker, Fire Service Programs Librarian

Ben Brock, Graphics Assistant
Don Burull, Graphics Assistant
Stephanie Guthrie, Graphics Assistant
Shelley Hollrah, Production Assistant
Scott Burke, Research Technician
Tim Frankenberg, Research Technician
Todd Haines, Research Technician
Jack Krill, Research Technician
Ryan Lewis, Research Technician
Dustin Stokes, Research Technician

Lynne C. Murnane

Lynne Murnane
Managing Editor

Fire Protection Publications would like to thank everyone who worked on or provided assistance for the full color version of IFSTA Essentials of Fire Fighting 4th edition. The following individuals and organizations were instrumental in locating and acquiring color photographs for this new version of this manual:

Jeff Fortney
Tara Gladden
Fred Stowell
Carl Goodson
Tom Ruane
Kyle Fortney
Clayton Stewart
Troy Williamson
Katie Sloan
Ben Brock
Jenny Brock
Craig Hannan
Susan Walker
Mike Wieder
Ann Moffat
Foster Cryer
Ben Brock

Stillwater (OK) Fire Department

Assistant Chief Rex Mott
Captain Rick Hauf
Lieutenant Bill Bunch
FF Todd Jones
FF Zach Logan
FF Rick Lozier
FF Tom Low
FF Kelly Williams
FF David Luckey
Captain Jim Morgan
FF Steve Sylvester

Tulsa (OK) Fire Department

Captain R.B. Ellis
FEO Jerry Roberts
FF Clayton Newell
FF Doug Carner
Captain Ray Evins
FEO Kevin McLurty
FF Abren Williams
FF Kevin Horner
Captain Dannie Caldwell
Captain Mike Thompson
VCO Frank Mason
FF Steve Johnston
FF Jimmy Ober
FF Kevin Fry

Oklahoma State Fire Service Training

Bryan West
Dan Knott
Harlan Giles
Phillip Pope
Jerimiah Hoffstatter

OSU Department of Environmental Health and Safety

Roy Mason
Cody Spybuck

Yukon (OK) Fire Department

Chief Jeff Lara
Captain Curtis Ogle
FF Billy Wilcher
FF Dustin Noel
FF Brad Pappe
Driver Kevin Jones

Also...

Diamond Auto Body, Stillwater, OK
Lowe's, Stillwater, OK
Donovan Davila
Richard Duda, Akron Brass Company
Frank Bateman
Gary Fortney
Julie Fortney
Sheldon Levi, IFPA
Sean Fortney
Thomas Locke
Dave Marshall, Yavapai College, AZ
Linda Miller
Rick Montemorra, Mesa (AZ) Fire Department
South Union (PA) Volunteer Fire Company
Michael Van Dyke, Montezuma-Rimrock (AZ) Fire Department

Introduction

In the early 1970s, the leaders of the various major fire service organizations saw the need for a standard of professional competence for firefighters. A committee was formed to develop this standard, the results of which became known to the fire service as National Fire Protection Association (NFPA) Standard 1001, *Standard for Fire Fighter Professional Qualifications*. The purpose of this standard is to specify, in terms of performance-based objectives, the minimum requirements of competence required of a person who wishes to serve as a firefighter, whether paid or volunteer.

NFPA 1001 has become widely accepted as the standard of measurement for all firefighters in North America and beyond. It must be pointed out that NFPA 1001 is a **minimum** standard. Local jurisdictions may desire to exceed the minimum requirements in any or all areas of the standard. This is acceptable. It is not acceptable for local jurisdictions to weaken any portion of the standard to suit their needs.

The original edition of NFPA 1001 contained three levels of competence: Fire Fighter I, Fire Fighter II, and Fire Fighter III. These three levels were maintained through the first several revisions of the standard. (In keeping with NFPA principles, all standards are revised every 3 to 5 years.) The last edition of NFPA 1001, adopted in May of 1992, saw a major change. After reviewing the existing (1987) edition of the standard and realigning information, the NFPA 1001 committee determined that the information left in the Fire Fighter III section was repetitive of many of the objectives in the lower levels of NFPA 1021, *Standard for Fire Officer Professional Qualifications*. Thus, it was decided that NFPA 1001 would be limited to two levels: Fire Fighter I and Fire Fighter II.

The 1997 edition of NFPA 1001 brought with it another major change: a Job Performance Requirement (JPR) format. A *Job Performance Requirement* reflects either what a firefighter actually does on the job or should be expected to do. This JPR format is being applied to all new NFPA professional qualifications standards.

The JPRs contained in NFPA 1001 are the result of a job task analysis that the NFPA 1001 committee conducted during this latest revision. Although the revision of the standard was extensive in terms of format, the technical content changes were minor. However, this revision significantly reorganized the Fire Fighter I and Fire Fighter II objectives contained in the previous edition of the standard. To help those interested in using the standard, a comparison chart of the previous and new editions is contained in Appendix C of the standard. Although the chapters within **Essentials of Fire Fighting,** 4th edition, have been reordered, the revision of NFPA 1001 did not result in a major reorganization of the information within those chapters.

Specific JPRs within NFPA 1001 may consist of prerequisite skills and knowledges that are unrelated and may not lend themselves to being addressed in the same chapter of **Essentials of Fire Fighting.** For example, JPR 3-3.11 requires the firefighter to "Perform vertical ventilation on a structure . . . given . . . personal protective equipment, ground and roof ladders . . . ." These subjects — vertical ventilation, personal protective equipment, and ladders — are also required knowledges and skills in other JPRs within the standard; therefore, it is not possible to address all of these subjects in the same chapter without repeating them in another chapter as well. Information for each of the subjects covered in 3-3.11 (vertical ventilation, personal protective equipment, and ladders) are contained within three separate chapters of **Essentials of Fire Fighting.** The reader must study the information in the various chapters to gain a complete understanding of the knowledge and skills required to fulfill the requirements of the JPR. The table on page xxiii (NFPA 1001 References Within Essentials of Fire Fighting) contains a matrix of the NFPA 1001 objective numbers and the **Essentials of Fire Fighting** chapter reference(s). This matrix is designed to assist the reader in finding the information necessary to fulfill the requirements of specific JPRs within the standard.

Due to the interrelatedness of the information, it was not possible to separate Fire Fighter I and Fire Fighter II information within the **Essentials of Fire Fighting** manual. However, each major heading within each chapter contains one or more numbered JPR references from the standard. The reader can find information related to those references under these major headings.

For ease of organization and presentation, **Essentials of Fire Fighting,** 4th edition, is divided into what the IFSTA **Essentials** committee believed was a logical progression of firefighter training. It should be noted that the standard itself does not require the objectives to be mastered in the order in which they appear. Local jurisdictions may decide exactly the order they wish the material to be presented.

Another change within this edition of **Essentials of Fire Fighting** is the use of Skill Sheets. The IFSTA **Essentials** committee believed that separating the written text from the step-by-step procedures would make the manual easier to read. Therefore, Skill Sheets describing the step-by-step procedures for many of the skills covered in the text are found at the end of that chapter.

It is important to understand the IFSTA **Essentials** committee's philosophy on the delineation between a Fire Fighter I and Fire Fighter II. A *Fire Fighter I* is a person who is minimally trained to function safely and effectively as a member of a fire fighting team under direct supervision. A person meeting the requirements of Level I is by no means considered a "complete" firefighter. This is not accomplished until the objectives of both Levels I and II have been satisfied. A *Fire Fighter II* may operate under general supervision and may be expected to lead a group of equally or lesser trained personnel through the performance of a specified task.

The acceptance and recognition of a national standard provides a baseline for professionalism in the international fire service. It is the intent of Fire Protection Publications and the International Fire Service Training Association (IFSTA) to promote this professionalism by providing excellence in the materials used to prepare fire personnel to meet the objectives of these standards.

SCOPE AND PURPOSE

The **Essentials of Fire Fighting** manual is designed to provide the firefighter candidate with the information needed to meet the fire-related performance objectives in NFPA 1001, Levels I and II. The methods shown throughout this text have been approved by the International Fire Service Training Association as accepted methods for accomplishing each task. However, they are **not** to be interpreted as the only methods to accomplish a given task. Other specific methods for achieving any performance task may be specified by a local authority having jurisdiction. For guidance in seeking additional methods for performing a given task, the student or instructor may consult any of the IFSTA expanded-topic manuals (such as **Hose Practices** or **Fire Service Ground Ladders**) for more in-depth information.

GENDER USAGE

The English language has historically given preference to the male gender. Among many words, the pronouns, "he" and "his" are commonly used to describe both genders. Society evolves faster than language, and the male pronouns still predominate our speech. IFSTA/Fire Protection Publications has made great effort to treat the two genders equally, recognizing that a significant percentage of fire service personnel are female. However, in some instances, male pronouns are used to describe both males and females solely for the purpose of brevity. This is not intended to offend readers of the female gender.

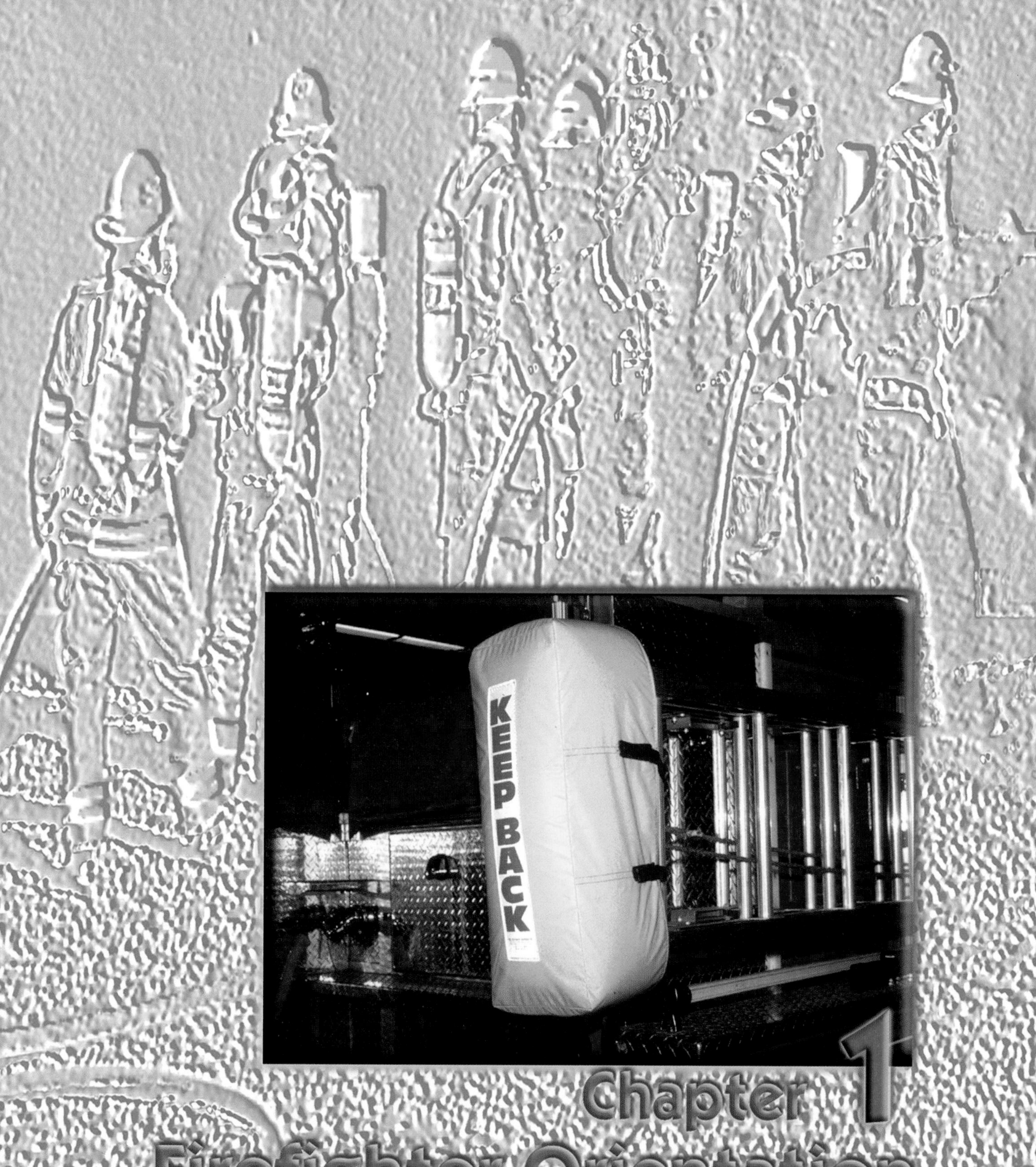

Chapter 1
Firefighter Orientation and Safety

Chapter 1
Firefighter Orientation and Safety

INTRODUCTION

[NFPA 1001: 3-1.1.1]

Fire fighting is one of the world's most honored but hazardous occupations. It is the duty of every fire department to practice life safety, incident stabilization, and property conservation. The firefighter's job is not comfortable or easy; it is a profession that exposes an individual to a high level of personal stress and danger. Fire fighting requires a high sense of personal dedication, a genuine desire to help people, and a high level of skill.

New firefighters enter into one of three categories: career, paid on call, or volunteer. Fire departments with career personnel (salaried firefighters) primarily protect larger towns and cities. Fire departments may also use paid-on-call firefighters to support their departments. These firefighters receive reimbursement for each call they attend. Volunteer fire departments are found in all sizes of communities. Volunteer fire departments and their firefighters greatly outnumber career departments and their firefighters.

Whenever there is an emergency, the fire department is one of the first entities called to the scene. Emergencies involve not only fires but incidents such as cave-ins, building collapses, motor vehicle accidents, aircraft crashes, tornadoes, hazardous materials incidents, civil disturbances, rescue operations, explosions, water incidents, and medical emergencies (Figure 1.1). The emergency list is unlimited.

Firefighters are involved with all types of people and are appreciated by some and scorned by others. Because firefighters are public employees, they are expected to calmly evaluate the problem and bring it to a successful conclusion. Firefighters are not extraordinary — they are ordinary people who often find themselves in extraordinary situations. They cannot do everything at once, and they and the public must accept this fact. Bringing any emergency situation to a safe conclusion requires knowledge, ability, and skill.

The purpose of this chapter is to acquaint the reader with the organization of the fire department, including the various positions and jobs found in a fire department. The chapter also covers the regulations governing the activities of firefighters and the Incident Management System (IMS). A discussion of interacting with other organizations will familiarize the reader with the types of agencies they may work with at an emergency scene. Finally, the chapter covers firefighter safety in the fire station, in training, on the apparatus, and at emergency scenes.

FIRE DEPARTMENT ORGANIZATION

[NFPA 1001: 3-1.1.1; 4-1.1.1]

An organizational chart shows the structure of the fire department and its chain of command. Small fire departments have a relatively simple chain of command, while large departments have a considerably more complex chart. Figure 1.2 shows an organizational chart for a medium-sized fire department. It is meant to serve only as a reference. Charts for local municipalities vary.

Organizational Principles

The firefighter should be aware of four basic organizational principles in order to operate effectively as a team member:

Figure 1.1 Firefighters respond to many different types of emergencies. *Photos courtesy of Chris Mickal, Steve George, Robert J. Bennet, Carl Goodson, and the Airline Pilots Association.*

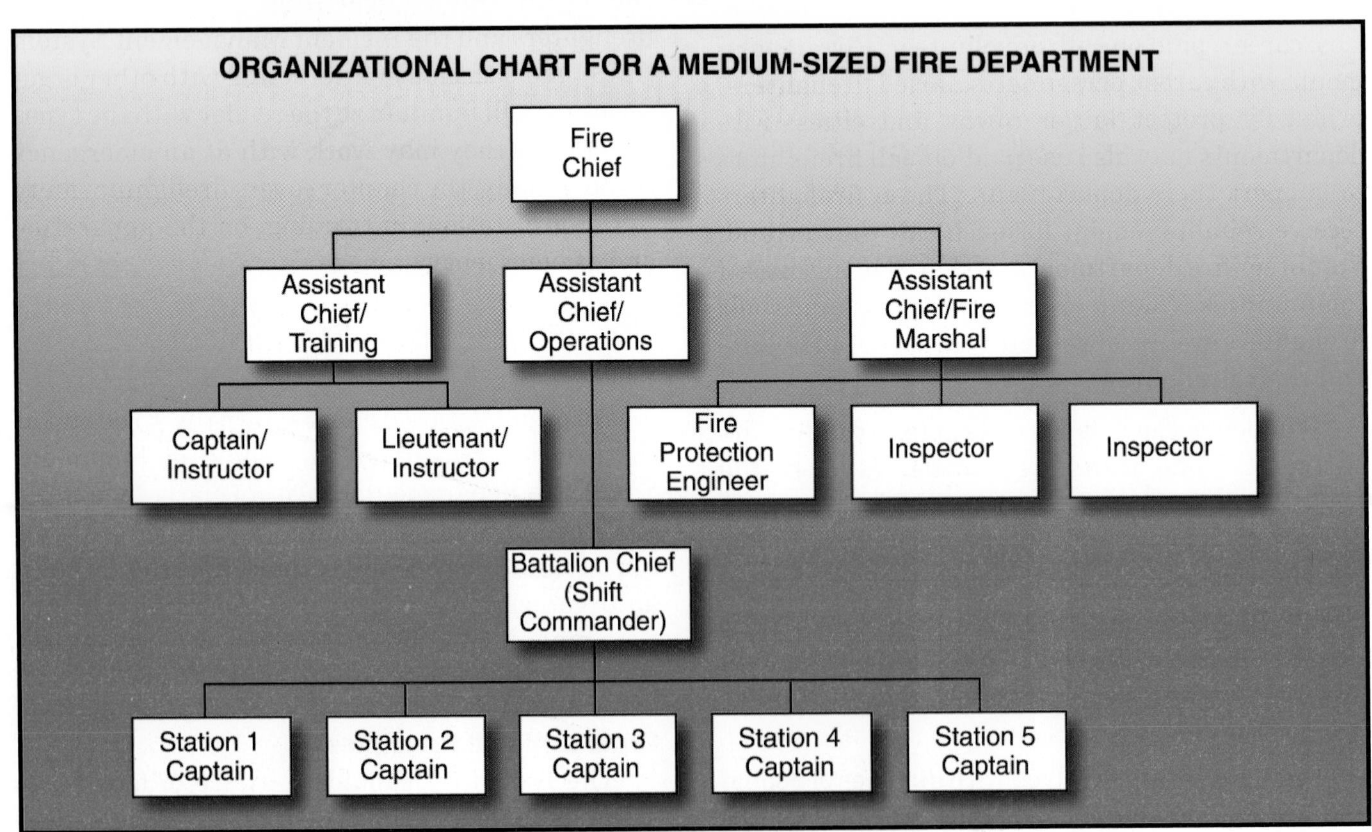

Figure 1.2 An organizational chart shows the structure of the fire department and its chain of command.

- Unity of command
- Span of control
- Division of labor
- Discipline

UNITY OF COMMAND

Unity of command is the principle that a person can report to only one supervisor. Directly, each subordinate reports to one boss; however, indirectly, everyone reports to the fire chief through the chain of command (Figure 1.3). The chain of command is the pathway of responsibility from the highest level of the department to the lowest.

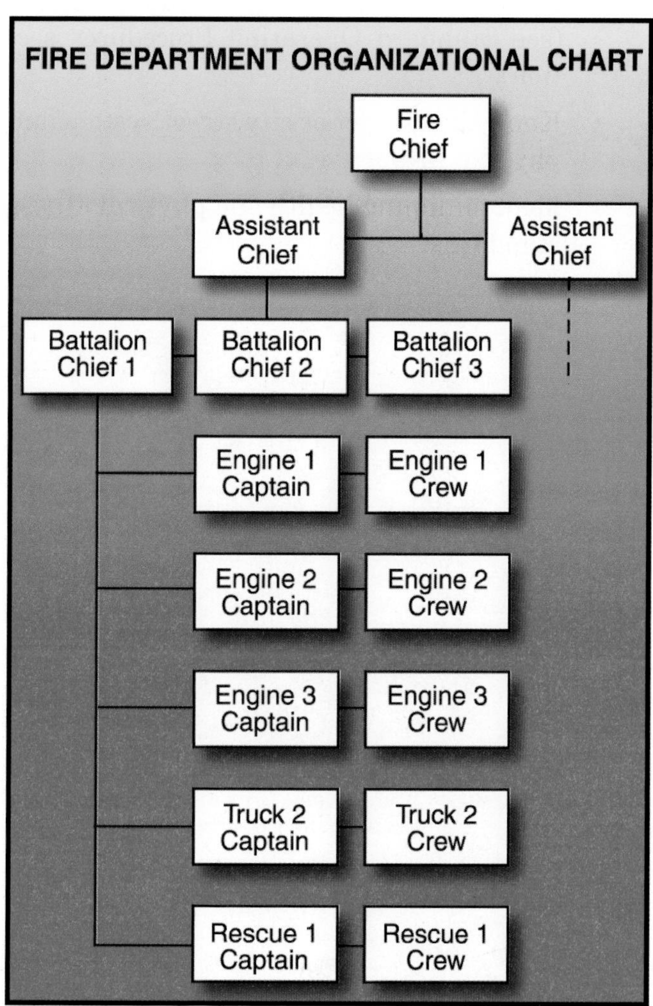

Figure 1.3 This simple organizational chart shows the department's chain of command.

SPAN OF CONTROL

Span of control is the number of personnel one individual can effectively manage. A rule of thumb in the fire service is that an officer can directly supervise three to seven firefighters effectively, but the actual number varies with the situation.

DIVISION OF LABOR

Division of labor is dividing large jobs into small jobs. These small jobs are then assigned to specific individuals. Division of labor is necessary in the fire service for the following reasons:

- To assign responsibility
- To prevent duplication of effort
- To make specific and clear-cut assignments

DISCIPLINE

Traditionally, discipline as applied to organizations has been understood to mean a well organized, adequately structured, uniform operation. However, in this instance *discipline* refers to an organization's responsibility to provide the direction needed to satisfy the goals and objectives it has identified. In other words, discipline is setting the limits or boundaries for expected performance and enforcing them. This direction may come in the form of rules, regulations, or policies, but regardless of the term used, it must define how the department plans to operate. The rules of the organization must be clearly written and presented.

Fire Companies

The standard operating unit of a fire department is the *company,* a group of firefighters assigned to a particular piece of fire apparatus or to a particular station. A company consists of a company officer(s), a driver/operator(s), and one or more firefighters (Figure 1.4).

A fire company is organized, equipped, and trained for definite functions. The functions and duties of similar fire companies may vary in different localities because of the inherent hazards of the area, the size of the department, and the scope of the department's activities. A small fire department may have only one fire company to carry out the functions that normally would be performed by several companies in a larger city. The following lists the general descriptions of fire companies:

- ***Engine company*** — Deploys hoselines for fire attack and exposure protection

Figure 1.4 The members of a company must work together.

- ***Truck (ladder) company*** — Performs forcible entry, search and rescue, ventilation, salvage and overhaul, and provides access to upper levels of a structure
- ***Rescue squad/company*** — Typically is responsible for the removal of victims from areas of danger or entrapment
- ***Brush company*** — Extinguishes wildland fires and protects structures in the urban-interface
- ***Hazardous materials company*** — Responds to and mitigates hazardous materials incidents
- ***Emergency medical company*** — Provides emergency medical care and support to patients

Fire Department Personnel

Fire fighting requires skill in preventing, combating, and extinguishing fires; answering emergency calls; and operating and maintaining fire department equipment, apparatus, and quarters. The work involves extensive training in fire fighting, rescue activities, hazardous materials, and emergency medical care. Firefighters must operate apparatus and perform dangerous assignments under emergency conditions, all of which require strenuous exertion amid hazards such as smoke and cramped surroundings (Figure 1.5). Although fire fighting and rescue work are the most demanding tasks, a significant portion of time is spent on inspections, training, and station duties.

FIREFIGHTER I AND FIREFIGHTER II

A firefighter must be an individual who can perform many functions. To function effectively, a firefighter must have certain knowledge and skills including:

- Meet the requirements of National Fire Protection Association (NFPA) Standard 1001, *Standard for Fire Fighter Professional Qualifications* (Figure 1.6).
- Know department organization, operation, and standard operating procedures (SOPs) (see Standard Operating Procedures section).
- Know the district or city street system and physical layout.
- Meet minimum health and physical fitness standards.

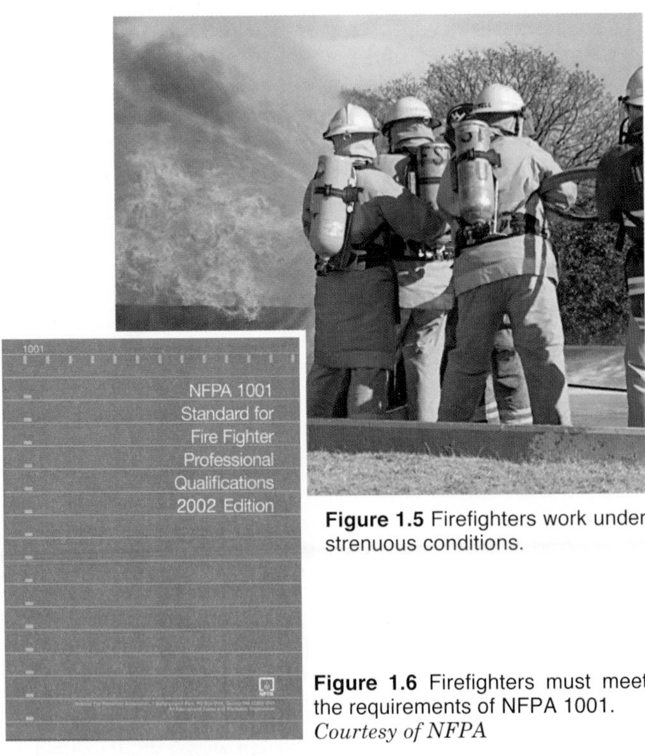

Figure 1.5 Firefighters work under strenuous conditions.

Figure 1.6 Firefighters must meet the requirements of NFPA 1001. *Courtesy of NFPA*

The following are some of the typical duties of a firefighter I and a firefighter II:

- Attend training courses; read and study assigned materials related to fire fighting, fire prevention, hazardous materials, and emergency medical care (Figure 1.7).

- Respond to medical emergencies and other patient care requests.
- Respond to fire alarms with the company, operate fire fighting equipment, lay and connect hose, maneuver nozzles and direct fire streams, raise and climb ladders, and use extinguishers and all fire fighting hand tools.
- Ventilate burning buildings by opening windows and skylights or by cutting holes in roofs or floors.
- Remove people from danger and administer first aid.
- Perform salvage operations, which include placing salvage covers, sweeping water, and removing debris.
- Complete overhaul operations with the goal of ensuring total fire extinguishment.
- Relay instructions, orders, and information, and give locations of alarms received from the telecommunicator.
- Exercise precautions to avoid injury while performing duties.
- Exercise loss control measures (reducing or eliminating loss and damage during and after a fire) to avoid unnecessary damage to or loss of property.
- Ensure safekeeping and proper care of all fire department property.
- Perform assigned fire inspections and checks of buildings and structures for compliance with fire prevention ordinances.

Figure 1.7 Training is a very important aspect of the firefighter's career.

OTHER FIRE DEPARTMENT PERSONNEL

Depending on local requirements and customs, other specialized fire service personnel may be used. Their duties and requirements vary depending on local needs and procedures. Following is a list of other positions among fire suppression personnel, their primary roles, and the NFPA standard covering their professional qualifications:

- ***Fire apparatus driver/operator*** — Safely drives assigned fire apparatus to and from fire and emergency scenes, operates pumps, aerial devices, or other mechanical equipment as required (NFPA 1002, *Standard for Fire Department Vehicle Driver / Operator Professional Qualifications*)
- ***Fire department officer*** — May fulfill any of the following responsibilities, depending upon the size and structure of the fire department (NFPA 1021, *Standard for Fire Officer Professional Qualifications*):
  - — The fire chief is ultimately responsible for all operations within the fire department.
  - — Fire department officers supervise a fire company in the station and at the fire scene. They may also supervise a group of fire companies in a specified geographical region of the city.
  - — Other roles assigned include operations, personnel/administration, public information, fire prevention, resources, and planning (Figure 1.8).

Figure 1.8 Fire department officers have both administrative and operational responsibilities.

- ***Fire department safety officer*** — Oversees a fire department's occupational safety and health program and monitors the operational safety of emergency incidents (NFPA 1521, *Standard for Fire Department Safety Officer*) (Figure 1.9)

Figure 1.9 On the emergency scene, the safety officer serves as an advisor to the incident commander.

In order to carry out the mission of the fire department, other personnel are also required. The following list describes some of these personnel:

- ***Communications/telecommunications personnel*** — Take emergency and nonemergency phone calls, process the information, dispatch units, maintain and provide communications link to companies that are in service, and complete incident reports (NFPA 1061, *Standard for Professional Qualifications for Public Safety Telecommunicators*)
- ***Fire alarm maintenance personnel*** — Maintain municipal fire alarm systems
- ***Apparatus and equipment maintenance personnel*** — Maintain all fire department apparatus and portable equipment
- ***Fire police personnel*** — Assist regular police officers in emergency operations with traffic control, crowd control, and scene security
- ***Information systems personnel*** — Manage the collection, entry, storage, retrieval, and dissemination of electronic databases such as fire reporting

Special Operations Personnel

If fire departments only provided standard structural fire protection to their communities, the positions discussed to this point in the chapter would cover all the bases. However, many fire departments today provide a wide variety of services to their jurisdictions. These special services require specially trained personnel. In many cases, these individuals serve as both regular firefighters and specialists in a particular discipline. The following list addresses some of the special operations found in many fire departments:

- ***Airport firefighter*** — Protects life and property, controls fire hazards, and performs general duties related to airport operations and aircraft safety (known as aircraft rescue and fire fighting [ARFF]) (NFPA 1003, *Standard for Airport Fire Fighter Professional Qualifications*) (Figure 1.10)

Figure 1.10 Airport firefighters wear special equipment and are trained to handle incidents involving all types of aircraft. *Courtesy of Robert Lindstrom.*

- ***Hazardous materials technician*** — Handles hazardous materials and nuclear, biological, and chemical (NBC) emergencies (NFPA 472, *Standard on Professional Competence of Responders to Hazardous Materials Incidents*)

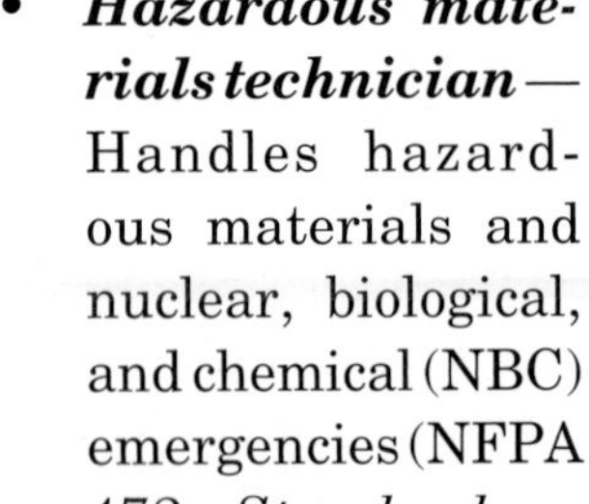

- ***Self-contained underwater breathing apparatus (SCUBA) diver*** — Performs both topside and underwater rescues and recoveries

- ***Special rescue technician*** — Handles special rescue situations such as high-angle (rope) rescue, trench and structural collapse, confined space entry, extrication operations, and cave or mine rescues

Fire Prevention Personnel

An effective fire prevention program decreases the need for suppression activities and thereby reduces the costs and risks of extinguishing fire. The fire prevention division of a fire department is typically headed by an assistant chief of the department. Depending on local customs, this person may be called the assistant chief in charge of fire prevention or the fire marshal. This individual has subordinate officers to fill the various roles within the division. The fire prevention division generally includes four major positions:

- ***Fire prevention officer/inspector*** — Conducts technical and supervisory work in the fire prevention program (Figure 1.11) (NFPA 1031, *Standard for Professional Qualifications for Fire Inspector*)
- ***Fire and arson investigator*** — Conducts the investigation of the fire area and makes analytical judgments based on the remains at the fire scene to determine the origin and cause of a fire (NFPA 1033, *Standard for Professional Qualifications for Fire Investigator*)
- ***Public fire and life safety educator*** — Informs the public about fire hazards, fire causes, precautions, and actions to take during a fire (NFPA 1035, *Standard for Professional Qualifications for Public Fire and Life Safety Educator*)
- ***Fire protection engineer/specialist*** — Acts as a consultant to the upper administration of the department in the areas of fire department operations and fire prevention

Figure 1.11 Inspectors may work with companies or work on their own.

Figure 1.12 Fire companies are commonly assigned to medical calls.

Emergency Medical Services (EMS) Personnel

Departments that provide first response to EMS incidents have trained first-aid responders on regular fire companies such as engines, trucks, or squads (Figure 1.12). These personnel may be trained to the first responder, emergency medical technician (EMT), or paramedic levels. The ambulance that responds to transport the victim also has trained crew members on board.

The following list highlights the roles of personnel who are trained to the first responder, emergency medical technician, or paramedic levels. In most cases these duties are in addition to those of a firefighter.

- ***First responder*** — Sustains the patient's life until more competent medical personnel arrive
- ***Emergency medical technician*** — Is trained to provide basic life support (BLS) for those whose lives are in danger

- ***Paramedic*** — Handles incidents similar to those handled by EMTs, but is able to provide advanced life support (ALS)

Training Personnel

The training that new firefighters receive is one of the most important aspects of job indoctrination. *A firefighter's training never ends.* New ideas, equipment, and tactics present new methods that must be learned. New materials and technology present challenges that never before existed. It is imperative that the fire service remain abreast of these changes. This is accomplished by the following training personnel who constantly improve and update the training program (Figure 1.13):

- ***Training officer/chief of training/drillmaster*** — Administers all fire department training activities (NFPA 1041, *Standard for Fire Service Instructor Professional Qualifications*)
- ***Instructor*** — Delivers training courses to the other members of the department (NFPA 1041, *Standard for Fire Service Instructor Professional Qualifications*)

Figure 1.13 Instructors prepare firefighters to respond to emergencies.

FIRE DEPARTMENT REGULATIONS

[NFPA 1001: 3-1.1.1; 4-1.1.1]

A fire department is composed of individuals with different backgrounds and different ideas about life. The success of a fire department depends on the willingness of its members to put aside their differences and work for the benefit of the department. To ensure that department members cooperate effectively, the methods of doing so are outlined in policies and procedures. When a firefighter joins a fire department, he is familiarized with the department's policies and procedures. If a firefighter has questions about these activities, he should contact a supervisor to clarify any misunderstanding that could cause trouble later. This section introduces some of the policies and procedures that most departments generally follow.

Policies and Procedures

It is important to understand the difference between policy and procedure. *Policy* is a guide to decision making within an organization. Policy originates mostly with top management in the fire department and points to the kinds of decisions that must be made by fire officers or other management personnel in specified situations.

Procedure is a kind of formal communication closely related to policy. Whereas policy is a guide to thinking or decision making, a *procedure* is a detailed guide to action. A procedure describes in writing the steps to be followed in carrying out organizational policy for some specific, recurring problem or situation.

Both orders and directives are essential for implementing the formal procedures of the department. They may be either written or verbal. An *order* is based upon a policy or procedure, whereas a *directive* is not based on a policy or procedure. On the fireground, fire officers issue many instructions, directives, and requests (Figure 1.14). However, because of the seriousness of the situation, all of these utterances are generally considered orders.

Standard Operating Procedures

Some fire departments have a predetermined plan for nearly every type of emergency that they can conceive of occurring. These plans are known as the department's *standard operating procedures (SOPs).* These procedures provide a standard set of actions that are the core of every fire fighting incident plan. The SOP may vary considerably in different localities, but the principle is usually the same.

Figure 1.14 Officers must not hesitate to give orders on the emergency scene.

Even though there are obvious variations in fires, most fires have more similarities than differences. These similarities are the basis for standard operating procedures. The incident commander (person in overall command of an incident) knows the SOPs and can base a plan of action upon them. Procedures have a built-in flexibility that allows, with reasonable justification, adjustments when unforeseen circumstances occur. The SOP is usually initiated by the first fire companies that reach the scene. The SOP is primarily a means to start the fire attack. It does not replace size-up, decisions based on professional judgment, evaluation, or command. In addition, there may be several SOPs from which to choose depending on fire severity, location, and the ability of first-in units to achieve control.

Examples of SOPs are as follows. These SOPs are performed with crews wearing complete protective clothing and self-contained breathing apparatus (SCBA).

1. The first unit on the scene assumes command.
2. The first-arriving engine attacks the fire.
3. The second-arriving engine lays a supply line(s) to the first engine.
4. The first-arriving ladder truck performs necessary forcible entry, search, rescue, and ventilation.

The SOP should follow the most commonly accepted order of fireground priorities:

- Life safety
- Incident stabilization
- Property conservation

The need to save lives in danger is always the first consideration. Once all possible victims have been rescued, attention is turned to stabilizing the incident. Last, firefighters should make all possible efforts to minimize damage to property. This can be accomplished through proper fire fighting tactics and good loss control techniques.

Following standard operating procedures reduces chaos on the fire scene. All resources can be used in a coordinated effort to rescue victims, stabilize the incident, and conserve property. Operational procedures that are standardized, clearly written, and mandated to each department member establish accountability and increase command and control effectiveness. When the firefighters of individual units are trained properly in SOPs, confusion is lessened. Firefighters will understand their duties and require a minimum of direction. SOPs also help prevent duplication of effort and uncoordinated operations because all positions are assigned and covered. The assumption and transfer of command, communications procedures, and tactical procedures are other areas that must be covered by the SOPs.

Safety is a top priority when designing SOPs. Requiring SCBA for all crews is an example of a safety consideration. SOPs should be applied to all situations, including medical responses. They should be designed to limit personnel exposure to contagious diseases. For example, SOPs may require personnel to use pocket masks when performing mouth-to-mouth resuscitation. They may also require personnel to wear rubber gloves and safety glasses to prevent contact with patients' body fluids during medical emergencies (Figure 1.15).

Standard operating procedures do not have to be limited to the emergency scene. Many departments prefer to carry out the administrative and personnel functions of the department through SOPs. SOPs may include regulations on dress, conduct, vacation and sick leave, station life and duties, and other departmental policies.

Figure 1.15 An SOP may dictate that firefighters responding to an EMS call wear protective rubber gloves.

INCIDENT MANAGEMENT SYSTEM

[NFPA 1001: 4-1.1.1; 4-1.1.2]

The Incident Management System (IMS) is designed to be applicable to incidents of all sizes and types. It applies to small, single-unit incidents that may last a few minutes and to complex, large-scale incidents involving several agencies and many mutual aid units that possibly can last for days or weeks.

Components of the IMS

The Incident Management System has a number of interactive components that provide the basis for clear communication and effective operations:

- Common terminology
- Modular organization
- Integrated communications
- Unified command structure
- Consolidated action plans
- Manageable span of control
- Predesignated incident facilities
- Comprehensive resource management

Overview

To understand the application of IMS, firefighters should know the major operational position descriptions within the IMS structure (Figure 1.16). These include Command, Operations, Planning, Logistics, and Finance/Administration.

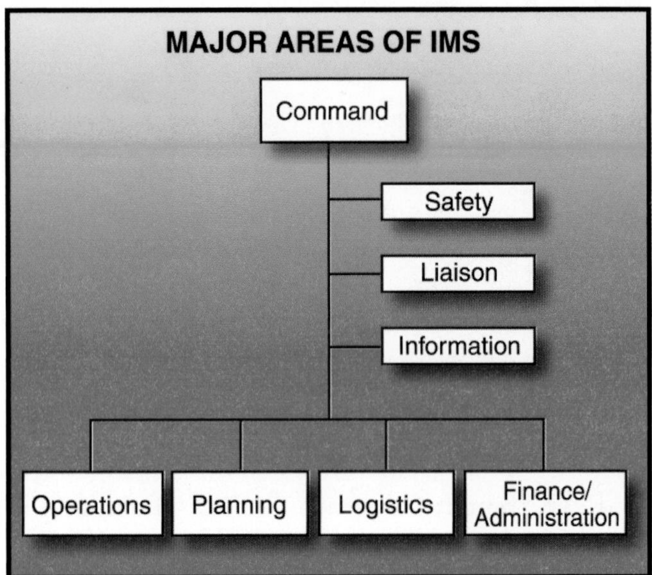

Figure 1.16 There are five major areas — command, operations, planning, logistics, and finance/administration — within the IMS system.

COMMAND

The person in overall command of an incident is the *Incident Commander (IC)* (Figure 1.17). The IC is ultimately responsible for all incident activities, including the development and implementation of a strategic plan. This process may include making a number of critical decisions and being responsible for the results of those decisions. The IC has the authority both to call resources to the incident and to release them from it. If the size and complexity of the incident requires, the IC may delegate authority to others, who together with the IC form the Command Staff. Positions within the Command Staff in-

Figure 1.17 The incident commander is located at a clearly identified command post.

Figure 1.18 Canteen units provide food and drinks at extended incidents.

clude the *Safety Officer, Liaison Officer,* and *Public Information Officer.*

OPERATIONS

The *Operations Officer* reports directly to the IC and is responsible for managing all operations that directly affect the primary mission of eliminating the problem. The Operations Officer directs the tactical operations to meet the strategic goals developed by the IC. Operations may be subdivided into as many as five branches if necessary.

PLANNING

Planning is responsible for the collection, evaluation, dissemination, and use of information concerning the development of the incident. Planning is also responsible for tracking the status of all resources assigned to the incident. Command uses the information compiled by Planning to develop strategic goals and contingency plans. Specific units under Planning include the *Resource Unit, Situation Status Unit, Demobilization Unit,* and any technical specialists whose services are required.

LOGISTICS

Logistics is responsible for providing the facilities, services, and materials necessary to support the incident. There are two branches within Logistics: the support branch and the service branch. The *service branch* includes medical, communications, and food services (Figure 1.18). The *support branch* includes supplies, facilities, and ground support (vehicle services).

FINANCE/ADMINISTRATION

Finance/Administration has the responsibility for tracking and documenting all costs and financial aspects of the incident. Generally, Finance/Administration will be activated only on large-scale, long-term incidents. Day-to-day mutual aid responses are usually considered to be reciprocal and do not require interagency reimbursement.

IMS Terms

The IMS uses several terms that all firefighters should understand.

COMMAND

Command is the function of directing, ordering, and controlling resources by virtue of explicit legal, agency, or delegated authority. It is important that lines of authority be clear to all involved. Lawful commands by those in authority should be followed immediately and without question.

DIVISION

Division is a geographic designation assigning responsibility for all operations within a defined area. Divisions are assigned clockwise around an outdoor incident with Division A at the front (street address side) of the incident. In buildings, divisions are usually identified by the floor or area to which they are assigned: First floor is Division 1, second floor is Division 2, etc. In a one-story building, the entire interior may be assigned as a division (Interior Division) (Figure 1.19). All groups or functional sectors operating within that specific geographic area report to that division supervisor. Organizationally, the division level is between a strike team or other operational unit and a branch.

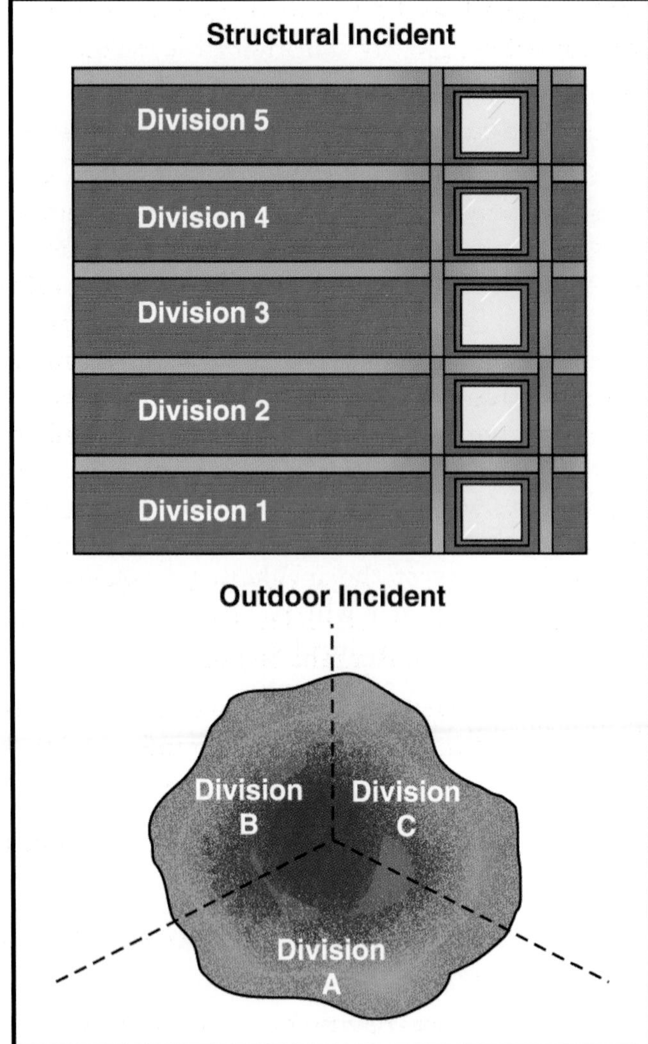

Figure 1.19 Examples of IMS divisions.

GROUP

Groups are functional designations (forcible entry, salvage, ventilation, etc.). When their assigned function has been completed, they are available for reassignment.

SECTOR

A *sector* is a geographic *or* functional assignment that is equivalent to a division or a group or both.

SUPERVISOR

A *supervisor* is someone in command of a division, a group, or a sector.

INCIDENT ACTION PLAN

The written or unwritten plan for managing the emergency is the *Incident Action Plan* (IAP). A plan should be formulated for *every* incident. Small, routine incidents usually do not require a written plan, but large, complex incidents do. The plan identifies the strategic goals and tactical objectives that must be achieved to eliminate the problem.

INCIDENT COMMANDER

The *Incident Commander (IC)* is the officer at the top of the incident chain of command and is in overall charge of the incident. The IC is ultimately responsible for everything that takes place at the emergency scene. The Incident Commander is primarily responsible for formulating the Incident Action Plan and for coordinating and directing all incident resources to implement the plan and meet its goals and objectives.

RESOURCES

Resources are all personnel and major pieces of apparatus on scene or en route on which status is maintained. Resources may be individual companies, task forces, strike teams, or other specialized units. Resources are considered to be *available* when they have checked in at the incident and are not currently committed to an assignment. It is imperative that the status of these resources be tracked so that they may be assigned when and where needed without delay.

Implementing the System

IMS should be initiated by the first person arriving on the scene of an emergency. This individual begins to evaluate the situation in order to answer the following questions:

- What has occurred?
- What is the current status of the emergency?
- Is anyone injured or trapped?
- Can the emergency be handled with the resources on scene or en route?
- Does the emergency fall within the scope of the individual's training?

If no life-threatening situation demands immediate action, the IC should begin to formulate an Incident Action Plan. The plan should reflect the following priorities:

1. Ensuring personnel safety and survival
2. Rescuing or evacuating endangered occupants

3. Eliminating the hazard
4. Conducting loss control
5. Cleaning up and protecting the environment

Whenever the IMS is implemented, there should be only ***ONE*** Incident Commander except in a multijurisdictional incident when a *unified command* is appropriate. A multijurisdictional incident involves agencies beyond the jurisdiction of one department or agency. Even when a unified command is used, the chain of command must be clearly defined. All orders should be issued by one person through the chain of command to avoid the confusion caused by conflicting orders.

With advice from the Operations Officer, the IC will gather enough resources to handle the incident and organize information to ensure that orders can be carried out promptly, safely, and efficiently. Having sufficient resources on scene will help to ensure the safety of all involved. The organization must be structured so that all available resources can be utilized to achieve the goals of the Incident Action Plan. If necessary, the Incident Commander can appoint a Command Staff to help gather, process, and disseminate information.

All incident personnel must function according to the Incident Action Plan. Company officers or sector officers should follow standard operating procedures, and every action should be directed toward achieving the goals and objectives specified in the plan. When all members (from the Incident Commander to the lowest ranking member of the team) understand their positions, roles, and functions in the Incident Management System, the system can serve to safely, effectively, and efficiently use resources to accomplish the plan.

TRANSFER OF COMMAND

The first-arriving fire department member must be prepared to transfer command to the next-arriving person with a higher level of expertise or authority. If the transfer cannot take place face-to-face, it can be accomplished over the radio, but *command can only be transferred to someone who is on scene*. As an incident grows larger, command may be transferred several times before the problem is brought under control. A smooth and efficient transfer of command will contribute greatly to bringing the incident to a timely and successful conclusion.

The person relinquishing command must provide the person assuming command as clear a picture of the situation as possible. This can be accomplished by giving a *situation status report,* which is an updated version of the incident evaluation performed on arrival. The person assuming command should acknowledge receipt of the information by repeating it back to the other person. If the reiteration is accurate, the recipient is ready to accept control of and responsibility for the management of the incident. The former IC can then be reassigned to an operating unit or retained at the Command Post (CP) as an aide or as a member of the Command Staff. The IC can call for any additional resources that might be needed.

Situation status report. The situation status report should include the following information:

- Description of what happened
- Whether anyone was/is injured or trapped
- What has been done so far
- Whether the problem has stabilized or is getting worse
- What resources are on scene or en route
- Whether it appears that current resources are adequate for the situation or that more resources need to be called

It is imperative that the information given be current and that it be clearly understood by the recipient. If a report of a fire in a trash container causes the new IC to picture a fire in a wastebasket when it is actually in a fully loaded Dumpster®, the resulting orders might be inappropriate. An example of a complete situation status report might be as follows:

> *A fire is burning in a Dumpster® inside the warehouse, at the northeast corner. No one is hurt, and the fire appears to be confined to the container, but it is close to the wall, and the smoke is pretty heavy.*

Acknowledge information. The new IC repeats the information back to verify that he understands the situation. For example:

I understand that the fire is confined to a Dumpster® in the northeast corner. It hasn't spread but has the potential to, and it's putting out a lot of smoke. No one has been hurt.

At this point, any miscommunication can be corrected, or the person being relieved of command can simply reply, *"That's correct."*

COMMAND AND CONTROL OF THE INCIDENT

Command and control of the incident does not transfer automatically when the information has been exchanged. If the problem does not exceed the level of training of the first IC and the senior member is satisfied with the manner in which the first IC is handling the situation, he may choose to leave the first IC in command. If not, the senior member assumes command and control of the incident.

NOTIFICATION THAT COMMAND HAS TRANSFERRED

If command is transferred, the former IC should announce the change to avoid any possible confusion caused by others hearing a different voice acknowledging messages and issuing orders. If everyone involved follows the chain of command and uses correct radio protocols, they should not be calling anyone by name, rank, or job title, so it should not matter who answers their radio messages. Because the early stages of an emergency can be chaotic, anything that can should be done to reduce the confusion. Announcing a transfer of command is one way of accomplishing that objective.

BUILDING THE ORGANIZATION

Emergency situations can be either as simple as a fire in a trash can or as complex as an explosion and fire of massive proportions. Depending on both the nature and the scope of the incident, different levels of incident management will be needed. The IMS should only be as large as is necessary to handle the incident safely and efficiently (Figure 1.20). When a complex emergency occurs, command may be transferred several times as the organization grows. It is important that the transitions be made as smoothly and as efficiently as possible.

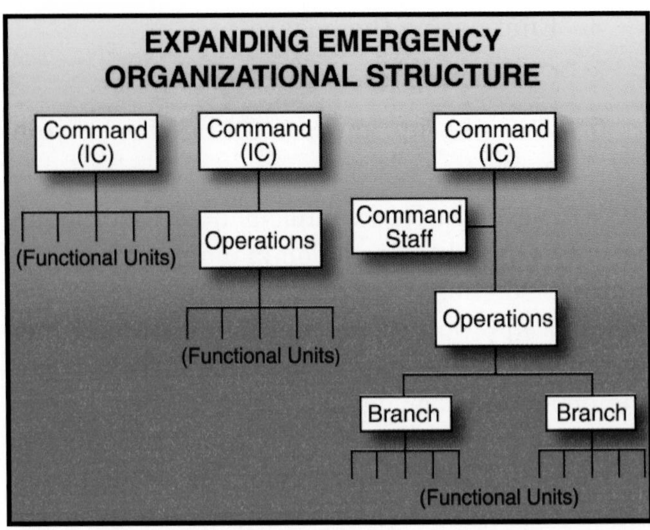

Figure 1.20 How a typical organizational structure may expand to deal with a growing incident.

TRACKING RESOURCES

One of the most important functions of an IMS is to provide a means of tracking all personnel and equipment assigned to the incident. Most units responding to an incident arrive fully staffed and ready to be assigned an operational objective; other personnel may have to be formed into units at the scene. To handle these and other differences in the resources available, the Incident Action Plan must contain a tracking and accountability system with the following elements:

- Procedure for checking in at the scene
- Way of identifying the location of each unit and all personnel on scene
- Procedure for releasing units no longer needed

TERMINATING THE INCIDENT

Once the incident has been brought under control and the size and complexity of the situation diminishes, the resources that are no longer needed should be released to return to their respective locations. This release is especially important when mutual aid units have been called, perhaps from considerable distances. Having an IMS in place will greatly assist in demobilizing methodically and efficiently. Adhering to a formal demobilization plan helps to recover loaned equipment, such as portable radios, and to identify and document any damaged or lost equipment.

INTERACTING WITH OTHER ORGANIZATIONS

[NFPA 1001: 3-1.1.1]

During the course of his career, a firefighter will be affected by and exposed to many different organizations that are a part of, or related to, the fire service. The purpose of this section is to acquaint the reader with these organizations.

Emergency Medical Services

Fire department personnel should establish a close working relationship with emergency medical services (EMS) personnel. In some areas, fire departments work very closely with private ambulance companies. Because one of the major functions of the fire department is the removal (and sometimes the initial treatment) of people trapped in wrecked vehicles and similar situations, it is important for firefighters to have an appropriate level of first-aid training (Figure 1.21). The level of training needed depends on the local EMS system and the department's standard operating procedures. In many jurisdictions, the main purpose of fire department personnel is to perform rescue and extrication functions. Beyond that, they provide only first responder medical treatment.

In most jurisdictions, once fire and EMS units are on the scene, EMS personnel are responsible for treating patients, and rescue personnel are responsible for freeing trapped victims and for scene safety. Close coordination between the two groups is very important to avoid working against each other, wasting valuable time, and perhaps further endangering victims and rescuers.

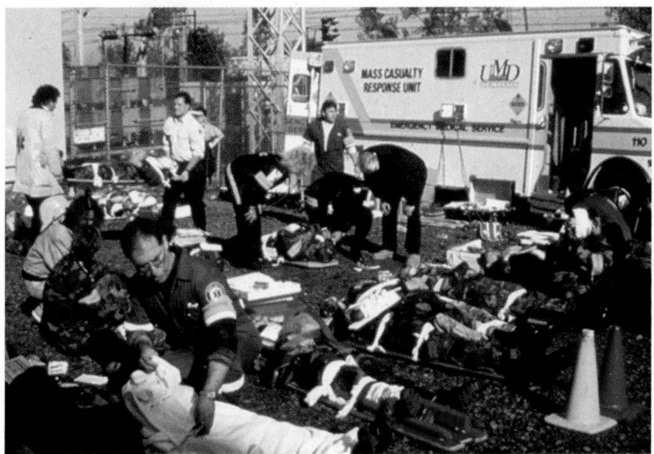

Figure 1.21 Firefighters can be an essential part of the EMS delivery system.

Hospitals

In some unusual incidents, hospital personnel may be called to the scene of an emergency (Figure 1.22). This situation is most likely to occur during a mass casualty incident. In such cases, hospital personnel are needed on-scene to assist in performing triage (sorting victims by the severity of their injuries) or conducting primary treatment of more seriously injured victims. An incident does not necessarily have to be large to require hospital personnel on the scene. Although quite rare, in some areas where EMS personnel are not trained to provide advanced life support, hospital personnel may be called to the scene to perform such functions as starting intravenous (IV) solutions while extrication operations are in progress. Another example where hospital personnel are needed is a serious industrial or agricultural equipment entrapment where major medical procedures (such as amputation of a limb) may be the only way to free the victim.

Figure 1.22 Hospital personnel may sometimes respond to the scene. *Courtesy of Mike Wieder.*

Law Enforcement

It is important for law enforcement and fire personnel to understand each other's functions and what to expect from each other at the scene. Firefighters may be called upon to assist law enforcement agencies in a variety of ways. These may include forcible entry to assist an investigation, emergency lighting to illuminate a crime scene, or a body recovery operation.

Law enforcement personnel may be present at the fire scene and may be part of the operation; however, they have their responsibilities just as any other unit does. They are responsible for maintaining the flow of traffic during rescue operations on highways, roads, and streets and for investigating traffic accidents on public roadways (Figure 1.23). When victims are either unconscious or otherwise unable to provide needed information, law enforcement personnel can often secure the information by using resources such as computer databases.

Utility Companies

Many incidents involve utilities (electric, gas, and water) in some way, so it is important for fire personnel to have a good working relationship with local utility company personnel. It is also important for responding fire units to coordinate with the utilities on mutual responses and to know what to do until the utility crews arrive. In addition, utility companies may have specially trained and equipped emergency response teams that can greatly assist in rescue efforts (Figure 1.24).

Figure 1.23 Traffic control is usually left to law enforcement.

Figure 1.24 A public utility emergency response team.

Other Agencies

In addition to the agencies mentioned, the fire department may come in contact with a number of other entities. These include public health departments, coroner/medical examiner's offices, and the Environmental Protection Agency (EPA), to name a few. Any possible contacts should be identified and a relationship established so that these agencies will be able to work more effectively with the fire department when an incident occurs.

FIREFIGHTER SAFETY

[NFPA 1001: 3-1.1.1; 4-1.1.1]

Fire fighting is one of the world's most dangerous jobs, and accidents in this profession can result in costly losses — the greatest loss being the death of a firefighter (Figure 1.25). Other losses may include lost manpower (due to injuries), damaged equipment (which is expensive to repair or replace), and legal expenses. In order to prevent these losses, it is necessary to prevent the accidents that cause them. Reducing accidents will save lives and money.

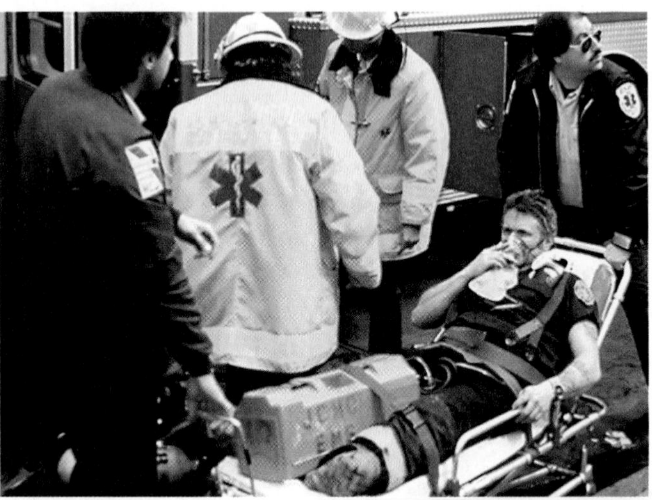

Figure 1.25 A firefighter receiving medical attention. *Courtesy of Chris Tompkins.*

Two basic factors motivate accident control efforts within the fire fighting profession: life safety and economy (Figure 1.26). The *life safety factor,* while interrelated with economics, stems from the natural desire to prevent needless suffering from physical pain or emotional stress. The *economic factor* includes legal expenses and expenses caused by the loss of manpower, apparatus, equipment, tools, property, or systems.

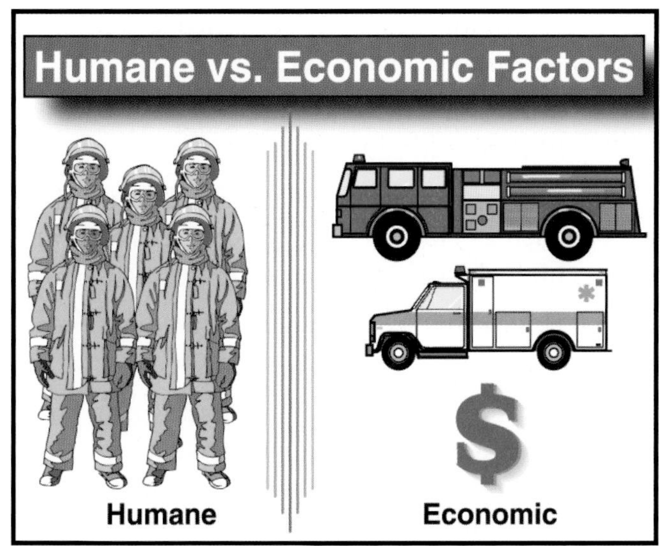

Figure 1.26 Safety is motivated by humane and economic factors.

Firefighters have traditionally accepted injuries and related losses as part of their vocation. Knowing their job to be one of the most hazardous, many firefighters are resigned to occupational accidents, injuries, and fatalities; this is compounded by the stereotypical image of the firefighter as heroic and fearless in the face of danger. Most firefighter injuries, however, are a direct result of preventable accidents. The firefighter should be too smart and too professional to take unnecessary risks.

Safety Standards for the Fire Service

NFPA 1500, *Standard on Fire Department Occupational Safety and Health Program,* contains the minimum requirements and procedures for a safety and health program. The standard may be applied to a fire department or similar organization, public or private. It calls upon the fire department to recognize safety and health as official objectives of the department and to provide as safe and healthy a work environment as possible. The basic concept of NFPA 1500 is to apply the same degree of safety throughout the fire service regardless of individual status or type of organization. Because it is a minimum standard, none of the objectives are intended to restrict a department or jurisdiction from exceeding the requirements specified.

Employee Interest

The success of a safety program will begin at the top of the fire department administrative chain (Figure 1.27). The administration's attitude toward safety is invariably reflected in the attitude of the supervising officers, which in turn affects firefighters. The main goals of any good safety program should be as follows:

- Prevent human suffering, deaths, injuries, illnesses, and exposures to hazardous atmospheres and contagious diseases.
- Prevent damage/loss of equipment.
- Reduce the incidence and severity of accidents and hazardous exposures.

Figure 1.27 Safety begins at the top of the fire department command structure.

An effective safety program becomes a matter of developing, promoting, and practicing an ongoing attitude of involvement throughout the organization.

Safety requires effort on the part of everyone. If one person does not participate or abide by the rules of the program, chances are that others will follow this bad example. Because of their leadership role, officers must provide a good example and must follow all safety rules. It is not enough to teach safety practices; they must be practiced and enforced. Breaking bad habits will not be easy for some, and once the new procedures are established, they must be maintained by *everyone*. If not, people will revert back to the old procedures.

Firefighter Health Considerations

Fire fighting is one of the most physically demanding and dangerous of all professions. Firefighters must be in good physical condition to handle the physical demands of the profession. Firefighters need strength to perform such tasks as rescuing victims, placing ladders, handling hoselines, and forcing entry with heavy tools (Figure 1.28). Aerobic endurance is required to move rapidly down hallways, climb ladders, or combat fires. Flexibility is needed to reach for equipment, tilt a ladder, and move a victim onto a ladder. The following list contains information essential to a firefighter's personal health:

- Stay informed about job-related health issues.
- Follow recommendations for vaccination against hepatitis B.
- Use precautions to avoid exposure to the human immunodeficiency virus (HIV).
- Learn proper lifting techniques to avoid muscle strains and other related injuries.
- Use tools designed to assist in lifting heavy objects.
- Properly clean and store tools and equipment used in patient care.

Figure 1.28 Moving a victim onto a ladder is an extremely physically demanding task.

- Maintain a regular exercise program to sustain physical fitness.
- Maintain a diet that reduces cholesterol, fat, and sodium intake.
- Be aware of cardiovascular and cancer risks such as smoking, high blood pressure, and high cholesterol levels.
- Have regular physicals and medical checkups.

Firefighter physical fitness should be an ongoing maintenance program. The department is responsible for ensuring that measures are taken to limit the number of stress-related accidents and illnesses. Physical fitness and health programs are a good way of fulfilling this responsibility. For more information on firefighter fitness and health considerations, refer to IFSTA's **Fire Department Occupational Safety** manual.

Employee Assistance Programs

An Employee Assistance Program (EAP) is one way the fire department can help its members and their families. An EAP offers confidential assistance with problems that could adversely affect job performance. Some of the areas in which an EAP can assist are as follows:

- Alcohol abuse
- Drug abuse
- Personal and interpersonal problems
- Stress
- Depression
- Anxiety
- Divorce
- Career development
- Nutrition
- Hypertension
- Smoking cessation
- Weight control

The program should be readily available to all members and their families. It should provide referrals to appropriate health care services, alcohol treatment services (i.e., Alcoholics Anonymous), community services, self-help groups, and other professionals. The program should provide counseling and education on health concerns. It should allow members easy, yet confidential, access to counseling and professional help for any problems or concerns that may be interfering with their daily well-being. Pamphlets and flyers detailing services can be distributed to make the program and information about the program accessible. Any service provided for departmental personnel should also be made available to family members.

Another significant area of employee assistance is debriefing individuals suffering from critical incident stress. Because the injuries suffered by the victims in fire and rescue incidents sometimes can be extremely gruesome and horrific, firefighters and any others who had to deal directly with the victims should participate in a critical incident stress debriefing (CISD) process (Figure 1.29). Because individuals react to and deal with extreme stress in different ways — some more successfully than others — and because the effects of unresolved stresses tend to accumulate, participation in this type of process should not be optional.

The process should actually start *before* firefighters enter the scene if it is known that conditions exist there that are likely to produce psychological or emotional stress for the firefighters involved. This is accomplished through a prebriefing process wherein the firefighters who are about to enter the scene are told what to expect so that they can prepare themselves.

Figure 1.29 Firefighters in a CISD session.

If firefighters are required to work more than one shift in these conditions, they should go through a minor debriefing, sometimes called *defusing,* at the end of each shift. They should also participate in the full debriefing process within 72 hours of completing their work on the incident.

SAFETY ON THE APPARATUS

[NFPA 1001: 3-1.1.1; 4-1.1.1]

The most common danger that the firefighter experiences is riding the apparatus to and from emergency calls. Passengers and driver/operators of emergency vehicles shall not dress while the apparatus is in motion. Preferably, all firefighters should ride within a fully enclosed portion of the cab (Figure 1.30). Firefighters who are not riding in enclosed seats should wear helmets and eye protection. If sirens and noise levels exceed 90 decibels, firefighters should also wear hearing protection (Figure 1.31). All firefighters must be seated with their seat belts fastened when the vehicle is in motion. Fire apparatus should have seat belts large enough to accommodate a firefighter in full protective clothing. Firefighters should NOT stand anywhere on the apparatus.

WARNING

Do not ride on the tailboard. Many firefighters have been killed falling from tailboards. This practice must be discontinued.

If it is absolutely necessary to ride in an unenclosed jump seat, safety bars are available that may prevent a firefighter from falling (Figure 1.32). These bars are not a substitute for safety procedures that require firefighters to ride in safe, enclosed positions wearing their seat belts; however, they do have value as an additional barrier between the firefighter and the road.

Firefighters should always use handrails when mounting or dismounting the apparatus (Figure

Figure 1.30 All new apparatus must have fully enclosed cabs.

Figure 1.31 Firefighters riding in open jump seats should wear appropriate ear and eye protection as well as their seat belts.

Figure 1.32 Safety bars provide very limited protection to occupants of the jump-seat area.

Figure 1.33 Use handrails when entering or exiting the apparatus.

1.33). Using handrails reduces the chance of firefighters accidentally slipping and falling from the apparatus. One exception to this rule is that firefighters should not use the handrail when dismounting an apparatus that has an aerial device extended close to electrical wires. If the aerial device contacts the charged lines and the firefighter is in contact with the apparatus and the ground at the same time, the firefighter might be electrocuted. Always jump clear of an apparatus that might be electrically energized (Figure 1.34).

Figure 1.34 Jump clear of aerial apparatus that may be in contact with energized electrical wires.

SAFETY IN THE FIRE STATION

[NFPA 1001: 3-1.1.1; 4-1.1.1]

Most firefighters' duties and activities center around the station, and a significant portion of their on-duty time is spent there. Hazards in the fire station not only endanger firefighters but can also endanger visitors who enter the station. Visitors are the responsibility of the fire department while they are in the building. Therefore, safe conditions must exist to limit the possibility of accidents and injuries (Figure 1.35).

Figure 1.35 Safety is a primary concern when visitors are in the station.

Personal Safety

Certain safety hazards are common to any fire station. Also, certain types of accidents are not limited to any specific location within a station. Improper lifting techniques and slip-and-fall accidents are two of the most common causes of injury.

Although back strains are the most common injuries related to improper lifting and carrying techniques, bruises, sprains, and fractures can also result. Improper lifting techniques not only cause personal injury, but they may also end in damage to equipment if it is dropped or improperly handled. Back injuries have been statistically proven to be the most expensive single type of accident in terms of worker's compensation, and they occur with surprising frequency.

Every firefighter should be instructed in the correct method of lifting. A firefighter should not attempt to lift or carry an object that is too bulky or heavy for one person to safely handle but should get help to lift or carry it (Figure 1.36). Lifting and carrying heavy or bulky objects without help can result in unnecessary strains and injuries.

Other common accidents are slips, trips, and/or falls. Numerous factors contribute to these types of accidents. A slip, trip, or fall generally results from poor footing. This can be caused by slippery surfaces, objects or substances on surfaces, inattention to footing on stairs, uneven surfaces, and similar hazards. These accidents can easily result in minor and serious injuries as well as damaged equipment. To prevent such accidents, it is important to stress good housekeeping. For example, floors must be kept clean and free from slipping hazards such as loose items and spills (Figure 1.37). Aisles must be unobstructed and stairs should be well lighted. In addition to walking surfaces (floors, stairs, and aisles), items such as handrails, slide poles, and slides must also be maintained in a safe condition.

Figure 1.36 When working around the station, use plenty of help to lift heavy loads.

Figure 1.37 Eliminate tripping hazards around the station.

Tool and Equipment Safety

Tools and equipment are vital to a firefighter's job. However, accidents can happen if the firefighter is not properly trained in the use and care of tools and equipment. Poorly maintained tools and equipment can be very dangerous and can result in costly accidents to firefighters in the station and at the emergency scene. NFPA 1500 stresses the importance of safety in every aspect of tool and equipment design, construction, purchase, usage, maintenance, inspection, and repair.

When working in a station shop or on the emergency scene, firefighters must use appropriate personal protective equipment (PPE). Using PPE is fundamental for safe work practices. Although PPE does not take the place of good tool engineering, design, and use practices, it does provide personal protection against hazards.

The most widely used tools in the station shop are hand tools and small power tools. Observe the following procedures when using hand and power tools:

- Wear appropriate personal protective equipment.
- Remove jewelry, including rings and watches.
- Select the appropriate tool for the job.
- Know the manufacturer's instructions and follow them.
- Inspect tools before use to determine their condition. If a tool has deteriorated or is broken, replace it.
- Provide adequate storage space for tools, and always return them promptly to storage after use.
- Inspect and clean tools before storing.
- Consult with and secure the approval of the manufacturer before modifying the tool.
- Use spark-resistant tools when working in flammable atmospheres such as around a vehicle's fuel system.

HAND TOOLS

Inspect all tools before each use to ensure that they are in good condition. This inspection may prevent an accident caused by tool failure. Homemade tool-handle extensions, or "cheaters," are sometimes incorrectly used to provide extra leverage for wrenches, pry bars, and similar tools (Figure 1.38). The use of a cheater can overload the tool beyond its designed capabilities. This overloading is unsafe and can cause the tool to break suddenly, not only while the cheater is attached but also later when the weakened tool is being used normally.

Figure 1.38 Misusing tools, such as adding a cheater bar to this wrench, may result in tool failure and/or injury to the user.

POWER TOOLS

Grinders, drills, saws, and welding equipment are commonly found in fire stations (Figure 1.39). Improperly used, these tools can cause a serious or life-threatening injury. Whether the tool is driven by air or electricity, it has a specific, safe method of operation that must be understood and followed. Only those firefighters who have read and who understand the tool manufacturer's instructions should be allowed to use power tools. It is important for instructions to be accessible to firefighters.

Repairs should always be made by someone trained and authorized to properly repair the damaged tool. Depending on the department, this person may be someone within the fire department or an outside equipment dealer and repair agent. Keeping accurate records of repairs can help spot misuse before the tool causes an accident.

Any electrical tool not marked "double insulated" should have a three-prong plug (Figure 1.40). For firefighter safety, the third prong must connect

Figure 1.39 Power tools, such as a grinder, are found in most fire stations.

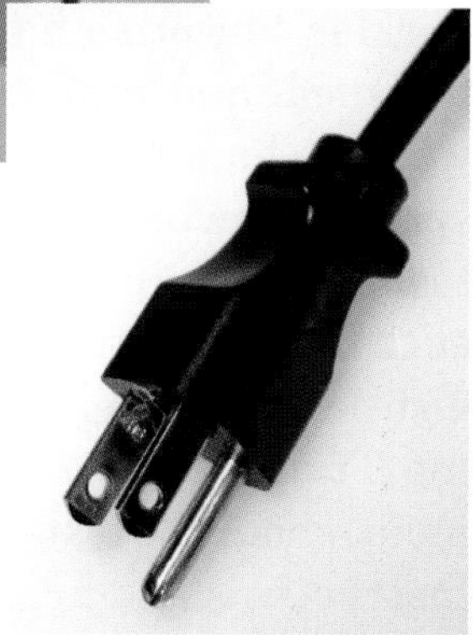

Figure 1.40 A three-prong plug should be used when a tool is not double insulated.

Figure 1.41 Rotary saws used in the station are usually mounted on a table or bench.

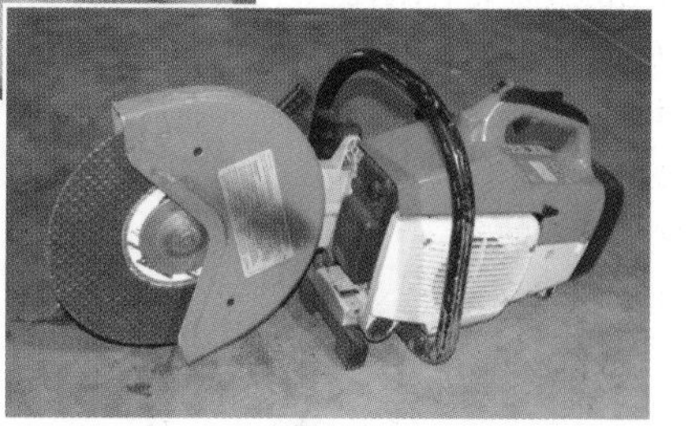

Figure 1.42 A rotary rescue saw.

to a ground while the tool is in use. Bypassing the ground plug in any way opens the door to injuries or deaths from unpredictable electrical shorts.

POWER SAWS

The most common types of power saws used by firefighters are rotary saws and chain saws. When improperly operated or maintained, a power saw can be the most dangerous type of tool a firefighter will use.

Rotary saws may be found in the station or on the emergency scene. Rotary saws found in the station are usually bench- or table-mounted (Figure 1.41). Rotary saws used on the emergency scene are generally of the rescue or forcible entry design (Figure 1.42).

Following a few simple safety rules when using power saws will prevent most typical accidents:

- Match the saw to the task and the material to be cut. Never push a saw beyond its design limitations.
- Wear proper protective equipment, including gloves and eye protection. Avoid wearing loose, dangling clothing that may become entangled in the saw.
- Have hoselines in place when forcing entry into an area where fire is suspected or when performing vertical ventilation. Hoselines are also essential when cutting materials that generate sparks.
- Avoid the use of all saws when working in a flammable atmosphere or near flammable liquids. Sparks generated by the saw or the saw's hot muffler are ignition sources for the vapors.
- Keep unprotected and unessential people out of the work area.
- Follow manufacturer's procedures for proper saw operation.
- Use caution to avoid igniting gasoline vapors when refueling a hot, gasoline-powered saw. It is best to allow the saw to cool before refueling.
- Keep blades and chains well sharpened. A dull saw is far more likely to cause an accident than a sharp one.

SAFETY IN TRAINING

[NFPA 1001: 3-1.1.1; 4-1.1.1]

NFPA 1500 requires that all personnel who may engage in structural fire fighting participate in training at least monthly. Ideally, this monthly training will reinforce safe practices until they become automatic. Other types of training are required on an "as-needed" basis. For example, training is required when new procedures or equipment are introduced. There should be at least two training sessions of this type per year.

Maintaining Personal Safety

All personnel participating in training at a drill site should be fully clothed in protective gear. Raising ladders, laying hose, or performing any other activity that simulates actual fire scene conditions requires the use of protective gear.

If trainees have colds, severe headaches, or other symptoms indicating physical discomfort or illness, they should not continue training until a medical examination can determine their fitness. Some trainees might feel uncomfortable telling an instructor they are physically unable to continue training. Older trainees and trainees who apparently are not in good physical condition should be watched closely for signs of fatigue, chest pains, or unusually labored breathing during heavy exercise. Physical discomfort or illness can lead to accidents; however, accidents can be prevented by determining the physical condition of all participating personnel before training.

Horseplay during training must be forbidden because it can lead to accidents and injuries (Figure 1.43). If the trainees want to play pranks, the instructor needs to find out why. Boredom often causes trainees to fidget and release their energy in horseplay. Boredom can develop easily in a training program that does not allow everyone to see the demonstration or to participate in the activities because of the size of the group.

Figure 1.43 Horseplay must not be tolerated.

Maintaining and Servicing Equipment

Equipment used for fire training evolutions must be in excellent condition. Items used frequently for training often wear out sooner than those used routinely in the fire station. Examples of frequently used training items are ropes, straps, buckles, and other harness parts that must be tied, fastened, and unfastened repeatedly. Tools with wooden handles can quickly become worn and splintered when used repeatedly by trainees. All tools and equipment should be inspected before each drill to ensure their reliability. Records must also be maintained on all equipment used for training. Training equipment, like all fire fighting equipment, should be tested according to manufacturer's instructions and applicable standards.

EMERGENCY SCENE SAFETY

[NFPA 1001: 3-3.4(a); 4-4.2(b)]

Upon arriving at the scene, the officer in charge has to decide whether it is safe and/or feasible to attempt an emergency operation. The IC must decide whether the operation about to be undertaken requires rescuing victims or stopping property loss. The IC must decide when the risks involved in either are great enough to warrant limiting the actions of fire fighting personnel. This may be a difficult decision to make, and the firefighters may feel frustrated because they cannot help a victim as much as they would like. It is necessary to weigh these feelings against the potential for firefighters becoming additional victims and the likelihood of the operation being successful.

All firefighters must remember that they did not cause the emergency incident; they are not responsible for the victim being in that situation; and they are not obligated to sacrifice themselves in a heroic attempt to save the victim — especially not in an attempt to recover a body. In fact, it is irresponsible and unprofessional for firefighters to take unnecessary risks that might result in their being incapacitated by an injury and therefore unable to perform the job for which they have been trained. The function of the fire/rescue service is not to add victims to the situation. The IC's first priority must be firefighter safety; the second priority is the victim's safety. The IC should never choose a course of action that requires firefighters to take unnecessary risks.

Crowd Control

Proper scene management reduces congestion and confusion by reducing the number of personnel within the perimeter of the emergency scene. Crowd control is essential to proper scene management. This function is usually the responsibility of the law enforcement agency on the scene, but it may sometimes have to be performed by firefighters or other rescue personnel. It is the responsibility of the IC to ensure that the scene is secured and properly managed.

Even in the most remote locations, bystanders or spectators are often drawn to the scene. Some may be people who were involved in the accident but are not injured. They are often quite curious and try to get as close to the scene as possible. All bystanders should be restrained from getting too close to the incident for their own safety and for that of victims and emergency personnel.

Emergency scenes tend to involve emotional situations that should be handled with care. This is particularly true when friends or relatives of the victims are at the scene. These particular bystanders are often difficult to deal with, and firefighters must treat them with sensitivity and understanding. Relatives and friends of victims should be gently but firmly restrained from getting too close. They should be kept some distance from the actual incident but still within the cordoned area. While they may console each other, they should not be left entirely on their own. A firefighter or other responsible individual should stay with them until the victims have been removed from the scene.

Cordoning off the area will keep bystanders a safe distance from the scene and out of the way of emergency personnel. There is no specific distance from the scene or area that should be cordoned off. The zone boundaries should be established taking into account the area needed by emergency personnel to work, the degree of hazard presented by elements involved in the incident, and the general topography of the area. Cordoning can be done with rope or fireline tape tied to signs, utility poles, parking meters, or any other objects readily available (Figure 1.44). Once the area has been cordoned off, the boundary should be monitored to make sure people do not cross the line.

Figure 1.44 The hazardous areas are cordoned off.

Personnel Accountability System

Each department must develop its own system of accountability that identifies and tracks all personnel working at an incident. The system should be standardized so that it is used at every incident. All personnel must be familiar with the system and participate when operating at an emergency incident. The system must also account for those individuals who respond to the scene in vehicles other than fire department apparatus.

Accountability is vital in the event of a serious accident or structural collapse. If the IC does not know who is on the fireground and where they are located, it is impossible to determine who and how many may be trapped inside or injured. Flashover (simultaneous ignition of room contents) and backdraft (fire explosion) may trap or injure firefighters (see Chapter 2, Fire Behavior). SCBAs can malfunction or run out of air. Firefighters can get lost in mazes of rooms and corridors. Too many firefighters have died because they were not discovered missing until it was too late.

TAG SYSTEM

A simple tag system can aid in accounting for personnel within the fireground perimeter. Personnel can be equipped with a personal identification tag (Figure 1.45). Upon entering the fireground perimeter, firefighters leave their tags at a given location or with a designated person (command post, apparatus compartment, company officer, control officer, or sector officer). Tags can be attached to a control board or personnel identification (ID) chart for quick reference. Upon leaving the fireground perimeter, the firefighters collect their tags. This system enables officers to know exactly who is operating on the fireground.

Figure 1.45 Identification badges may be worn on both the station uniform and the turnout clothing.

SCBA TAG SYSTEM

An SCBA tag system provides closer accountability for personnel inside a structure. All personnel entering a hazardous atmosphere must be required to wear full protective clothing with SCBA. These firefighters must be trained and certified for SCBA use. Each SCBA is provided with a tag containing the name of the user and the air pressure. Having individually assigned breathing apparatus ensures that the user is familiar with the apparatus and has a properly fitting facepiece. Upon entering a building, personnel give their tags to a designated supervisor (Figure 1.46). The supervisor records time of entry and expected time of exit. This supervisor also does a brief check to ensure that all protective equipment is properly used and in place. This provides complete accountability for those inside the structure and ensures that they are in proper gear. Firefighters leaving the danger area take back their tags so that the control officer knows who is safely outside and who is still inside the structure or danger area. Relief crews are sent in before the estimated time of the sounding of the low-pressure alarms.

Figure 1.46 A firefighter checks in with the Accountability Officer.

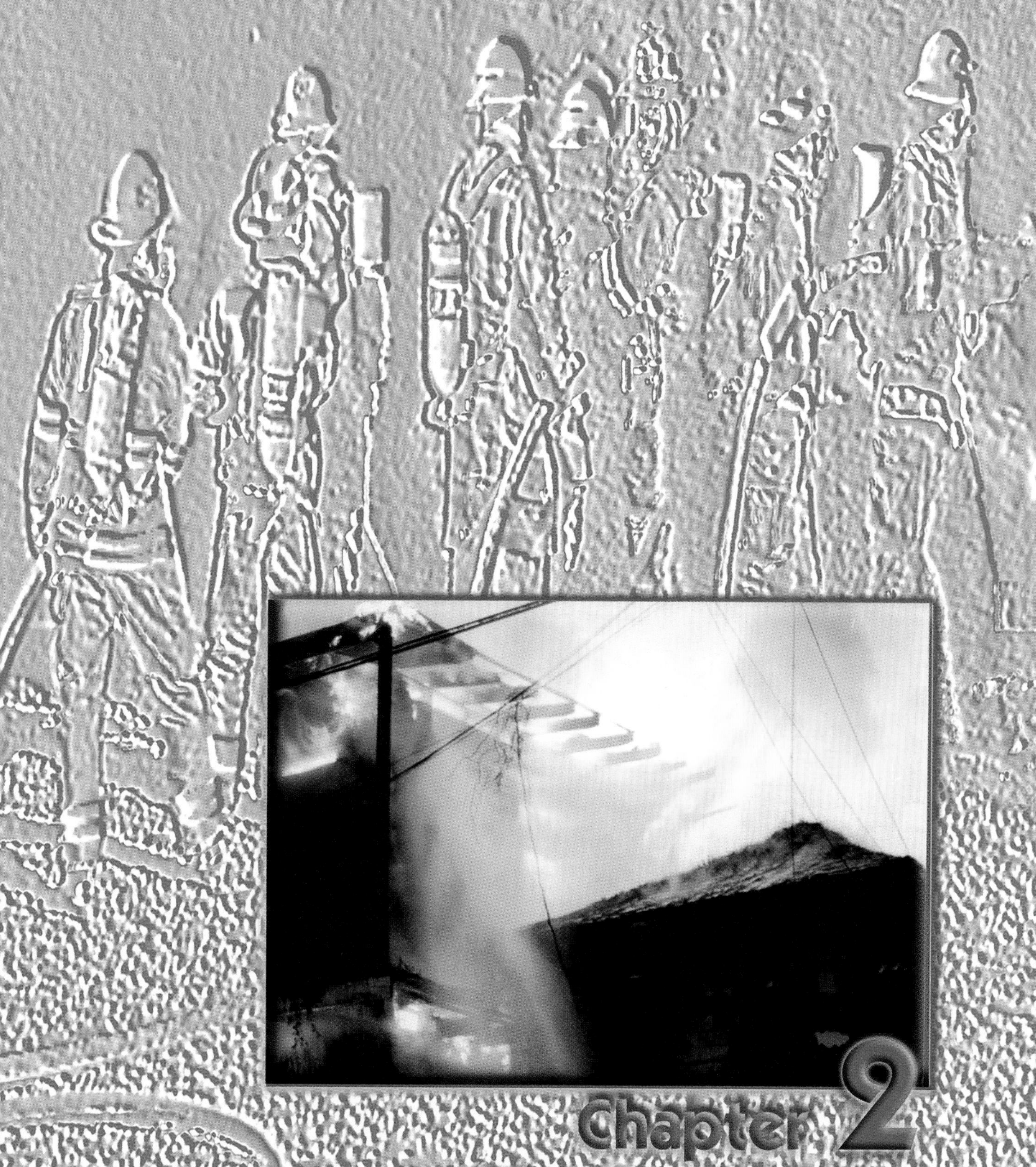

Chapter 2
Fire Behavior

Chapter 2
Fire Behavior

INTRODUCTION

Firefighters responding to a fire may have to cope rapidly with a variety of conditions. The fire may be "exposing" (endangering) another structure or groups of structures as in wildland/interface fires. The smoke and flames may be creating a "life hazard" (danger to survival) to occupants. The room of fire origin may be close to "flashover" (simultaneous ignition of room contents). If a building is not ventilated, there may be a "backdraft" (fire explosion) potential. All of these conditions result from fire and the way it behaves. To perform safely and effectively in any fire fighting function, firefighters should have a basic understanding of the science of fire and the factors that affect its ignition, growth, and spread (fire behavior).

This chapter deals primarily with the types of fire a structural firefighter encounters. Many of the concepts discussed in this chapter hold true for wildland fires, but a number of additional factors must be addressed in those incidents. Wildland fires are dealt with in a separate manual.

Fire has been both a help and a hindrance to mankind throughout history. Fire has heated our homes, cooked our food, and helped us to become technologically advanced. Fire, in its hostile mode, has also endangered us for as long as we have used it. Technically, fire is a chemical reaction that requires fuel, oxygen, and heat to occur. Over the last 30 years, scientists and engineers have learned a vast amount about fire and its behavior.

This chapter introduces several basic concepts from physical science that affect the ignition and development of a fire. Firefighters can use the information in this chapter to interpret what they see on the fireground and to develop methods to prevent, extinguish, and investigate fires. An understanding of fire behavior and the phases a fire passes through as it grows, also discussed in this chapter, assists firefighters in selecting the proper tactics to attack and extinguish fires. This knowledge also helps firefighters in recognizing potential hazards to themselves and others while they work on the fireground.

PHYSICAL SCIENCE

[NFPA 1001: 3-3.9(a); 3-3.11(a)]

Fire is a rapid chemical reaction that gives off energy and products of combustion that are very different in composition from the fuel and oxygen that combined to produce them. To understand the reaction we call fire, how it grows, and its products (products of combustion), we need to look at some basic concepts from physical science. *Physical science* is the study of the physical world around us and includes the sciences of chemistry and physics and the laws related to matter and energy. The basic science information in this section is referred to throughout this chapter.

Measurement Systems

Any scientific discussion presents information using numbers. Firefighters use numbers frequently while performing their jobs. They regularly use a numerical system to describe the hoselines they use to attack fires — 1¾-inch (45 mm) — or the capacity of the pump on an engine — 1,500 gpm (5 678 L/min) — or the length of a ladder — 24 feet (7.3 m). For these numbers to make any sense, they must be used with some unit of measurement that describes what is being measured — distance, mass, or time, for instance. The units are based on a measurement system. In the United States the *English* or *Customary System* is commonly used. Most other nations and the scientific community use a form of the metric system called the *International System of Units* or *SI* (after the French Systeme International).

Each system defines specific units of measure. Table 2.1 shows some of the base units used in each system.

TABLE 2.1
The Base Units of Measurement

Quantity	Customary System	SI System
Length	Foot (ft)	Meter (m)
Mass		Kilogram (kg)
Time	Second (s)	Second (s)
Temperature	Fahrenheit (°F)	Celsius (°C)
Electric current	Ampere (A)	Ampere (A)
Amount of a substance		Mole (mol)
Luminous intensity		Candela (cd)

A large variety of derived units are generated from these base units. For example, the base unit for length in SI is the *meter (m)*. From this base unit, you can derive measurements for area in square meters (m^2) and volume in cubic meters (m^3). Measurements for speed can be derived from length and time and described in feet per second or meters per second (fps or m/s). In the discussion of fire behavior that follows, the derived units for heat, energy, work, and power will be introduced and discussed. Other units used in the SI are *hour (h)* and *liter (L)*. While they are not considered base units, they are widely accepted and used. While mass is considered a base unit, weight is not. *Weight* is the measurement of the gravitational attraction on a specific mass. In the Customary system, the unit for weight is the *pound (lb)*. In the SI, weight is considered to be a force and is measured in *newtons (N)*. Both Customary and SI units are provided throughout the chapter.

One reason why the scientific community uses the SI is that it is a very logical and simple system based on powers of 10. This allows for the manipulation and conversion of units without the fractions needed with the Customary System. For example, in the Customary System the unit of length is the foot. The other recognized units are the inch (1/12th of a foot), the yard (36 inches or 3 feet), and the mile (5,280 feet or 1,760 yards). In the SI, the unit of length is the meter. To express length in larger or smaller terms, the system uses the prefixes shown in Table 2.2. Thus, a centimeter is 1/100th of a meter, and a kilometer is 1,000 meters.

TABLE 2.2
Names and Symbols for SI Prefixes

Prefix	Symbol	Multiply By
Tera	T	10^{12}, 1 trillion
Giga	G	10^{9}, 1 billion
Mega	M	10^{6}, 1 million
Kilo	k	10^{3}, 1,000
Deci	d	10^{-1}, one tenth
Centi	c	10^{-2}, one hundredth
Milli	m	10^{-3}, one thousandth
Micro	μ	10^{-6}, one millionth
Nano	n	10^{-9}, one billionth
Pico	p	10^{-12}, one trillionth

Energy and Work

In any science, energy is one of the most important concepts. *Energy* is simply defined as the capacity to perform work. Work occurs when a force is applied to an object over a distance (Figure 2.1). In other words, *work* is the transformation of energy from one form to another. The SI unit for work is the *joule (J)*. The joule is a derived unit based on a force in expressed newtons (also a derived unit — kg m/s^2) and distance in meters. In the Customary System the unit for work is the *foot-pound (ft lb)*.

The many types of energy found in nature include the following:

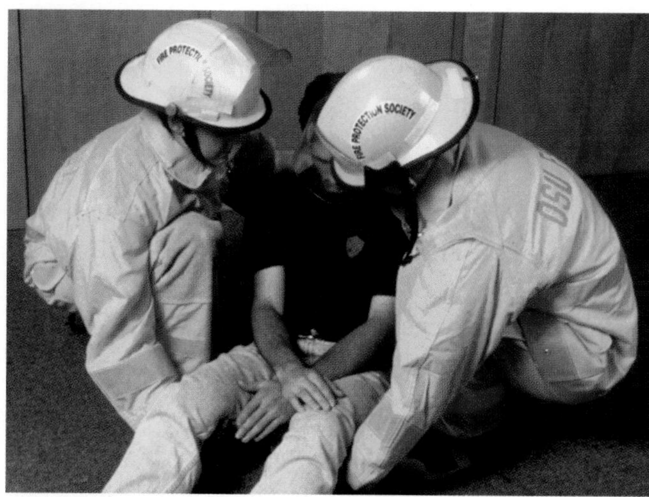

Figure 2.1 The firefighters moving a victim in a rescue is an example of work.

Figure 2.2 The water in the hose with the nozzle shut has potential energy. When the nozzle is opened, the water is converted to kinetic energy.

- ***Chemical*** — Energy released as a result of a chemical reaction such as combustion
- ***Mechanical*** — Energy an object in motion possesses such as a rock rolling down a hill
- ***Electrical*** — Energy developed when electrons flow through a conductor
- ***Heat*** — Energy transferred between two bodies of differing temperature such as the sun and the earth
- ***Light*** — Visible radiation produced at the atomic level such as a flame produced during the combustion reaction
- ***Nuclear*** — Energy released when atoms are split (fission) or joined together (fusion); nuclear power plants generate power as a result of the fission of uranium-235

Energy exists in two states: kinetic and potential. *Kinetic energy* is the energy possessed by a moving object. *Potential energy* is the energy possessed by an object that can be released in the future (Figure 2.2). A rock on the edge of a cliff possesses potential mechanical energy. When the rock falls from the cliff, the potential energy is converted to kinetic energy. In a fire, fuel has potential chemical energy. As the fuel burns, the chemical energy is converted to kinetic energy in the form of heat and light.

Power

Power is an amount of energy delivered over a given period of time. In our earlier example for work (Figure 2.1), firefighters are shown moving a victim over a distance in a rescue. The firefighters were expending energy over a distance and thus performing work. If the time to complete the rescue were known, then the amount of power required to perform the rescue could be determined.

Throughout history, people have used fire to generate power in many ways. A fuel's potential energy is released during combustion and converted to kinetic energy to run a generator or turn a shaft that "powers" a machine. The derived units for power are *horsepower (hp)* in the Customary System and *watts (W)* in SI.

In the study of fire behavior, researchers frequently address power when they consider the rate at which various fuels or fuel packages (groups of fuels) release heat as they burn. During the last several decades, researchers at the National Institute of Standards and Technology (NIST) have compiled a great deal of information on the heat release rates (HRRs) of many fuels and fuel packages. This information is very useful in the study of fire behavior because it provides data on just how much energy is released over time when various types of fuels are burned. Heat release rates for specific fuel packages are discussed in more detail in the Fire Development section later in the chapter.

Heat and Temperature

Anyone who has ever fought or even watched a fire fighting operation knows a tremendous amount of heat is generated. *Heat* is the energy transferred from one body to another when the temperatures of the bodies are different. Heat is the most common form of energy encountered on earth. *Temperature* is an indicator of heat and is the measure of the warmth or coldness of an object based on some standard. In most cases today, the standard is based on the freezing (32°F and 0°C) and boiling points (212°F and 100°C) of water. Temperature is measured using *degrees Celsius (°C)* in SI and *degrees Fahrenheit (°F)* in the Customary System (Figure 2.3).

The approved SI unit for all forms of energy including heat is the *joule*. While joules are used to describe heat in current literature, heat was described in terms of calories (Cal) or British thermal units (Btu) for many years. A *calorie* is the amount of heat required to raise the temperature of 1 gram of water 1 degree Celsius. The *British thermal unit* is the amount of heat required to raise the temperature of 1 pound of water 1 degree Fahrenheit. The calorie and the Btu are not approved SI units but are still frequently used. The relationship between the calorie and the joule is called the *mechanical equivalent of heat,* where 1 calorie equals 4.187 joules and a Btu equals 1,055 joules.

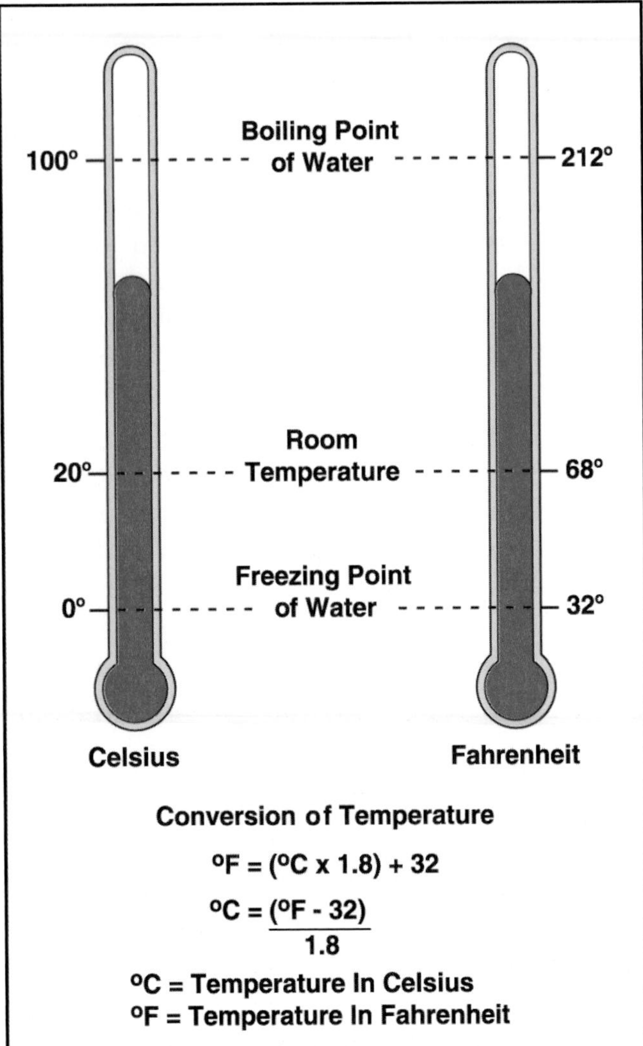

Figure 2.3 Comparison of Celsius and Fahrenheit scales.

Transmission of Heat

The transfer of heat from one point or object to another is a basic concept in the study of fire. The transfer of heat from the initial fuel package to other fuels in and beyond the area of fire origin controls the growth of any fire. Firefighters use their knowledge of heat transfer to estimate the size of a fire before attacking it and to evaluate the effectiveness of an attack. The definition of heat makes it clear that for heat to be transferred from one body to another, the two bodies must be at different temperatures. Heat moves from warmer objects to those that are cooler. The rate at which heat is transferred is related to the temperature differential of the bodies. The greater the temperature difference between the bodies, the greater the transfer rate. The transfer of heat from body to body is measured as energy flow (heat) over time. In the SI, heat transfer is measured in *kilowatts (kW)*. In the Customary System, the units are *Btu per second (Btu / s)*. Both units (kW and Btu/s) are expressions that relate to power (see the discussion of mechanical equivalent of heat in the preceding Heat and Temperature section).

Heat can be transferred from one body to another by three mechanisms: *conduction, convection,* and *radiation*. Each of these is discussed in some detail in the following sections.

CONDUCTION

When a piece of metal rod is heated at one end with a flame, the heat travels throughout the rod (Figure 2.4). This transfer of energy is due to the increased activity of atoms within the object. As heat is applied to one end of the rod, atoms in that area begin to move faster than their neighbors. This activity causes an increase in the collisions between the atoms. Each collision transfers energy

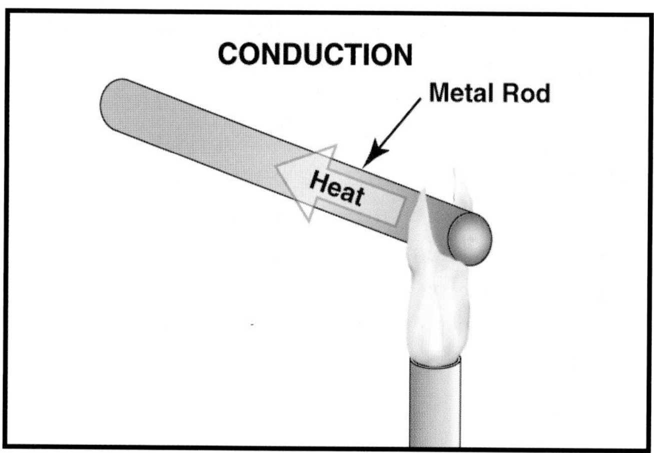

Figure 2.4 The temperature along the rod rises because of the increased movement of molecules from the heat of the flame. This is an example of conduction.

to the atom being hit. The energy, in the form of heat, is transferred throughout the rod.

This type of heat transfer is called conduction. *Conduction* is the point-to-point transmission of heat energy. Conduction occurs when a body is heated as a result of direct contact with a heat source. Heat cannot be conducted through a vacuum because there is no medium for point-to-point contact.

In general, heat transfer early in the development of all fires is almost entirely due to conduction. Later, as the fire grows, hot gases begin to flow over objects some distance away from the point of ignition and conduction again becomes a factor. The heat from the gases in direct contact with structural components or other fuel packages is transferred to the object by conduction.

Heat insulation is closely related to conduction. Insulating materials do their jobs primarily by slowing the conduction of heat between two bodies. Good insulators are materials that do not conduct heat well because of their physical makeup and thus disrupt the point-to-point transfer of heat energy. The best commercial insulators used in building construction are those made of fine particles or fibers with void spaces between them filled with a gas such as air.

CONVECTION

As a fire begins to grow, the air around it is heated by conduction. The hot air and products of combustion rise. If you hold your hand over a flame, you are able to feel the heat even though your hand is not in direct contact with the flame. The heat is being transferred to your hand by convection. *Convection* is the transfer of heat energy by the movement of heated liquids or gases. When heat is transferred by convection, there is movement or circulation of a fluid (any substance — liquid or gas — that will flow) from one place to another. As with all heat transfer, the flow of heat is from the warmer area to the cooler area (Figure 2.5).

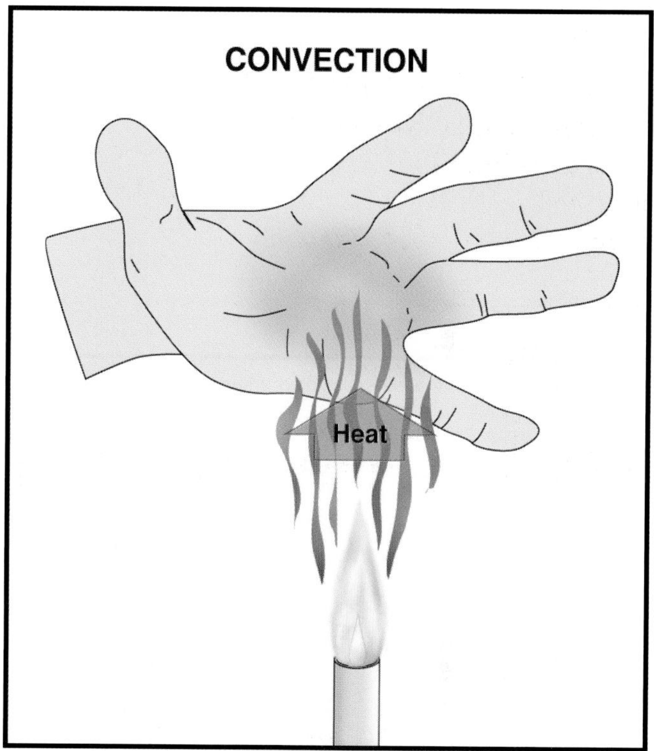

Figure 2.5 Convection is the transfer of heat energy by the movement of heated liquids or gases.

RADIATION

If you hold your hand a few inches (millimeters) to the side of the small fire used as an example in the preceding section, you would also be able to feel heat. This heat reaches your hand by radiation. *Radiation* is the transmission of energy as an electromagnetic wave (such as light waves, radio waves, or X rays) without an intervening medium (Figure 2.6). Because it is an electromagnetic wave, the energy travels in a straight line at the speed of light. All warm objects will radiate heat. The best example of heat transfer by radiation is the sun's heat. The energy travels at the speed of light from the sun through space (a vacuum) and warms the earth's

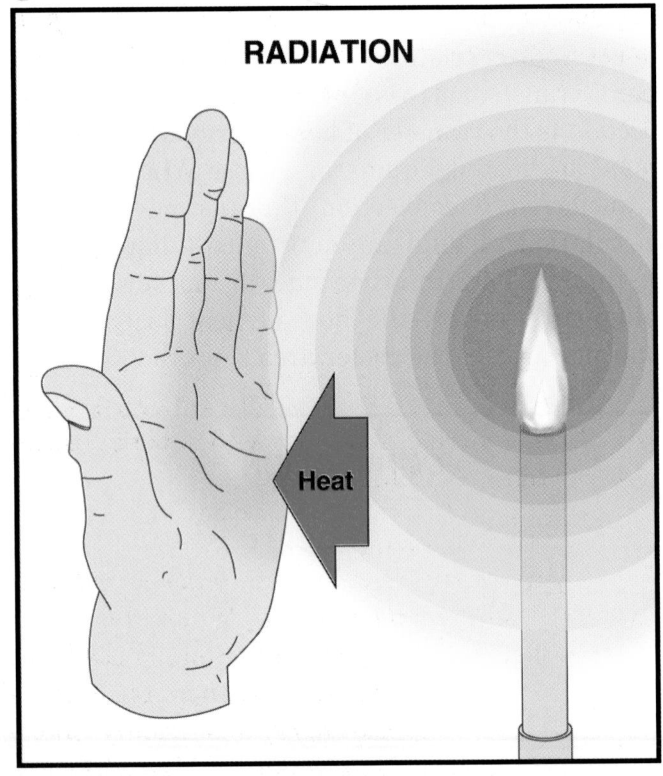

Figure 2.6 Radiation is the transmission of energy as an electromagnetic wave without an intervening medium.

surface. Radiation is the cause of most exposure fires (fires ignited in fuel packages or buildings that are remote from the fuel package or building of origin). As a fire grows, it radiates more and more energy in the form of heat. In large fires, it is possible for the radiated heat to ignite buildings or other fuel packages some distance away (Figure 2.7). Heat energy being transmitted by radiation travels through vacuums and substantial air spaces that would normally disrupt conduction and convection. Materials that reflect radiated energy will disrupt the transmission of heat.

Matter

As you look at the world around you, the physical materials you see are called *matter*. It is said that matter is the "stuff" that makes up our universe. *Matter* is anything that occupies space and has mass. Matter can be described by its physical appearance or more technically by its physical properties such as mass, size, or volume.

Figure 2.7 Radiated heat is one of the major sources of fire spread to exposures.

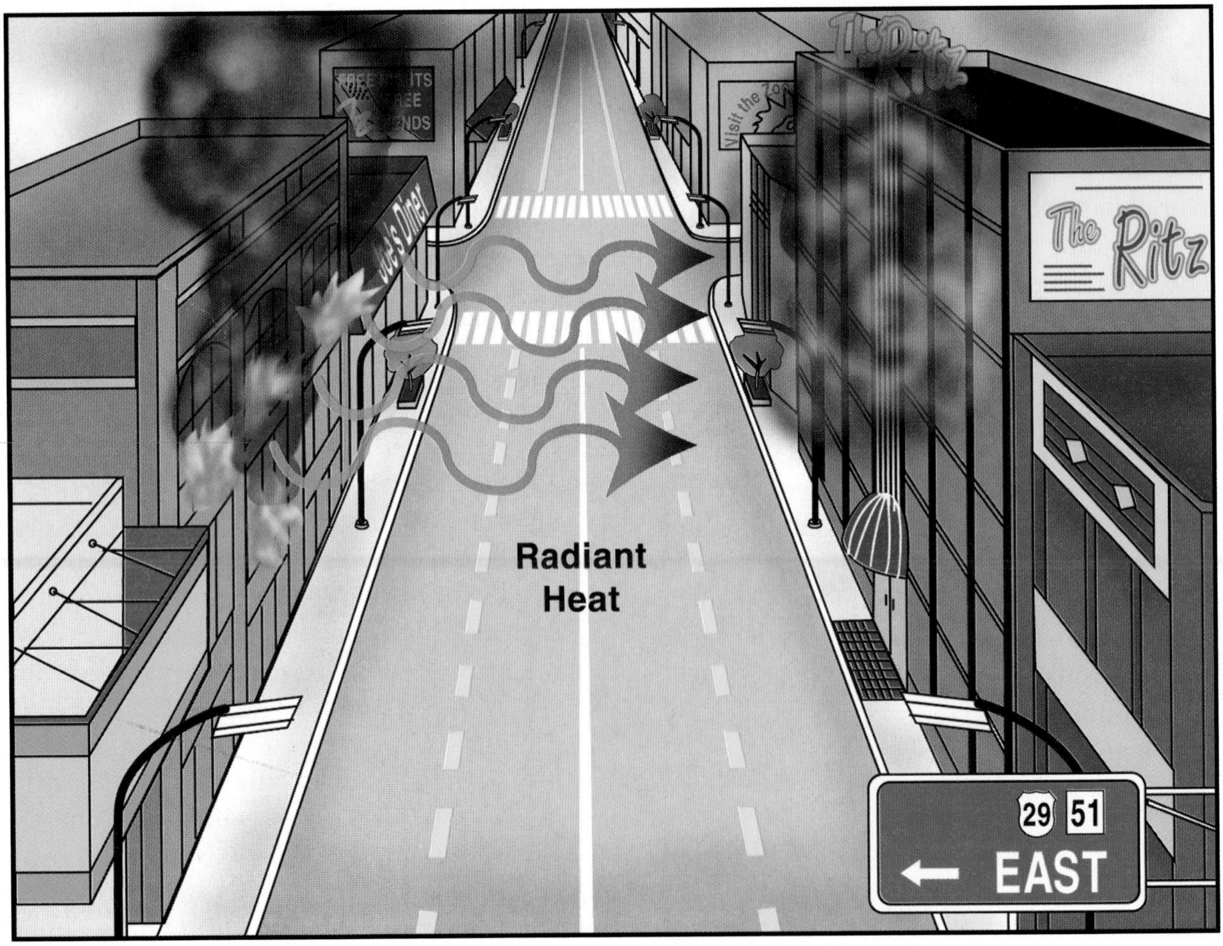

In addition to those properties that can be measured, matter also possesses properties that can be observed such as its physical state (solid, liquid, or gas), color, or smell. One of the best and most common examples of the physical states of matter is water. At normal atmospheric pressure (the pressure exerted by our atmosphere on all objects) and temperatures above 32°F (0°C), water is found as a liquid. At sea level *atmospheric pressure* is defined as 760 mm of mercury measured on a barometer. When the temperature of water falls below 32°F (0°C) and the pressure remains the same, water changes state and becomes a solid called *ice.* At temperatures above its boiling point, water changes state to a gas called *steam.*

Temperature, however, is not the only factor that determines when a change of state will occur. The other factor is pressure. As the pressure on the surface of a substance decreases, so does the temperature at which it boils. The opposite is also true. If the pressure on the surface increases, so will the boiling point. This is the principle used in pressure cookers. The boiling point of the liquid increases as the pressure inside the vessel increases. Thus, foods cook faster in the device because the temperature of the boiling water is greater than 212°F (100°C).

Matter can also be described using terms derived from its physical properties of mass and volume. *Density* is a measure of how tightly the molecules of a solid substance are packed together (Figure 2.8). Density is determined by dividing the mass of a substance by its volume. It is expressed as kg/m^3 in SI and lb/ft^3 in the Customary System.

The common description for liquids is specific gravity. *Specific gravity* is the ratio of the mass of a given volume of a liquid compared with the mass of an equal volume of water. Thus, water has a specific gravity of 1. Liquids with a specific gravity less than 1 are lighter than water, while those with a specific gravity greater than 1 are heavier than water.

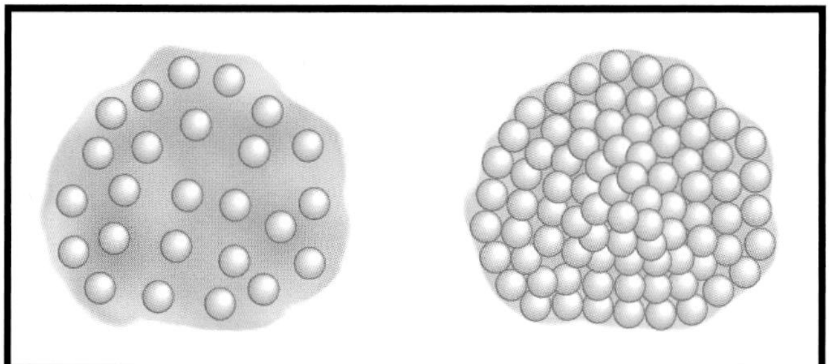

Figure 2.8 The molecules on the right are more dense than those on the left.

The description for gases is vapor density. *Vapor density* is defined as the density of gas or vapor in relation to air. Since air is used for the comparison, it has a vapor density of 1 (as with specific gravity and liquids). Gases with a vapor density of less than 1 will rise, and those with vapor densities greater than 1 will fall.

Conservation of Mass and Energy

As fire consumes a fuel, its mass is reduced. What happens to this material? Where does it go? The answer to these questions is one of the basic concepts of modern physical science: the *Law of Conservation of Mass-Energy* (commonly shortened to the *Law of Conservation of Mass*). The law states: *Mass and energy may be converted from one to another, but there is never any net loss of total mass-energy*. In other words, mass and energy are neither created nor destroyed. The law is fundamental to the science of fire. The reduction in the mass of a fuel results in the release of energy in the form of light and heat. This principle enables researchers to calculate the heat release rate of materials by using instruments that determine mass loss and temperature gain when a fuel is burned.

The firefighter should be aware of this concept during preplanning operations and size-up (initial evaluation of a situation) at fires. The more fuel available to burn, the more potential there is for greater amounts of energy being released as heat during a fire. The more heat released, the more extinguishing agent needed to control a fire.

Chemical Reactions

Before we begin the discussion of combustion and fire growth, it is important to understand the concept of chemical reactions. Whenever matter is transformed from one state to another or a new substance is produced, chemists describe the transformation as a *chemical reaction.* The simplest of these reactions occurs when matter changes state, which is called a *physical change*. In a physical change the chemical makeup of the substance is not altered. The change of state that occurs when water freezes is a physical change.

A more complex reaction occurs when substances are transformed into new substances with different physical and chemical properties. These changes are defined as *chemical changes*. The change that occurs when hydrogen and oxygen are combined to form water is a chemical change. In this case, the chemical and physical properties of the materials being combined are altered. Two materials that are normally gases (hydrogen and oxygen) at room temperature are converted into a substance that is a clear liquid (water) at room temperature.

Chemical and physical changes almost always involve an exchange of energy. Reactions that give off energy as they occur are called *exothermic*. Reactions that absorb energy as they occur are called *endothermic*. When fuels are burned in air, the fuel vapors chemically react with the oxygen in the air, and both heat and light energies are released in an exothermic reaction. Water changing state from liquid to gas (steam) requires the input of energy, thus the conversion is endothermic.

One of the more common chemical reactions on earth is oxidation. *Oxidation* is the formation of a chemical bond between oxygen and another element. Oxygen is one of the more common elements on earth (our atmosphere is composed of 21 percent oxygen) and reacts with almost every other element found on the planet. The oxidation reaction releases energy or is exothermic. The most familiar example of an oxidation reaction is rusting of iron. The combination of oxygen and iron produces a flaky red compound called *iron oxide* or, more commonly, *rust*. Because this is an exothermic process, it always produces heat. Normally, the process is very slow, and the heat dissipates before it is noticed. If the material that is rusting is in a confined space and the heat is not allowed to dissipate, the oxidation process will cause the temperature in the space to increase.

One of the most common examples of heat production in confined spaces is in cargo ships loaded with iron filings. Oxidization of the filings confined within the hold of the ship generates heat that cannot be dissipated because of its location. This heat is conducted to the hull and subsequently to the water outside the ship. When the vessel is in motion, the heat is transferred to the water the ship is moving through and goes unnoticed. When the ship is stationary, however, the fact that heat is being conducted to the surrounding water becomes apparent when the water near the ship begins to boil. While the temperature does not usually increase to the point that flaming ignition (fire) occurs, the condition can be quite dramatic.

COMBUSTION

[NFPA 1001: 3.3.10(a); 4-3.2(b)]

Fire and combustion are terms that are often used interchangeably. Technically, however, fire is a form of combustion. *Combustion is a self-sustaining chemical reaction yielding energy or products that cause further reactions of the same kind.*[1] Combustion is, using the term discussed earlier, an exothermic reaction. *Fire is a rapid, self-sustaining oxidization process accompanied by the evolution of heat and light of varying intensities.*[2] The time it takes a reaction to occur determines the type of reaction that is observed. At the very slow end of the time spectrum is oxidation, where the reaction is too gradual to be observed. At the faster end of the spectrum are explosions that result from the very rapid reaction of a fuel and an oxidizer. These reactions release a large amount of energy over a very short time (Figure 2.9).

Fire Tetrahedron

For many years, the fire triangle (oxygen, fuel, and heat) was used to teach the components of fire. While this simple example is useful, it is not technically correct. For combustion to occur, four components are necessary:

- Oxygen (oxidizing agent)
- Fuel
- Heat
- Self-sustained chemical reaction

These components can be graphically described as the *fire tetrahedron* (Figure 2.10). Each component of the tetrahedron must be in place for com-

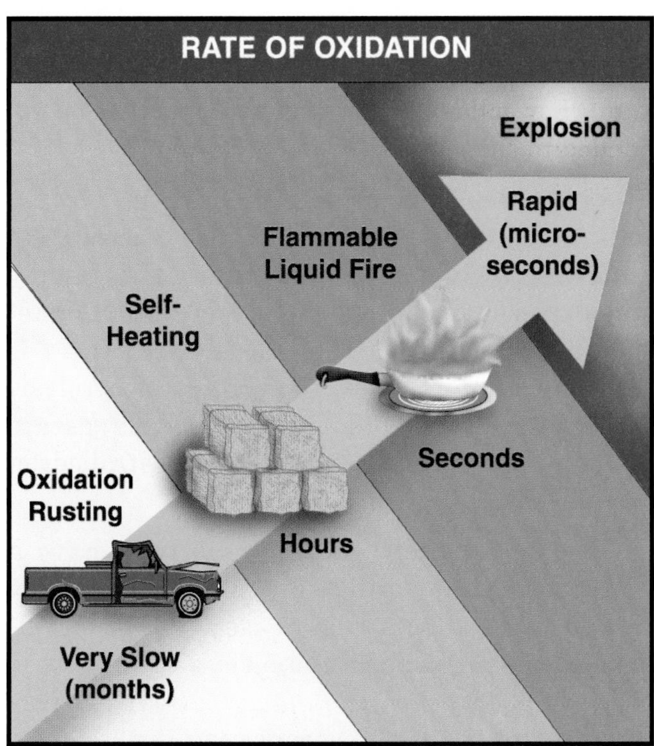

Figure 2.9 Combustion, a self-sustaining chemical reaction, may be very slow or very rapid.

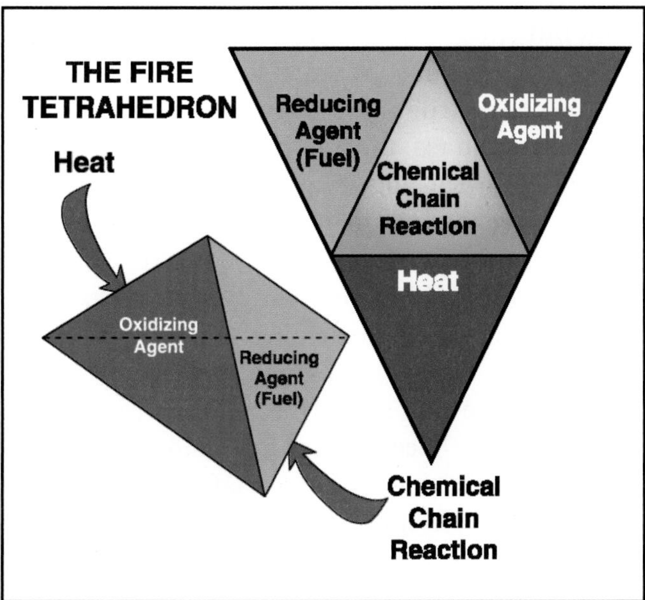

Figure 2.10 The components of the fire tetrahedron.

bustion to occur. This concept is extremely important to students of fire suppression, prevention, and investigation. Remove any one of the four components and combustion will not occur. If ignition has already occurred, the fire is extinguished when one of the components is removed from the reaction.

To better explain fire and its behavior, each of the components of the tetrahedron is discussed in the following sections.

OXYGEN (OXIDIZING AGENT)

Oxidizing agents are those materials that yield oxygen or other oxidizing gases during the course of a chemical reaction. Oxidizers are not themselves combustible, but they support combustion when combined with a fuel. While oxygen is the most common oxidizer, other substances also fall into the category. Table 2.3 lists other common oxidizers.

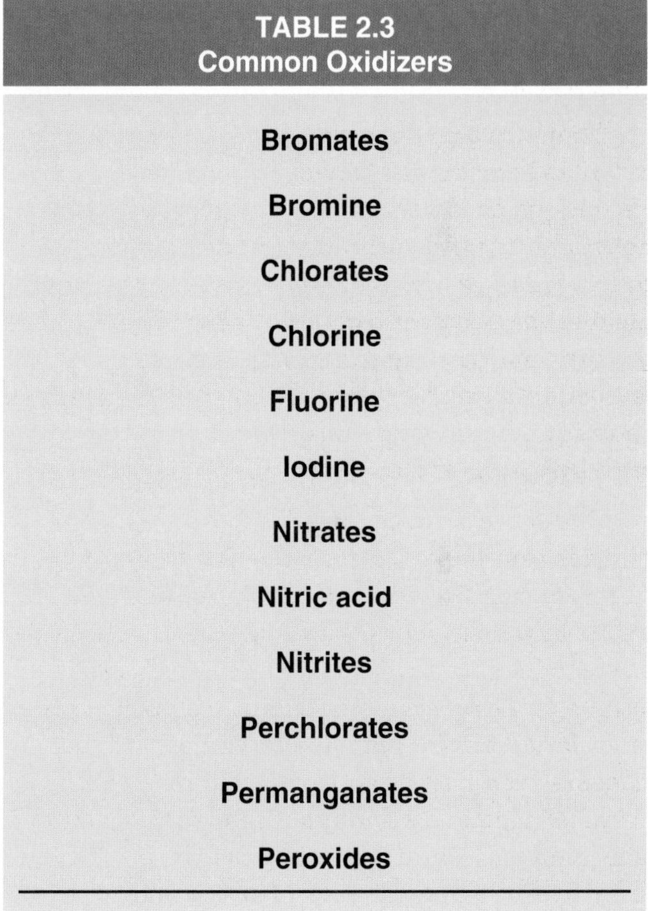

TABLE 2.3 Common Oxidizers
Bromates
Bromine
Chlorates
Chlorine
Fluorine
Iodine
Nitrates
Nitric acid
Nitrites
Perchlorates
Permanganates
Peroxides

For the purposes of this discussion, the oxygen in the air around us is considered the primary oxidizing agent. Normally, air consists of about 21 percent oxygen. At room temperature (70°F or 21°C), combustion is supported at oxygen concentrations as low as 14 percent. Research shows, however, that as temperatures in a compartment fire increase, lower concentrations of oxygen are needed to support flaming combustion. In studies of compartment fires, flaming combustion has been observed at post-flashover temperature conditions (the fully developed and decay stages) when oxygen concentrations have been very low (see the Fire Development section). Some research indicates the concentration can be less than 2 percent.

When oxygen concentrations exceed 21 percent, the atmosphere is said to be *oxygen enriched.* Under these conditions, materials exhibit very different burning characteristics. Materials that burn at normal oxygen levels burn more rapidly in oxygen-enriched atmospheres and may ignite much easier than normal. Some petroleum-based materials will autoignite in oxygen-enriched atmospheres. Many materials that do not burn at normal oxygen levels burn readily in oxygen-enriched atmospheres. One such material is Nomex® fire-resistant material, which is used to construct much of the protective clothing worn by firefighters. At normal oxygen levels, Nomex® does not burn. When placed in an oxygen-enriched atmosphere of approximately 31 percent oxygen, however, Nomex® ignites and burns vigorously. Fires in oxygen-enriched atmospheres are more difficult to extinguish and present a potential safety hazard to firefighters operating in them. These conditions can be found in health care facilities, industrial occupancies, and even private homes where occupants use oxygen breathing equipment.

FUEL

Fuel is the material or substance being oxidized or burned in the combustion process. In scientific terms, the fuel in a combustion reaction is known as the *reducing agent.* Most common fuels contain carbon along with combinations of hydrogen and oxygen. These fuels can be further broken down into hydrocarbon-based fuels (such as gasoline, fuel oil, and plastics) and cellulose-based materials (such as wood and paper). Other fuels that are less complex in their chemical makeup include hydrogen gas and combustible metals such as magnesium and sodium. The combustion process involves two key fuel-related factors: the physical state of the fuel and its distribution. These factors are discussed in the following paragraphs.

From the earlier discussion on matter, it should be understood that a fuel may be found in any of three states of matter: solid, liquid, or gas. To burn, however, fuels must normally be in the gaseous state. For solids and liquids, energy must be expended to cause these state changes.

Fuel gases are evolved from solid fuels by pyrolysis. *Pyrolysis* is the chemical decomposition of a substance through the action of heat. Simply stated, as solid fuels are heated, combustible materials are driven from the substance. If there is sufficient fuel and heat, the process of pyrolysis generates sufficient quantities of burnable gases to ignite if the other elements of the fire tetrahedron are present.

Because of their nature, solid fuels have a definite shape and size. This property significantly affects their ease of ignition. Of primary consideration is the surface-to-mass ratio of the fuel. The *surface-to-mass ratio* is the surface area of the fuel in proportion to the mass. One of the best examples of the surface-to-mass ratio is wood. To produce usable materials, a tree must be cut into a log. The mass of this log is very high, but the surface area is relatively low, thus the surface-to-mass ratio is low. The log is then milled into boards. The result of this process is to reduce the mass of the individual boards as compared to the log, but the resulting surface area is increased, thus increasing the surface-to-mass ratio. The sawdust that is produced as the lumber is milled has an even higher surface-to-mass ratio. If the boards are sanded, the resulting dust has the highest surface-to-mass ratio of any of the examples. As this ratio increases, the fuel particles become smaller (more finely divided — for example, sawdust as opposed to logs), and their ignitability increases tremendously (Figure 2.11). As the surface area increases, more of the material is exposed to the heat and thus generates more burnable gases due to pyrolysis.

A solid fuel's actual position also affects the way it burns. If the solid fuel is in a vertical position, fire spread will be more rapid than if it is in a horizontal position. For example, if you were to ignite a sheet of ⅛-inch plywood paneling that was laying horizontally on two saw horses, the fire would consume the fuel at a relatively slow rate.

Figure 2.11 Materials with a high surface-to-mass ratio require less energy to ignite.

The same type of paneling in the vertical position burns much more rapidly. The rapidity of fire spread is due to increased heat transfer through convection as well as conduction and radiation (Figure 2.12).

For liquids, fuel gases are generated by a process called vaporization. In scientific terms, *vaporization* is the transformation of a liquid to its vapor or gaseous state. The transformation from liquid to vapor or gas occurs as molecules of the substance escape from the liquid's surface into the surrounding atmosphere. In order for the molecules to break free of the liquid's surface, there must be some energy input. In most cases, this energy is provided in the form of heat. For example, water left in a pan eventually evaporates. The energy required for this process comes from the sun or surrounding environment. Water in the same pan placed on a stove and heated to boiling vaporizes much more rapidly because there is more energy being applied to the system. The rate of vaporization is determined by the substance and the amount of heat energy applied to it.

Vaporization of liquid fuels generally requires less energy input than does pyrolysis for solid fuels. This is primarily caused by the different densities of substances in solid and liquid states and by the fact that molecules of a substance in the liquid state have more energy than when they are in the solid state. Solids also absorb more of the energy because of their mass. The volatility or ease with which a liquid gives off vapor influences its ignitability. All liquids give off vapors to a greater or lesser degree in the form of simple evaporation. Liquids that easily give off quantities of flammable or combustible vapors can be dangerous.

Like the surface-to-mass ratio for solid fuels, the surface-to-volume ratio of liquids is an important factor in their ignitability. A liquid assumes the shape of its container. Thus, when a spill or release occurs, the liquid assumes the shape of the ground (flat), flows, and accumulates in low areas. When contained, the specific volume of a liquid has a relatively low surface-to-volume ratio. When it is released, this ratio increases significantly as does the amount of fuel vaporized from the surface.

Gaseous fuels can be the most dangerous of all fuel types because they are already in the natural state required for ignition. No pyrolysis or vaporization is needed to ready the fuel and less energy is required for ignition.

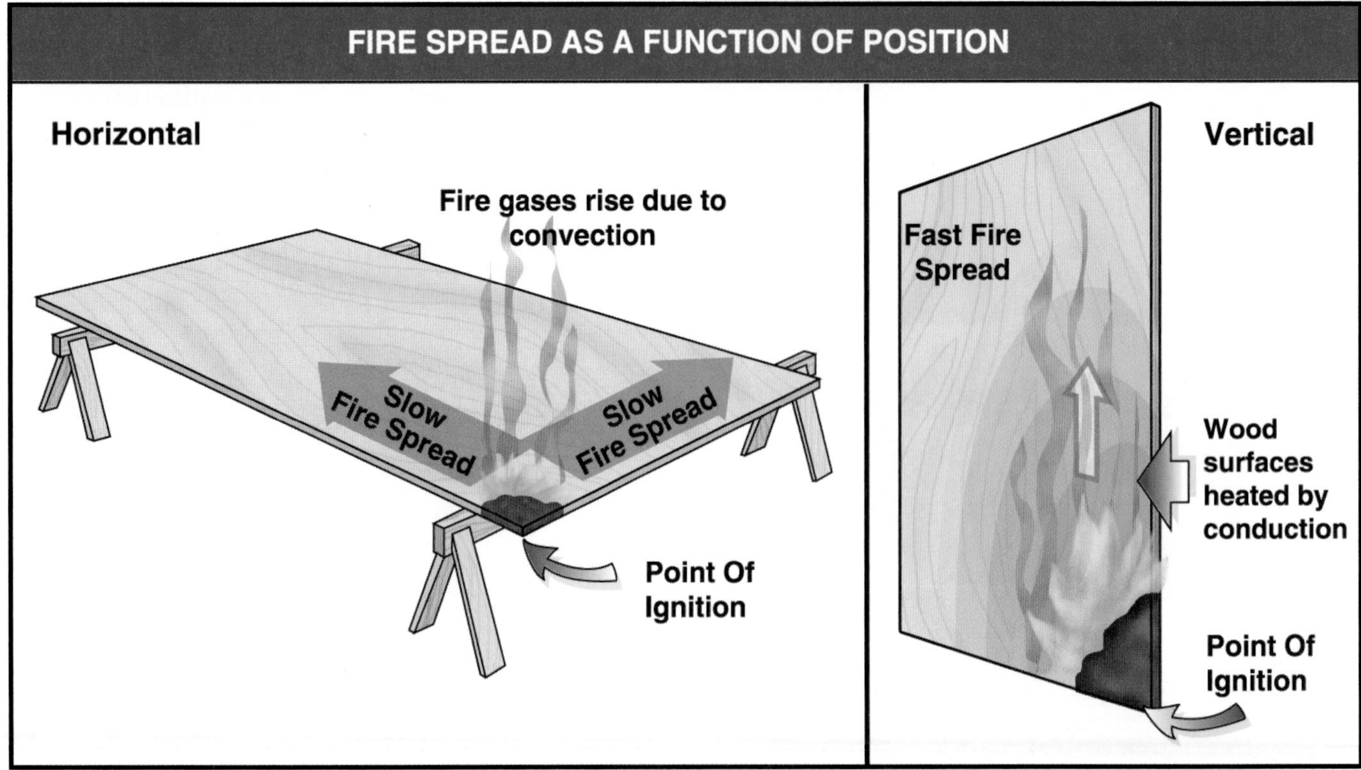

Figure 2.12 The actual position of a solid fuel affects the way it burns.

For combustion to occur after a fuel has been converted into a gaseous state, it must be mixed with air (oxidizer) in the proper ratio. The range of concentrations of the fuel vapor and air (oxidizer) is called the *flammable (explosive) range.* The flammable range of a fuel is reported using the percent by volume of gas or vapor in air for the lower flammable limit (LFL) and for the upper flammable limit (UFL). The *lower flammable limit* is the minimum concentration of fuel vapor and air that supports combustion. Concentrations that are below the LFL are said to be *too lean* to burn. The *upper flammable limit* is the concentration above which combustion cannot take place. Concentrations that are above the UFL are said to be *too rich* to burn.

Table 2.4 presents the flammable ranges for some common materials. The flammable limits for combustible gases are presented in chemical handbooks and documents such as National Fire Protection Association (NFPA) Standard 49, *Hazardous Chemicals Data,* and NFPA 325, *Fire Hazard Properties of Flammable Liquids, Gases, and Volatile Solids.* The limits are normally reported at ambient temperatures and atmospheric pressures. Variations in temperature and pressure can cause the flammable range to vary considerably. Generally, increases in temperature or pressure broaden the range and decreases narrow it.

TABLE 2.4
Flammable Ranges for Selected Materials

Material	Lower Flammable Limit (LFL)	Upper Flammable Limit (UFL)
Acetylene	2.5	100.0
Carbon Monoxide	12.5	74.0
Ethyl Alcohol	3.3	19.0
Fuel Oil No. 1	0.7	5.0
Gasoline	1.4	7.6
Hydrogen	4.0	75.0
Methane	5.0	15.0
Propane	2.1	9.5

Source: NFPA 325, Fire Hazard Properties of Flammable Liquids, Gases, and Volatile Solids, 1994 edition.

It is often helpful to the firefighter to identify groups of fuels or fuel packages in an area or building compartment. In outside areas, fuel packages could be clusters of underbrush or trees growing close together. In an outside storage area, a fuel package might be a tank of fuel oil or flammable liquid or a pile of lumber or building materials (Figure 2.13). In buildings, combustible components (both structural and interior finish) and the contents can be considered fuel packages. A foam-padded upholstered chair or couch in a living room, a mattress and box spring unit in a bedroom, or a computer and office furniture in a business office would be considered fuel packages.

Figure 2.13 The lumber stacked in front of this building as well as the structural components of the building are examples of fuel packages.

The total amount (mass) of fuel in a compartment or specific location multiplied by the heat of combustion of the materials is called the *fuel load* or *fire load.* The term is commonly used to describe the maximum heat that would be released if all of the materials in an area or compartment burned. The concept of fire load was the basis for many of the fire-resistance requirements found in today's building codes. In one study, researchers found the fire loads of typical basement recreation rooms to average about 5.8 pounds per square foot (psf) or 28.3 kg/m^2. Fire loading is normally expressed in terms of the heat of combustion of wood. Materials with different heats of combustion are converted to be equivalent to wood. The available fuel in a space and the proximity of fuel packages to each other have a significant impact on the growth and development of fires. As the amount of available fuel increases, the potential heat release rate of a fire in the compartment increases. If fuel packages are very close together, the amount of energy required for a fire in one package to generate enough heat to ignite the nearby (target) fuel package is less than that for target packages at greater distances.

HEAT

Heat is the energy component of the fire tetrahedron. When heat comes into contact with a fuel, the energy supports the combustion reaction in the following ways (Figures 2.14 a and b):

- Causes the pyrolysis or vaporization of solid and liquid fuels and the production of ignitable vapors or gases
- Provides the energy necessary for ignition
- Causes the continuous production and ignition of fuel vapors or gases so that the combustion reaction can continue

Most of the energy types discussed earlier in the chapter produce heat. For our discussion of fire and its behavior, however, chemical, electrical, and mechanical energy are the most common sources of heat that result in the ignition of a fuel. Each of these sources is discussed in depth in this section.

Chemical. Chemical heat energy is the most common source of heat in combustion reactions. When any combustible is in contact with oxygen, oxidation occurs. This process almost always results in the production of heat. The heat generated when a common match burns is an example of chemical heat energy.

Self-heating (also known as spontaneous heating) is a form of chemical heat energy that occurs when a material increases in temperature without the addition of external heat. Normally the heat is produced slowly by oxidation and is lost to the surroundings almost as fast as it is generated. In order for self-heating to progress to spontaneous ignition, the material must be heated to its ignition temperature (minimum tempera-

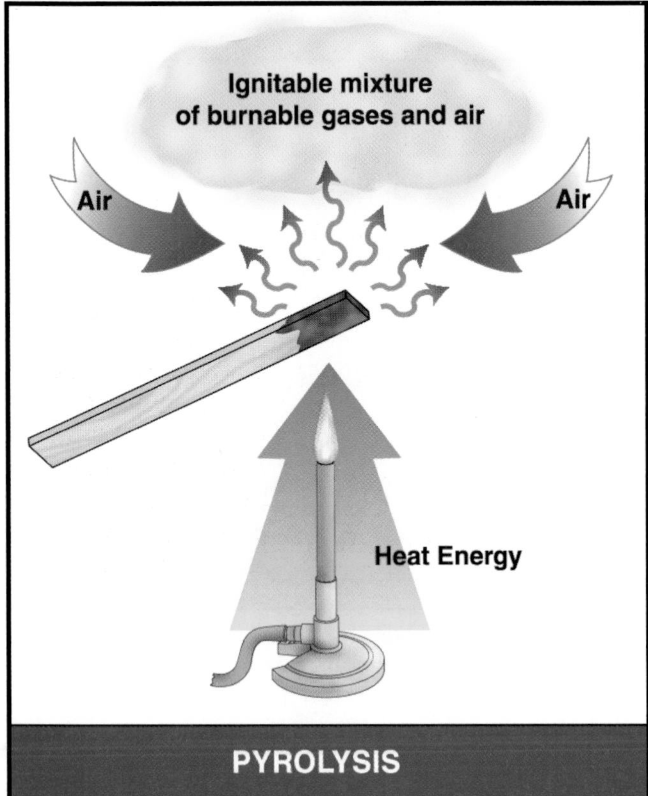

Figure 2.14a Pyrolysis takes place as the wood decomposes from the action of the heat-generating vapors. These vapors then mix with air, producing an ignitable mixture.

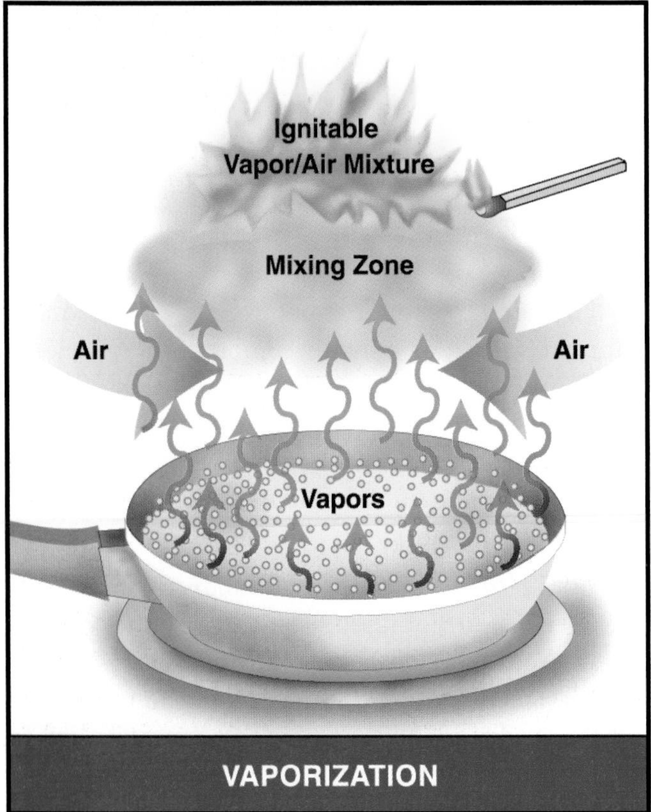

Figure 2.14b Vaporization occurs as fuel gases are generated from the action of heat. These vapors then mix with air, producing an ignitable mixture.

ture at which self-sustained combustion occurs for a specific substance). For spontaneous ignition to occur, the following events must occur:

- The rate of heat production must be great enough to raise the temperature of the material to its ignition temperature.
- The available air supply (ventilation) in and around the material being heated must be adequate to support combustion.
- The insulation properties of the material immediately surrounding the fuel must be such that the heat being generated does not dissipate.

An example of a situation that could lead to spontaneous ignition would be a number of oil-soaked rags that are rolled into a ball and thrown into a corner. If the heat generated by the natural oxidation of the oil and cloth is not allowed to dissipate, either by movement of air around the rags or some other method of heat transfer, the temperature of the cloth will eventually become sufficient to cause ignition.

The rate of the oxidation reaction, and thus the heat production, increases as more heat is generated and held by the materials insulating the fuel. In fact, the rate at which most chemical reactions occur doubles with each 18°F increase in the temperature of the reacting materials. The more heat generated and absorbed by the fuel, the faster the reaction causing the heat generation. When the heat generated by a self-heating reaction exceeds the heat being lost, the material may reach its ignition temperature and ignite spontaneously. Table 2.5 lists some common materials that are subject to self-heating.

Electrical. Electrical heat energy can generate temperatures high enough to ignite any combustible materials near the heated area. Electrical heating can occur in several ways, including the following:

- Current flow through a resistance
- Overcurrent or overload
- Arcing
- Sparking

TABLE 2.5
Materials Subject to Spontaneous Heating

Material	Tendency
Charcoal	High
Fish meal/fish oil	High
Linseed oil rags	High
Brewers grains/feed	Moderate
Fertilizers	Moderate
Foam rubber	Moderate
Hay	Moderate
Manure	Moderate
Iron metal powder	Moderate
Waste paper	Moderate
Rags (bales)	Low to moderate

Source: Fire Protection Handbook, NFPA 18th edition, Table A-10, 1997.

- Static
- Lightning

Mechanical. Mechanical heat energy is generated by friction and compression. *Heat of friction* is created by the movement of two surfaces against each other. This movement results in heat and/or sparks being generated. *Heat of compression* is generated when a gas is compressed. Diesel engines use this principle to ignite fuel vapor without a spark plug. The principle is also the reason that self-contained breathing apparatus (SCBA) cylinders feel warm to the touch after they have been filled (Figure 2.15).

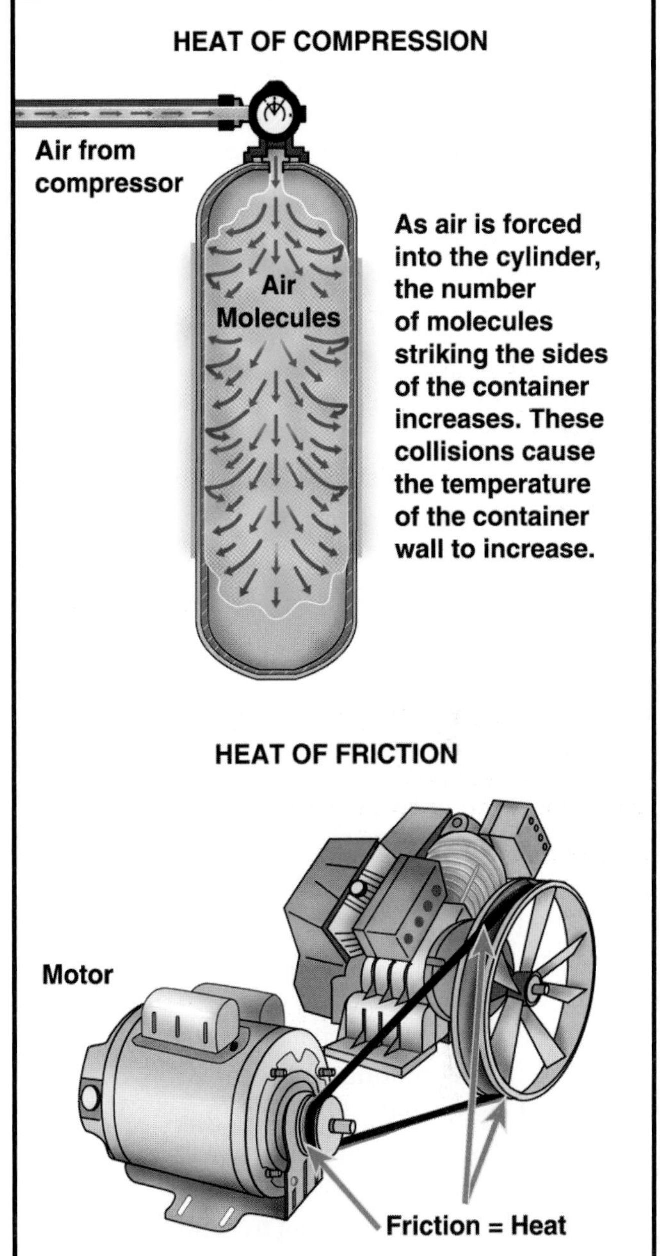

Figure 2.15 Examples of mechanical heat energy.

Nuclear. Nuclear heat energy is generated when atoms either split apart (fission) or combine (fusion). In a controlled setting, fission heats water to drive steam turbines and produce electricity. Fusion reactions cannot be contained at this time and have no commercial use. The sun's heat (solar energy) is a product of a fusion reaction and thus is a form of nuclear energy.

SELF-SUSTAINED CHEMICAL REACTION

Combustion is a complex reaction that requires a fuel (in the gaseous or vapor state), an oxidizer, and heat energy to come together in a very specific way. Once flaming combustion or fire occurs, it can only continue when enough heat energy is produced to cause the continued development of fuel vapors or gases. Scientists call this type of reaction a chain reaction. A *chain reaction* is a series of reactions that occur in sequence with the results of each individual reaction being added to the rest. An excellent illustration of a chain reaction is given by Faughn, Chang, and Turk in their textbook *Physical Science:*

> *An example of a chemical chain reaction is a forest fire. The heat from one tree may initiate*

the reaction (burning) of a second tree, which, in turn ignites a third, and so on. The fire will then go on at a steady rate. But if one burning tree ignites, say, two others, and each of these two ignite two more, for a total of four, and so on, the rate of burning speeds rapidly. Such uncontrolled, runaway chain reactions are at the heart of nuclear bombs.[3]

The self-sustained chemical reaction and the related rapid growth are the factors that separate fire from slower oxidation reactions. Slow oxidation reactions do not produce heat fast enough to reach ignition, and they never generate sufficient heat to become self-sustained. Examples of slow oxidation are the rusting of iron (mentioned earlier) and the yellowing of paper.

Fire Development

When the four components of the fire tetrahedron come together, ignition occurs. For a fire to grow beyond the first material ignited, heat must be transmitted beyond the first material to additional fuel packages. In the early development of a fire, heat rises and forms a plume of hot gas. If a fire is in the open (outside or in a large building), the fire plume rises unobstructed, and air is drawn (entrained) into it as it rises. Because the air being pulled into the plume is cooler than the fire gases, this action has a cooling effect on the gases above the fire. The spread of fire in an open area is primarily due to heat energy that is transmitted from the plume to nearby fuels. Fire spread in outside fires can be increased by wind and sloping terrain that allow exposed fuels to be preheated (Figure 2.16).

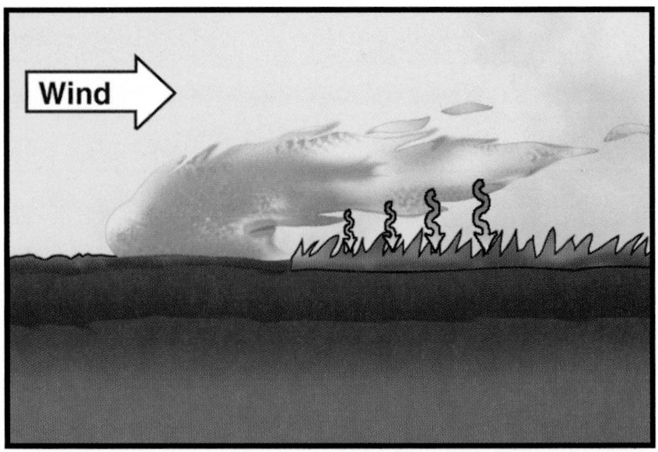

Figure 2.16 Outdoor fire spread is affected by wind and terrain.

The development of fires in a compartment is more complex than those in the open. For the purposes of this discussion, a *compartment* is an enclosed room or space within a building. The term *compartment fire* is defined as a fire that occurs within such a space. The growth and development of a compartment fire is usually controlled by the availability of fuel and oxygen. When the amount of fuel available to burn is limited, the fire is said to be *fuel controlled.* When the amount of available oxygen is limited, the condition is said to be *ventilation controlled.*

Recently, researchers have attempted to describe compartment fires in terms of stages or phases that occur as the fire develops. These stages are as follows:

- Ignition
- Growth
- Flashover
- Fully developed
- Decay

Figure 2.17 shows the development of a compartment fire in terms of time and temperature. It should be noted that the stages are an attempt to describe the complex reaction that occurs as a fire develops in a space with no suppression action taken. The ignition and development of a compartment fire is very complex and influenced by many variables. As a result, all fires may not develop through each of the stages described. The information is presented to depict fire as a dynamic event that is dependent on many factors for its growth and development.

IGNITION

Ignition describes the period when the four elements of the fire tetrahedron come together and combustion begins. The physical act of ignition can be *piloted* (caused by a spark or flame) or *nonpiloted* (caused when a material reaches its ignition temperature as the result of self-heating) such as spontaneous ignition. At this point, the fire is small and generally confined to the material (fuel) first ignited. All fires — in an open area or within a compartment — occur as a result of some type of ignition.

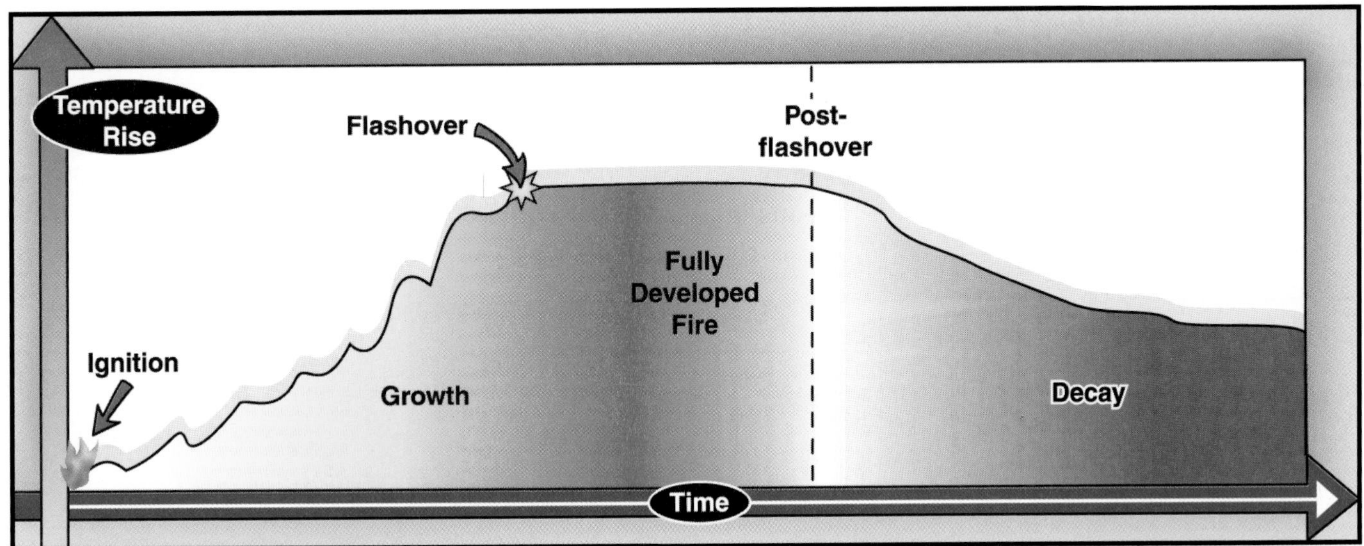

Figure 2.17 Stages of fire development in a compartment.

GROWTH

Shortly after ignition, a fire plume begins to form above the burning fuel. As the plume develops, it begins to draw or entrain air from the surrounding space into the column. The initial growth is similar to that of an outside unconfined fire, with the growth a function of the fuel first ignited. Unlike an unconfined fire, however, the plume in a compartment is rapidly affected by the ceiling and walls of the space. The first impact is the amount of air that is entrained into the plume. Because the air is cooler than the hot gases generated by the fire, the air has a cooling effect on the *temperatures within the plume.* The location of the fuel package in relation to the compartment walls determines the amount of air that is entrained and thus the amount of cooling that takes place. Fuel packages near walls entrain less air and thus have higher plume temperatures. Fuel packages in corners entrain even less air and have the highest plume temperatures. This factor significantly affects the temperatures in the developing hot-gas layer above the fire. As the hot gases rise, they begin to spread outward when they hit the ceiling. The gases continue to spread until they reach the walls of the compartment. The depth of the gas layer then begins to increase.

The temperatures in the compartment during this period depend on the amount of heat conducted into the compartment ceiling and walls as the gases flow over them and on the location of the initial fuel package and the resulting air entrainment. Research shows that the gas temperatures decrease as the distance from the centerline of the plume increases. Figure 2.18 shows the plume in a typical compartment fire and the factors that affect the temperature of the developing hot-gas layer.

The growth stage will continue if enough fuel and oxygen are available. Compartment fires in the growth stage are generally fuel controlled. As the fire grows, the overall temperature in the compartment increases (see Figure 2.17, stages of fire) as does the temperature of the gas layer at the ceiling level (Figure 2.19).

FLASHOVER

Flashover is the transition between the growth and the fully developed fire stages and is not a specific event such as ignition. During flashover, conditions in the compartment change very rapidly as the fire changes from one that is dominated by the burning of the materials first ignited to one that involves all of the *exposed combustible surfaces* within the compartment. The hot-gas layer that develops at the ceiling level during the growth stage causes radiant heating of combustible materials remote from the origin of the fire (Figure 2.20). Typically, radiant energy (heat flux) from the hot-gas layer exceeds 20 kW/m^2 when flashover occurs. This radiant heating causes pyrolysis in the combustible materials in the compartment.

The gases generated during this time are heated to their ignition temperature by the radiant energy from the gas layer at the ceiling (Figure 2.21).

While scientists define flashover in many ways, most base their definition on the temperature in a compartment that results in the simultaneous ignition of all of the combustible contents in the space. While no exact temperature is associated with this occurrence, a range from approximately 900°F to 1,200°F (483°C to 649°C) is widely used. This range correlates with the ignition temperature of carbon monoxide (CO) (1,128°F or 609°C), one of the most common gases given off from pyrolysis.

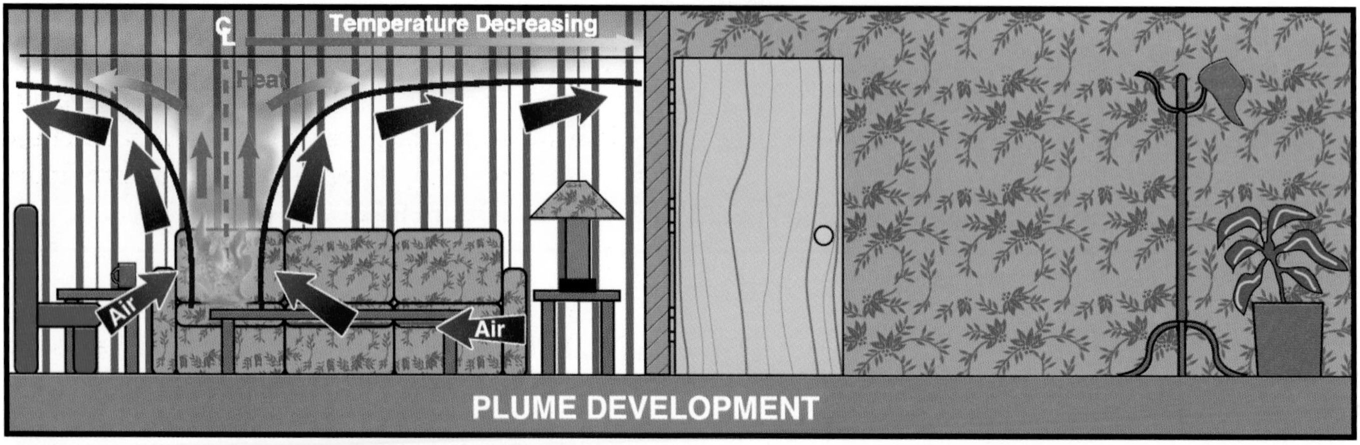

Figure 2.18 Initially, the temperature of the fire gases decreases as they move away from the centerline of the plume.

Figure 2.19 As the fire grows, the overall temperature in the compartment increases as does the temperature of the gas layer at the ceiling level.

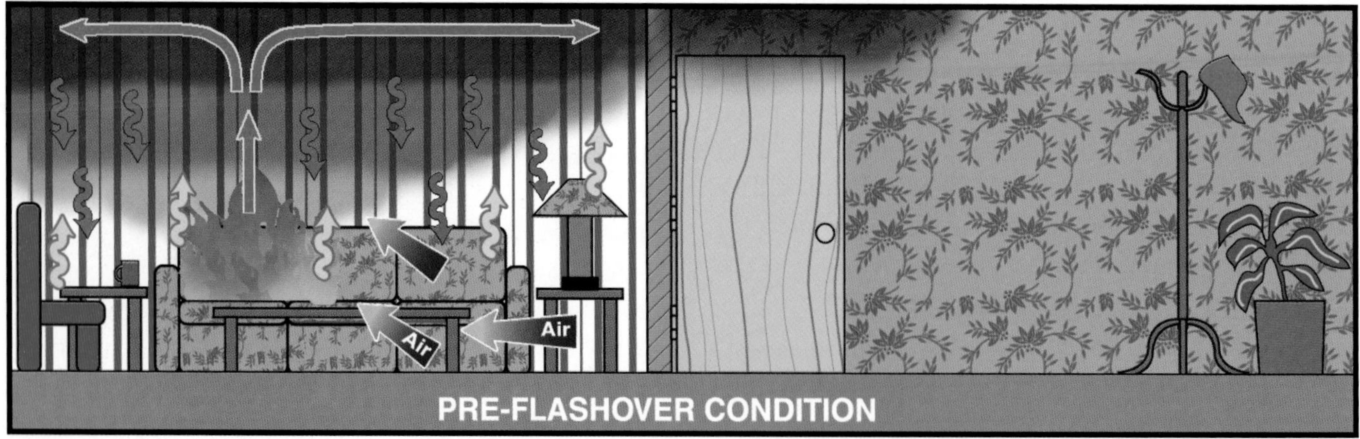

Figure 2.20 The radiant heat (red arrows) from the hot-gas layer at the ceiling heats combustible materials, which produces vapors (green arrows).

Just prior to flashover, several things are happening within the burning compartment: The temperatures are rapidly increasing, additional fuel packages are becoming involved, and the fuel packages in the compartment are giving off combustible gases as a result of pyrolysis. As flashover occurs, the combustible materials in the compartment and the gases given off from pyrolysis ignite. The result is full-room involvement. The heat release from a fully developed room at flashover can be on the order of 10,000 kW or more.

Occupants who have not escaped from a compartment before flashover occurs are not likely to survive. Firefighters who find themselves in a compartment at flashover are at extreme risk even while wearing their personal protective equipment.

FULLY DEVELOPED

The fully developed fire stage occurs when all combustible materials in the compartment are involved in fire. During this period of time, the burning fuels in the compartment are releasing the maximum amount of heat possible for the available fuel packages and producing large volumes of fire gases. The heat released and the volume of fire gases produced depend on the number and size of the ventilation openings in the compartment. The fire frequently becomes ventilation controlled, and thus large volumes of unburned gases are produced. During this stage, hot unburned fire gases are likely to begin flowing from the compartment of origin into adjacent spaces or compartments. These gases ignite as they enter a space where air is more abundant (Figure 2.22).

DECAY

As the fire consumes the available fuel in the compartment, the rate of heat release begins to decline. Once again the fire becomes fuel controlled, the amount of fire diminishes, and the temperatures within the compartment begin to

Figure 2.21 An example of flashover.

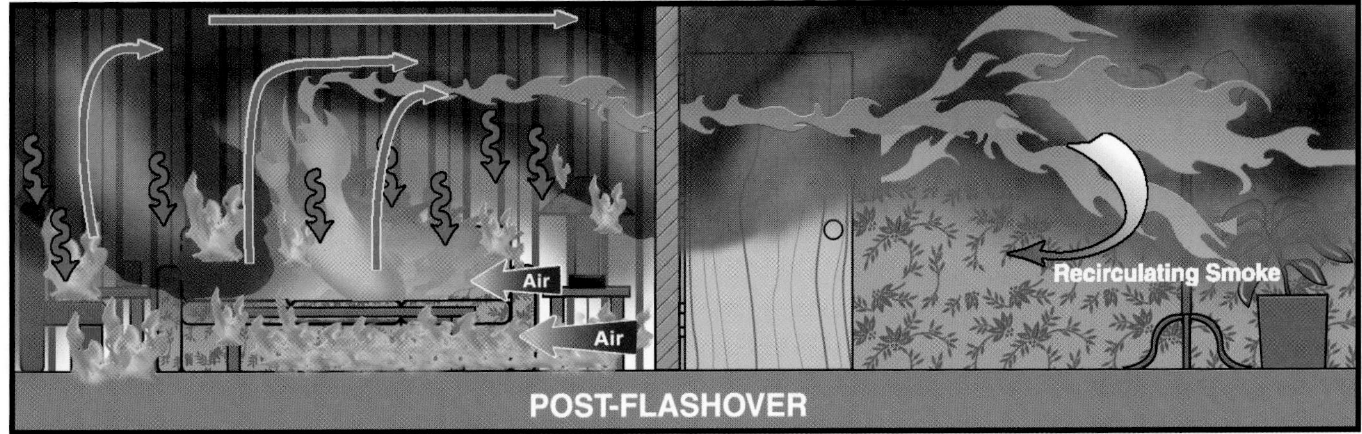

Figure 2.22 A fully developed fire.

decline. The remaining mass of glowing embers can, however, result in moderately high temperatures in the compartment for some time.

FACTORS THAT AFFECT FIRE DEVELOPMENT

As the fire progresses from ignition to decay, several factors affect its behavior and development within the compartment:

- Size, number, and arrangement of ventilation openings
- Volume of the compartment
- Thermal properties of the compartment enclosures
- Ceiling height of the compartment
- Size, composition, and location of the fuel package that is first ignited
- Availability and locations of additional fuel packages (target fuels)

For a fire to develop, enough air to support burning beyond the ignition stage must be available. The size and number of ventilation openings in a compartment determine how the fire develops within the space. The compartment's size and shape and ceiling height determine if a significant hot-gas layer will form. The location of the initial fuel package is also very important in the development of the hot-gas layer. The plumes of burning fuel packages in the center of a compartment entrain more air and are cooler than those against walls or in corners of the compartment.

The temperatures that develop in a burning compartment are the direct result of the energy released as the fuels burn. Because matter and energy are conserved, any loss in mass caused by the fire is converted to energy. In a fire, the resulting energy is in the form of heat and light. The amount of heat energy released over time in a fire is called the *heat release rate* (HRR). HRR is measured in Btu/s or kilowatts (kW). The heat release rate is directly related to the amount of fuel being consumed over time and the heat of combustion (the amount of heat a specific mass of a substance gives off when burned) of the fuel being burned. See Table 2.6 for maximum heat release rates for several common items. This information gives representative numbers for typical fuel items.

TABLE 2.6
Heat Release Rates for Common Materials

Material	Maximum Heat Release Rate	
	kW	Btu/s
Wastebasket (0.53 kg) with milk cartons (0.40 kg)	15	14.2
Upholstered chair (cotton padded) (31.9 kg)	370	350.7
Four stacking chairs (metal frame, polyurethane foam padding) (7.5 kg each)	160	151.7
Upholstered chair (polyurethane foam) (28.3 kg)	2,100	1,990.0
Mattress (cotton and jute) (25 kg)	40	37.9
Mattress (polyurethane foam) (14 kg)	2,630	2,492.9
Mattress and box springs (cotton and polyurethane foam) (62.4 kg)	660	626.0
Upholstered sofa (polyurethane foam) (51.5 kg)	3,200	3,033.0
Gasoline/kerosene (2 sq ft pool)	400	379.0
Christmas tree (dry) (7.4 kg)	500	474.0

Source: NFPA 921, Guide for Fire and Explosion Investigations, Table 3-4; NBSIR 85-3223 Data Sources for Parameters Used in Predictive Modeling of Fire Growth and Smoke Spread; and NBS Monograph 173, Fire Behavior of Upholstered Furniture.

Firefighters should be able to recognize potential fuel packages in a building or compartment and use this information to estimate the fire growth potential for the building or space. Materials with high heat release rates such as polyurethane foam-padded furniture, polyurethane foam mattresses, or stacks of wooden pallets, for example, would be expected to burn rapidly once ignition occurs. Fires in materials with lower heat release rates would be expected to take longer to develop. In general, low-density materials (such as polyurethane foam) burn faster (have a higher HRR) than higher density materials (cotton padding) of similar makeup.

One final relationship between the heat generated in a fire and fuel packages is the ignition of additional fuel packages that are remote from the first package ignited. The heat generated in a compartment fire is transmitted from the initial fuel package to other fuels in the space by all three modes of heat transfer. The heat rising in the initial fire plume is transported by convection. As the hot gases travel over surfaces of other fuels in the compartment, heat is transferred to them by conduction. Radiation plays a significant role in the transition from a growing fire to a fully developed fire in a room. As the hot-gas layer forms at the ceiling, hot particles in the smoke begin to radiate energy to the other fuel packages in the compartment. These remote fuel packages are sometimes called *target fuels.* As the radiant energy increases, the target fuels begin pyrolysis and start to give off ignitable gases. When the temperature in the compartment reaches the ignition temperature of these gases, the entire room becomes involved in fire (flashover).

SPECIAL CONSIDERATIONS

[NFPA 1001:3-3.10(a); 3-3.11(a)]

Several conditions or situations that occur during a fire's growth and development should be discussed. This section provides an overview of these conditions and the potential safety concerns for each.

Flameover/Rollover

The terms *flameover* and *rollover* describe a condition where flames move through or across the unburned gases during a fire's progression. Flameover is distinguished from flashover by its involvement of only the fire gases and not the surfaces of other fuel packages within a compartment. This condition may occur during the growth stage as the hot-gas layer forms at the ceiling of the compartment. Flames may be observed in the layer when the combustible gases reach their ignition temperature. While the flames add to the total heat generated in the compartment, this condition is not flashover. Flameover may also be observed when unburned fire gases vent from a compartment during the growth and fully developed stages of a fire's development. As these hot gases vent from the burning compartment into the adjacent space, they mix with oxygen; if they are at their ignition temperature, flames often become visible in the layer (Figure 2.23).

Thermal Layering of Gases

The *thermal layering* of gases is the tendency of gases to form into layers according to temperature. Other terms sometimes used to describe this tendency are *heat stratification* and *thermal balance*. The hottest gases tend to be in the top layer, while the cooler gases form the lower layers (Figure 2.24). Smoke, a heated mixture of air, gases, and particles, rises. If a hole is made in a roof, smoke will rise from the building or room to the outside. Thermal layering is critical to fire fighting activities. As long as the hottest air and gases are allowed to rise, the lower levels will be safer for firefighters.

This normal layering of the hottest gases to the top and out the ventilation opening can be dis-

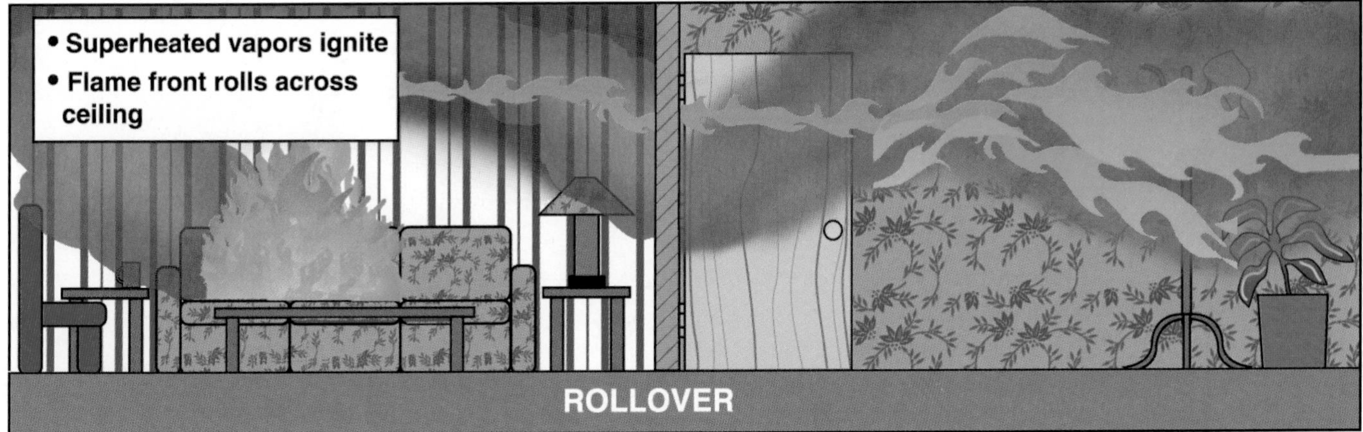

Figure 2.23 An example of rollover.

rupted if water is applied directly into the layer. When water is applied to the upper level of the layer, where the temperatures are highest, the rapid conversion to steam can cause the gases to mix rapidly. This swirling mixture of smoke and steam disrupts normal thermal layering, and hot gases mix throughout the compartment (Figure 2.25). This process is sometimes referred to as *disrupting the thermal balance* or *creating a thermal imbalance.* Many firefighters have been burned when thermal layering was disrupted. Once the normal layering is disrupted, forced ventilation procedures (such as using fans) must be used to clear the area. The proper procedure under these conditions is to ventilate the compartment, allow the hot gases to escape, and direct the fire stream at the base of the fire, keeping it out of the hot upper layers of gases.

Backdraft

Firefighters operating at fires in buildings must use care when opening a building to gain entry or to provide horizontal ventilation (opening doors or windows). As the fire grows in a compartment, large volumes of hot, unburned fire gases can collect in unventilated spaces. These gases may be at or above their ignition temperature but have insufficient oxygen available to actually ignite. Any action during fire fighting operations that

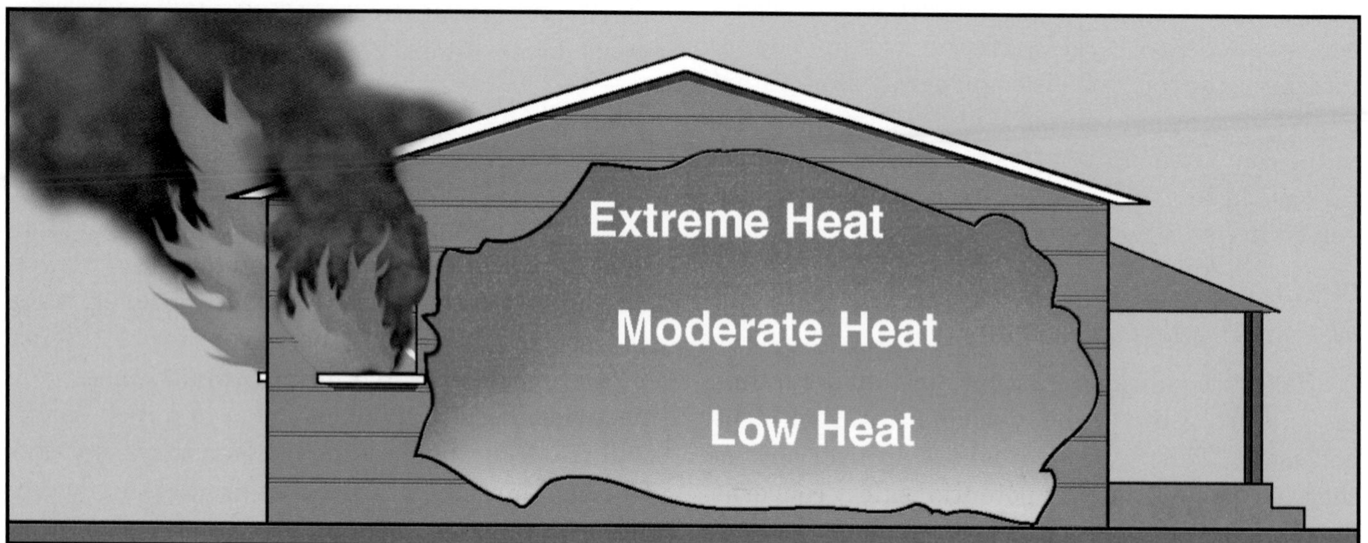

Figure 2.24 Under normal fire conditions in a closed structure, the highest levels of heat will be found at ceiling level, and the lowest level of heat will be found at floor level.

Figure 2.25 Applying water to the upper level of the thermal layer creates a thermal imbalance.

allows air to mix with these hot gases can result in an explosive ignition called *backdraft* (Figure 2.26). Many firefighters have been killed or injured as a result of backdrafts. The potential for backdraft can be reduced with proper vertical ventilation (opening at highest point) because the unburned gases rise. Opening the building or space at the highest possible point allows them to escape before entry is made.

The following conditions may indicate the potential for a backdraft:

- Pressurized smoke exiting small openings
- Black smoke becoming dense gray yellow
- Confinement and excessive heat
- Little or no visible flame
- Smoke leaving the building in puffs or at intervals (appearance of breathing)
- Smoke-stained windows

Products of Combustion

As a fuel burns, the chemical composition of the material changes. This change results in the production of new substances and the generation of energy (Figure 2.27). As a fuel is burned, some of it is actually consumed. The Law of Conservation of Mass tells us that any mass lost converts to energy. In the case of fire, this energy is in the form of light and heat. Burning also results in the generation of airborne fire gases, particles, and liquids. These materials have been referred to throughout this chapter as products of combustion or smoke. The heat generated during a fire is one of the products of combustion. In addition to being responsible for the spread of a fire, heat also causes burns, dehydration, heat exhaustion, and injury to a person's respiratory tract.

While the heat energy from a fire is a danger to anyone directly exposed to it, smoke causes most deaths in fires. The materials that make up smoke

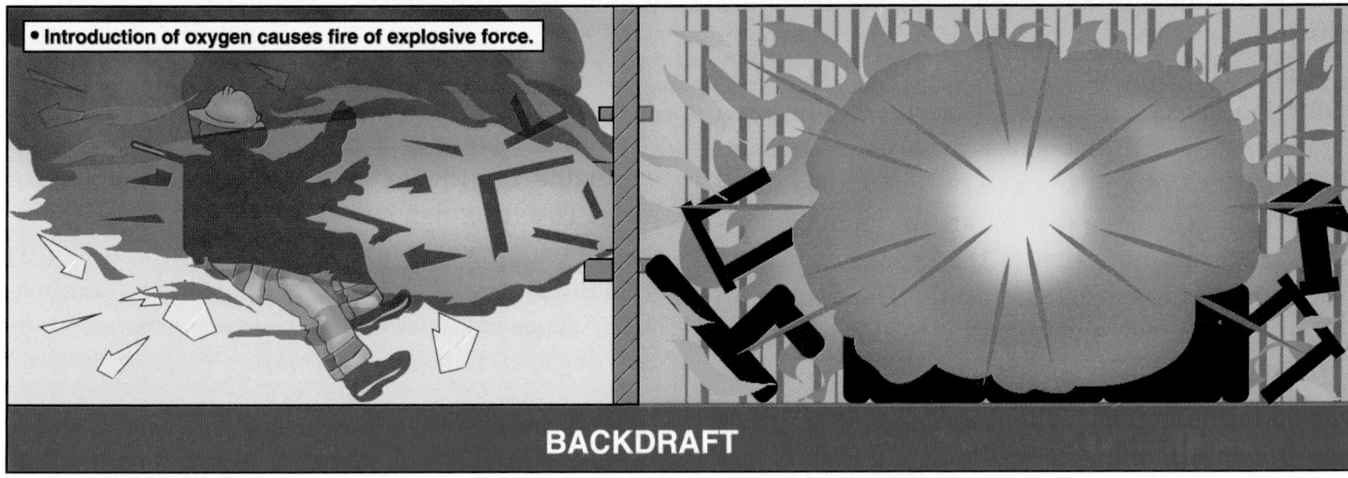

Figure 2.26 Improper ventilation during fire fighting operations may result in a backdraft.

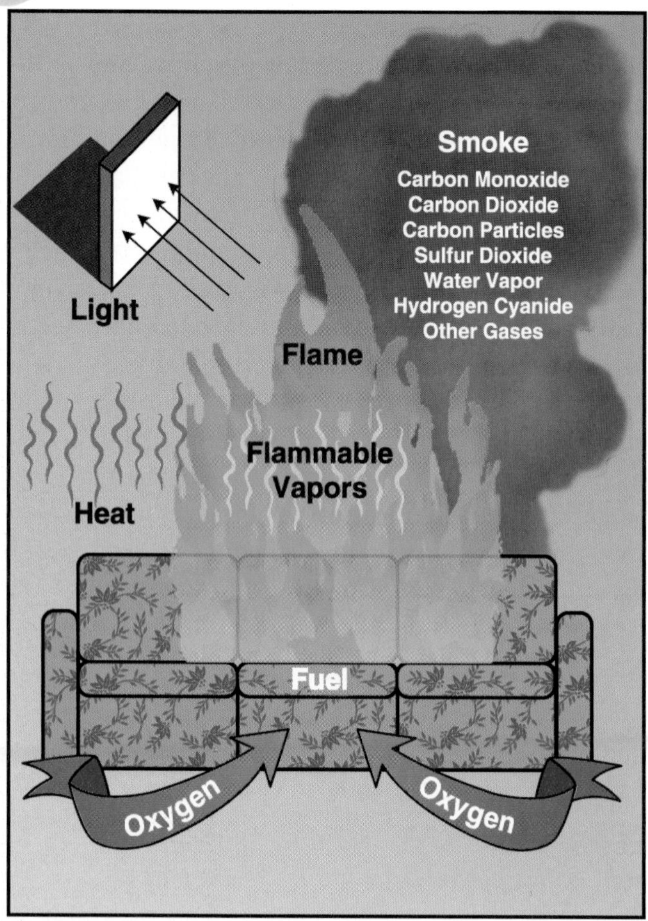

Figure 2.27 The four products of combustion are heat, light, smoke, and fire gases.

vary from fuel to fuel, but generally all smoke can be considered toxic. The smoke generated in a fire contains narcotic (asphyxiant) gases and irritants. Narcotic or asphyxiant gases are those products of combustion that cause central nervous system depression, which results in reduced awareness, intoxication, and can lead to loss of consciousness and death. The most common narcotic gases found in smoke are carbon monoxide (CO), hydrogen cyanide (HCN), and carbon dioxide (CO_2). The reduction in oxygen levels as a result of a fire in a compartment will also cause a narcotic effect in humans. Irritants in smoke are those substances that cause breathing discomfort (pulmonary irritants) and inflammation of the eyes, respiratory tract, and skin (sensory irritants). Depending on the fuels involved, smoke will contain numerous substances that can be considered irritants.

The most common of the hazardous substances contained in smoke is carbon monoxide. While CO is not the most dangerous of the materials found in smoke, it is almost always present when combustion occurs. While someone may be killed or injured by breathing a variety of toxic substances in smoke, carbon monoxide is the one that is most easily detected in the blood of fire victims and thus most often reported. Because the substances in smoke from compartment fires are deadly (either alone or in combination), firefighters must use SCBA for protection when operating in smoke.

Flame is the visible, luminous body of a burning gas. When a burning gas is mixed with the proper amounts of oxygen, the flame becomes hotter and less luminous. The loss of luminosity is caused by a more complete combustion of the carbon. For these reasons, flame is considered to be a product of combustion. Of course, it is not present in those types of combustion that do not produce a flame such as smoldering fires.

FIRE EXTINGUISHMENT THEORY

[NFPA 1001: 3-3.10(a)]

Fire is extinguished by limiting or interrupting one or more of the essential elements in the combustion process (fire tetrahedron). A fire may be extinguished by reducing its temperature, eliminating available fuel or oxygen, or stopping the self-sustained chemical chain reaction (Figure 2.28).

Temperature Reduction

One of the most common methods of extinguishment is cooling with water. This process depends on reducing the temperature of a fuel to a point where it does not produce sufficient vapor to burn. Solid fuels and liquid fuels with high flash points can be extinguished by cooling. However, cooling with water cannot sufficiently reduce vapor production to extinguish fires involving low flash point liquids and flammable gases. The use of water for cooling is also the most effective method available for the extinguishment of smoldering fires. To extinguish a fire by reducing its temperature, enough water must be applied to the burning fuel to absorb the heat being generated by combustion. Types of streams and extinguishing methods are discussed later in the manual.

Fuel Removal

Removing the fuel source effectively extinguishes some fires. The fuel source may be removed

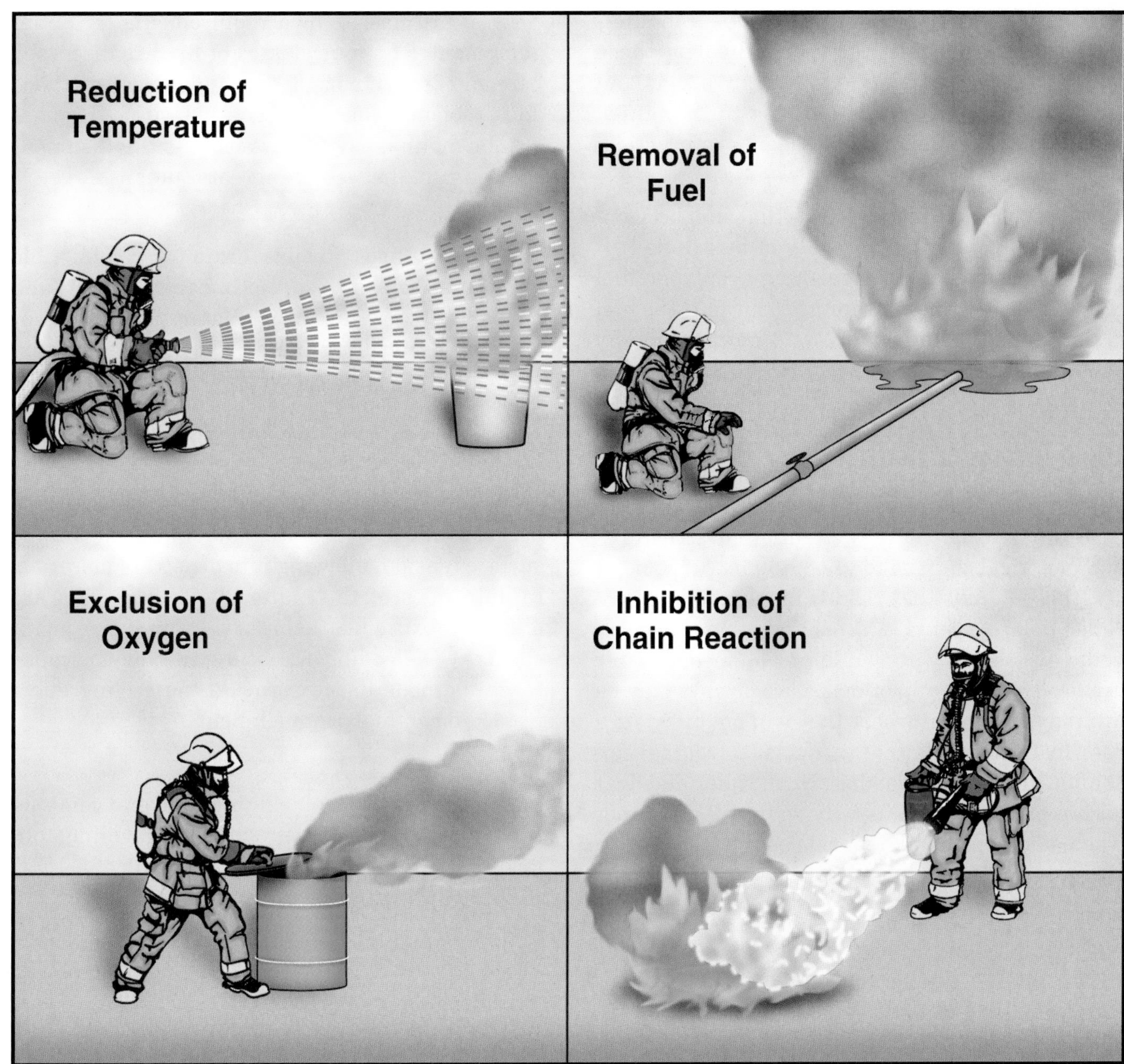

Figure 2.28 Four methods of fire extinguishment.

by stopping the flow of liquid or gaseous fuel or by removing solid fuel in the path of a fire. Another method of fuel removal is to allow a fire to burn until all fuel is consumed.

Oxygen Exclusion

Reducing the oxygen available to the combustion process reduces a fire's growth and may totally extinguish it over time. In its simplest form, this method is used to extinguish cooking stove fires when a cover is placed over a pan of burning food. The oxygen content can be reduced by flooding an area with an inert gas such as carbon dioxide, which displaces the oxygen and disrupts the combustion process. Oxygen can also be separated from fuel by blanketing the fuel with foam. Of course, neither of these methods works on those rare fuels that are self-oxidizing.

Chemical Flame Inhibition

Extinguishing agents such as some dry chemicals and halogenated agents (halons) interrupt the combustion reaction and stop flaming. This method of extinguishment is effective on gas and liquid fuels because they must flame to burn. Smoldering fires are not easily extinguished by these agents. The very high agent concentrations and extended periods of time necessary to extinguish smoldering fires make these agents impractical in these cases.

Most ignitable liquids (those that support combustion) have a specific gravity of less than 1. If water is used as an extinguishing agent, the fuel can float on it while continuing to burn. If the fuel is unconfined, using water could unintentionally spread a fire.

The solubility (ability of a substance to mix with water) of a liquid fuel in water is also an important factor in extinguishment. Liquids of similar molecular structure tend to be soluble in each other while those with different structures and electrical charges tend not to mix. In chemistry, those liquids that readily mix with water are called *polar solvents.* Alcohol and other polar solvents dissolve in water. If large volumes of water are used, alcohol and other polar solvents may be diluted to the point where they will not burn. As a rule, hydrocarbon liquids (nonpolar solvents — not soluble in water) do not dissolve in water and float on top of water. This is why water alone cannot wash oil off our hands; the oil does not dissolve in the water. Soap must be added to water to dissolve the oil.

Vapor density also affects extinguishment of both ignitable liquids and gaseous fuels. Gases that are less dense than air (vapor densities of less than 1) tend to rise and dissipate when released. Gases or vapors with vapor densities greater than 1 tend to hug the ground and travel as directed by terrain and wind. Common hydrocarbon gases such as ethane and propane have vapor densities greater than air and tend to collect near the surface when released. Natural gas (methane) is an example of a hydrocarbon gas with a vapor density less than air. When released, methane tends to rise and dissipate.

CLASSIFICATION OF FIRES

[NFPA 1001: 3-3.15(a)]

The classification of a fire is important to the firefighter when discussing extinguishment. Each class of fire has its own requirements for extinguishment. The four classes of fire are discussed here, along with normal extinguishment methods and problems. These classes will be used throughout the manual when the various extinguishment methods are discussed in greater detail.

Class A Fires

Class A fires involve ordinary combustible materials such as wood, cloth, paper, rubber, and many plastics (Figure 2.29). Water is used to cool or quench the burning material below its ignition temperature. The addition of Class A foams (sometimes referred to as *wet water*) may enhance water's ability to extinguish Class A fires, particularly those that are deep seated in bulk materials (such as piles of hay bales, sawdust piles, etc.). This is because the Class A foam agent reduces the water's surface tension, allowing it to penetrate more easily into piles of the material. Class A fires are difficult to extinguish using oxygen-exclusion methods like CO_2 flooding or coating with foam because those methods do not provide the cooling effect needed for total extinguishment.

Class B Fires

Class B fires involve flammable and combustible liquids and gases such as gasoline, oil, lacquer, paint, mineral spirits, and alcohol (Figure 2.30). The smothering or blanketing effect of oxygen exclusion is most effective for extinguishment and also helps reduce the production of additional vapors. Other extinguishing methods include removal of fuel, temperature reduction when possible, and the interruption of the chain reaction with dry chemical agents such as Purple K®.

Class C Fires

Fires involving energized electrical equipment are Class C fires (Figure 2.31). Household appliances, computers, transformers, and overhead transmission lines are examples. These fires can sometimes be controlled by a nonconducting extinguishing agent such as halon, dry chemical, or carbon dioxide. The fastest extinguishment procedure is to first de-energize high-voltage circuits and then fight the fire appropriately depending upon the fuel involved.

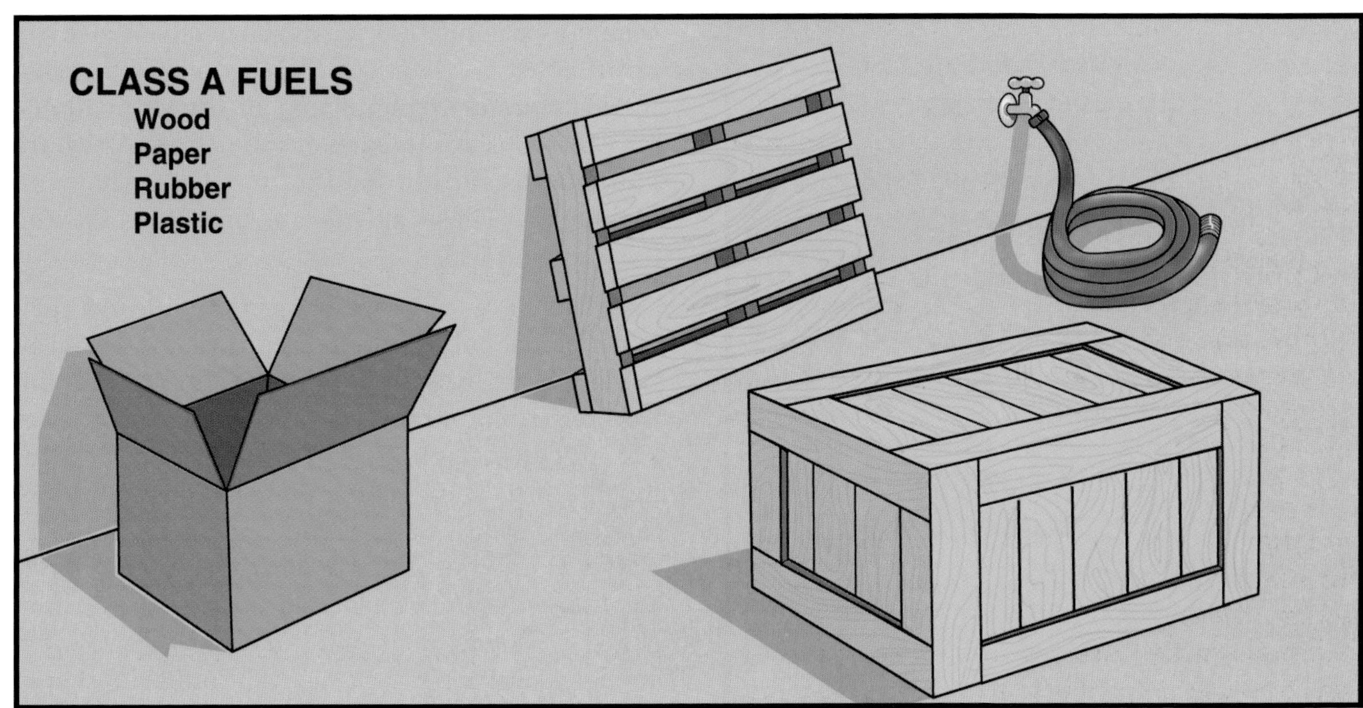

Figure 2.29 Examples of Class A fuels.

Figure 2.30 Examples of Class B fuels.

Figure 2.31 Examples of Class C fuels.

Class D Fires

Class D fires involve combustible metals such as aluminum, magnesium, titanium, zirconium, sodium, and potassium (Figure 2.32). These materials are particularly hazardous in their powdered form. Proper airborne concentrations of metal dusts can cause powerful explosions, given a suitable ignition source. The extremely high temperature of some burning metals makes water and other

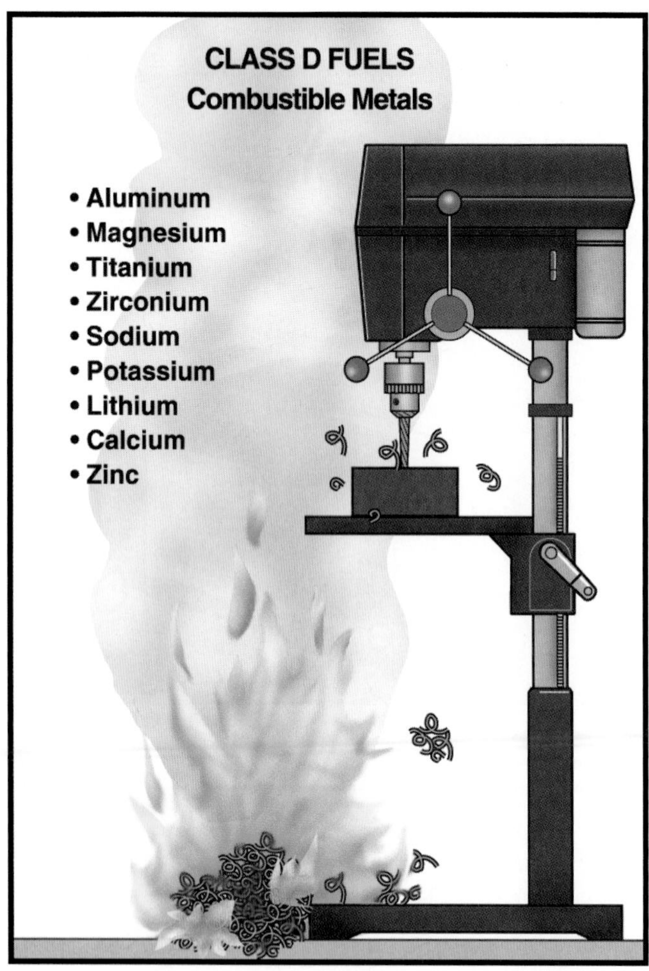

Figure 2.32 Examples of Class D fuels.

common extinguishing agents ineffective. No single agent effectively controls fires in all combustible metals. Special extinguishing agents are available for control of fire in each of the metals. They are marked specifically for the metal fire they can extinguish. These agents are used to cover the burning material.

Firefighters may find these materials in a variety of industrial or storage facilities. It is essential to use caution in a Class D materials fire. Information regarding a material and its characteristics should be reviewed prior to attempting to extinguish a fire. The burning material should be isolated and treated as recommended in its Material Safety Data Sheet (MSDS) or in the *North American Emergency Response Guidebook* (NAERG) from the U.S. Department of Transportation. All personnel operating in the area of the material should be in full protective equipment, and those exposed should be limited to only the people necessary to contain or extinguish the fire.

ENDNOTES

[1] Richard L. Tuve, *Principles of Fire Protection Chemistry,* Boston: National Fire Protection Association, 1976, p. 125.

[2] Ibid.

[3] Jerry S. Faughn, Raymond Chang, and Jon Turk, *Physical Science,* Second Edition, Orlando: Harcourt Brace, 1995, p. 317.

OTHER REFERENCES

Flammable Ranges, NFPA 325, *Fire Hazard Properties of Flammable Liquids, Gases, and Volatile Solids,* 1994.

Chapter 3, "Basic Fire Science," NFPA 921, *Guide for Fire and Explosion Investigations,* 1995.

Chapter 3
Building Construction

Chapter 3
Building Construction

INTRODUCTION

From a safety standpoint, all firefighters should have a basic knowledge of the principles of building construction. Knowledge of the various types of building construction and how fires react in each type give the firefighter and fire officer an edge in planning for a safe and effective fire attack. History has shown that failure to recognize the potential dangers presented by a particular type of construction and the effects a fire has on it can lead to deadly results.

New technologies and designs are being used for building construction every day. Therefore, it is impossible to highlight every conceivable situation in this chapter. The purpose of this chapter is to introduce the firefighter to some of the most basic and common types of building construction and their fire protection characteristics. The chapter will also introduce the firefighter to common building construction terms and components. This chapter also discusses some of the indicators that signify danger during fire fighting operations.

TYPES OF BUILDING CONSTRUCTION

[NFPA 1001: 3-3.11(a)]

Each of the model building codes classifies building construction in different terms. In general, construction classifications are based on the type of materials used in the construction and on the fire-resistance rating requirements of certain structural components. Most building codes have the same five construction classifications, but use different terms to name the classifications. The five types of building construction include:

- Type I (fire-resistive) construction
- Type II (noncombustible or limited combustible) construction
- Type III (ordinary) construction
- Type IV (heavy timber) construction
- Type V (wood-frame) construction

Type I (Fire-Resistive) Construction

Fire resistance provides structural integrity during a fire. Fire-resistive construction has structural members, including walls, columns, beams, floors, and roofs, made of noncombustible or limited combustible materials (Figure 3.1). The fire-resistive compartmentation provided by partitions and floors tends to retard the spread of fire through the building. These features allow time for occupant evacuation and interior fire fighting. Because of the limited combustibility of the materials of construction, the primary fire hazards are the contents of the structure. In a

Figure 3.1 Type I Construction.

fire-resistive structure, firefighters are able to launch an interior attack with greater confidence than in a building that is not fire resistant. The ability of fire-resistive construction to confine the fire to a certain area can be compromised by openings made in partitions and by improperly designed and dampered heating and air-conditioning systems.

Type II (Noncombustible or Limited Combustible) Construction

Noncombustible or limited combustible construction is similar to fire-resistive construction except that the degree of fire resistance is lower. Noncombustible construction has a fire-resistance rating on all parts of the structure (exterior and interior load-bearing walls and building materials). Materials with no fire-resistance ratings, such as untreated wood, may be used in limited quantities (Figure 3.2). Again, one of the primary fire protection concerns is the contents of the building. The heat buildup from a fire in the building can cause structural supports to fail. Another potential problem is the type of roof on the building. Noncombustible or limited combustible construction buildings often have flat, built-up roofs. These roofs contain combustible felt, insulation, and roofing tar (Figure 3.3). Fire extension to the roof can eventually cause the entire roof to become involved and fail.

Type III (Ordinary) Construction

Ordinary construction features exterior walls and structural members constructed of noncombustible or limited combustible materials. Interior structural members, including walls, columns, beams, floors, and roofs, are completely or partially constructed of wood (Figure 3.4). The wood used in these members is of smaller dimensions than that required for heavy timber construction. See the Type IV (Heavy Timber) Construction section that follows. The primary fire concern specific to ordinary construction is the problem of fire and smoke spreading through concealed spaces. These spaces are between the walls, floors, and ceiling. Heat from a fire may be conducted to these concealed spaces through

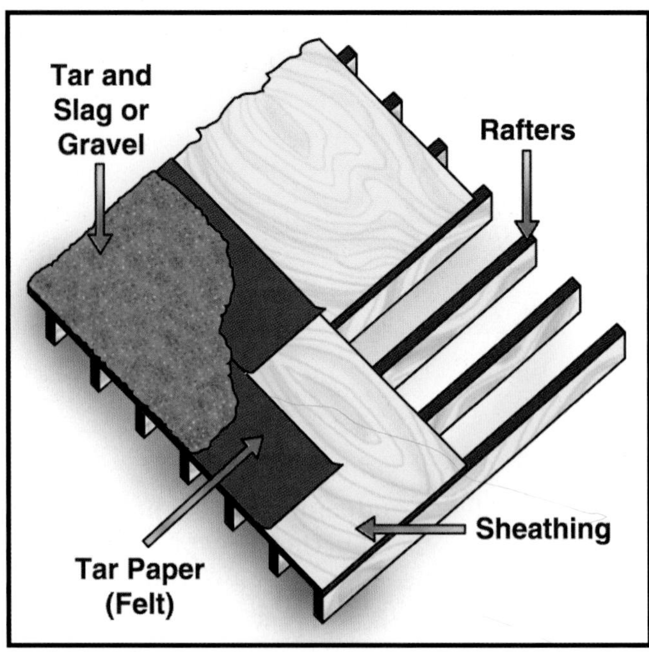

Figure 3.3 The components of a built-up roof.

Figure 3.2 Type II Construction.

Figure 3.4 Type III Construction.

finish materials, such as drywall, gypsum board, or plaster, or the heat can enter the concealed spaces through holes in the finish materials. From there, the heat, smoke, and gases may spread to other parts of the structure. If enough heat is present, the fire may actually burn within the concealed spaces and feed on the combustible construction materials in the space. These hazards can be reduced considerably by placing firestops inside these spaces to limit the spread of the combustion by-products (heat, smoke, etc.).

Type IV (Heavy Timber) Construction

Heavy timber construction features exterior and interior walls and their associated structural members made of noncombustible or limited combustible materials. Other interior structural members, including beams, columns, arches, floors, and roofs, are made of solid or laminated wood with no concealed spaces (Figure 3.5). This wood must have dimensions large enough to be considered heavy timber. These dimensions vary depending on the particular code being used.

Figure 3.5 Type IV Construction.

Heavy timber construction was used extensively in old factories, mills, and warehouses. It is rarely used today in new construction, other than occasionally in churches. The primary fire hazard associated with heavy timber construction is the massive amount of combustible contents presented by the structural timbers in addition to the contents of the building. Though the heavy timbers remain stable for a long period under fire conditions, they give off tremendous amounts of heat and pose serious exposure protection problems for firefighters.

Type V (Wood-Frame) Construction

Wood-frame construction has exterior walls, bearing walls, floors, roofs, and supports made completely or partially of wood or other approved materials of smaller dimensions than those used for heavy timber construction. Wood-frame construction is the type commonly used to construct the typical single-family residence. This type of construction presents almost unlimited potential for fire extension within the building of origin and to nearby structures, particularly if the nearby structures are also wood-frame construction (Figure 3.6). Firefighters must be alert for fire coming from doors or windows extending to the exterior of the structure.

Figure 3.6 Type V Construction.

EFFECTS OF FIRE ON COMMON BUILDING MATERIALS

[NFPA 1001: 3-3.11(a); 4-3.2(a)(b)]

All materials react differently when exposed to heat or fire. Knowledge of how these materials react gives fire suppression personnel an idea of what to expect during fire fighting operations at a particular occupancy. This part of the chapter reviews the common materials found in building construction and explains how they react to fire involvement.

Wood

Wood is used in various structural support systems. It may be used in *load-bearing walls* (those that support structural weight) or *nonload-bearing walls* (those that do not support structural weight). Most exterior walls are load-bearing walls. A *party wall* that supports two adjacent structures is a load-bearing wall (Figure 3.7). A *partition wall* that simply divides two areas within a structure is an example of a nonload-bearing wall (Figure 3.8). Some interior walls may also be load bearing, although this is often difficult to tell by just looking at them. This information should be obtained during pre-incident planning trips.

The reaction of wood to fire conditions depends mainly on two factors: the size of the wood and the moisture content of the wood. The smaller the wood size, the more likely it is to lose structural integrity. Large pieces of wood, such as those used in heavy timber construction, retain much of their original structural integrity, even after extensive fire exposure. Smaller pieces of wood can be protected by drywall or gypsum to increase their resistance to heat or fire.

Figure 3.7 A diagram of a party wall.

Figure 3.8 Walls that separate offices or rooms are commonly nonload-bearing walls.

The moisture content of the wood affects the rate at which it burns. Wood with a high moisture content (sometimes referred to as *green* wood) does not burn as fast as wood that has been cured or dried. In some cases, fire retardants may be added to wood to reduce the speed at which it ignites or burns. However, fire retardants are not always totally effective in reducing fire spread.

Water used during extinguishing operations does not have a substantial negative effect on the structural strength of wood construction materials. Applying water to burning wood minimizes damage by stopping the charring process, which reduces wood's strength. Firefighters should check wood studs and structural members for charring to ascertain their structural integrity.

Newer construction often contains composite building components and materials that are made of wood fibers, plastics, and other substances joined by glue or resin binders. Such materials include plywood, particleboard, fiberboard, and paneling. Some of these products may be highly combustible, can produce significant toxic gases, or can rapidly deteriorate under fire conditions.

Masonry

Masonry includes bricks, stones, and concrete masonry products (Figure 3.9). Masonry in a variety of wall types is commonly used for *fire wall assemblies,* which consist of all the components needed to provide a separating fire wall that meets the requirements of a specified fire-resistance rating. The components include the wall structure, doors, windows, and any other opening protection meeting the required protection-rating criteria. Fire wall assemblies may be used to separate two connected structures and prevent the spread of fire from one structure to the next. Fire wall assemblies can also divide large structures into smaller portions and contain a fire to that particular portion of the structure. *Cantilever walls* are free-standing fire walls commonly found on large churches and shopping centers (Figure 3.10).

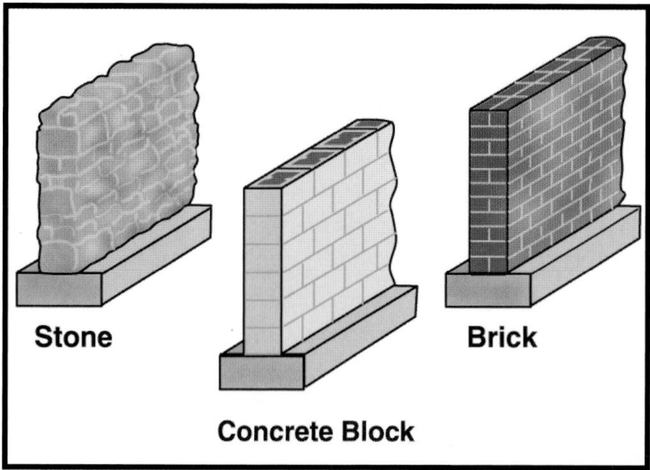

Figure 3.9 The main types of masonry construction are stone, concrete masonry products, and brick.

Figure 3.10 A cantilever wall.

Block walls may be load-bearing walls; however, most brick and stone walls are *veneer walls,* which are decorative and usually attached to the outside of some type of load-bearing frame structure.

Masonry is minimally affected by fire and exposure to high temperatures. Bricks rarely show any signs of loss of integrity or serious deterioration. Stones may spall or lose small portions of their surface when heated. Blocks may crack, but they usually retain most of their strength and basic structural stability. The mortar between the bricks, blocks, and stone may be subject to more deterioration and should be checked for signs of weakening (Figure 3.11).

Figure 3.11 Mortar is used between bricks, blocks, or stone.

Rapid cooling, which can occur when water is used to extinguish fire, may cause bricks, blocks, or stone to spall and crack. This is a common problem when water is used to extinguish chimney flue fires. The water causes the flue liner or firebricks to crack. Masonry products should be inspected for signs of this damage after extinguishment has been completed.

Cast Iron

Cast iron is rarely used in modern construction; it typically is found only in older buildings. It was commonly used as an exterior surface covering (veneer wall). These large sections were fastened to the masonry on the front of the building. The cast iron stands up well to fire and intense heat situations, but it may crack or shatter when rapidly cooled with water. A primary concern from a fire fighting standpoint is that the bolts or other connections that hold the cast iron to the building can fail, causing these large, heavy sections of metal to come crashing down.

Steel

Steel is the primary material used for structural support in modern building construction (Figure 3.12). Steel structural members elongate when heated. A 50-foot (15 m) beam may elongate by as much as 4 inches (100 mm) when heated from room temperature to about 1,000°F (538°C). If the steel is restrained from movement at the ends, it buckles and fails somewhere in the middle. For all purposes, the failure of steel structural members can be anticipated at temperatures near or above 1,000°F (538°C). The temperature at which a specific steel member fails depends on many variables, including the size of the member, the load it is under, the composition of the steel, and the geometry of the member. For example, a lightweight, open-web truss will fail much quicker than a large, heavy I beam.

Figure 3.12 A typical steel superstructure.

From a fire fighting perspective, firefighters must be aware of the type of steel members used in a particular structure. Firefighters also need to determine how long the steel members have been exposed to heat; this gives an indication of when the members might fail. Another possibility for firefighters to consider is that elongating steel can actually push out load-bearing walls and cause a collapse (Figure 3.13). Water can cool steel structural members and reduce the risk of failure, which reduces the risk of a structural collapse.

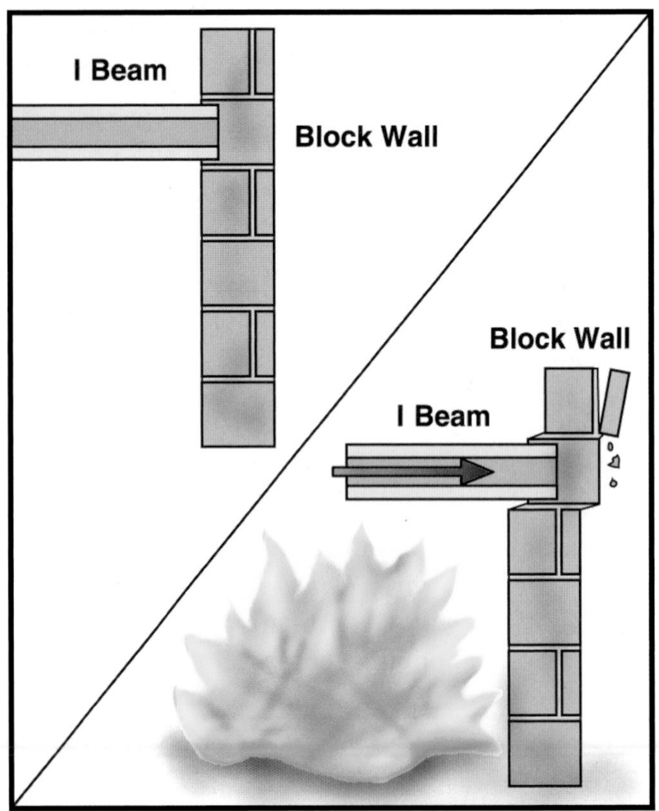

Figure 3.13 As beams expand, they can push a wall outward, forcing a collapse.

Reinforced Concrete

Reinforced concrete is concrete that is internally fortified with steel reinforcement bars or mesh (Figure 3.14). This gives the material the compressive strength of concrete along with the tensile strength of steel. Reinforced concrete does not perform particularly well under fire conditions; it loses strength and spalls. Heating may cause a failure of the bond between the concrete and the steel reinforcement. Firefighters should look for cracks and

Figure 3.14 Reinforcing rods increase structural integrity.

spalls in reinforced concrete surfaces. This is an indication that damage has occurred and that strength may be reduced.

Gypsum

Gypsum is an inorganic product from which plaster and plasterboards are constructed (Figure 3.15). It is unique because it has a high water content, and the evaporation of this water requires a great deal of heat. The water content gives gypsum excellent heat-resistant, fire-retardant properties. Gypsum is commonly used to provide insulation to steel and wood structural members that are less adapted to high heat situations because it breaks down gradually under fire conditions. In areas where the gypsum has failed, the structural members behind it will be subjected to higher temperatures and could fail as a result.

Figure 3.15 Gypsum board is used to cover interior walls.

Glass/Fiberglass

Glass is not typically used for structural support; it is used in sheet form for doors and windows (Figure 3.16). Wire-reinforced glass may provide some thermal protection as a separation, but for the most part conventional glass is not an effective barrier to fire extension. Heated glass may crack and shatter when it is struck by a cold fire stream.

Fiberglass is typically used for insulation purposes. The glass component of fiberglass is not a significant fuel, but the materials used to bind the fiberglass may be combustible and can be difficult to extinguish.

Figure 3.16 Some structures have large quantities of glass.

FIREFIGHTER HAZARDS RELATED TO BUILDING CONSTRUCTION

[NFPA 1001: 3-3.9(a); 3-3.11(a); 4-3.2(a)(b)]

The primary objective of understanding building construction and materials principles is to apply that information to the fireground. Firefighters should use their knowledge of these principles to monitor building conditions for signs of structural instability. Any problems that are noted should be reported to incident command personnel as quickly as possible. Even though a specific safety officer may be designated at the scene, it is the obligation of all personnel to constantly monitor for unsafe conditions. The following sections highlight some of the more critical issues related to building construction that affect firefighter safety.

Dangerous Building Conditions

Firefighters must be aware of the dangerous conditions created by a fire, as well as dangerous conditions that may be created by firefighters trying to extinguish a fire. A potentially serious situation can be compounded if firefighters fail to recognize the seriousness of the situation and take actions that only make the situation worse.

There are two primary types of dangerous conditions that may be posed by a particular building:

- Conditions that contribute to the spread and intensity of the fire
- Conditions that make the building susceptible to collapse

These two conditions are obviously related — conditions that contribute to the spread and intensity of the fire increase the likelihood of structural collapse. The following sections describe some of these conditions.

FIRE LOADING

Fire load is the maximum heat that can be produced if all the combustible materials in a given area burn. Heavy *fire loading* is the presence of large amounts of combustible materials in an area of a building. The arrangement of materials in a building directly affects fire development and severity and must be considered when determining the possible duration and intensity of a fire.

Heavy content fire loading is perhaps one of the most critical hazards in commercial and storage facilities because the fire may quickly override the capabilities of a fire sprinkler system (if present) and cause access problems for fire fighting personnel during fire fighting operations (Figure 3.17). Proper inspection and code enforcement procedures are the most effective defense against these hazards.

Figure 3.17 Commercial occupancies, such as this tire warehouse, may have heavy content fire loading.

COMBUSTIBLE FURNISHINGS AND FINISHES

Combustible furnishings and finishes contribute to fire spread and smoke production (Figure 3.18). These two elements have been identified as major factors in the loss of many lives in fires. Proper inspection and code enforcement procedures are the most effective defense against these hazards.

ROOF COVERINGS

Roof coverings are the final outside layer that is placed on top of a roof assembly. Common roof

Figure 3.18 Furniture stores have considerable amounts of combustible furnishings.

coverings include wood and composite shingles, tile, slate, tin, and asphaltic tar paper. The combustibility of a roof's surface is a basic concern to the fire safety of an entire community. Some of the earliest fire regulations ever imposed in the United States hundreds of years ago related to combustible roof-covering issues because these roof coverings caused several conflagrations from flaming embers flying from roof to roof.

History has shown time and time again that wood shake shingles in particular, even when treated with fire retardant, can significantly contribute to fire spread. This is a particular problem in wildland/urban interface fire situations where wood shake shingle roofs have contributed to large fires (Figure 3.19). Firefighters must use aggressive exposure protection tactics when faced with this type of fire.

Figure 3.19 Wood shake shingles are major contributors to fire spread.

WOODEN FLOORS AND CEILINGS

Combustible structural components such as wood framing, floors, and ceilings also contribute to the fire loading in a building. Prolonged exposure to fire may weaken them and increase the chances of collapse.

LARGE, OPEN SPACES

Large, open spaces in buildings contribute to the spread of fire throughout. Such spaces may be found in warehouses, churches, large atriums, common attics or cocklofts, and theaters (Figure 3.20). In these facilities, proper vertical ventilation (channeling smoke from a building at its highest point) is essential for slowing the spread of fire (see Chapter 10, Ventilation).

Figure 3.20 Buildings with large open spaces burn fast, and in many cases, they collapse quickly.

BUILDING COLLAPSE

Many firefighters have been seriously injured or killed by a structural collapse at fire fighting operations. The collapse results from damage to the structural system of the building caused by the fire or by fire fighting operations. Knowledge of the types of construction and the ability to recognize them is important to firefighters. Some buildings, because of their construction and age, are more inclined to collapse than others. For example, buildings featuring lightweight or truss construction will succumb to the effects of fire much quicker than a heavy timber building. Older buildings that have been exposed to weather and have been poorly maintained are more likely to collapse than newer, well-maintained buildings (Figure 3.21). Wooden structural components in older structures may also dehydrate to the point that their ignition temperature decreases and their flame spread characteristics increase. Information on building age and construction type should be obtained when conducting inspections and documented in pre-incident plans.

Figure 3.21 Old buildings can become very dangerous during fires.

The longer a fire burns in a building, the more likely it will collapse. Fire weakens the structural support system until it becomes incapable of holding the weight of the building. The time it takes for this to happen varies with the fire severity, the type of construction, the presence or absence of heavy industrial machinery on upper floors or on the roof, and the general condition of the building. Firefighters should be aware of the following indicators of building collapse and be on the lookout for them at every fire:

- Cracks or separations in walls, floors, ceilings, and roof structures (Figure 3.22)
- Evidence of existing structural instability such as the presence of tie rods and stars that hold walls together (Figure 3.23)
- Loose bricks, blocks, or stones falling from buildings
- Deteriorated mortar between the masonry
- Walls that appear to be leaning
- Structural members that appear to be distorted
- Fires beneath floors that support heavy machinery or other extreme weight loads
- Prolonged fire exposure to the structural members

Figure 3.22 Firefighters should look for cracks and repaired cracks in the walls of buildings that may affect structural integrity.

Figure 3.23 Reinforcement stars on a building are an indication that the building was already in bad shape before the fire started.

- Unusual creaks and cracking noises
- Structural members pulling away from walls
- Excessive weight of building contents

Fire fighting operations also increase the risk of building collapse. Improper vertical ventilation techniques can result in the cutting of structural supports that could weaken the structure. The water used to extinguish a fire adds extra weight to the structure and can weaken it. Water only a few inches (millimeters) deep over a large area can add many tons (tonnes) of weight to an already weakened structure.

Immediate safety precautions must be taken if fire personnel believe that the collapse of a building is imminent or even possible. First, all personnel who are operating within the building should immediately evacuate. Second, a collapse zone should be set up around the perimeter of the building (Figure 3.24). The collapse zone should be equal to

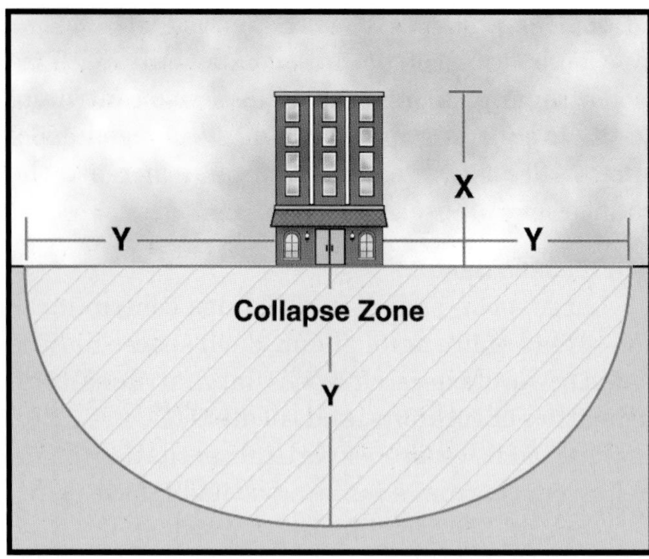

Figure 3.24 A collapse zone should be equal to one and a half times the height of the building.

one and a half times the height of the building. No personnel or apparatus should be allowed to operate in the collapse zone except to place unmanned master stream devices (see Chapter 14, Fire Control). Once these devices have been placed, personnel should immediately retreat to an area outside the collapse zone. Firefighters must always be aware of any evacuation or emergency signals used by their department.

Lightweight and Truss Construction Hazards

One of the most serious building construction hazards facing firefighters today is the increased use of lightweight and trussed support systems (Figure 3.25). Lightweight construction is most commonly found in houses, apartments, and small commercial buildings. The two most common types are lightweight metal and lightweight wood trusses. *Lightweight steel trusses* are made from a long steel

Figure 3.25 Lightweight truss construction can be extremely hazardous to firefighters during fire fighting operations.

bar that is bent at a 90-degree angle with flat or angular pieces welded to the top and bottom (Figure 3.26). *Lightweight wood trusses* are constructed of 2- x 3- or 2- x 4-inch boards that are connected together by gusset plates (Figure 3.27). *Gusset plates* are small metal plates (usually 18 to 22 gauge metal) with prongs that penetrate about ¾-inch (10 mm) into the wood.

Experience has shown that lightweight metal and wood trusses will fail after 5 to 10 minutes of exposure to fire. For steel trusses, 1,000°F (538°C) is the critical temperature. Gusset plates in wood trusses will fail early when exposed to heat. Although the trusses may be protected with fire-retardant treatments to give longer protection, most are not protected at all.

Figure 3.26 Unprotected lightweight steel trusses fail quickly when exposed to high heat.

Figure 3.27 Wood trusses connected together with gusset plates.

Wooden I beams are also used in lightweight construction. They have fire characteristics similar to wood trusses and similar precautions should be used when they are found in a structure (Figure 3.28).

Other types of trusses, such as bowstrings, are found in virtually every community. They are used in buildings that have large open spaces such as car dealerships, bowling alleys, factories, and supermarkets. Bowstrings are often easily denoted by their rounded appearance, though many appear otherwise (Figure 3.29).

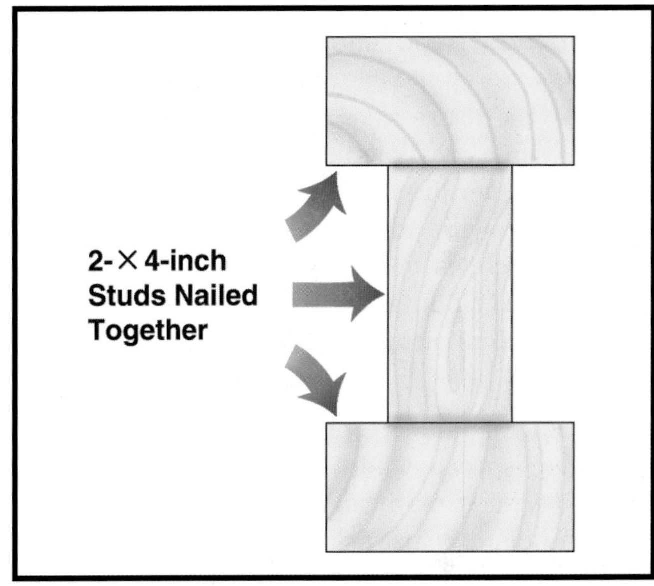

Figure 3.28 Lightweight wooden I beams are made by nailing 2- x 4-inch boards together.

Figure 3.29 The telltale rounded shape of a bowstring truss roof may be hidden by square-end parapets.

All trusses are designed to work as an integral unit. Some members are in *tension* (vertical and horizontal stresses that tend to pull things apart), and others are in *compression* (vertical and horizontal stresses that tend to press things together). One thing common to all types of trusses is that if one member fails, the entire truss is likely to fail. Once an entire truss fails, usually the truss next to it fails, and the domino principle soon takes over until a total collapse occurs.

It is important that firefighters know what buildings in their district have truss roofs or floors. Truss-containing buildings exposed to fire conditions for 5 to 10 minutes (which is usually how long they have been exposed before the fire department arrives) should not be entered, and crews should not go onto the roofs.

Construction, Renovation, and Demolition Hazards

The risk of fire rises sharply for a number of reasons when construction, renovation, or demolition is being performed on a structure. One contributing factor is the additional fire load and ignition sources (such as open flames from torches and sparks from grinding or cutting processes) brought by building contractors and their associated equipment.

Buildings under construction are subject to rapid fire spread when they are partially completed because many of the protective features such as plasterboard are not yet in place (Figure 3.30). The exposed wood framing can be likened to a vertical lumberyard. The lack of doors or other measures that would normally slow fire spread are also contributing factors to rapid fire growth.

Figure 3.30 "Skeleton" of a wood building under construction.

Buildings that are being renovated, demolished, or abandoned are also subject to faster than normal fire growth. Breached walls, open stairwells, missing doors, and disabled fire protection systems are all potential problems. The potential for a sudden building collapse during fire conditions is also a serious consideration. Arson is also a factor at construction or demolition sites because of easy access into the building.

Due to the rising costs of new construction, renovating old buildings is becoming more popular. Hazardous situations may arise during renovation construction because occupants and their belongings may remain in the building while construction continues. Fire detection or alarm systems may be taken out of service or damaged during renovation. With the accumulation of debris, new construction materials, and equipment, exits can easily be blocked if good housekeeping is not maintained, preventing the egress of persons from the building in an emergency.

Chapter 4
Firefighter Personal Protective Equipment

Chapter 4
Firefighter Personal Protective Equipment

INTRODUCTION

Firefighters require the best personal protective equipment available because of the hostile environment in which they perform their duties (Figure 4.1). All equipment discussed in this chapter is required by NFPA 1500, *Standard on Fire Department Occupational Safety and Health Program*. Providing and using quality protective equipment will not necessarily guarantee firefighter safety; however, injuries can be reduced and prevented if protective clothing and breathing apparatus are used properly. All protective equipment has inherent limitations that must be recognized so that firefighters do not overextend the item's range of protection. Extensive training in the use and maintenance of equipment is required to ensure that the equipment provides optimum protection.

All firefighters operating at an emergency scene must wear full protective equipment (which includes personal protective clothing and self-contained breathing apparatus) suitable to that incident (Figure 4.2). Personal protective clothing refers to the garments firefighters must wear while performing their jobs. Full protective equipment for structural fire fighting consists of the following:

Figure 4.1 Fire fighting exposes personnel to a hostile environment.

Figure 4.2 A firefighter working at a structural fire should always wear full personal protective equipment, which includes an SCBA and a PASS device.

- ***Helmet*** — Protects the head from impact and puncture injuries as well as from scalding water

- ***Protective hood*** — Protects portions of the firefighter's face, ears, and neck not covered by the helmet or coat
- ***Protective coat and trousers*** — Protect trunk and limbs against cuts, abrasions, and burn injuries (resulting from radiant heat), and provide limited protection from corrosive liquids
- ***Gloves*** — Protect the hands from cuts, wounds, and burn injuries
- ***Safety shoes or boots*** — Protect the feet from burn injuries and puncture wounds
- ***Eye protection*** — Protects the wearer's eyes from flying solid particles or liquids
- ***Hearing protection*** — Limits noise-induced damage to the firefighter's ears when loud noise situations cannot be avoided
- ***Self-contained breathing apparatus (SCBA)*** — Protects the face and lungs from toxic smoke and products of combustion
- ***Personal alert safety system (PASS)*** — Provides life-safety protection by emitting a loud shriek if the firefighter should collapse or remain motionless for approximately 30 seconds

The first part of this chapter discusses overall protective clothing, including eye protection, hearing protection, work uniforms, standard protective gear, and wildland fire fighting gear. The second part of this chapter gives an extensive overview of protective breathing equipment. Included is information on the different types of protective breathing apparatus and personal alert safety systems (PASS). Reasons why protective breathing equipment must be worn and the general procedures for donning, doffing, inspecting, and maintaining breathing equipment are given. Changing and refilling air cylinders are also covered. The last portion of the chapter covers safety precautions and using self-contained breathing apparatus during emergency situations.

PERSONAL PROTECTIVE CLOTHING

[NFPA 1001: 3-1.1.2; 3-3.1; 3-3.2; 3-3.2(a); 3-3.2(b)]

NFPA 1971, *Standard on Protective Ensemble for Structural Fire Fighting,* includes coats, trousers, coveralls, helmets, gloves, footwear, and interface components (protective hoods and wristlets) as parts of the multiple elements of clothing and equipment designed to provide protection for firefighters during structural fire fighting and certain other operations. All components of the protective ensemble must have an appropriate product label for that component permanently and conspicuously attached (Figure 4.3). This label contains the following information:

"THIS . . . MEETS THE . . . REQUIREMENTS OF NFPA 1971, *STANDARD ON PROTECTIVE ENSEMBLE FOR STRUCTURAL FIRE FIGHTING,* 1997 EDITION."

- Manufacturer's name, identification, or designation
- Manufacturer's address
- Country of manufacture
- Manufacturer's . . . identification number or lot number or serial number

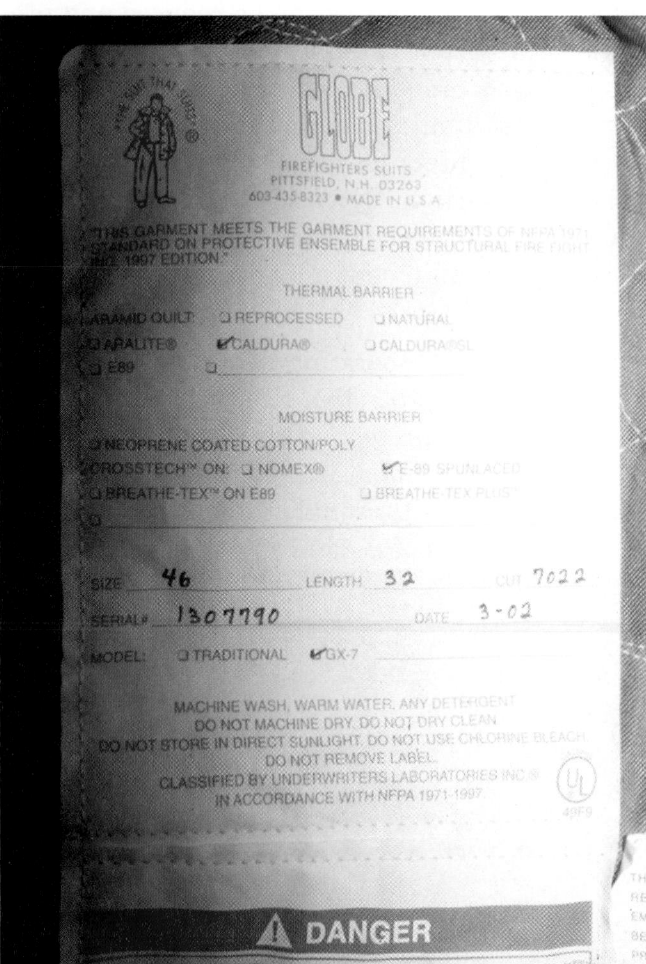

Figure 4.3 An information label required for ensemble items.

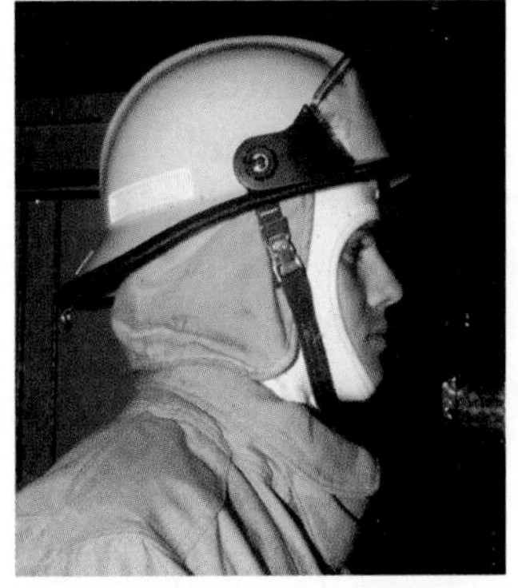

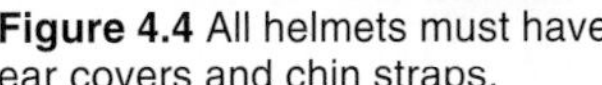

Figure 4.4 All helmets must have ear covers and chin straps.

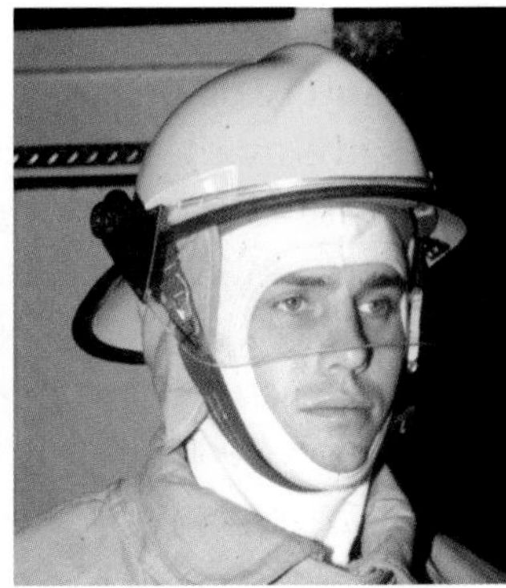

Figure 4.5 A typical helmet faceshield.

- Month and year of manufacture (not coded)
- Model name, number, or design
- Size or size range
- Garment materials (coats, trousers, coveralls, hoods)
- Footwear size and width (boots)
- Cleaning precautions

All equipment worn by the firefighter should meet current applicable standards. The firefighter must understand the design and purpose of the various types of protective clothing and be especially aware of each garment's inherent limitations. The following sections highlight some of the important features of specific types of firefighter personal protective clothing.

Helmets

Head protection was one of the first concerns for firefighters. The traditional function of the helmet was to shed water, not to protect from heat, cold, or impact. The wide brim, particularly where it extends over the back of the neck, was designed to prevent hot water and embers from reaching the ears and neck. Newer helmet designs perform this function as well as provide the following additional benefits:

- Protect the head from impact
- Provide protection from heat and cold
- Provide faceshields for secondary protection of the face and eyes when SCBA is not required

Helmets must have ear covers, which should always be used during fire fighting. Chin straps ensure that helmets stay in place upon impact (Figure 4.4).

For secondary face and eye protection, faceshields are provided that attach to the helmet (Figure 4.5). Most of these flip up and out of the field of vision fairly easily and are generally acceptable to firefighters. Most assemblies do not interfere with protective breathing equipment.

Eye Protection

Perhaps one of the most common injuries on the fireground is injury to the eyes. Eye injuries are not always reported because they are not always debilitating. Eye injuries can be serious, but they are fairly easy to prevent. It is important to protect the eyes on the fireground and while performing duties around the station. Eye protection for the firefighter comes in many forms such as safety glasses, safety goggles, helmet faceshields, and SCBA masks (Figure 4.6). Faceshields provide secondary protection and may not provide the eye protection required against flying particles or splashes. NFPA 1500 requires that goggles or other appropriate primary eye protection be worn when participating in operations where protection from flying particles or chemical splashes is necessary.

Firefighters may encounter a variety of situations where eye protection, other than that afforded by a helmet faceshield or SCBA mask, is required. Other situations where more eye protection is needed include fireground and station operations (such as welding, grinding, or cutting), vehicle extrications or brush fires, and inspections in industrial occupancies.

Figure 4.6 Various types of eye protection available to the firefighter.

Safety glasses and goggles protect against approximately 85 percent of all eye hazards. Several styles are available, including some that fit over prescription glasses. Firefighters who wear prescription safety eyeglasses should select frames and lenses that meet ANSI Standard Z87.1, *Practice for Occupational and Educational Eye and Face Protection,* for severe exposure to impact and heat.

Warning signs should be posted near operations requiring eye protection. Use of eye protection must be required through departmental standard operating procedures and enforced by supervisory personnel.

Hearing Protection

Firefighters are exposed to a variety of sounds in the station, en route to the scene, and on the fireground. Often, exposure to these sounds or a combination of sounds can produce permanent hearing loss. To prevent exposure to unacceptably high levels of noise, it is necessary for a department to initiate a hearing protection program to identify, control, and reduce potentially harmful noise and/or provide protection from it. Eliminating or reducing noise level is the best solution; however, this is often not possible. Therefore, acceptable hearing protection should be provided to firefighters, and it should be used in accordance with standard operating procedures.

The most common use of hearing protection is for firefighters who ride apparatus that exceed maximum noise exposure levels. Intercom/ear protection systems provide a dual benefit because of their ability to reduce the amount of noise the ear is exposed to and at the same time allow the crew to communicate or monitor the radio (Figure 4.7).

Earplugs or earmuffs may be used for hearing protection (Figure 4.8). If earplugs are used, each firefighter should be issued a personal set. However, there are some potential hazards associated with earplugs and earmuffs. For example, in a structural fire fighting situation, earmuffs can compromise protection of the face by making it awkward to use SCBA and protective hoods. Earplugs may melt when exposed to intense heat. For these reasons, wearing hearing protection during structural fire fighting is impractical.

Figure 4.7 Intercom/ear protection allows the crew to communicate and at the same time reduces the amount of noise to which they are exposed.

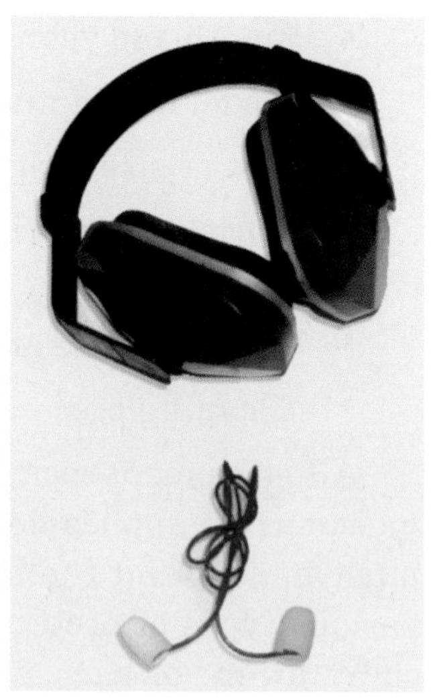

Figure 4.8 Earmuffs and earplugs provide the firefighter with hearing protection.

Protective Hoods

Protective hoods are designed to protect the firefighter's ears, neck, and face from exposure to extreme heat. These hoods also cover areas not otherwise protected by the SCBA facepiece, ear covers, or coat collar. Hoods are typically made of fire-resistant material and are available in long or short styles (Figures 4.9 a and b). Protective hoods used in conjunction with the SCBA facepiece provide effective protection. However, care must be taken to ensure that the hood does not interfere with the facepiece-to-face seal (Figure 4.10).

Figure 4.9a Longer protective hoods extend over the shoulders and chest.

Figure 4.9b The short protective hood covers the ears, neck, and face from exposure to extreme heat.

Figure 4.10 Proper placement of the protective hood will not interfere with the facepiece-to-face seal.

Firefighter Protective Coats

Firefighter protective coats are used for protection in structural fire fighting and other fire department activities (Figure 4.11). NFPA 1971 requires that all protective coats be made of three components: outer shell, moisture barrier, and thermal barrier (Figure 4.12). These barriers serve to trap insulating air that inhibits the transfer of heat from the outside to the firefighter's body. They also protect the firefighter from direct flame contact, hot water and vapors, cold temperatures, and any number of other environmental hazards. Clearly, the construction and function of each component is important to the firefighter's safety.

Figure 4.11 Structural fire fighting protective coat.

Figure 4.12 The moisture barrier protects the firefighters from water, steam, hot vapors, or corrosive liquids.

WARNING

All inner liners of the protective coat must be in place during any fire fighting operation. Failure to wear the entire coat and liner system during fire conditions may expose the firefighter to severe heat that could result in serious injury or death.

Firefighter protective coats have many features that provide additional protection and convenience to the wearer. Collars must be turned up to protect the wearer's neck and throat (Figure 4.13). Wristlets prevent water, embers, and other foreign debris from rolling down inside the sleeves (Figure 4.14). The closure system on the front of protective coats prevents water or fire products from entering through gaps between the snaps or clips (Figure 4.15).

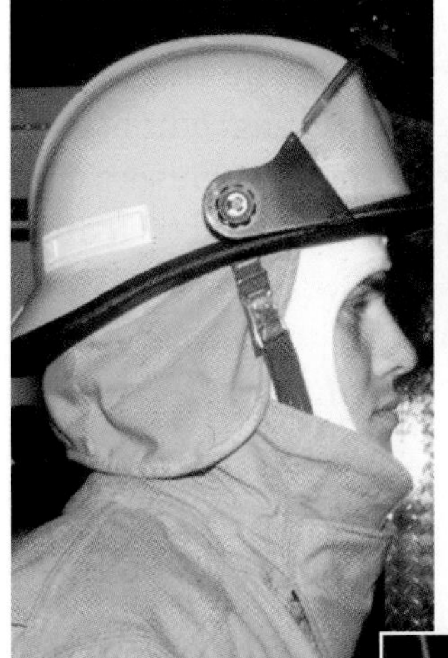

Figure 4.13 The collar protects the firefighter's neck and throat.

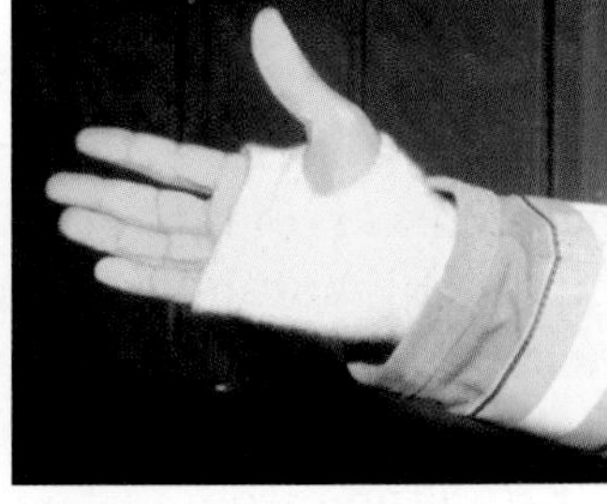

Figure 4.14 The wristlet that extends over the hand should have a thumbhole to prevent it from sliding up the wrist.

Figure 4.15 The storm flap covers the closure area and prevents steam, water, and fire products from entering the gaps between the closures.

Protective coats that meet NFPA standards are designed to be cleaned according to manufacturer's specifications. Reflective trim should be maintained according to NFPA standards. Trim should not be obscured by pockets, patches, or storm flaps.

Firefighter Protective Trousers

Protective trousers are an integral part of the firefighter's protective ensemble. Three-quarter boots and long coats alone do not provide adequate protection for the lower torso or extremities and are no longer permissible according to NFPA 1500. When selecting protective trousers, consider the same concepts of fabric selection, moisture barriers, and other considerations used to select protective coats. The layering principles that apply to coats also apply to trousers. Options, such as reinforced knees and leather cuffs, may increase the durability of protective trousers (Figure 4.16). Suspenders should be the heavy-duty type so that pants do not sag when they become wet (Figure 4.17). Protective trousers that meet NFPA standards are designed to be cleaned according to manufacturer's specifications. Reflective trim should be maintained according to NFPA standards.

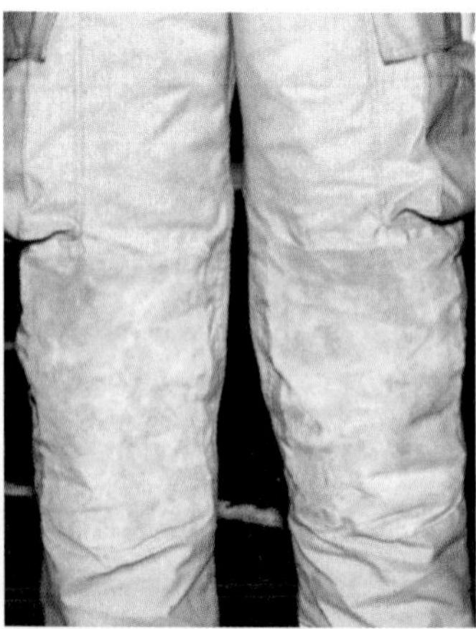

Figure 4.16 Reinforced knees prolong the life of protective trousers.

Figure 4.17 Heavy-duty suspenders keep trousers from sagging when they become wet.

Hand Protection

The most important characteristics of gloves are the protection they provide against heat or cold penetration and their resistance to cuts, punctures, and liquid absorption. Gloves must allow enough dexterity and tactile feel for the firefighter to perform the job effectively. If the gloves are too awkward and bulky, the firefighter may not be able to do fine manipulative work (Figure 4.18). Gloves must fit properly and be designed to provide protection as well as to allow dexterity. Unfortunately, in order to provide protection, dexterity is often reduced.

Figure 4.18 Gloves must have enough flexibility to allow the firefighter to perform fire fighting tasks.

Foot Protection

There are numerous hazards to the feet on the fire scene. Embers, falling objects, and nails are examples of commonly encountered hazards (Figure 4.19). Appropriate foot protection should be selected to ensure that the risk of injury from these hazards is minimized. Because of the nature of their work, firefighters will need to have the following two kinds of foot protection:

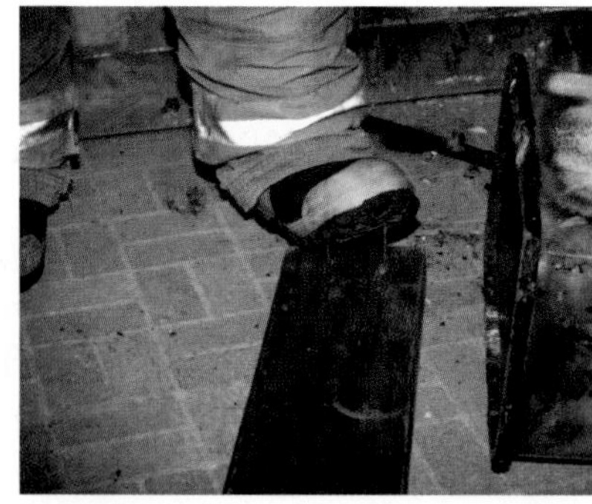

Figure 4.19 Fire scenes have numerous foot hazards.

- Protective boots for fire fighting and emergency activities (Figure 4.20)
- Safety shoes for station wear and other fire department activities that include inspections, emergency medical responses, and similar activities (Figure 4.21)

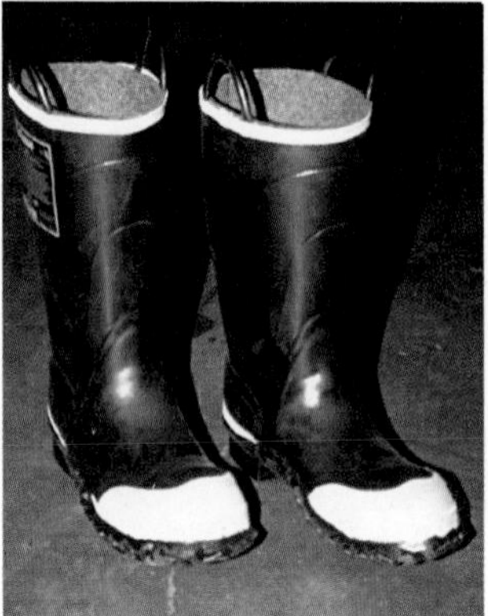

Figure 4.20 The rubber boot is one style of protective boot.

Figure 4.21 These leather protective boots can also serve as station safety shoes. *Courtesy of the Warrington Group, LTD.*

Puncture resistance should be provided by a stainless steel midsole plate. If there is doubt about midsole protection, X-ray the boot. Some fire departments require insulation laminated into the rubber. The only disadvantage to this requirement is that the added weight tends to increase firefighter fatigue. Select a boot lining that will not break up and cause blisters and discomfort. There are also protective boots with shin pads to reduce the strain from leg locks and crawling. Boots should have well-secured pull loops.

Each firefighter should be fitted as accurately as possible. Half sizes are available in both men's and women's boots. Firefighters should not share protective boots because this practice is unsanitary. When boots are reissued, sanitize them with procedures recommended by an industrial hygienist.

Safety shoes or boots should be worn while conducting inspections or while doing work around the station. Some departments require safety shoes or boots as part of their daily uniform. Safety shoes usually have safety toes, puncture-resistant soles, or special inserts. These shoes provide good support for climbing, give increased physical agility, and are generally less fatiguing than protective boots. Leather fire fighting boots can be used for work at the station, for conducting inspections, and for fire fighting operations.

Wildland Personal Protective Clothing

Personal protective clothing used for structural fire fighting is generally too bulky, too hot, and too heavy to be practical for use in wildland fire fighting. Specifications for wildland fire fighting personal protective clothing (often called *brush gear*) and equipment are contained in NFPA 1977, *Standard on Protective Clothing and Equipment for Wildland Fire Fighting.* Wildland personal protective clothing includes gloves, goggles, brush jackets/pants or one-piece jumpsuits, head and neck protection, and footwear (Figure 4.22). Different forms of respiratory protection for wildland firefighters are also available.

Wildland fire fighting gloves are made of leather or other suitable materials and must provide wrist protection. They should be comfortable and sized correctly to prevent abrasions and blisters.

The cuffs of the sleeves and the pants legs of protective clothing are closed snugly around the wrists and ankles. The fabric is treated cotton or some other inherently flame-resistant material. Underwear of 100 percent cotton, including a long-sleeved T-shirt, should be worn under brush gear. Socks should be made of natural fiber.

CAUTION: Firefighters should *never* wear synthetic materials at a fire; these materials melt when heated and stick to the wearer's skin. This greatly increases the likelihood of major burn injuries.

Hard hats or helmets with chin straps must be worn for head protection. Lightweight wildland helmets are preferred to structural helmets. They should be equipped with a protective shroud for face and neck protection. Goggles with clear lenses should also be worn.

Figure 4.22 Wildland firefighters should be equipped with personal protective equipment designed especially for wildland fire conditions.

What is deemed acceptable in footwear for wildland fire fighting varies in different geographical regions, but some standard guidelines apply in all areas. Lace-up or zip-up safety boots with lug or grip-tread soles are most often used. Boots should be at least 8 to 10 inches (200 mm to 250 mm) high to protect the lower leg from burns, snakebites, and cuts and abrasions.

Station/Work Uniforms

Firefighter accident statistics show that certain types of clothing can contribute to on-the-job injuries. Certain synthetic fabrics, such as polyester, can be especially hazardous because they can melt during exposure to high temperatures. Some of the materials that have high temperature resistance are as follows:

- Organic fibers such as wool and cotton
- Synthetic fibers such as Kevlar® aramid fibers, Nomex® fire-resistant material, PBI® polybenzimidazole fiber, Kynol® phenolic resins, Gore-Tex® water repellent fabric, Orlon® acrylic fiber, neoprene, Teflon® fluorocarbon resins (nonstick coatings), silicone, and panotex

All firefighter station and work uniforms should meet the requirements set forth in NFPA 1975, *Standard on Station / Work Uniforms for Fire Fighters*. The purpose of the standard is to provide minimum standards for work wear that will not contribute to firefighter injury and negate the effects of the outer protective clothing. Garments falling under this standard include trousers, shirts, jackets, and coveralls, but not underwear (Figure 4.23). Underwear made of 100 percent cotton is recommended.

Figure 4.23 Station uniforms should provide additional protection to the firefighter.

The main part of the standard requires that no components of garments ignite, melt, drip, or separate when exposed to heat at 500°F (260°C) for 5 minutes. A garment meeting all requirements of the standard will have a notice to that effect permanently attached. It is important to note that while this clothing is designed to be fire resistant, it is not designed to be worn for fire fighting operations. Standard structural fire fighting clothing must always be worn over these garments when a firefighter is engaged in structural fire fighting activities. Wildland protective clothing, depending on design and local preference, may be worn over station uniforms or directly over undergarments.

Care of Personal Protective Clothing

In order for personal protective clothing to perform properly, it must be maintained within the manufacturer's specifications. Each piece of protective clothing has a particular manufacturer's recommended maintenance procedure that should be followed to ensure it is ready for service.

Helmets should be properly cleaned and maintained to ensure their durability and maximum life expectancy. The following are guidelines for their proper care and maintenance.

- Remove dirt from the shell. Dirt absorbs heat faster than the shell itself, thus exposing the wearer to more severe heat conditions.
- Remove chemicals, oils, and petroleum products from the shell as soon as possible (Figure 4.24). These agents may soften the shell material and reduce its impact and dielectric protection. See the manufacturer's instructions for suggested cleansers.
- Repair or replace helmets that do not fit properly (Figure 4.25). A poor fit reduces the helmet's ability to resist the transmission of force.
- Repair or replace helmets that are damaged. This includes leather helmets that have become cracked or brittle with age.

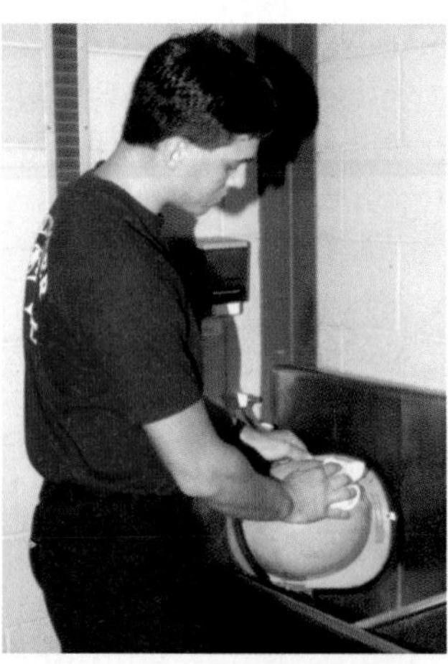

Figure 4.24 Clean dirt and chemicals from the helmet shell.

Figure 4.25 The firefighter should try on the helmet to ensure a proper fit. Some may be adjusted to correct a loose or tight fit.

- Inspect suspension systems frequently to detect deterioration. Replace if necessary.
- Consult the helmet manufacturer if a helmet needs repainting. Manufacturers can inform the department about the choice of paints available for a particular shell material.
- Remove polycarbonate helmets that have come into contact with hydraulic oil from service and check them. Some oils attack the polycarbonate material and weaken the helmet.

Cleanliness also affects the performance of protective coats, trousers, and hoods. The outer shells should be cleaned regularly. Clean outer shells have better fire resistance; dirty protective clothing absorbs more heat. Follow the manufacturer's directions for cleaning. The directions are usually contained on a tag sewn to the garment. NFPA 1500 requires that protective clothing be cleaned through either a cleaning service or fire department facility that is equipped to handle contaminated clothing.

Gloves and boots should also be cleaned according to the manufacturer's instructions. NFPA 1581, *Standard on Fire Department Infection Control Program*, further requires that personal protective clothing be cleaned and dried at least every six months in accordance with the manufacturer's recommendations.

SELF-CONTAINED BREATHING APPARATUS

[NFPA 1001: 3-3.1; 3-3.1(a); 3-3.4(b); 3-3.9(a); 3-3.10(a)]

Protective breathing apparatus is extremely crucial to the well-being of the firefighter. Failure to use this equipment could lead to failed rescue attempts, firefighter injuries, or firefighter fatalities. The well-trained firefighter should be knowledgeable of respiratory hazards, the requirements for wearing protective breathing apparatus, the procedures for donning and doffing the apparatus, and the proper care and maintenance of the equipment.

Respiratory Hazards

The lungs and respiratory tract are more vulnerable to injury than any other body areas, and the gases encountered in fires are, for the most part, dangerous in one way or another. It should be a fundamental rule in fire fighting that no one be permitted to enter any potentially toxic atmosphere, such as an interior or exterior fire attack, below-grade rescue, or hazardous materials emergency, unless equipped with protective breathing apparatus (Figure 4.26). All situations should be monitored for firefighter safety.

There are four common hazardous atmospheres associated with fires or other emergencies. These atmospheres include the following:

- Oxygen deficiency
- Elevated temperatures
- Smoke
- Toxic atmospheres (with and without fire)

Figure 4.26 Firefighters should always wear their SCBA while performing a fire attack.

OXYGEN DEFICIENCY

The combustion process consumes oxygen while producing toxic gases that either physically displace oxygen or dilute its concentration. When oxygen concentrations are below 18 percent, the human body responds by increasing its respiratory rate. Symptoms of oxygen deficiency by percentage of available oxygen are shown in Table 4.1. Oxygen deficiency can also occur in below-grade locations, chemical storage tanks, grain bins, silos, and other confined spaces. Another area of potential hazard would be a room protected by a total-flooding carbon dioxide extinguishing system after discharge.

TABLE 4.1
Physiological Effects of Reduced Oxygen (Hypoxia)

Oxygen in Air (Percent)	Symptoms
21	None — normal conditions
17	Some impairment of muscular coordination; increase in respiratory rate to compensate for lower oxygen content
12	Dizziness, headache, rapid fatigue
9	Unconsciousness
6	Death within a few minutes from respiratory failure and concurrent heart failure

NOTE: These data cannot be considered absolute because they do not account for difference in breathing rate or length of time exposed.

These symptoms occur only from reduced oxygen. If the atmosphere is contaminated with toxic gases, other symptoms may develop.

Some departments have the ability to monitor atmospheres and measure these hazards directly. When this capability exists, it should be used. Where monitoring is not possible or monitor readings are in doubt, self-contained breathing apparatus should be worn.

ELEVATED TEMPERATURES

Exposure to heated air can damage the respiratory tract, and if the air is moist, the damage can be much worse. Excessive heat taken quickly into the lungs can cause a serious decrease in blood pressure and failure of the circulatory system. Inhaling heated gases can cause pulmonary edema (accumulation of fluids in the lungs and associated swelling), which can cause death from asphyxiation. The tissue damage from inhaling hot air is not immediately reversible by introducing fresh, cool air.

SMOKE

The smoke at a fire is a suspension of small particles of carbon, tar, and dust floating in a combination of heated gases (Figure 4.27). The particles provide a means for the condensation of some of the gaseous products of combustion, especially aldehydes and organic acids formed from carbon. Some of the suspended particles in smoke are merely irritating, but others may be lethal. The size of the particle determines how deeply into the unprotected lungs it will be inhaled.

Figure 4.27 A common structure fire gives off large volumes of smoke.

TOXIC ATMOSPHERES ASSOCIATED WITH FIRE

The firefighter should remember that a fire means exposure to combinations of irritants and toxicants whose toxicity cannot be predicted accurately. In fact, the combination can have a synergistic effect in which the combined effect of two or more substances is more toxic or more irritating than the total effect would be if each were inhaled separately.

Inhaled toxic gases may have several harmful effects on the human body (Figure 4.28). Some of

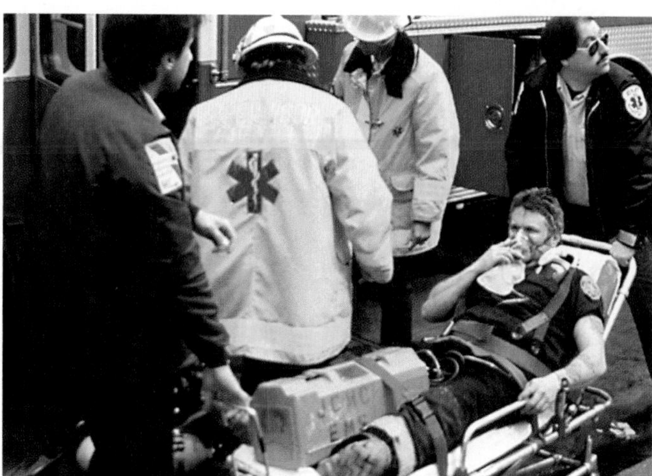

Figure 4.28 Oxygen therapy being given to a downed firefighter who was exposed to a toxic atmosphere. *Courtesy of Ron Jeffers.*

the gases directly cause disease of the lung tissue and impair its function. Other gases have no directly harmful effect on the lungs but pass into the bloodstream and to other parts of the body and impair the oxygen-carrying capacity of the red blood cells.

The particular toxic gases given off at a fire vary according to four factors:

- Nature of the combustible
- Rate of heating
- Temperature of the evolved gases
- Oxygen concentration

Table 4.2 addresses some of the most commonly found gases on the fire scene. The immediately dangerous to life and health (IDLH) concentrations are from the *National Institute for Occupational Safety and Health (NIOSH) Pocket Guide to Chemical Hazards*. The current NIOSH definition for an IDLH exposure condition is one *"that poses a threat of exposure to airborne contaminants when that exposure is likely to cause death or immediate or delayed permanent adverse health effects or prevent escape from such an environment."* These values were established to ensure that a worker could escape without injury or irreversible health effects from an IDLH exposure in the event of the failure of respiratory protection equipment.

Because more fire deaths occur from carbon monoxide (CO) poisoning than from any other toxic

TABLE 4.2
Toxic Atmospheres Associated With Fire

Toxic Atmospheres	Sensibility	IDLH*	Caused By	Miscellaneous
Carbon Dioxide (CO_2)	Colorless; odorless	40,000 ppm**	Free-burning	End product of complete combustion of carboniferous materials
Carbon Monoxide (CO)	Colorless; odorless	1,200 ppm	Incomplete combustion	Cause of most fire-related deaths
Hydrogen Chloride (HCl)	Colorless to slightly yellow; pungent odor	50 ppm	Burning plastics (e.g., polyvinyl chloride [PVC])	Irritates eyes and respiratory tract
Hydrogen Cyanide (HCN)	Colorless; bitter almond odor	50 ppm	Burning of wool, nylon, polyurethane foam, rubber, and paper	Chemical asphyxiate; hampers respiration at the cellular and tissue level
Nitrogen Dioxide (NO_2)	Reddish-brown; pungent, acrid odor	20 ppm	Given off around silos or grain bins; also liberated when pyroxylin plastics decompose	Irritates nose and throat
Phosgene ($COCl_2$)	Colorless; odor of musty hay; tasteless	2 ppm	Produced when refrigerants such as Freon contact flame	Forms hydrochloric acid in lungs due to moisture

* Immediately dangerous to life and health — any atmosphere that poses an immediate hazard to life or produces immediate irreversible, debilitating effects on health

** Parts per million — ratio of the volume of contaminants (parts) compared to the volume of air (million parts)

product of combustion, a greater explanation of this toxic gas is necessary. This colorless, odorless gas is present with every fire. The poorer the ventilation and the more inefficient the burning, the greater the quantity of carbon monoxide formed. A rule of thumb, although subject to much variation, is that the darker the smoke, the higher the carbon monoxide levels. Black smoke is high in particulate carbon and carbon monoxide because of incomplete combustion.

The blood's hemoglobin combines with and carries oxygen in a loose chemical combination called *oxyhemoglobin*. The most significant characteristic of carbon monoxide is that it combines with the blood's hemoglobin so readily that the available oxygen is excluded. The loose combination of oxyhemoglobin becomes a stronger combination called *carboxyhemoglobin (COHb)*. In fact, carbon monoxide combines with hemoglobin, creating carboxyhemoglobin, about 200 times more readily than does oxygen. The carbon monoxide does not act on the body, but crowds oxygen from the blood and leads to eventual hypoxia of the brain and tissues, followed by death if the process is not reversed.

Concentrations of carbon monoxide in air above five-hundredths of one percent (0.05 percent) (500 parts per million [ppm]) can be dangerous. When the level is more than 1 percent, unconsciousness and death can occur without physiological signs. Even at low levels, the firefighter should not use signs and symptoms for safety factors. Headaches, dizziness, nausea, vomiting, and cherry-red skin can occur at many concentrations, based on an individual's dose and exposure. Therefore, these signs and symptoms are not good indicators of safety. Table 4.3 shows the toxic effects of different levels of carbon monoxide in air. These effects are not absolute because they do not take into account variations in breathing rate or length of exposure. Such factors could cause toxic effects to occur more quickly.

Measurements of carbon monoxide concentrations in air are not the best way to predict rapid physiological effects because the actual reaction is from the concentration of carboxyhemoglobin in the blood, causing oxygen starvation. High oxygen user organs, such as the heart and brain, are damaged early. The combination of carbon monoxide with the blood is greater when the concentration in air is greater. An individual's general physical condition, age, degree of physical activity, and length of exposure all affect the actual carboxyhemoglobin level in the blood. Studies have shown that it takes years for carboxyhemoglobin to dissipate from the bloodstream. People frequently exposed to carbon monoxide develop a tolerance to it, and they can function asymptomatically (without symptoms) with residual levels of serum carboxyhe-

TABLE 4.3
Toxic Effects of Carbon Monoxide

Carbon Monoxide (CO) (ppm*)	Carbon Monoxide (CO) in Air (Percent)	Symptoms
100	0.01	No symptoms — no damage
200	0.02	Mild headache; few other symptoms
400	0.04	Headache after 1 to 2 hours
800	0.08	Headache after 45 minutes; nausea, collapse, and unconsciousness after 2 hours
1,000	0.10	Dangerous — unconsciousness after 1 hour
1,600	0.16	Headache, dizziness, nausea after 20 minutes
3,200	0.32	Headache, dizziness, nausea after 5 to 10 minutes; unconsciousness after 30 minutes
6,400	0.64	Headache, dizziness, nausea after 1 to 2 minutes; unconsciousness after 10 to 15 minutes
12,800	1.26	Immediate unconsciousness; danger of death in 1 to 3 minutes

*ppm — parts per million

moglobin that would produce significant discomfort in the average adult. The bottom line is that firefighters may be suffering the effects of CO exposure even though they are asymptomatic.

Experiments have provided some comparisons relating air and blood concentrations to carbon monoxide. A 1-percent concentration of carbon monoxide in a room will cause a 50-percent level of carboxyhemoglobin in the bloodstream in 2½ to 7 minutes. A 5-percent concentration can elevate the carboxyhemoglobin level to 50 percent in only 30 to 90 seconds. A person previously exposed to a high level of carbon monoxide may react later in a safer atmosphere because the newly formed carboxyhemoglobin may be traveling through the body. A person so exposed should not be allowed to use breathing apparatus or resume fire control activities until the danger of toxic reaction has passed. Even with protection, a toxic condition could be endangering consciousness.

A hardworking firefighter may be incapacitated by a 1-percent concentration of carbon monoxide. The stable combination of carbon monoxide with the blood is only slowly eliminated by normal breathing. Administering pure oxygen is the most important element in first aid care. After an uneventful convalescence from a severe exposure, signs of nerve or brain injury may appear any time within three weeks. This is why an overcome firefighter who quickly revives should not be allowed to reenter a smoky atmosphere.

TOXIC ATMOSPHERES NOT ASSOCIATED WITH FIRE

Hazardous atmospheres can be found in numerous situations in which fire is not involved. Many industrial processes use extremely dangerous chemicals to make ordinary items (Figure 4.29). For example, quantities of carbon dioxide would be stored at a facility where wood alcohol, ethylene, dry ice, or carbonated soft drinks are manufactured. Any other specific chemical could be traced to numerous, common products.

Figure 4.29 Hazardous atmospheres can be found in numerous situations such as this large factory.

Many refrigerants are toxic and may be accidentally released, causing a rescue situation to which firefighters may respond. Ammonia and sulfur dioxide are two dangerous refrigerants that irritate the respiratory tract and eyes. Sulfur dioxide reacts with moisture in the lungs to form sulfuric acid. Other gases also form strong acids or alkalies on the delicate surfaces of the respiratory system.

An obvious location where a chlorine gas leak may be encountered is at a manufacturing plant; a not-so-obvious location is at a swimming pool or water park (Figure 4.30). Incapacitating concentrations can be found at either location. Chlorine is also used in manufacturing plastics, foam, rubber, and synthetic textiles and is commonly found at water and sewage treatment plants.

Figure 4.30 Swimming pools and water parks may have large quantities of chlorine on the property.

Sometimes the leak is not at the manufacturing plant but occurs during transportation of the chemical. Train derailments have resulted in container failures, exposing the public to toxic chemicals and gases. The large quantities of gases released can travel long distances.

Because of the likelihood of the presence of toxic gas, rescues in sewers, storm drains, caves, trenches, storage tanks, tank cars, bins, silos, manholes, pits, and other confined places require the use of self-contained breathing apparatus (Figure 4.31). Workers have been overcome by harmful gases in large tanks during cleaning or repairs; unprotected personnel have also been overcome while attempting a rescue. In addition, the atmosphere in many of these areas is oxygen deficient and will not support life even though there may be no toxic gas. For more information on confined spaces, see IFSTA's **Fire Service Search & Rescue** manual.

The manufacture and transport of hazardous materials have made virtually every area a potential site for a hazardous materials incident. Hazardous materials are routinely transported by vehicle, rail, water, air, and pipeline (Figure 4.32). A firefighter needs to be able to recognize when a chemical spill or incident is hazardous and know when to wear protective breathing apparatus. The

Figure 4.31 Sewers and storm drains are confined spaces that require SCBA.

Courtesy of Howard Chatterton.

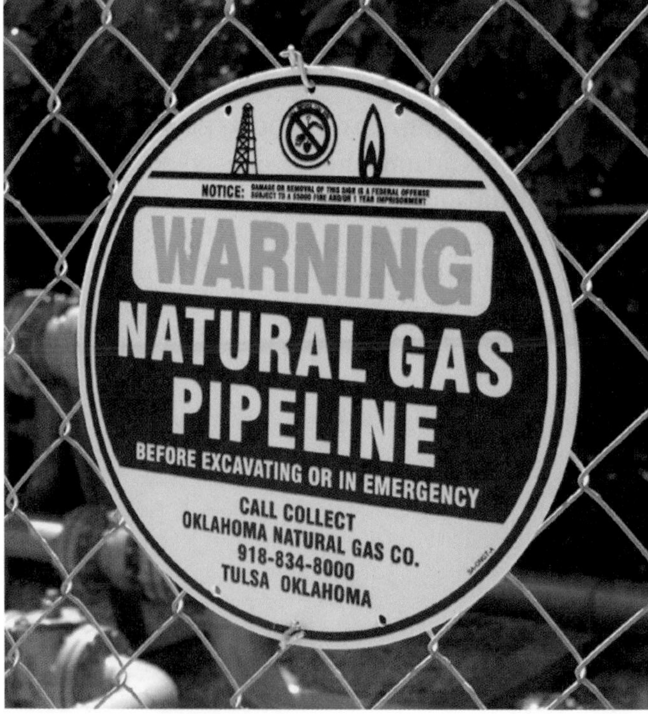

Figure 4.32 There are a number of transportation modes that are used to move hazardous materials.

United States Department of Transportation (DOT) defines a hazardous material as "*any substance which may pose an unreasonable risk to health and safety of operating or emergency personnel, the public, and/or the environment if it is not properly controlled during handling, storage, manufacture, processing, packaging, use, disposal, or transportation.*" Hazardous materials can range from chemicals in liquid or gas form to radioactive materials to etiologic (disease-causing) agents. Fire may complicate the hazards and pose an even greater danger. Many times a response to an industrial site may deal with hazardous materials. Self-contained breathing apparatus should be a mandatory piece of protective equipment when dealing with hazardous materials situations.

When responding to a vehicle accident involving a truck, the placard on the truck should serve as a warning that the atmosphere may be toxic and that self-contained breathing apparatus should be worn. In industrial facilities, placards and labels placed on containers provide warning of the dangerous materials inside. It is safer to attempt to view these placards and labels through binoculars from a distance before moving in close to them.

Do not limit using self-contained breathing apparatus to transportation hazardous materials incidents only. Common calls, such as natural gas leaks or carbon monoxide poisonings, may also require the use of self-contained breathing apparatus. *When in doubt, wear self-contained breathing apparatus!*

For more information on hazardous materials, see IFSTA's **Hazardous Materials for First Responders** manual or Fire Protection Publications' **Hazardous Materials: Managing the Incident** manual.

Protective Breathing Apparatus Limitations

To operate effectively, the firefighter must be aware of the limitations of protective breathing apparatus. These include limitations of the wearer, equipment, and air supply.

LIMITATIONS OF WEARER

Several factors affect the firefighter's ability to use SCBA effectively. These factors include physical, medical, and mental limitations.

Physical

- ***Physical condition*** — The wearer must be in sound physical condition in order to maximize the work that can be performed and to stretch the air supply as far as possible.
- ***Agility*** — Wearing a protective breathing apparatus restricts the wearer's movements and affects his balance. Good agility will help overcome these obstacles.
- ***Facial features*** — The shape and contour of the face affects the wearer's ability to get a good facepiece-to-face seal.

One issue of frequent debate is the use of contact lenses while wearing a protective breathing apparatus facepiece. The Occupational Safety and Health Administration (OSHA) standard for respiratory protection, 29 CFR 1910.134, prohibits firefighters from wearing contact lenses while using a respirator. However, this regulation has been repeatedly challenged by users.

Based on the results of an OSHA-funded research project assessing the hazards associated with the wearing of contact lenses with full-facepiece respirators and the review of other reports and studies, OSHA has adopted a policy that states:

> *Violations of the respirator standard involving the use of rigid gas-permeable or soft (hydrophilic) contact lenses with any type of respirator shall be characterized as de minimis. A violation is characterized as de minimis if it has no direct or immediate relationship to employee safety or health. Citations are not issued for de minimis violations, and there is no monetary penalty or requirement for abatement.*

NOTE: This policy does not apply to hard, nonpermeable lenses.

NFPA 1500 allows the firefighter to wear soft contact lenses while using SCBA if the firefighter has demonstrated successful long-term (at least 6 months) use of contact lenses without any problems.

Medical

- ***Neurological functioning*** — Good motor coordination is necessary for operating in protective breathing equipment. The firefighter must be of sound mind to handle emergency situations that may arise.

- ***Muscular/skeletal condition*** — The firefighter must have the physical strength and size required to wear the protective equipment and to perform necessary tasks.
- ***Cardiovascular conditioning*** — Poor cardiovascular conditioning can result in heart attacks, strokes, or other related problems during strenuous activity.
- ***Respiratory functioning*** — Proper respiratory functioning will maximize the wearer's operation time in a self-contained breathing apparatus.

Mental

- ***Adequate training in equipment use*** — The firefighter must be knowledgeable in every aspect of protective breathing apparatus use (Figure 4.33).
- ***Self-confidence*** — The firefighter's belief in his ability will have an extremely positive overall effect on the actions that are performed.
- ***Emotional stability*** — The ability to maintain control in an excited or high stress environment will reduce the chances of a serious mistake being made.

Figure 4.33 The firefighter must be thoroughly familiar with the use of SCBA.

LIMITATIONS OF EQUIPMENT

In addition to being concerned about the limitations of the wearer, firefighters must also be cognizant of the limitations of the equipment.

- ***Limited visibility*** — The facepiece reduces peripheral vision, and facepiece fogging can reduce overall vision.
- ***Decreased ability to communicate*** — The facepiece hinders voice communication.
- ***Increased weight*** — The protective breathing equipment adds 25 to 35 pounds (11 kg to 16 kg) of weight to the firefighter, depending on the model.
- ***Decreased mobility*** — The increase in weight and the splinting effect of the harness straps reduce the firefighter's mobility (Figure 4.34).

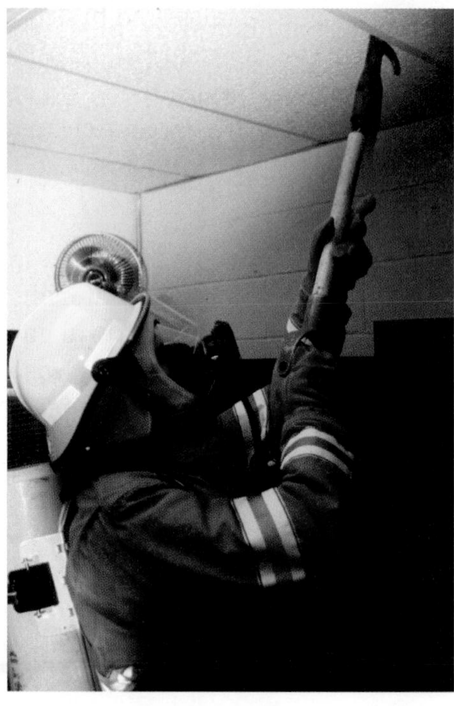

Figure 4.34 SCBA decreases the firefighter's mobility; for example, working above one's head can become difficult.

LIMITATIONS OF AIR SUPPLY

Air supply is another factor to consider when discussing protective breathing apparatus limitations. Some limitations are based on the apparatus user whereas others are based on the actual supply of air in the cylinder.

- ***Physical condition of user*** — The poorer the firefighter's physical condition, the faster the air supply is expended.

- ***Degree of physical exertion*** — The higher the physical exertion, the faster the air supply is expended.
- ***Emotional stability of user*** — The firefighter who becomes excited increases respiratory rate and uses air faster than a calm firefighter.
- ***Condition of apparatus*** — Minor leaks and poor adjustment of regulators result in excess air loss.
- ***Cylinder pressure before use*** — If the cylinder is not filled to capacity, the amount of working time is reduced proportionately.
- ***Training and experience of user*** — Properly trained and highly experienced personnel are able to draw the maximum air supply from a cylinder.

Types of Breathing Apparatus

There are two types of self-contained breathing apparatus used in the fire service: *open-circuit* and *closed-circuit.* Open-circuit SCBA is used much more frequently than closed-circuit SCBA. In fact, closed-circuit breathing apparatus is rarely used in today's fire service. Open-circuit SCBA uses compressed air; closed-circuit uses compressed or liquid oxygen. The exhaled air in open-circuit SCBA is vented to the outside atmosphere. Closed-circuit SCBA is also known as *rebreather* apparatus because the user's exhaled air stays within the system for reuse. Closed-circuit SCBA and open-circuit airline equipment are only used in some extended hazardous materials and rescue operations. Regardless of the type of SCBA used, training in its use is essential.

OPEN-CIRCUIT SELF-CONTAINED BREATHING APPARATUS

Several companies manufacture open-circuit SCBA, each with different design features or mechanical construction (Figure 4.35). Certain parts, such as cylinders and backpacks, are interchangeable; however, such substitution voids NIOSH and Mine Safety and Health Administration (MSHA) certification and is not a recommended practice. Substituting different parts may also void warranties and leave the department or firefighter liable for any injuries incurred.

Figure 4.35 Most commonly used SCBA in the fire service are the open-circuit type.

There are four basic SCBA component assemblies:

- ***Backpack and harness assembly*** — Holds the air cylinder on the firefighter's back
- ***Air cylinder assembly*** — Includes cylinder, valve, and pressure gauge
- ***Regulator assembly*** — Includes high-pressure hose and low-pressure alarm
- ***Facepiece assembly*** — Includes facepiece lens, an exhalation valve, and a low-pressure hose (breathing tube) if the regulator is separate; also includes head harness or helmet mounting bracket

Backpack and harness assembly. The backpack assembly is designed to hold the air cylinder on the firefighter's back as comfortably and securely as possible. Adjustable harness straps provide a secure fit for whatever size the individual requires. The waist straps are designed to help properly distribute the weight of the cylinder or pack to the hips (Figure 4.36). One problem is that waist straps are often not used or are removed. NIOSH and MSHA certify the entire SCBA unit, and removal of waist straps could void warranties.

Air cylinder assembly. Because the cylinder must be strong enough to safely contain the high pressure of the compressed air, it constitutes the main weight of the breathing apparatus (Figure 4.37). The weight of air cylinders varies with each manufacturer and depends on the material used to fabricate the cylinder. Manufacturers offer cylinders of various sizes, capacities, and features to correspond to their varied uses in responses. Commonly found sizes of air cylinders used in fire, rescue, and hazardous materials responses include the following:

- 30-minute, 2,216 psi (15 290 kPa), 45 ft^3 (1 270 L) cylinders
- 30-minute, 4,500 psi (31 000 kPa), 45 ft^3 (1 270 L) cylinders
- 45-minute, 3,000 psi (21 000 kPa), 66 ft^3 (1 870 L) cylinders
- 45-minute, 4,500 psi (31 000 kPa), 66 ft^3 (1 870 L) cylinders
- 60-minute, 4,500 psi (31 000 kPa), 87 ft^3 (2 460 L) cylinders

Regulator assembly. Air from the cylinder travels through the high-pressure hose to the regulator. The regulator reduces the pressure of the cylinder air to slightly above atmospheric pressure and controls the flow of air to meet the respiratory requirements of the wearer (Figure 4.38). When the firefighter inhales, a pressure differential is created in the regulator. The apparatus diaphragm moves inward, tilting the admission valve so that low-pressure air can flow into the facepiece. The diaphragm is then held open, which creates the positive pressure. Exhalation moves the diaphragm back to the "closed" position. Some SCBA units have regulators that fit into the facepiece (Figure 4.39). On other units, the regulator is on the firefighter's chest or waist strap.

Depending on the SCBA model, it will have control valves for normal and emergency operations. These are the mainline valve and the bypass valve (Figure 4.40). During normal operation, the

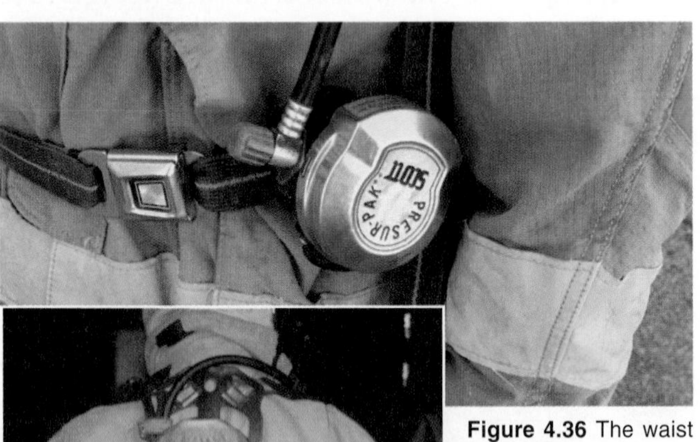

Figure 4.36 The waist strap helps to distribute the weight of the cylinder.

Figure 4.37 The air cylinder constitutes the main weight of the SCBA.

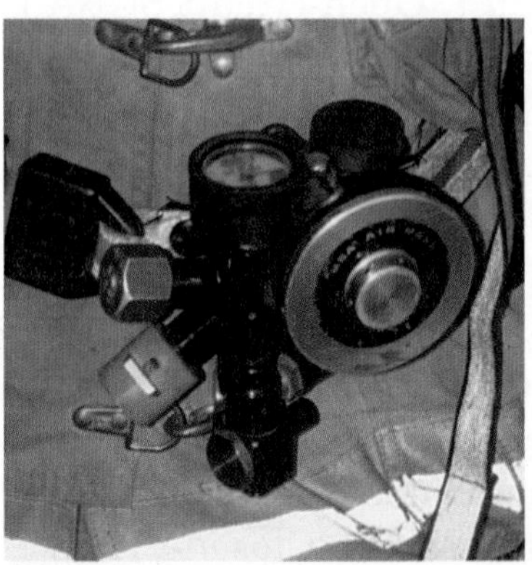

Figure 4.38 The regulator controls the flow of air to meet the respiratory requirements of the user.

Figure 4.39 This regulator connects directly to the facepiece.

Figure 4.40 The mainline valve (bottom) and the bypass valve (top).

mainline valve is fully open and locked if there is a lock. The bypass valve is closed. On some SCBA, the bypass valve controls a direct airline from the cylinder in the event that the regulator fails. Once the valves are set in their normal operating position, they should not be changed unless the emergency bypass is needed (Figure 4.41).

Figure 4.41 A firefighter operating the bypass valve.

A remote pressure gauge that shows the air pressure remaining in the cylinder is mounted in a position visible to the wearer (Figure 4.42). This remote pressure gauge should read within 100 psi (700 kPa) of the cylinder gauge if increments are in psi (kPa). If increments are shown in other measurements, such as percents or fractions, both measurements should be the same (Figure 4.43). These pressure readings are most accurate at or near the upper range of the gauge's rated working pressures. Low pressures in the cylinder may cause inconsistent readings between the cylinder and regulator gauges. If they are not consistent, rely on the lower reading and check the equipment for any needed repair before using it again. All units have an audible alarm that sounds when the cylinder pressure decreases to approximately one fourth of the maximum rated pressure of the cylinder, depending on the manufacturer. SCBA teams should leave the fire area *immediately* after the first firefighter's alarm sounds (Figure 4.44).

Figure 4.42 Remote pressure gauge.

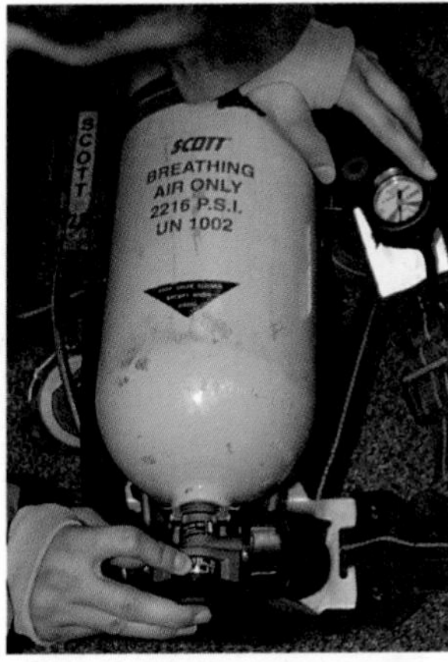

Figure 4.43 The cylinder pressure gauge reading and the remote gauge reading should be compared.

Figure 4.44 This alarm sounds when the firefighter needs to leave the area because of a low air supply.

Facepiece assembly. A facepiece provides some protection from facial and respiratory burns and holds in the cool breathing air. The facepiece assembly consists of the facepiece lens, an exhalation valve, and a low-pressure hose to carry the air from the regulator to the facepiece if the regulator is separate (Figure 4.45). The facepiece lens is made of clear safety plastic and is connected to a flexible rubber mask. The facepiece is held snugly against the face by a head harness with adjustable straps, net, or some other arrangement (Figure 4.46). Some helmets have a face mask bracket that connects directly to the

Figure 4.45 This low-pressure hose carries the air from the regulator to the facepiece.

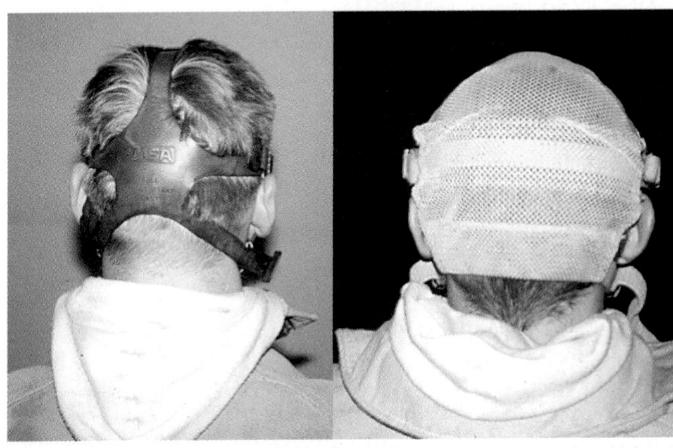

Figure 4.46 A head harness with adjustable straps and a mesh or hairnet model.

helmet instead of using a head harness (Figure 4.47). The lens should be protected from scratches during use and storage. Some facepieces have a speech diaphragm to make communication clearer.

The facepiece for an SCBA with a harness-mounted regulator has a low-pressure hose, or breathing tube, attached to the facepiece with a clamp or threaded coupling nut. The low-pressure hose brings air from the regulator into the facepiece; therefore, it must be kept free of kinks and away from contact with abrasive surfaces (Figure 4.48). The hose is usually corrugated to prevent collapse when a person is working in close quarters, breathing deeply, or leaning against a hard surface. Some units have no low-pressure hose because the regulator is attached directly to the facepiece.

Figure 4.47 This helmet has an SCBA facepiece bracket that connects directly to the helmet.

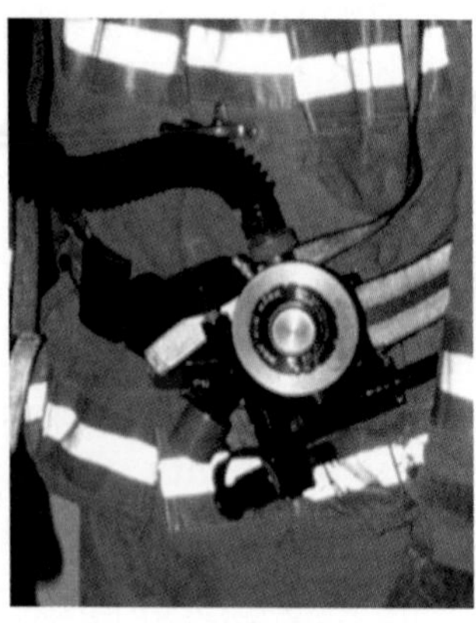

Figure 4.48 The low-pressure hose must be kept free of kinks.

Figure 4.49 The exhalation valve is a simple, one-way valve that releases an exhaled breath without admitting any of the contaminated outside atmosphere.

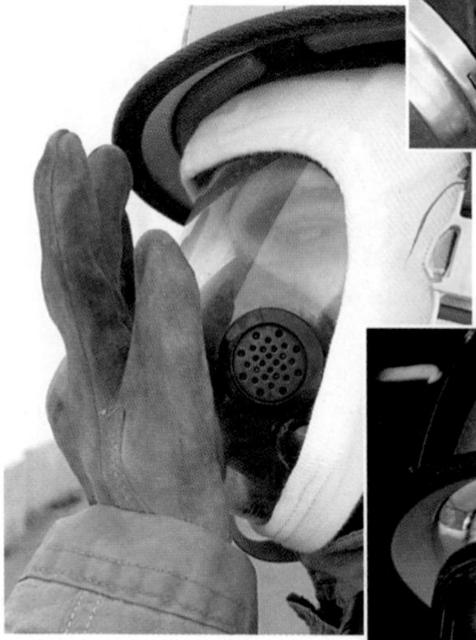

Figure 4.50 Test the exhalation valve before entering a hazardous atmosphere.

Figure 4.51 A fogged facepiece can hamper a firefighter's vision.

The exhalation valve is a simple, one-way valve that releases an exhaled breath without admitting any of the contaminated outside atmosphere (Figure 4.49). Dirt or foreign materials can cause the valve to become partially opened, which may permit excess air from the tank to escape the facepiece. Therefore, it is important that the valve be kept clean and free of foreign material. It is also important that the exhalation valve be tested by the firefighter during facepiece-fit tests and before entering a hazardous atmosphere (Figure 4.50).

An improperly sealed facepiece or a fogged lens can cause problems for the wearer. The different temperatures inside and outside the facepiece where the exhaled air or outside air is moist can cause the facepiece lens to fog, which hampers vision (Figure 4.51). Internal fogging occurs when the lens is cool, causing the highly humid exhaled breath to condense. As the cooler, dry air from the cylinder passes over the facepiece lens, it often removes the condensation. External fogging occurs when condensation collects on the relatively cool lens during interior fire fighting operations.

External fogging can be removed by wiping the lens. One of the following methods can be used to prevent or control internal fogging of a lens.

- ***Use a nosecup*** — Facepieces can be equipped with a nosecup that deflects exhalations away from the lens (Figure 4.52). However, if the nosecup does not fit well, it will permit exhaled air to leak into the facepiece and condense on the lens.
- ***Apply an antifogging chemical*** — Special antifogging chemicals recommended by the manufacturer can be applied to the lens

of the facepiece. Some SCBA facepieces are permanently impregnated with an antifogging chemical.

When storing the facepiece, it may be packed in a case or stored in a bag or coat pouch (Figure 4.53). Wherever it is stored, the straps should be left fully extended for donning ease and to keep the facepiece from becoming distorted.

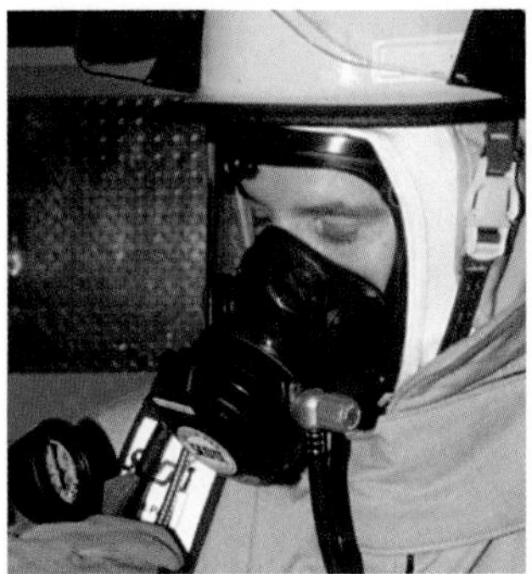

Figure 4.52 Facepiece nosecups help reduce mask fogging.

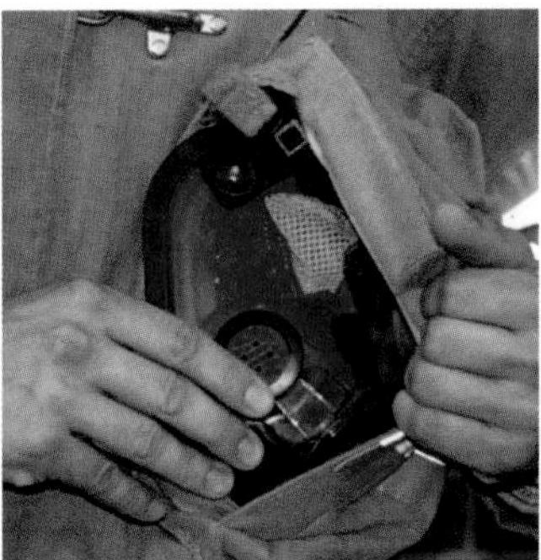

Figure 4.53 Facepiece stored in a coat pouch.

OPEN-CIRCUIT AIRLINE EQUIPMENT

Incidents involving hazardous materials or rescues often require a longer air supply than can be obtained from standard open-circuit SCBA. In these situations, an airline attached to one or several large air cylinders can be connected to an open-circuit facepiece, regulator, and egress cylinder (Figure 4.54). Airline equipment enables the firefighter to travel limited distances from the *regulated* air supply source, allowing the firefighter to work for several hours without the encumbrance of a backpack. For more information on open-circuit airline equipment, see IFSTA's **Self-Contained Breathing Apparatus** manual.

Figure 4.54 Airline breathing equipment is useful for extended duration operations.

CLOSED-CIRCUIT BREATHING APPARATUS

Closed-circuit breathing apparatus are not used in the fire service as commonly as open-circuit breathing apparatus. However, they are sometimes used for hazardous materials incidents because of their longer air supply duration (Figure 4.55). Closed-circuit SCBA are available with durations of 30 minutes to 4 hours and usually weigh less than open-circuit units of similarly rated service time. They weigh less because a smaller cylinder containing pure oxygen is used. For more information on closed-circuit breathing apparatus, see IFSTA's **Self-Contained Breathing Apparatus** manual.

Figure 4.55 A typical closed-circuit SCBA.

Mounting Protective Breathing Apparatus

Methods of storing self-contained breathing apparatus vary from department to department. Each department should use the most appropriate method to facilitate quick and easy donning (Figure 4.56). SCBA can be placed on the apparatus in seat mounts, side mounts, and compartment mounts, and stored in cases. If placed in seat mounts, the SCBA should be arranged so that it may be donned without the firefighter having to remove the seat belt.

Personal Alert Safety Systems

The use of personal alert safety system (PASS) devices by all firefighters and rescuers is mandatory under NFPA 1500. (The acronym PAD [personal alert device] is also used). A downed or disoriented firefighter inside a structure poses a

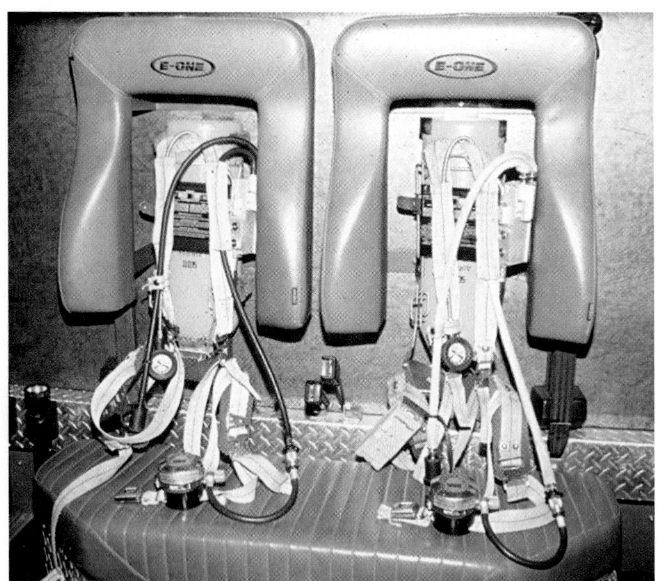

Figure 4.56 Seat-mounted SCBA are shown in the fire fighting vehicle.

severe rescue problem. PASS devices are designed to assist rescuers attempting to locate the firefighter, even in dense smoke. The device, about the size of a portable transistor radio, is worn on the firefighter's self-contained breathing apparatus or coat, and a switch is turned on before entering a structure (Figure 4.57). If the firefighter should collapse or remain motionless for approximately 30 seconds, the PASS device will emit a loud, pulsating shriek. It can also be activated manually. Either way, rescuers can follow the sound to locate the lost or downed firefighter. Some SCBA manufacturers have integrated a distress alarm system into the SCBA air circuit. Once the cylinder valve is opened, the distress alarm system is automatically activated. This type of system can also be activated manually without opening the cylinder valve.

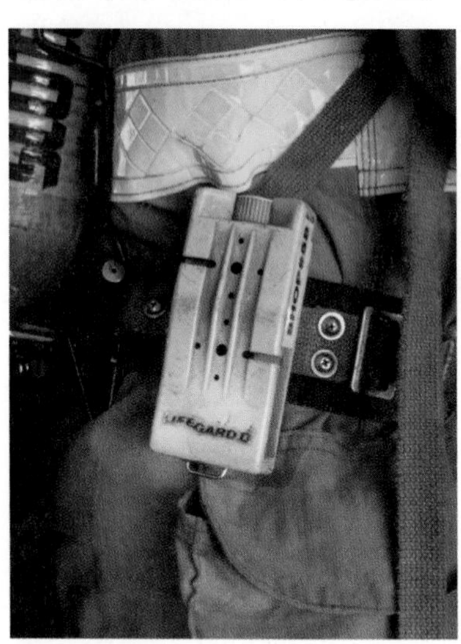

Figure 4.57 PASS devices can save firefighters' lives.

PASS devices can save lives, but they must be used and maintained properly. The user must remember to turn on and test the device before entering a structure. Training classes should be conducted on techniques to be used when attempting rescue of a lost firefighter. Locating even the loud shriek of a PASS device in poor visibility conditions can be more difficult than expected because the sound reflects off walls, ceilings, and floors. Rescuers have a tendency to sidestep established search procedures when they think they can tell the location of the alarm sound. Noise from SCBA operation and muffled hearing because of protective hoods also adds to the difficulty. Recommendations for use of PASS devices include the following:

- Make sure that the system selected meets the requirements of NFPA 1982, *Standard on Personal Alert Safety Systems (PASS) for Fire Fighters*.
- Test the PASS at least weekly, and maintain in accordance with manufacturer's instructions.
- Conduct practical training with the PASS under realistic conditions to teach firefighters how to react appropriately to PASS alarm activations.
- Retrain every six months with PASS devices.
- Train firefighters to always turn on and test the device before entering a hazardous atmosphere.
- Train rescuers to listen for the distress sound by stopping in unison, controlling breathing, and lifting hood or earflaps away from ears.
- Turn the PASS device off to facilitate communications when a downed firefighter is located.

DONNING AND DOFFING PROTECTIVE BREATHING APPARATUS

[NFPA 1001: 3-1.1.2; 3-3.1; 3-3.1(a); 3-3.1(b)]

Several methods can be used to don self-contained breathing apparatus, depending on how the SCBA is stored. The methods used in the fire service include the over-the-head method, the coat method, donning from a seat, and donning from a rear mount or compartment mount. The steps needed to get the SCBA onto the firefighter vary somewhat with each method. Also, there are different steps for securing different makes and models of self-contained breathing apparatus. Due to the variety of SCBA, it is impossible to list step-by-step procedures for each manufacturer's model. Therefore, the information in this section is intended only as a general description of the different donning techniques. The wearer should follow manufacturer's instructions and local standard operating procedures for donning and doffing their particular SCBA.

General Donning Considerations

Regardless of the SCBA model or method of donning, several precautionary safety checks should be made prior to donning the SCBA. For departments that have daily shift changes, these checks may occur at shift change. The apparatus is then placed in the apparatus-mounted storage racks or placed back in the storage case. For departments that are unable to inspect their breathing apparatus daily, these checks should be made immediately prior to donning the SCBA no matter how it is stored.

- Check the air cylinder gauge to ensure that the cylinder is full. NFPA 1404, *Standard for a Fire Department Self-Contained Breathing Apparatus Program,* recommends no less than 90 percent of cylinder capacity (Figure 4.58).
- Check the remote gauge and cylinder gauge to ensure that they read within 100 psi (700 kPa) of the same pressure. Gauges not marked in increments of 100 psi (700 kPa) should read relatively close to each other.
- Check the harness assembly and facepiece to ensure that all straps are fully extended (Figure 4.59).
- Check all valves to ensure that they are in the proper position.

Once these checks are complete, the protective breathing apparatus may be donned using the most appropriate method.

Donning From a Storage Case

The following donning methods require the SCBA to be positioned in front of the firefighter, ready to don.

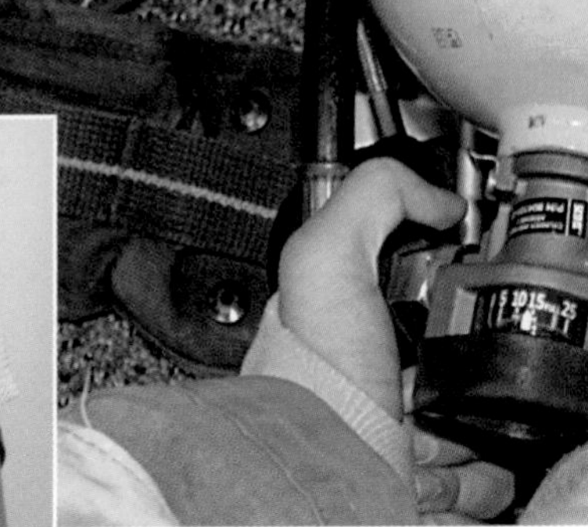

Figure 4.58 Check the cylinder gauge to make sure that the air cylinder is full.

Figure 4.59 The harness assembly and facepiece straps should be fully extended.

- ***Over-the-head-method*** — The harness assembly is raised overhead. As the SCBA slides down the wearer's back, the arms slide into their respective harness shoulder strap loops (Figure 4.60).
- ***Coat method*** — The SCBA is donned like a coat, putting one arm at a time through the shoulder strap loops. The unit should be arranged so that *either* shoulder strap can be grasped for lifting (Figure 4.61).

Skill Sheet 4-1 describes the general procedures for donning full protective clothing and SCBA.

Figure 4.60 The over-the-head method.

Figure 4.61 The coat method.

Donning From a Seat Mount

Valuable time can be saved if the SCBA is mounted on the back of the firefighter's seat in the vehicle (Figure 4.62). By having a seat mount, firefighters can don SCBA while en route to an incident. Donning from a seat mount should only be done, however, if it can be safely accomplished without the firefighter having to remove his seat belt.

Figure 4.62 An SCBA seat-mounting bracket.

Seat-mounting hardware comes in three main types: lever clamp, spring clamp, or flat hook. Part of this hardware is a hold-down device for securing the SCBA to the bracket. A drawstring or other quick-opening bag should enclose the facepiece to keep it clean and to protect it from dust and scratches. (**NOTE:** Do not keep the facepiece connected to the regulator during storage. These parts must be separate to check for proper facepiece seal.)

Donning en route is accomplished by releasing the hold-down device, inserting the arms through the straps while sitting with the seat belt on, then adjusting the straps for a snug fit (Figures 4.63 and 4.64).

Figure 4.63 When donning en route, the first step is to insert the arms through the straps.

Figure 4.64 The second step is to adjust the straps for a snug fit.

WARNING

Never stand to don SCBA while the vehicle is moving. Standing places both you and other firefighters in danger of serious injury in the event of a fall. NFPA 1500 requires firefighters to remain seated and belted at all times while the emergency vehicle is in motion.

The cylinder's position should match the proper wearing position for the firefighter. The visible seat-mounted SCBA reminds and even encourages personnel to check the equipment more frequently. Because it is exposed, checks can be made more conveniently. When exiting the fire apparatus, be sure to adjust the straps for a snug and comfortable fit.

Side or Rear Mount

Although it does not permit donning en route, the side- or rear-mounted SCBA may be desirable (Figure 4.65). This type of mount saves time because the following steps are eliminated: removing the equipment case from the fire apparatus, placing it on the ground, opening the case, and picking up the unit. However, because the unit is exposed to weather and physical damage, a canvas cover is desirable (Figure 4.66).

If the mounting height is right, firefighters can don SCBA with little effort. Having the mount near the running boards or near the tailboard allows the firefighter to don the equipment while sitting. The donning steps are essentially the same as those for seat-mounted SCBA.

Figure 4.65 Some apparatus have SCBA mounted on the sides for easy donning. *Courtesy of Ron Bogardus.*

Figure 4.66 Canvas covers help protect the SCBA from excess dirt and moisture. *Courtesy of Ron Bogardus.*

Compartment or Backup Mount

SCBA stored in a closed compartment can be ready for rapid donning by using any number of methods (Figure 4.67). A mount on the inside of a compartment presents the same advantages as does side-mounted equipment. Some compartment doors, however, may not allow a firefighter to stand fully while donning SCBA. Other compartments may be too high for the firefighter to don the SCBA properly.

Certain compartment mounts feature a telescoping frame that holds the equipment out of the way inside the compartment when it is not needed (Figure 4.68). One type of compartment mount telescopes outward, then upward or downward to proper height for quick donning.

The backup mount provides quick access to SCBA (some high-mounted SCBA must be removed from the vehicle and donned using the over-the-head or coat method). The procedure for donning SCBA using the backup method is similar to the method used for mounts from which the SCBA can be donned while seated.

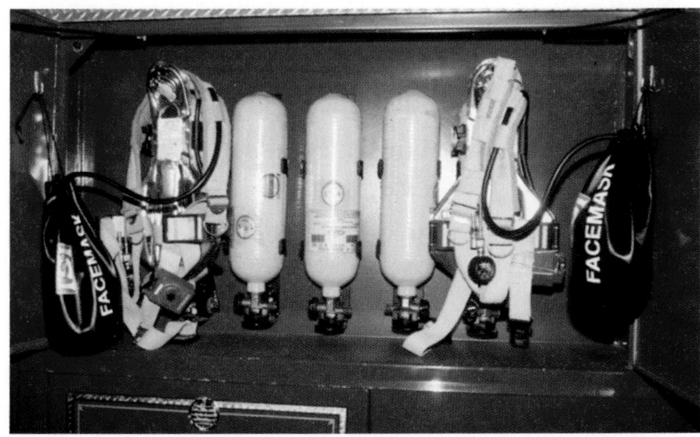

Figure 4.67 A compartment-mount installation.

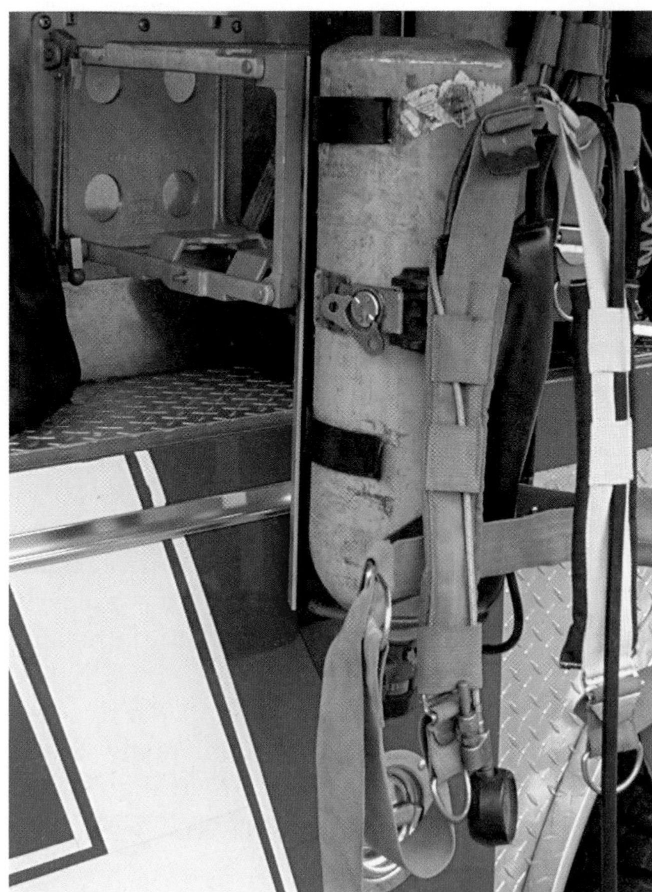

Figure 4.68 A compartment mount featuring a telescoping frame to hold the equipment inside the compartment provides the proper height for donning.

Donning the Facepiece

The facepieces for most SCBA are donned similarly. One important difference in facepieces is the number of straps used to tighten the head harness. Different models from the same manufacturer may have a different number of straps. Another important difference is the location of the regulator. The regulator may be attached to the facepiece or mounted on the waist belt. The shape and size of facepiece lenses may also differ. Despite these variations, the uses and donning procedures for facepieces are essentially the same.

NOTE: Interchanging facepieces, or any other part of the SCBA, from one manufacturer's equipment to another makes any warranty and certification void.

An SCBA facepiece cannot be worn loosely, or it will not seal against the face properly. An improper seal may permit toxic gases to enter the facepiece and be inhaled. Firefighters shall not let long hair, sideburns, beards, or facial hair interfere with the seal of the facepiece, thus preventing contact and a proper seal with the skin. Most fire departments simplify this policy by insisting that firefighters be clean shaven. Temple pieces of glasses and missing dentures can also affect facepiece fit.

A firefighter should not rely solely on tightening facepiece straps to ensure proper facepiece fit. A facepiece tightened too much will be uncomfortable or may cut off circulation to the face. Each firefighter must be fitted with a facepiece that conforms properly with the face shape and size (Figure 4.69). For this reason, many SCBA are available with different sized facepieces. Nosecups, if used, must also properly fit the firefighter.

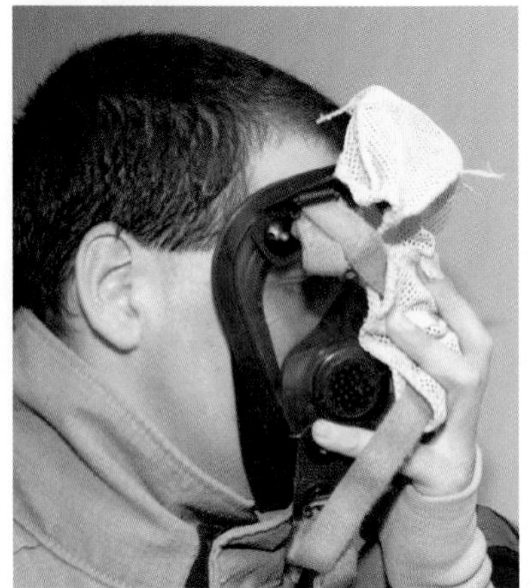

Figure 4.69 A firefighter trying on a facepiece for proper fit.

The following are general considerations for donning all SCBA facepieces.

- No hair should come between the skin and the sealing surface of the facepiece.
- The chin should be centered in the chin cup, and the harness is centered at the rear of the head.
- Facepiece straps should be tightened by pulling them evenly and simultaneously to the rear. Pulling the straps outward, to the sides, may damage them and prevent proper engagement with the adjusting buckles. Tighten the lower straps first, then the temple straps, and finally the top strap if there is one.
- The facepiece should be checked for proper seal and operation (exhalation valve functioning properly, all connections secure, and donning mode switch in proper position if present).
- Positive pressure should be checked by gently breaking the facepiece seal. This can be done by inserting two fingers under the edge of the facepiece (Figure 4.70). You should be able to feel air moving past your fingers. If you cannot feel air movement, remove the unit and have it checked.
- The hood is worn over the facepiece harness or straps. All exposed skin must be covered, and vision must not be obscured (Figure 4.71). No portion of the hood should be located between the facepiece and the face.
- The helmet should be worn with all straps secured.

NOTE: Helmets are provided with adjustable chin straps to ensure helmets remain on firefighters' heads during fire fighting operations. This is especially important while operating inside structures.

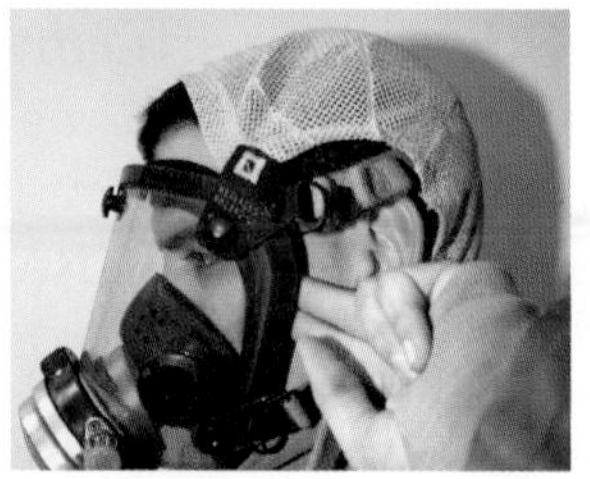

Figure 4.70 Check for positive pressure by inserting two fingers under the edge of the facepiece.

Figure 4.71 The protective hood is worn over the facepiece.

Doffing SCBA

Doffing techniques differ for different types of SCBA. Generally, there are certain actions that apply to all SCBA when doffing.

- Make sure you are out of the contaminated area and that SCBA is no longer required.
- Discontinue the flow of air from the regulator to the facepiece.
- Disconnect the low-pressure hose from the regulator or remove the regulator from the facepiece, depending upon type of SCBA.
- Remove the facepiece.
- Remove the backpack assembly while protecting the regulator.
- Close cylinder valve.
- Relieve pressure from the regulator in accordance with manufacturer's instructions.
- Extend all straps.
- Refill and replace the cylinder.
- Clean and disinfect the facepiece.

INSPECTION AND MAINTENANCE OF PROTECTIVE BREATHING APPARATUS

[NFPA 1001: 3-5.3; 3-5.3(a); 3-5.3(b)]

NFPA 1404 and NFPA 1500 require all SCBA to be inspected after each use, weekly, monthly, and annually (Figure 4.72).

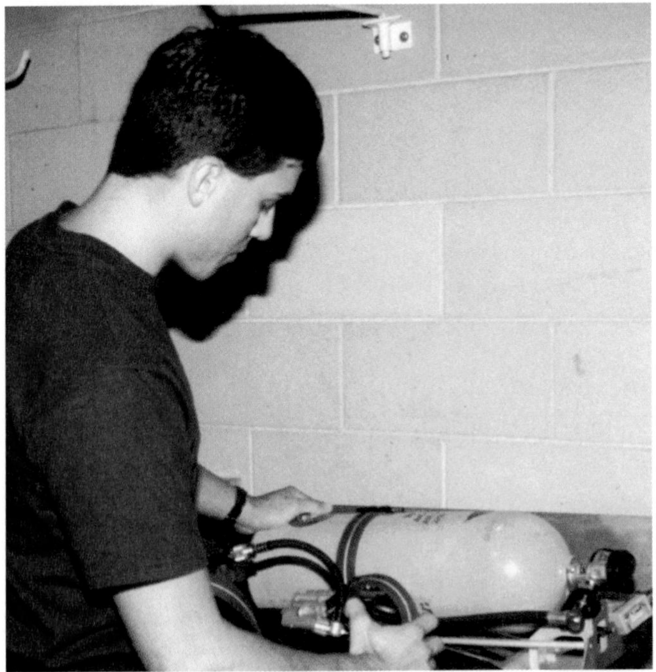

Figure 4.72 Personnel should inspect SCBA after each use, weekly, monthly, and annually.

Daily/Weekly Inspections

Self-contained breathing apparatus requires proper care and inspection before and after each use to provide complete protection. Proper care should include making a daily inspection as soon as possible after reporting for duty. Some organizations may not be able to check the units every day. In this case, the SCBA should be checked at least weekly and after each use. The following is a list of things to check.

- Cylinder is full.
- All gauges work. The cylinder gauge and the remote gauge should read within 100 psi (700 kPa) of each other. Gauges not marked in increments of 100 psi (700 kPa) should read relatively close to each other.
- Low-pressure alarm is in working condition. The alarm should sound briefly when the cylinder valve is turned on and again as the pressure is relieved.
- All hose connections are tight and free of leaks.
- Facepiece is clean and in good condition.
- Harness system is in good condition and straps are in the fully extended position.
- All valves are operational. After checking the bypass valve, make sure that it is fully closed.

Breathing apparatus should be cleaned and sanitized immediately after each use. Moving parts that are not clean may malfunction. A facepiece that has not been cleaned and sanitized may contain an unpleasant odor and can spread germs to other department members who may wear the mask at a later time. An air cylinder with less air than prescribed by the manufacturer renders the apparatus inefficient if not useless.

The facepiece should be thoroughly washed with warm water containing any mild commercial disinfectant, and then it should be rinsed with clear, warm water (Figure 4.73). Special care should be given to the exhalation valve to ensure proper operation. The air hose should be inspected for cracks or tears. The facepiece should be dried with a lint-free cloth or air dried.

CAUTION: Do not use paper towels to dry the lens as the paper towel will scratch the plastic lens.

Many departments now issue personal facepieces to each firefighter. This eliminates the

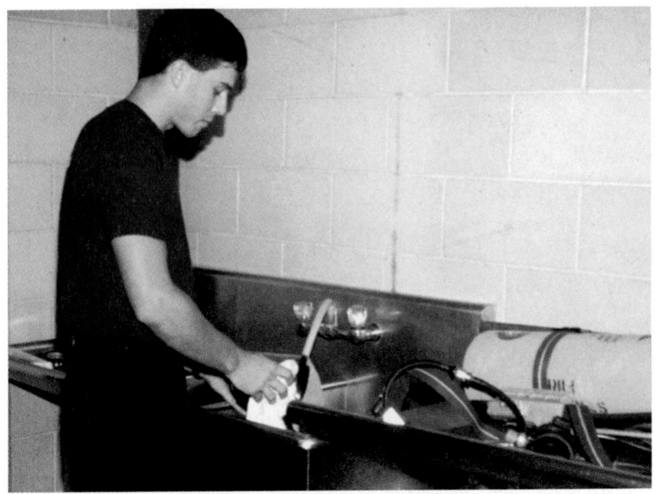

Figure 4.73 The facepiece should be thoroughly washed.

risk of spreading germs from one wearer to the next. Even though each firefighter has his own facepiece, it is still important that it be cleaned after each use.

Monthly Inspection and Maintenance

Monthly inspections should include removing the equipment from service and checking the following:

- All components for deterioration
- Leaks around valves and air cylinder connections
- Operation of all gauges, valves, regulator, exhalation valve, and low-air alarm

Annual Inspection and Maintenance

Annual maintenance, testing, and repairs requiring the expertise of factory certified technicians should be done in accordance with manufacturer's recommendations. This level of maintenance requires specialized training. The service provider must be able to disassemble the apparatus into its basic components and conduct tests using specialized tools and equipment generally not available to all fire departments.

Air cylinders must be stamped or labeled with the date of manufacture and the date of the last hydrostatic test (Figure 4.74). Steel and aluminum cylinders must be tested every five years; composite cylinders every three years. This procedure is necessary to meet the requirements of the United States Department of Transportation. Always empty cylinders before returning them for servicing and testing.

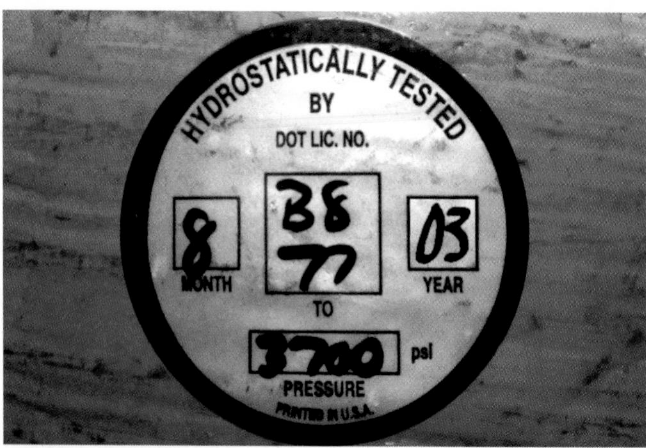

Figure 4.74 Each cylinder must be stamped with the date of the most recent hydrostatic test.

Reservicing Self-Contained Breathing Apparatus Cylinders

Air cylinders for self-contained breathing apparatus are filled from either a cascade system (a series of at least three, 300 cubic-foot [8 490 L] cylinders) or directly from a compressor purification system (Figures 4.75 a and b). No matter how the cylinders are filled, the same safety precautions

Figure 4.75a Cylinders can be refilled from a cascade system.

Figure 4.75b Cylinders can also be filled from a compressor purification system.

apply: Put the cylinders into a shielded fill station, prevent cylinder overheating by filling slowly, and be sure that the cylinder is completely full but not overpressurized. Skill Sheet 4-2 provides a sample procedure for filling an SCBA cylinder from a cascade system. Skill Sheet 4-3 provides a sample procedure for filling an SCBA cylinder from a compressor/purifier.

USING SELF-CONTAINED BREATHING APPARATUS

[NFPA 1001: 3-3.1; 3-3.1(a); 3-3.1(b); 3-3.4; 3-3.4(a); 3-3.4(b); 3-3.8(a); 3-3.8(b)]

Firefighters have to wear SCBA in many different types of incidents. In addition to being familiar with the donning, operation, and doffing of protective breathing apparatus, the firefighter must also be trained in the safe use of the apparatus. The preceding sections of this chapter discussed why and how to operate self-contained breathing apparatus. This section covers safety precautions when using SCBA, emergency situations that may arise when using SCBA, and the use of SCBA in areas of obscured vision and restricted openings.

Safety Precautions for SCBA Use

Fire fighting is a strenuous, demanding activity, so firefighters need to be in good physical condition. Although protective gear is designed to protect firefighters, it can also work against them at the same time. The basic required protective coat can be a virtual sweatbox. It builds up body heat and hinders movement that increases firefighter exhaustion. This condition is intensified when self-contained breathing apparatus is used under emergency conditions. The difference between the weight of ordinary street clothes and fire fighting gear plus the SCBA unit has been measured at an extra 47 pounds (21 kg). The breathing unit alone can weigh from 25 to 35 pounds (11 kg to 16 kg), depending on size and type. All firefighters should be aware of the signs and symptoms of heat-related conditions that can occur under these situations. *Know your own limitations and abilities!*

When using self-contained breathing apparatus, the following items should be remembered and observed for maximum safety.

- All firefighters who wear SCBA should be certified as physically fit by a physician using criteria established by the fire department.
- Firefighters should closely monitor how they are feeling while wearing SCBA and rest when they become fatigued (Figure 4.76).

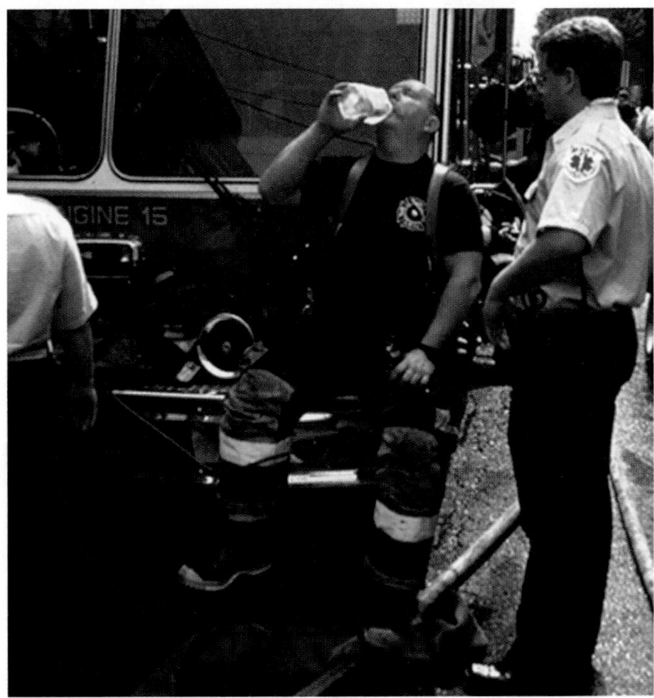

Figure 4.76 Firefighters who work in protective clothing are susceptible to heat-stroke.

- Air-supply duration will vary with the following:
  - — Firefighter conditioning
  - — Task performed
  - — Level of training
  - — Operational environment
  - — Degree of excitement
  - — Other variables
- Once entering a contaminated area, firefighters should not remove their breathing apparatus until they have left the contaminated area. Improved visibility does not ensure that the area is free from contamination.
- When wearing SCBA, firefighters should work in groups of two or more.

Emergency Situations

The emergencies created by the malfunction of protective breathing apparatus can be overcome in several ways. In all of these emergencies, the conservation of air and immediate withdrawal from the hazardous atmosphere are of the utmost importance. The following is a list of suggestions that can effectively resolve an emergency situation:

- Do not panic! Panicking causes rapid breathing that uses more valuable air.
  — Control breathing while crawling.
  — Communicate with other team members.
- Stop and think. How did you get to where you are? Downstairs? Upstairs? Left turns?
- Listen
  — For noise from other personnel.
  — For hose and equipment operation.
  — For sounds that indicate the location of fire.
- Use the portable radio to announce your last known location.
- Activate your PASS device.
- Place a flashlight on the floor with the light shining toward the ceiling.
- Remember the different methods to find a way out:
  — Follow the hoseline out if possible (male coupling is closest to exit, female is closest to the fire) (Figure 4.77).
  — Crawl in a straight line (hands on floor, move knee to hand).
  — Crawl in one direction (all left-hand turns, all right-hand turns) once in contact with the wall.
  — Call for directions, call out, or make noise for other firefighters to assist you.
  — Break a window or breach a wall to escape if possible.
- Lie flat on the floor close to a wall so that you will be easier to find if you are exhausted or feel you may lose consciousness.

Firefighters should practice controlled breathing when using SCBA. However, when air supply is low, they may practice skip breathing. *Skip breathing* is an emergency breathing technique used to extend the use of the remaining air supply. To use this technique, the firefighter inhales (as during regular breathing), holds the breath as long as it would take to exhale, and then inhales once again before exhaling. The firefighter should take normal breaths and exhale slowly to keep the carbon dioxide in the lungs in proper balance.

Although a regulator usually works as designed, it can malfunction. One method of using SCBA when the regulator becomes damaged or malfunctions is to open the bypass valve to provide a flow of air into the facepiece (Figure 4.78). The bypass valve should be closed after the firefighter takes a breath and then opened each time the next breath is needed.

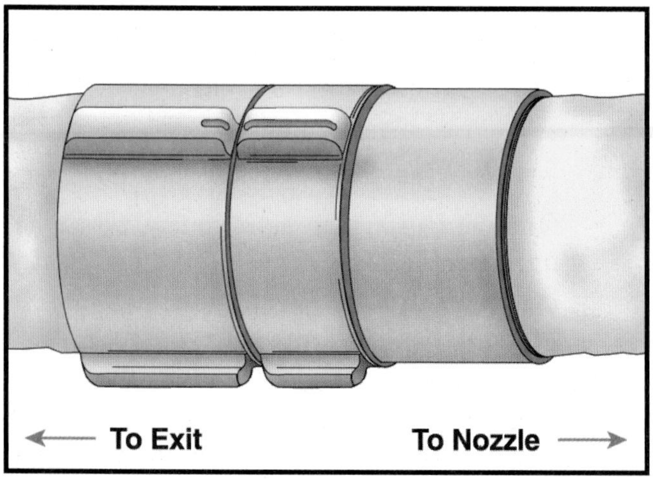

← To Exit
To Nozzle →

Figure 4.77 Hose couplings will indicate the direction toward the exit.

Figure 4.78 This firefighter is using the bypass valve because of regulator malfunction or damage.

If the facepiece fails, various extreme techniques may be available as emergency measures. A thorough knowledge of your department's SCBA is necessary. Firefighters should obtain this type of training based on their particular SCBA, following the manufacturer's recommendations and department's standard operating procedures.

Evacuation signals are used when command personnel decide that all firefighters should be pulled from a burning building or other hazardous area because conditions have deteriorated beyond the point of reasonable safety. All firefighters should be familiar with their department's method of sounding an evacuation signal. There are several ways this communication may be done. The two most common ways are to broadcast a radio message ordering them to evacuate and to sound the audible warning devices on the apparatus at the fire scene for an extended period of time. The radio broadcast of an evacuation signal should be handled in a manner similar to that described for emergency traffic. The message should be broadcast several times to make sure that everyone hears it. The use of audible warning devices on apparatus, such as sirens and air horns, works in small structures but may not be heard by everyone when working in a large building.

Special Uses of SCBA

In order to operate at maximum efficiency, the firefighter must be able to operate effectively in areas of obscured vision and negotiate tight passages without having to completely shed the breathing apparatus. The following sections address techniques for accomplishing these tasks.

OPERATING IN AREAS OF OBSCURED VISIBILITY

In many instances where protective breathing apparatus is required, firefighters will be operating in an area of obscured visibility. Most interior fire attacks and many exterior attacks present firefighters with heavy smoke conditions that may reduce visibility to zero. Firefighters must learn techniques for moving about and performing critical tasks when vision is diminished.

The primary method of moving about in areas of obscured visibility is by crawling. Crawling is beneficial for several reasons. First, it allows firefighters to remain close to the floor and avoid the higher heat found closer to ceiling level. Second, crawling allows firefighters to feel in front of themselves as they move along. This prevents them from falling through holes burned in the floor, falling through stair or elevator shafts, or running into objects in front of them. Crawling also allows firefighters to feel for victims who may be lying on the floor or feel for furniture. If firefighters can see the floor, they may be able to move about using a crouched or "duck" walk (Figure 4.79). This method is slightly faster than crawling but is more dangerous unless firefighters can clearly see the floor in front of them.

When entering an area of obscured visibility, firefighters must always operate in teams of at least two, and they should always have some sort of guideline that leads them back to the point of entrance if necessary. The guideline may be a hoseline, rope, or electrical cord. In the event it becomes necessary to evacuate the structure in a hurry, the firefighters can turn around and follow the guideline to safety. If for some reason the team does not have a guideline, or becomes separated from it, they should proceed to a wall and follow it until a door or window is found.

Figure 4.79 Firefighter using the "duck" walk.

EXITING AREAS WITH RESTRICTED OPENINGS UNDER EMERGENCY CIRCUMSTANCES

In an emergency, firefighters may need to exit an opening that is too small to allow them to pass through in a normal manner while wearing SCBA.

It may be necessary to loosen parts of the SCBA harness or remove the backpack completely, exit the restricted area, and resecure or redon the backpack. Removing the SCBA should be limited only to the extent to which it is necessary to exit the area. The procedures for removing the SCBA and maneuvering through an opening will be determined by the type of apparatus. Firefighters should be familiar with their particular SCBA. Some things to keep in mind include the following:

- Maintain contact with the regulator.
- Loosen straps as necessary to reduce your profile.
- Reduce your profile further by removing one or both backpack harness straps if absolutely necessary.
- Push the SCBA in front of you as necessary, maintaining control of the SCBA at all times.

Changing Cylinders

With care and caution, a firefighter can change an air cylinder at the scene of an emergency so that the equipment can be used again as soon as possible. A tarp can be placed on the ground to help protect cylinders that are not in use. Cylinders that are out of service should be marked and moved away from any cylinders that are serviced and ready for use. Changing cylinders can be either a one- or two-person job. Skill Sheet 4-4 describes the one-person method for changing an air cylinder. When there are two people, the firefighter with an empty cylinder simply positions the cylinder so that it can be easily changed by the other firefighter (Figures 4.80 and 4.81).

Figure 4.80 One firefighter slides a full cylinder into the backpack assembly while the other firefighter braces to remain steady.

Figure 4.81 The firefighter receiving a full cylinder may choose to kneel while the cylinder is being replaced.

SKILL SHEET 4-1 DONNING PERSONAL PROTECTIVE EQUIPMENT

Step 1: Don protective coat, trousers, and boots.

Step 2: Pull protective hood down around the neck.

Step 3: Place gloves in a readily accessible location.

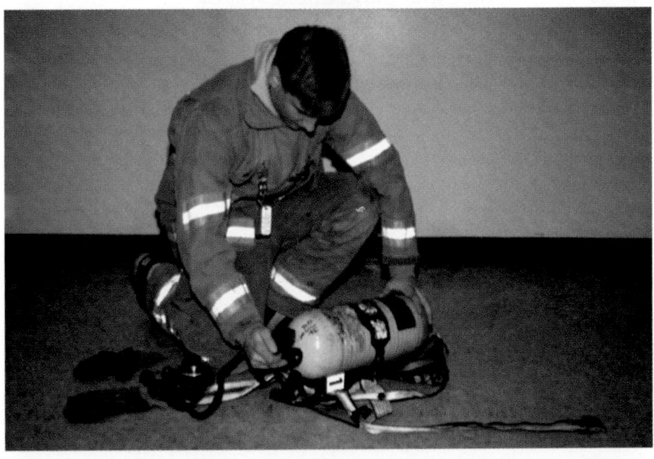

Step 4: Position SCBA in front of you ready for donning.

Step 5: Check the cylinder pressure (should be at least 90 percent full).

Step 6: Open the cylinder valve slowly and listen for the audible alarm as the system pressurizes.

Step 7: Verify the operation of the low air supply warning alarm.

NOTES:

- If the audible alarm does not sound or if it sounds but does not stop, place the unit out of service.
- On some styles of SCBA, the audible alarm does not sound when the cylinder valve is opened. Each firefighter must know the operation of his particular unit.

Step 8: Check the regulator gauge (remote gauge on some SCBA) and the cylinder gauge to ensure that they read within 100 psi (700 kPa) of the same pressure.

Step 9: Don the SCBA in accordance with manufacturer's recommendations (secure all straps, properly position the facepiece, check the exhalation valve, check facepiece seal, and attach low-pressure tube to the regulator or attach the regulator and air line to the facepiece depending on the style of SCBA; check donning mode switch if present, and activate the airflow; activate the PASS device).

Step 10: Place the hood and helmet in proper position for fireground operations.

Step 11: Don gloves.

SKILL SHEET 4-2 FILLING AN SCBA CYLINDER

From a Cascade System

NOTE: This skill sheet is for sample purposes only. The procedures outlined here may not be applicable to your cascade system. Always check the manufacturer's instructions before attempting to fill any cylinders.

Step 1: Check the hydrostatic test date of the cylinder.

Step 2: Inspect the SCBA cylinder for damage such as deep nicks, cuts, gouges, or discoloration from heat.

NOTE: If the cylinder is damaged or is out of hydrostatic test date, remove the cylinder from service and tag it for further inspection and hydrostatic testing.

CAUTION: Never attempt to fill a cylinder that is damaged or that is out of hydrostatic test date.

Step 3: Place the SCBA cylinder in a fragment-proof fill station.

Step 4: Connect the fill hose to the cylinder.

NOTE: If the fill hose has a bleed valve, make sure that it is closed.

Step 5: Open the SCBA cylinder valve.

Step 6: Open the valve at the fill hose, the valve at the cascade system manifold, or the valves at both locations if the system is so equipped.

NOTE: Some cascade systems may have a valve at the fill hose, at the manifold, or at both places.

Step 7: Open the valve of the cascade cylinder that has the least pressure but that has more pressure than the SCBA cylinder.

NOTE: The airflow from the cascade cylinder must be slow enough to avoid "chatter" or excessive heating of the cylinder being filled.

Step 8: Watch to see that the cylinder gauge needle rises slowly by about 300 to 600 psi (2 100 kPa to 4 200 kPa) per minute.

NOTE: Your hand should be able to rest on the SCBA cylinder without undue discomfort from the heating of the cylinder.

Step 9: Close the cascade cylinder valve when the pressures of the SCBA and the cascade cylinder equalize.

NOTE: If the SCBA cylinder is not yet completely full, open the valve on the cascade cylinder with the next highest pressure.

Step 10: Repeat Step 9 until the SCBA cylinder is completely full.

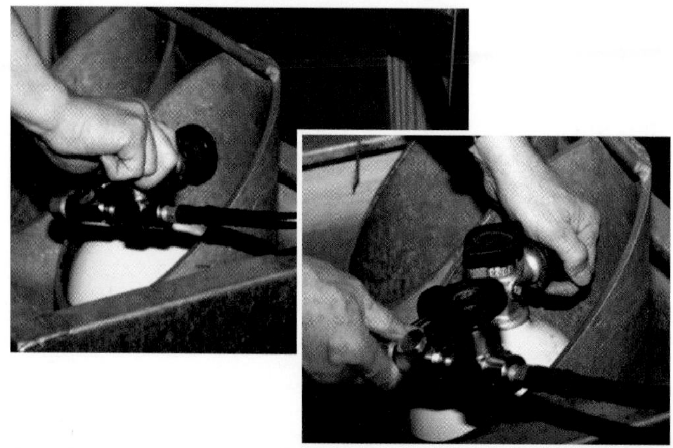

Step 11: Close the valve or valves at the cascade system manifold and/or fill line if the system is so equipped.

Step 12: Close the SCBA cylinder valve.

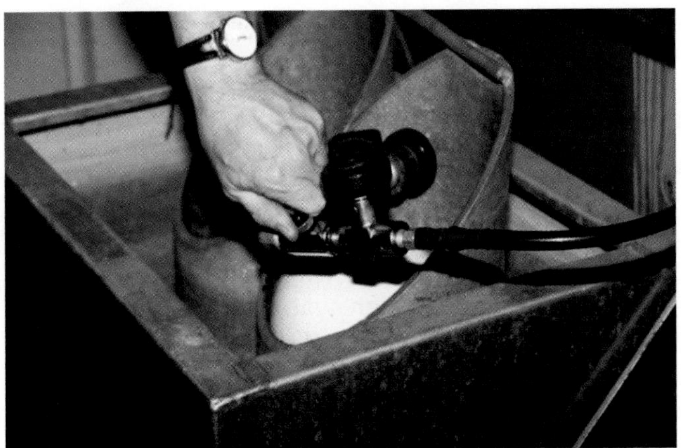

Step 13: Open the hose bleed valve to bleed off excess pressure between the cylinder valve and the valve on the fill hose.

CAUTION: Failure to do so could result in O-ring damage.

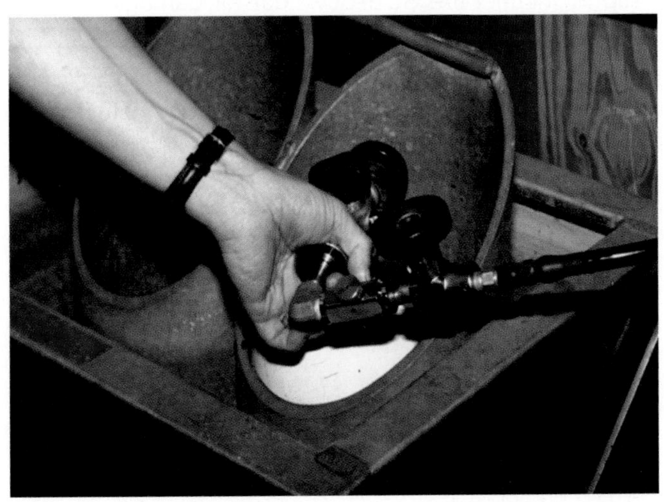

Step 14: Disconnect the fill hose from the SCBA cylinder.

Step 15: Remove the SCBA cylinder from the fill station.

Step 16: Return the cylinder to proper storage.

SKILL SHEET 4-3 FILLING AN SCBA CYLINDER

From a Compressor/Purifier

NOTE: This skill sheet is for sample purposes only. The procedures outlined here may not be applicable to your compressor/purifier system. Always check the compressor/purifier manufacturer's instructions before attempting to fill any cylinders.

Step 1: Check the hydrostatic test date of the cylinder.

Step 2: Inspect the SCBA cylinder for damage such as deep nicks, cuts, gouges, or discoloration from heat.

NOTE: If the cylinder is damaged or out of hydrostatic test date, remove the cylinder from service and tag it for further inspection and hydrostatic testing.

CAUTION: Never attempt to fill a cylinder that is damaged or that is out of hydrostatic test date.

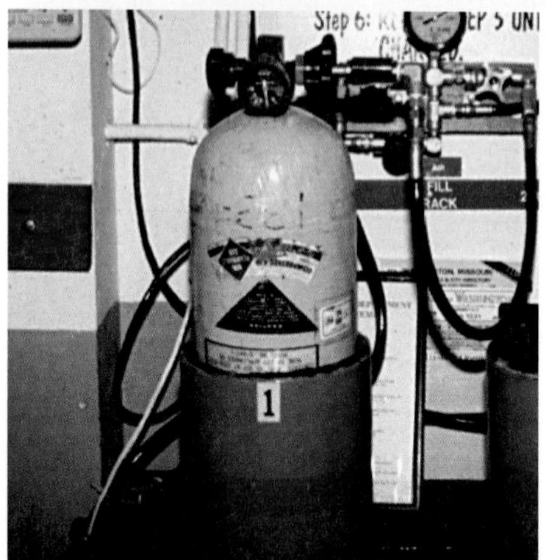

Step 3: Place the SCBA cylinder in a fragment-proof fill station.

Step 4: Connect the fill hose to the cylinder.

Step 5: Make sure that the hose bleed valve is closed.

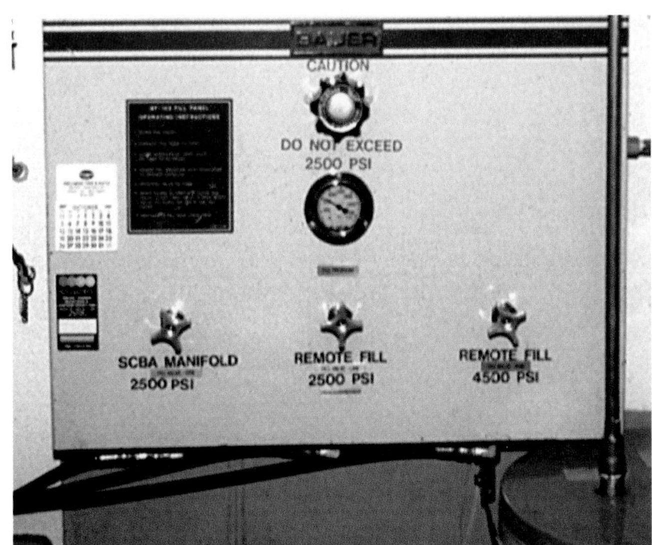

Step 6: Open the SCBA cylinder valve.
Step 7: Turn on the compressor/purifier and open the outlet valve.

Step 8: Set the cylinder pressure adjustment on the compressor (if applicable) or manifold to the desired full-cylinder pressure.

NOTE: If there is no cylinder pressure adjustment, you must watch the pressure gauge on the cylinder during filling to determine when it is full.

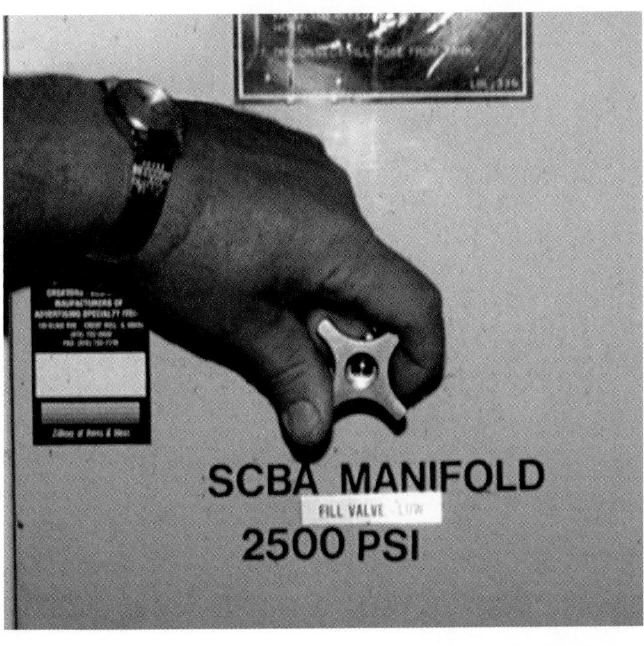

Step 9: Open the manifold valve (if applicable), and again check the fill pressure.

Step 10: Open the fill station valve and begin filling the SCBA cylinder.

NOTE: Airflow should be slow (300 to 600 psi [2 100 kPa to 4 200 kPa] per minute) to avoid excessive heating of the cylinder.

Step 11: Close the fill station valve when the cylinder is full.
Step 12: Close the SCBA cylinder valve.

Step 13: Open the hose bleed valve to bleed off excess pressure between the cylinder valve and the valve on the fill station.

CAUTION: Failure to do so could result in O-ring damage.

Step 14: Disconnect the fill hose from the SCBA cylinder.

Step 15: Remove the SCBA cylinder from the fill station, and return the cylinder to proper storage.

SKILL SHEET 4-4 CHANGING AN AIR CYLINDER

One Person

Step 1: Disconnect the regulator from the facepiece or disconnect the low-pressure hose from the regulator.

Step 2: Doff the unit.

Step 3: Obtain a full air cylinder and have it ready.

Step 4: Close the cylinder valve on the used bottle.

Step 5: Release the pressure from the high-pressure hose.

NOTES:

- On some units, the pressure must be released by breathing down the regulator or opening the mainline valve. Refer to the manufacturer's instructions for the correct method for the particular unit.
- If the pressure is not released, the high-pressure coupling will be difficult to disconnect.

Step 6: Disconnect the high-pressure coupling from the cylinder.

NOTE: If more than hand force is required to disconnect the coupling, repeat Step 5 and then again attempt to disconnect the coupling.

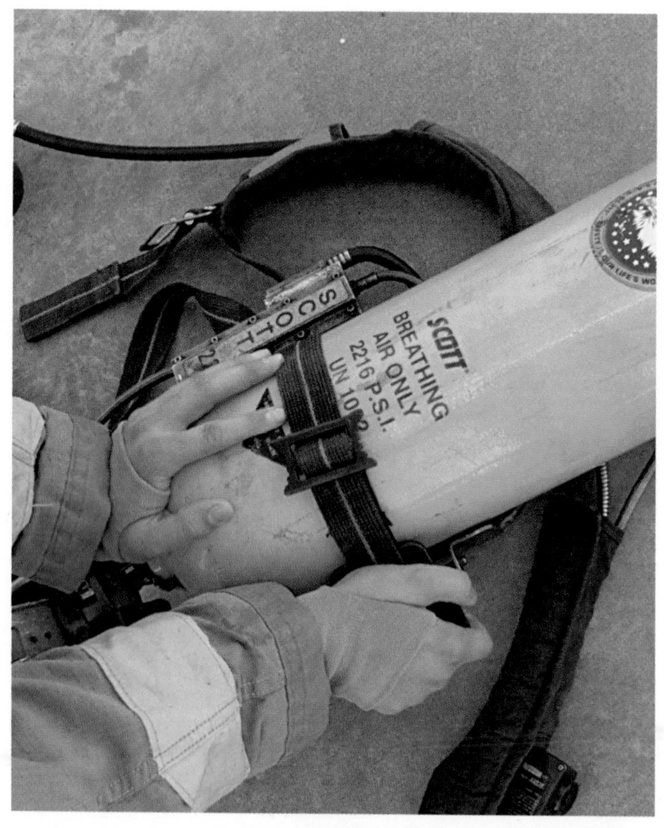

Step 7: Lay the hose coupling on the ground, directly in line with the cylinder outlet.

NOTE: This will serve as a reminder so that the replacement cylinder can be aligned correctly and easily.

Step 8: Be sure that grit or liquids do not enter the end of the unprotected high-pressure hose prior to attaching it to the cylinder outlet valve.

Step 9: Release the cylinder clamp and remove the empty cylinder.

Step 10: Place the new cylinder into the backpack.

Step 11: Position the cylinder outlet.

Step 12: Lock the cylinder into place.

NOTE: For some cylinders, it may be necessary to rotate the cylinder one-eighth turn to the left; this protects the high-pressure hose by lessening the angle of the hose and prevents twisting.

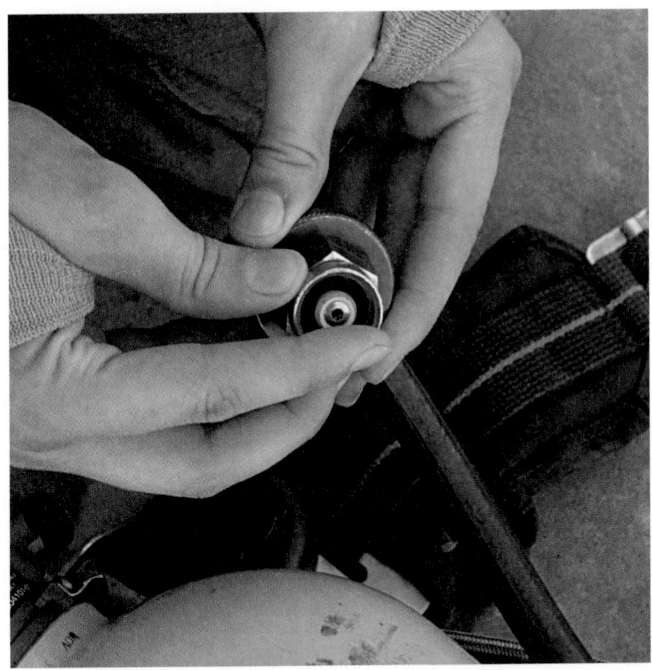

Step 13: Check the cylinder valve opening and the high-pressure hose fitting for debris.

Step 14: Check the condition of the O-ring.

Step 15: Clear any debris from the cylinder valve opening by quickly opening and closing the cylinder valve or by wiping the debris away.

Step 16: Replace the O-ring if it is distorted or damaged.

Step 17: Connect the high-pressure hose to the cylinder valve opening.

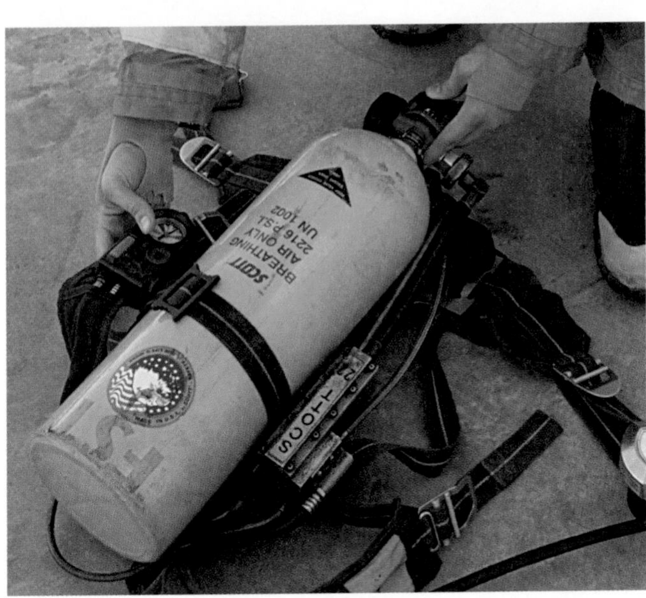

NOTE: Do not overtighten; hand tightening is sufficient.

Step 18: Open the cylinder valve.

Step 19: Check the gauges on the cylinder and the regulator.

NOTES:

- Both gauges should register within 100 psi [700 kPa] of each other — if increments are in psi [kPa] — when the cylinder is pressurized to its rated capacity. If increments are in other measurements, such as fractions or minutes, they should correspond.
- Some units require that the mainline valve on the regulator be opened in order to obtain a gauge reading. Seal the regulator outlet port by placing a hand over it. On a positive-pressure regulator, the port must be sealed for an accurate regulator gauge reading.

Chapter 5 Portable Extinguishers

Chapter 5
Portable Extinguishers

INTRODUCTION

The portable fire extinguisher, one of the most common fire-protection appliances in use today, is found in fixed facilities and on fire apparatus (Figure 5.1). A portable fire extinguisher is excellent to use on incipient fires. In many cases, a portable extinguisher can extinguish a small fire in much less time than it would take to deploy a hoseline.

It is important that firefighters be knowledgeable about the different types of portable fire extinguishers and their correct use. This chapter covers the various types of portable fire extinguishers that firefighters are likely to encounter. Also covered is information on the rating, selection, and inspection of portable fire extinguishers. NFPA 10, *Standard for Portable Fire Extinguishers*, provides additional information on rating, placement (location), and use of portable extinguishers.

Figure 5.1 Fire extinguishers may be located at fixed positions in an occupancy or mounted on the apparatus.

TYPES OF PORTABLE FIRE EXTINGUISHERS

[NFPA 1001: 3-3.15(a)]

There are many different types of portable fire extinguishers. This section highlights some of the common extinguishers encountered by fire service personnel. Table 5.1 shows the operational characteristics of different types of portable fire extinguishers discussed in the following subsections.

Firefighters should not rely on extinguishers found in occupancies. These extinguishers may not have been maintained properly or they may be obsolete. Responding firefighters should rely on the extinguishers carried on their pumping apparatus. NFPA 1901, *Standard for Automotive Fire Apparatus*, requires that pumping apparatus have two approved portable fire extinguishers with mounting brackets. These must be suitable for use on Class B and Class C fires. The stated minimum size requirement for a dry chemical extinguisher is one with a rating of 80 B:C; the required rating for a carbon dioxide (CO_2) extinguisher is 10 B:C. Ratings represent the type of fire plus performance capability (see Extinguisher Rating System section). NFPA 1901 also requires pumping apparatus to carry one 2½ gallon (10 L) or larger water extinguisher with a mounting bracket for use on Class A fires.

NOTE: Water-type extinguishers should be protected against freezing if they are going to be

TABLE 5.1
Operational Characteristics of Portable Fire Extinguishers

Extinguisher	Type	Agent	Fire Class	Size	Stream Reach	Discharge Time
Pump-Tank Water	Hand-carried; backpack	Water	A only	1½–5 gal (6 L to 20 L)	30–40 ft (9.1 m to 12.2 m)	45 sec to 3 min
Stored-Pressure Water	Hand-carried	Water	A only	1¼–2½ gal (5 L to 10 L)	30 ft–40 ft (9.1 m to 12.2 m)	30–60 sec
Aqueous Film Forming Foam (AFFF)	Hand-carried	Foam	A & B	2½ gal (10 L)	20–25 ft (6.1 m to 7.6 m)	Approximately 50 sec
Halon 1211*	Hand-carried; wheeled	Halon	B & C	Hand-carried: 2½–20 lb (1 kg to 9 kg)	8–18 ft (2.4 m to 5.5 m)	8–18 sec
				Wheeled: to 150 lb (68 kg)	20–35 ft (6.1 m to 10.7 m)	30–44 sec
Halon 1301	Hand-carried	Halon	B & C	2½ lb (1 kg)	4–6 ft (1.2 m to 1.8 m)	8–10 sec
Carbon Dioxide	Hand-carried	Carbon dioxide	B & C	2½–20 lb (1 kg to 9 kg)	3–8 ft (1 m to 2.4 m)	8–30 sec
Carbon Dioxide	Wheeled	Carbon dioxide	B & C	50–100 lb (23 kg to 45 kg)	8–10 ft (2.4 m to 3 m)	26–65 sec
Dry Chemical	Hand-carried stored-pressure; cartridge-operated	Sodium bicarbonate, potassium bicarbonate, ammonium phosphate, potassium chloride	B & C	2½–30 lb (1 kg to 14 kg)	5–20 ft (1.5 m to 6.1 m)	8–25 sec
Multipurpose Dry Chemical	Hand-carried stored-pressure; cartridge-operated	Monoammonium phosphate	A, B, & C	2½–30 lb (1 kg to 14 kg)	5–20 ft (1.5 m to 6.1 m)	8–25 sec
Dry Chemical	Wheeled; ordinary or multipurpose		A, B, & C	75–350 lb (34 kg to 159 kg)	Up to 45 ft (13.7 m)	20 sec to 2 min
Dry Powder	Hand-carried; wheeled	Various, depending on metal fuel (this description for sodium chloride plus flow enhancers)	D only	Hand-carried: to 30 lb (14 kg) Wheeled: 150 lb & 350 lb (68 kg & 159 kg)	4–6 ft (1.2 m to 1.8 m)	28–30 sec

* Rating: Those larger than 9 lb (4 kg) capacity have small Class A ratings (1-A to 4-A).

exposed to temperatures lower than 40°F (4°C). Freeze protection may be provided by adding antifreeze to the water or by storage in warm areas.

Pump-Tank Water Extinguishers

Pump-tank water extinguishers are intended for use on small Class A fires only (Figure 5.2). There are several kinds of pump-tank water extinguishers, but all operate in a similar manner. They are generally equipped with a double-acting pump. The procedures for operating a pump-tank water extinguisher are shown in Skill Sheet 5-1.

Figure 5.2 Typical pump-tank fire extinguishers.

Figure 5.3 A typical stored-pressure water extinguisher. *Courtesy of Ansul, Inc., Marinette, Wisconsin.*

Figure 5.4 A pressure gauge clearly shows operable range.

Stored-Pressure Water Extinguishers

Stored-pressure water extinguishers, also called *air-pressurized water (APW) extinguishers,* are useful for all types of small Class A fires and are often used for extinguishing confined hot spots during overhaul operations, as well as for extinguishing chimney flue fires (Figure 5.3).

Water is stored in a tank along with either compressed air or nitrogen. A gauge located on the side of the valve assembly shows when the extinguisher is properly pressurized (Figure 5.4). When the operating valve is activated, the water is forced up the siphon tube and out through the hose (Figure 5.5).

Class A foam concentrate is sometimes added to a water extinguisher to enhance its effectiveness. The addition of Class A foam serves as a wetting agent that aids in extinguishing deep-seated fires, vehicle fires, and wildland fires. The procedures for operating stored-pressure extinguishers are listed in Skill Sheet 5-2.

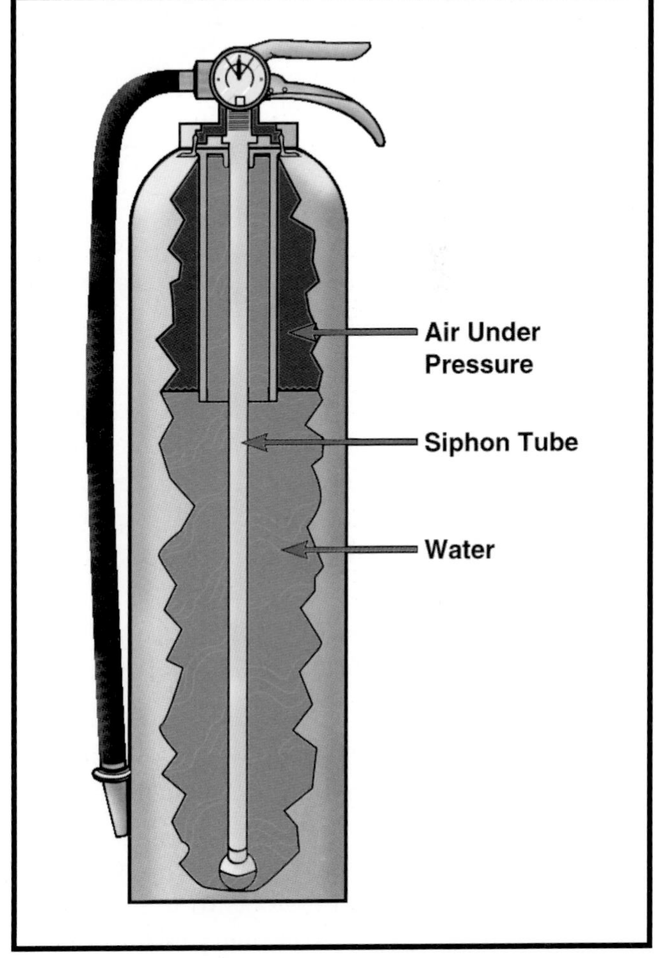

Figure 5.5 Cutaway of a stored-pressure water extinguisher.

Aqueous Film Forming Foam (AFFF) Extinguishers

Aqueous film forming foam (AFFF) extinguishers are suitable for use on Class A and Class B fires. They are particularly useful in combating fires or suppressing vapors on small liquid fuel spills.

AFFF extinguishers are different from stored-pressure water extinguishers in two ways. The AFFF extinguisher tank contains a specified amount of AFFF concentrate mixed with the water, and it has an air aspirating nozzle that aerates the foam solution, producing a better quality foam than a standard extinguisher nozzle provides (Figure 5.6).

Figure 5.6 A typical AFFF fire extinguisher. *Courtesy of Amerex Corp.*

The water/AFFF solution is expelled by compressed air or nitrogen stored in the tank with the solution. To prevent the disturbance of the foam blanket when applying the foam, it should not be applied directly onto the fuel; it should be allowed to either gently rain down onto the fuel surface or deflect off an object (Figure 5.7).

When AFFF and water are mixed, the resulting finished foam floats on the surface of fuels that are lighter than water. The vapor seal created by the film of water extinguishes the flame and prevents reignition (Figure 5.8). The foam has both good wetting and good penetrating properties on Class A fuels but is ineffective on flammable liquids that are water-soluble (polar solvents) such as alcohol and acetone.

AFFF extinguishers are not suitable for fires in Class C or Class D fuels. They are not suitable for three-dimensional fires such as in fuel flowing down from an elevated point and fuel under pressure spraying from a leaking flange. They are most effective on static pools of flammable liquids.

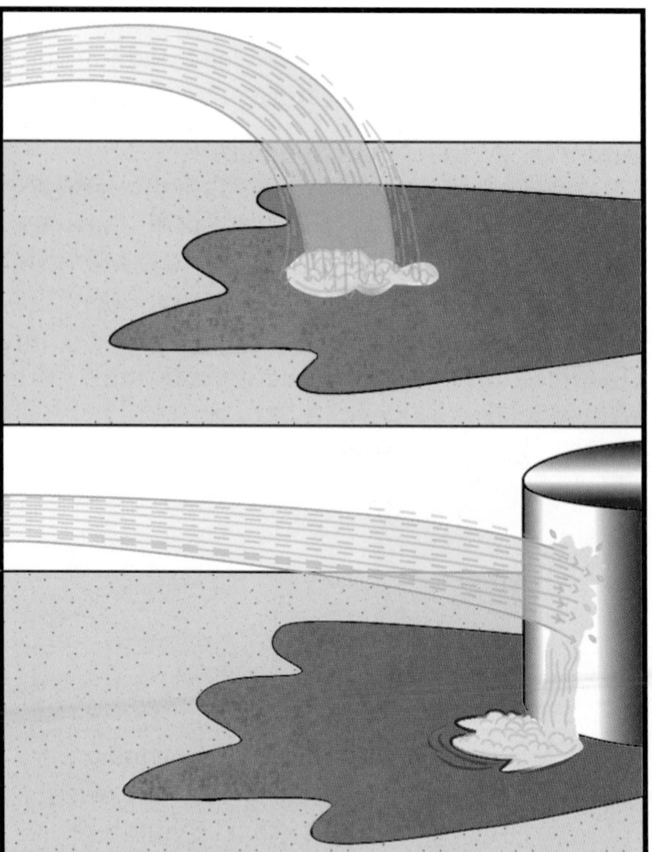

Figure 5.7 Two ways in which AFFF can be applied.

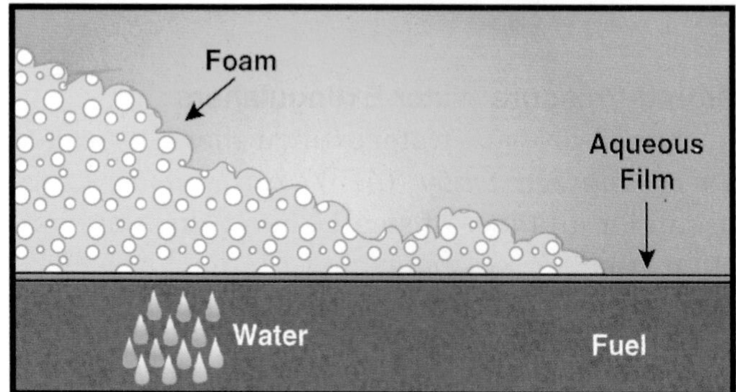

Figure 5.8 The film of AFFF floats ahead of the foam blanket.

Halon Extinguishers

Because of their ozone-depletion potential, halogenated extinguishing agents are included in the *Montreal Protocol on Substances that Deplete the Ozone Layer*. This international agreement requires a complete phaseout of the production of halogens by the year 2000. The only exceptions allowed under the agreement are for essential uses where no suitable alternatives are available. The United States unilaterally decided to stop producing halogens at the end of 1993, and research was begun on possible alternative extinguishing agents.

The following information on halon extinguishers is included because these units may still be found.

Halon is a generic term for halogenated hydrocarbons and is defined as *a chemical compound that contains carbon plus one or more elements from the halogen series (fluorine, chlorine, bromine, or iodine)*. While a large number of halogenated compounds exist, only a few are used to a significant extent as fire extinguishing agents. The two most common ones are Halon 1211 (bromochlorodifluoromethane) and Halon 1301 (bromotrifluoromethane).

Halogenated vapor is nonconductive and is effective in extinguishing surface fires in flammable and combustible liquids and electrical equipment. However, these agents are not effective on fires in self-oxidizing fuels such as combustible metals, organic peroxides, and metal hydrides. Although the halons have long been used for the protection of internal combustion engines, their primary modern-day application is for the protection of sensitive electronic equipment such as computers.

Figure 5.9 A typical hand-carried halon extinguisher.

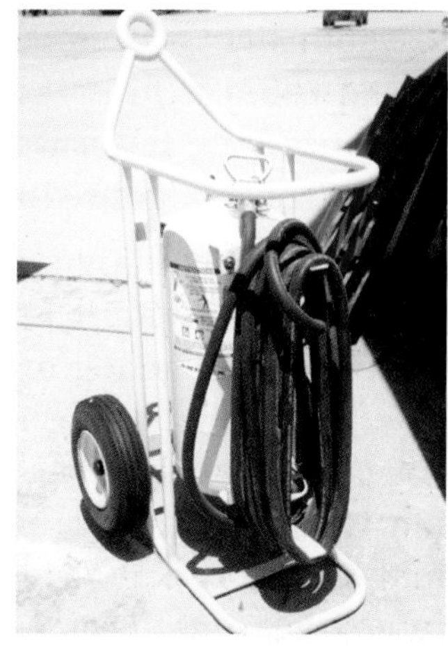

Figure 5.10 A typical wheeled halon extinguisher. *Courtesy of Ansul, Inc., Marinette, Wisconsin.*

HALON 1211

How Halon 1211 or any other halogenated agent extinguishes fire is not definitely known, but research suggests that it interrupts the chain reaction of the combustion process. Halon 1211 extinguishers are intended primarily for use on Class B and Class C fires (Figure 5.9); however, Halon 1211 extinguishers greater than 9 pounds (4 kg) in capacity also have a low Class A rating (1-A to 4-A, depending on size) (see Extinguisher Rating System section). Larger Halon 1211 extinguishers are found as wheeled units up to 150 pounds (68 kg) (Figure 5.10).

Halon 1211 is stored in the extinguisher as a liquefied compressed gas, but nitrogen is added to the tank to increase discharge pressure and stream reach. Halon 1211 is discharged from an extinguisher in a clear liquid stream, giving it greater reach than a gaseous agent; however, the stream may be affected by wind when operated outside.

HALON 1301

Halon 1301 is normally not used by itself in portable fire extinguishers because the agent is discharged as a nearly invisible gas that is highly susceptible to being affected by wind. In a confined space, such as a computer room, the agent's volatility allows it to disperse faster than Halon 1211. For this reason, and because it is effective at a lower concentration than Halon 1211, Halon 1301 is the agent of choice in most total-flooding systems using halogenated agents (Figure 5.11).

Figure 5.11 A Halon 1301 tank in a fixed system.

WARNING

When halon is used to extinguish a fire, it decomposes and liberates toxic components, so these agents should not be used in unventilated, confined spaces.

Carbon Dioxide Extinguishers

Carbon dioxide (CO_2) extinguishers are found as both handheld units and wheeled units. CO_2

extinguishers are effective in extinguishing Class B and Class C fires (Figure 5.12). Because their discharge is in the form of a gas, they have a limited reach. They do not require freeze protection.

Carbon dioxide is stored under its own pressure as a liquefied compressed gas ready for release at anytime. The agent is discharged through a plastic or rubber horn on the end of either a short hose or tube. The gaseous discharge is usually accompanied by little dry ice crystals or carbon dioxide "snow" (Figure 5.13). This snow sublimes — changes into a gaseous form — shortly after discharge. When released, the carbon dioxide gas displaces available oxygen and smothers the fire. CO_2 produces no vapor-suppressing film on the surface of the fuel; therefore, reignition of the fuel is always a danger.

Figure 5.12 A typical handheld CO_2 fire extinguisher. *Courtesy of Ansul, Inc., Marinette, Wisconsin.*

Figure 5.13 CO_2 "snow" formed with the moisture in the air. *Courtesy of Ansul, Inc., Marinette, Wisconsin.*

Carbon dioxide wheeled units are similar to the handheld units except that they are considerably larger (Figure 5.14). These units are intended to be used only on Class B and Class C fires. Wheeled units are most commonly used in airports and industrial facilities. After being wheeled to the fire, the hose (usually less than 15 feet [5 m] long) must be deployed or unwound from the unit before use. The principle of operation is the same as in the smaller handheld units.

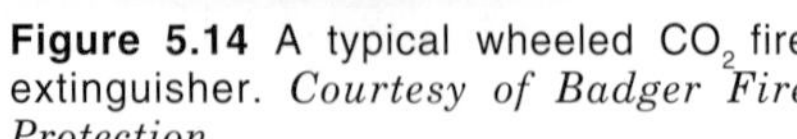

Figure 5.14 A typical wheeled CO_2 fire extinguisher. *Courtesy of Badger Fire Protection.*

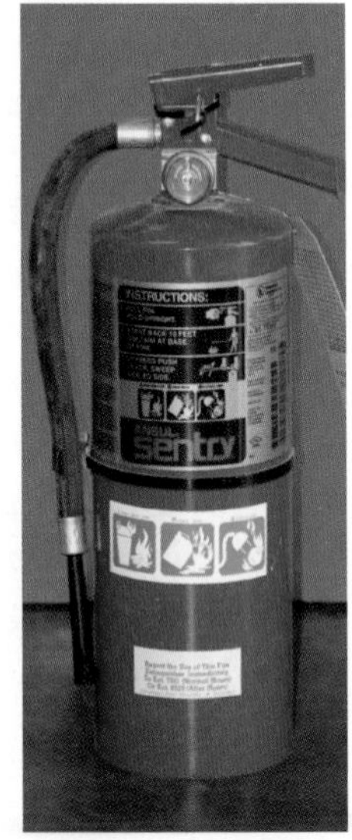

Figure 5.15 A hand-carried dry chemical fire extinguisher rated for Classes A, B, and C fires.

Dry Chemical Extinguishers

The terms *dry chemical* and *dry powder* are often incorrectly used interchangeably. Dry chemical agents are for use on Class A-B-C fires and/or Class B-C fires. Dry powder agents are for Class D fires only. Dry chemical extinguishers are among the most common portable fire extinguishers in use today. There are two basic types of dry chemical extinguishers: (1) regular B:C-rated and (2) multipurpose and A:B:C-rated (see Extinguisher Rating System section) (Figure 5.15). Unless specifically noted in this section, the characteristics and operation of both types are exactly the same. The following are commonly used dry chemicals.

- Sodium bicarbonate
- Potassium bicarbonate
- Urea-potassium bicarbonate
- Potassium chloride
- Monoammonium phosphate

During manufacture, these agents are mixed with small amounts of additives that make the agents moisture resistant and prevent them from caking. This process keeps the agents ready for use even after being undisturbed for long periods, and it makes them free flowing.

CAUTION: Never mix or contaminate dry chemicals with any other type of agent because they may chemically react and cause a dangerous rise in pressure inside the extinguisher.

The dry chemical agents themselves are nontoxic and generally considered quite safe to use. However, the cloud of chemicals may reduce visibility and create respiratory problems like any airborne particulate. Some dry chemicals are compatible with foam, but others will degrade the foam blanket. On Class A fires, the discharge should be directed at whatever is burning in order to cover it with chemical. When the flames have been knocked down, the agent should be applied intermittently as needed on any smoldering hot spots. Many dry chemical agents are corrosive to metals, so it may be better to use another agent such as carbon dioxide on them.

HANDHELD UNITS

There are two basic designs for handheld dry chemical extinguishers: *stored-pressure* and *cartridge-operated* (Figure 5.16). The stored-pressure type is similar in design to the air-pressurized water extinguisher, and a constant pressure of about 200 psi (1 400 kPa) is maintained in the agent storage tank. Cartridge-operated extinguishers employ a pressure cartridge connected to the agent tank (Figure 5.17). The agent tank is not pressurized until a plunger is pushed to release the gas from the cartridge. Both types of extinguishers use either nitrogen or carbon dioxide as the pressurizing gas. Cartridge-operated extinguishers use a carbon dioxide cartridge unless the extinguisher is going to be subjected to freezing temperatures; in such cases, a dry nitrogen cartridge is used. The procedures for operating cartridge-operated extinguishers are listed in Skill Sheet 5-3.

Figure 5.16 Cartridge-operated and stored-pressure dry chemical extinguishers. *Courtesy of Ansul, Inc., Marinette, Wisconsin.*

Figure 5.17 Cutaway of a cartridge-operated dry chemical extinguisher.

WHEELED UNITS

Dry chemical wheeled units are similar to the handheld units but are on a larger scale (Figure 5.18). They are rated for Class A, Class B, and Class C fires based on the dry chemical in the unit (see Extinguisher Rating System section).

Figure 5.18 A typical wheeled dry chemical extinguisher. *Courtesy of Ansul, Inc., Marinette, Wisconsin.*

Operating the wheeled dry chemical extinguisher is similar to operating the handheld, cartridge-type dry chemical extinguisher. The extinguishing agent is kept in one tank, and the pressurizing gas is stored in a separate cylinder. When the extinguisher is in position at a fire, the hose should be stretched out completely first (Figure 5.19). This procedure is recommended

Figure 5.19 The hose must be deployed before being pressurized.

because removing the hose can be more difficult after it is charged and the powder can sometimes pack in any sharp bends in the hose. The pressurizing gas should be introduced into the agent tank and allowed a few seconds to fully pressurize the tank before the nozzle is opened. The agent is applied the same as that described for the handheld, cartridge-type dry chemical extinguishers.

CAUTION: The top of the extinguisher should be pointed away from the firefighter or other personnel when pressurizing the unit. Because of the size of the nozzle, the firefighter should be prepared for a significant nozzle reaction when it is opened (Figure 5.20).

Figure 5.20 The operator should prepare for a significant nozzle reaction when the nozzle is opened.

Extinguishers and Powder Extinguishing Agents for Metal Fires

The extinguishing agents discussed so far in this chapter generally should not be used on Class D (combustible metal) fires. Special extinguishing agents and application techniques have been developed to control and extinguish metal fires. No single agent will control or extinguish fires in all combustible metals. Some agents are effective against fires in several metals; others are effective on fires in only one type of metal. Some powdered agents can be applied with portable extinguishers, but others must be applied by either a shovel or a scoop. The appropriate application technique for any given dry powder is described in the manufacturer's technical sales literature. Firefighters should be thoroughly familiar with the information that applies to any agent carried on their emergency response vehicles.

Portable extinguishers for Class D fires come in both handheld and wheeled models. (Figure 5.21). Whether a particular dry powder is applied with an extinguisher or with a scoop, it must be applied in sufficient depth to completely cover the area that is burning to create a smothering blanket (Figure 5.22). The agent should be applied gently to avoid breaking any crust that may form over the burning metal. If the crust is broken, the fire may flare up and expose more uninvolved material to combustion. Care should be taken to avoid scattering the burning metal.

Figure 5.21 Handheld and wheeled Class D extinguishers. *Wheeled extinguisher courtesy of Amerex Corp.*

Additional applications may be necessary to cover any hot spots that develop.

If the burning metal is on a combustible surface, the fire should first be covered with powder. Then, a 1- to 2-inch (25 mm to 50 mm) layer of powder should be spread nearby and the burning metal shoveled onto this layer with more powder added as needed (Figure 5.23). After extinguishment, the material should be left undisturbed. Disposal should not be attempted until the mass has cooled completely.

EXTINGUISHER RATING SYSTEM

[NFPA 1001: 3-3.15(a)]

Portable fire extinguishers are classified according to the types of fire (A, B, C, or D) for which they are intended (Figure 5.24). In addition to the classification represented by the letter, Class A

Figure 5.22 The dry powder must completely cover the area that is burning to create a smothering blanket.

Figure 5.23 Class D agent may be shoveled onto the fire.

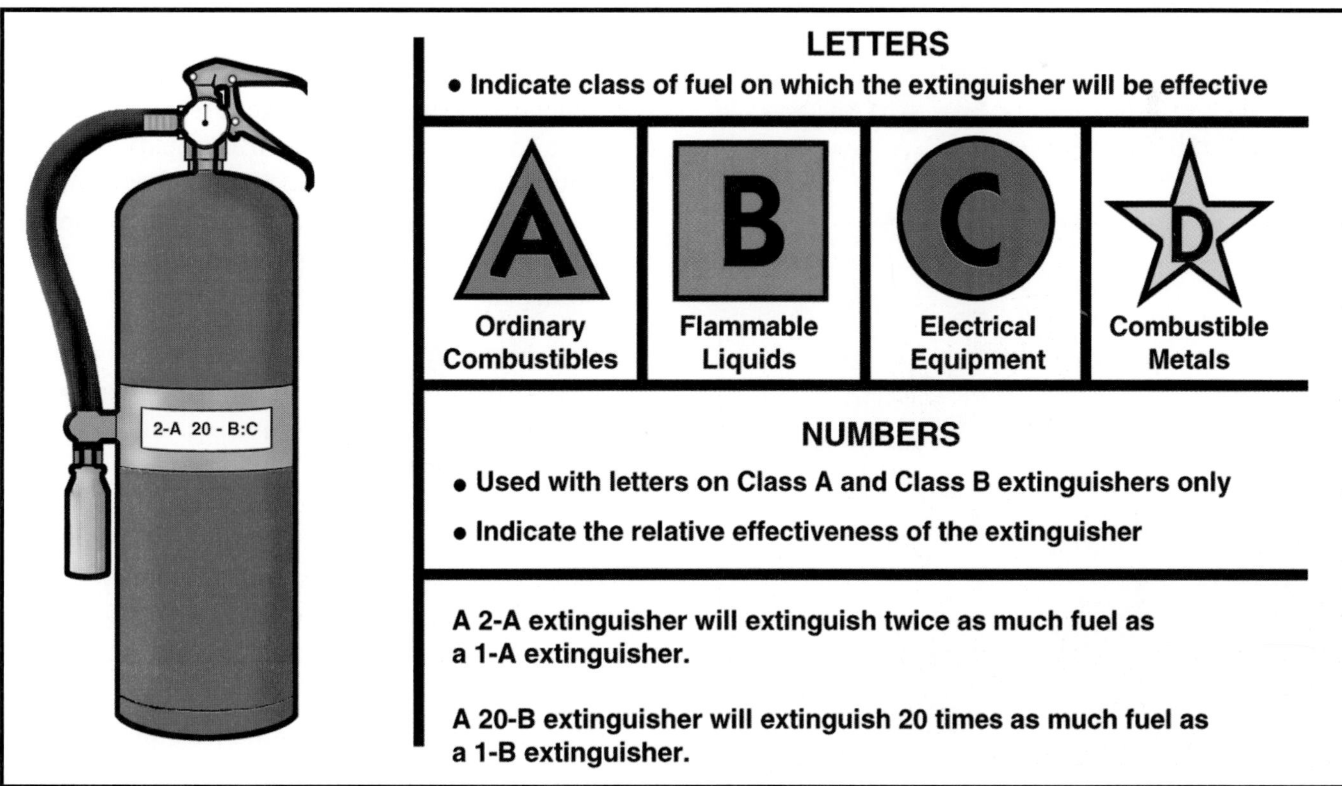

Figure 5.24 Extinguishers are classified according to their intended use.

and Class B extinguishers are also rated according to performance capability, which is represented by a number. The classification and numerical rating system is based on tests conducted by Underwriters Laboratories Inc. (UL) and Underwriters Laboratories of Canada (ULC). These tests are designed to determine the extinguishing capability for each size and type of extinguisher.

Class A Ratings

Class A portable fire extinguishers are rated from 1-A through 40-A. The Class A rating of water extinguishers is primarily based on the amount of extinguishing agent and the duration and range of the discharge used in extinguishing test fires. For a 1-A rating, 1¼ gallons (5 L) of water are required. A 2-A rating requires 2½ gallons (10 L) or twice the 1-A capacity (Figure 5.25).

Class B Ratings

Extinguishers suitable for use on Class B fires are classified with numerical ratings ranging from 1-B through 640-B (Figure 5.26). The rating is based on the approximate square foot (square meter) area of a flammable liquid fire that a nonexpert operator can extinguish. The nonexpert operator is expected to extinguish 1 square foot (0.09 m^2) for each numerical rating or value of the extinguisher rating.

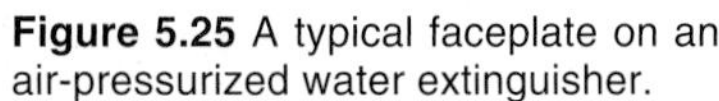

Figure 5.25 A typical faceplate on an air-pressurized water extinguisher.

Figure 5.26 Class B extinguishers are available in a variety of sizes.

Class C Ratings

There are no fire extinguishing capability tests specifically conducted for Class C ratings. Extinguishers for use on Class C fires receive only the letter rating because Class C fires are essentially Class A or Class B fires involving energized electrical equipment. The extinguishing agent is tested for nonconductivity. The Class C rating confirms the extinguishing agent will not conduct electricity. The Class C rating is assigned in addition to the rating for Class A and/or Class B fires.

Class D Ratings

Test fires for establishing Class D ratings vary with the type of combustible metal being tested. The following factors are considered during each test.

- Reactions between the metal and the agent
- Toxicity of the agent
- Toxicity of the fumes produced and the products of combustion
- Time to allow metal to burn out without fire suppression efforts versus time to extinguish

When an extinguishing agent is determined to be safe and effective for use on a combustible metal, the details of instruction are included on the faceplate of the extinguisher, although no numerical rating is given. Class D agents cannot be given a multipurpose rating for use on other classes of fire.

Multiple Markings

Extinguishers suitable for more than one class of fire are identified by combinations of the letters A, B, and/or C or the symbols for each class. The three most common combinations are Class A-B-C, Class A-B, and Class B-C (Figure 5.27). All new portable fire extinguishers must be labeled with their appropriate markings. Any extinguisher not properly marked is not a listed unit and should not be used.

The ratings for each separate class of extinguisher are independent and do not effect each other. To better understand the rating system, a common-sized extinguisher, such as the multipurpose extinguisher rated 4-A 20-B:C, can be reviewed. This extinguisher should extinguish a Class A fire that is 4 times larger than a 1-A fire, extinguish approximately 20 times as much Class B fire as a 1-B extinguisher, and extinguish a deep-layer flammable liquid fire of a 20 square-foot (2 m^2) area. It is also safe to use on fires involving energized electrical equipment.

Figure 5.27 An extinguisher intended for Class A, Class B, and Class C fires.

Figure 5.28 The alphabetical/geometric symbols for the four classes of fire.

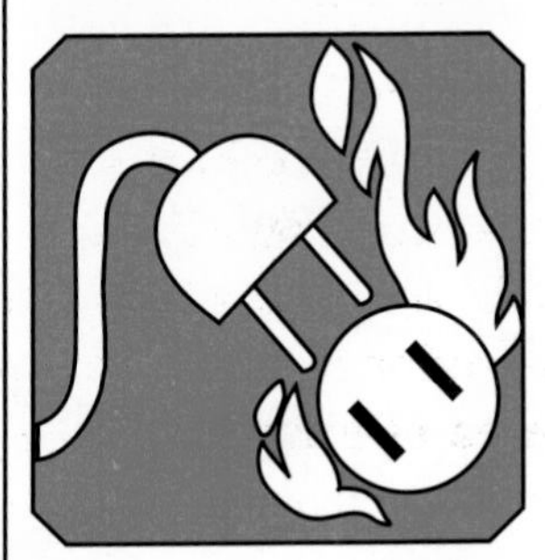

Figure 5.29 The pictographs representing Class A, Class B, and Class C fires.

There are two systems of labeling portable fire extinguishers. One system uses geometric shapes of specific colors with the class letter shown within the shape (Figure 5.28). The other system, currently recommended in NFPA 10, uses pictographs to make the selection of the most appropriate fire extinguishers easier (Figure 5.29). This system also shows the types of fires on which the extinguisher should *not* be used (Figure 5.30). Regardless of which system is used, it is important that the markings are clearly visible.

Figure 5.30 Pictographs showing the classes of fires for which an extinguisher is not suitable.

SELECTING PORTABLE FIRE EXTINGUISHERS

[NFPA 1001: 3-3.15; 3-3.15(b)]

Selection of the proper portable fire extinguisher depends on numerous factors:

- Classification of the burning fuel

- Rating of the extinguisher
- Hazards to be protected
- Severity of the fire
- Atmospheric conditions
- Availability of trained personnel
- Ease of handling extinguisher
- Any life hazard or operational concerns

Select extinguishers that minimize the risk to life and property but are effective in extinguishing the fire. It is unwise to use dry chemical extinguishers with a corrosive agent in areas where highly sensitive computer equipment is located. The residue left after use could potentially do more damage to the sensitive electronic equipment than a fire. In these particular areas, halon or carbon dioxide extinguishers are better choices (Figure 5.31).

Figure 5.31 Halon extinguishers are used in computer areas.

USING PORTABLE FIRE EXTINGUISHERS

[NFPA 1001: 3-3.15; 3-3.15(a)(b)]

Portable extinguishers come in many sizes and types. While the operating procedures of each type of extinguisher are similar, firefighters should become familiar with the detailed instructions found on the label of the extinguisher. Fire extinguishers on the emergency response vehicles must be inspected regularly to ensure that they are accessible and operable (see Inspecting Fire Extinguishers section).

A firefighter should quickly check an extinguisher before attempting to use it in an emergency (Figure 5.32). This check is necessary to assure the firefighter that the extinguisher is charged and operable. Such a check may protect the firefighter from injury caused by a defective extinguisher. If the extinguisher appears to be in working order, the firefighter can then use it to suppress a fire.

When inspecting an extinguisher immediately before use, a firefighter should check the following:

- External condition — no apparent damage
- Hose/nozzle — in place
- Weight — feels as though it contains agent
- Pressure gauge (if available) — in operable range

Figure 5.32 The firefighter should quickly check the extinguisher before using it.

After selecting the appropriate size and type of extinguisher for the situation, a firefighter approaches a fire from the windward side (wind at the firefighter's back). The firefighter must be sure the extinguishing agent reaches the fire — if it cannot, the agent is wasted (Figure 5.33). Smaller extinguishers require a closer approach to the fire than larger units, thus radiant heat may prevent the firefighter from getting close enough for the agent to reach the fire. Adverse winds also can limit the reach of an agent. However, operating an extinguisher close to the fire can cause another problem. Discharging agent directly onto the fuel can sometimes scatter lightweight solid fuels or penetrate the surface of liquid fuels (Figure 5.34). Apply agent from a point where it reaches but does not disturb the fuel (Figure 5.35). The general operating procedures shown in Skill Sheets 5-1 through 5-3 should be followed when operating fire extinguishers.

After the fire is knocked down, the firefighter may move closer to achieve final fire extinguishment. However, if extinguishment is not achieved after an entire extinguisher has been discharged onto the fire, the firefighter should withdraw and reassess the situation. If the fire is in solid fuel that has been reduced to the smoldering phase, it should be overhauled using an appropriate tool to pull it apart and a charged hoseline to achieve complete extinguishment. If the fire is in a liquid fuel, it may be necessary to either apply the appropriate type of foam through a hoseline or simultaneously attack the fire with more than one extinguisher. If more than one extinguisher is used simultaneously, firefighters must work in unison and maintain a constant awareness of each other's actions and positions. Firefighters should lay empty fire extinguishers on their sides after use. This procedure signals that they are empty and reduces the chance of someone taking one and approaching a fire with an empty extinguisher.

Figure 5.33 Apply agent to the base of the flames.

Figure 5.34 Sweep the nozzle from side to side.

Figure 5.35 Agent being applied from an appropriate distance.

INSPECTING PORTABLE FIRE EXTINGUISHERS

Fire extinguishers must be inspected regularly to ensure that they are accessible and operable. Verify that extinguishers are in their designated locations, that they have not been activated or tampered with, and that there is no obvious physical damage or condition present that prevents their operation. Servicing of portable fire extinguishers (or any other privately owned fire suppression or detection equipment) is the responsibility of the property owner or building occupant.

Although it is usually performed by the building owner or the owner's designate, fire inspectors should include extinguisher inspections in their building inspection and pre-incident planning programs (Figure 5.36). During inspections, inspectors should remember that there are three important factors that determine the value of a fire

Figure 5.36 Pre-incident planning should include the inspection of fire extinguishers.

extinguisher: its serviceability, its accessibility, and the user's ability to operate it.

NFPA 10 requires and explains the procedures for hydrostatic testing of extinguisher cylinders. The test results must be recorded on the extinguisher. The hydrostatic test results on high- and low-pressure cylinders are recorded differently. Maintenance personnel should refer to NFPA 10 for specific information on extinguisher testing and recording of results.

The following procedures should be part of every fire extinguisher inspection.

- Check to ensure that the extinguisher is in a proper location and that it is accessible (Figure 5.37).
- Inspect the discharge nozzle or horn for obstructions. Check for cracks and dirt or grease accumulations.
- Inspect extinguisher shell for any physical damage.
- Check to see if the operating instructions on the extinguisher nameplate are legible.
- Check the lock pins and tamper seals to ensure that the extinguisher has not been tampered with.
- Determine if the extinguisher is full of agent and fully pressurized by checking the pressure gauge, weighing the extinguisher, or inspecting the agent level. If an extinguisher is found to be deficient in weight by 10 percent, it should be removed from service and replaced.
- Check the inspection tag for the date of the previous inspection, maintenance, or recharging.
- Examine the condition of the hose and its associated fittings.

Figure 5.37 Make sure that access to the extinguisher is not blocked.

If any of the items listed are deficient, the extinguisher should be removed from service and repaired in accordance with department policies. The extinguisher should be replaced with an extinguisher that has an equal or greater rating.

DAMAGED PORTABLE EXTINGUISHERS

Leaking, corroded, or otherwise damaged extinguisher shells or cylinders should be discarded or returned to the manufacturer for repair (Figure 5.38). Damaged extinguishers can fail at any time and could result in serious injury to the user or people standing nearby.

CAUTION: Never attempt to repair the shell or cylinder of a defective fire extinguisher. Contact the manufacturer for instructions on where to have it repaired or replaced.

Figure 5.38 Damaged extinguishers should be removed from service until repaired and tested.

If an extinguisher shows only slight damage or corrosion but it is uncertain whether or not the unit is safe to use, it should be hydrostatically tested by either the manufacturer or a qualified testing agency. Leaking hoses, gaskets, nozzles, and loose labels can be replaced by firefighters.

OBSOLETE PORTABLE EXTINGUISHERS

American manufacturers stopped making inverting-type fire extinguishers in 1969. These included soda-acid, foam, internal cartridge-operated water and loaded stream, and internal cartridge dry chemical extinguishers (Figure 5.39). Manufacturing of extinguishers made of copper or brass with cylinders either soft soldered or riveted together was also discontinued at that time. Because of the toxicity of carbon tetrachloride and chlorobromomethane, extinguishers using these agents were prohibited in the workplace. OSHA regulations 29 CFR 1910.157, Subpart L (c)(5) (dated September 12, 1980), required employers to permanently remove all of these obsolete extinguishers from service by January, 1982. Occasionally, firefighters encounter old fire extinguishers in old buildings, particularly factories and educational facilities. If the occupant asks firefighters to dispose of an extinguisher, they should dispose of it in accordance with fire department policies and procedures.

Figure 5.39 Obsolete extinguishers should be removed from service.

SKILL SHEET 5-1 OPERATING A PORTABLE FIRE EXTINGUISHER

Pump-Tank Extinguishers

Step 1: Make sure the unit is full of water before attempting to extinguish a fire. This can usually be determined by the weight of the unit. If not, open the cap and check the water level.

Step 2: Carry the pump tank to the fire.

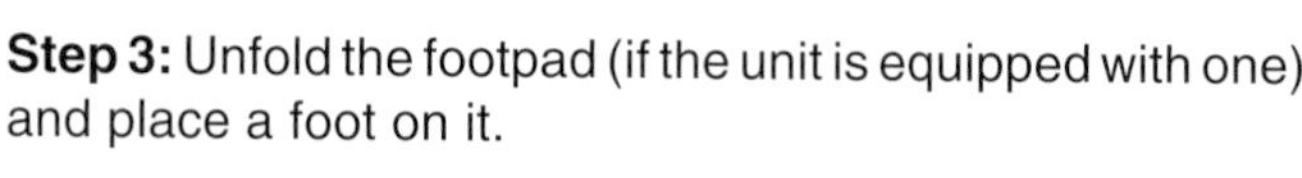

Step 3: Unfold the footpad (if the unit is equipped with one) and place a foot on it.

Step 4: Pump the extinguisher with one hand and direct the stream with the other. (**NOTE:** Short, brisk strokes will produce a continuous stream.)

Step 5: Sweep the nozzle to spread the water over the entire fire area.

Step 6: Move closer and complete extinguishment after the fire has been knocked down.

SKILL SHEET 5-2 OPERATING A PORTABLE FIRE EXTINGUISHER

Stored-Pressure Extinguishers
Water, Halon, Dry Chemical, CO_2

Step 1: Select the appropriate extinguisher based on the size and type of fire.

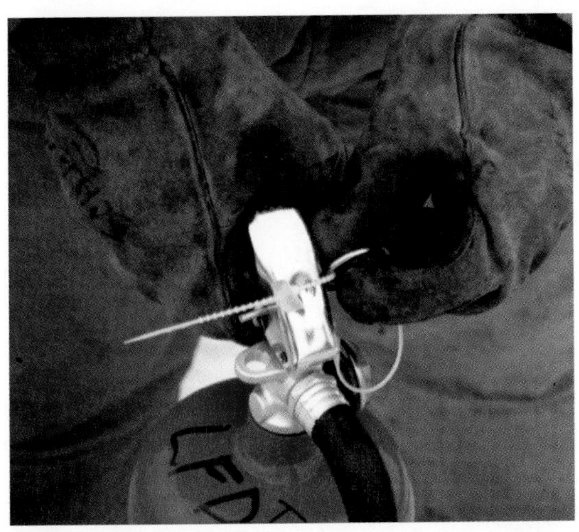

Step 2: Pull the safety pin at the top of the extinguisher, breaking the plastic or thin wire seal in the process.

Step 3: Point the nozzle or horn in a safe direction and discharge a very short test burst to ensure proper operation.

NOTE: If the hose is clipped to the extinguisher body, release it before discharging the agent.

Step 4: Carry the extinguisher to within stream reach of the fire.

Step 5: Aim the nozzle or horn toward the material that is burning.

CAUTION: The nozzles of Halon 1211 and CO_2 extinguishers may freeze during discharge, so touching the nozzle with bare skin may result in frostbite.

Step 6: Squeeze the carrying handle and the discharge handle together to start the flow of agent. Release the handle to stop the flow.

Step 7: Sweep the nozzle back and forth at the base of the flames to ensure full coverage by the extinguishing agent until fire is extinguished.

NOTES:

- Do not plunge **dry chemical** into flammable liquid fires.
- **Halon** and **CO_2** applications should continue after the flames are extinguished to reduce the possibility of reignition.

Step 8: Watch for smoldering hot spots or possible reignition of flammable liquids. Make sure that the fire is out.

SKILL SHEET 5-3 OPERATING A PORTABLE FIRE EXTINGUISHER

Cartridge-Operated Extinguishers
Dry Chemical, Dry Powder

Step 9: Back away from the fire area.
Step 1: Select the appropriate extinguisher based on the size and type of fire.

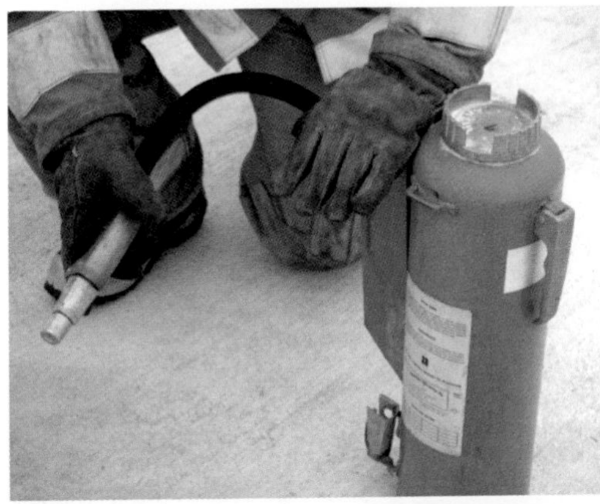

Step 2: Remove hose from its storage position.

Step 3: Position to one side of the extinguisher and depress the activation plunger.

Step 4: Point the nozzle or horn in a safe direction, and discharge a very short test burst to ensure proper operation.

Step 5: Carry the extinguisher to within stream reach of the fire.

Step 6: Aim the nozzle or horn toward the material that is burning.

Step 7: Squeeze the carrying handle and the discharge handle together to start the flow of agent. Release the handle to stop the flow.

Step 8: Sweep the nozzle back and forth. Start at the near edge of the fire and move forward while sweeping the nozzle from side to side until fire is extinguished.

NOTE: Do not plunge **dry chemical** into flammable liquid fires.

Step 9: Watch for smoldering hot spots or possible reignition of flammable liquids. Make sure that the fire is out. Be prepared to reapply agent if reignition occurs.

Step 10: Back away from the fire area.

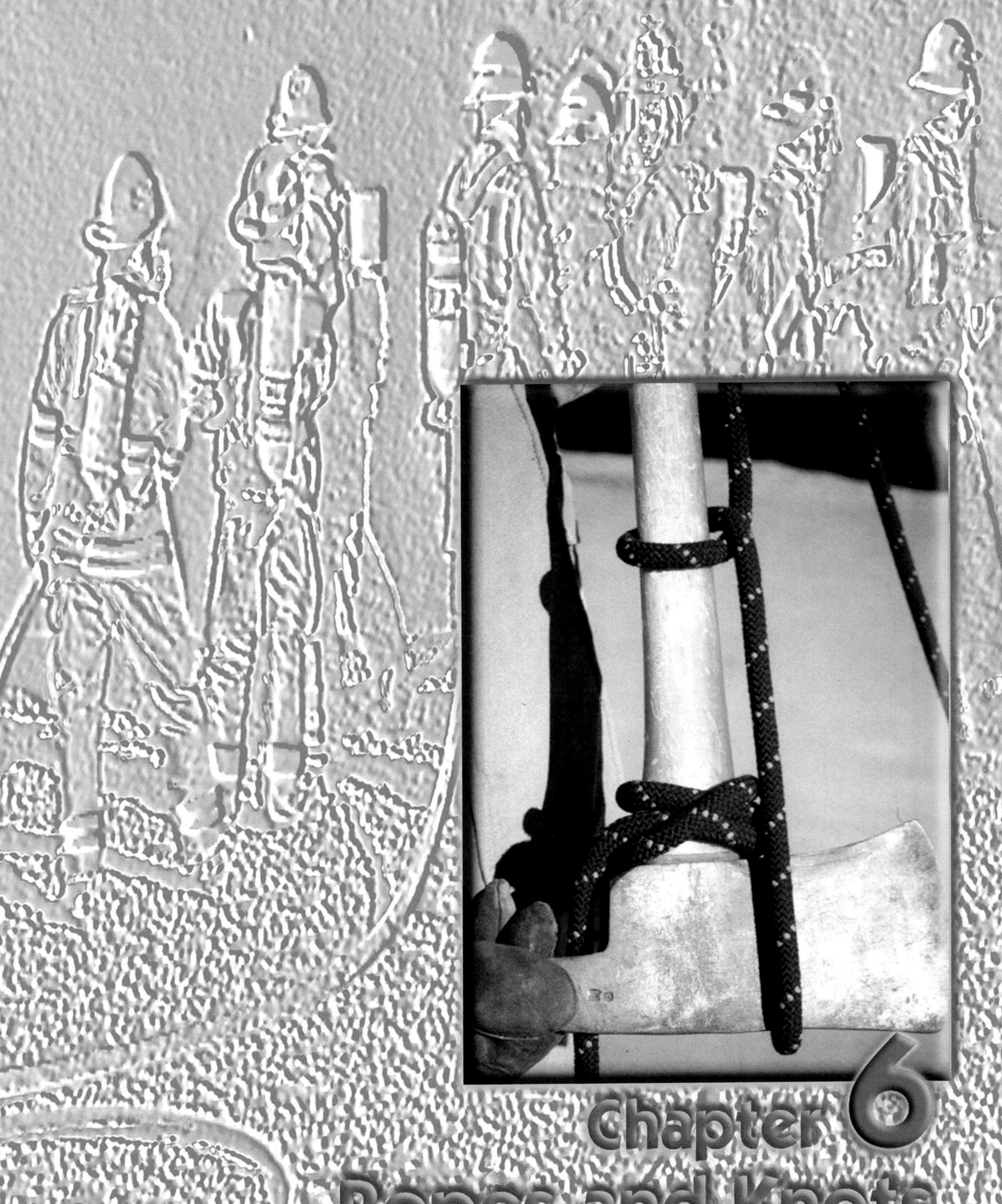

Chapter 6
Ropes and Knots

Chapter 6
Ropes and Knots

INTRODUCTION

Rope is one of the oldest tools used by the fire service. Rope is very valuable for applications such as hauling tools, accomplishing rescues from areas of different elevations, stabilizing vehicles, and cordoning off areas. Firefighters must be knowledgeable of the different types of rope so that the correct one is chosen for the required job. The ability to tie proper knots is crucial to the safety of rope maneuvers. The knots discussed in this chapter are limited to only those basic knots that NFPA 1001, *Standard for Fire Fighter Professional Qualifications*, requires firefighters to know. Local policies may require firefighters to know additional knots or to use different methods than are shown in this chapter — this knowledge is not discouraged. However, any knots that deviate from the standard should be thoroughly tested under controlled conditions before use in life safety applications.

This chapter covers the types of rope and their usage along with rope construction and materials. The proper care, inspection, record keeping, and storage of fire service ropes are also discussed. Several types of knots commonly used in the fire service are described as well. Finally, the methods for hoisting tools and equipment are reviewed. An overview of rope rescue is provided as an introduction to the subject. For expanded information on ropes and knots and rope rescue, consult IFSTA's **Fire Service Search & Rescue** manual.

TYPES OF ROPE AND THEIR USAGE

[NFPA 1001: 3-1.1.1]

Fire service rope falls into two classifications: life safety rope and utility rope. *Life safety rope* is used to support rescuers and/or victims during actual incidents or training (see Rope Rescue section) (Figure 6.1). *Utility rope* is used in any instance, excluding life safety applications, where the use of a rope is required (see Hoisting Tools and Equipment section).

Figure 6.1 It is important that the proper rope be used for life safety applications. *Courtesy of Laura Mauri.*

All life safety (rescue) rope must conform to NFPA 1983, *Standard on Fire Service Life Safety Rope and System Components,* which defines life safety rope as *"rope dedicated solely for the purpose of supporting people during rescue, fire fighting, or other emergency operations, or during training evolutions.*" Only rope of block creel construction (without knots or splices in fibers) using continuous filament virgin fiber for load-bearing elements is suitable for life safety applications. Rope made of any other material or construction should not be used.

According to NFPA 1983, the rope manufacturer must supply the purchaser with information regarding use criteria, inspection procedures, maintenance procedures, and criteria for retiring life safety rope from service. The manufacturer is further required to supply criteria to consider before a life safety rope is reused in life safety situations. Included in these criteria are the following conditions that must be met.

- Rope has not been visibly damaged.
- Rope has not been exposed to heat, direct flame impingement, or abrasion.
- Rope has not been subjected to any impact load (a force suddenly applied to a rope from a falling load).
- Rope has not been exposed to liquids, solids, gases, mists, or vapors of any chemical or other material that can deteriorate rope.
- Rope passes inspection when inspected by a qualified person following the manufacturer's inspection procedures both before and after each use.

The manufacturer must also provide the user with information about removing a life safety rope from service when it does not meet all of the previously stated conditions or if there is any reason to doubt its safety or serviceability. Any life safety rope that fails to pass inspection or has been impact loaded should be destroyed immediately. In this context, "destroy" means that it is altered in such a manner that it cannot be mistakenly used as a life safety rope again. This could include disposing of the rope or removing the manufacturer's label and cutting the rope into shorter lengths to be used as utility rope. A rope that has been subjected to impact loading must have an entry made in its logbook, otherwise, there is no way to determine by inspection if the rope has been impact loaded (see Maintaining a Rope Logbook section).

Utility rope can be used to hoist equipment, secure unstable objects, or cordon off an area. Although there are industry standards concerning the physical properties of utility rope, there are no standards set forth for utility rope applications; however, common sense should prevail in its use. Regularly inspect utility rope to see if it is damaged.

ROPE MATERIALS

[NFPA 1001: 3-1.1.1]

For many years natural fiber rope was the primary type of rope used for rescue. Natural fiber ropes are made of manila, sisal, and cotton (Figure 6.2). However, after extensive testing and evaluation, natural fiber rope is no longer accepted for use in life safety applications. It is acceptable to use natural fiber rope for utility purposes; however, it must not be used for specific rescue purposes.

Figure 6.2 Utility rope can be used for hoisting operations.

Advances in synthetic rope construction have made its use preferable to natural fiber rope for life safety applications. Synthetic fiber rope has excellent resistance to mildew and rotting, has excellent strength, and is easy to maintain. Unlike natural fiber rope, which is made of short overlapping strands of fiber, synthetic rope may feature continuous fibers running the entire length of the rope. Table 6.1 provides the general characteristics of rope fibers. Firefighters should become familiar with the manufacturers' specifications and limitations for the ropes used in their department.

ROPE CONSTRUCTION

[NFPA 1001: 3-1.1.1]

Two types of rope are used in life safety situations: dynamic rope and static rope. Each type has its advantages and disadvantages because of different design and performance criteria.

Dynamic (high-stretch) rope is used when long falls are a possibility such as in rock climbing. To reduce the shock of impact on both climbers and

TABLE 6.1
Rope Fiber Characteristics

Characteristics	Nylon	Polyester	Polypropylene	Polyethylene	Manila	Cotton	Kevlar® Aramid	H. Spectra ® Polyethylene
Strength	3*	4*	5*	6*	7*	8*	2*	1*
Wet Strength vs. Dry Strength	85%	100%	100%	100%	115%	115%	90%	100%
Shock Load Ability	1*	3*	2*	4*	5*	6*	7*	7*
Floats or Sinks in Water (Specific Gravity)	Sinks (1.14)	Sinks (1.38)	Floats (0.92)	Floats (0.95)	Sinks (1.38)	Sinks (1.54)	Sinks (1.45)	Floats (0.97)
Elongation at Break (Approximately)	20–34%	15–20%	15–20%	10–15%	10–15%	5–10%	2–4%	< 4%
Melting Point	480°F (249°C)	500°F (260°C)	330°F (166°C)	275°F (135°C)	Does not melt; chars at 350°F (177°C)	Does not melt; chars at 300°F (149°C)	Does not melt; chars at 800°F (427°C)	275°F (135°C)
Abrasion Resistance	3*	2*	7*	6*	4*	8*	5*	1*
Resistance:								
Sunlight	Good	Excellent	Poor	Fair	Good	Good	Good	Good
Rot	Excellent	Excellent	Excellent	Excellent	Poor	Poor	Excellent	Excellent
Acids	Poor	Good	Good	Good	Poor	Poor	Poor	Excellent
Alkalis	Good	Poor	Good	Good	Poor	Poor	Good	Excellent
Oil & Gas	Good	Good	Good	Good	Poor	Poor	Good	Excellent
Electrical Conductivity Resistance	Poor	Good	Good	Good	Poor	Poor	Poor	Good
Storage Requirements	Wet or Dry	Wet or Dry	Wet or Dry	Wet or Dry	Dry Only	Dry Only	Wet or Dry	Wet or Dry

*** Scale: Best = 1; Poorest = 8**

Source: Wellington Leisure Products, Inc.

their anchor systems in falls, dynamic rope is designed for high stretch without breaking. However, this elasticity is a disadvantage when trying to raise or lower heavy loads, so dynamic rope is *not* considered practical for hauling applications.

Static (low-stretch) rope is the rope of choice for most rescue incidents. It is designed for low stretch without breaking, and this makes it better suited for raising and lowering heavy loads. Static rope is used for hauling, rescue, rappelling, and where no falls are likely to occur or only very short falls are possible.

The most common types of rope construction are laid, braided, braid-on-braid, and kernmantle (Figure 6.3).

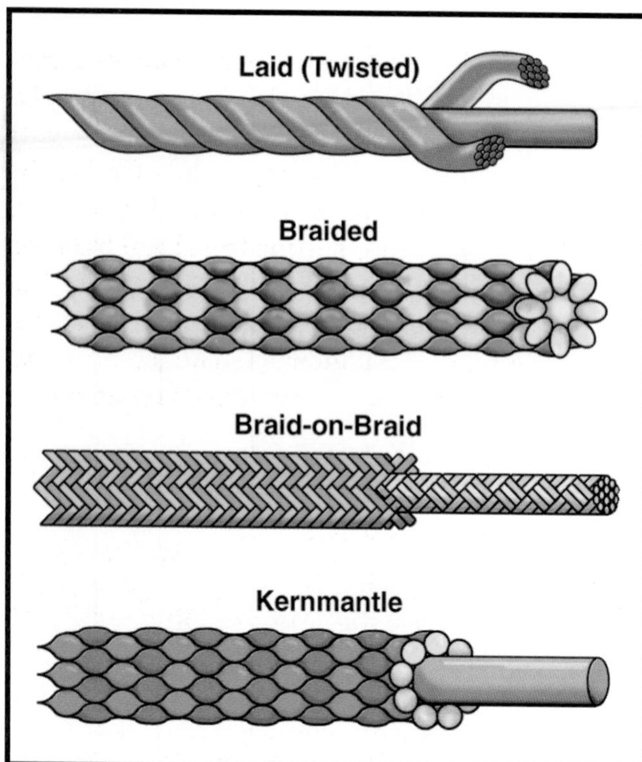

Figure 6.3 Common types of rope construction.

Laid (Twisted) Natural or Synthetic Rope

Laid ropes are constructed by twisting yarns together to form strands. Generally, three strands are twisted together to make the final rope. How tightly these ropes are twisted and the type of fiber used determine the rope's properties. Twisted rope is susceptible to abrasion and other types of physical damage. Twisting a rope leaves all three load-bearing strands exposed at various points along the rope. Although this exposure allows for easy inspection, it also means that any damage immediately affects the rope's strength.

Braided Rope

Although some braided ropes are made from natural fibers, most are of the synthetic variety. *Braided rope* is constructed by uniformly intertwining strands of rope together (similar to braiding a person's hair). Braided rope reduces or eliminates the twisting common to laid ropes. Because of its construction characteristics, the load-bearing fibers are subject to direct abrasion and damage.

Braid-On-Braid Rope (Double Braid)

Because braid-on-braid is a jacketed rope, it is often confused with kernmantle rope (see following section). *Braid-on-braid rope* is just what the name implies: It is constructed with both a braided core and a braided sheath. The sheath has a herringbone-pattern appearance.

Braid-on-braid rope is very strong. Half of its strength is in the sheath and the other half is in the core. A disadvantage of braid-on-braid rope is that it does not resist abrasion as well as kernmantle rope. Another disadvantage is that the sheath may slide along the inner core of the rope.

Kernmantle Rope

Kernmantle, a jacketed rope, is composed of a braided covering or sheath (mantle) over the main load-bearing strands (kern). The core runs parallel with the covering, which increases the rope's stretch resistance and load characteristics. The load characteristics are also affected by the method of manufacture. The core (kern) is made of high-strength fibers; these account for most of the total strength of the rope. The sheath provides a small amount of the rope's overall strength, absorbs most of the abrasion, and protects the load-bearing core. Kernmantle rope comes in both dynamic and static types. Dynamic kernmantle is most commonly used as a sport rope for rock or ice climbing. Static kernmantle is most commonly used as rescue rope where stretch is an undesirable characteristic.

ROPE MAINTENANCE

[NFPA 1001: 3-1.1.1; 3-5.3; 3-5.3(a)(b)]

In order for rescue rope to be ready and safe for use when needed, it must be properly maintained.

This maintenance includes inspecting and cleaning the rope as well as maintaining a logbook of its use and maintenance.

Inspecting Rope

Inspect all types of rope after each use. Inspect it visually and tactilely (by touch). When making inspections, use the following methods described for the various ropes and note any observations. They should be inspected for shards of glass, metal shavings, wood splinters, or other foreign objects that could cause damage. If any of these are found, the rope should be taken out of service.

LAID ROPE

Inspect synthetic laid ropes for soft, crusty, stiff, or brittle spots; for areas of excessive stretching; for cuts, nicks, or abrasions; for chemical damage; for dirt; and for other obvious flaws. Laid rope should be untwisted and checked internally for these flaws (Figure 6.4). The presence of mildew does not necessarily indicate a problem; however, the rope should be cleaned and reinspected. A foul smell might indicate rotting or mildew in manila rope.

Figure 6.4 Inspect laid rope by untwisting the sections.

BRAIDED ROPE

Visually inspect braided rope for exterior damage such as heat sears (caused by friction or fire), nicks, and cuts. Also, visually inspect for excess or unusual fuzziness. Tactilely inspect for permanent mushy spots or other deformities.

BRAID-ON-BRAID ROPE

Figure 6.5 The seared end of a braid-on-braid rope.

Inspect braid-on-braid rope for heat sears, nicks, and cuts. Also inspect for the sheath sliding on the core. If sliding is found, cut the end of the rope and pull off the excess material; then sear the end (Figure 6.5). Inspect for lumps that indicate core damage. A reduction in the rope's diameter may indicate that the core has broken. Carefully examine any type of damage or questionable wear to the sheath.

KERNMANTLE ROPE

Inspecting kernmantle rope for damage is somewhat difficult because the damage may not be obvious. The inspection can be performed by putting a slight tension on the rope while feeling for any lumps, depressions, or soft spots (Figure 6.6). A temporary soft spot resulting from hard knots or sharp bends in the rope may be felt; however, the fibers within the core may realign themselves over time if the rope is undamaged. The only way to determine whether such a soft spot is damage or

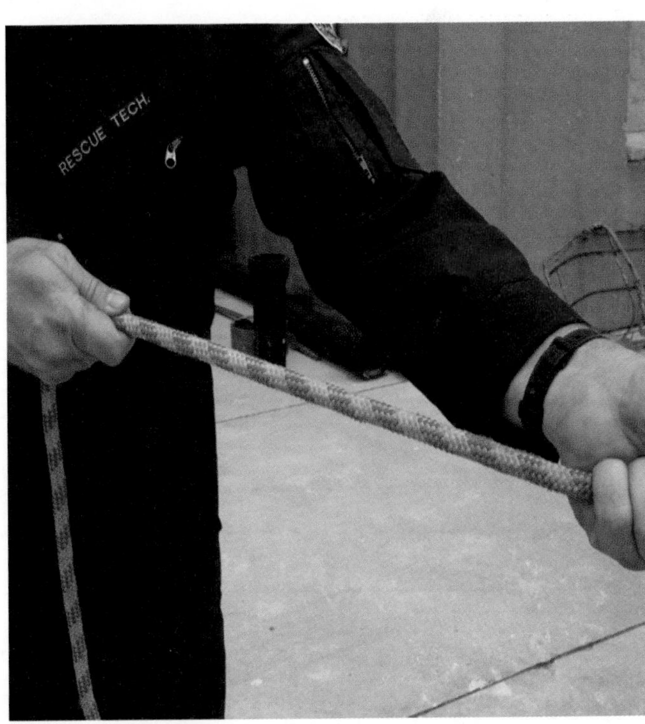

Figure 6.6 A firefighter inspects a kernmantle rope.

just temporarily misaligned core fibers is by carefully inspecting the outer sheath. Any damage to the outer sheath indicates probable damage to the core. The core of a kernmantle rope can be damaged without visible evidence on the outer sheath. If there is any doubt about the rope's integrity, it should be downgraded to utility status.

The rope should also be inspected for irregularities in shape or weave, foul smells, discoloration from chemical contamination, roughness, abrasions, or fuzziness. A certain amount of fuzziness is normal and is not necessarily a cause for concern. If there is a great amount of fuzziness in one spot or if the overall amount is excessive based upon the inspector's judgment and experience, the rope should be downgraded.

Maintaining a Rope Logbook

When a piece of rescue rope is purchased, it should be permanently identified and a record (rope logbook) started and kept throughout its working life. The date of each use and the inspection/maintenance records of the rope should be entered into the logbook. This information helps determine when the rope should be retired (Figure 6.7). The log should be kept in a waterproof envelope, usually placed in a pocket sewn on the side of the rope's storage bag (see Storage of Life Safety Ropes section) (Figure 6.8).

Cleaning Rope

Methods of washing and drying rope vary with each manufacturer, so it is always advisable to contact them for specific cleaning and drying in-

Oklahoma State University
Fire Service Training
Rope Log

Rope Type: ______________ Rope Size: ______________ Rope #: ______________
Manufacturer: ______________ Model: ______________ Rope Color: ______________
Purchased From: ______________ Date: ______________ Bag Color: ______________

Date	Sign-Out	Use	Possible Damage/Comments	Sign-In

Figure 6.7 A typical rope log.

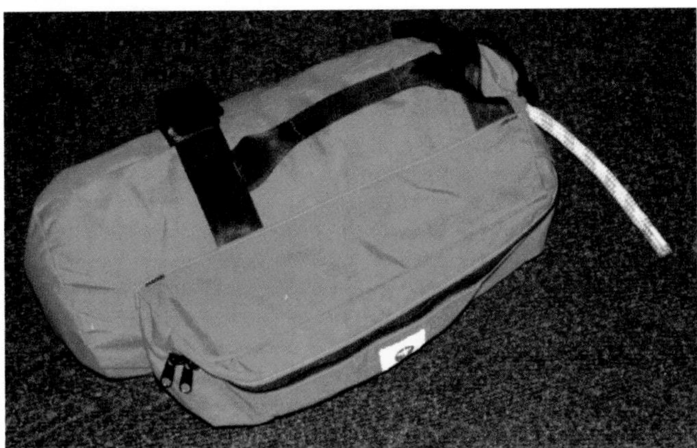

Figure 6.8 The pouch containing the rope logbook. *Courtesy of Laura Mauri*

structions for the type of rope or ropes in use. The following sections give some general guidelines for cleaning rope.

NATURAL FIBERS

Natural fiber rope cannot be cleaned effectively because water cannot be used in the cleaning process. Although water initially strengthens natural fiber rope, after continual exposure to wetting and drying, water weakens and damages the fiber. Wipe or gently brush the rope to remove as much of the dirt and grit as possible.

SYNTHETIC FIBERS

Cool water and mild soap are least likely to damage synthetic fiber ropes. Bleaches or strong cleaners should not be used. Some synthetic rope may feel stiff after washing, but this is not a cause for concern. There are three principal ways to clean synthetic rope: hand washing, special rope-washing device, or regular clothes-washing machine.

Washing by hand consists of wiping the rope with a cloth or scrubbing it with a brush and then thoroughly rinsing with clean water (Figure 6.9). Commercial rope-washing devices that can be connected to a standard faucet or garden hose are available (Figure 6.10). Rope is fed manually through the device, and multidirectional streams of water clean all sides of the rope at the same time. These devices do an adequate job of cleaning mud and other surface debris from rope, but for a more thorough cleaning, rope should be washed in a clothes-washing machine.

Figure 6.9 Rope may need to be scrubbed.

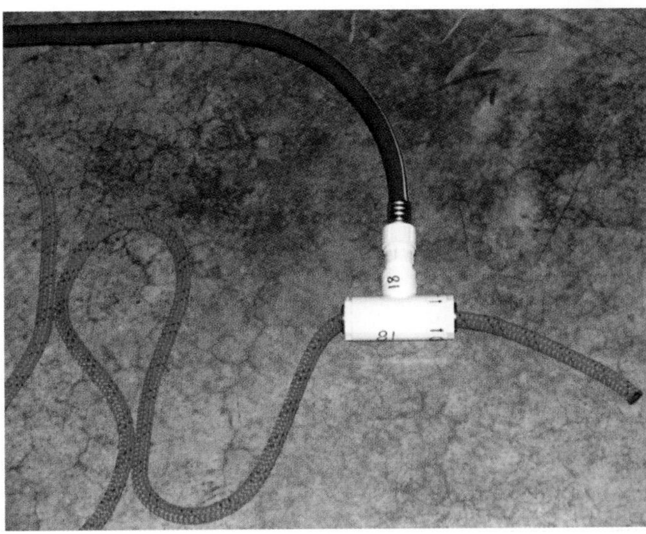

Figure 6.10 A typical hose-bib rope washer. *Courtesy of Laura Mauri.*

Front-loading washing machines without plastic windows are the best type to use for washing synthetic rope. Front-loaders that have a plastic window are not recommended because the plastic can cause enough friction with the rope during the spin cycle to damage the rope. Top-loaders may also damage the rope during agitation. The washer should be set on the coolest wash/rinse temperature available, and only a small amount of mild soap, if anything, added. The rope can be further protected by putting it in a cloth bag before placing it in the washer, or it can be coiled first into a bird's-nest coil (Figure 6.11).

Figure 6.11 A rope is loaded into a washing machine.

Once the rope has been washed, it should be dried. It can be spread out on a hose rack *out of direct sunlight*, suspended in a hose tower, or loosely coiled in a hose dryer.

STORAGE OF LIFE SAFETY ROPES

[NFPA 1001: 3-5.3]

Rescue (life safety) ropes can be stored in various coils, but most rescue units store their ropes in rope bags. Regardless of how it is stored, where it is stored is of critical importance. Rescue rope should be stored in spaces or compartments that are clean and dry but have adequate ventilation. It should not be exposed to chemical contaminants, such as battery acid or hydrocarbon fuels, or the fumes or vapors of these substances. Rescue rope should not be stored in the same compartments where gasoline-powered rescue tools or the spare fuel for these tools are stored.

Bagging a Rope

The best method for storing kernmantle rope and other life safety rope is to place it into a storage bag (Figure 6.12). The bag allows easy carrying of the rope and keeps dirt and grime from the rope as well. The bag may have a drawstring and shoulder straps for ease in carrying. Nylon bags can be used as well as canvas.

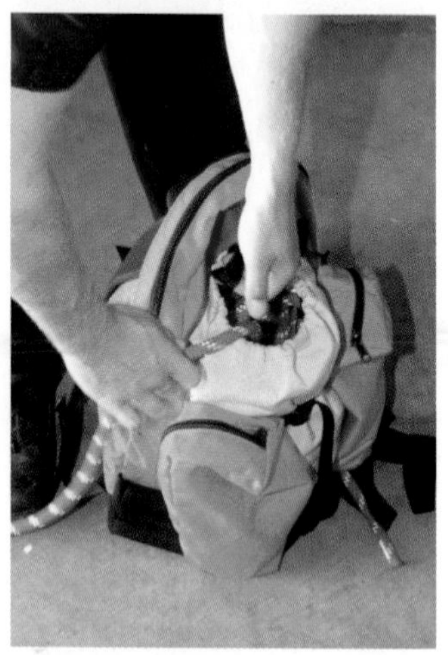

Figure 6.12 Rescue rope can be stored in stuff bags.

Coiling/Uncoiling a Rope

Coiling rope so that it may be placed into service with a minimum of delay is very essential in the fire service. An improperly coiled rope may result in the failure of an evolution. Refer to Skill Sheets 6-1 and 6-2 for the procedures for coiling and uncoiling a rope.

KNOTS

[NFPA 1001: 3-1.1.1; 3-1.1.2]

Knots are used to join or connect objects or to form loops. The ability to tie knots is a vital part of fire and rescue operations. Improperly tied knots can be extremely hazardous to both rescuers and victims. The descriptions of how to tie knots include terms for parts of a rope (Figure 6.13). The *running end* is the part used for hoisting, pulling, or belaying. The *working end* is the part used in forming the knot (commonly referred to as the *loose end* or *bitter end*). The *standing part* is the part between the working end and the running end.

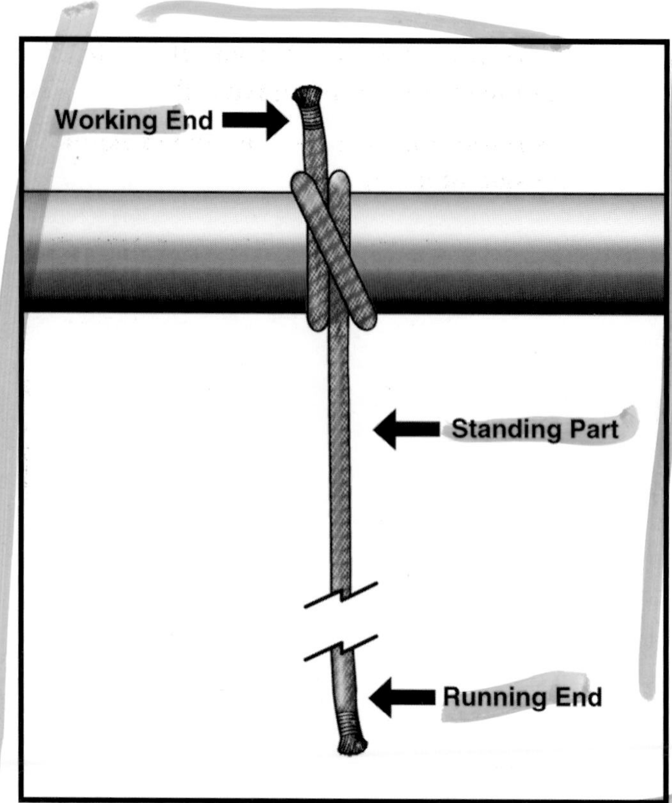

Figure 6.13 Parts of a rope.

All knots should be dressed after they are tied; that is, they should be tightened until snug with all slack removed. In order to prevent slipping, a safety knot should be applied to the tail of the working end of the rope. Safety knots include the single- and double-overhand knots. Other knots used in the fire service are the bowline, half hitch, clove hitch, figure-eight family, and becket bend (sheet bend).

Elements of a Knot

To be suitable for use in rescue, a knot must be easy to tie and untie, be secure under load, and reduce the rope's strength as little as possible. A rope's strength is reduced whenever it is bent. The tighter the bend, the more strength is lost. Some knots create tighter bends than others and thereby reduce the rope's strength to a greater degree. Bight, loop, and round turn are names for the bends that a rope undergoes in the formation of a knot or hitch. Knots and hitches are formed by combining these elements in different ways so that the tight part of the rope bears on the free end to hold it in place. Each of these formations is shown in the following figures:

- The *bight* is formed by simply bending the rope back on itself while keeping the sides parallel (Figure 6.14).

- The *loop* is made by crossing the side of a bight over the standing part (Figure 6.15).
- The *round turn* consists of further bending one side of a loop (Figure 6.16).

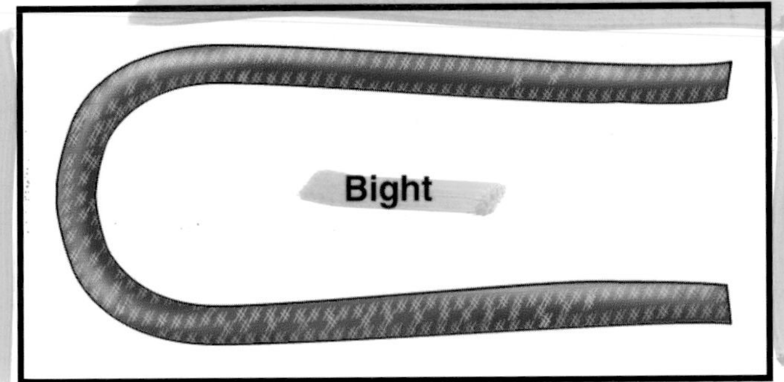

Figure 6.14 A bight.

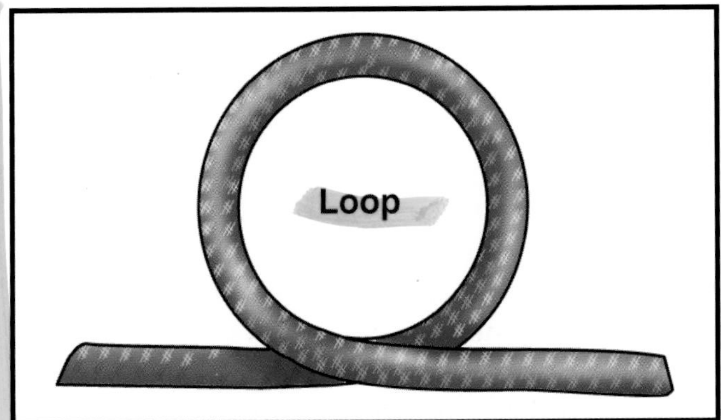

Figure 6.15 A loop.

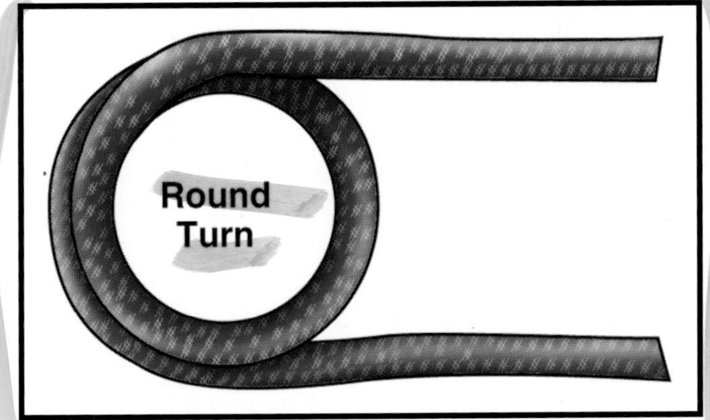

Figure 6.16 A round turn.

Single/Double Overhand Safety Knots

As an added measure of safety, use an overhand safety knot whenever tying any type of knot (Figure 6.17). Although any properly tied knot should hold, it is always desirable to provide the highest level of safety possible. Use of the overhand safety knot eliminates the danger of the end of the rope slipping back through the knot and causing the knot to fail. Skill Sheet 6-3 describes the procedure for tying the single overhand knot. The double overhand knot is tied in the same manner but the rope is doubled before tying.

Figure 6.17 An overhand safety knot.

Bowline

The bowline is an important knot in the fire service, sharing a degree of acceptance in both life safety and other fire service applications. The bowline is easily untied and is a good knot for forming a single loop that will not constrict the object it is placed around. Firefighters should be able to tie the bowline in the open as well as around an object. The method in Skill Sheet 6-4 is one way of tying the bowline, although other methods may be just as effective.

Half Hitch

The half hitch is particularly useful in stabilizing tall objects that are being hoisted. The half hitch is always used in conjunction with another knot or hitch. For example, when hoisting a pickhead axe, a half hitch is used around the handle. The half hitch is formed by making a round turn around the object. The standing part of the rope is passed under the round turn on the side opposite the intended direction of pull (Figure 6.18). Several half hitches can be applied in succession if required.

Figure 6.18 A half hitch.

Clove Hitch

The clove hitch may be formed by several methods. It consists essentially of two half hitches. Its principal use is to attach a rope to an object such as a pole, post, or hoseline. The clove hitch is not regarded as suitable for use in anchoring a life safety rope (or in a life safety application). The clove hitch may be formed anywhere in the rope from one end to the middle. When properly applied, it withstands a pull in either direction without slipping. If the knot will be subjected to repeated loading and unloading, it should be backed up with an overhand safety knot. Skill Sheet 6-5 describes the procedure for tying a clove hitch.

The clove hitch, when formed by the method described in Skill Sheet 6-5, cannot be placed over an object that has no free end such as the center of a hoseline. Therefore, it is necessary to know how to tie the clove hitch around an object. Skill Sheet 6-6 describes the procedure.

Figure-Eight Family of Knots

The figure-eight family of knots has gained increased acceptance and popularity for fire and rescue service applications. There are several variations of the figure eight that are commonly used such as the figure-eight follow through and the figure-eight on a bight. The figure eight is the foundation knot for these knots. Refer to Skill Sheet 6-7 for the procedure for tying the figure-eight knot.

FIGURE-EIGHT FOLLOW THROUGH

The figure-eight follow through is used to tie ropes of equal diameters together or to tie a rope around an object, the end of which is not available. Refer to Skill Sheet 6-8 for procedures on tying the figure-eight follow through knot.

FIGURE-EIGHT ON A BIGHT

The figure-eight on a bight is a good way to tie a loop in either the middle or the end of a rope. It is tied by forming a bight in either the end of the rope or at any point along its length and then tying a simple figure eight with the doubled part of the rope (bight). Refer to Skill Sheet 6-9 for procedures on tying the figure-eight on a bight.

Becket Bend (Sheet Bend)

The becket bend (or sheet bend) is used for joining two ropes of unequal diameters or joining a rope and a chain. It is also unlikely to slip when the rope is wet. These advantages make it useful and dependable in fire service rope work. However, the becket bend is not suitable in life safety applications. Refer to Skill Sheet 6-10 for procedures on tying the becket bend.

HOISTING TOOLS AND EQUIPMENT

[NFPA 1001: 3-1.1.1; 3-1.1.2; 3-3.11(b)]

A common activity at large-scale fire or rescue incidents is to use ropes to haul various pieces of equipment from one elevation to another (Figure 6.19). Almost any piece of equipment can be hauled with rope. Common sense and knowledge of proper knots and hitches aid in securing these objects. For example, anything with a closed-type D-ring handle can be raised or lowered using a bowline or figure eight. The hoisting of pressurized cylinders, such as fire extinguishers or SCBA bottles, is not recommended.

Use the proper knots and securing procedures to prevent dropping the equipment. This prevents damage to the equipment and serious injury to anyone standing below. Depending on local policy, a separate tag line may also be tied to any of these pieces of equipment or the object may be tied in the center of the rope so that the hoisting rope also serves as the tag line. When used, the tag line is guided by firefighters on the ground who prevent the equipment from coming in contact with the structure or other objects as it is raised (Figure 6.20). When one rope serves as both the tag line and the hoisting line, the knot or hitch-tying methods and methods of hoisting may vary. Keep safety in mind first, then select the method of hoisting. Practice improves performance. The following sections discuss how to hoist various pieces of equipment and give hoisting safety considerations.

Figure 6.19 Rope being used to hoist equipment.

Figure 6.20 A tag line is used to prevent the tool from hitting against the side of the building.

Hoisting Safety Considerations

- Have solid footing and make necessary preparations before starting a hoisting operation.
- Use the hand-over-hand method to maintain control of the rope during a hoisting operation.
- Protect rope from physical damage when rope must be pulled over sharp edges such as cornices or parapet walls. Edge rollers can be used for this purpose (Figure 6.21).
- Work in teams to ensure firefighter safety when working from heights.
- Look to ensure all personnel are clear of the hoisting area.
- Avoid hoisting operations near electrical hazards if possible. If this is not possible, use extreme caution.
- Ensure a charged hoseline's nozzle handle is secure to prevent accidental discharge when hoisting charged hoselines.

Figure 6.21 Typical edge rollers. *Courtesy of Laura Mauri.*

Axe

The procedure for attaching and hoisting an axe is the same for either a pick-head axe or flat-head axe. Using the following method, the same rope can be used as both the hoist rope and as the tag line. Refer to Skill Sheet 6-11 for the procedure on hoisting an axe.

Pike Pole

To raise a pike pole (with the head up), place a clove hitch toward the end of the handle, followed by a half hitch in the middle of the handle and another half hitch around the head (Figure 6.22).

Ladder

Use a bowline or figure-eight on a bight and slip it first through two rungs of the ladder about one-third of the way down from the top. After pulling that loop through, slip it over the top of the ladder (Figure 6.23).

Figure 6.22 Raise a pike pole (with the head up) by placing a clove hitch toward the end of the handle followed by a half hitch in the middle of the handle and another around the head.

Figure 6.23 To hoist a ladder, use a bowline or figure-eight on a bight and slip it first through two of the rungs about one-third of the way down from the top. After pulling that loop through, slip it over the top of the ladder.

Hoselines

Hoisting hose is possibly the safest way of getting hoselines to upper levels. As with advancing hose up a ladder, it is easier and safer to hoist a dry hoseline (see Skill Sheet 6-12); however, charged lines may also be hoisted. It is most desirable to bleed the pressure from a charged hoseline before hoisting. If it is not possible to do so, use the procedures in Skill Sheet 6-13 to hoist the charged hoseline. Care should be taken to reduce the possibility of damaging the coupling or the nozzle as the hoseline is being raised.

Portable Fans

To securely hoist a portable fan, tie a bowline or figure-eight follow through around two of the connecting rods between the front and back plates.

This will be the hauling line. A tag line should be attached to the bottom of the unit (Figure 6.24). This line is controlled by personnel on the ground who keep the fan from bouncing against the side of the building while it is being raised.

Figure 6.24 Use a figure-eight follow through or a bowline to hoist a portable fan. Notice the tag line prevents the fan from bouncing against the building.

ROPE RESCUE

[NFPA 1001: 3-1.1.1)

When victims are located above or below grade in rescue situations, the most efficient and sometimes the only means of reaching them and getting them to ground level may be by the use of ropes, knots, and rope systems (Figure 6.25). Rope rescue is a technical skill that requires specialized training. It may be necessary to lower rescuers into a confined space and to hoist a victim out with a mechanical advantage system made of rescue rope. Victims stranded on a rock ledge or on an upper floor of a partially collapsed building may have to be lowered to the ground with rescue ropes.

Rescue rope, webbing, and appropriate hardware are used for a variety of purposes. Rescue rope and harnesses are used to protect rescuers and victims as they move and/or work in elevated locations where a fall could cause injury or death. Ropes in combination with webbing are the primary tools for raising and lowering rescuers, equipment, and victims. Rope and appropriate hardware are used to create a variety of mechanical advantage and safety systems. For more information on rope rescue, see IFSTA's **Fire Service Search & Rescue** manual.

Figure 6.25 Rope rescue is a technical skill that requires specialized training.

SKILL SHEET 6-1 COILING A ROPE

Step 1: Measure off and reserve a length of rope (about three times the distance between standards, the structures used as a support) at the front of the rope to secure the coil when completed.

NOTE: A compact, finished coil can also be prepared by using the beams of a ladder as standards.

Step 2: Drape this length of rope over one of the standards.

Step 3: Wrap the remaining rope around the standards until sufficient width is developed. It may be necessary to make two layers to coil all rope.

NOTE: Avoid making the coils too tight, which would make removal of the finished coil difficult.

Step 4: Wrap the last portion of the rope around the loops.

Step 5: Fasten the end securely by tucking the end of the rope under the last wrap.

Step 6: Form a bight with the length measured in Step 1.

Step 7: Insert the bight through the end of the coil as shown.

Step 8: Place the end of the rope through the opposite end of the coil.

Step 9: Insert the end of the rope through the bight.

Step 10: Finish the coil by tucking the end of the rope next to the end in Step 5.

Step 11: Remove the coil from the standards.

SKILL SHEET 6-2 UNCOILING A ROPE

Step 1: Release the tie.

Step 2: Grasp the inside of the coil.

Step 3: Pull out two or three loops to loosen the coil.

Step 4: Look below for other people or potentially hazardous situations or obstructions.

Step 5: Drop the coil. If the coil is not carefully prepared, it may not pay out to the ground.

NOTE: Another method of uncoiling the rope is to uncoil it by hand and then lower it hand over hand.

SKILL SHEET 6-3 TYING THE SINGLE OVERHAND KNOT

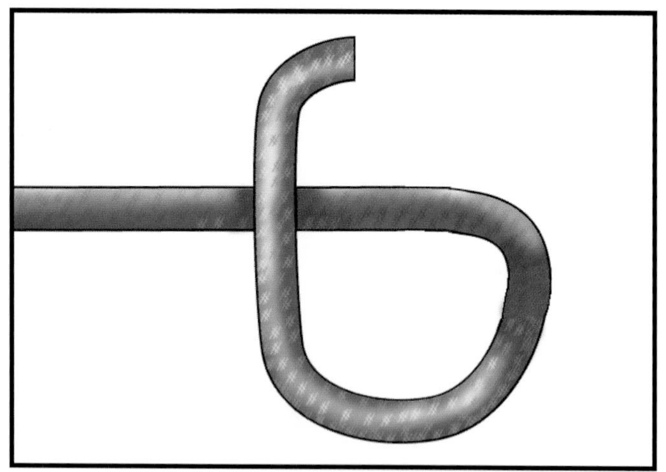

Step 1: Form a loop in the rope.

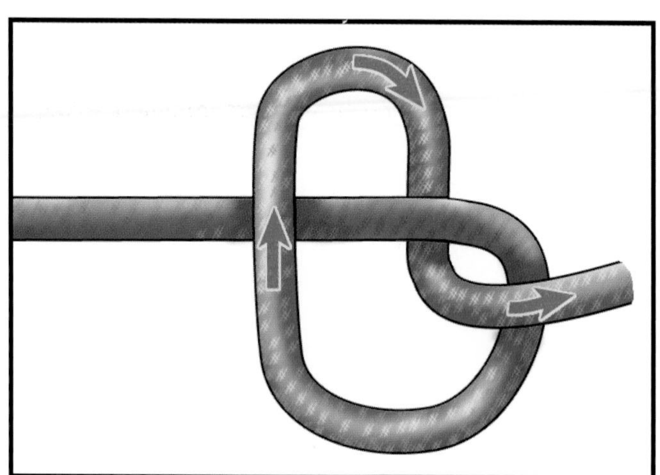

Step 2: Insert the end of the rope through the loop.

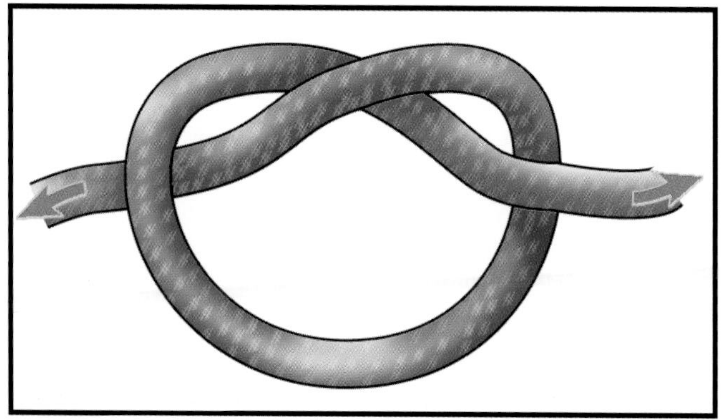

Step 3: Dress the knot by pulling on both ends of the rope at the same time.

SKILL SHEET 6-4 TYING A BOWLINE

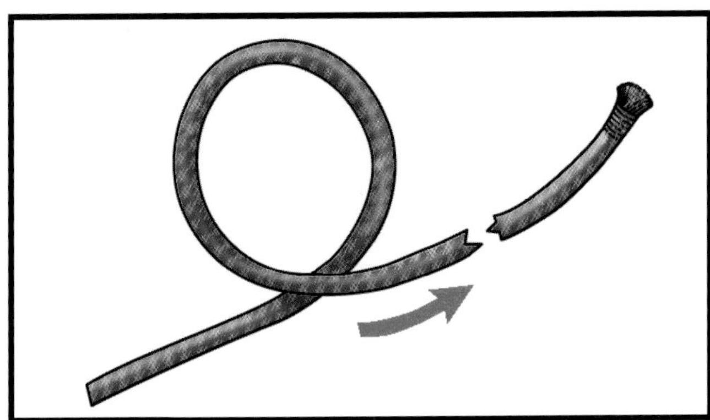

Step 1: Select enough rope to form the size of the knot desired.

Step 2: Form an overhand loop in the standing part.

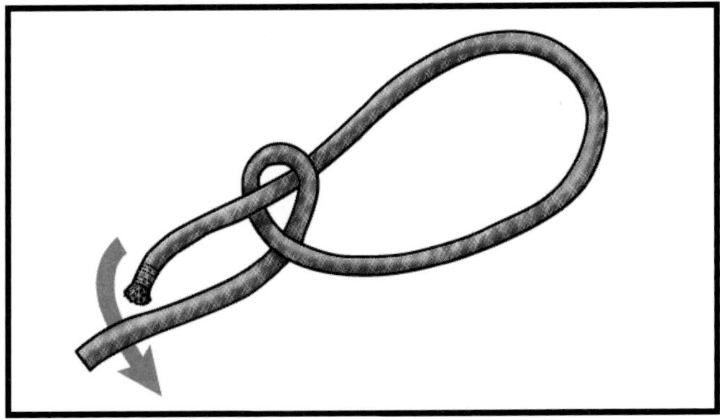

Step 3: Pass the working end upward through the loop.

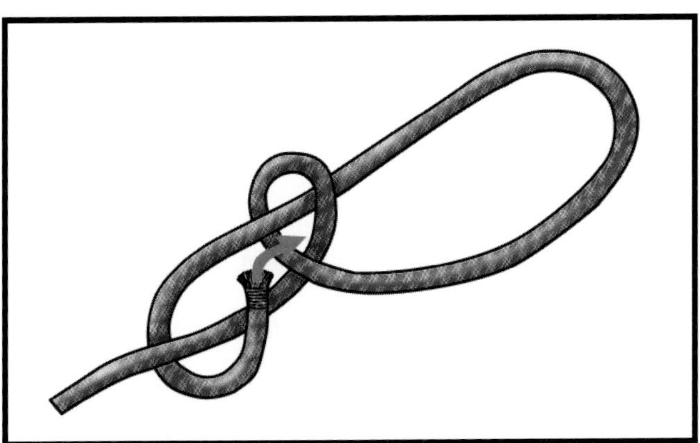

Step 4: Pass the working end over the top of the loop under the standing part.

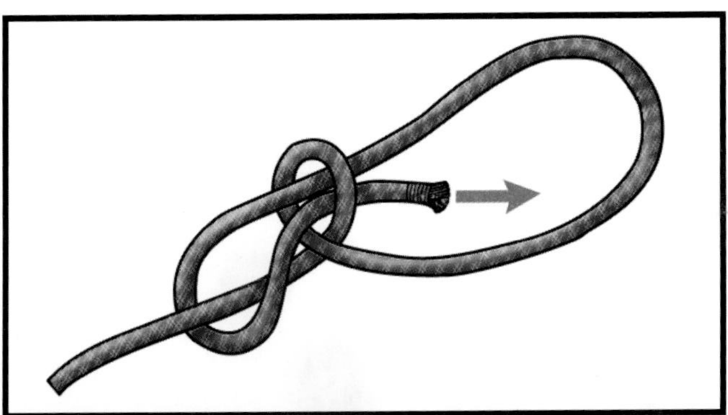

Step 5: Bring the working end completely around the standing part and down through the loop.

Step 6: Pull the knot snugly into place, forming an *inside* bowline with the working end on the inside of the loop.

SKILL SHEET 6-5 TYING A CLOVE HITCH

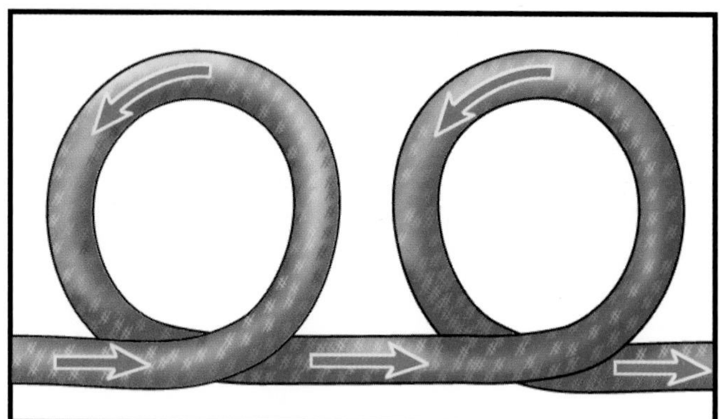

Step 1: Form a loop in your left hand with the working end to the right crossing under the standing part.

Step 2: Form another loop in your right hand with the working end crossing under the standing part.

Step 3: Slide the right-hand loop on top of the left-hand loop.

NOTE: This is the important step in forming the clove hitch knot.

Step 4: Hold the two loops together at the rope forming the clove hitch.

Step 5: Slide the knot over the object.

Step 6: Pull the ends in opposite directions to tighten.

SKILL SHEET 6-6 TYING A CLOVE HITCH AROUND AN OBJECT

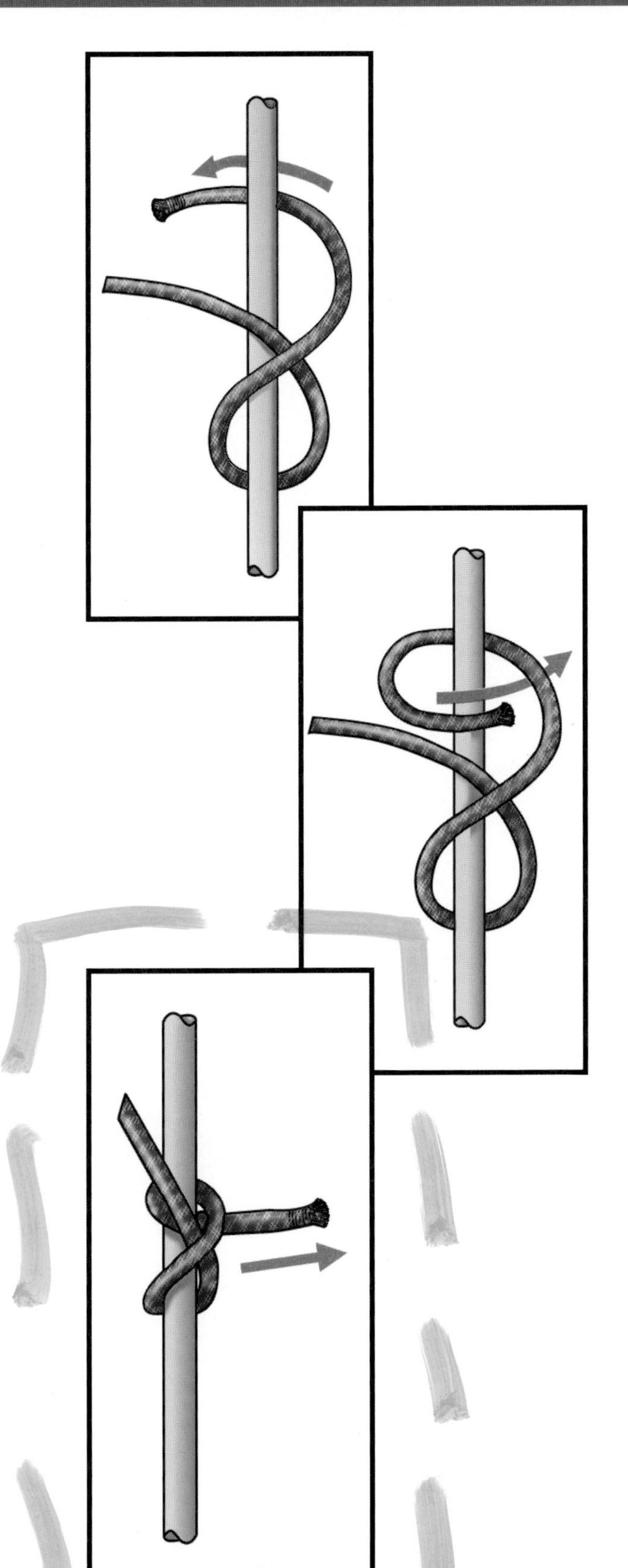

Step 1: Make one complete loop around the object, crossing the working end over the standing part.

Step 2: Complete the *round turn* about the object just above the first loop as shown.

Step 3: Pass the working end under the upper wrap, just above the cross.

Step 4: Set the hitch by pulling.

SKILL SHEET 6-7 TYING A FIGURE EIGHT

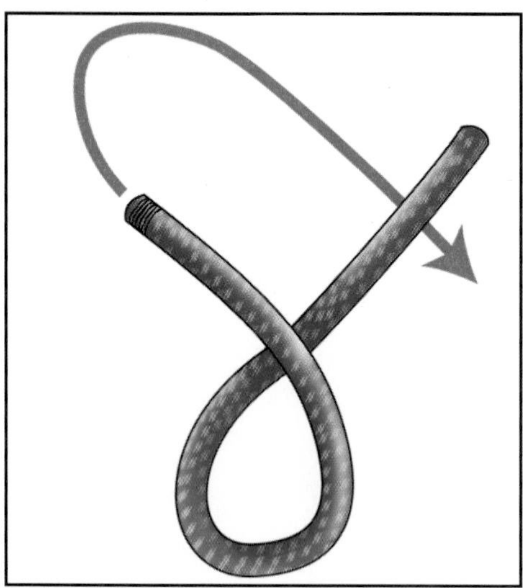

Step 1: Make a loop in the rope.

Step 2: Pass the working end completely around the standing part.

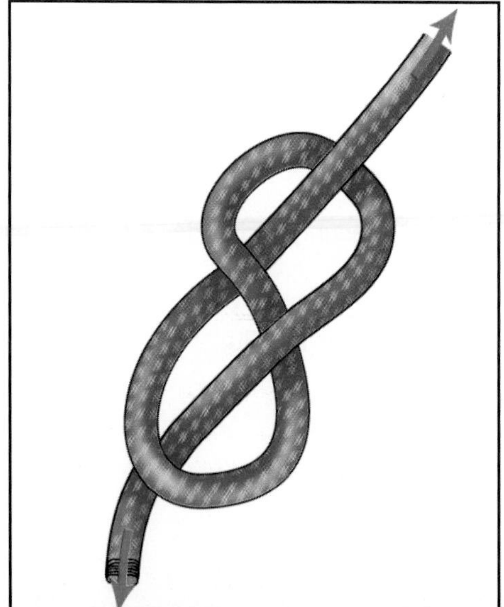

Step 3: Insert the end of the rope back through the loop.

Step 4: Dress the knot by pulling on both the working end and standing part of the rope at the same time.

SKILL SHEET 6-8 TYING A FIGURE-EIGHT FOLLOW THROUGH

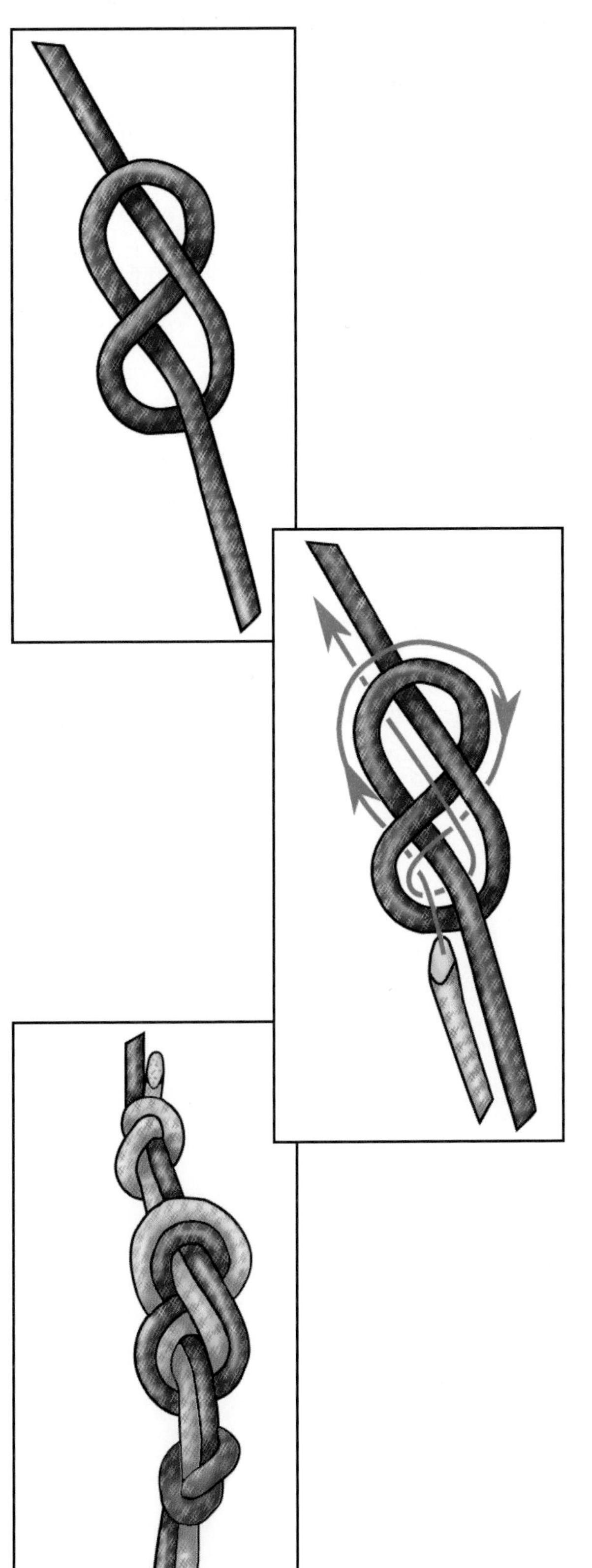

Step 1: Tie a figure-eight knot on one end of the rope.

Step 2: Feed the end of the other rope through the figure-eight knot in reverse. It should follow (hence the name) the exact path of the original knot.

Step 3: Use a safety knot, such as the overhand, with this knot.

NOTE: The figure-eight follow through can be tied in the middle of the rope by placing the figure eight at a point far enough along the rope to allow sufficient rope to go around the object.

SKILL SHEET 6-9 TYING A FIGURE-EIGHT ON A BIGHT

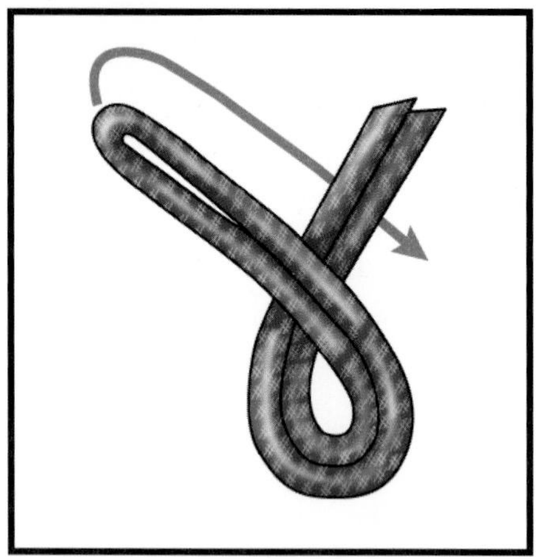

Step 1: Form a bight in the working end of the rope.

Step 2: Pass it over the standing part to form a loop.

Step 3: Pass the bight under the standing part and then over the loop and down through it; this forms the figure eight.

Step 4: Extend the bight through the knot to whatever size working loop is needed.

Step 5: Dress the knot.

SKILL SHEET 6-10 TYING A BECKET BEND

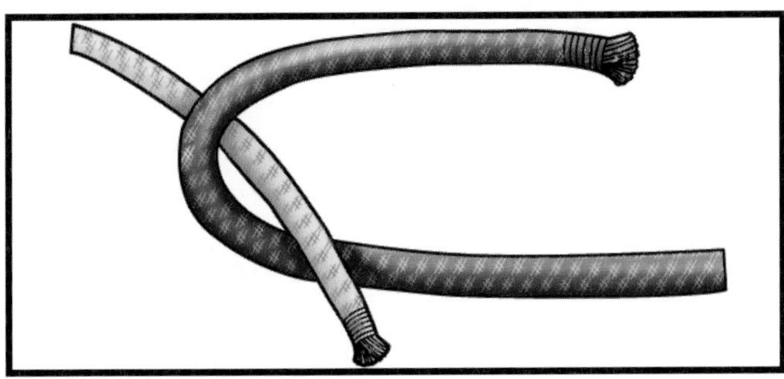

Step 1: Form a bight in one of the ends to be tied (if two ropes of unequal diameter are being tied, the bight always goes in the larger of the two).

Step 2: Pass the end of the second rope through the bight.

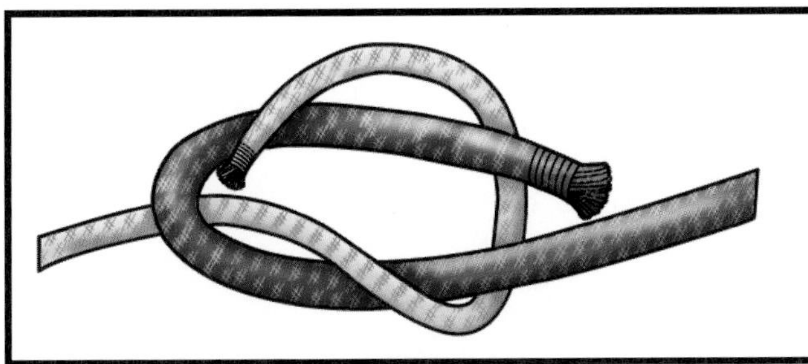

Step 3: Bring the loose end around both parts of the bight.

Step 4: Tuck this end under its own standing part and over the bight.

Step 5: Pull the knot snug.

SKILL SHEET 6-11 HOISTING AN AXE

Step 1: Tie a clove hitch using the method shown in Skill Sheet 6-5.

Step 2: Slide the clove hitch down the axe handle to the axe head.

NOTE: The excess running end of the rope becomes the tag line.

Step 3: Loop the working end of the rope around the head of the axe and back up the handle.

Step 4: Tie a half hitch on the handle a few inches (millimeters) above the clove hitch.

Step 5: Tie another half hitch at the butt end of the handle.

SKILL SHEET 6-12 HOISTING A DRY HOSELINE

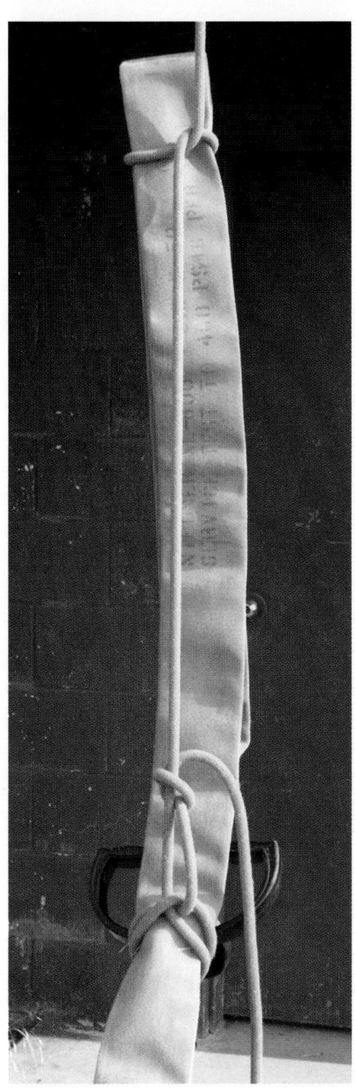

Step 1: Lower a rope of appropriate length from the intended destination of the hoseline.

Step 2: Fold the nozzle end of the hoseline back over the rest of the hose so that an overlap of 4 to 5 feet (1.2 m to 1.5 m) is formed.

Step 3: Tie a clove hitch, with an overhand safety knot, around the tip of the nozzle and the hose it is folded against so that they are lashed together.

Step 4: Place a half hitch on the doubled hose about 12 inches (300 mm) from the loop end.

NOTE: With the ties properly placed, the hose will turn on the hose roller so that the coupling and nozzle will be on top as the hose passes over the roller.

SKILL SHEET 6-13 HOISTING A CHARGED HOSELINE

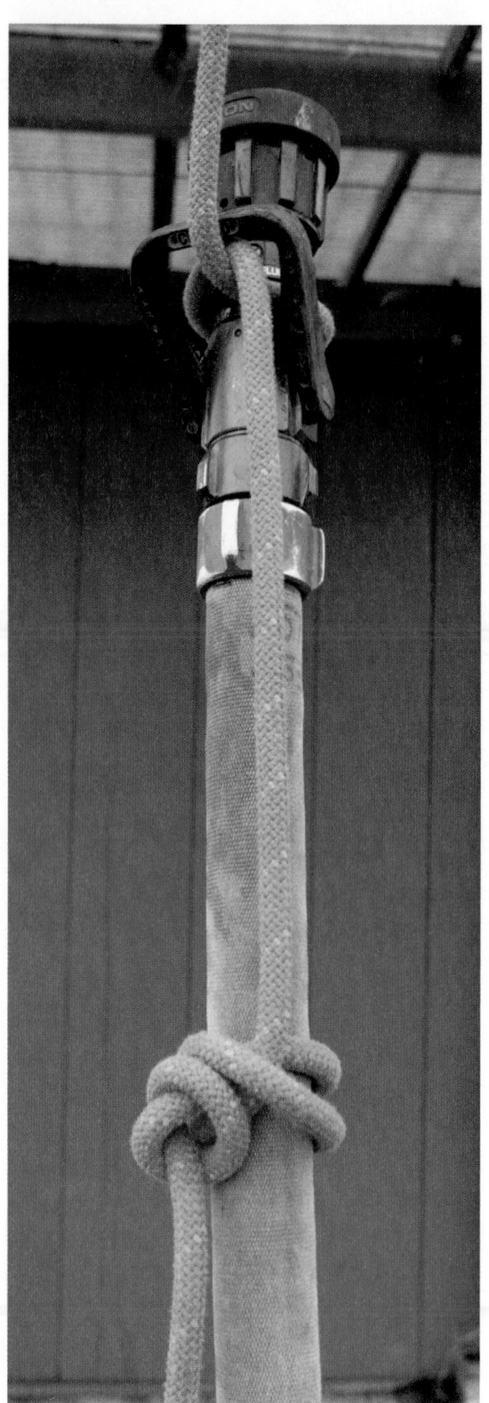

Step 1: Lower a rope of appropriate size from the intended destination of the hoseline.

Step 2: Tie a clove hitch, with an overhand safety knot, around the hose about 1 foot (0.3 m) below the coupling and nozzle.

Step 3: Tie a half hitch through the nozzle handle and around the nozzle itself in a manner that allows the rope to hold the nozzle shut while it is being hoisted.

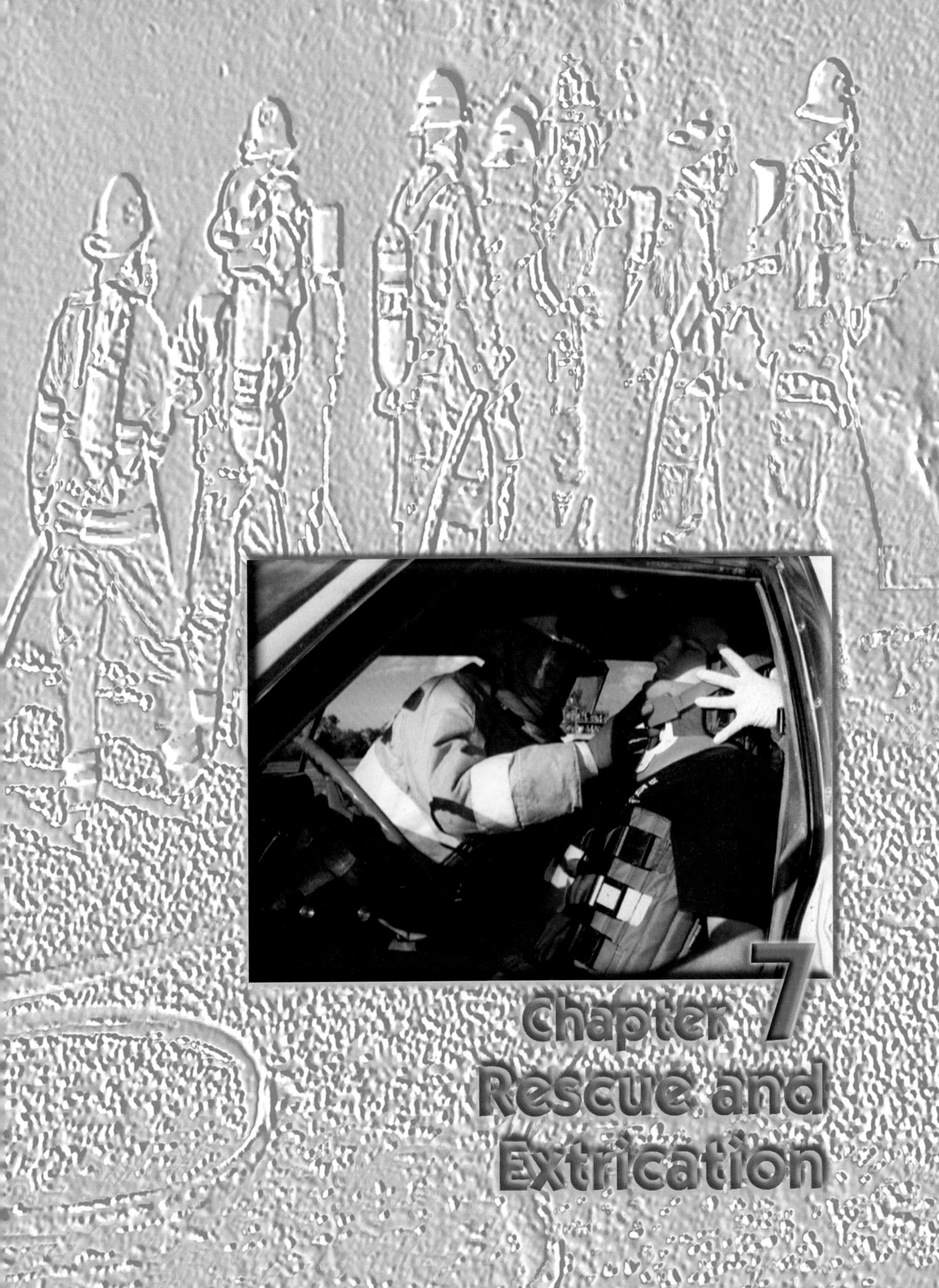

Chapter 7
Rescue and Extrication

Chapter 7
Rescue and Extrication

INTRODUCTION

While the entire fire service is dedicated to saving lives and property, rescue and extrication deal exclusively with life-threatening situations. Because they are life-threatening situations, firefighters must be thoroughly prepared for any potential rescue and/or extrication situation they encounter. IFSTA makes a definite distinction between rescue and extrication. *Rescue* incidents involve the removal and treatment of victims from situations involving natural elements, structural collapse, elevation differences, or any other situation not considered to be an extrication incident. *Extrication* incidents involve the removal and treatment of victims who are trapped by some type of man-made machinery or equipment.

This chapter covers the basics of rescue and extrication equipment and techniques as required by NFPA 1001. For more extensive information on extrication and rescue, see the IFSTA manuals **Principles of Extrication** and **Fire Service Search and Rescue.**

FIREGROUND SEARCH AND RESCUE

[NFPA 1001: 3-3.8; 3-3.8(a); 4-3.2; 4-3.2(a); 4-3.2(b); 4-4.2; 4-4.2(b)]

Fire departments were originally organized to protect life and property from fire. However, the mission of most fire departments has been expanded to include rescuing people from a wide range of hazardous environments. The vast majority of search and rescue operations conducted by firefighters are on the fireground. Even though thousands of people die in fires each year in the United States and Canada, many more are successfully rescued by firefighters.

Building Search

Regardless of how small a structure fire may look upon arrival, the fire department must always do a thorough search of the building. Even in relatively minor fires, there may be occupants in the building who are incapable of exiting on their own. Not locating a victim until after a "minor" fire is extinguished or, worse yet, missing a victim entirely is unacceptable.

While size-up is initially the responsibility of the first-arriving officer, all firefighters should look at the entire building and its surroundings as they approach. Careful observation will give them some indication as to the size of the fire, whether or not the building is likely to be occupied, the probable structural integrity of the building, and some idea of the amount of time it will take to effectively search the structure. Their initial exterior size-up will help them maintain their orientation within the building. They should identify their alternate escape routes (windows, doors, fire escapes) *before* they enter the building. Once inside, their specific location can sometimes be confirmed by looking out windows.

To obtain information about those who might still be inside and where they might be found, as well as to obtain information about the location and extent of the fire, firefighters should first question occupants who have escaped the fire (Figure 7.1). If possible, all information should be verified; in any case, firefighters should not assume that all occupants are out until the building has been searched by fire department personnel. Because neighbors may be familiar with occupants' habits and room locations, they may be able to suggest where occupants are likely to be found.

Figure 7.1 A firefighter questions an occupant about the fire.

They may also have seen an occupant near a window prior to the fire department's arrival. Information on the number and location of victims should be relayed to the incident commander (IC) and all incoming units.

Conducting a Search

There are two objectives of a building search: finding victims (searching for life) and obtaining information about the extent of the fire (searching for fire extension). In most structure fires, the search for life requires two types of searches: primary and secondary.

A *primary search* is a rapid but thorough search that is performed either before or during fire suppression operations. It is often carried out under extremely adverse conditions, but it must be performed expeditiously. During the primary search, members must be sure to search the known or likely locations of victims as rapidly as conditions allow, moving quickly to search all affected areas of the structure as soon as possible. The search team(s) can verify that the fire conditions are as they appeared from the outside or report any surprises they may encounter.

A *secondary search* is conducted *after* the fire is under control and the hazards are somewhat abated. It should be conducted by personnel other than those who conducted the primary search. It is a very thorough, painstaking search that attempts to ensure that any remaining occupants have been found.

WARNING

Neither interior nor exterior fire attacks should be attempted unless firefighters are wearing appropriate personal protective equipment.

PRIMARY SEARCH

During the primary search, rescuers should always use the buddy system — working in teams of two or more. By working together, two rescuers can conduct a search quickly while maintaining their own safety.

Primary search personnel should always carry forcible entry tools with them whenever they enter a building and throughout the search (Figure 7.2). Valuable time is lost if rescuers have to return to their apparatus to obtain this equipment. Also, tools used to force entry may be needed to force a way out of the building if rescuers become trapped.

Figure 7.2 Search/rescue personnel should always carry forcible entry tools with them.

Depending on conditions within the fire building, rescuers may be able to search while walking in an upright position, or they may have to crawl on their hands and knees (Figure 7.3). If there is only light smoke and little or no heat, walking is the most rapid means of searching a building. Searching on hands and knees (beneath the smoke) can increase visibility and reduce the chances of tripping or falling into stairways or holes in floors. Move up and down stairs on hands and knees; when ascending, proceed head first, and when descending, proceed feet first. Movement in this position is much slower than when walking, but it is usually noticeably cooler near the floor.

Figure 7.3 Search teams may have to proceed on all fours.

When searching within a structure, rescuers should move systematically from room to room, searching each room completely, while constantly listening for sounds from victims. On the fire floor, firefighters should start their search as close to the fire as possible and then search back toward the entrance door. This procedure allows the search team to reach those in the most danger first — those who would be overtaken by any fire extension that might occur while the rest of the search was in progress. Because those who are a greater distance from the fire are in less immediate danger, they can wait to be reached as the team moves back toward safety.

It is very important for rescuers to search all areas such as bathrooms, bathtubs, shower stalls, closets, under beds, behind furniture, attics, basements, and any areas where children may hide and where either infirm or disoriented victims may be found (Figure 7.4). Rescuers should search the perimeter of each room, and they should extend their arms or legs or use the handle of a tool to reach completely under beds and other furniture (Figure 7.5). When the perimeter has been searched, they should then search the middle of the room.

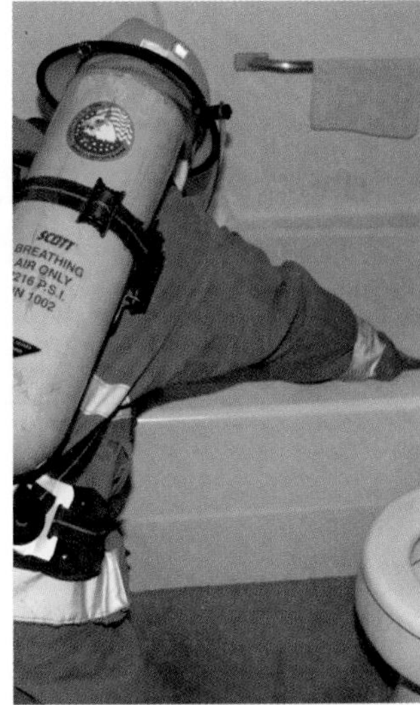

Figure 7.4 Every area must be searched.

During the primary search, visibility may be extremely limited, so rescuers may have to identify objects by touch — touch may provide the only clue to what type of room the team is in. Visibility being obscured by smoke should be reported through channels to the IC because it may indicate a need for additional ventilation.

Rescue teams should maintain radio contact with their supervisor and periodically report their progress and their needs in accordance with departmental procedures (Figure 7.6). Informing the IC of any areas that have not been completely

Figure 7.5 A tool handle helps in searching under furniture.

Figure 7.6 Search team leader reports progress.

searched is especially important so that additional search teams can be assigned to these areas if necessary.

During the primary search, negative information is just as important as positive information to ensure a complete search. If the search has to be aborted for any reason, the officer in charge should be notified immediately and the search resumed as soon as possible.

SECONDARY SEARCH

After the initial fire suppression and ventilation operations have been completed, personnel other than those who conducted the primary search are assigned to conduct a secondary search of the fire building. During the secondary search, speed is not as critical as thoroughness (Figure 7.7). The secondary search is conducted just as systematically as the primary search to ensure that no rooms or spaces are missed. As in the primary search, any negative information, such as the fire beginning to rekindle in some area, is reported immediately.

Figure 7.7 Firefighters must conduct a thorough secondary search.

Multistory Buildings

When searching in multistory buildings, the most critical areas are the fire floor, the floor directly above the fire, and the topmost floor (Figure 7.8). These floors should be searched immediately because this is where any remaining occupants will be in the greatest jeopardy due to rising smoke, heat, and fire. The majority of victims are likely to be found in these areas. Once these floors have been searched, the intervening floors should be checked.

During the primary search, doors to rooms not involved in fire should be closed to prevent the spread of fire into these areas. The exits, hallways, and stairs should be kept as clear as possible of unused hoselines and other equipment to facilitate the egress of occupants and to reduce the tripping hazard (Figure 7.9).

While still the source of much debate within the fire service, some departments insist that search and rescue personnel have a charged hoseline with them on all floors. Because advancing a charged hoseline during a search is a time-consuming process that may unnecessarily delay and impede the primary search, other departments make this an option based on conditions. Firefighters must be guided by their department's policy.

Figure 7.8 These areas have the highest search priority in multistory buildings.

Figure 7.9 Exit stairways should be kept clear of trip hazards.

Search Methods

When rooms, offices, or apartments extend from a center hallway, teams should be assigned to search both sides of the hallway. If two teams are available, each can take one side of the hallway. If there is only one search team, they search down one side of the hallway and back up the other side (Figure 7.10).

Entering the first room, the searchers turn right or left and follow the walls around the room until they return to the starting point. As rescuers leave the room, they turn in the same direction they used to enter the room and continue to the next room to be searched (Figure 7.11). For example, if they turned left when they entered the room, they turn left when they leave the room. When removing a victim to safety or to exit the building, rescuers must turn opposite the direction used to enter the room. It is important that rescuers exit each room through the same doorway they entered to ensure a complete search. This technique may be used to search most buildings, from a one-story, single-family residence to a large high-rise building.

In most cases, the best method of searching small rooms is for one member to stay at the door while another member searches the room. The searcher remains oriented by maintaining a more-or-less constant dialogue with the member at the door. The searcher keeps the member at the door informed of the progress of the search. When the room search is completed, the two rejoin at the doorway, close and mark the door (see Marking Systems section), and proceed to the next room. When searching the next room, the partners exchange their roles of searching the room and waiting at the door.

Figure 7.10 Following a wall helps searchers remain oriented.

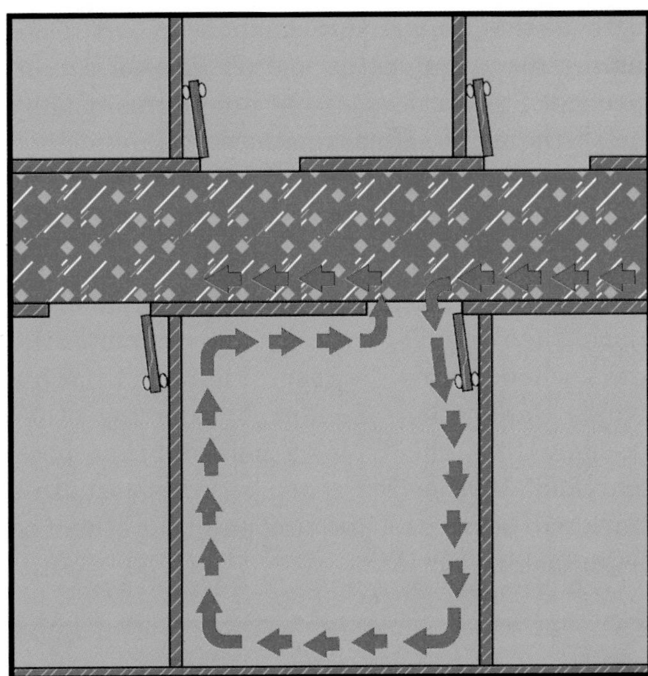

Figure 7.11 Searchers should always turn in the same direction when entering and leaving rooms.

This last method reduces the likelihood of rescuers becoming lost within the room, which reduces some of the stress of the situation. When searching relatively small rooms, this technique is often quicker than when both members search together because the searcher can move along more quickly without the fear of becoming disoriented.

Marking Systems

Several methods of marking searched rooms are used by the fire service: chalk or crayon marks, masking tape, specially designed door markers, and latch straps over doorknobs (Figure 7.12). Latch straps also serve the secondary function of preventing a rescuer from being locked in a room. Methods that might contribute to fire spread, such as blocking doors open with furniture, or

Figure 7.12 This latch strap indicates that the room has been searched.

methods that require subsequent searchers to enter the room to find the marker are not recommended. Standard operating procedures usually dictate the method of marking; however, any method used must be known to and clearly understood by all personnel who may participate in the search.

It is a good idea for search teams to use a two-part marking system. The team affixes half of the mark when entering the room and completes the mark when exiting the room (Figure 7.13). This avoids duplication of effort by alerting other rescuers that the room is being or has been searched. If a search team becomes lost, this mark will serve as a starting point for others to begin looking for it.

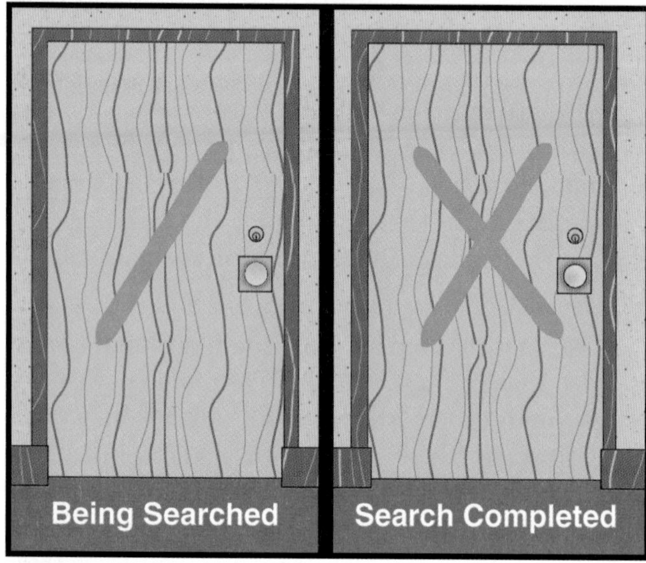

Figure 7.13 A typical search marking system.

SAFETY

[NFPA 1001: 3-3.4(a); 3-3.4(b); 3-3.8; 3-3.8(b); 4-4.2(b)]

While searching for victims in a fire, rescuers must always consider their own safety. Incident commanders also must consider the hazards to which rescuers may be exposed while performing search and rescue. Safety is the primary concern of rescuers because hurried, unsafe rescue attempts may have serious consequences for rescuers as well as victims.

Personnel must be properly trained and equipped with the necessary tools to accomplish a rescue in the least possible time. A rope is a typical search and rescue tool. It may be used as a guideline when conducting search and rescue operations in the dark or under extremely hazardous conditions. Other search and rescue tools include marking devices (to indicate which rooms have already been searched) and forcible entry tools (to aid in entry and egress and to enlarge the sweep area when searching) (Figure 7.14). An informed decision should be made about whether the search teams need to take protective hoselines with them.

Figure 7.14 Searchers keep tools with them throughout the search.

Safety During Building Searches

Every time a firefighter or rescuer responds to a fire, a human life may be in jeopardy. In order to assess the degree to which someone may be threatened, a search is initiated as soon as possible. While rescuers must work quickly, they must also operate safely and with sound judgment if they are to fulfill their assignment and avoid becoming victims themselves.

As personnel search a multistory building, especially when visibility is limited because of smoke and/or darkness, they must always be alert for weakened or hazardous structural conditions, especially the floors. They should continually feel the floor in front of them with their hands or a tool to ensure that the floor is still intact (Figure 7.15). Otherwise, they may blindly crawl into an open elevator shaft, a stairway, an arsonist's trap, or a hole that may have burned through the floor. Personnel on or directly below the fire floor should also be alert for signs that the floor/ceiling assembly above them has weakened.

When searching within a fire building, personnel should be very cautious when opening doors. They should feel the top of the door and the doorknob to determine the heat level (Figure 7.16). If the door is excessively hot, it should not be opened until a charged hoseline is in position. Firefighters should not remain in front of the door while opening it. They should stay to one side, keep low, and slowly open the door. If there is fire behind the door, staying low allows the heat and combustion products to pass over their heads.

NOTE: Some departments insist that their firefighters keep their gloves (and all other parts of the protective ensemble) on when in a burning building; others allow them to remove a glove to feel a door for heat. Firefighters should be guided by local protocols.

If an inward-opening door is difficult to open, firefighters should not kick the door to force it open because a victim may have collapsed just inside the door. Kicking the door may injure the victim further, and it is neither a safe nor a very professional way to force a door. The door should be slowly pushed open and the area behind it checked for possible victims.

Figure 7.15 A searcher checks the floor ahead.

Figure 7.16 A firefighter checks a door for heat.

Trapped or Disoriented Firefighters

Even with the best incident command or accountability system in place, unusual circumstances can lead to a firefighter, or a group of firefighters, becoming trapped or disoriented within a burning structure. Unexpected structural collapse, doors closing behind crews, or firefighters straying from a hoseline or search rope are all ways that this scenario may evolve.

Firefighters who become disoriented should try to remain calm. Becoming overly excited reduces a firefighter's ability to think and react quickly. Excitement or disorientation also causes firefighters to expend their air supplies faster than normal. If possible, firefighters should try to retrace their steps to their original locations. If retracing is not possible, firefighters should try to seek an exit from the building or at least from the area that is on fire. Firefighters should shout for help periodically so that other personnel who may be in the area will hear them. If they are not having any success finding their way out, they should find a place of relative safety and activate their PASS devices. If disoriented firefighters can locate a hoseline, they can crawl along it and feel the first set of couplings they come to. The female coupling is toward the nozzle and the male is toward the water source. The male coupling has lugs on its shank; the female does not (Figure 7.17). Following the hoseline will lead them either to an exit or to the nozzle team.

If they find a window, they can signal for assistance by straddling the windowsill and turning on their PASS devices, by using their flashlights, by

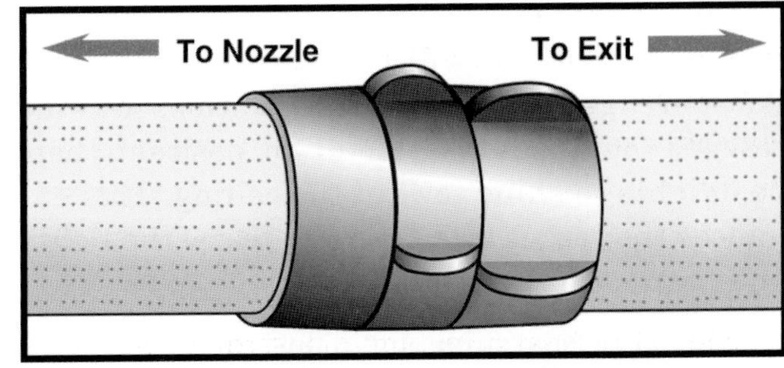

Figure 7.17 Hose couplings will indicate the direction toward the exit.

yelling and waving their arms, or by throwing objects out the window. However, under no circumstances should firefighters throw out their helmets or any other parts of their protective ensemble.

Firefighters who become trapped by a structural collapse or suffer some sort of injury that prevents them from moving about do not have all the options that a disoriented firefighter has. These firefighters should immediately activate their PASS devices and try to maintain their composures to maximize their air supplies (Figure 7.18).

Figure 7.18 PASS devices can save firefighters' lives.

If either trapped or disoriented firefighters have radios, they should try to make radio contact as quickly as possible with other personnel on the emergency scene. They should try to describe their location as accurately as possible to narrow down the search area for rescuers.

If lost firefighters cannot find their way out of a building, they should attempt to stay close to a wall as rescuers normally search around the walls before making sweeps of large interior areas. If firefighters become exhausted or close to losing consciousness, they should assume horizontal positions on the floor next to an exterior wall, hallway, or doorway; this maximizes the audible effects of the PASS devices. This position also maximizes quick discovery by rescue crews. If lost firefighters assume these positions to await rescue, they should attempt to position their flashlights to shine toward the ceiling. This enhances the rescue crew's ability to see the lights and locate the firefighters.

Rescuers searching for a lost or disoriented firefighter should first try to quickly obtain an idea of the last location of the firefighter. When performing the search, the rescuers should stop every so often and become perfectly quiet. This may allow the rescuers to hear calls for help or the downed firefighter's PASS device tone.

If it becomes necessary to remove a downed firefighter, the rescuers should use any safe means possible. In most cases, the need to exit the hostile atmosphere overrides the need to stabilize injuries before moving the firefighter (Figure 7.19). If the firefighter has a functioning SCBA, carefully move the firefighter so as not to dislodge the mask. If the firefighter does not have a functioning SCBA, either connect the mask to the buddy breathing connection on a rescuer's SCBA or simply quickly remove the victim from the hazardous atmosphere.

WARNING

At no time should rescuers remove their facepieces or in any way compromise the proper operation of their SCBA in an attempt to share them with another firefighter or victim.

Safety Guidelines

The following is a list of safety guidelines that should be used by search and rescue personnel in any type of search operation within a building.

Figure 7.19 Move the downed firefighter from the building as soon as possible.

- Do not enter a building in which the fire has progressed to the point where viable victims are not likely to be found.
- Attempt entry only after ventilation is accomplished when backdraft conditions exist.
- Work from a single operational plan. Crews should not be allowed to freelance.
- Maintain contact with command, which has control over search/rescue teams.
- Monitor constantly fire conditions that might affect search teams and individual firefighters.
- Have a rapid intervention team constantly available to help firefighters or teams in need of assistance.
- Use the established personnel accountability system without exception.
- Be aware of the secondary means of egress established for personnel involved in the search.
- Wear full personal protective equipment, including SCBA and PASS device.
- Work in teams of two or more and stay in constant contact with each other. Rescuers are responsible for themselves and each other.
- Search systematically to increase efficiency and to reduce the possibility of becoming disoriented.
- Stay low and move cautiously while searching.
- Stay alert — use all senses.
- Monitor continually the structure's integrity.
- Feel doors for excessive heat before opening them.
- Mark entry doors into rooms and remember the direction turned when entering the room. To exit the building, turn in the opposite direction when exiting the room.
- Maintain contact with a wall when visibility is obscured. Working together, search team members can extend their reach by using ropes or straps.
- Have a charged hoseline at hand whenever possible when working on the fire floor (or the floor immediately below or above the fire) because it may be used as a guide for egress as well as for fire fighting.
- Coordinate with ventilation teams before opening windows to relieve heat and smoke during search.
- Close the door, report the condition, and be guided by the group/sector supervisor's orders if fire is encountered during a search.
- Inform the group/sector supervisor immediately of any room(s) that could not be searched, for whatever reason.
- Report promptly to the supervisor once the search is complete. Besides giving an "all clear" search report, also report the progress of the fire and the condition of the building.

VICTIM REMOVAL

[NFPA 1001: 3-3.8; 3-3.8(a); 3-3.8(b)]

An ambulatory or semiambulatory victim may only require help to walk to safety —walking being probably the least laborious of all transportation methods. One or two rescuers may be needed, depending on how much help is available and the size and condition of the victim (Figure 7.20).

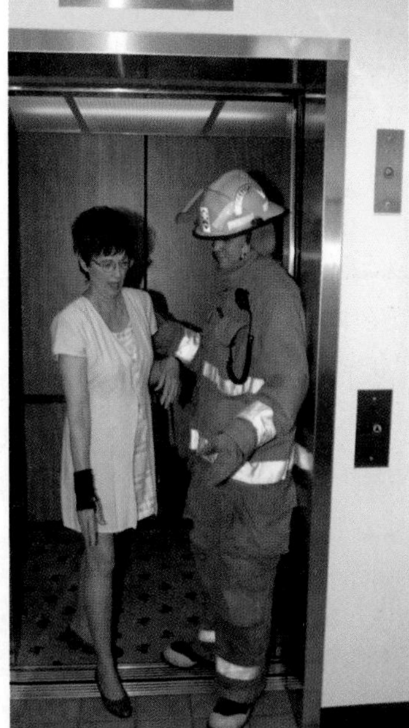

Figure 7.20 Occupants may only need to be escorted to safety.

The victim is not moved before treatment is provided unless there is an immediate danger to the victim or to rescuers. Emergency moves are necessary under the following conditions:

- There is fire or danger of fire in the immediate area.
- Explosives or other hazardous materials are involved.
- It is impossible to protect the accident scene.
- It is impossible to gain access to other victims who need immediate life-saving care.
- The victim is in cardiac arrest and must be moved to a different area (a firm surface for instance) so that rescuers can administer cardiopulmonary resuscitation (CPR).

The chief danger in moving a victim quickly is the possibility of aggravating a spinal injury. In an extreme emergency, however, the possible spinal injury becomes secondary to the goal of preserving life.

If it is necessary to perform an emergency move, the victim should be pulled in the direction of the long axis of the body — not sideways. Jackknifing the victim should also be avoided. If the victim is on the floor, pull on the victim's clothing in the neck or shoulder area (Figure 7.21). It may be easier to pull the victim onto a blanket and then drag the blanket.

It is always better to have two or more rescuers when attempting to lift or carry an adult. One rescuer can safely carry a small child, but two, three, or even four rescuers may be needed to safely lift and carry a large adult. An unconscious victim is always more difficult to lift; the person is unable to assist in any way, and a relaxed body becomes "dead weight" (Figure 7.22).

It is not easy for inexperienced people to lift and carry a victim correctly. Their efforts may be uncoordinated, and they usually need close supervision to avoid further injury to the victim. Rescuers helping to carry a victim should guard against losing their balance. They should lift as a team and with proper technique to avoid jostling the victim unnecessarily.

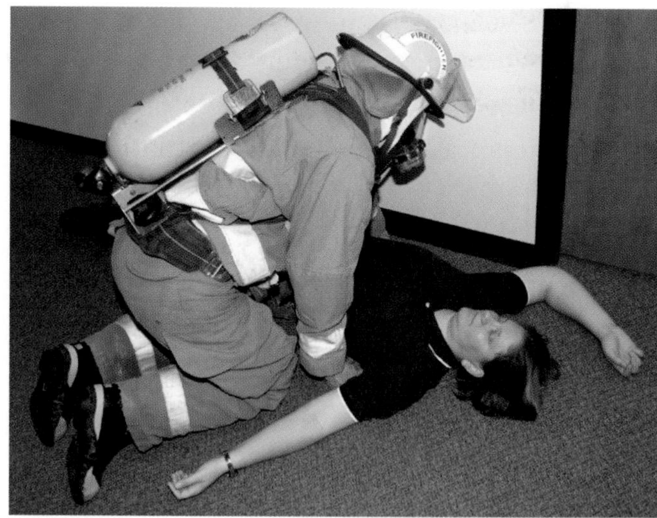

Figure 7.22 An unconscious adult is very difficult to lift.

Lifting incorrectly is also one of the most common causes of injury to rescuers. Rescuers should always remember to keep their backs straight and lift with their legs, not their backs (Figure 7.23). If immobilization of a fracture is not feasible until the victim has been moved a short distance, one rescuer should support the weight of the injured part while others move the victim (Figure 7.24). There are a number of carries and drags that may be used to move a victim from an area quickly; these are described in the following sections.

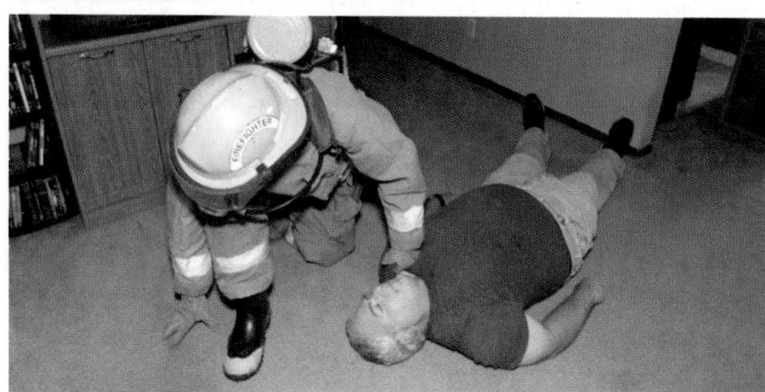

Figure 7.21 One way to move an unconscious victim in an emergency.

Figure 7.23 Rescuers should keep their backs straight when lifting.

Figure 7.24 Close coordination between rescuers is necessary to avoid aggravating the injury.

Cradle-in-Arms Lift/Carry

This lift/carry is effective for carrying children or very small adults if they are conscious. It is usually not practical for carrying an unconscious adult because of the weight and relaxed condition of the body. Skill Sheet 7-1 describes the procedure for the cradle-in arms lift/carry.

Seat Lift/Carry

This lift/carry can be used with a conscious or an unconscious victim and is performed by two rescuers. Skill Sheet 7-2 describes the procedure for the seat lift/carry.

Two- or Three-Person Lift/Carry

Many victims are more comfortable when left in a supine position, and this lift/carry is an effective way to lift a victim who is lying down. The two- or three-person lift/carry is often used for moving a victim from a bed to a gurney, especially when the victim is in cramped quarters. If the victim is small, two rescuers may be sufficient for the carry; if the victim is large, three rescuers may be needed. Skill Sheet 7-3 describes the procedure for the two- or three-person lift/carry.

Moving a Victim Onto a Long Backboard or Litter

Occasionally, rescuers will have the advantage of being able to use some type of litter to remove a victim. There are many different types of litters such as the standard ambulance cot, army litter, scoop stretcher, basket litter, and long backboard. The long backboard is one of the most common types of litters used by fire service personnel. This section highlights the proper techniques for moving a victim onto a long backboard. Similar techniques should be used for moving people onto stretchers and basket litters.

Immobilizing a victim who is suspected of having a spinal injury on a long backboard requires four rescuers. One rescuer is needed to maintain in-line stabilization throughout the process, and three rescuers are needed to actually move the victim to the board. It is critical that the victim with a suspected spinal injury be moved in such a way to avoid any unnecessary jolting or twisting of the spinal column. For this reason, the rescuer who applies and maintains in-line stabilization directs the other rescuers in their actions to ensure that the victim's head and body are moved as a unit. When dangers at the scene are life-threatening to the victim and rescuers or the victim is not suspected of having a cervical spine injury and is just being relocated, this process may be performed with only two rescuers — one to maintain in-line stabilization and one to move the victim. Skill Sheet 7-4 describes the procedure for moving a victim suspected of having a cervical spine injury onto a long backboard or litter.

Extremities Lift/Carry

The extremities lift/carry is used on either a conscious or an unconscious victim. This technique requires two rescuers. Skill Sheet 7-5 describes the procedure for the extremities lift/carry.

Chair Lift/Carry

The chair lift/carry is used for either a conscious or an unconscious person. Be sure that the chair used is sturdy; do not attempt this carry using a folding chair. Skill Sheets 7-6 and 7-7 describe two methods of performing the chair lift/carry.

Incline Drag

This drag is used by one rescuer to move a victim down a stairway or incline and is very useful for moving an unconscious victim. Skill Sheet 7-8 describes the method of performing the incline drag.

Blanket Drag

This drag is implemented by one rescuer using a blanket, rug, or sheet. Skill Sheet 7-9 describes the procedure for the blanket drag.

RESCUE AND EXTRICATION TOOLS AND EQUIPMENT

[NFPA 1001: 3-3.16; 3-3.16(a); 3-3.16(b); 4-4.1(a); 4-4.1(b); 4-4.2; 4-4.2(b); 4-5.2; 4-5.2(a); 4-5.2(b)]

The skills and techniques required for rescue and extrication work can be learned only through complete training. Although it is impossible to anticipate every extrication situation, rescue personnel will be best prepared if they are proficient with their equipment. The following sections highlight some of the tools that are more commonly used by firefighters who perform or assist in rescue and extrication functions.

Emergency Power and Lighting Equipment

It would certainly make operations easier if all emergency situations occurred during daylight hours. Unfortunately, this does not always happen. Many incidents occur in poor lighting conditions such as during the hours of darkness and in windowless buildings. These conditions create the need to artificially light the scene, which provides a safer, more efficient atmosphere in which to work. Firefighters must be knowledgeable of when and how to properly and safely operate the emergency power and lighting equipment.

POWER PLANTS

An *inverter* is a step-up transformer that converts a vehicle's 12- or 24-volt DC current into 110- or 220-volt AC current. Inverters are used on emergency vehicles when small amounts of power are needed or when small electrically operated tools need to be used (Figure 7.25). Advantages of inverters are fuel efficiency and low or nonexistent noise during operation. Disadvantages include limited power supply capability and limited mobility from the vehicle.

Generators are the most common power source used for emergency services; they can be portable or vehicle-mounted. Portable generators are powered by small gasoline or diesel engines and generally have 110- and/or 220-volt capacities (Figure 7.26). Most portable generators are light enough to be carried by two people. They are extremely useful when electrical power is needed in an area that is not accessible to the vehicle-mounted system or when less power is needed.

Vehicle-mounted generators usually have a larger power-generating capacity than portable units (Figure 7.27). In addition to providing power for portable equipment, vehicle-mounted generators provide power for the floodlighting system on the vehicle. Vehicle-mounted generators can be powered by gasoline, diesel, or propane gas engines or by hydraulic or power take-off systems. Fixed floodlights are usually wired directly to the unit through a switch, and outlets are also provided for other equipment. These power plants generally have 110- and 220-volt output capabilities with capacities up to 50 kilowatts — occasionally greater. However, mounted generators with a separate engine are noisy, making communication difficult near them.

Figure 7.25 Some inverters are mounted in the compartment of the apparatus.

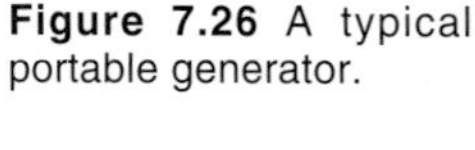

Figure 7.26 A typical portable generator.

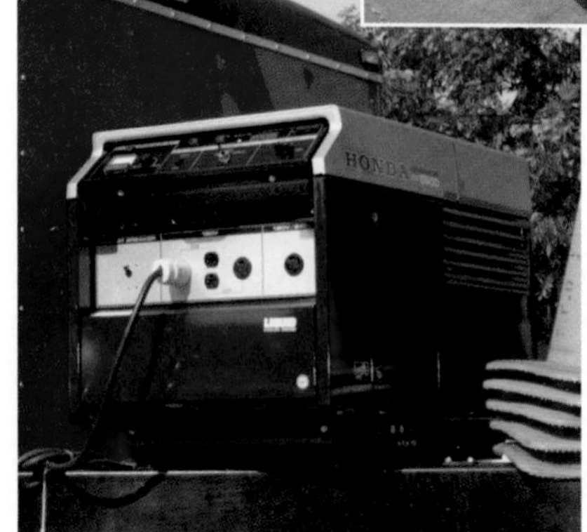

Figure 7.27 A vehicle-mounted generator.

LIGHTING EQUIPMENT

Lighting equipment can be divided into two categories: portable and fixed. Portable lights are used in areas where fixed lights are not able to illuminate because of opaque obstructions or when additional lighting is necessary. Portable lights generally range from 300 to 1,000 watts (Figure 7.28). They may be supplied with power by a cord from either a vehicle-mounted power plant or from a self-contained power unit. The lights usually have handles for ease of carrying and large bases for stability. Some portable lights are mounted on telescoping stands, which allow them to be directed more effectively.

Fixed lights are mounted to a vehicle, and their main function is to provide overall lighting of the emergency scene. Fixed lights are usually mounted so that they can be raised, lowered, or turned to provide the best possible lighting. Often, these lights are mounted on telescoping poles that allow both vertical and rotational movement (Figure 7.29). More elaborate designs include hydraulically operated booms with a bank of lights (Figure 7.30). These banks of lights generally have a capacity of 500 to 1,500 watts per light. The amount of lighting should be carefully matched with the amount of power available from the power plant. Overtaxing the power plant gives poor lighting, may damage the power generating unit or the lights, and restricts the operation of other electrical tools using the same power supply.

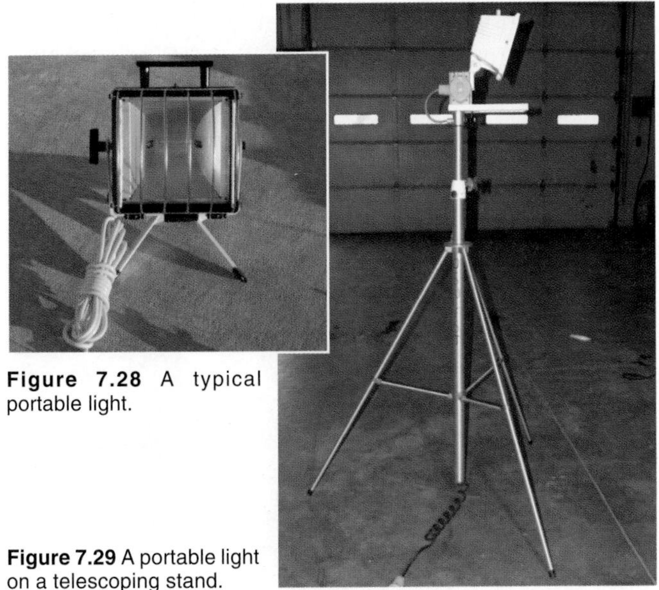

Figure 7.28 A typical portable light.

Figure 7.29 A portable light on a telescoping stand.

Figure 7.30 A typical lighting unit. *Courtesy of Mike Wieder.*

AUXILIARY ELECTRICAL EQUIPMENT

A variety of other equipment may be used in conjunction with power plants and lighting equipment. Electrical cables or extension cords are necessary to conduct electric power to portable equipment. Cords may be stored in coils, on portable cord reels, or on fixed automatic rewind reels (Figure 7.31). Twist-lock receptacles provide secure, safe connections (Figure 7.32). Electrical cable should be waterproof, explosionproof,

Figure 7.31 Typical power-cord reel.

Figure 7.32 A twist-lock adapter.

and have adequate insulation with no exposed wires. Junction boxes may be used when multiple connections are needed (Figure 7.33). The junction box has several outlets and is supplied through one inlet from the power plant. All outlets should be equipped with ground-fault circuit interrupters and conform to NFPA 70E, *Standard for Electrical Safety Requirements for Employee Workplaces*.

Figure 7.33 A typical junction box.

In situations where mutual aid departments frequently work together and have either different sizes or different types of receptacles (for example, one has two prongs, the other has three), adapters should be carried so that equipment can be interchanged (Figure 7.34). Adapters should also be carried to allow rescuers to plug their equipment into standard electrical outlets.

Figure 7.34 A variety of electrical adapters.

MAINTAINING POWER PLANTS AND LIGHTING EQUIPMENT

Servicing and maintaining portable power plants and lighting equipment are essential for reliable operation. The following are guidelines for the servicing and maintenance of this equipment; however, they do not replace the equipment's owner's manual.

- Run power plants once a week while testing electrical devices for operating status.
- Check gas and oil levels weekly and after every use.
- Wear gloves when changing quartz bulbs. Normal hand oil can cause a bulb to explode when it is energized.
- Inspect electrical cords for damage at weekly intervals.
- Inspect the spark plug, spark plug wire, and carburetor weekly. A spare spark plug should be readily accessible.
- Change extra gasoline approximately every three weeks to ensure freshness.

Hydraulic Tools

Rescue tools can be operated manually or by powered hydraulics. The development of powered hydraulic rescue tools has revolutionized the process of removing victims from various types of entrapments. The wide range of uses, speed, and superior power of these tools has made them the primary tools used in many rescue situations. Manual hydraulic tools operate on the same principles as powered hydraulic tools except that the hydraulic pump is manually powered by a rescuer operating a pump lever.

POWERED HYDRAULIC TOOLS

Powered hydraulic rescue tools receive their power from hydraulic fluid pumped through special high-pressure hoses. Although there are a few pumps that are operated by compressed air, most are powered by electric motors or by two- or four-cycle gasoline engines. These units may be portable and carried with the tool, or they may be mounted on the vehicle and may supply power to the tool through a hose reel line (Figure 7.35). Manually operated pumps are also available in case of a power

Figure 7.35 A typical hydraulic rescue tool power unit.

unit failure (Figure 7.36). Four basic types of powered hydraulic tools are used in rescue incidents: *spreaders, shears, combination spreader/shears,* and *extension rams.*

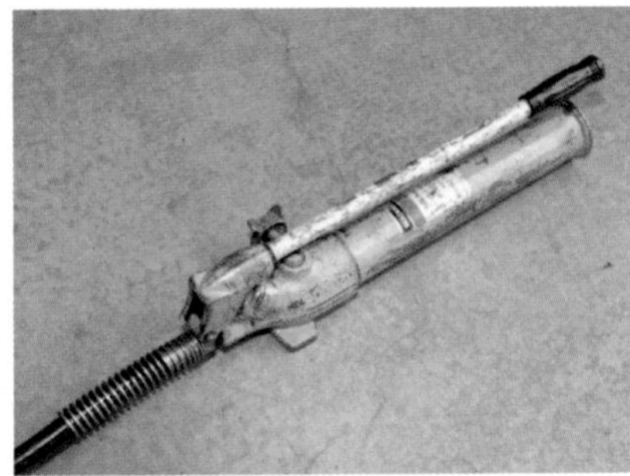

Figure 7.36 A manually operated hydraulic pump.

Spreaders. Powered hydraulic spreaders were the first powered hydraulic tools to become available to the fire/rescue service (Figure 7.37). They are capable of either pushing or pulling. Depending on the brand, this tool can produce up to 22,000 psi (154 000 kPa) of force at the tips of the tool. The tips of the tool may spread as much as 32 inches (813 mm) apart.

Shears. Hydraulic shear tools are capable of cutting almost any metal object that can fit between their blades, although some models cannot cut case-hardened steel (Figure 7.38). The shears may also be used to cut other materials such as plastics or wood. Shears are typically capable of producing up to 30,000 psi (206 850 kPa) of cutting force and have an opening spread of approximately 7 inches (180 mm).

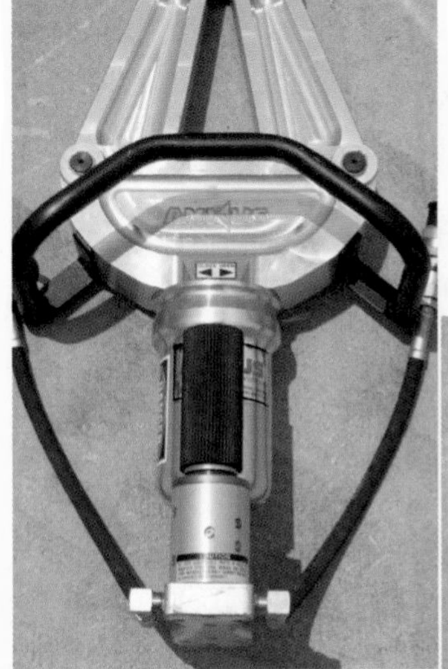

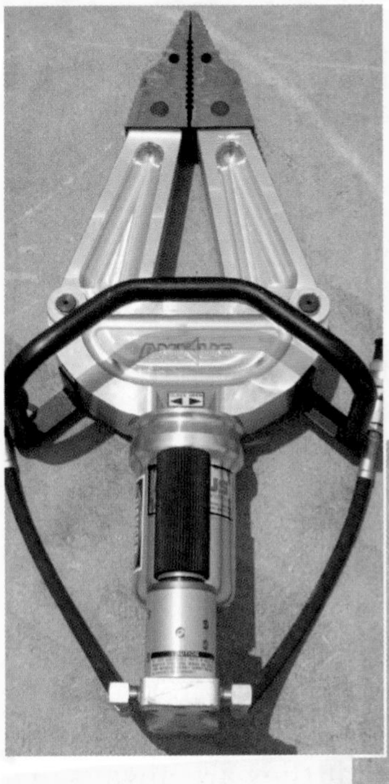

Figure 7.37 Typical hydraulic spreaders.

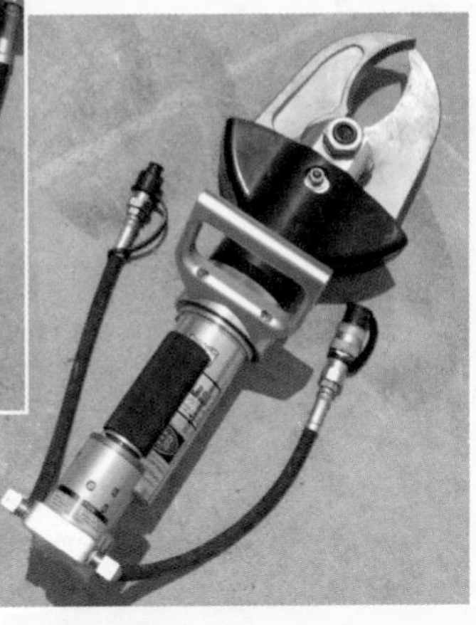

Figure 7.38 Typical hydraulic shears.

Combination spreader/shears. Most manufacturers of powered hydraulic rescue equipment offer a combination spreader/shears tool (Figure 7.39). This tool consists of two arms equipped with spreader tips that can be used for pulling or pushing. The inside edges of the arms are equipped with cutting shears similar to those described in the previous paragraph. This combination tool is excellent for a small rapid-intervention vehicle or for departments where limited resources prevent the purchase of larger and more expensive individual spreader and cutting tools. However, the combination tool's spreading and cutting capabilities are somewhat less than those of the individual units.

Extension rams. Extension rams are designed primarily for straight pushing operations, although they are effective at pulling as well. These tools are especially useful when it is necessary to push objects farther than the maximum opening distance of the hydraulic spreaders (Figure 7.40). The largest of these extension rams can extend from a closed length of 36 inches (914 mm) to an extended length of nearly 63 inches (1 600 mm). They open with a pushing force of about 15,000 psi (104 000 kPa). The closing force is about one-half that of the opening force.

Figure 7.39 Combination spreader/shears.

Figure 7.40 Hydraulic rams.

MANUAL HYDRAULIC TOOLS

Two manual hydraulic tools are used frequently in extrication work: the *porta-power tool system* and the *hydraulic jack*. The primary disadvantage of manual hydraulic tools is that they operate slower than powered hydraulic tools, and they are labor-intensive.

Porta-power tool system. The porta-power tool system is basically a commercial shop tool that has been adopted by the fire/rescue service (Figure 7.41). It is operated by transmitting pressure from a manual hydraulic pump through a hydraulic hose to a tool assembly. A number of different tool accessories allows the porta-power tool system to be used on a variety of applications.

The primary advantage of the porta-power tool over the hydraulic jack is that the porta-power tool has accessories that allow it to be operated in narrow places where the jack either will not fit or cannot be operated. The primary disadvantage of the porta-power tool is that assembling complex combinations of accessories and the actual operation of the tool is time-consuming.

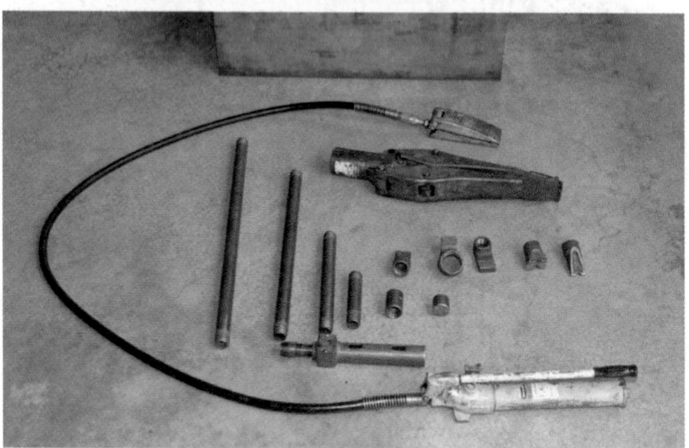

Figure 7.41 A typical porta-power set.

Hydraulic jack. The hydraulic jack is designed for heavy lifting applications (Figure 7.42). It is also an excellent compression device for shoring or stabilizing operations (see Shoring section). Most hydraulic jacks have lifting capacities up to 20 tons (20.3 tonnes [t]), but units with a higher capacity are available.

Any kind of jack, hydraulic or otherwise, should have flat, level footing and should be used in conjunction with cribbing (Figure 7.43). On a soft surface, a flat board or steel plate with wood on top should be put under the jack to distribute the force placed on the jack.

Figure 7.42 One type of hydraulic jack.

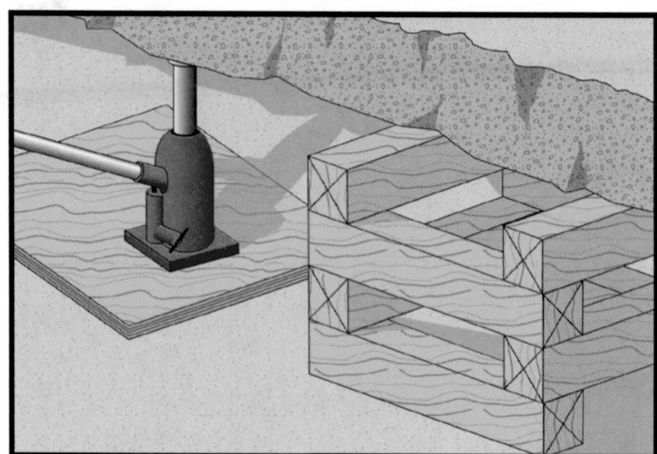

Figure 7.43 Cribbing is used in conjunction with a jack.

Nonhydraulic Jacks

There are several kinds of jacks that can be considered hand tools because they do not operate with hydraulic power. Although these tools are effective for their designed purposes, they do not have the same amount of power as hydraulic jacks. The following sections describe several of the nonhydraulic types of jacks. See the Hydraulic jack section for safety guidelines when using any type of jack.

SCREW JACKS

Screw jacks can be extended or retracted by turning the shaft. Jacks should be checked for wear after each use so that they are always in a state of readiness. They should also be kept clean and lightly lubricated, with particular attention paid to the screw thread. Footplates should also be checked for wear or damage. Footplates make contact with whatever is being stabilized by the jacks.

The two types of screw jacks are the *bar screw jack* and the *trench screw jack.* Both jacks have a male-threaded core similar to a bolt and a means to turn the core.

Bar screw jacks. Bar screw jacks are excellent for supporting collapsed structural members (Figure 7.44). These jacks are normally not used for lifting; their primary use is to hold an object in place, not to move it. The jacks are extended or retracted as the shaft is rotated in the base. The shaft is turned by pushing a long bar that is inserted through a hole in the top of the shaft.

Figure 7.44 A typical bar screw jack.

Trench screw jacks. Because of their ease of application, durability, and relatively low cost, trench screw jacks are sometimes used to replace wooden cross braces in trench rescue applications. These devices consist of a swivel footplate with a stem that is inserted into one end of a length of 2-inch (50 mm) steel pipe (not to exceed 6 feet [2 m] in length) and a swivel footplate with a threaded stem that is inserted into the other end of the pipe (Figure 7.45). An adjusting nut on the threaded stem is turned to vary the length of the jack and to tighten it between opposing members in a shoring (stabilizing) system.

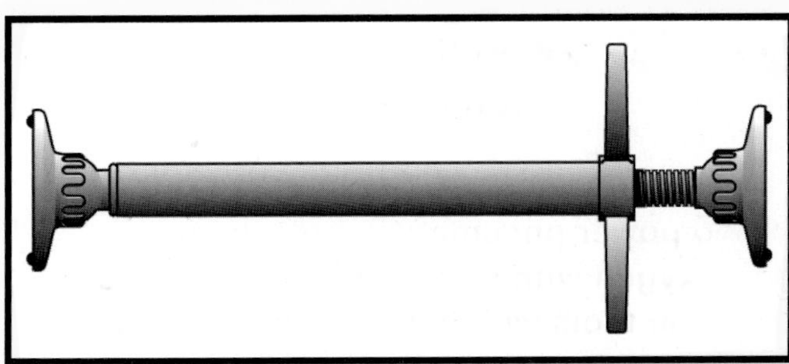

Figure 7.45 A typical trench screw jack.

RATCHET-LEVER JACK

Also known as *high-lift* jacks, these medium-duty jacks consist of a rigid I beam with perforations in the web and a jacking carriage with two ratchets on the geared side that fits around the I beam (Figure 7.46). One ratchet holds the carriage underneath. The second ratchet is combined with a lever that is pushed down to force the carriage upward. The ratchets can be reversed to move the carriage down.

Ratchet jacks can be dangerous because they are the least stable of all the various types of jacks. If the load being lifted shifts, ratchet-lever jacks may simply fall over, allowing the load to suddenly drop to its original position. Also, the ratchets can fail under a heavy load.

> **WARNING**
>
> **Rescuers should never work under a load supported only by a jack. If the jack fails or the load shifts, severe injury or death may result. The load should also be supported by properly placed cribbing.**

Figure 7.46 Ratchet-lever jacks.

Cribbing

Rescue vehicles should carry an adequate amount of appropriately sized cribbing. Cribbing is essential in many rescue operations. It is most commonly used to stabilize objects but also has many other uses. Large wedges may be used to shim up loose cribbing (Figure 7.47). The wedges may be driven in with a mallet or a piece of cribbing.

Figure 7.47 Wedges can be used to shim up cribbing.

When wood is selected for cribbing, it should be solid, straight, and free of such major flaws as large knots or splits. Various sizes of wood can be used, but the most popular are 2- x 4-inch and 4- x 4-inch hardwood lumber. The length of the pieces may vary, but 16 to 18 inches (400 mm to 450 mm) is standard. The ends of the blocks may be painted different colors for easy identification by length. Other surfaces of the cribbing should be free of paint or any other finish because they can make the wood slippery, especially when it is wet. Individual pieces of cribbing may have a hole drilled through the end with a loop of rope tied through the hole for easy carrying and safe removal from under objects (Figure 7.48). Other commercially manufactured synthetic materials are used in crib construction.

Cribbing can be stored in numerous ways. It can be stacked in a compartment with the grab handles facing out for easy access (Figure 7.49). It can also be placed on end inside a storage crate (Figure 7.50).

Figure 7.48 Cribbing with rope handles attached.

Figure 7.49 Cribbing stored in a compartment.

Figure 7.50 Cribbing stored in crates is easy to move.

Pneumatic (Air-Powered) Tools

Pneumatic tools use compressed air for power. The air can be supplied by vehicle-mounted air compressors, apparatus brake system compressors, SCBA cylinders, or cascade system cylinders. Air chisels and pneumatic nailers are two types of pneumatic tools.

WARNING

Never use compressed oxygen supplies to power pneumatic tools. Mixing pure oxygen with grease and oils found on the tools will result in fire or violent explosion.

AIR CHISELS

Pneumatic-powered chisels (also called *air chisels, pneumatic hammers,* or *impact hammers*) are useful for rescue and extrication work. Most air chisels operate at air pressures between 100 and 150 psi (700 kPa and 1 050 kPa). These tools come with a variety of interchangeable bits to fit the needs of almost any situation (Figure 7.51). In addition to cutting bits, special bits for such operations as breaking locks or driving in plugs are also available. Often used in vehicle extrication situations, these tools are good for cutting medium- to heavy-gauge sheet metal and for popping rivets and bolts. Cutting heavier-gauge metal requires more air at higher pressures.

CAUTION: The sparks produced while cutting metal with pneumatic chisels may provide an ignition source for flammable vapors.

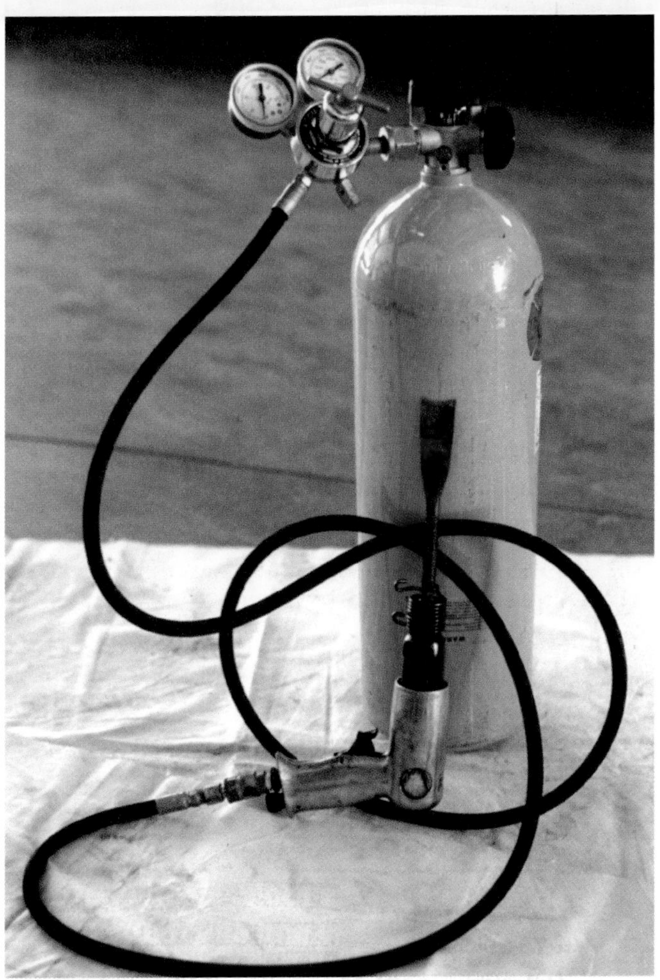

Figure 7.51 A typical air chisel. *Courtesy of Vespra (ONT) Fire Department.*

PNEUMATIC NAILERS

Air-operated nailers can be used to drive nails into wood or masonry. They are especially useful for nailing wedges and other wooden components of shoring systems into place (Figure 7.52).

Figure 7.52 A pneumatic nailer.

Lifting/Pulling Tools

Rescuers must sometimes lift or pull an object to free a victim. Several rescue tools have been developed to assist in this task. These include *tripods, winches, come-alongs, chains, air bags,* and *block and tackle systems.*

TRIPODS

Rescue tripods are needed to create an anchor point above a utility cover or other opening. This allows rescuers to be safely lowered into confined spaces and rescuers and victims to be raised out of them (Figure 7.53).

Figure 7.53 A rescue tripod.

WINCHES

Vehicle-mounted winches are excellent pulling tools. They can usually be deployed faster than other lifting/pulling devices, generally have a greater travel or pulling distance, and are much stronger. Winches are usually mounted on the front bumper, but some are located at the rear of the vehicle (Figure 7.54). The three most common drives for winches are electric, hydraulic, and power take-off. Either chain or steel cables are used for pulling.

Winches should be equipped with handheld, remote-control operating devices (Figure 7.55). These devices allow the operator to get a better view of the operation and to stand away from the winch since being near the winch can be dangerous if the cable breaks. Rescuers should position the winch as close to the object being pulled as possible so that if the cable breaks, there will be less cable to suddenly recoil and less chance of injury.

Figure 7.54 A vehicle-mounted winch.

Figure 7.55 The winch operator uses the remote-control operating device.

CAUTION: Whenever possible, a winch operator should stay farther away from the winch than the length of the cable from the winch to the load (Figure 7.56).

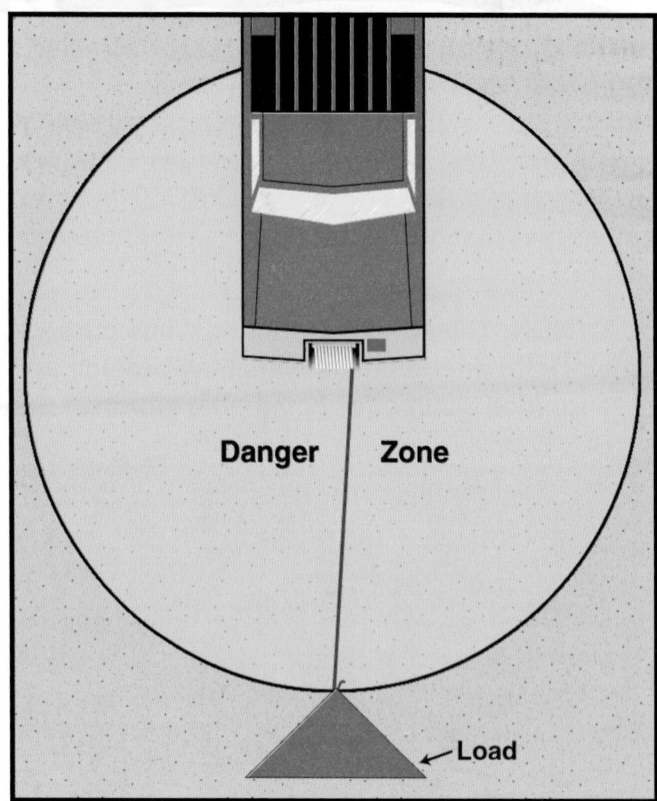

Figure 7.56 The danger zone in a winch operation.

COME-ALONGS

Another lifting/pulling tool used in rescue is the come-along (Figure 7.57). It is a portable cable winch operated by a manual ratchet. In use, the come-along is attached to a secure anchor point, and the cable is run out to the object to be moved. Once both ends are attached, the lever is operated to rewind the cable, which pulls the movable object toward the anchor point. The most common sizes or ratings of come-alongs are 1 to 10 tons (1.02 t to 10.2 t).

CHAINS

Winches and come-alongs may use chains as part of a lifting/pulling system. Only alloy steel chains of the correct size should be used in rescue work (Figure 7.58). Alloy steel chains are highly

Figure 7.57 One type of come-along.

Figure 7.58 Typical rescue chains.

resistant to abrasion, making them ideal for rescue and extrication work. Special alloys are available that are resistant to corrosive or hazardous atmospheres. Proof coil chain, also called *common* or *hardware chain,* is not suitable for emergency situations.

AIR BAGS

Air bags give rescuers the ability to lift or displace objects that cannot be lifted with other rescue equipment (Figure 7.59). There are three basic types of lifting bags: *high-pressure, medium-pressure,* and *low-pressure.* A fourth type of bag is used for sealing leaks but has little, if any, rescue application.

High-pressure bag. High-pressure bags consist of a tough, neoprene rubber exterior reinforced with steel wire or Kevlar® aramid fiber. Deflated, the bags lie completely flat and are about 1 inch (25 mm) thick (Figure 7.60). They come in various sizes that range in surface area from 6 x 6 inches (150 mm by 150 mm) to 36 x 36 inches (900 mm by 900 mm). Depending on the size of the bags, they may inflate to a height of 20 inches (500 mm).

Figure 7.59 Air bags come in a variety of shapes and sizes. *Courtesy of Safety Corporation of America.*

Figure 7.60 A deflated high-pressure air bag.

Low- and medium-pressure bags. Low- and medium-pressure bags are considerably larger than high-pressure bags and are most commonly used to lift or stabilize large vehicles or objects (Figure 7.61). Their primary advantage over high-pressure air bags is that they have a much greater lifting distance. Depending on the manufacturer, a lifting bag may be capable of lifting an object 6 feet (2 m) above its original position.

Air bag safety rules. Operators should follow these safety rules when using air bags:

- The lifting operation should be planned before starting.
- Operators should be thoroughly familiar with the equipment — its operating principles and methods — and its limitations.

- Operators should follow the manufacturer's recommendations for the specific system used.
- All components should be kept in good operating condition with all safety seals in place.
- Operators should have available an adequate air supply and sufficient cribbing before beginning operations.
- The bags should be positioned on or against a solid surface.
- The bags should never be inflated against sharp objects.
- The bags should be inflated slowly and monitored continually for any shifting.
- Rescuers should never work under a load supported only by air bags.
- The load should be continuously shored up with enough cribbing to adequately support the load in case of bag failure.
- When box cribbing is used to support an air bag, the top layer should be solid; leaving a hole in the center may cause shifting and collapse (Figure 7.62).
- Bags should not be allowed to contact materials hotter than 220°F (104°C).
- Bags should never be stacked more than two high. With the smaller bag on top, the bottom bag should be inflated first (Figure 7.63). A single multicell bag is preferred.

Figure 7.61 Low-pressure bags in operation.

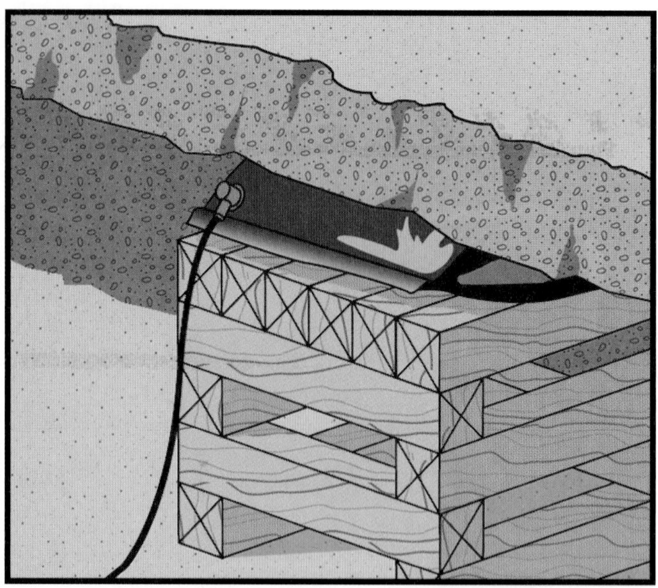

Figure 7.62 Air bags can be supported by cribbing.

Figure 7.63 Air bags may be stacked.

CAUTION: Air bags should be inspected regularly and should be removed from service if any evidence of damage or deterioration is found.

BLOCK AND TACKLE SYSTEMS

Because of their mechanical advantage in converting a given amount of pull to a working force greater than the pull, block and tackle is useful for lifting or pulling heavy loads. A *block* is a wooden or metal frame containing one or more pulleys called *sheaves*. *Tackle* is the assembly of ropes and blocks through which the line passes to multiply the pulling force (Figure 7.64). Block and

Figure 7.64 A double block and tackle system.

tackle is covered in greater detail in IFSTA's **Fire Service Search and Rescue** manual.

> **WARNING**
> Block and tackle systems are not life safety devices and should only be used to lift or stabilize objects and NOT for lifting people.

VEHICLE EXTRICATION

[NFPA 1001: 4-4.1; 4-4.1(a); 4-4.1(b); 4-4.2]

Frequently encountered rescue situations involve automobile accidents with victim entrapment (Figure 7.65). These accidents are the result of collisions with other automobiles, larger motor vehicles, or stationary objects. Because a victim who is trapped may be seriously injured, proper extrication procedures are essential to prevent further injury and to speed the victim's removal. It is also critical that firefighters coordinate with emergency medical personnel who are providing first aid to the victim.

Figure 7.65 Vehicle extrications are the most common rescue incidents.

Scene Size-Up

Scene size-up is essential in accomplishing an efficient extrication. Size-up should begin as soon as the first emergency vehicle approaches the accident scene. Taking several minutes to carefully assess the scene can help avoid confusion, clarify required tasks, prevent further injuries to victims, and prevent injury to personnel.

When arriving on the scene of a motor vehicle accident, firefighters should carefully select where to park their vehicle. It should be close enough to the scene for the equipment and supplies to be readily available but not so close that it might interfere with on-scene activities. The emergency vehicle is positioned to provide a barrier to protect the scene from oncoming traffic, but it is safer and more desirable that at least one traffic lane be kept open for other traffic, including other emergency vehicles. Despite the advantages of obtaining advance information from telecommunicators, firefighters face many unknowns when arriving at an emergency. Rescue personnel should be observant as they approach the scene and consider the following:

- What are the traffic hazards?
- How many and what type(s) of vehicle(s) are involved?
- Where and how are the vehicles positioned?
- Is there a fire or potential for a fire?
- Are there any hazardous materials involved?
- Are there any utilities, such as gas or electricity, that may have been damaged? If so, are they posing a hazard to the victims and rescue personnel?
- Is there a need for additional resources?

Assessing the Need for Extrication Activities

At the scene, personnel should make a more thorough assessment of the situation before taking any action. Personnel should assess the immediate area around each vehicle and assess the entire scene in more detail. The rescuer who assesses each vehicle should be concerned with the number of victims in or around the vehicle and the severity of their injuries. The rescuer should also assess the condition of the vehicle, extrication tasks that may be required, and any hazardous conditions that might exist. Ideally, there will be one rescuer to assess each vehicle involved in the incident, but this may not be possible (Figure 7.66). If there is only one rescuer available and more than one vehicle to survey, the rescuer must check each one separately and report the conditions in each vehicle to the incident commander.

While each vehicle is being checked, another rescuer should be assigned to survey the entire

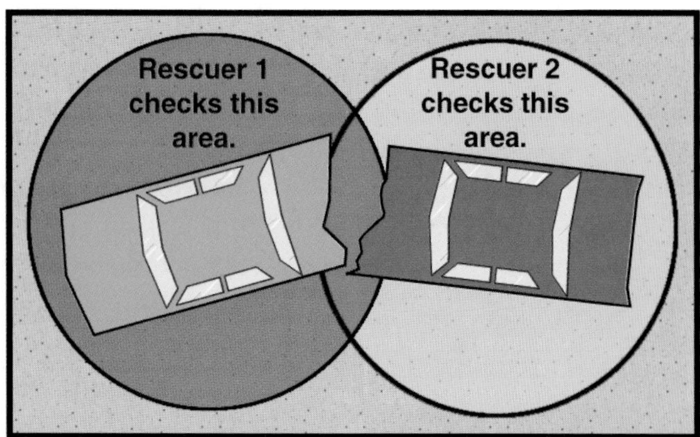

Figure 7.66 Rescuers should search around each of the involved vehicles.

area around the scene (Figure 7.67). This person should check to see if there are any other vehicles involved that may not be readily apparent (over an embankment, for example), any victims who have been thrown clear of the vehicles, any damage to structures or utilities that present a hazard, or any other circumstances that warrant special attention.

Rescue personnel who are trained in first aid or more advanced emergency medical techniques should triage the victims to determine the extent of injury and entrapment (Figure 7.68). This information aids the incident commander in determining the order in which victims should be removed. Of course, more seriously injured victims must receive higher priority than those with minor injuries. Victims who are not trapped should be removed first to make more working room for rescuers who are trying to remove those entrapped. As each assessment is completed, the rescuer should report his findings to the incident commander.

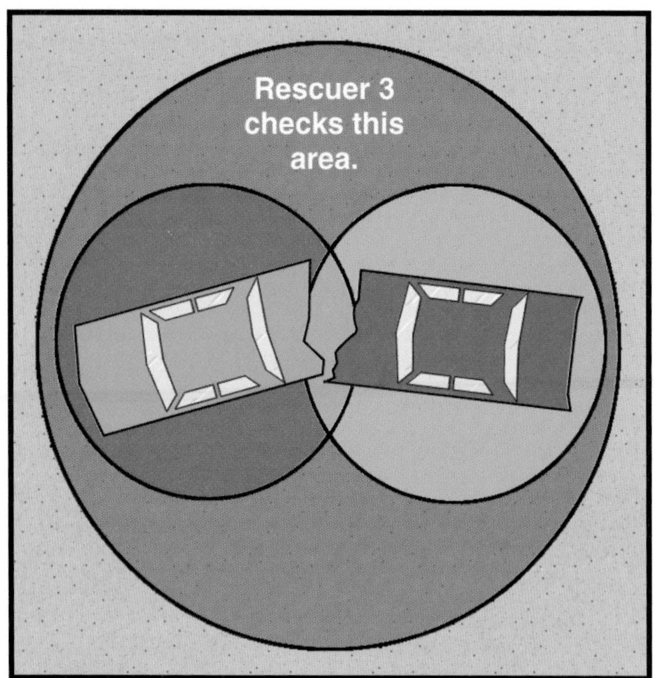

Figure 7.67 A third rescuer should make a general sweep and search of the entire scene. This person should be on the lookout for victims that may have been thrown or staggered clear of the vehicles.

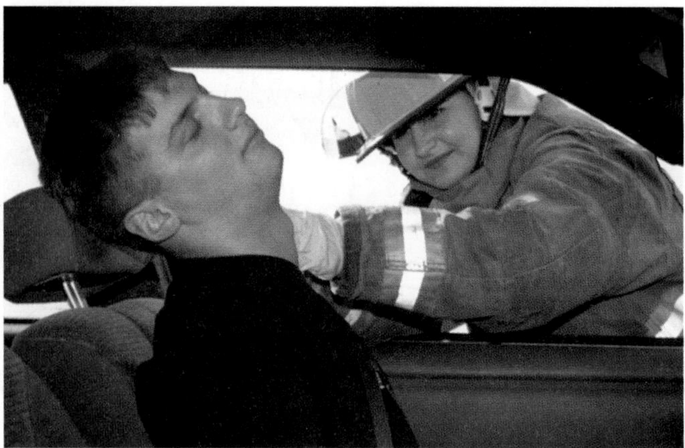

Figure 7.68 A trained rescuer should assess the victim's condition as soon as possible.

Stabilizing the Vehicle

Following scene assessment, rescuers must stabilize the vehicle(s). This is vital to prevent further damage to the vehicle(s), further injury to the victim(s), or possible injuries to emergency personnel. Proper stabilization refers to the process of providing additional support to key places between the vehicle and the ground or other solid anchor points. The primary goal of stabilization is to maximize the area of contact between the vehicle and the ground to prevent any further movement of the vehicle.

Vehicles can be found in a number of different positions following a collision. Rescuers are often tempted to test the stability of the vehicle in the position in which it is found. Rescuers must be trained to resist this temptation because the slightest push in the wrong place may cause the vehicle to move. This is particularly true of vehicles that are on their sides or resting partially over a cliff or embankment (Figure 7.69).

Most vehicles involved in collisions remain upright. Rescuers must realize that even though the vehicle still has all its wheels on the ground, some stabilization is required to ensure maximum

Figure 7.69 Crib the vehicle on both sides to prevent accidental flipping.

Figure 7.71 When a vehicle is on level ground, chock the wheels in both directions.

stability for extrication operations. The vehicle should be stabilized to prevent both vertical and horizontal movement.

Several methods can be used to prevent horizontal motion. The most common method is to chock the vehicle's wheels. It is most important to chock the wheels on the downhill side of a vehicle that is sitting on a grade (Figure 7.70). If the vehicle is on level ground, chock the wheels in both directions (Figure 7.71). Chocking can be accomplished with standard wheel chocks, pieces of cribbing or other wood, or other appropriately sized objects.

It may also be possible to use one or more of the vehicle's own mechanical systems to assist in stabilization. This will depend on whether or not these systems are still operable. If possible, place automatic transmissions in the park position; place manual transmissions in gear. Set parking or emergency brakes.

CAUTION: Do not rely on mechanical systems, even if they are operable, as the sole source of stabilization. They should be used only with other stabilization procedures.

There are numerous ways to prevent a vehicle from moving vertically. Jacks, air-lifting bags, and cribbing are used most frequently for this purpose. Different types of jacks can be used to support the frame of the vehicle. The advantage of jacks is that they can be adjusted to the required height; their disadvantage is that they are time-consuming to place. Air-lifting bags can also be used for support. To be effective, at least two air-lifting bags are needed. They should be positioned either one on each side of the vehicle or one in the front and one in the rear (Figure 7.72).

Figure 7.70 When a vehicle is resting on an incline, chock the wheels in the downhill direction.

Figure 7.72 Air bags can be used to support overturned vehicles; however, they do permit some bouncing movement that a solid box crib would not.

Standard wooden cribbing is also an effective stabilizing tool (Figure 7.73). Cribbing may be built up in a box formation until enough is used to support the vehicle. It may be necessary to use wedges as the top pieces to ensure solid contact between the cribbing and the vehicle (Figure 7.74). Special step blocks can also be used to provide rapid stabilization of the vehicle (Figure 7.75). At least one and preferably two step blocks should be placed on each side of the vehicle.

When using any of these methods, rescuers must take care to avoid placing any part of their bodies under the vehicle while placing the stabilizing device. There is always the possibility that the vehicle may drop unexpectedly, injuring or killing the person beneath it. Handle cribbing on the sides to prevent any crushing hand injuries should a sudden drop occur (Figures 7.76 a and b).

On occasion, vehicles will be found in positions other than upright such as upside down, on their side, or on an embankment. Under these circumstances, rescuers should use whatever means available to stabilize the vehicle. Generally, a combination of cribbing, ropes, webbing, and chains are used to accomplish these types of stabilization tasks.

Figure 7.73 It may be necessary to construct tall box cribs to stabilize overturned vehicles.

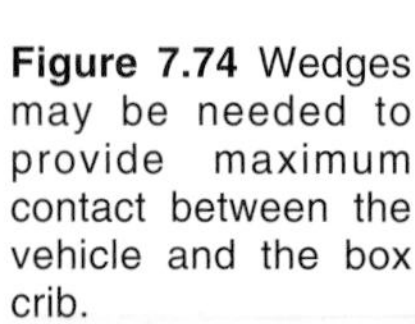

Figure 7.74 Wedges may be needed to provide maximum contact between the vehicle and the box crib.

Figure 7.75 Insert the step block until it makes solid contact with a portion of the vehicle's undercarriage.

Figure 7.76a Push rear portions of the box crib into place with another piece of cribbing to keep from having to reach under the vehicle.

Figure 7.76b To prevent hand injuries, hold cribbing and wedges on the sides while inserting them.

Gaining Access to Victims

In general, there are three methods of gaining access to victims in vehicles:

- Through a normally operating door
- Through a window
- By compromising the body of the vehicle

The simpler the required operation, the better for all concerned. When complex maneuvers are required to gain access into a vehicle, extrications become long, complicated, and ultimately more dangerous. For example, if a vehicle is not badly damaged, access may be obtained by simply opening an undamaged or operable door. However, when there is severe structural damage, when the roof is collapsed, or when foreign materials are crushing the passenger compartment, gaining access can be a lengthy and complex process.

Supplemental Restraint System (SRS) and Side-Impact Protection System (SIPS)

Modern technology has added increased collision protection for vehicle occupants by means of Supplemental Restraint Systems (SRSs) and Side-Impact Protection Systems (SIPSs), also called *air bags* (Figure 7.77). These systems can be either electrically or mechanically operated. Although air bags have saved many lives, they have also added a potential rescuer safety hazard: accidental activation of the SRS or SIPS during extrication operations. These air bags can deploy with a speed of 200 mph (322 kmph) and exert a tremendous force.

An electrically operated restraint system receives its energy from the vehicle's battery and is designed to activate through a system of electronic sensors installed on the vehicle. These systems have a reserve energy supply that is capable of deploying an air bag even if the battery is disconnected or destroyed in the accident. When the battery is disconnected, the reserve energy supply will drain, disarming the restraint system. Vehicle manufacturers have different time estimates on how long it takes for the reserve to deplete entirely. According to one manufacturer's service manual, the system can maintain sufficient voltage to deploy an air bag for up to 10 seconds after the battery is disconnected; another says the reserve can last up to 10 minutes.

Fire suppression or extrication activities are capable of accidentally activating electrically or mechanically operated restraint systems. For electrically operated systems, an electrical impulse during the extrication process may cause the air bag to deploy. There have been reports of rescuers being physically ejected from a vehicle with a connected battery when the "loaded" SRS was accidentally deployed during extrication operations. Personal protective equipment must be worn and extreme care taken when performing extrication operations on vehicles with SRS or SIPS.

On many vehicle models, the only method to prevent the accidental firing of electrical-type systems is to turn the ignition switch to the "off" position, disconnect both battery cables, and wait for the reserve power supply to drain down. However, some vehicle models are equipped with a key-operated switch that disables and drains the reserve power to passenger-side air bags (Figure 7.78).

Mechanically operated systems are sometimes used in SIPS design and do not require power from the vehicle's electrical system to activate. Therefore, these air bags may be deployed even if the battery has been disconnected. In these systems, disarming or preventing deployment of the air bag may require that the connection between the sensor and the air bag inflation unit be separated. How and where this is done is specific to each vehicle make and model.

Disentanglement and Patient Management

Rescuers should choose the easiest route available to gain access to a vehicle. They should try to

Figure 7.77 A Supplemental Restraint System in place and unactivated.

Figure 7.78 Some vehicles are equipped with an SRS key-operated switch.

open the doors normally, but if they are jammed, the windows would be the next logical choice. Once access to the vehicle is gained, at least one rescuer with appropriate emergency medical training should be placed in the vehicle to begin stabilization of the patient and to protect the patient while disentanglement procedures are in progress. Initial assessment and treatment should be done in accordance with local EMS protocols.

Once the patient's injuries have been assessed, treatment can begin simultaneously with preparation for removal from the vehicle. The most important point to remember is that the vehicle is removed from around the patient and not the reverse. Various parts of the vehicle, such as the steering wheel, seat, pedals, and dashboard, may trap the occupant. The situation should be assessed with the patient's safety foremost in the rescuer's mind.

PATIENT REMOVAL

Packaging means wounds have been dressed and bandaged, fractures have been splinted, and the patient's body has been immobilized to reduce the possibility of further injury (Figure 7.79). Proper packaging protects the patient and facilitates the patient's removal. Once the path has been cleared and the patient has been properly packaged for removal, rescuers should cover sharp edges to prevent cutting themselves or the patient. Openings should be widened and edges padded with blankets or fire hose that has been split and prepared beforehand. Openings should be wide enough so that the patient can be removed as smoothly as possible with no jerking or sudden movements (Figure 7.80).

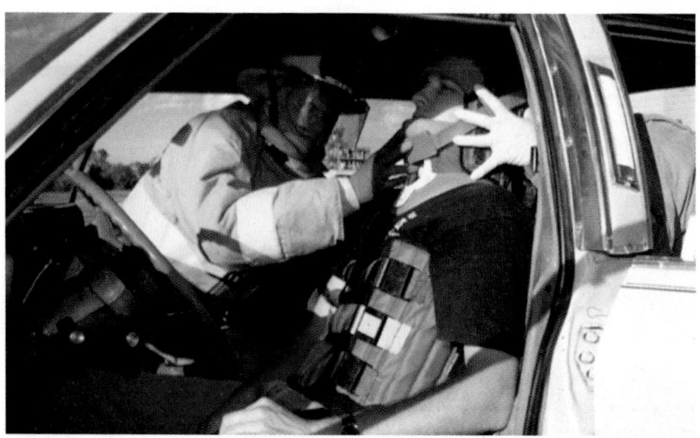

Figure 7.79 Firefighters package a victim for removal.

Figure 7.80 Completely removing both doors and the post between them will give maximum access to the passenger compartment.

REMOVING GLASS

A common task required of rescuers at the scene of a vehicle extrication is removing glass from the vehicle. Glass may need to be removed to facilitate access to the passenger compartment or to lessen the injury hazard posed by remaining fragments of glass. Before discussing glass-removal techniques, it is important to understand the two primary types of glass used in vehicles: *safety (laminated) glass* and *tempered glass.*

Safety (laminated) glass. Safety or laminated glass is manufactured from two sheets of glass that are bonded to a sheet of plastic sandwiched between them (Figure 7.81). This type of glass is most commonly used for windshields and some rear windows. Impact produces many long, pointed shards with sharp edges. The plastic laminate sheet retains most of these shards and fragments in place. When broken, glass stays attached to the laminate and moves as a unit. This facilitates windshield removal. Some manufacturers have laminated an additional layer of plastic

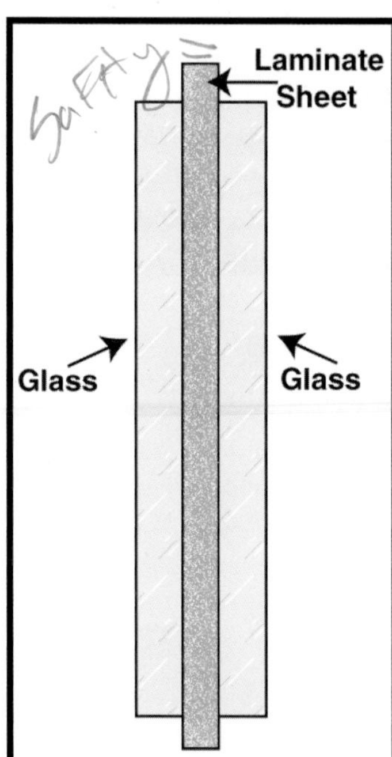

Figure 7.81 This illustration shows the basic construction of laminated glass.

to the passenger-compartment side of the windshield. This provides greater protection from lacerations when impacted.

Tempered glass. Tempered glass is most commonly used in side windows and some rear windows. When struck, tempered glass is designed so that small lines of fracture are spread throughout the entire plate. This results in the glass separating into many small pieces. This lessens the hazard of long, pointed pieces of glass, but presents new problems, among them small nuisance lacerations to unprotected body parts and the entrance of small pieces of glass into open wounds or the eyes.

REMOVING LAMINATED GLASS

Removing windshields and laminated rear windows is somewhat more complicated and time-consuming than removing tempered side or rear windows. This is mainly because of the difference in glass types. Windshields and rear windows that are constructed of safety or laminated glass will not disintegrate and fall out like tempered glass windows. Since more laminates are being added to windshields, it may not be as easy to chop through the windshields of newer vehicles. In this case, the best method for removing glass is with a saw. The following common hand tools can be used to cut laminated glass:

- Air chisel
- Axe (standard or aircraft crash axe)
- Reciprocating saw
- Handsaw with a coarse blade such as those used in commercially produced tools

Total windshield removal is performed before the roof is laid back or removed. This method requires two rescuers, one on each side of the vehicle for cutting the windshield. The passengers and rescuers inside the vehicle should be covered with a tarp or protective blanket. Two rescuers should be used to hold the cover over the passenger(s) and rescuer inside the vehicle. A backboard can be added to protect people inside the vehicle from being struck by tools or loose glass.

Total windshield removal is accomplished in the following manner (Figure 7.82): An opening is made in both upper corners of the windshield. The opening should penetrate all layers of lamination. A saw or other cutting tool is then used to cut down the two short sides of the windshield to the lower corners. Another cut is made across the bottom edge of the windshield to connect with the cuts on the short sides. Once all cuts are made, the bottom of the windshield is gently pulled outward and upward to begin separating the window from the upper mount. The window is then folded rearward over the roof. The windshield can then be placed underneath the vehicle or removed from the area entirely.

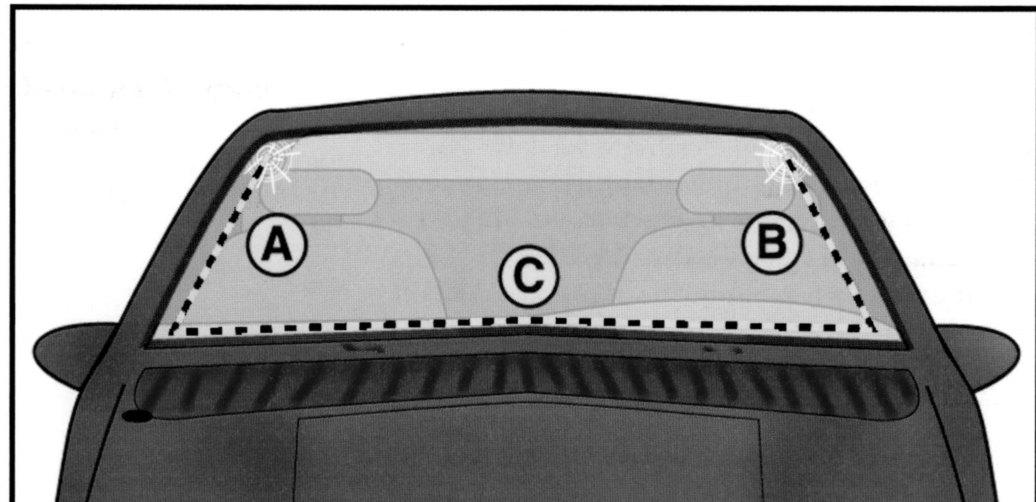

Figure 7.82 Cuts necessary for total windshield removal.

REMOVING TEMPERED GLASS

Removing side and rear windows constructed of tempered glass is a fairly simple task. These windows can easily be broken by either striking them with a sharp, pointed object in the lower corner of the window or by using a spring-loaded center punch pressed into the lower corner of the window. When using a center punch, the hand holding the punch should be braced by the opposite hand (Figure 7.83). This prevents the rescuer from

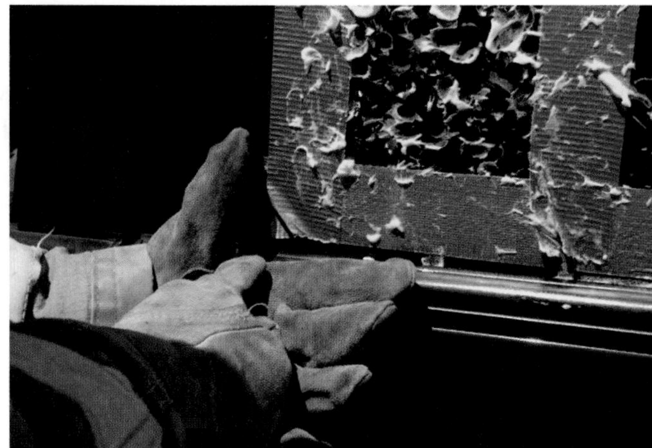

Figure 7.83 Note the hand position of the rescuer using the spring-loaded center punch. The left hand is positioned to keep the right hand from going through the window as the glass breaks.

Figure 7.84 Spray-on adhesive controls glass once it shatters. Use duct tape with loops formed on it to act as a handle for removing the glass.

sticking his hand into the glass when it breaks and also prevents the center punch from coming in contact with a victim who may be close to the window. A standard center punch or Phillips™ screwdriver may also be used. It will need to be driven into the window with a hammer or mallet. The pick end of a pick-head axe or Halligan tool will also work if nothing else is available.

When glass is broken using these methods, most of it will usually drop straight to the floor. To protect against any injuries from the loose glass, rescuers should wear full protective equipment, including eye protection. If it is necessary to break a window to gain primary access to the victim, choose one as far away from the victim as possible.

One method commonly used to control broken glass is to apply a sheet of self-adhering contact paper to the window before breaking the glass. This gives the window basically the same properties as laminated glass. Once the paper is applied, the window can be broken as previously described and most of the pieces of glass will stick to the paper, allowing the window to be removed as a unit.

Another method of controlling glass is to apply a commercially marketed spray aerosol that forms a laminated-type coating on the glass (Figure 7.84). This coating sets up in a matter of seconds and allows the glass to be broken and retained in a sheet (Figure 7.85). Then the glass can be removed in sheets instead of in little pieces (Figure 7.86).

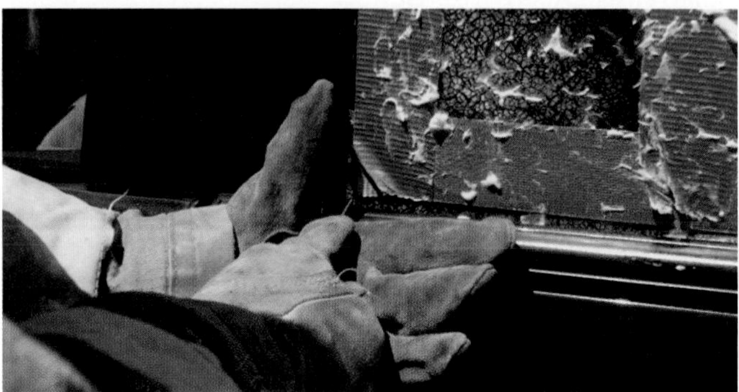

Figure 7.85 When the glass is broken, the adhesive retains it in a sheet.

Figure 7.86 Once the glass is broken, remove it carefully toward the outside of the vehicle.

When working with rear windows, rescuers must remember that some rear windows will be tempered and some will be laminated. If the window is not responding to removal techniques for tempered glass, it is probably laminated glass and will have to be removed in a manner similar to that for windshields.

REMOVING THE ROOF AND DOORS

The disentanglement procedures used for any particular accident vary depending on the circumstances. A common evolution that is required is the removal of the vehicle's roof. A-, B-, and C-posts are designations given to vehicle door posts from front to back (Figure 7.87). The A-post is the front post area where the front door is connected to the body. The B-post is the post between the front and rear doors on a four-door vehicle or the door handle end post on a two-door vehicle. The C-post is the post nearest the handle on the rear door of a four-door vehicle. On a two-door vehicle, the rear roof post may be considered the C-post. Removal can be done by either cutting all the roof posts and removing the roof entirely or by cutting only the front posts and folding the roof back over the trunk (Figure 7.88). New materials such as plastics used in vehicle construction may prevent the roof from bending. In this case, the best method is to cut all roof posts and remove the entire roof. Unibody vehicles have features that affect them when their roofs are removed. Vehicles should be well supported before compromising the body of the vehicle. A third step block should be placed under the B-post of the vehicle.

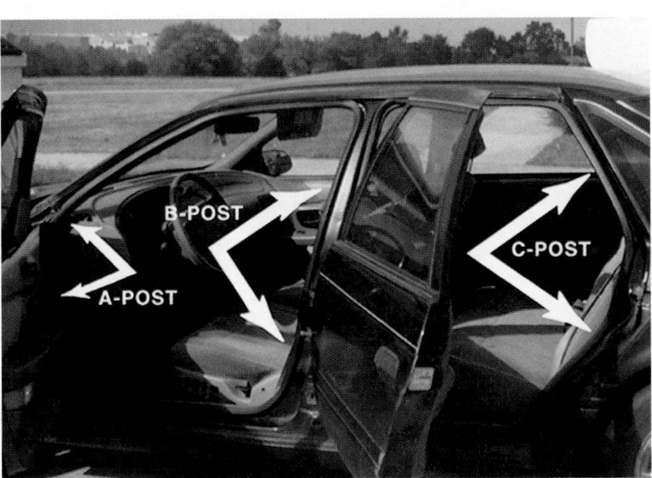

Figure 7.87 This photograph highlights the locations of the A-, B, and C-posts on a four-door automobile.

Figure 7.88 A long bar placed at the point where the fold is to be made will facilitate the folding process.

Doors can be opened from the handle side or removed completely by inserting the rescue tool in the crack on the hinge side (Figure 7.89). Here again, the outer door panel may be made of plastic. The rescuer may have to remove this outer skin to gain access to the metal frame.

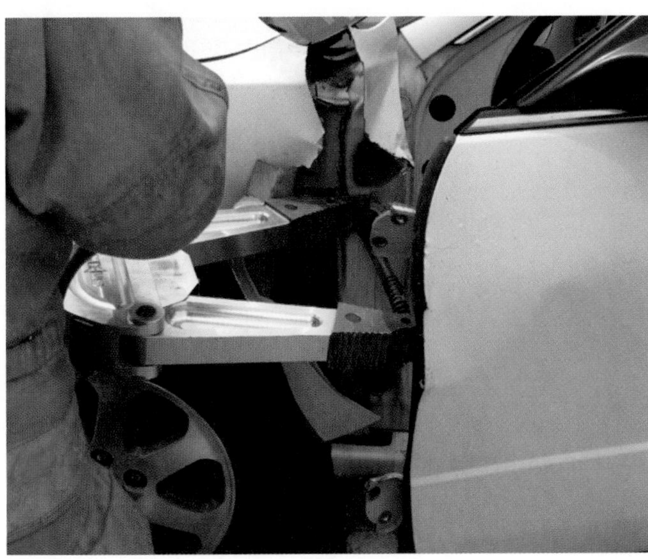

Figure 7.89 When the bottom hinge breaks, that side of the door should be free.

DISPLACING THE DASHBOARD

Often, the victim is trapped by the steering wheel or the dashboard. The dashboard-displacement method is the best method to remove dashboard wreckage from the patient after a front-end collision. The dashboard displacement method is accomplished by removing the windshield, cutting the front roof posts, and folding back the roof. Cut a relief notch in both A-posts as close to the rocker panel as possible. Place a hydraulic ram on each side of the vehicle and push the dashboard assembly up and away from the front seat area (Figure 7.90). Inserting cribbing into the cuts on the A-posts keeps the dashboard from settling back into place. The rams can then be removed (Figure 7.91).

Figure 7.90 As the rams are extended, the entire front portion of the passenger compartment is opened.

Figure 7.91 A piece of cribbing can be inserted into the relief cuts to hold the dashboard when the extension rams are removed.

SPECIAL RESCUE SITUATIONS

[NFPA 1001: 3-3.7(a); 4-4.2; 4-4.2(a)]

Firefighters may encounter many different scenarios involving rescue. Specialized rescues can include rescue from collapsed buildings, trench cave-ins, caves or tunnels, electrical contact, water and ice, industrial machinery, and elevators. These rescue operations require advanced training and equipment. Firefighters should be educated on special rescue situations so that they can identify the need for a special rescue team. Firefighters may also be used to assist rescue personnel and retrieve necessary tools and equipment. Firefighters should be familiar with their department's capabilities for handling special rescue situations. The following sections provide information to assist the firefighter in determining the need for specialized rescue assistance. For additional information on these types of rescue operations see IFSTA's **Fire Service Search and Rescue** manual.

Rescue From Collapsed Buildings

Building collapse may occur as a result of fire, weather conditions, earthquake, explosions, or simply because an old or otherwise weak structural component fails (Figure 7.92). The difficulty encountered in reaching a victim in a collapsed building depends upon conditions that are found. Immediate rescue of surface and lightly trapped victims should be accomplished first. Rescue of a heavily trapped victim is a more complicated endeavor and requires more time. This type of rescue depends upon the services of specially trained rescue workers who have a knowledge of building construction and collapse and who are proficient in the use of special rescue tools, equipment, and techniques.

Figure 7.92 Parking garages are collapse-prone structures. *Courtesy of Mehmet Celebi, US Geological Survey, NOAA National Graphics Data Center.*

TYPES OF COLLAPSE

Structures collapse in predictable patterns. Knowing and recognizing these patterns can help rescuers make more informed decisions about the likelihood of finding viable victims in the rubble and about the need for shoring and tunneling (see Shoring and Tunneling sections). The four most common patterns of structural collapse are *pancake, V-shaped, lean-to,* and *cantilever.*

Pancake collapse. This pattern of collapse is possible in any building where simultaneous failure of two opposing exterior walls results in the upper floors and the roof collapsing on top of each other such as in a stack of pancakes — thus, the name of this collapse pattern (Figure 7.93). The pancake collapse is the pattern least likely to contain voids in which live victims may be found, but it must be assumed that there are live victims in the rubble until it is proven otherwise.

Figure 7.93 Pancake collapse.

V-shaped collapse. This pattern of collapse occurs when the outer walls remain intact and the floor(s) and/or roof structure fail in the middle (Figure 7.94). This pattern offers a good chance of habitable void spaces being created on both sides of the collapse.

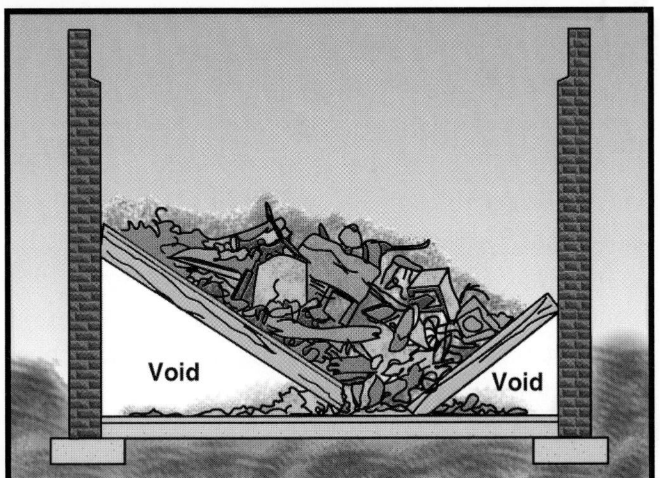

Figure 7.94 V-shaped collapse.

Lean-to collapse. This pattern of collapse occurs when one outer wall fails while the opposite wall remains intact. The side of the roof assembly that was supported by the failed wall drops to the floor forming a triangular void beneath it (Figure 7.95).

Cantilever collapse. This pattern of collapse occurs when one sidewall of a multistory building collapses leaving the floors attached to and supported by the remaining sidewall (Figure 7.96). This pattern also offers a good chance of habitable

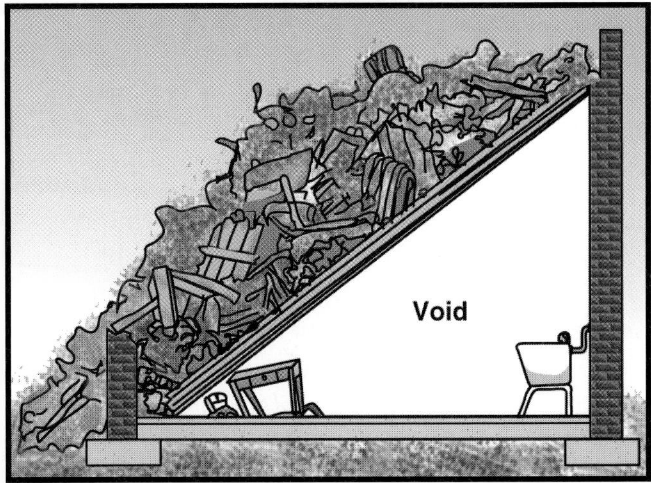

Figure 7.95 Lean-to collapse.

Figure 7.96 Cantilever collapse.

voids being formed under the supported ends of the floors. This collapse pattern is perhaps the least stable of all the patterns and is the most vulnerable to secondary collapse.

HAZARDS

There are many actual and potential hazards involved in structural collapse rescue, and they may take any of a wide variety of forms. However, most of the hazards associated with this type of operation fall into one or both of two categories: *environmental* and *physical*.

Environmental. Before rescuers can begin to search the rubble of a collapsed structure for victims, they may have to contend with a number of environmental problems — those that are in and around the collapse. Many of the secondary hazards — those that were created by the collapse or that developed after it — are environmental in nature. Most of the potential environmental hazards involve damaged utilities, atmospheric contamination, hazardous materials contamination, darkness, temperature extremes, noise, fire, or adverse weather.

Physical. Physical hazards are those hazards associated with working in and around piles of heavy, irregularly shaped pieces of rubble that may suddenly shift or fall without warning. The primary physical hazards are those related to secondary collapse, working in unstable debris, working in confined spaces (some of them below grade), working around exposed wiring and rebar, and dealing with heights.

SHORING

Shoring is a general term used to describe any of a variety of means by which unstable structures or parts of structures can be stabilized (Figure 7.97). It is the process of preventing the sudden or unexpected movement of objects that are too large to be moved in a timely manner and that may pose a threat to victims and/or rescuers. Shoring is not intended to move heavy objects but is just intended to stabilize them. Stabilizing objects with shoring may involve applying air bags, applying cribbing, using jacks, constructing a system of wooden braces, or using a combination of these methods.

Figure 7.97 A shoring system in place.

TUNNELING

Tunneling primarily involves removing smaller rubble and debris to create a path to a victim whose location is known (Figure 7.98). Tunneling may involve shoring large pieces of overhanging rubble, but shoring is not its main function. Because it is a slow and dangerous process, tunneling should be used only when all other means of reaching a victim have proven ineffective.

If a victim is known to be under tons (tonnes) of rubble and debris and time does not allow for working down to the victim by removing layers of debris from above, tunneling through the debris may be the only option. Rescuers must be very careful when they begin tunneling because when a piece of debris is moved, there is a chance that it will start a chain reaction of falling debris. This could undo all of the work accomplished to that point and/or bury the rescuers under tons (tonnes) of debris.

Rescue From Trench Cave-Ins

Trench construction occurs in virtually every city and town; in many jurisdictions, it occurs almost daily somewhere within their boundaries. With all of this excavation going on, cave-ins are bound to happen, and they do. Many people killed in trench incidents are would-be rescuers who fail to stabilize the trench before they enter it, and they become additional victims when the trench caves in on them. Knowing how to make a trench safe to enter and taking the time to do it give both the victim and the rescuer the best chance for survival.

Rescue operations depend on making the site as safe as possible by using shoring or cribbing to hold back other weakened earth formations (Figure 7.99). Rescuers should not be sent into a trench unless their safety can be reasonably ensured and they have been trained. Meanwhile, rescue apparatus, nonessential personnel, heavy equipment, and spectators should be moved back to avoid causing secondary cave-ins.

Several safety precautions firefighters and officers must remember when they are involved in cave-ins and excavation rescues are as follows:

- Only rescuers with advanced trench rescue skills should enter a trench.

Figure 7.98 Typical debris tunnels.

Figure 7.99 Aluminum hydraulic shores stabilizing a trench.

- A trench should not be entered until it has been safely shored.
- Rescuers entering a trench should have on proper protective equipment to protect them from physical, atmospheric, and environmental hazards associated with working in and around trenches.
- If a trench is found to be either oxygen-deficient or contaminated, rescuers will have to wear self-contained breathing apparatus or the trench will have to be mechanically ventilated before rescuers are allowed to enter.
- Exit ladders should be placed in trenches. Ladders should extend at least 3 feet (1 m) above the top of the trench (Figure 7.100).
- Firefighters should be careful with the tools they use in a trench to avoid injuring each other or the victim(s).
- Unnecessary fire department personnel and bystanders should be kept out of a trench and away from its edge.

Figure 7.100 Ground ladders are a critical part of trench rescue.

- Rescuers should be aware of any other hazards that might exist at the scene such as underground electrical wiring, water lines, explosives, or toxic or flammable gases.

Rescue From Caves and Tunnels

Although firefighters may be called when someone is lost or injured in a cave, they are usually not trained or equipped to perform these rescues. Rescue from caves must be done by those who are familiar with the uniquely hostile environment of a cave and who have the training and equipment needed. Unless they are specially trained to operate in these environments, fire/rescue personnel usually confine their activities to aboveground support of other cave-rescue personnel.

Rescues Involving Electricity

Rescues involving energized electrical lines or equipment are some of the most common situations to which firefighters are called (Figure 7.101). But the frequency with which these situations occur should not lull rescuers into a false sense of security — these situations can be extremely dangerous. Improper actions by rescue personnel can result in their being injured or killed instantly. Whenever rescuers respond to any situation involving electricity, they should *always* do the following:

- Assume that electrical lines or equipment are energized.
- Call for the power provider to respond. Let only power company personnel cut electrical wires.
- Control the scene.

Electrical wires on the ground can be dangerous without even being touched. Downed electrical lines can energize wire fences or other metal objects with which they come in contact. When an energized electrical wire comes in contact with the ground, current flows outward in all directions from the point of contact. As the current flows away from the point of contact, the voltage drops progressively (Figure 7.102). Depending upon the voltage involved and other variables, such as ground moisture, this energized field can extend for sev-

Figure 7.101 Typical overhead electric power lines.

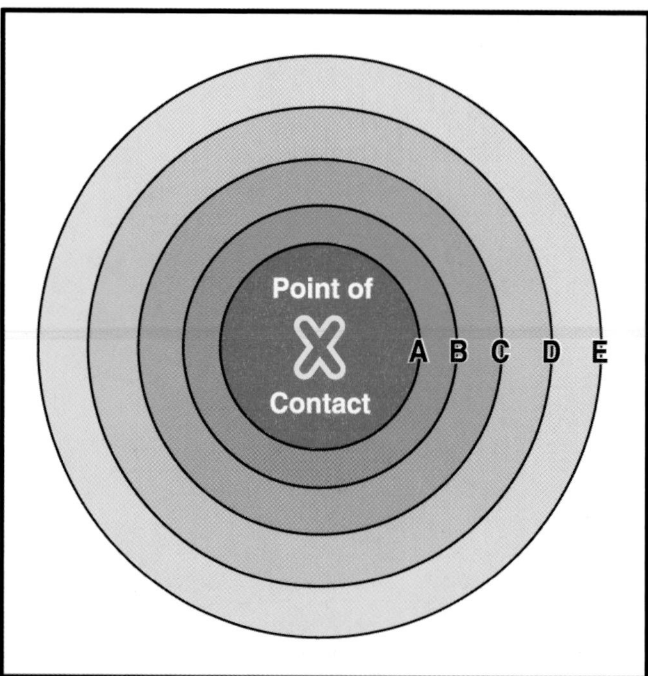

Figure 7.102 Voltage drops as it spreads away from the source.

eral feet (meters) from the point of contact. A rescuer walking into this field can be electrocuted (Figure 7.103). To avoid this hazard, rescuers should stay away from downed wires a distance equal to one span between poles until they are certain that the power has been shut off (Figure 7.104).

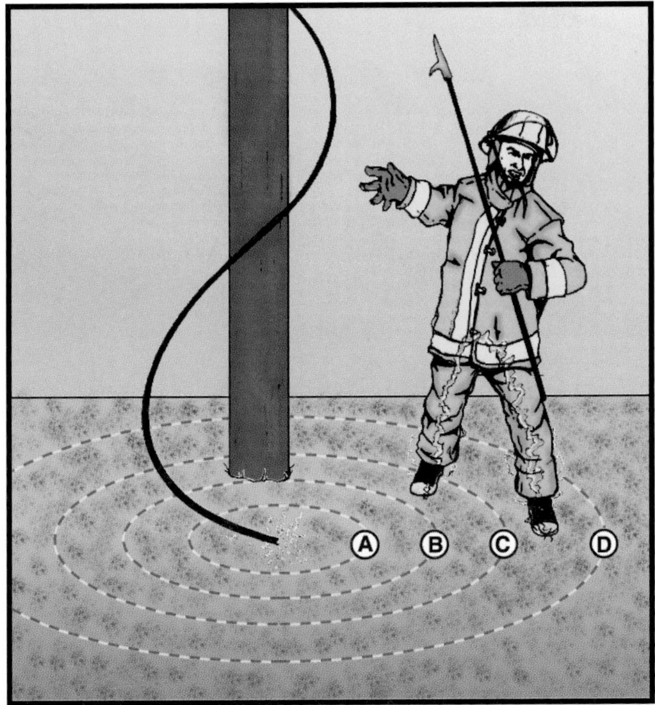

Figure 7.103 Rescuers must approach downed wires with caution.

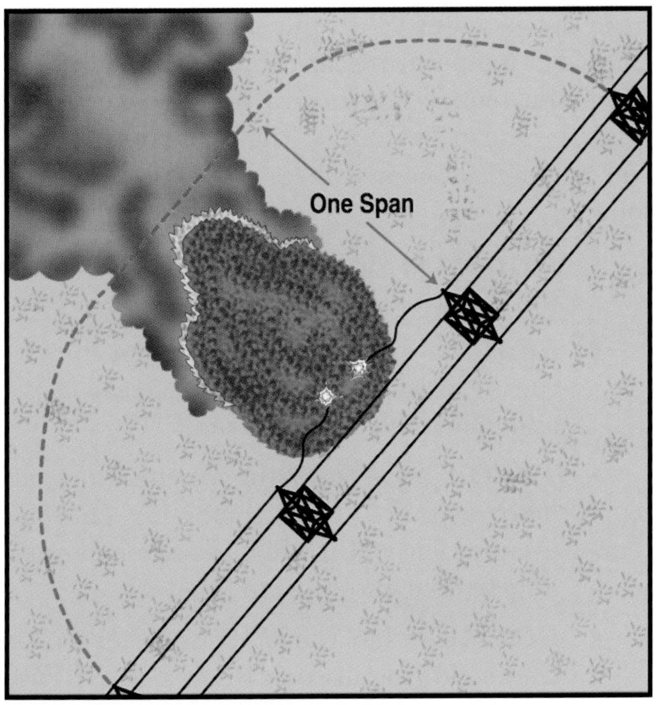

Figure 7.104 Firefighters should stay well clear of energized power lines.

Water and Ice Rescue

All jurisdictions have the potential for water rescue and recovery operations. These situations can occur in swimming pools, lakes, ponds, rivers, streams, other bodies of water, and at low-head dams and water treatment facilities (Figure 7.105). Areas subject to freezing temperatures also provide the potential for ice emergencies. It is important to denote the distinction between rescues and recoveries. *Rescues* are situations where a victim is stranded, floundering, or has been submerged for a short period of time (usually less than half an hour). In these cases, the potential for saving the victim is real. *Recoveries* are situations where a victim has been submerged for such a long period of time that he is most probably dead, and the goal of the operation is to recover the body.

All rescue personnel should wear appropriate personal protective equipment when operating at water and ice incidents. Standard firefighter turnout clothing is not acceptable. Proper personal protective equipment includes a water rescue helmet and an appropriate personal flotation device (PFD). When working in or around ice or cold water, thermal protective suits should also be worn (Figure 7.106).

Figure 7.105 A typical low-head dam.

Figure 7.106 Proper protective equipment must be worn at water and ice rescues.

WATER RESCUE METHODS

The following methods can be used in order to rescue a victim during a water emergency.

- ***REACH*** — Extend a long-handled tool to the victim (Figure 7.107).
- ***THROW***—Throw a rope or flotation device with an attached rope to the victim (Figure 7.108).
- ***ROW*** — Use a boat to retrieve the victim.
- ***GO*** — Swim to the victim and drag the victim to safety.

WARNING

The ROW and GO rescue techniques should be attempted only by those who have been specifically trained in their application.

Figure 7.107 A rescuer extends a tool handle to a victim.

Figure 7.108 Throwing a lifeline to a victim may be all that is needed.

ICE RESCUE METHODS

The steps in performing an ice rescue are designed to be as simple as possible because the rescuer has other factors to consider. One of those other factors is the unpredictability of the ice. Just because ice is thick does not mean that it is strong; *the victim in the water has demonstrated that the ice is weak.*

WARNING

Until they have donned life jackets/PFDs or environmental/thermal protection suits (dry suits), rescuers should stay off the ice.

Rescue personnel must contend with the weather and its effect on those involved in the incident and on the scene. The victim will almost certainly be suffering the effects of hypothermia, so having an advanced life support unit on scene to start immediate patient care is critical. Another factor for ice rescuers to consider is that the victim may not be able to help in his own rescue. With frozen hands, the victim may not be able to grasp a rope or other aid, and with heavy, wet clothing, the victim may even have difficulty keeping his head above water. With immersion in ice water, the body's temperature can drop dramatically, and the victim's chances of survival may depend on how quickly he can get out of the water and into a warmer environment. The ice rescue protocol is as follows:

- Instruct the victim ***not*** to try to get out of the water until a rescuer says to.
- ***REACH*** — Implement only when the victim is close to solid ground and is responsive and able to hold onto an aid.
- ***THROW*** — Allows the rescuer to span more distance while remaining on solid ground. The victim must be responsive and able to hold onto the aid.
- ***GO*** — Use when the victim is either too far from solid ground to use REACH or THROW or is incapable of grasping an aid (Figure 7.109).

Figure 7.109 Only those trained for ice rescue should attempt the GO technique. *Courtesy of Steve Taylor.*

WARNING

The GO rescue technique should be attempted only by those who have been specifically trained in this application.

Industrial Extrication

Industrial extrications are among the most challenging rescue situations that firefighters will ever face (Figure 7.110). Because there is an endless number of machines that have the potential to entrap victims, it is impossible to list specific techniques for victim removal. When surveying the situation, personnel should take into account the following:

- Medical condition and degree of entrapment of the victim
- Number of rescue personnel required
- Type and amount of extrication equipment needed
- Need for special personnel, equipment, or expert assistance
- Level of fire or hazardous material hazard that is present

These observations are critical to the rest of the incident. For example, if a victim is seriously entangled and in danger of bleeding to death, amputation by a doctor brought to the scene may be required to save the person's life.

Figure 7.110 Industrial machinery causes special rescue situations for firefighters.

If it becomes obvious during the initial survey that the problem is beyond the capability of the rescue team, outside expertise is required. In most cases, this will be plant personnel on site who are more familiar with the involved machinery. Plant maintenance personnel are usually good sources of information. In rare cases, it may be necessary to go to off-site sources, such as machinery manufacturers, for help. Ideally, these outside sources are identified during pre-incident planning.

Elevator Rescue

Most elevator emergencies involve elevators that are stuck between floors because of a mechanical or power failure. Upon arrival at the scene of an elevator emergency, firefighters should have an elevator mechanic dispatched to the scene (Figure 7.111). Unless there is a medical emergency in the elevator car, the best approach is to reassure the occupants that help is on the way and then wait for the elevator mechanic to arrive and handle the problem.

Figure 7.111 A firefighter consults with an elevator mechanic.

An elevator mechanic is trained to make mechanical adjustments to the elevator that may enable passengers to exit from the elevator car in a normal manner. Under no circumstances should firefighters alter the elevator's mechanical system in an attempt to move the elevator. Adjustments to the mechanical system of the elevator installation should be performed only by the elevator mechanic.

If there is an emergency situation requiring immediate action or if the mechanical problem cannot be immediately fixed, it may be necessary to conduct an elevator rescue. These rescues require training in the use of proper rescue techniques. Only trained personnel should attempt elevator rescues.

Figure 7.112 An emergency phone can be used to calm the occupants.

Regardless of the type of situation, communication must be established with the passengers to assure them of their safety and that work is being done to release them. If a telephone or intercom is not available, shouting through the door near the stall location may be sufficient for passing messages back and forth. Communication with the passengers is essential for their morale and mental state and should be established and maintained throughout the operation (Figure 7.112).

Escalator Rescue

Escalators, also called *moving stairways,* are stairways with electrically powered steps that move continuously in one direction. (Figure 7.113). Each individual step rides a track. The steps are linked together and move around the frame by a step chain. The handrails move at the same rate as the stairs. The driving unit is most commonly located under the upper landing and is covered by a landing plate.

Many escalators have manual stop switches located on a nearby wall, at the base of the escalator, or at a point close to where the handrail goes into the newel base (Figure 7.114). Operation of the switch stops the stairs and sets an emergency brake. The stairs should be stopped during rescues or when firefighters are advancing hoselines up or down the moving stairway. As with the elevator, an escalator technician should be requested to assist in removing victims.

Figure 7.113 Escalators are found in many occupancies.

Figure 7.114 Most escalators have emergency stop controls.

SKILL SHEET 7-1 CRADLE-IN-ARMS LIFT/CARRY

One Rescuer

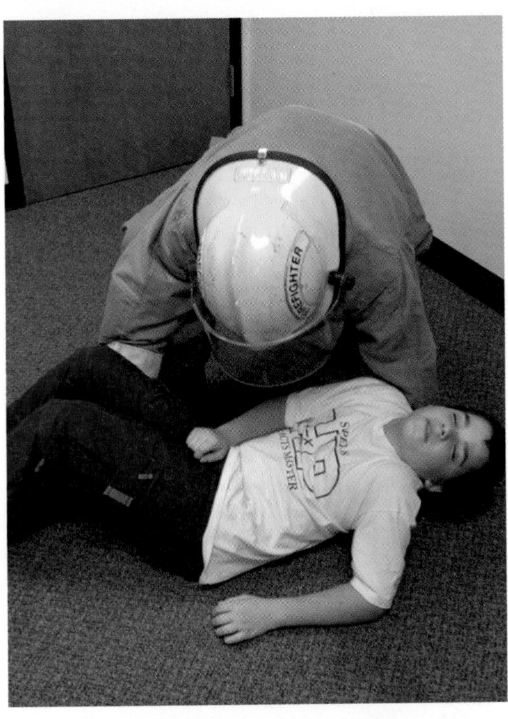

Step 1: Place one arm under the victim's arms and across the back.

Step 2: Place the other arm under the victim's knees.

Step 3: Keep the back straight while preparing to lift.

Step 4: Lift the victim to about waist height.

Step 5: Carry the victim to safety.

SKILL SHEET 7-2 SEAT LIFT/CARRY

Two Rescuers

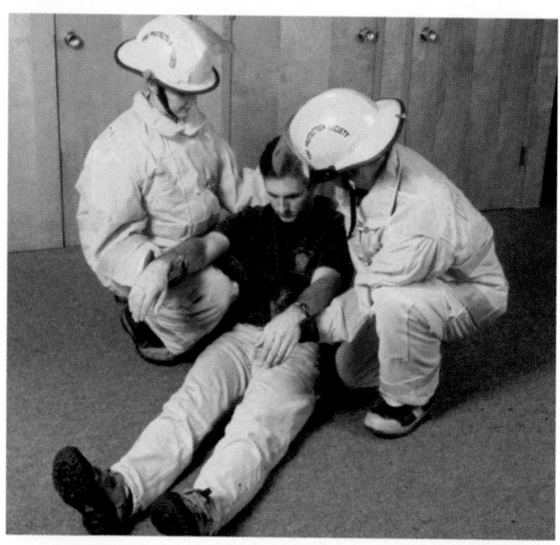

Step 1: Raise the victim to a sitting position.

Step 2: Link arms across the victim's back.

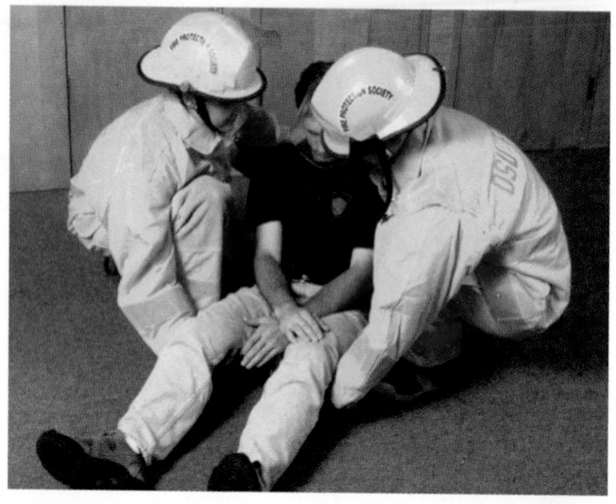

Step 3: Reach under the victim's knees to form a seat.

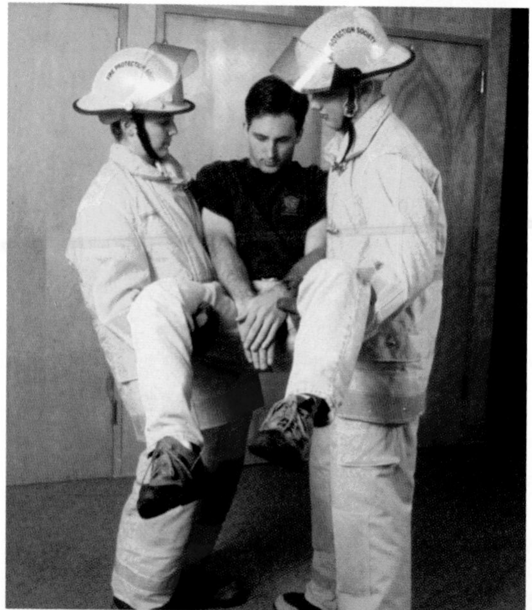

Step 4: Stand.

Step 5: Lift the victim (use your legs).

Step 6: Move the victim to safety.

SKILL SHEET 7-3 TWO- OR THREE-PERSON LIFT/CARRY

To a Gurney

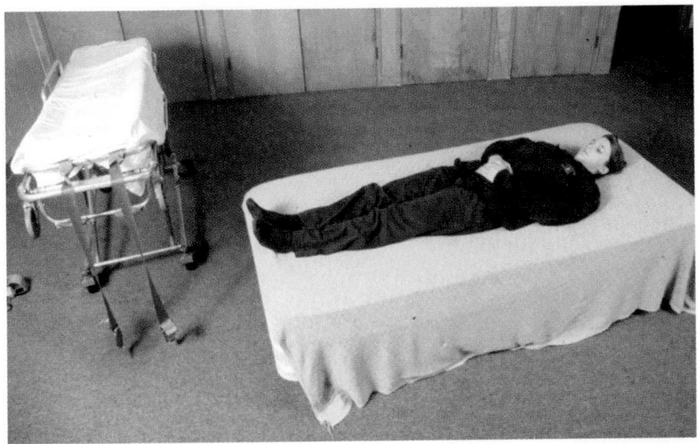

NOTE: **All victim movements are carried out under the direction of Rescuer #1.**

Step 1: Position the gurney so that the victim can be carried to it and placed on it with the least amount of movement. This may require leaving the gurney in the fully raised position.

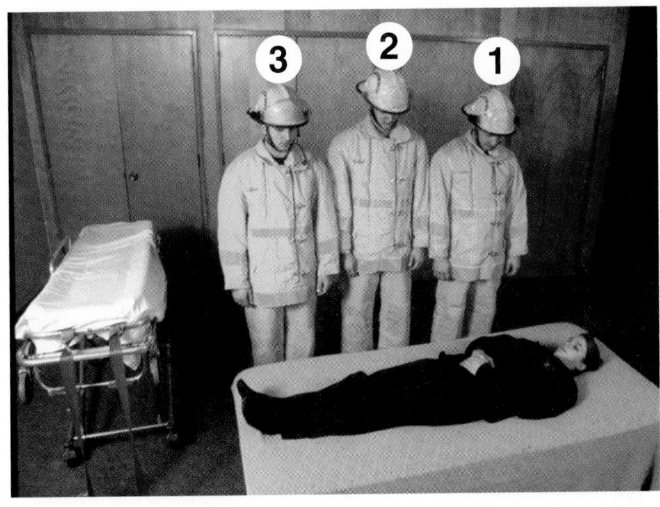

Step 2: Position rescuers on the side of the victim that is easiest to reach and/or that will facilitate placing the victim on the gurney.

Step 3: ***All Rescuers:*** Crouch or kneel as close to the victim as possible, keeping backs straight.

Step 4: ***Rescuer #1:*** Place one hand under the victim's head and the other hand and arm under the victim's upper back.

Step 5: ***Other Rescuers:*** Place arms under the victim at rescuers' respective positions.

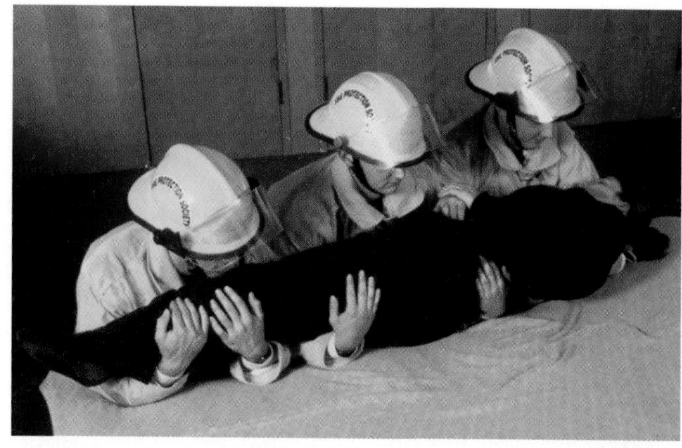

Step 6: ***All Rescuers:*** Roll the victim carefully toward rescuers' chests.

Step 7: ***All Rescuers:*** Stand while holding the victim against rescuers' chests.

Step 8: ***All Rescuers:*** Carry the victim to the desired location.

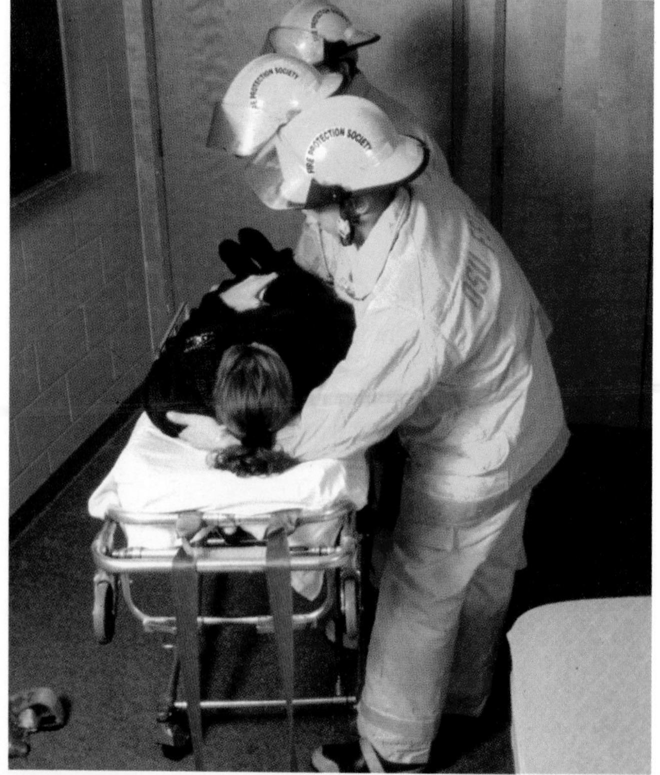

Step 9: Reverse the above procedures on the signal of Rescuer #1 to place the victim on the gurney.

NOTE: **With a smaller victim, two rescuers can perform this lift. One rescuer supports the victim's head and upper back, and the other rescuer supports the victim's torso and legs.**

SKILL SHEET 7-4 MOVING A VICTIM ONTO A LONG BACKBOARD OR LITTER

Four Rescuers

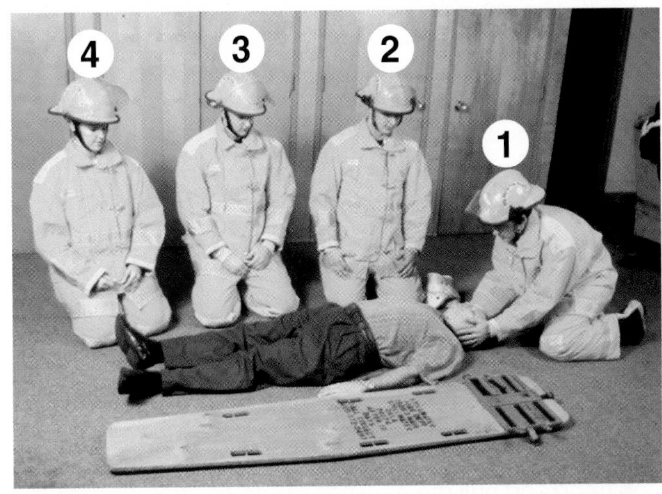

Step 1: ***Rescuer #1:*** Apply in-line stabilization.

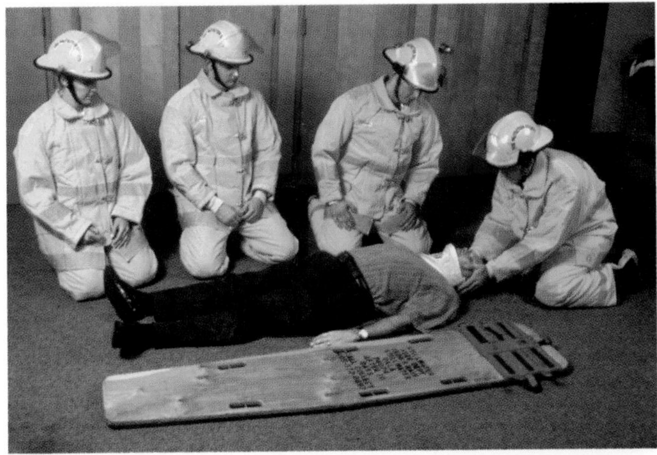

Step 2: ***Rescuer #2:*** Apply a cervical collar.

Step 3: ***Rescuers #3 and #4:*** Place the backboard alongside and parallel to the victim.

Step 4: ***Rescuers #2, #3, and #4:*** Kneel on one side of the victim.

Step 5: ***Rescuer #1:*** Continue to maintain in-line stabilization throughout the lift. Give lifting directions to other rescuers throughout the procedure.

Step 6: ***Rescuer #2:*** Raise the patient's arm over the patient's head on the side the patient will be rolled toward.

Step 7: ***Rescuer #2:*** Grasp the victim's opposite shoulder and upper arm.

Step 8: ***Rescuer #3:*** Grasp the victim's waist and buttocks on the opposite side.

Step 9: ***Rescuer #4:*** Grasp the victim's lower thigh and calf on the opposite side.

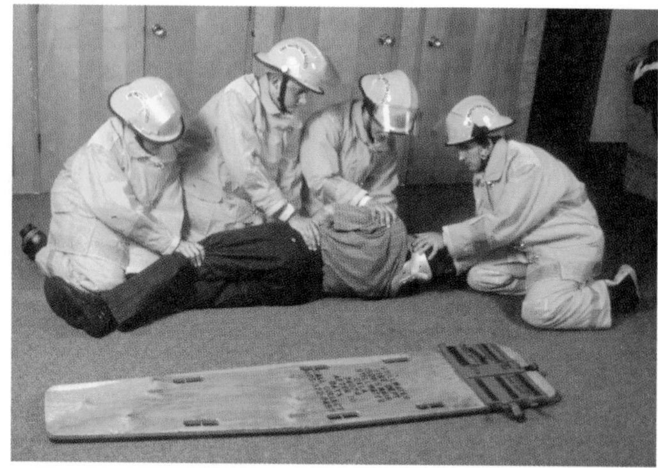

Step 10: ***Rescuers #2, #3, and #4:*** Roll the victim gently toward rescuers as a unit at the direction of Rescuer #1.

Step 11: ***Rescuer #3:*** Reach across the victim's body with one hand and pull the backboard into position against the victim.

Step 12: ***Rescuers #2, #3, and #4:*** Roll the victim onto the board at the direction of Rescuer #1, again making sure that the victim's head and body are rolled as a unit.

NOTE: The victim will not be completely on the backboard at this point.

Step 13: ***Rescuers #2, #3, and #4:*** Move the victim gently so that the victim is centered on the backboard. Move only at the command of Rescuer #1, who continues to maintain in-line stabilization.

CAUTION: This step must be carefully coordinated in order to move the victim's head and body as a unit.

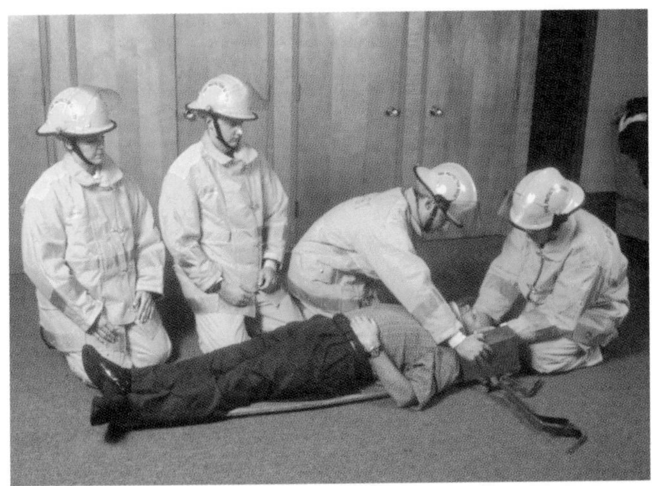

Step 14: ***Rescuer #2:*** Place rolled towels, blankets, or specially designed immobilization devices on both sides of the victim's head.

Step 15: ***Rescuer #2:*** Secure these items and the victim's head to the board with a cravat or tape that passes over the forehead. If an immobilizer is used, place the sides in position and secure the chin and forehead straps.

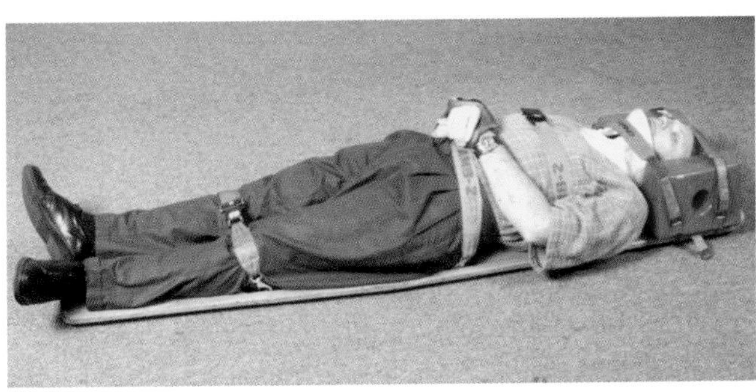

Step 16: ***Rescuers #2, #3, and #4:*** Fasten the victim to the board with the appropriate strap — one across the chest, one above the hips, and one above the knees.

Step 17: ***Rescuers #2, #3, and #4:*** Pad any void areas between the patient and the board.

SKILL SHEET 7-5 EXTREMITIES LIFT/CARRY

Two Rescuers

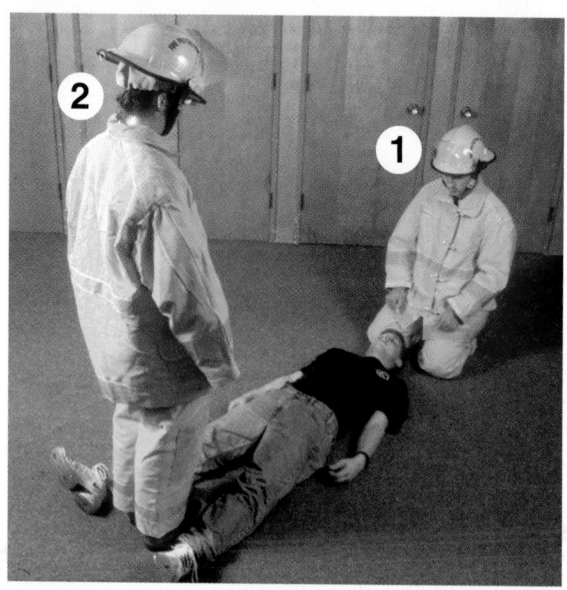

Step 1: ***Both Rescuers:*** Turn the victim (if necessary) so that the victim is supine.

Step 2: ***Rescuer #1:*** Kneel at the head of the victim.

Step 3: ***Rescuer #2:*** Stand between the victim's knees.

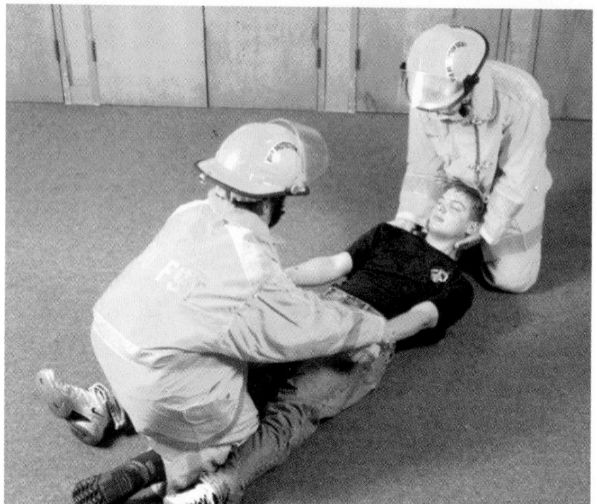

Step 4: ***Rescuer #1:*** Support the victim's head and neck with one hand and place the other hand under the victim's shoulders.

Step 5: ***Rescuer #2:*** Grasp the victim's wrists.

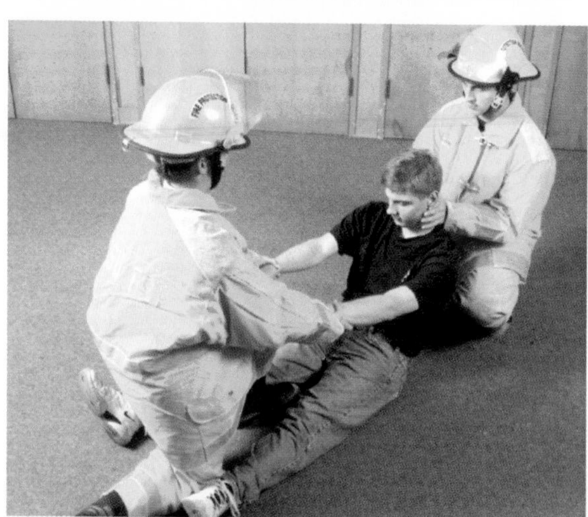

Step 6: ***Rescuer #2:*** Pull the victim to a sitting position.

Step 7: ***Rescuer #1:*** Push gently on the victim's back.

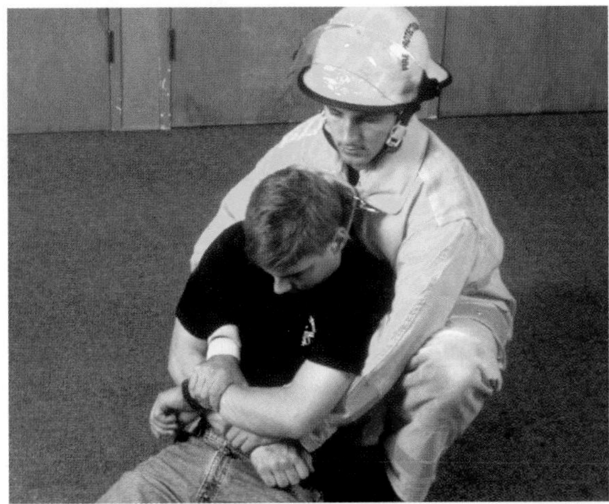

Step 8: ***Rescuer #1:*** Reach under the victim's arms and grasp the victim's wrists as Rescuer #2 releases them.

NOTE: Grasp the victim's left wrist with the right hand and right wrist with the left hand.

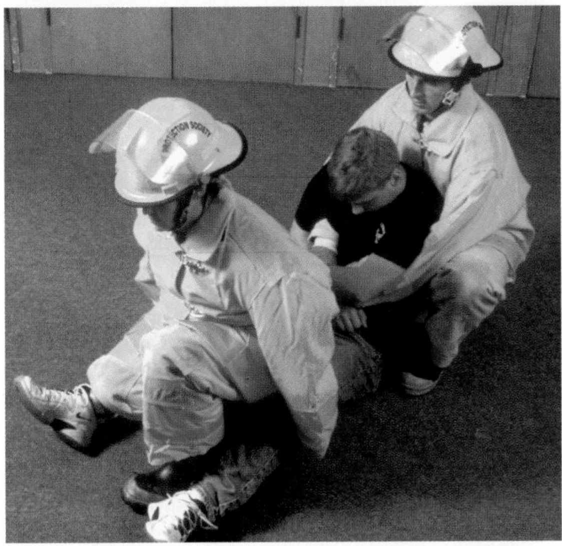

Step 9: ***Rescuer #2:*** Turn around, kneel down, and slip hands under the victim's knees.

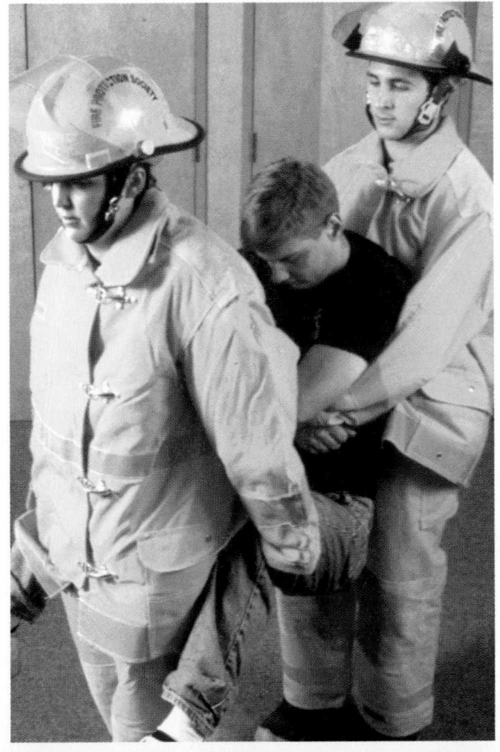

Step 10: ***Both Rescuers:*** Stand and move the victim on a command from Rescuer #1.

SKILL SHEET 7-6 CHAIR LIFT/CARRY

Method 1 — Two Rescuers

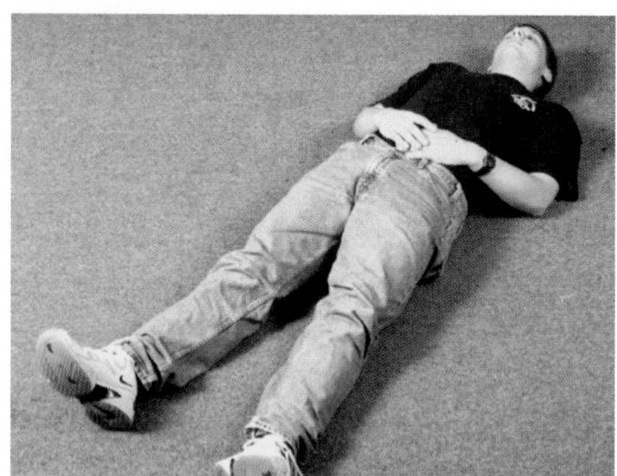

Step 1: ***Both Rescuers:*** Turn the victim (if necessary) so that the victim is supine.

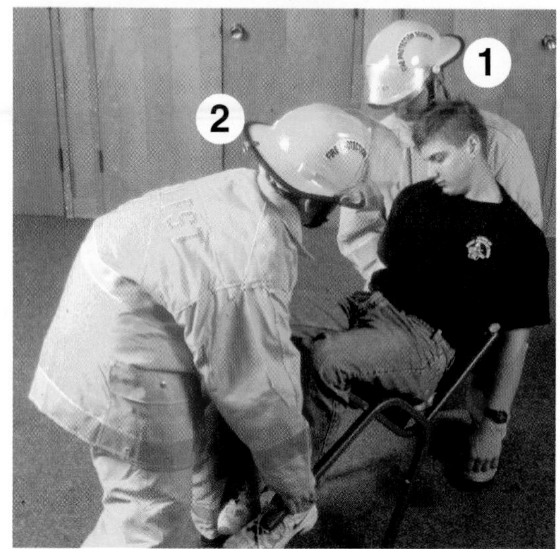

Step 2: ***Rescuer #1:*** Lift the victim's knees until the knees, buttocks, and lower back are high enough to slide a chair under the victim.

Step 3: ***Rescuer #2:*** Slip a chair under the victim.

Step 4: ***Both Rescuers:*** Raise the victim and chair to a 45-degree angle.

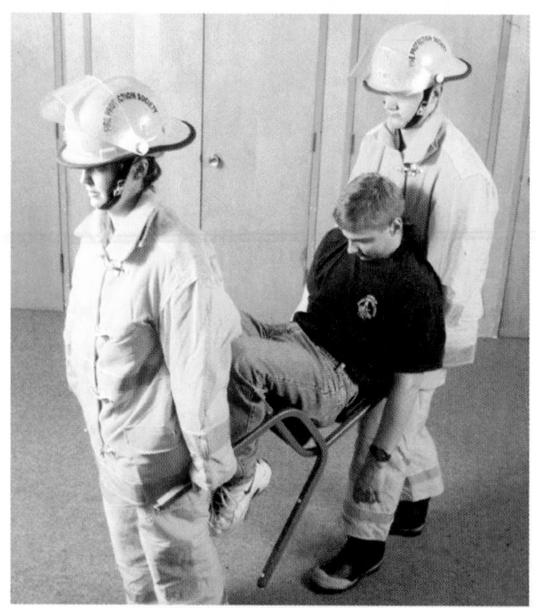

Step 5: ***Both Rescuers:*** Lift the seated victim with one rescuer carrying the legs of the chair and the other carrying the back of the chair.

SKILL SHEET 7-7 CHAIR LIFT/CARRY

Method 2 — Two Rescuers

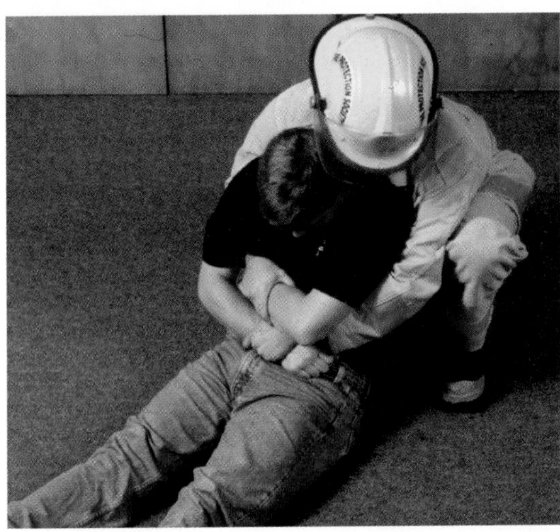

Step 1: ***Rescuer #1:*** Place the victim in a sitting position.

Step 2: ***Rescuer #1:*** Reach under the victim's arms and grasp the victim's wrists.

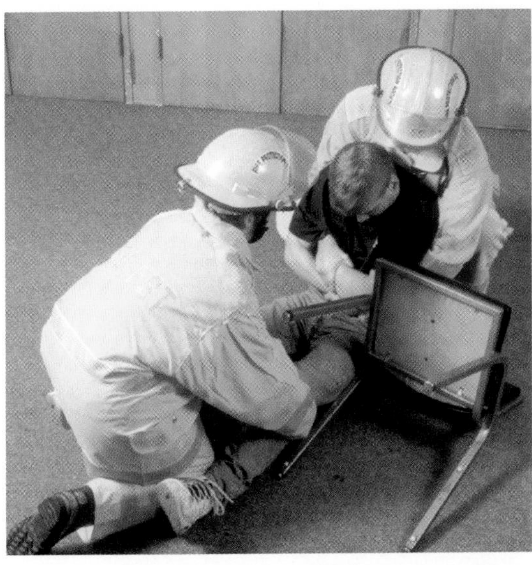

Step 3: ***Rescuer #2:*** Position the chair next to the victim.

Step 4: ***Rescuer #2:*** Grasp the victim's legs under the knees.

Step 5: ***Both Rescuers:*** Lift gently and place the victim onto the chair.

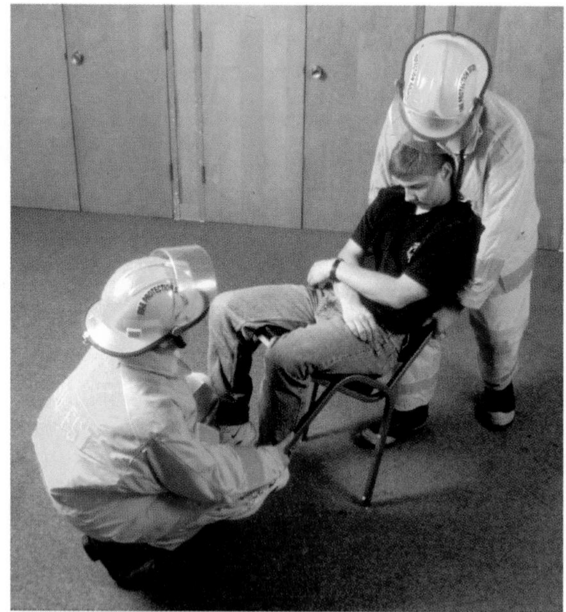

Step 6: ***Both Rescuers:*** Raise the victim and chair to a 45-degree angle.

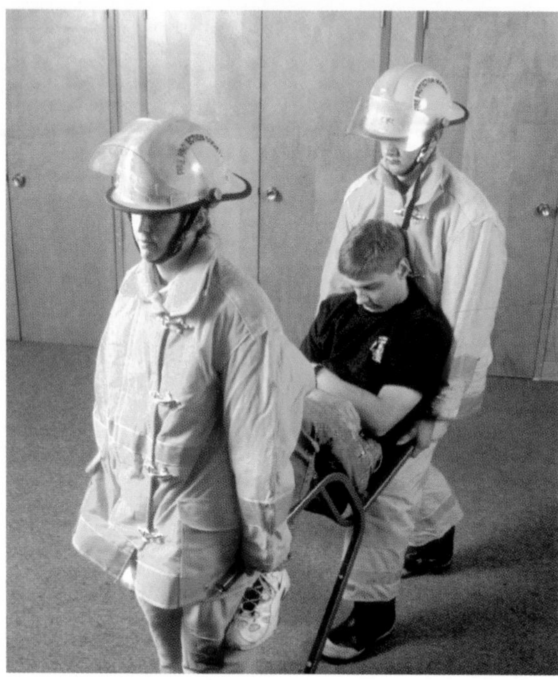

Step 7: ***Both Rescuers:*** Lift the seated victim with one rescuer carrying the legs of the chair and the other carrying the back of the chair.

SKILL SHEET 7-8 INCLINE DRAG

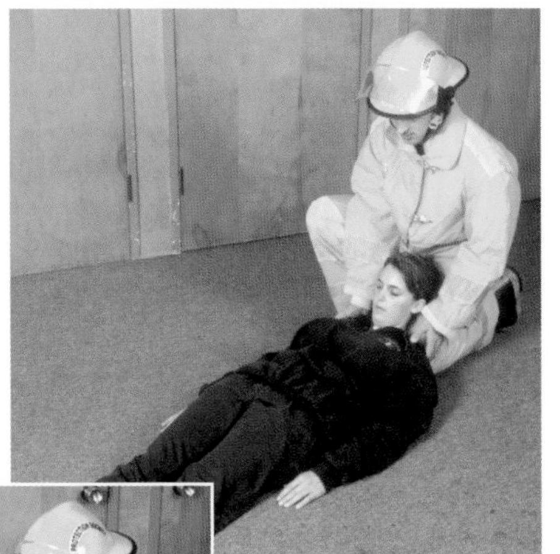

Step 1: Turn the victim (if necessary) so that the victim is supine.

Step 2: Kneel at victim's head.

Step 3: Support the victim's head and neck.

Step 4: Lift the victim's upper body into a sitting position.

Step 5: Reach under the victim's arms.

Step 6: Grasp the victim's wrists.

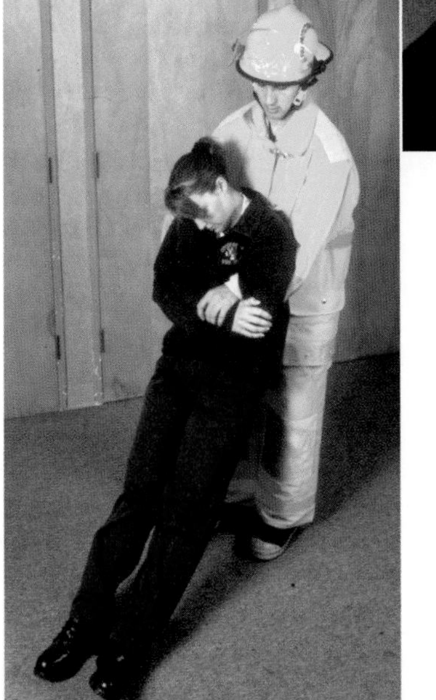

Step 7: Stand. The victim can now be eased down a stairway or ramp to safety.

SKILL SHEET 7-9 BLANKET DRAG

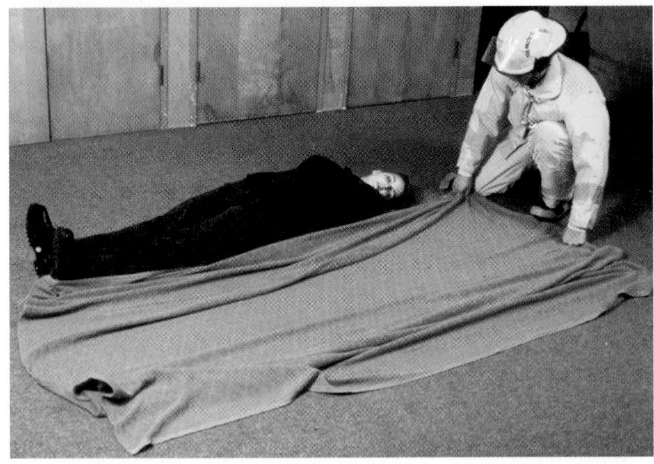

Step 1: Spread a blanket next to the victim, making sure that it extends above the victim's head.

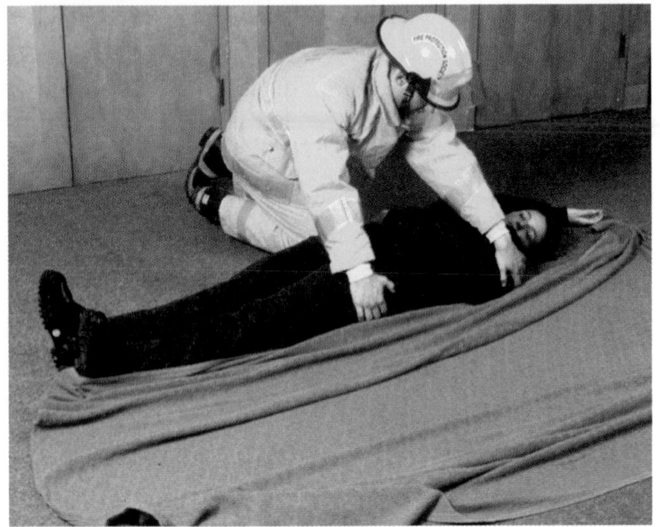

Step 2: Kneel on both knees at the victim's side opposite the blanket.

Step 3: Extend the victim's arm above the victim's head.

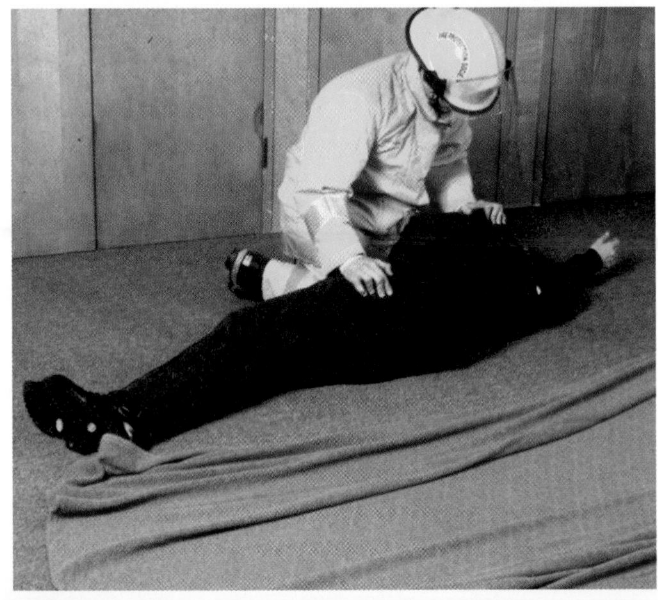

Step 4: Roll victim against your knees.

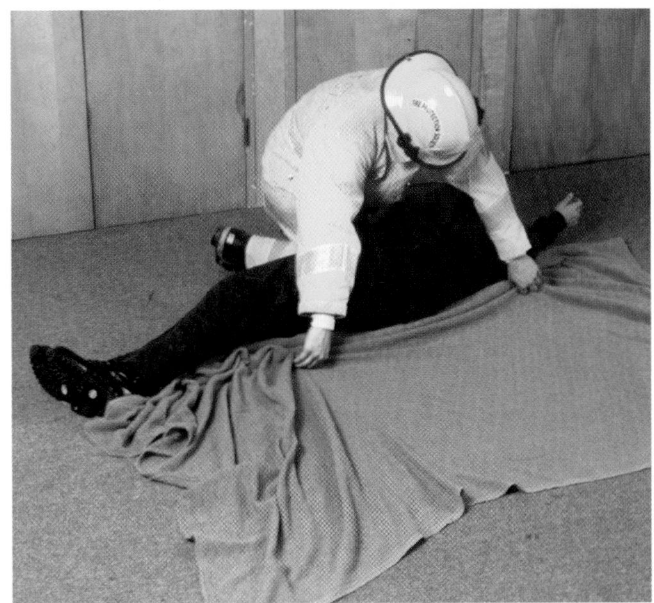

Step 5: Pull the blanket against the victim, gathering it slightly against the victim's back.

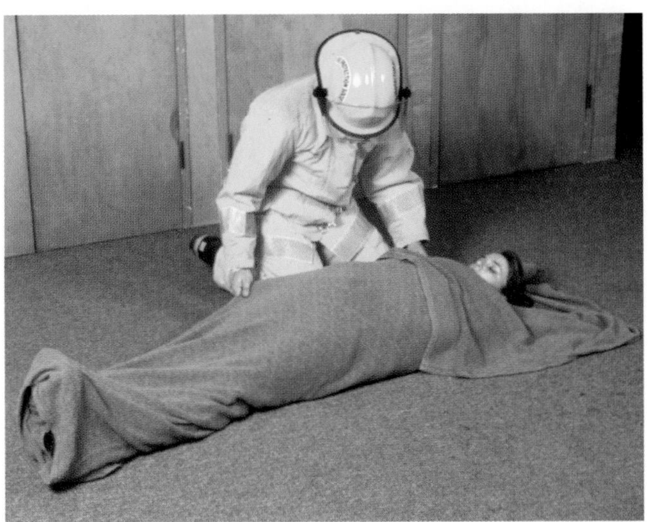

Step 6: Allow victim to roll gently onto the blanket.

Step 7: Straighten the blanket on both sides.

Step 8: Wrap the blanket around the victim.

Step 9: Tuck the lower ends around the victim's feet.

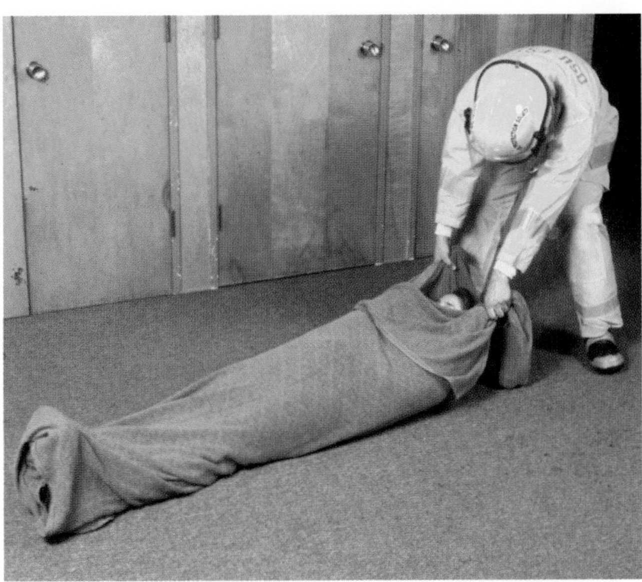

Step 10: Pull the end of the blanket at the victim's head.

Step 11: Drag the victim to safety.

Chapter 8
Forcible Entry

Chapter 8
Forcible Entry

INTRODUCTION

Modern society is security conscious. Private homes, commercial occupancies, and vehicles are all more heavily secured than in previous times (Figure 8.1). Firefighters must be able to get past security measures during fires, rescues, and sometimes even during odor investigations or alarm malfunctions. Forcible entry may be necessary to accomplish this task.

Forcible entry is the technique used by fire department personnel to gain access to a structure whose normal means of access is locked, blocked, or nonexistent. Forcible entry techniques, when properly used, do a minimal amount of damage to the structure or structural components and provide quick access for firefighters. Forcible entry should not be used when normal means of access are readily available. Additionally, forcible entry techniques may be required to open means of egress (exit) from structures.

Figure 8.1 Heavily secured occupancies make forcible entry difficult.

A knowledge of forcible entry techniques increases a firefighter's effectiveness. Knowing the construction features of doors, windows, and other barriers, knowing proper tool selection, and knowing forcible entry techniques greatly enhances a firefighter's effort on the fireground. The ability to use forcible entry techniques quickly and effectively also demonstrates firefighters' professionalism to the community they serve.

Forcible entry is a learned skill. It requires up-to-date knowledge of the construction features of the types of barriers that will be encountered. This includes doors, walls, floors, locks, padlocks, windows, and fences. Firefighters must remember that the purpose of these security devices is to keep out people. Forcible entry is not easy and must be practiced often. Selection of the appropriate tool or set of tools is imperative in forcible entry. A complete and thorough understanding of the basic types of tools used in forcible entry also ensures that the firefighter performing the task will be efficient and safe.

This chapter highlights the many tools that can be used for forcible entry operations. Their proper use, care, and maintenance are crucial to the success of the forcible entry operation. Characteristics of the various types of barriers that may have to be forced open, such as doors, floors, walls, fences, and windows, are also covered. Skill sheets are included to demonstrate actual forcible entry techniques. The opening of roofs is covered in Chapter 10, Ventilation.

FORCIBLE ENTRY TOOLS

[NFPA 1001: 3-3.3; 3-3.3(b); 3-3.7(a); 3-3.7(b); 3-3.10(b); 3-3.11(b); 3-3.12(a); 3-5.3; 3-5.3(a); 3-5.3(b); 4-3.2(a); 4-3.2(b); 4-4.1(b); 4-4.2(a); 4-4.2(b)]

Before any type of forcible entry technique can be discussed, a firefighter must have a complete working knowledge of the tools available to perform the task. Selection of the proper tool may make the difference in whether the barrier faced is successfully forced. This section begins by highlighting the various categories of tools used for forcible entry operations. Also included in this section is information on the proper use, care, and maintenance of tools, all of which are crucial to the success of the forcible entry operation.

Forcible entry tools can be divided into four basic categories:

- Cutting tools
- Prying tools
- Pushing/pulling tools
- Striking tools

Cutting Tools

There are many different types of cutting tools. These tools are often specific to the types of materials they can cut and how fast they can cut them. There is no such thing as a single cutting tool that will efficiently cut all materials. Using a cutting tool on materials for which it was not designed can destroy the tool and endanger the operator. Cutting tools may be either manual or powered. The following sections discuss the different types of cutting tools.

AXES AND HATCHETS

The axe is the most common type of cutting tool available in the fire service. There are two basic types of axe configurations in use today: *pick-head axe* and *flat-head axe* (Figure 8.2). Smaller axes and hatchets are also available to the firefighter, but often these tools are too lightweight for effective use during forcible entry operations. Smaller versions of either the pick-head or flat-head axe are fine for use in overhaul and salvage, but they are inefficient for forcible entry.

Pick-head axe. The pick-head axe comes with either a 6-pound head or an 8-pound head (3 kg or 3.6 kg). Handle sizes vary according to specifications, but they are made of either wood or

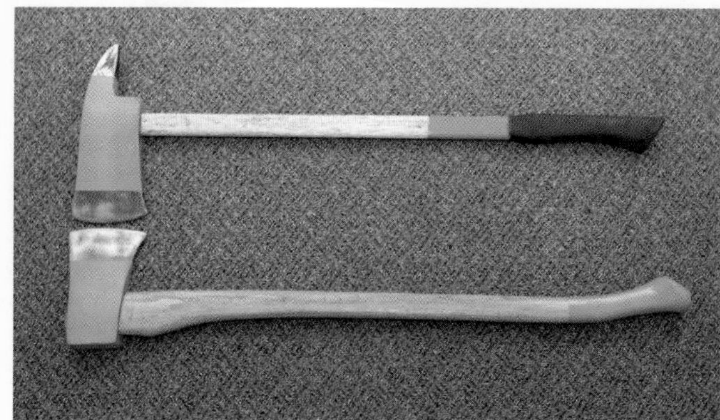

Figure 8.2 The pick-head axe and the flat-head axe are common cutting tools.

fiberglass. The tool is very effective for cutting through wood, shingles, and other natural and lightweight materials. The pick end serves to give the firefighter an opportunity to make a starting point to begin cutting or to pierce materials.

Flat-head axe. Like the pick-head axe, the flat-head axe comes in either 6- or 8-pound (3 kg or 3.6 kg) head weights with either a wood or a fiberglass handle. It also cuts through a variety of natural materials. When paired with a prying tool, the flat-head axe becomes a vital addition to the forcible entry team because the flat head can be used as a striking tool.

HANDSAWS

There are times when the handsaw is necessary because of a small work space. Handsaws that are commonly used by firefighters include the *carpenter's handsaw* (both rip cut and crosscut), *keyhole saws, hacksaws,* and *coping saws* (Figure 8.3). Handsaws are extremely slow. The knowl-

Figure 8.3 Various types of handsaws

edge of which saw is required, good handsaw maintenance, and practice in handsaw use will make a firefighter proficient when the handsaw is the tool of choice for the job.

POWER SAWS

Power saws are the "heavy hitters" of the fire service. These machines make fast and efficient cuts in a variety of materials. However, like any other tool in the toolbox, there are times when these saws should and should not be used. Power saws can be divided into several categories including *rotary (circular) saw, reciprocating saw, chain saw,* and *ventilation saw*.

CAUTION

- **Do not push a saw (or any tool) beyond the limits of its design and purpose; two things may occur: tool failure (including breakage) and/or injury to the operator.**
- **Never use a power saw in a flammable atmosphere. The saw's motor or sparks from the cutting operation can ignite a fire or cause an explosion.**
- **Always use eye protection when operating any power saw.**

Rotary (circular) saw. The fire service version of this device is most often gasoline powered and has changeable blades (Figure 8.4). The blades often spin more than 6,000 rpm. Blades range from large-toothed blades for quick rough cuts to fine teeth for a more precise cut. Carbide-tipped teeth are available and are far superior to standard blades because they are less prone to dulling with heavy use. Blades specifically designed for cutting metal are also available, and these are the types of blades most often used in forcible entry. There are many manufacturers of rotary-type saws. The firefighter should be familiar with the type purchased by the department. Following both manufacturers' recommendations and departmental operating procedures are imperative to maintaining a firefighter's personal safety when operating saws.

Reciprocating saw. The reciprocating saw is a very powerful, versatile, and highly controllable saw (Figure 8.5). It can use a variety of blades for cutting different materials. This saw has a short, straight blade that moves forward and backward with an action similar to that of a handsaw. Its major drawback, however, is that most all reciprocating saws require electricity, which may not be readily available on the fireground. Do not discard the idea of using this type of saw for only that reason. A reciprocating saw can be very beneficial in a number of forcible entry situations.

Chain saw. The chain saw has been used for years by the logging industry (Figure 8.6). This handy, wood-cutting saw has found a place in the fire service, especially during natural disasters, such as tornadoes and ice storms, when trees and limbs must be cleared from streets and access routes.

Ventilation saw. The ventilation saw is a relative

Figure 8.5 A reciprocating saw.

Figure 8.4 A rotary saw.

Figure 8.6 Typical chain saws.

newcomer to the fire service as a forcible entry tool (Figure 8.7). It is sometimes more efficient than the rotary saw. It is important that the ventilation saw be powerful enough to penetrate dense material yet lightweight enough to be easily handled in awkward positions. When equipped with a carbide-tipped chain, depth gauge, and kickback protection, the saw makes fast cuts through natural materials. It should not be used as a metal cutting saw. Lightweight and capable of being held at various angles, the ventilation saw should not be overlooked when considering certain forcible entry situations.

Figure 8.7 A ventilation saw with depth gauge. *Courtesy of Cutters Edge.*

METAL CUTTING DEVICES AND CUTTING TORCHES

Bolt cutters are metal cutting devices used in forcible entry to cut bolts, iron bars, pins, cables, hasps, chains, and some padlock shackles (Figure 8.8). The continual advancement in security technology is limiting the use of the bolt cutter as a viable entry tool. High-security chains, hasps, and padlock shackles cannot be cut with bolt cutters. These materials shatter the cutting surface of the bolt cutter or cause the handles to fail due to the tremendous pressures that must be exerted by the firefighter. Bolt cutters should not be used to cut case-hardened materials found in locks and other security devices.

Figure 8.8 Bolt cutters are excellent for cutting chains, iron bars, cables, and other materials that are not case-hardened.

In the instances where high-security devices are found, it may be necessary to use a cutting torch (Figure 8.9). The cutting torch operates by burning away the material being cut. Cutting torches use a mixture of flammable gases and generate a flame with a temperature of more than 5,700°F (3 149°C). The cutting torch cuts through almost all materials with ease; however, the use of a cutting torch is a very technical skill that requires training and much practice. Only firefighters well versed in its use and limitations should attempt to use a torch on the fireground. Specific manufacturers' recommendations for the torch must be followed as well as department operating procedures.

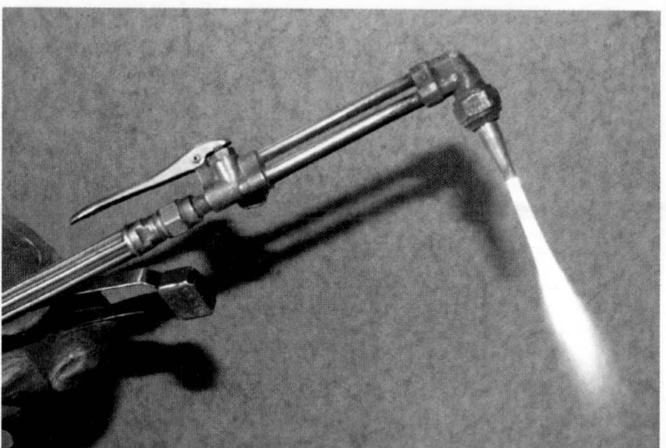

Figure 8.9 Cutting torches are effective for cutting metal that is too thick to be easily cut with saws.

Prying Tools

Prying tools provide an advantage to the firefighter for opening doors, windows, locks, and moving heavy objects. Hand (manual) prying tools use the basic principle of the lever to provide a mechanical advantage. This means that when properly using the prying tool, a firefighter is able to generate more force on an object with the tool than without it. Leverage applied incorrectly works against the firefighter. The correct tool must be selected first. If an object cannot be forced with one tool, a different tool should be selected.

Hydraulic prying tools can be either powered hydraulic or manual hydraulic. Powered hydraulic tools receive their power from hydraulic fluid pumped through special high-pressure hoses. Although there are a few pumps that are operated by compressed air, most are powered by either electric motors or by two- or four-cycle gasoline engines. Manual hydraulic tools operate slower than powered hydraulic tools, and they are labor-intensive. The hydraulic door opener is operated by transmitting pressure from a manual hydraulic pump through a hydraulic hose to a tool assembly.

MANUAL PRYING TOOLS

A large variety of hand (manual) prying tools is available to the fire service (Figure 8.10):

- Crowbar
- Halligan-type bar
- Pry (pinch) bar
- Hux bar
- Claw tool
- Kelly tool
- Pry axe
- Flat bar

Many departments have names for tools other than the names found in this manual. Firefighters should become familiar with the types and names of the tools carried on their apparatus. Firefighters need to be familiar with other important aspects of hand prying tools such as which surfaces may be used for striking, which are prying surfaces, etc. Efficiency in the use of a tool under emergency situations is directly affected by the firefighter's familiarity with the tool's functions. Some prying tools can also be used effectively as striking tools, although most cannot. For safe and efficient use of a tool, it should be used for its intended purpose.

HYDRAULIC PRYING TOOLS

Hydraulic prying tools that one person can operate have proven very effective in extrication rescues. They are also useful in forcible entry situations. These tools are useful for a variety of different operations involving prying, pushing, or pulling. The rescue tools and hydraulic door opener are examples of hydraulic prying tools.

Rescue tools. The hydraulic rescue spreader tool, most often associated with vehicle extrication, has some uses in forcible entry. Depending on the manufacturer, the tips on these tools can spread as much as 32 inches (813 mm). Their capability to exert force in either spreading or pulling makes them a valuable tool in some instances. The hydraulic ram is another hydraulic rescue tool. Although designed primarily for vehicle extrication, hydraulic rams have spreading capabilities ranging from 36 inches (900 mm) to an extended length of nearly 63 inches (1 600 mm). In certain forcible entry situations, these tools may be invaluable. One use is to place the ram in between either side of a door frame to spread the frame apart far enough to allow the door to swing open (Figure 8.11).

Figure 8.11 A hydraulic ram can be used to spring a door open.

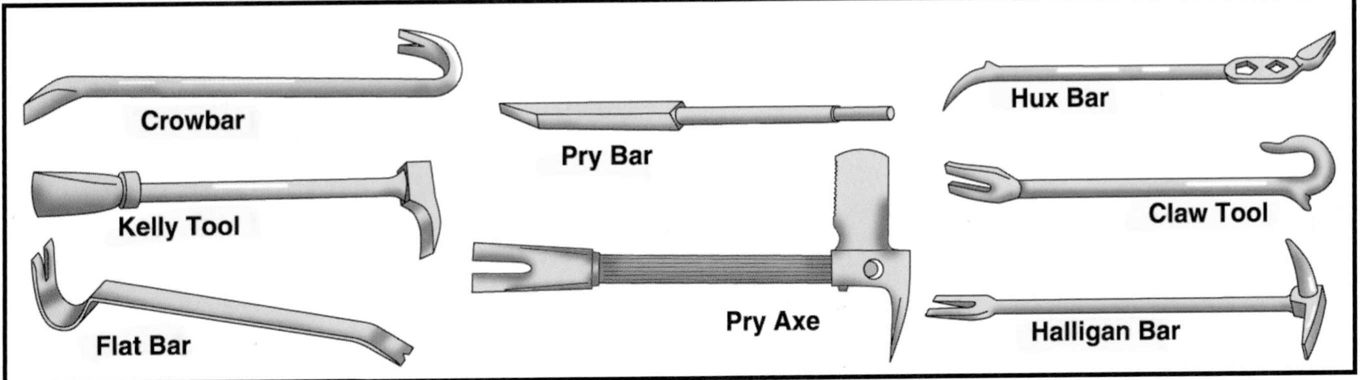

Figure 8.10 Various types of manual prying tools.

Hydraulic door opener. This hand-operated spreader device is relatively lightweight. It consists of a hand pump and spreader device (Figure 8.12). The spreader device has intermeshed teeth that can be easily slipped into a narrow opening such as between a door and door frame. A few pumps of the handle causes the jaws of the spreader device to open, exerting pressure on the object to be moved. The pressure usually causes the locking mechanism or door to fail. These are extremely valuable tools when more than one door must be forced such as in apartments or hotels. Although mobile and lightweight, these devices can place the firefighter in a dangerous position during their use if they are not used according to the manufacturers' recommendations.

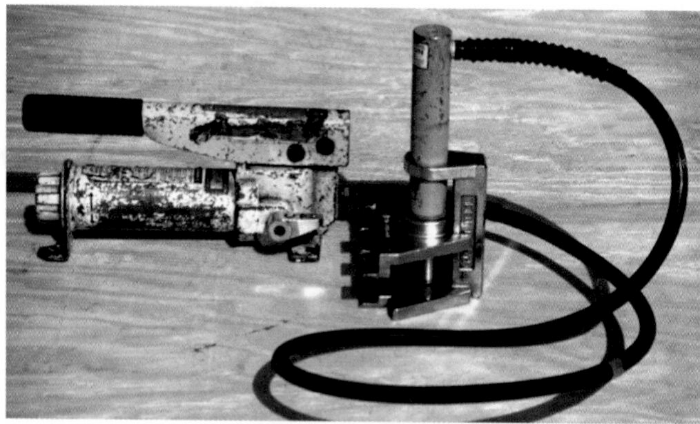

Figure 8.12 A hydraulic door opener is used to open doors that swing away from the firefighter.

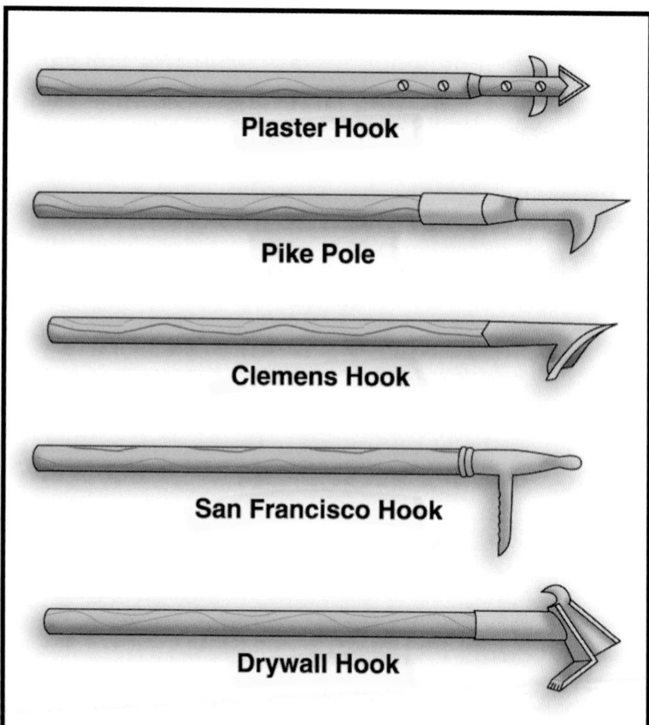

Figure 8.13a Various types of pushing/pulling tools.

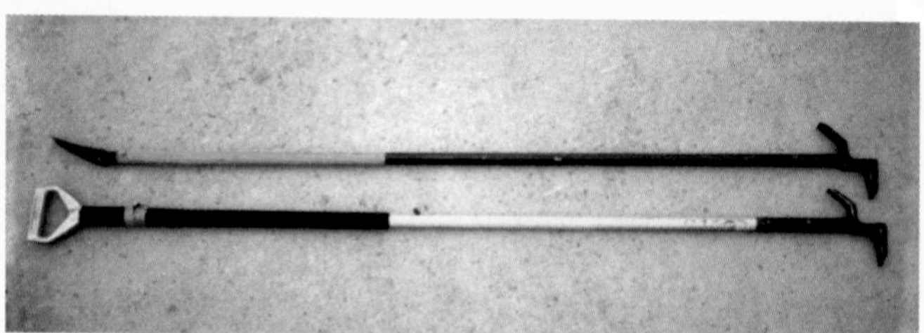

Figure 8.13b The roofman's hook (top) is an all-metal hook designed for heavy prying and pulling. The multipurpose hook (bottom) has a wood handle that cannot withstand the same type of use.

Pushing/Pulling Tools

Another category of tools available for forcible entry use is the push/pull category (Figures 8.13 a and b). These tools have limited use in forcible entry, but in certain instances, such as breaking glass and opening walls or ceilings, they are the tools of choice. This category of tools includes the following:

- Standard pike pole
- Clemens hook
- Plaster hook
- Drywall hook
- San Francisco hook
- Multipurpose hook
- Roofman's hook

Pike poles and hooks give the firefighter a reach advantage when performing certain tasks. By using a pike pole to break a window, the firefighter is able to stay out of the way of falling glass. The pike pole also allows the firefighter to remove shards and the window frame from a safer distance. The plaster hook has two knifelike wings that depress as the head is driven through an obstruction and reopen or spread outward under the pressure of self-contained springs (Figure 8.14). With the exception of the roofman's hook, which is all metal, pike poles and hooks should not be depended

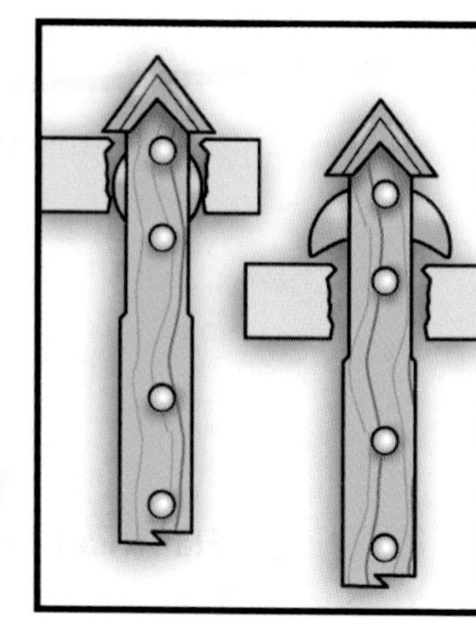

Figure 8.14 This illustrates the operation of a plaster hook.

on for leverage. Their strength is in pushing or pulling, not prying. If a lever is needed, select the appropriate prying tool. Handles of pike poles and hooks are easily broken by the application of inappropriate force.

Striking Tools

A striking tool is a very basic hand tool consisting of a weighted head attached to a handle (Figures 8.15 a and b). Some examples are as follows:

- Sledgehammer (8, 10, and 16 pounds [3.6 kg, 5 kg, and 7.3 kg])
- Maul
- Battering ram
- Pick
- Flat-head axe
- Mallet
- Hammer
- Punch
- Chisel

In certain instances, a striking tool is the only tool required. However, in most forcible entry situations, the striking tool is used in conjunction with another tool to effect entry. As common as they are, striking tools are dangerous when improperly used, carried, or maintained. Striking tools can crush fingers, toes, and other body parts. Improperly maintained striking surfaces may cause chips or splinters of metal to fly into the air. Proper eye protection must be used when using striking tools.

Figure 8.15b Various hammers, mallets, punches, and chisels should be included in the forcible entry tool kit.

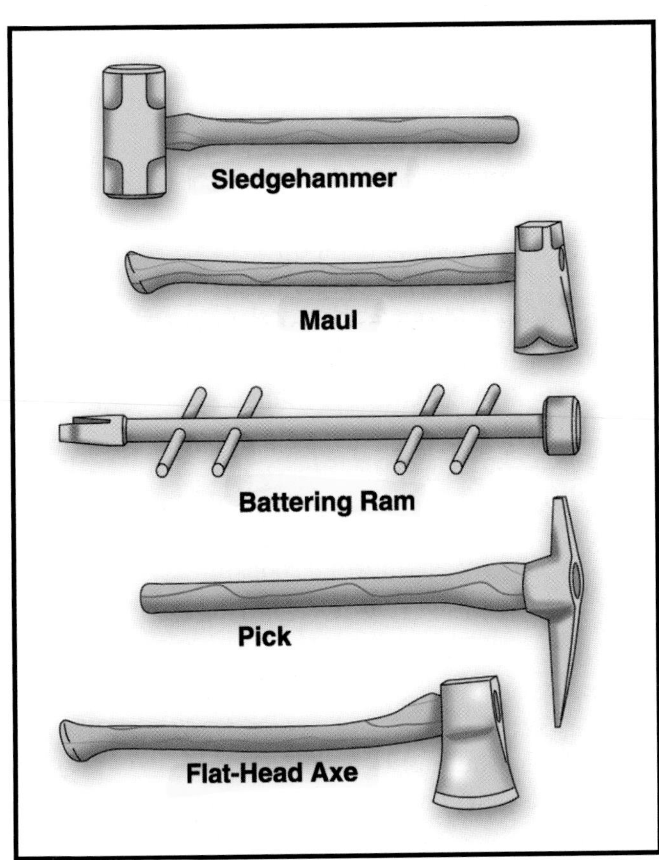

Figure 8.15a Various types of striking tools.

Tool Combinations

There is no single forcible entry tool that provides the firefighter with the needed force or leverage to handle all forcible entry situations. To effectively conduct forcible entry techniques, firefighters should choose combinations of tools to make a tool set. The types of tool sets carried vary, depending on building construction, security concerns, tool availability, and other factors within a fire department and the area served. The most important factor to consider is selecting the proper tools to do the job. Using tools for situations for which they are not designed is an extremely dangerous practice. Pre-incident surveys will help the firefighter determine what tools are required.

Tool Safety

Hand and power tools used in the fire service can be extremely dangerous if misused or used carelessly. Firefighters must become familiar with all the tools they will use, which includes reading and following all manufacturers' guidelines as well as individual department standard operating procedures on tool safety. In atmospheres that could be explosive, extreme caution should be taken in the use of power and hand tools that may cause arcs or sparks. When tools are not in use, they should be kept in properly designated places on the apparatus (Figure 8.16). Check the location of tools carried on the apparatus and make sure they are secured in their holders. The following sections contain information concerning prying tool safety, safety information particular to circular saws, and safety when using power saws in general.

Figure 8.16 When not in use, all tools should be stored in their proper places on the apparatus.

PRYING TOOL SAFETY

As with other tools, using prying tools incorrectly creates a safety hazard. For example, it is not acceptable to use a "cheater bar" or to strike the handle of a pry bar with other tools. A *cheater bar* is a piece of pipe added to a prying tool to lengthen the handle, thus providing additional leverage. Use of a cheater bar can put forces on the tool that are greater than the tool was designed to handle. This can cause serious injury if the tool slips, breaks, or shatters, and it can destroy the tool. If a job cannot be done with one tool, use another. Do not use a prying tool as a striking tool unless it has been designed for that purpose.

CIRCULAR SAW SAFETY

The circular saw must be used with extreme care to prevent injury from the high-speed rotary blade. Blades from different manufacturers may look alike, but may not be interchangeable. Always store blades in a clean, dry environment free of hydrocarbon fumes (such as gasoline). Blades should not be stored in any compartment where gasoline fumes accumulate (such as where spare saw fuel is kept) because the hydrocarbons will attack the bonding material in the blades and make them subject to sudden disintegration during use.

SAFETY WHEN USING POWER SAWS

Following a few simple safety rules when using any type of power saw will prevent most typical accidents.

- Match the saw to the task and the material to be cut. Never push a saw beyond its design limitations.
- Wear proper protective equipment always, including gloves and eye protection.
- Do not use any power saw when working in a flammable atmosphere or near flammable liquids.
- Keep unprotected and nonessential people out of the work area.
- Follow manufacturer's guidelines for proper saw operation.
- Keep blades and chains well sharpened. A dull saw is more likely to cause an accident than a sharp one.
- Be aware of hidden hazards such as electrical wires, gas lines, and water lines.

Carrying Tools

Firefighters must carry tools and tool combinations in the safest manner possible. Precautions should be taken to protect the carrier, other firefighters, and bystanders. Some recommended safety practices for carrying tools are as follows:

- ***Axes*** — Carry the axe with the blade away from the body. With pick-head axes, grasp the pick with a hand to cover it. Axes should never be carried on the shoulder (Figure 8.17).

Figure 8.17 Two methods of carrying an axe.

- ***Prying tools*** — Carry these tools with any pointed or sharp edges away from the body. With multiple surfaces, this will be somewhat difficult.
- ***Combinations of tools*** — Strap tool combinations together (Figure 8.18). Halligan-type bars and flat-head axes can be married together and strapped. Short sections of old hose can be slipped over the handles of some tools and smaller prying tools inserted into the hose.
- ***Pike poles and hooks*** — Carry these tools with the tool head down, close to the ground, and ahead of the body when outside a structure. When entering a building, carefully invert the tool and carry it with the head upright close to the body (Figures 8.19 a and b). These tools are especially dangerous because they can severely injure anyone accidentally poked with the working end of the tool.
- ***Striking tools*** — Keep the heads of these tools close to the ground. Maintain a firm grip. Mauls and sledgehammers are heavy and may slip.
- ***Power tools*** — Never carry a power tool that is running. Transport the tool to the area where the work will be performed and start it there. Running power tools are lethal weapons.

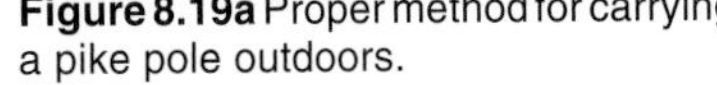

Figure 8.19a Proper method for carrying a pike pole outdoors.

Figure 8.19b Proper method for carrying a pike pole indoors.

Figure 8.18 Some tools such as the Halligan bar and the flat-head axe can be tied together for ease of carrying.

Care and Maintenance of Forcible Entry Tools

Proper care and maintenance of all forcible entry tools are essential ingredients of any forcible entry operation. Forcible entry tools will function as designed if they are properly maintained and kept in the best of condition. Tool failure on the fireground may have harsh consequences, including severe injury or death. Always read manufacturers' recommended maintenance guidelines for all tools, especially power tools. The following sections describe some basic maintenance procedures for various forcible entry tools.

WOOD HANDLES

- Inspect the handle for cracks, blisters, or splinters (Figure 8.20).
- Sand the handle to minimize hand injuries.
- Wash the handle with mild detergent, rinse, and wipe dry. Do not soak the handle in water because it will cause the wood to swell.
- Apply a coat of boiled linseed oil to the handle to prevent roughness and warping. Do not paint or varnish the handle.
- Check the tightness of the tool head.
- Limit tool marking (such as company identification, department name). A small stripe painted on the handle for identification is used by some departments.

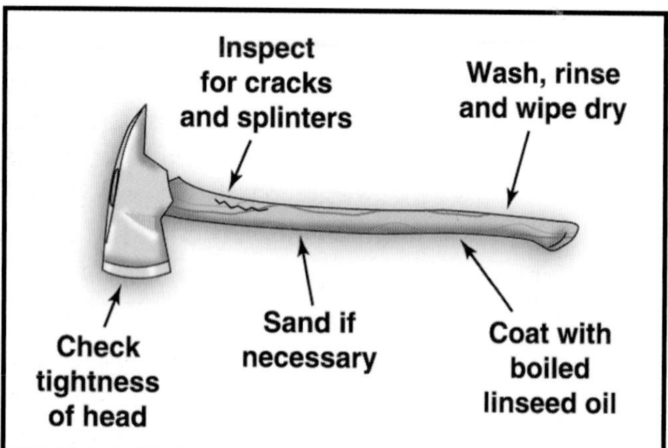

Figure 8.20 These areas should be inspected on wood handles.

FIBERGLASS HANDLES

- Wash the handle with mild detergent, rinse, and wipe dry.
- Check the tightness of the tool head.

CUTTING EDGES

- Inspect the cutting edge for nicks, tears, or metal spurs.
- Replace cutting edges when required.
- File the cutting edges by hand; grinding weakens the tool (Figure 8.21).

PLATED SURFACES

- Inspect for damage.
- Wipe plated surfaces clean, or wash with mild detergent and water.

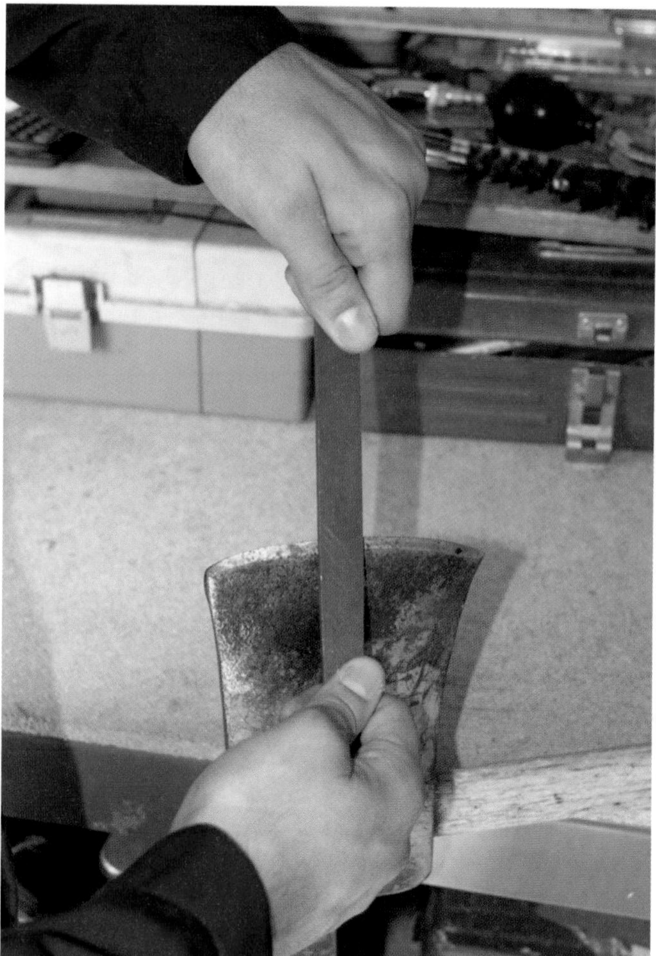

Figure 8.21 Cutting tools must be kept sharp. *Courtesy of Kyle Fortney.*

UNPROTECTED METAL SURFACES

- Keep free of rust.
- Oil the metal surface lightly. Light machine oil works best. Avoid using any metal protectant that contains 1-1-1-trichloroethane. This chemical may cause the material of the handle to decompose.
- Avoid painting. Paint hides defects.
- Inspect the metal for spurs, burrs, or sharp edges, and file them off when found.

AXE HEADS

The manner in which the axe head is maintained directly affects how well it works. If the blade is extremely sharp and its body is ground too thin, pieces of the blade may break when cutting gravel roofs or striking nails and other materials in flooring. If the body of the blade is too thick, regardless of its sharpness, it is difficult to drive the axe head through ordinary objects.

NOTE: DO NOT PAINT AXE HEADS! Painting hides faults in the metal. Paint also may cause the cutting surface to stick and bind.

POWER EQUIPMENT

- Read and follow manufacturers' instructions.
- Inspect and ensure power tools will start manually.
- Check blades for completeness and readiness.
- Replace blades that are worn.
- Check all electrical components (cords, etc.) for cuts and frays.
- Ensure that all guards are functional and in place.
- Ensure that fuel is fresh. A fuel mixture may deteriorate over time.

DOOR SIZE-UP AND CONSTRUCTION FEATURES

[NFPA 1001: 3-3.3; 3-3.3(a); 3-3.3(b)]

The primary obstacle firefighters face in gaining access to a building is a locked or blocked door. Forcible entry is required in these situations. Size-up of the door is an essential part of the forcible entry task. Recognizing how the door functions, how it is constructed, and how it is locked are critical issues to successful forcible entry. From a forcible entry standpoint, doors function in one of the following ways:

- Swinging (either inward or outward)
- Sliding
- Revolving
- Overhead

Regardless of the type of door, firefighters should try the door to make sure that it is locked before force is used (Figure 8.22). Remember, "Try before you pry!" If the door opens, there is no need for forcible entry. If it is locked, begin additional size-up. Look at the door. Which way does it swing? In? Out? Does it slide left or right? Does it roll up? An easy way to recognize which way a door swings is to look for the door hinges. If you can see the hinges of the door, it swings toward you. If you cannot see

Figure 8.22 Remember to always "Try before you pry."

the hinges, the door swings away from you. Access doors to residences usually swing inward. Commercial, public assembly doors and industrial doors, according to building codes, swing outward. This is not the case in all instances, however, and a firefighter must do a size-up to determine the swing of the door.

There will be times that even the best size-up and forcible entry effort will not be successful. It is important for the firefighter to remember not to get focused on one effort and one technique. If the door does not force using the technique chosen, choose another. If the tool chosen is inadequate, choose another. Spending too much time forcing a door is counterproductive. If the door proves to be too well secured, find another means of access.

After determining how a door functions, a firefighter must understand how the door is constructed. Doors range in construction types from interior hollow core to high-security steel. Locks are a problem, but the door is also an obstacle to quick and efficient forcible entry. Building supply stores carry various types of doors that are available to the consumer. The most common door firefighters encounter is the wood swinging door, followed next by the steel swinging door. Others likely to be found are sliding doors, revolving doors, overhead doors, and fire doors.

Wood Swinging Doors

There are three general categories of wood swinging doors: *panel*, *slab*, and *ledge*. Entry doors on structures are usually panel or slab.

The door is only one component of a door assembly. Doorjambs are the sides of the opening into which the door is fitted. Wood swinging doors may have either rabbeted or stopped jambs (Figure 8.23). The rabbeted jamb is a shoulder milled into the casing that the door closes against to form a seal. Whole door assemblies that are bought by contractors or do-it-yourselfers usually have rabbeted jambs. The stopped jamb has a piece of molding added to the door frame for the door to close on. Unlike the rabbeted jamb, a stopped jamb can be easily removed with prying tools, allowing firefighters easier access to the door-lock assembly.

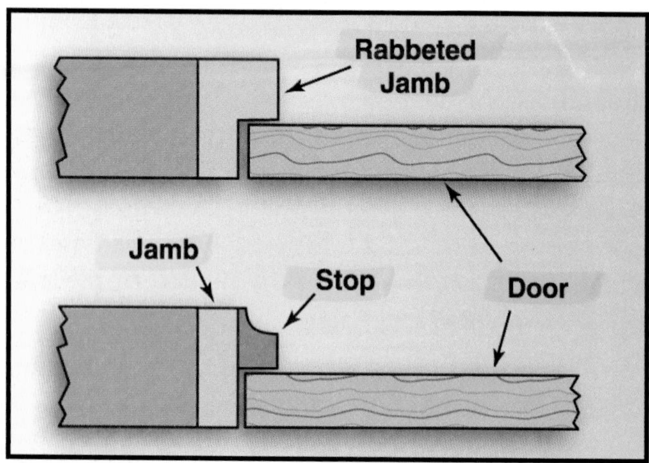

Figure 8.23 Typical rabbeted and stopped jambs.

PANEL DOORS

Wood panel doors are made of solid wood members inset with panels (Figure 8.24). The panels may be wood, plastic, or other such materials. Panel doors often have glass, Lexan® (polycarbonate) plastic, or Plexiglas® acrylic plastic panels fitted into the door to allow in light. These panels may be held in place by molding that can be removed for quick access.

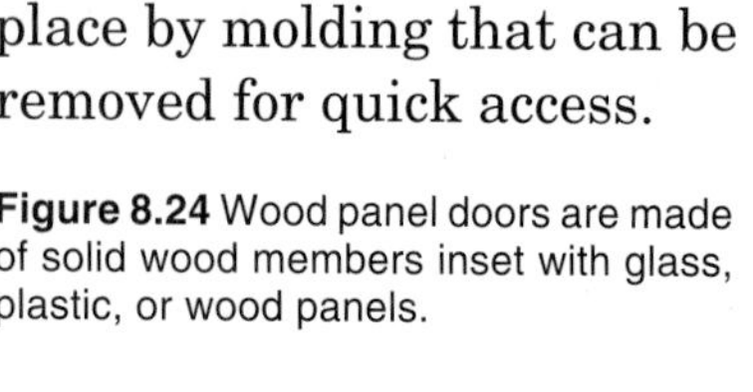

Figure 8.24 Wood panel doors are made of solid wood members inset with glass, plastic, or wood panels.

SLAB DOORS

The slab door, a very common door, is constructed in two ways: *solid core* and *hollow core* (Figure 8.25). Many interior doors in residences are hollow core. The name is misleading because it would lead you to believe that the entire core of the door is hollow, which is not true. The core or center portion of the door is made up of a web or grid of glued wood strips over which several layers of plywood veneer panels have been glued. The purposes of a hollow core door are to decrease its weight and lower its cost. Most exterior slab doors found on newly constructed residences are hollow core, but the exterior slab doors on older homes may be solid core. Slab doors are not pierced by windows or other openings. Panels on a slab door are purely decorative.

Solid core doors have a much more substantial construction than hollow core doors. The core of a solid core door is constructed of some type of solid material. In very old homes, the doors may be made of thick planks that have been tongue and grooved together. Modern solid core doors may be filled

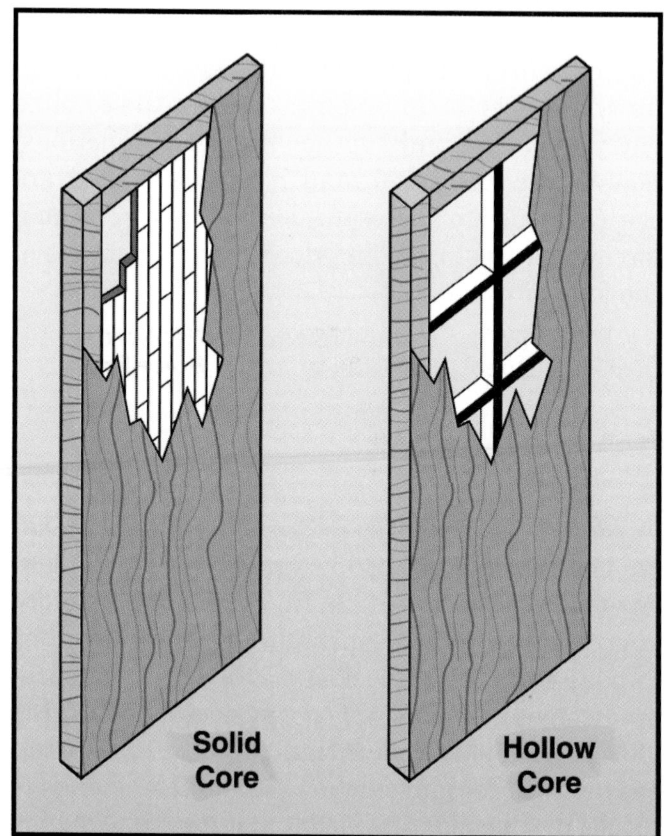

Figure 8.25 Doors may be of the solid or hollow core types.

with a material used for insulation or soundproofing. Other doors may be filled with a compressed mineral material for fire resistance. In either case, the solid core door is solid with a plywood veneer covering. The solid core door is much heavier and more expensive than the hollow core. In high crime areas, panel doors have been replaced with heavier solid core slab doors.

LEDGE DOORS

Ledge doors, also known as *batten doors,* are found on warehouses, storerooms, barns, sheds, and other structures (Figure 8.26). Although many are made commercially, firefighters will also find many homemade versions of this type of door. These doors are made of built-up materials, including boards, plywood sheeting, particleboard, etc. This type of door is generally locked with some type of surface lock, hasp, padlock, bolt, or bar. Hinges on this type of door are generally pin type, fastened with screws or bolts.

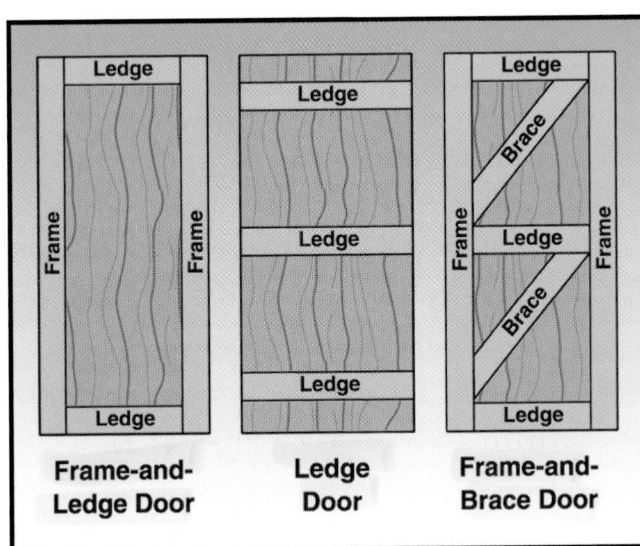

Figure 8.26 Types of ledge doors.

Metal Swinging Doors

Metal swinging doors are classified as *hollow metal, metal covered,* and *tubular.* Metal swinging doors are more difficult to force due to their construction and design. The metal door is most often set in a metal doorjamb (Figure 8.27). There is very little "spring" to the door. Add a few locks, and the metal door is a formidable enemy. It is generally considered impractical to force a metal door in a metal frame in masonry.

Metal doors vary greatly in their construction. Their end use designates their construction criteria. Metal-covered doors may have a solid wood door underneath the metal, or it may be a hollow metal door filled with fire-resistive materials. High-security metal doors are virtually impenetrable.

The structural design of tubular metal doors is of seamless rectangular tube sections (Figure 8.28). A groove is provided in the rectangular tube for glass or metal panels. The tube sections form a door with unbroken lines all in one piece. These doors are found on exterior openings of modern buildings. The tubular doors are hung with conventional hardware except that the balance principle of hanging is sometimes used. The operating hardware consists of an upper and a lower arm, each connected by a concealed pivot. The arms and pivots are visible from the exterior side only. From the interior side, the balanced door resembles any other door.

Figure 8.27 A typical metal swinging door.

Figure 8.28 A typical rectangular tubular metal door.

Tubular aluminum doors with narrow stiles are also quite commonly used. The panels of these doors are generally glass but some metal panels are used. Tubular aluminum doors are comparatively light in weight, are strong, and are not subject to much spring within the aluminum frame.

When faced with the need to force a metal door, firefighters should consider the use of power tools, especially rotary saws or hydraulic tools. Do not waste too much time trying to force the door. If the door will not open after a few tries, move on to another site. In some cases, it may be easier to breach the wall next to a steel door rather than try to force the door itself.

Sliding Doors

Sliding doors travel either left or right of their opening and in the same plane as the opening. This type of door generally is attached to a metal track by roller or guide wheels that make it easy to slide. These doors are often called *pocket doors* when used as an interior door. The door will slide into a "pocket" in a wall or partition, sliding out of sight (Figure 8.29).

Figure 8.29 A typical sliding pocket door.

The more common type of sliding door is the door assembly used in patio areas of residences or as doors to porches or balconies in houses, hotels, apartments, etc. Patio sliding glass doors usually slide either left or right of a stationary glass panel. The slider door is hung from guide wheels in a metal track. Usually, there is a lockable lightweight screened sliding door on the very outside of the assembly (Figure 8.30). The glass panels and sliding door are heavy glass window panels set in a metal or wood frame. These glass panels are normally double-thickness glass (Thermopane®), and newer doors may be triple-glass pane. Some door assemblies may have tempered (safety) glass (glass heat treated to increase its strength and flexibility), which make these doors very heavy and expensive.

Figure 8.30 Many residential occupancies have sliding doors that exit to a patio.

Patio sliding doors may sometimes be barred or blocked by a metal rod or a special device. These devices are commonly called *burglar blocks.* This feature can easily be seen from the outside, and it practically eliminates any possibility of forcing without causing excessive damage. If it is necessary to enter, the glass will have to be broken using the techniques described for tempered plate glass doors later in this chapter (see Tempered Plate Glass Doors section).

Revolving Doors

A *revolving door* is made up of quadrants (glass door panels) that revolve around a center shaft (Figure 8.31). The number of quadrants in the door varies with the manufacturer and how the door is used. The revolving door turns within a metal or glass housing assembly that is open on each side to allow users' entry and egress. The ends of the door panels are usually fitted with some type of large rubber weather stripping to help prevent the transfer of cold air into the building in winter or the loss of air conditioning in the summer.

Figure 8.31 A typical revolving door.

Revolving doors may be locked in various ways, and in general, they are considered difficult to force when locked. Usually, there are swinging doors on either side of the revolving door. It is more effective to force through the swinging door than trying to force open a locked revolving door (Figure 8.32).

Figure 8.32 It is more effective to force through a swinging door next to the revolving door.

All revolving doors are equipped with a mechanism that allows them to collapse during an emergency. A problem is that not all revolving doors collapse in the same way. Fire department preincident surveys must be conducted to locate revolving doors and to determine how their individual collapse mechanisms work. There are three basic types of mechanisms involved in making revolving doors collapse: *panic proof, drop arm,* and *metal braced*.

PANIC-PROOF TYPE

This mechanism has a ¼-inch (6 mm) cable holding the door quadrants apart. The collapse mechanism is triggered by forces pushing in opposite directions on the quadrants (Figure 8.33).

Figure 8.33 These firefighters are collapsing a panic-proof revolving door.

DROP-ARM TYPE

The drop-arm mechanism has a solid arm passing through one of the quadrants. A pawl is located on the quadrant the arm passes through. To collapse the system, press the pawl to disengage the arm, then push the quadrant to one side.

METAL-BRACED TYPE

This type mechanism resembles a gate hook and eye assembly. To collapse the mechanism, lift the hook and fasten it back against the fixed quadrant. Hooks are located on both sides of the quadrant. Generally, the pivots are cast iron and are easily broken by applying force to the quadrant at the pivot points.

Overhead Doors

Overhead doors have a wide variety of uses. They are generally constructed of wood, metal, or fiberglass. Overhead doors pose quite a forcible entry problem. These doors are heavily secured, sometimes motor driven, and are usually spring-loaded or balanced. Forcible entry may be difficult, but it is not impossible. Overhead doors are classified as follows: *sectional (folding), rolling steel,* and *slab.*

The sectional (folding) overhead door is not too difficult to force entry through unless it is either motor driven or remotely controlled. The latch mechanism is generally located in the center of the door. It controls two locks, one located on each side of the door. The lock and latch may also be located on only one side. These latches and locks are illustrated in Figure 8.34.

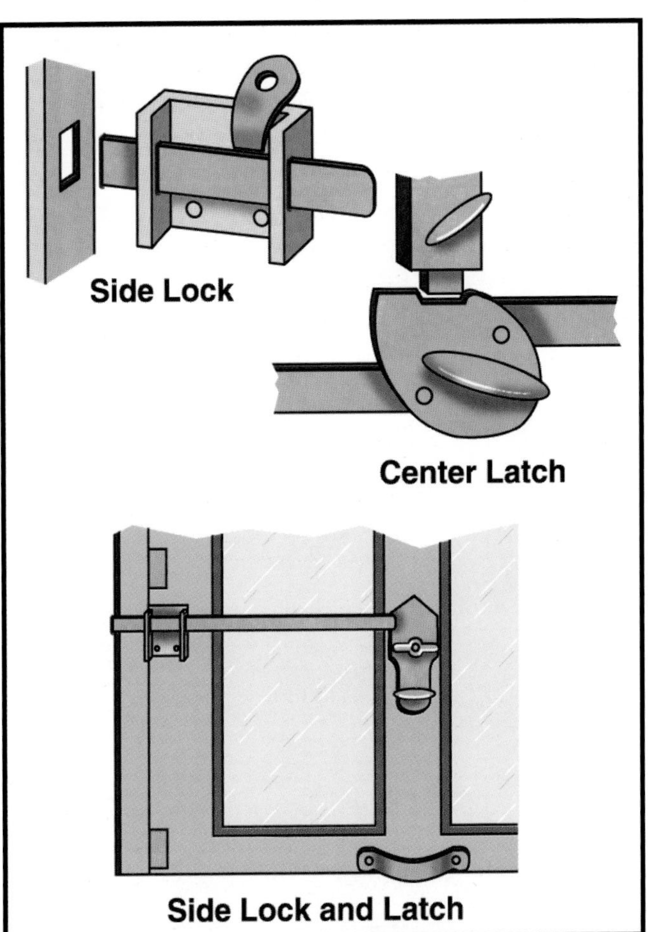

Figure 8.34 Types of overhead door latches.

Sectional overhead doors may be forced by prying upward at the bottom of the door with a good prying tool, but less damage will be done and time will be saved if a panel is removed and the latch is turned from the inside (Figure 8.35). Some overhead doors may be locked with a padlock through a hole at either end of the bar, or the padlock may even be in the track. These systems of locking may make it necessary to cut a hole in the door to gain access and remove the padlock.

Figure 8.36 A slab-type garage door. Note that during forcible-entry operations, this door might not be able to be opened all the way unless the automobile in front of it is moved.

Pivoting or overhead slab doors, sometimes called *awning doors,* are more difficult to force due to the nature of the door. Because the spring mechanism must pivot the door out and up, techniques must be used and care taken to not jam the door in its tracks or it will not open (Figure 8.36). Wood pivoting doors are very heavy. Pivoting or overhead slab doors are locked similarly to the sectional or folding door. Sometimes it is possible

Figure 8.35 After removing a panel or pane of glass, reach through to open the lock.

to pry outward with a bar at each side near the bottom (Figure 8.37). This action tends to bend the lock bar enough to pass the keeper.

Rolling steel doors, used as high-security doors, are designed to keep people out. They are normally locked with several padlocks and pins. The door can be manually operated, mechanically operated, or motor driven (Figure 8.38). If the door is motor or gear driven, it may be necessary to pull the manual-release chain or rope on the inside of the door. This release is generally hanging on either side of the door near the roller track. Rolling steel doors are among the toughest forcible entry challenges faced by firefighters. They are best accessed by cutting a triangle-shaped opening large enough for firefighters to crawl through. This can be done with a rotary rescue saw or a cutting torch.

Figure 8.37 Pry open a slab door by using a pry bar at each corner.

Figure 8.38 A rolling steel door.

CAUTION: All overhead doors should be blocked open (up position) to prevent injury to firefighters should the control device fail.

Fire Doors

Fire-door assemblies protect door openings in walls that are required to be rated as fire-barrier assemblies or fire walls. A fire-door assembly includes the door, frame, and associated hardware (Figure 8.39).

Types of standard fire doors include *horizontal and vertical sliding, single and double swinging,* and *overhead rolling* (Figures 8.40 and 8.41). They

Figure 8.39 Double swinging fire doors, common in the corridors of commercial buildings, automatically close when alarm systems are activated.

Figure 8.40 An overhead rolling steel fire door.

Figure 8.41 A horizontal sliding door in an elevator hoistway.

may or may not be counterbalanced. Counterbalanced doors are generally employed on openings to freight elevators, and they are mounted on the face of the wall inside the shaft (Figure 8.42). Fire doors may be mechanically, manually, or electrically operated.

There are two standard means by which fire doors operate: self-closing and automatic closing. When the *self-closing* type door is opened, it returns to the closed position on its own (Figure 8.43). *Automatic-closing* type doors, which normally remain open, close when the hold-open device releases the door upon activation of either a local smoke detector or a fire alarm system (Figure 8.44).

Figure 8.44 A magnetic hold-open device.

Figure 8.45 A typical swinging fire door.

Figure 8.42 A counterbalanced fire door.

Figure 8.43 Self-closing door mechanism.

Swinging fire doors are generally used on stair enclosures and in other areas where they must be opened and closed frequently in normal service (Figure 8.45). Vertical sliding fire doors are normally open and arranged to close automatically. They are employed where horizontal sliding or swinging fire doors cannot be used. Some vertical sliding models utilize telescoping sections that slide into position vertically on side-mounted tracks; the sections are operated by counterweights. Overhead rolling fire doors may be installed where space limitations prevent the installation of other types. Like vertical sliding doors, overhead rolling doors are arranged to close automatically.

Fire doors that slide horizontally are preferable to other types when floor space is limited. They operate on overhead tracks that are mounted in such a way that when a fusible link releases the door, a counterweight causes the door to move across the opening. Horizontal sliding fire doors also close automatically.

Overhead rolling doors have a barrel that is usually turned by a set of gears located near the top of the door on the inside of the building. This feature makes the door exceptionally difficult to force. Whenever possible, entrance to the building should be gained at some other point and the door operated from the inside.

Most interior fire doors do not lock when they close; they can be opened without using forcible entry techniques. Doors that are used on exterior openings may be locked; therefore, the lock must be forced.

A precautionary measure that firefighters should take when passing through an opening protected by a fire door is to block the door open to prevent its closing and trapping them. Fire doors have also been known to close behind firefighters and cut off the water supply in a hoseline.

LOCKS AND LOCKING DEVICES

[NFPA 1001: 3-3.3; 3-3.3(a)]

Locking devices vary from a simple lock to a series of very sophisticated locking devices. As a continuing part of the size-up procedure, the firefighter must have an understanding of the types of locks and locking devices that will be encountered during forcible entry. Although locks come in a variety of brand names, they can be divided into four basic types: *mortise lock, bored (cylindrical) lock, rim lock,* and *padlock.*

Mortise Lock

This lock mechanism is designed to fit into a cavity in the door (Figure 8.46). It usually consists of a latch mechanism and an opening device (doorknob, lever, etc.). Older mortise locks have just the latch to hold the door closed, while other mortise locks have a bolt or bar (tang). When the lock is in the locked position, the bolt protrudes from the lock into a keeper that is mortised into the jamb. Newer mortise locks may also have larger and longer dead-bolt features for added security. Mortise locks can be found on private residences, commercial buildings, and industrial buildings.

Bored (Cylindrical) Lock

Bored locks are so named because their installation involves boring two holes at right angles to one another: one through the face of the door to accommodate the main locking mechanism and the other in the edge of the door to receive the latch or bolt mechanism. One type of bored lock is the key-in-knob lock.

The *key-in-knob lock* has a keyway in the outside knob; the inside knob may contain either a keyway or a button (Figure 8.47). The button may be a push button or a push and turn button. Key-in-knob locks are equipped with a latch mechanism that is locked and unlocked by both the key and, if present, by the knob button. In the unlocked position, a turn of either knob retracts the spring-loaded beveled latch bolt, which is usually no longer than ¾ inch (19 mm). Because of the relatively short length of the latch, key-in-knob locks are some of the most vulnerable to prying operations. If the door and frame are pried far enough apart, the latch clears the strike and allows the door to swing open.

Figure 8.46 Mortise locks have not only latches but also dead bolts.

Figure 8.47 A key-in-knob lock has a keyway in the outside knob; the inside knob may contain either a keyway or a button.

Rim Lock

The rim lock is one of the most common locks in use today. It is best described as being surface-mounted and for this reason is used as an add-on lock for doors that already have other types of locks (Figure 8.48). This lock is found in all types of occupancies, including houses, apartments, and some commercial buildings. The rim lock can be identified from the outside by a cylinder that is recessed into the door in a bored latching mechanism fastened to the inside of the door and a strike mounted on the edge of the door frame.

Figure 8.48 The interlocking dead bolt is the most pry-resistant type of rim lock.

Padlock

Padlocks include portable or detachable locking devices (Figure 8.49). There are two basic types of padlocks: regular and heavy-duty. *Regular padlocks* have shackles of ¼ inch (6 mm) or less in diameter and are not case-hardened. *Heavy-duty padlocks* have shackles more than ¼ inch (6 mm) in diameter and are case-hardened. Many heavy-duty padlocks have what is called *toe and heel locking.* Both ends of the shackle are locked when depressed into the lock mechanism. These shackles will not pivot if one side of the shackle is cut. Both sides of the shackle must be cut in order to remove the lock.

Figure 8.49 Various types of regular and heavy-duty padlocks.

NONDESTRUCTIVE RAPID-ENTRY METHOD

The problem of gaining rapid entry without destruction has confronted fire departments for as long as locks have existed. In trying to find a solution, many departments have attempted to keep an inventory of keys to all the buildings in their areas. While this procedure does reduce damage from forcible entry, it also presents a problem of maintaining an inventory of keys and gaining quick access to the right key at the right time. The problems presented by locked doors can be eliminated through the use of a rapid-entry key box system (Figure 8.50). All necessary keys to the building, storage areas, gates, and elevators are kept in a key box mounted at a high-visibility location on the building's exterior. Only the fire department carries a master key that opens all boxes in its jurisdiction.

Figure 8.50 Exterior key boxes provide a means of nondestructive entry.

Proper mounting is the responsibility of the property owner. The fire department should indicate the desired location for mounting, inspect the completed installation, place the building keys inside, and lock the box with the department's master key. Unauthorized duplication of the master key is prevented because key blanks are not available to locksmiths and cannot be duplicated with conventional equipment.

CONVENTIONAL FORCIBLE ENTRY THROUGH DOORS

[NFPA 1001: 3-3.3; 3-3.3(a); 3-3.3(b); 3-3.10(b)]

Conventional forcible entry is the use of standard fire department tools to open doors and windows to gain access. Once a firefighter has made a size-up of a door, forcible entry, if needed, can be performed (Figure 8.51). In this section, various methods of opening doors are discussed. Forcible entry through windows is discussed later in the chapter (see Forcing Windows section). If there are

Figure 8.51 Firefighters conducting conventional forcible entry through a door.

no glass panels in the door to break and a door is definitely locked, the firefighter must force the door open. In conventional forcible entry, the best tool combination to use for a large variety of forcible entry situations is the 8-pound (3.6 kg) flathead axe and the Halligan-type bar.

Breaking Glass

The first technique of forcible entry is to break the glass near the door or in the door. Once the glass is broken, the firefighter reaches inside and operates the lock mechanism. During size-up of the door, take special note of the glass. It may be easier to break the glass, but will it cause more damage? Glass, especially tempered glass, is very expensive. Plexiglas® and Lexan® may also be found in and around doors for security and safety reasons. Many local building codes forbid the use of glass in areas where people may be injured by falling through them or having similar accidents.

Because glass will shatter into fragments with sharp cutting edges, the act of breaking glass must be done in a manner to ensure the safety of the firefighter (Figure 8.52). Wear full protective equipment, especially hand and eye protection. If breaking the glass to gain access into a fire building, SCBA should be worn and a charged hoseline should be in place, ready to attack the fire. The techniques used for breaking both door glass and window glass are similar. If breaking the glass is the most appropriate method of entry, *do it!* Use the techniques described in Skill Sheet 8-1 to make entry through the glass as safe as possible.

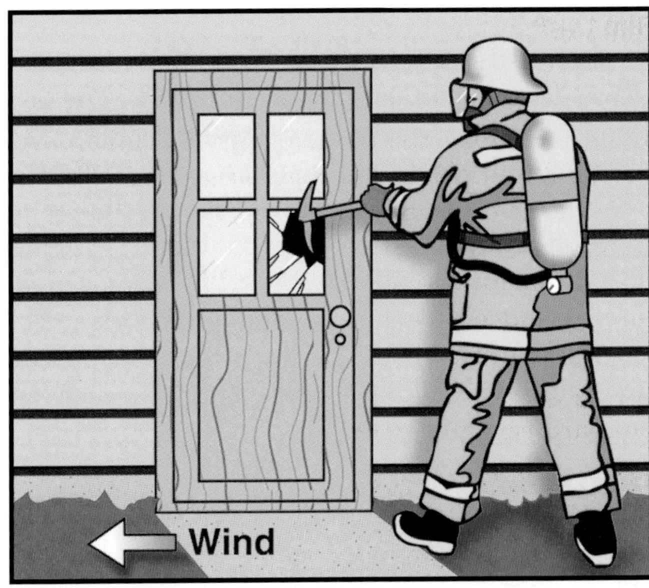

Figure 8.52 Stand to the windward side when breaking glass.

Forcing Swinging Doors

A common type of door is one that swings to open and close. Swinging doors have mounting hardware that permits them to pivot on one side of the opening. These doors can be either inward or outward swinging doors. Forcing entry through these types of doors are basic skills, but they require practice to master them.

INWARD SWINGING DOORS

Conventional forcible entry of inward swinging doors requires either one or two skilled firefighters (Figure 8.53). Skill Sheet 8-2 describes the technique for forcing locked inward swinging doors.

Figure 8.53 Firefighters forcing entry through an inward swinging door.

OUTWARD SWINGING DOORS

Outward swinging doors present a different set of problems for firefighters (Figure 8.54). The key issue in forcing an outward swinging door is to get a forcible entry tool into the space between the door and the doorjamb, open that space, and allow the lock bolt to slip from its keeper. Outward swinging doors, sometimes called *flush fitting doors,* can be forced using either the adze end or the fork end of the Halligan-type bar. Skill Sheet 8-3 describes the procedure for making forcible entry through an outward swinging door.

Figure 8.54 Firefighters forcing entry through an outward swinging door.

Special Circumstances

The basic techniques described earlier will work on most conventionally locked doors. There are circumstances where additional measures may need to be taken to force a door due to building construction features, door construction, or higher security. A few of the doors needing additional forcing measures are *double swinging doors, doors with drop bars,* and *tempered plate glass doors.*

DOUBLE SWINGING DOORS

These doors can present a problem depending on how they are secured (Figure 8.55). If they are secured only by a mortise lock, the doors can be pried apart far enough to let the bolt slip past the keeper. By inserting the adze end of a Halligan-type bar between the doors and pushing down and outward, the bolt should clear the keeper. Some sets of double doors have a security molding over the space between the two doors. This molding must be removed. On metal doors, this molding is steel and is very difficult to remove.

Figure 8.55 Typical double swinging doors.

DOORS WITH DROP BARS

Some double swinging and single entry doors may be secured by a drop-bar assembly (Figure 8.56). A *drop-bar assembly* is a bar, either wood or steel, that is dropped across the door and held in place by wood or metal stirrups. If this type of locked door must be entered, try one of the following methods:

- Insert a small narrow tool into the space between the double doors (if there is enough room), and try to lift the bar up and out of its stirrup.
- Cut a triangular hole into the door just below the bar. Reach in and push the bar up and out of the stirrup.
- Insert the blade of a rotary power saw into either the space between the jam and the door or between the doors in double doors, and cut the bar.

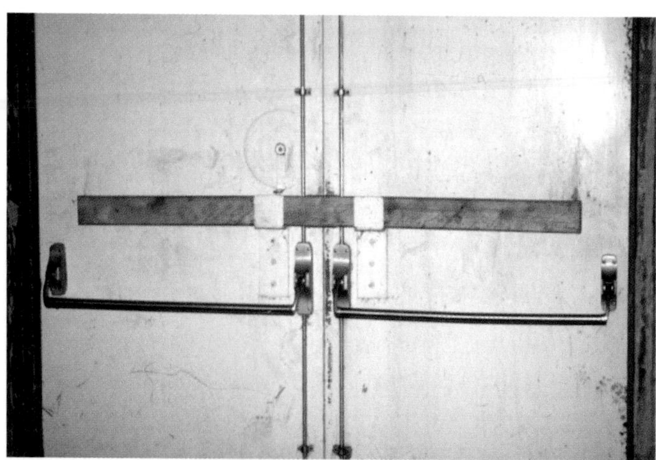

Figure 8.56 Sometimes doors are secured with metal or wood bars.

TEMPERED PLATE GLASS DOORS

In certain locations, especially in commercial stores, light industry, and institutional occupancies, firefighters may be faced with tempered plate glass doors (Figure 8.57). These doors are heavy and extremely expensive. The tempered glass that is mounted in the door frame is difficult to break. Unlike regular plate glass, tempered glass resists heat, and when broken, shatters into thousands of small cubelike pieces.

When it becomes necessary to break through a tempered plate glass door, the glass should be shattered at a bottom corner. To break the glass, use the pick end of a pick-head axe. The firefighter should wear a suitable faceshield to protect against eye injury, or turn away from the door as the glass is being broken. Some departments place a shield made from a salvage cover as close to the glass as possible, and the blow is struck through the cover (Figure 8.58). The remaining glass should then be removed from the frame.

Tempered plate glass doors should be broken only as a last resort for access. Another technique that firefighters can use to open tempered plate glass doors as well as other doors is the through-the-lock method (discussed in the next section).

Figure 8.57 Tempered plate glass doors typically are set in metal frames.

Figure 8.58 Before breaking the glass, cover the door with a salvage cover to protect the firefighters.

Through-the-Lock Forcible Entry

The through-the-lock method is the preferred method of entry for many commercial doors, residential security locks, padlocks, and high-security doors. This technique is very effective and does a minimal amount of damage to the door when performed correctly.

Through-the-lock forcible entry requires a good size-up of both the door and the lock mechanism. If the door and lock are suitable for conventional forcible entry, then it should be used. If the firefighter is unable to open the door through conventional forcible entry, the through-the-lock entry method should be used.

On many types of commercial doors, the lock cylinder can actually be unscrewed from the door. This is common on storefront doors because this technique is used by locksmiths to rekey locks after occupancy changes. If the lock is not protected by a collar or shield, use the procedure described in Skill Sheet 8-4.

Removing the lock cylinder is only half the battle. Through-the-lock forcible entry is really about operating the lock as though firefighters had the key to the lock. Once the lock cylinder is removed, firefighters use a key tool to operate the lock mechanism. The key tool is usually flat steel with a bent end on the cam end and a flat screwdriver shaped blade on the other end.

The through-the-lock method, like conventional forcible entry, requires patience and practice. Along with standard forcible entry striking and prying tools, special tools are also needed for this forcible entry technique. Some examples of these special tools are K-tool, A-tool, J-tool, and shove knife.

K-TOOL

The K-tool is useful in pulling all types of lock cylinders (rim, mortise, or tubular) (Figure 8.59). Used with a Halligan-type bar or other prying tool, the K-tool is forced behind the ring and face of the cylinder until the wedging blades take a bite into the cylinder (Figure 8.60). The front metal loop of the tool acts as a fulcrum for leverage and holds the fork end of the prying tool.

When a cylinder is found close to the threshold or jamb, the narrow blade side of the tool will usually fit behind the ring. A close-clearance situation will often be found on a sliding glass door, but only a ½-inch (13 mm) clearance is needed. Once the cylinder is removed, a key tool can be used in the hole to move the locking bolt to the open position. If the lock must be pulled by prying and the K-tool is capable of being slipped over the lock, use the technique described in Skill Sheet 8-5.

Figure 8.59 A key tool is designed to manipulate the interior locking mechanism after a lock cylinder is removed using the K-tool.

Figure 8.60 The K-tool jaws are designed to bite into a cylinder.

A-TOOL

The A-tool is a different tool that accomplishes the same job as the K-tool (Figure 8.61). The A-tool causes slightly more damage to the door than a K-tool, but it will rapidly pull the cylinder. Many locks are manufactured with collars or protective cone-shaped covers over them to prevent anyone from using a lock-pulling device. The A-tool was developed as a direct result of those lock design

Figure 8.61 The A-tool is designed for removing lock cylinders.

changes. The *A-tool* is a sharp notch with cutting edges machined into a prying tool. The notch resembles the letter *A*. This tool is designed to cut behind the protective collar of a lock cylinder and maintain a hold so the lock can be pried out.

The curved head and long handle are then used to provide the leverage for pulling the cylinder. The chisel head on the other end of the tool is used when necessary to gouge out the wood around the cylinder for a better bite of the working head. When pulling protected dead bolt lock cylinders and collared or tubular locks, use the A-tool and the procedure described in Skill Sheet 8-6.

J-TOOL

The *J-tool* is a wire-type device designed to fit through the space between double swinging doors equipped with panic hardware (Figure 8.62). By inserting the J-tool through the weather stripping between the doors, a firefighter can manipulate the panic bar. Panic bars operate with minimal pressure exerted on them.

Figure 8.62 The J-tool is used to manipulate the panic bar on double swinging doors with panic hardware.

SHOVE KNIFE

This flat steel tool is one of the oldest burglar tools, and it provides firefighters rapid access to outward swinging latch-type doors (Figure 8.63). This tool, when used properly, can slide a latch back past its keeper, allowing the door to open. It is an invaluable tool in locations where doors lead into smoke tower exit stairways that lock from the stairway side.

Figure 8.63 The shove knife is an excellent tool for use on outward swinging latch-type doors.

Forcible Entry Involving Padlocks

Padlocks are portable locking devices that are used to secure a door, window, or other access. Padlocks range from the very simple, easily broken type to the high-security, virtually impenetrable type. Firefighters must be capable of defeating either the padlock itself or the device to which it is fastened. Conventional forcible entry tools can be used to break padlocks and gain access. Additional tools are available to make forcible entry through padlocks easier. Some of these tools include the following (Figure 8.64):

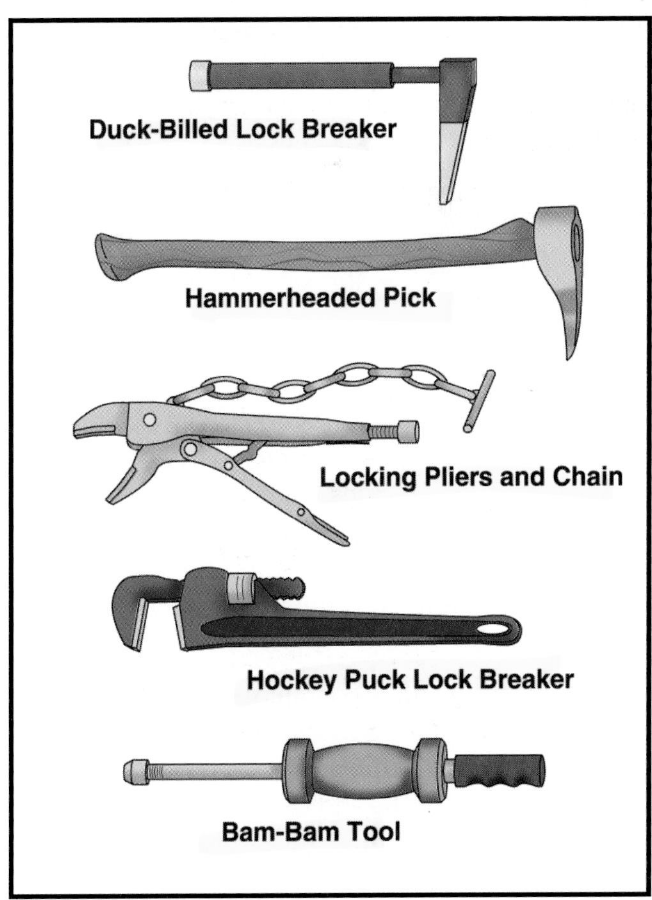

Figure 8.64 Various tools used for forcing entry through padlocks.

- Duck-billed lock breaker
- Hammerheaded pick
- Locking pliers and chain
- Hockey puck lock breaker
- Bam-bam tool

Size-up of the lock is important. If the lock is small with shackles of ¼ inch (6 mm) or less and not case-hardened, forcible entry can be made using the techniques described in Skill Sheets 8-7 through 8-9.

SPECIAL TOOLS AND TECHNIQUES FOR PADLOCKS

If the shackles of the padlock exceed ¼ inch (6 mm) and the lock, including body, is case-hardened, the firefighter faces a difficult forcible entry task. Conventional methods of forcing padlocks will not work effectively. Firefighters may need to select either the duck-billed lock breaker or the bam-bam tool (Figure 8.65).

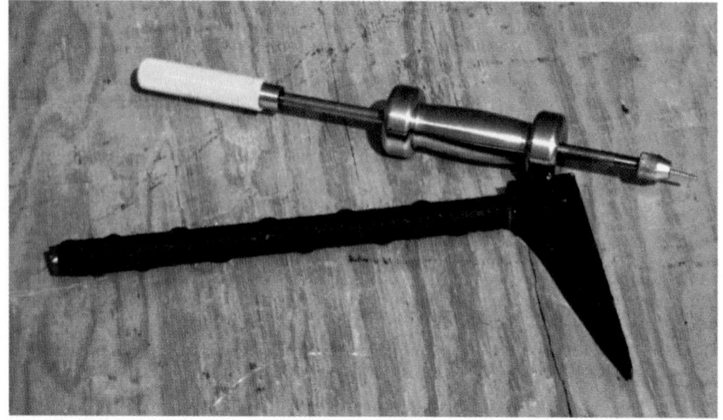

Figure 8.65 The duck-billed lock breaker and the bam-bam tool can be used to force entry through padlocks.

Duck-billed lock breaker. The *duck-billed lock breaker* is a wedge-shaped tool that will widen and break the shackles of padlocks, much like using the hook of a Halligan-type bar. This tool is inserted into the lock shackle and driven by a maul or flat-head axe until the padlock shackles break (Figure 8.66).

Bam-bam tool. This tool uses case-hardened screws that are driven into the actual keyway lock mechanism of the padlock. Once the screw is firmly set, a few hits with the sliding hammer will pull the lock tumbler out of the padlock body. The flat end of a key tool or a screwdriver can then be inserted to trip the lock mechanism (Figure 8.67).

Figure 8.66 Drive the duck-billed lock breaker into the shackle opening with a sledgehammer or flat-head axe.

Figure 8.67 Screw the case-hardened sheet metal screw of the bam-bam tool at least ¾ inch (19 mm) into the keyway, and maintain alignment to prevent the screw from breaking.

NOTE: This method will NOT work on Master Locks, American Locks, and other high-quality locks. These locks have a case-hardened retaining ring in the lock body that prevents the lock cylinder from being pulled out.

CUTTING PADLOCKS WITH SAWS OR CUTTING TORCHES

Using either a power rotary saw with a composite metal cutting blade or a cutting torch may be the quickest method of removing some padlocks. High-security padlocks are designed with heel and toe shackles. Heel and toe shackles will not pivot if only one side of the shackle is cut. Cutting padlocks with a power saw or torch is very dangerous work. Do not try to cut a loose padlock. Work with a partner. One firefighter should fasten a set of locking pliers and chain to the lock body. Pull the lock straight out from the staple. The second firefighter cuts both sides of the padlock shackle with the power saw or torch (Figure 8.68).

Figure 8.68 When cutting the shackle with a circular saw, a second person should hold tension on the padlock with locking pliers and chain.

FENCES

Property owners and occupants faced with a high risk of break-in often take measures beyond protecting their buildings with well-built and heavily locked doors and windows. One of these measures is to install fences, which present special problems to firefighters.

Fences can be made of wood, masonry, woven wire, or metal. They may be topped with barbed wire or razor wire. Fences may also be used to keep guard animals on the premises, so extreme care should be taken when entering into a fenced area.

Cutting metal fences with bolt cutters or removing wood boards are ways to gain access. Wire fences should be cut near posts to provide adequate space for fire apparatus and to lessen the danger of injury from the whip coil of loosened wires (Figure 8.69). Fence gates are often secured with padlocks or chains, so many of the techniques previously discussed will allow firefighters to gain access.

Using ladders to bridge fences, especially masonry fences, is another quick way of gaining access over a fence (Figure 8.70). Size-up is important in accessing areas through fences as well as in all aspects of forcible entry.

Figure 8.69 Wire fences should be cut near the posts.

Figure 8.70 An A-frame ladder can be used to bridge a fence.

FORCING WINDOWS

[NFPA 1001: 3-3.3; 3-3.3(a); 3-3.3(b); 3-3.10(b)]

Forcible entry can take place through windows, though they are not the preferred entry point into a fire building. Windows are sometimes easier to force than doors, and entry can be made to open a locked door from inside the structure. As with doors, size-up of windows is critical to a successful forced entry. Breaking the glass is the general technique, but often this slows entry into the structure while the glass and frame are being cleared. Breaking the glass of the wrong window may also intensify fire growth and draw fire to uninvolved sections of the building.

Breaking window glass on the fireground presents a multitude of hazards to both firefighters and civilians. Flying glass shards may travel great distances from windows on upper floors. Glass shards on floors, porches, etc., make movement for advancing hose teams or rescue crews difficult. Glass may shower victims inside the structure, causing additional harm. Wire glass requires great effort to break and remove because the wire prevents the glass from falling out of the frame. A sharp tool, such as the pick of an axe, hook of a Halligan-type bar, pike pole, or hook, is required to break this type of glass. Thermopane® windows or triple-glaze windows can cost the owner of the occupancy a large sum of money. Firefighters must determine if the benefits of breaking the window outweigh the damage that will be caused or will breaking the window cause more damage than necessary. Thermopane® windows will also slow firefighters because normally this glass is more difficult to break and is held into its sash by a butyl rubber glue, making shard removal difficult and time-consuming.

Windows come in a variety of types and sizes. The basic windows include double-hung (checkrail) windows, hinged (casement) windows, projected (factory) windows, and awning or jalousie windows. There are also various high-security windows and openings (Lexan®, barred, and screened).

Double-Hung (Checkrail) Windows

The *checkrail,* or more commonly *double-hung window,* has been an extremely popular window in building construction (Figure 8.71). Structures hundreds of years old are fitted with double-hung windows. Manufactured in either wood, metal, or vinyl clad, the window is made up of two sashes. The top and bottom sashes are fitted into the window frame and operate by sliding up or down. Often they are counterweighted for ease of movement. Newer double-hung windows, often referred to as *replacement windows,* not only move up and down, but tip inward for cleaning. Double-hung windows may contain ordinary glass (single-, double-, or triple-pane), Thermopane® glass, wire glass, or in certain circumstances, Plexiglas® acrylic plastic or Lexan® plastic.

Figure 8.71 A typical double-hung (checkrail) window.

Ordinarily, the double-hung window is secured by one or two thumb-operated locking devices located where the bottom of the top sash meets the top of the bottom sash (Figure 8.72). They may

Figure 8.72 Locking devices for a double-hung window.

also be more securely fastened by window bolts. Replacement windows have two side-bolt-type mechanisms located on each side of the sash that, when operated, allow the window sash to tip inward.

Forcing techniques for double-hung windows depend on how the window is locked and the sash frame material. The general technique for forcing a double-hung wood window is described in Skill Sheet 8-10.

NOTE: Metal windows are more difficult to pry. The lock mechanism will not pull out of the sash and may jam, creating further problems. Use the same technique given for a wood window, but if the lock does not give with a minimal amount of pressure, it may be quicker to break the window glass and open the lock manually.

In emergency situations where a window is the best means of access to a structure, valuable time can be saved and firefighter safety increased if the window glass is broken, the entire window area is cleared of all glass, and both top and bottom sashes are completely removed. This is especially true of metal double-hung windows. Glass and sash removal can be accomplished with an 8- or 10-foot (2.4 m or 3 m) pike pole or hook, allowing the firefighter to remain a safe distance from the window and falling glass (Figure 8.73). Removal of the sashes prevents any obstacles from snagging a firefighter's equipment or breathing apparatus.

Figure 8.73 The firefighter's hands should be higher than the point of impact to prevent shards of glass from falling onto them.

Hinged (Casement) Windows

Casement windows are hinged windows constructed of wood or metal. This type of window is often called a *crank out window,* but it should not be confused with an awning or jalousie window (see Awning and Jalousie Windows section). The casement window consists of two sashes mounted on side hinges that swing outward, away from the structure, when the window crank assembly is operated (Figure 8.74).

Figure 8.74 Casement windows may be found on all types of structures.

Locking devices vary for the casement window from simple thumb-operated devices to latch-type mechanisms. In addition to the locking devices, the casement window can only be opened by operating the crank mechanism. Casement windows are extremely difficult to force. Usually, they have at least four locking devices as well as two crank devices. This type of window is also very narrow and presents a more difficult entry for firefighters. If possible another means of entry should be sought. If it becomes necessary to force entry through a casement window, the most practical way is to use the following steps:

Step 1: Break the lowest pane of glass, and clean out the sharp edges.

Step 2: Force or cut the screen in the same area.

Step 3: Reach in and upward to unlock the latch.

Step 4: Operate the cranks or levers at the bottom.

Step 5: Completely remove the screen, and enter.

Projected (Factory) Windows

Projected windows are most often associated with factories, warehouses, and other commercial and industrial locations (Figure 8.75). These windows are most often metal sashes with wire glass. The most practical method of forcing factory-type windows is the same as that described for casement windows. Firefighters should not enter through projected windows unless it cannot be avoided. The metal frames and wire glass make it difficult to effectively accomplish rapid forcible entry. Often, these windows may have bars over the outside and inside to prevent entry. The best method of forcible entry for a projected window is to seek another entry point!

Factory windows often cover a large area, but the window openings themselves are very small. Factory windows are usually located several feet (meters) off the floor and present a very high risk to the firefighter. If entry must be made, consideration should be given to using a power saw or cutting torch to cut the frame of the window and enlarge the opening. These windows function by pivoting at either the top or bottom. They are classified by the way that they swing when opened: projected-in, projected-out, or pivoted-projected. If forcible entry must be made through a factory window, use the same procedures previously described for casement windows.

Figure 8.75 A projected (factory) window.

PROJECTED-IN

The bottom rail of the window swings into the occupancy toward the person who is opening it. The top rail of the window slides in a metal channel as the window is opened.

PROJECTED-OUT

The bottom rail of the window swings away from the building. The top rail slides into a metal channel as the window is opened.

PIVOTED-PROJECTED

Pivoted-projected windows are usually operated by a push bar that is notched to hold the window in place. Screens are seldom used with this type of window, but when present, they are on the side opposite the direction of the projection.

Awning and Jalousie Windows

Awning windows consist of large sections of glass about 1 foot (305 mm) wide and as long as the window width. They are constructed with a metal or wood frame around the glass panels, which are usually double-strength glass (Figure 8.76).

Figure 8.76 An awning window.

Jalousie windows consist of small sections about 4 inches (100 mm) wide and as long as the window width. They are usually constructed without frames, and the glass is heavy plate that has been ground to overlap when closed (Figure 8.77).

The glass sections of both awning and jalousie windows are supported on each end by a metal operating mechanism. This mechanical device may be exposed or concealed along the sides of the window. Each glass panel opens the same distance outward when the crank is turned. The operating crank and gear housing are located at the bottom of the window.

Awning or jalousie windows are the most difficult of all types to force. Even with the louvers open, it is obvious that there is not enough room between the louvers to permit a person to enter. Entrance through these windows requires the re-

Figure 8.77 A jalousie window.

moval of several panels. Because of the cost of jalousie windows and the danger to firefighters, these openings should be avoided unless absolutely necessary.

High-Security Windows

The world is becoming more and more security conscious, and the manufacturers of building supplies are striving to meet those consumer demands. Firefighters must be prepared when they encounter windows that are heavily secured. Some examples of high-security situations that are becoming common forcible entry problems are discussed in the following sections.

LEXAN® WINDOWS

In areas where window breakage is a recurring problem, firefighters may encounter window assemblies that have Plexiglas® or other acrylic thermoplastics in place of glass. Lexan® is one example of this replacement thermoplastic. Lexan® is 250 times stronger than safety glass, 30 times stronger than acrylic, and classified as self-extinguishing. Lexan® is virtually impossible to break with conventional forcible entry tools. The following are two recommended techniques for forcing entry through Lexan®.

- Cut the Lexan® using a rotary power saw with a carbide-tipped, medium-toothed blade (approximately 40 teeth). Large-toothed blades will skid off the surface, and smaller toothed blades will melt the Lexan® and cause the blade to bind.
- Discharge a carbon dioxide fire extinguisher on the Lexan® window, then immediately strike the Lexan® with the pick of a fire axe or other suitable sharp tool. The intense cold combined with the sharp blow will shatter the Lexan®.

BARRED OR SCREENED WINDOWS AND OPENINGS

Often, building owners will add metal bars or metal mesh screens over windows and sometimes door openings. Mesh guards may be permanently installed, hinged at the top or side, or fitted into brackets and locked securely (Figure 8.78). Regardless of how they are installed, forcing wire mesh guards involves considerable time and should be avoided.

Figure 8.78 Some standard windows are fitted with heavy mesh screening for additional security.

A more permanent security measure is to install heavy metal bars in the masonry above and below the window. These "burglar" bars vary in their types and construction, but their main feature is that they are difficult to force open (Figure 8.79). Some bars are attached directly to the building, while some are attached to the window frame. Forcible entry through burglar bars is a difficult and time-consuming task. Forcible entry considerations for burglar bar installations are as follows:

- Shear off the bolt heads for the mesh screen or bar assembly if they are visible and accessible using the flat-head axe and the

Figure 8.79 Several types of barred windows. *Courtesy of Edward Prendergast.*

Halligan-type bar. Place the axe blade behind the bolt head and strike it with the Halligan-type bar. Shear all the bolt heads off and remove the screen or bars.

- Cut the bar assembly or its attachments to the structure using a rotary power saw fitted with a metal cutting blade.
- Cut the bar assembly or screen from the building using an oxyacetylene torch.

Burglar bars and screen assemblies present a hazard to firefighters and occupants alike. Size-up of the bar or screen assembly is critical. Some assemblies are hinged and swing away from the window for cleaning or egress. Firefighters should look for locking devices that indicate that the burglar bars swing outward.

BREACHING WALLS

[NFPA 1001: 3-3.3; 3-3.3(a); 3-3.3(b); 3-3.12(b)]

During fire fighting operations, forcible entry situations may arise where it would be faster and more efficient to gain access through the wall of a structure rather than through a conventional opening, especially in buildings or other occupancies with high-security devices in place. Opening a hole in a wall to gain access is known as *breaching*. This action should be taken by experienced firefighters with a thorough knowledge of building construction and good size-up techniques.

A size-up of the situation must be conducted before any opening is made. Breaching load-bearing walls in a structure already weakened by fire can be a very dangerous task. The improper location of the breach or the removal of too many structural components could be disastrous. Walls conceal electrical wiring, plumbing, gas lines, and other components of the building utilities (Figure 8.80). The area selected for the breach must be clear of all these obstructions. The various techniques for breaching different types of walls are covered in the following sections.

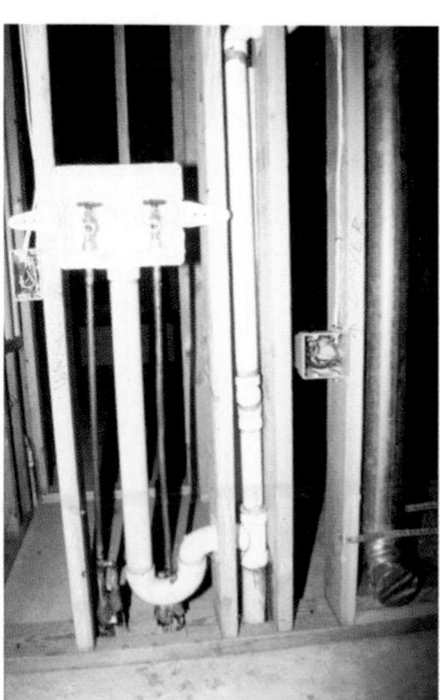

Figure 8.80 Avoid breaching walls containing obstructions such as plumbing, electrical wiring, and gas lines.

Plaster or Gypsum Partition Walls

Interior walls may or may not be load bearing, depending on the need of the wall to support the weight of overhead ceiling and roof spans. Fire resistance is another factor considered in partition walls. Fire-resistive partitions can be constructed from a wide variety of materials, including plaster and gypsum wallboard. Gypsum wallboard and plaster are relatively easy to penetrate with forcible entry tools. If an opening needs to be made in a partition, the following procedures can be followed:

Step 1: Select the location of the opening.

Step 2: Check the wall for electric wall plugs and switches.

Step 3: Have a wide variety of forcible entry tools available, including hand and power tools.

Step 4: Sound the wall to locate studs.

Step 5: Cut along the studs to make a large opening (at least three bays wide) (Figure 8.81).

Step 6: Remove one stud, if possible, from the center of the breach to enlarge the opening for firefighters to pass. **NOTE:** Use a hand or power saw to cut the stud level with the top of the opening.

Step 7: Use the breach to gain access to the area, and search to find the normal means of entry.

Figure 8.81 When making an opening in a partition, cut the wall covering along the studs.

Brick or Concrete Block Walls

Masonry walls can be the toughest type to breach. One appliance that may be used is the battering ram. The battering ram is made of iron with handles and hand guards. One end is jagged for breaking brick and stone, and the other end is rounded and smooth for battering walls and doors (Figure 8.82). The ram requires two to four firefighters to use. The firefighters work together to swing the ram back and forth into the wall (Figure 8.83). Each time the ram strikes the wall, a little more masonry material is knocked away.

Figure 8.82 This battering ram can be used to force holes in masonry walls.

Figure 8.83 At least two firefighters are required to breach a masonry wall.

Power tools such as air chisels, hydraulic spreaders, and rotary rescue saws with masonry blades prove to be the best methods for breaching masonry and concrete walls. They are faster and usually require only one person to operate. Use a power tool until a diamond- or triangular-shaped hole of desired size is formed. Open the breach large enough for firefighters to pass through. Use the breach to gain access to the area, and search to find the normal means of entry.

Metal Walls

Metal walls are found in many buildings (Figure 8.84). Prefabricated metal walls are common in both rural and urban settings, and they present difficult obstacles to firefighters when they must be breached.

Figure 8.84 Many new buildings have metal exterior walls.

Breaching a metal wall, like all other walls, should be a last resort. Size-up of the metal wall is critical. The walls are usually fastened to studs (which may be metal) by nails, rivets, bolts, screws, or other fasteners. Normal forcible entry hand tools are almost useless in this situation. If opening a metal wall cannot be avoided, a metal-cutting power saw is normally the best tool to use. Make sure no building utilities are located in the area selected for cutting. The metal should then be cut along the studding to provide stability for the saw and for ease of repair. After the metal is cut, it should be folded back out of the way where it will not endanger the firefighter.

If no studs can be located, it may be assumed that the metal wall bears the entire load of the structure. If that is so and the wall must be breached, cut a hole in the wall in the shape of a triangle. Make two cuts to form the triangle, and bend the metal outwards and fold it along the bottom or third side of the cut. Cutting a triangle in the wall distributes the wall's load more evenly and reduces the risk of collapse.

BREACHING FLOORS

[NFPA 1001: 3-3.9(b); 3-3.12(b)]

There are almost as many kinds of floors as there are buildings. The type of floor construction is, however, limited to the two basics: wood and concrete. Either of the two may be finished with a variety of floor-covering materials. Concrete slab floors over tested and rustproof plumbing are quite common. Generally speaking, the floors of upper stories of family dwellings are still wood joist with subfloor and finish construction.

It is not uncommon for a floor to be classified according to its covering instead of the material from which it is constructed. The feasibility of opening a floor during a fire fighting operation obviously depends upon how it was constructed and from what material. A wood floor does not in itself ensure that it can be penetrated easily. Many wood floors are laid over a concrete slab. The type of floor construction can be determined by pre-incident surveys of business and industrial structures, but similar information for residential structures is not easily obtained. Some accepted and recommended techniques for opening wood and concrete floors are offered in the following sections.

Wood Floors

The wood joists of wood floor construction are usually spaced a maximum of 16 inches (400 mm) apart. A subfloor consisting of either 1-inch boards or 4-foot by 8-foot sheets of plywood is first laid over the joists. The finish flooring, which may be linoleum, tile, hardwood, or carpeting, is laid last. The subfloor made of boards may be diagonal to the joists, and the finished floor may be at right angles to the joists. Plywood subflooring is generally laid at right angles to the joists; however, this practice varies widely. The procedure for opening a wood floor is shown in Skill Sheet 8-11.

Neat cuts in wood floors can be made with power saws. A wood-cutting blade can be provided for a circular saw, or a saber or chain saw may be used. It is better to supply power to electric saws from a portable generator carried on the fire apparatus than to depend upon domestic power during a fire. Carpets and rugs should also be removed or rolled to one side before a floor is cut.

Concrete/Reinforced Concrete Floors

The general construction of reinforced concrete floors makes them extremely difficult to force, and opening them should be bypassed if possible. If concrete floors must be opened, the most feasible means is to use a compressed-air or electric jackhammer (Figure 8.85). Unless a jackhammer is readily available, this process is extremely slow and may not prove beneficial for fire extinguishment; however, it might be the best means for rescue operations.

Concrete cutting blades are available for most portable power saws. There are also special-purpose nozzles that are designed to penetrate masonry and some concrete. Although these devices are primarily nozzles, they also qualify as forcible entry tools. They are sometimes called puncture or penetrating nozzles because of their ability to be driven into hard objects (Figure 8.86). It is best to first strike the masonry or concrete with a sledgehammer to shatter the concrete topping and provide a center for the tool. When wood or other materials are used as a finish over concrete, it often gives the appearance of a floor other than concrete.

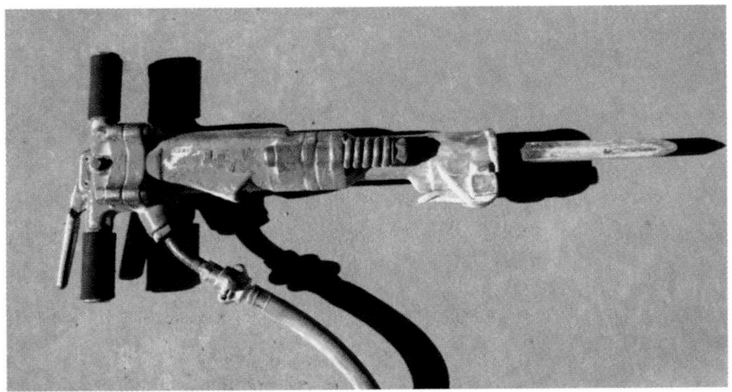

Figure 8.85 The jackhammer is a heavy-duty tool for breaking through masonry walls.

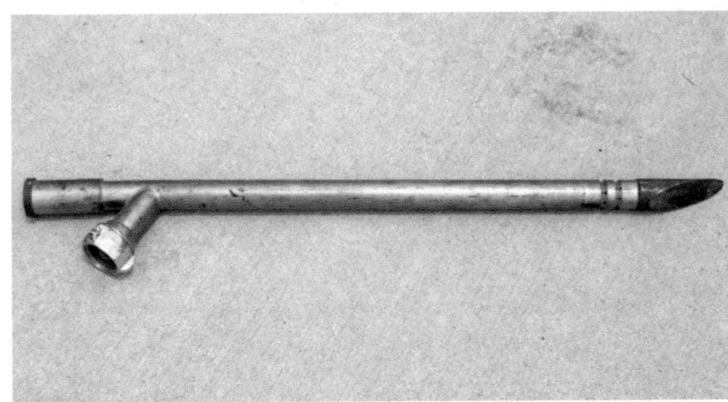

Figure 8.86 A piercing nozzle. *Courtesy of Superior Flamefighter Inc.*

SKILL SHEET 8-1 BREAKING DOOR GLASS

Step 1: Choose the most appropriate tool.

Step 2: Stand to the windward side of the glass panel or pane to be broken.

NOTE: *Windward* means to stand with the wind at the firefighter's back. Broken glass shards will move away from the body.

Step 3: Strike the glass as close to the top of the pane as possible.

NOTE: To avoid losing control of the tool, do not use excessive force.

Step 4: Keep hands above the point of impact or at an angle to the impact.

Step 5: Use the tool to clean all the broken glass out of the frame once the glass has been broken.

Step 6: Reach inside and find the door-lock mechanism.

NOTE: Be sure to use a gloved hand!

Step 7: Operate the lock.

Step 8: Open the door.

SKILL SHEET 8-2 CONVENTIONAL FORCIBLE ENTRY

Inward Swinging Door — Two Firefighters

Step 1: ***Firefighter #1:*** Place the fork of a Halligan-type bar just above or below the lock with the bevel side of the fork against the door.

Step 2: ***Firefighter #1:*** Angle the tool slightly up or down.

Step 3: ***Firefighter #2:*** Strike the tool with the back side of a flat-head axe.

NOTE: Strike the tool only when Firefighter #1 calls for the strike.

Step 4: ***Firefighter #2:*** Drive the forked end of the tool past the interior doorjamb.

Step 5: ***Firefighter #1:*** Move the bar slowly perpendicular to the door being forced to prevent the fork from penetrating the interior doorjamb.

NOTE: If unusual resistance is met, remove the bar, and turn it over. Begin again with the concave side of the fork now against the door.

Step 6: ***Firefighter #1:*** Make sure that the fork has penetrated between the door and the doorjamb.

Step 7: ***Firefighter #1:*** Exert pressure on the tool toward the door, forcing it open.

NOTE: If additional leverage is needed, Firefighter #2 can slide the head of the axe between the fork and the door.

CAUTION: The door may swing open uncontrollably when pressure is exerted on the Halligan-type bar. Maintain control of the door at all times. Placing locking pliers and chain or a utility rope on the doorknob will allow the forcible entry team to maintain control of the door.

SKILL SHEET 8-3 CONVENTIONAL FORCIBLE ENTRY

Outward Swinging Door — Adze End Method
Two Firefighters

Step 1: ***Firefighter #1:*** Place the adze of the Halligan-type bar just above or below the lock.

NOTE: If there are two locks, place the adze between the locks.

Step 2: ***Firefighter #2:*** Strike the tool using a flat-head axe on the surface behind the adze, driving the adze into the space between the door and the jamb.

NOTE: Strike the tool only when Firefighter #1 calls for it.

Step 3: ***Firefighter #1:*** Make sure the adze is sufficiently driven into the space.

Step 4: ***Firefighter #1:*** Pry down and out with the fork end of the tool.

SKILL SHEET 8-4 THROUGH-THE-LOCK FORCIBLE ENTRY

Unscrewing the Lock Cylinder

Step 1: Size-up the door and lock.

Step 2: Check the position of the keyway.

NOTE: The keyway is always in the 6 o'clock position.

Step 3: Place a set of locking pliers firmly on the lock cylinder.

NOTE: Make sure that the tool bites hard into the cylinder.

Step 4: Unscrew the lock cylinder from the door and remove it.

Step 5: Look inside the lock and identify the type of mechanism.

Step 6: Insert the appropriate key tool into the lock through the cylinder hole.

Step 7: Manipulate the locking mechanism.

Step 8: Open the door.

SKILL SHEET 8-5 THROUGH-THE-LOCK FORCIBLE ENTRY

Using the K-Tool

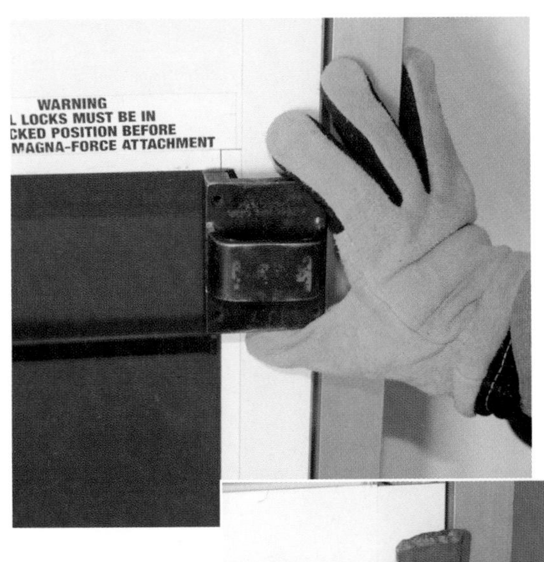

Step 1: Size-up the door and lock.

Step 2: Make sure the lock is not protected by a collar or shield.

Step 3: Check the position of the keyway.

NOTE: The keyway is always at the 6 o'clock position.

Step 4: Slide the K-tool over the lock cylinder face.

Step 5: Tap the K-tool down with a Halligan-type bar or the back of a flat-head axe.

Step 6: Insert the adze end of the pry tool into the strap on the top of the K-tool.

Step 7: Drive the K-tool further onto the cylinder.

NOTE: Make sure the K-tool has an adequate bite into the lock cylinder.

Step 8: Pry UP on the tool.

Step 9: Insert the key tool through the cylinder hole to manipulate the locking mechanism.

Step 10: Open the door.

SKILL SHEET 8-6 THROUGH-THE-LOCK FORCIBLE ENTRY

Using the A-Tool

Step 1: Size-up the door and lock.

Step 2: Check the position of the keyway.

NOTE: The keyway is always at the 6 o'clock position.

Step 3: Insert the opening of the A-tool between the lock cylinder and the door frame.

NOTE: The A-tool should be at an approximate 45-degree angle to the lock.

Step 4: Tap the A-tool firmly in place behind the lock cylinder.

NOTE: The firefighter may have to drive the A-tool into the frame of the door in order to get behind a tight lock.

Step 5: Pry up on the tool.

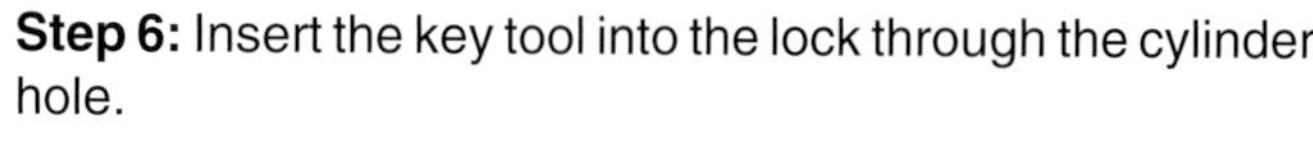

Step 6: Insert the key tool into the lock through the cylinder hole.

Step 7: Manipulate the locking mechanism.

Step 8: Open the door.

SKILL SHEET 8-7 CONVENTIONAL FORCIBLE ENTRY THROUGH PADLOCKS

Method One — Two Firefighters

Step 1: ***Firefighter #1:*** Insert the hook of a Halligan-type bar into the shackle of the lock.

Step 2: ***Firefighter #1:*** Pull the lock out away from the staple.

Step 3: ***Firefighter #2:*** Strike the Halligan-type bar sharply with a flat-head axe.

Step 4: ***Firefighter #2:*** Drive the hook of the bar through the lock shackle, breaking it.

SKILL SHEET 8-8 CONVENTIONAL FORCIBLE ENTRY THROUGH PADLOCKS

Method Two — Fork End

Step 1: Place the fork of the Halligan-type bar over the padlock shackles.

Step 2: Twist the lock until the shackles break.

NOTE: This will NOT work if the staple that the lock is attached to is weak. Twisting the padlock will result in a twisted hasp device that will require additional work to open. Use this technique only where the padlock is fastened to a high-security staple (one that will not twist or break easily).

SKILL SHEET 8-9 — CONVENTIONAL FORCIBLE ENTRY THROUGH PADLOCKS

Method Three — Bolt Cutters

Step 1: Cut the shackles of the padlock, the chain, or the staple with bolt cutters.

NOTE: Do NOT cut case-hardened metal with bolt cutters.

SKILL SHEET 8-10 — FORCIBLE ENTRY THROUGH A DOUBLE-HUNG WOOD WINDOW

Step 1: Insert the blade of an axe or a prying tool under the center of the bottom sash as much in line with the lock mechanism as possible.

Step 2: Pry upward to force the screws out of the lock.

Step 3: Open the window.

SKILL SHEET 8-11 BREACHING A WOOD FLOOR

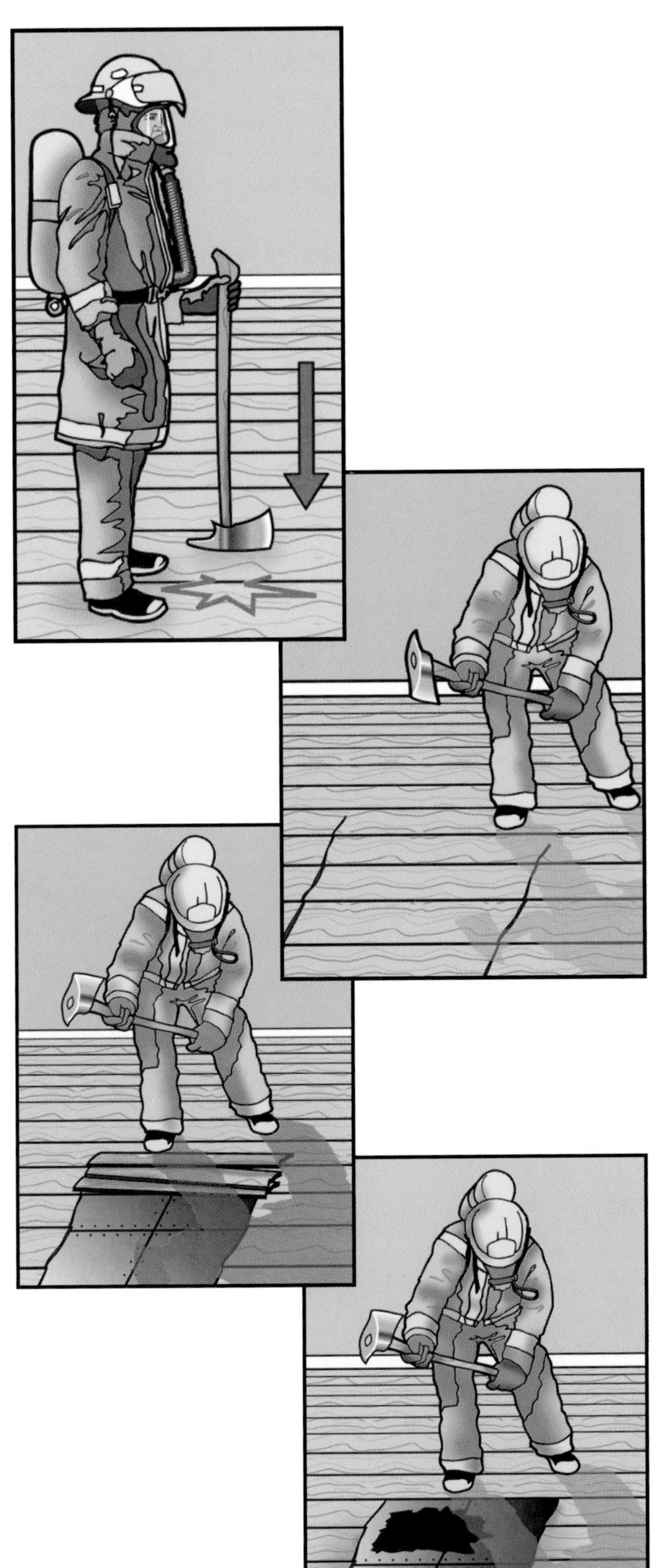

Step 1: Determine the approximate location for the hole based on need.

Step 2: Sound for floor joists to determine the exact location.

Step 3: Cut one side of the finished floor by using angle cuts.

Step 4: Cut the other side of the finished floor in like manner.

Step 5: Remove the flooring or floor coverings (including tile, linoleum, and carpet) with the pick of the axe.

Step 6: Cut the subfloor using the same technique and angle cuts.

NOTE: It is usually advisable to cut all sides of the subfloor before removing the boards. If just a few boards are removed before the others are cut, the heat and smoke conditions may prohibit completion of the job.

Chapter 9
Ground Ladders

Chapter 9
Ground Ladders

INTRODUCTION

Fire service ladders are essential in the performance of many fireground and rescue scene functions. From both the tactical and safety standpoints, it is crucial that firefighters be knowledgeable of the characteristics and proper uses of ground ladders. Fire service ladders are similar to any other ladder in shape and design; however, they tend to be built more rigidly and are capable of withstanding heavier loads than commercial ladders. Their use under adverse conditions requires that they provide a margin of safety not usually expected of commercial ladders. NFPA 1931, *Standard on Design of and Design Verification Tests for Fire Department Ground Ladders*, contains the requirements for the design and manufacturer's testing of ground ladders.

This chapter first introduces the reader to basic ladder parts and terms that are common to most ladders. Then, the various types of ladders in use by the fire service are reviewed. The chapter also details the proper care, carrying, deployment, and use of fire service ground ladders. For more information on the subjects covered in this chapter, see IFSTA's **Fire Service Ground Ladders** manual.

BASIC PARTS OF A LADDER

[NFPA 1001: 3-3.5 (a)]

In order to successfully continue this discussion of fire service ladders, the firefighter must have a basic understanding of the various parts of the ladder. Many of these terms apply to all types of ladders; others may be specific to a certain type of ladder.

- ***Beam*** — Main structural member of a ladder supporting the rungs or rung blocks (Figure 9.1)
- ***Bed section (base section)*** — Lowest or widest section of an extension ladder; this section always maintains contact with the ground or other supporting surface (Figure 9.2)

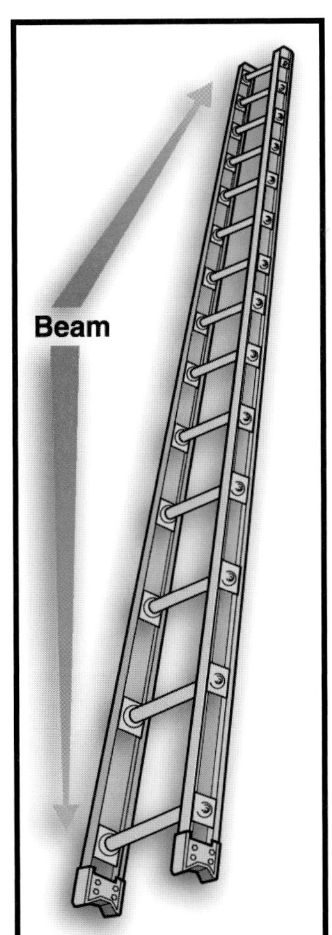

Figure 9.1 A ladder beam.

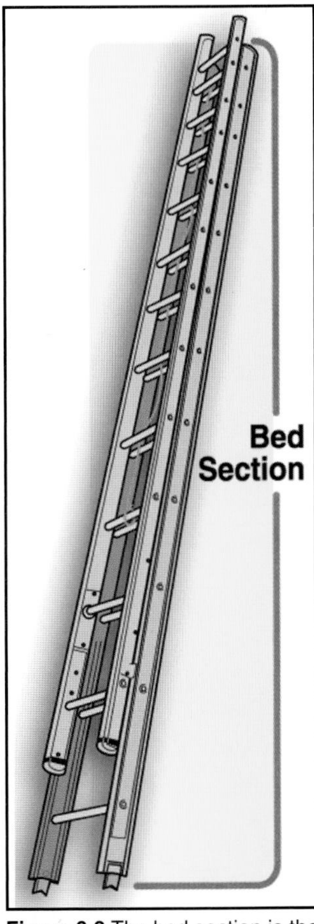

Figure 9.2 The bed section is the widest section of an extension ladder.

- ***Butt (also called heel)*** — Bottom end of the ladder; the end that is placed on the ground or other supporting surface when the ladder is raised (Figure 9.3)
- ***Butt spurs*** — Metal safety plates or spikes attached to the butt end of ground ladder beams to prevent slippage (Figure 9.4)
- ***Dogs*** — See pawls bullet

Figure 9.4 The butt spur helps keep the ladder anchored on soft surfaces.

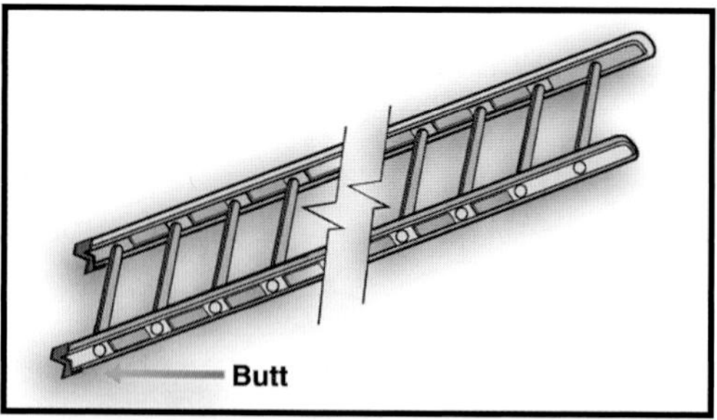

Figure 9.3 The bottom of the ladder is known as the butt.

- ***Fly*** — Upper section(s) of extension or some combination ladders (Figure 9.5)
- ***Footpads*** — Rubber or neoprene foot plates, usually of the swivel type, attached to the butt of the ladder (Figure 9.6)
- ***Guides*** — Wood or metal strips, sometimes in the form of slots or channels, on an extension ladder that guide the fly section while being raised

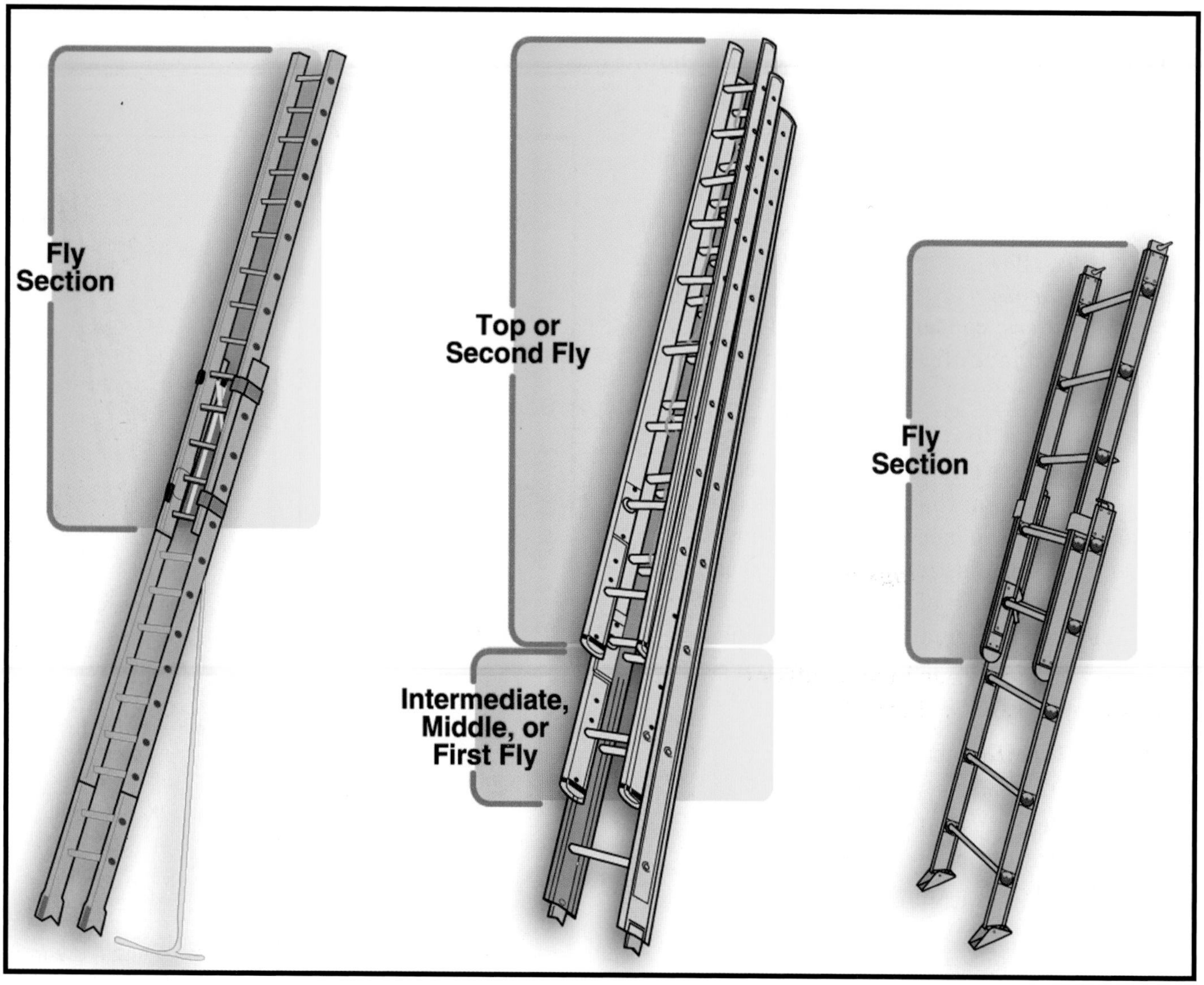

Figure 9.5 Fly sections of extension and combination ladders.

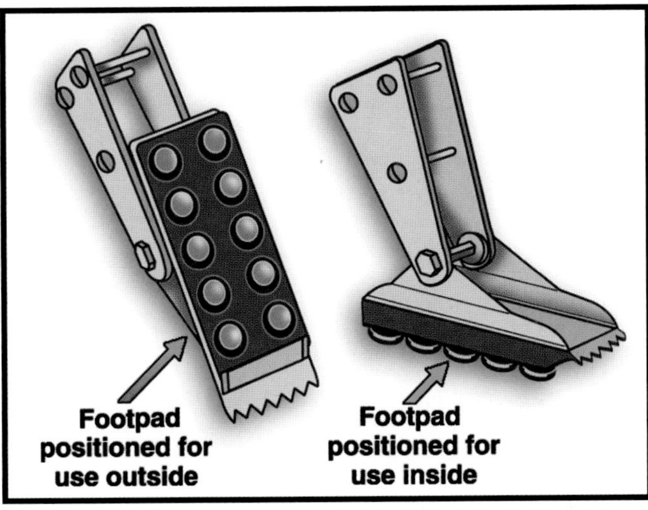

Figure 9.6 Some footpads may have steel toes that may be used on a soft surface.

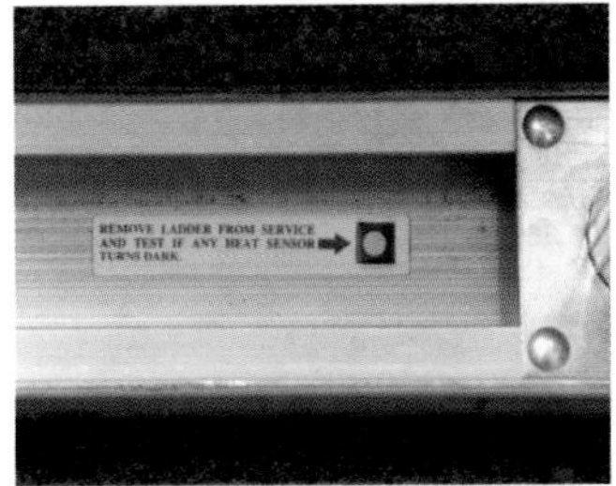

Figure 9.8 A heat sensor label includes directions on how to read it.

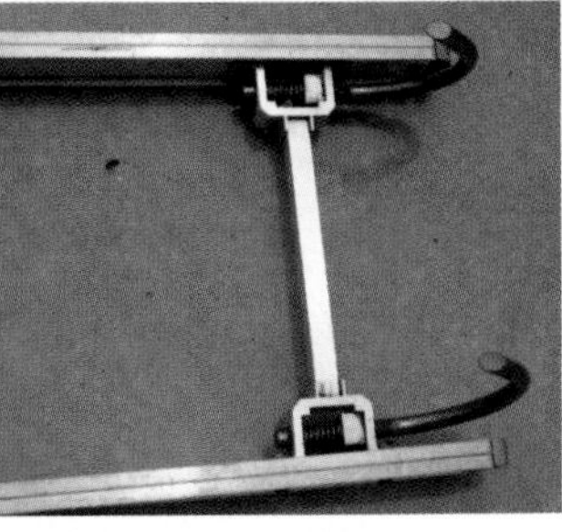

Figure 9.9 A roof ladder with hooks open.

- ***Halyard*** — Rope or cable used for hoisting and lowering the fly sections of an extension ladder; also called fly rope (Figure 9.7)
- ***Heat sensor label*** — Label affixed to the inside of each beam of each ladder section; a color change indicates that the ladder has been exposed to a sufficient degree of heat that it should be tested before further use (Figure 9.8)
- ***Hooks*** — Curved metal devices installed on the tip end of roof ladders to secure the ladder to the highest point on the roof of a building (Figure 9.9)
- ***Locks*** — See pawls bullet
- ***Pawls (also called dogs or ladder locks)***—Devices attached to the inside of the beams on fly sections used to hold the fly section in place after it has been extended (Figure 9.10)
- ***Protection plates*** — Strips of metal attached to ladders at chafing points, such as the tip, or at areas where it comes in contact with the apparatus mounting brackets (Figure 9.11)
- ***Pulley*** — Small, grooved wheel through which the halyard is drawn on an extension ladder (Figure 9.12)

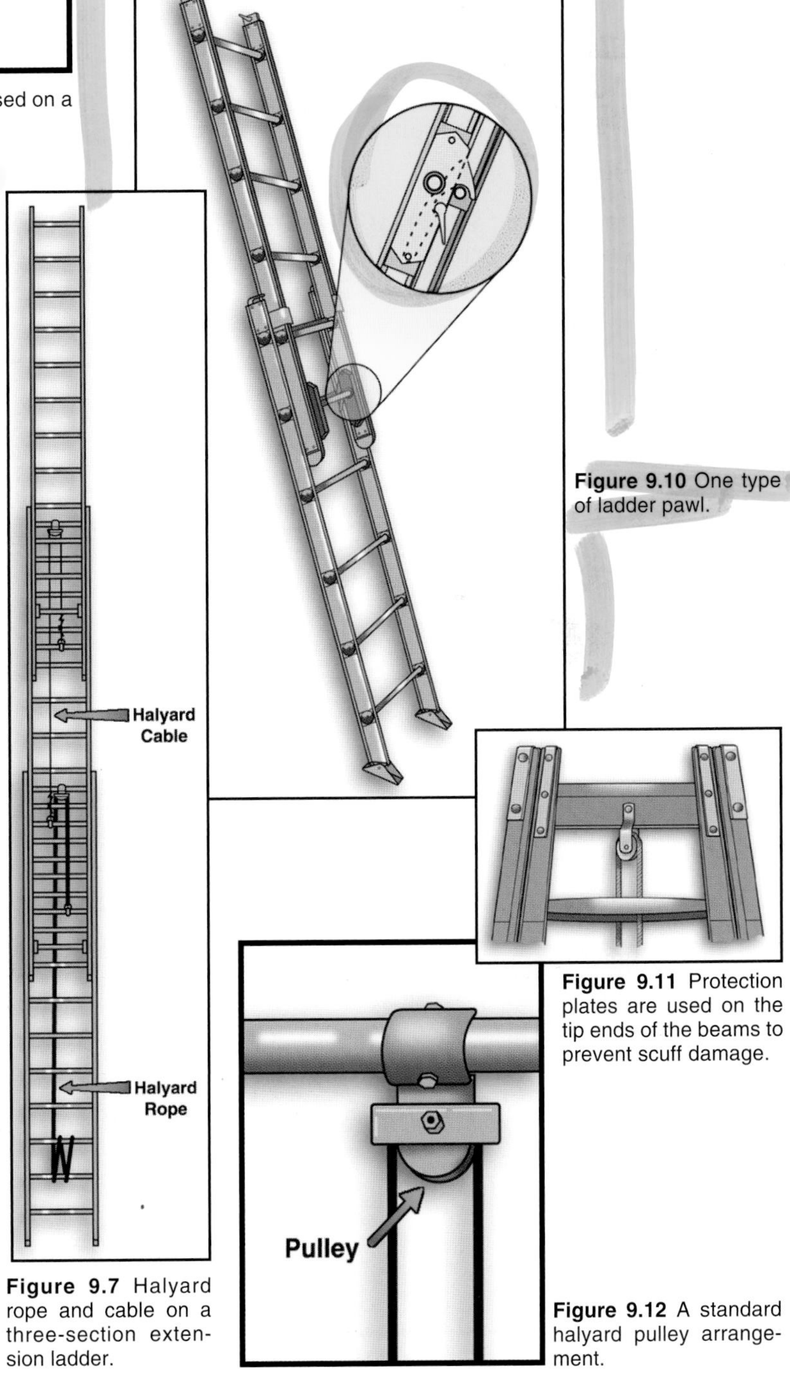

Figure 9.7 Halyard rope and cable on a three-section extension ladder.

Figure 9.10 One type of ladder pawl.

Figure 9.11 Protection plates are used on the tip ends of the beams to prevent scuff damage.

Figure 9.12 A standard halyard pulley arrangement.

- ***Rails*** — The two lengthwise members of a trussed ladder beam that are separated by truss or separation blocks (Figure 9.13)
- ***Rungs*** — Cross members that provide the foothold for climbing; the rungs extend from one beam to the other except on a pompier ladder the rungs pierce the single beam (Figure 9.14)
- ***Stops*** — Wood or metal pieces that prevent the fly section from being extended too far (Figure 9.15)
- ***Tie rods*** — Metal rods running from one beam to the other (Figure 9.16)
- ***Tip (top)*** — Extreme top of a ladder (Figure 9.17)
- ***Truss block*** — Separation pieces between the rails of a trussed ladder; sometimes used to support rungs (Figure 9.18)

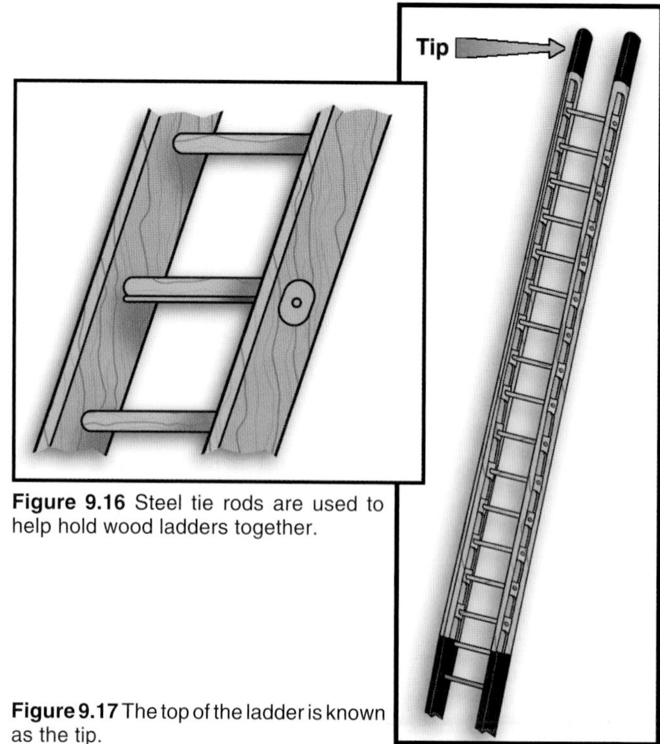

Figure 9.16 Steel tie rods are used to help hold wood ladders together.

Figure 9.17 The top of the ladder is known as the tip.

Figure 9.13 The rails of a trussed ladder beam.

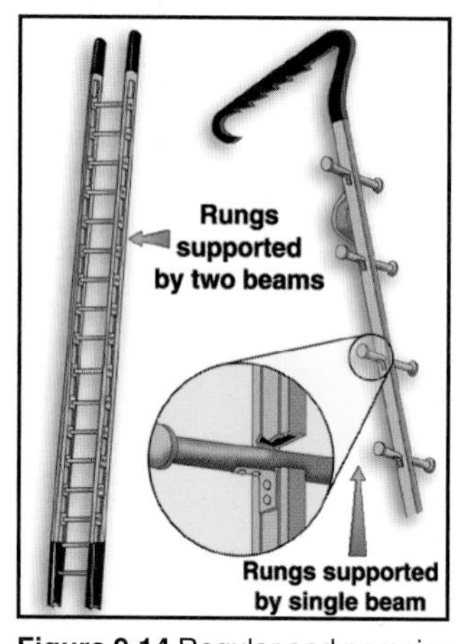

Figure 9.14 Regular and pompier ladder rungs.

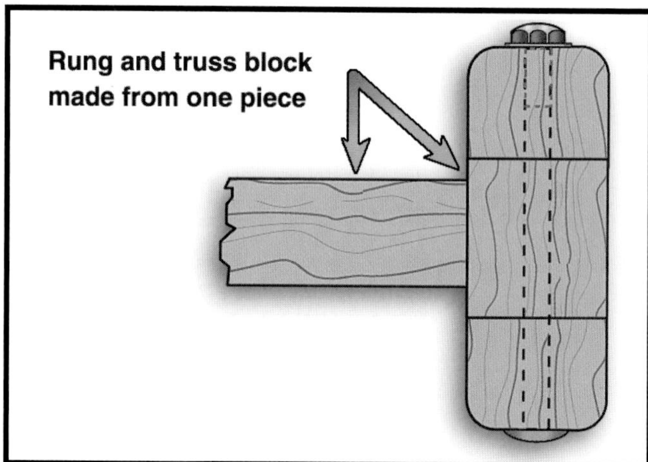

Figure 9.18 A typical truss block.

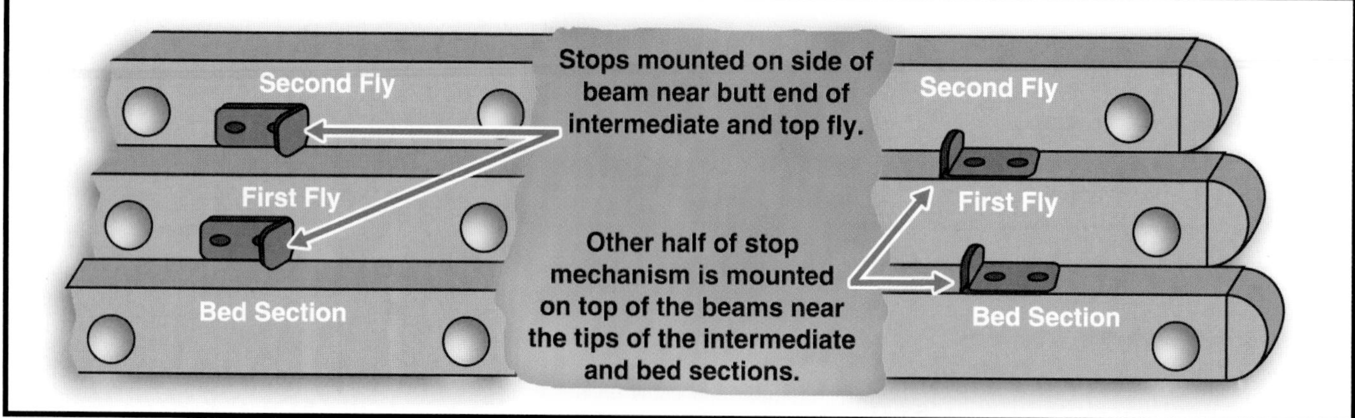

Figure 9.15 Common types of stops used for extension and pole ladders.

LADDER TYPES

[NFPA 1001: 3-3.5]

All of the various types of fire service ladders have a purpose. Many of them, however, are more adaptable to a specific function than they are to general use. Their identifying name is often significant regarding their use, and firefighters frequently make reference to them by association. The descriptions in the following sections more clearly identify fire service ladders.

Single (Wall) Ladders

A *single ladder* is nonadjustable in length and consists of only one section (Figure 9.19). Its size is designated by the overall length of the beams. The single ladder is used for quick access to windows and roofs on one- and two-story buildings. Single ladders must be constructed to have maximum strength and minimum weight. These ladders may be of the trussed type in order to reduce their weight. Lengths vary from 6 to 32 feet (2 m to 10 m) with the more common lengths ranging from 12 to 20 feet (4 m to 6 m).

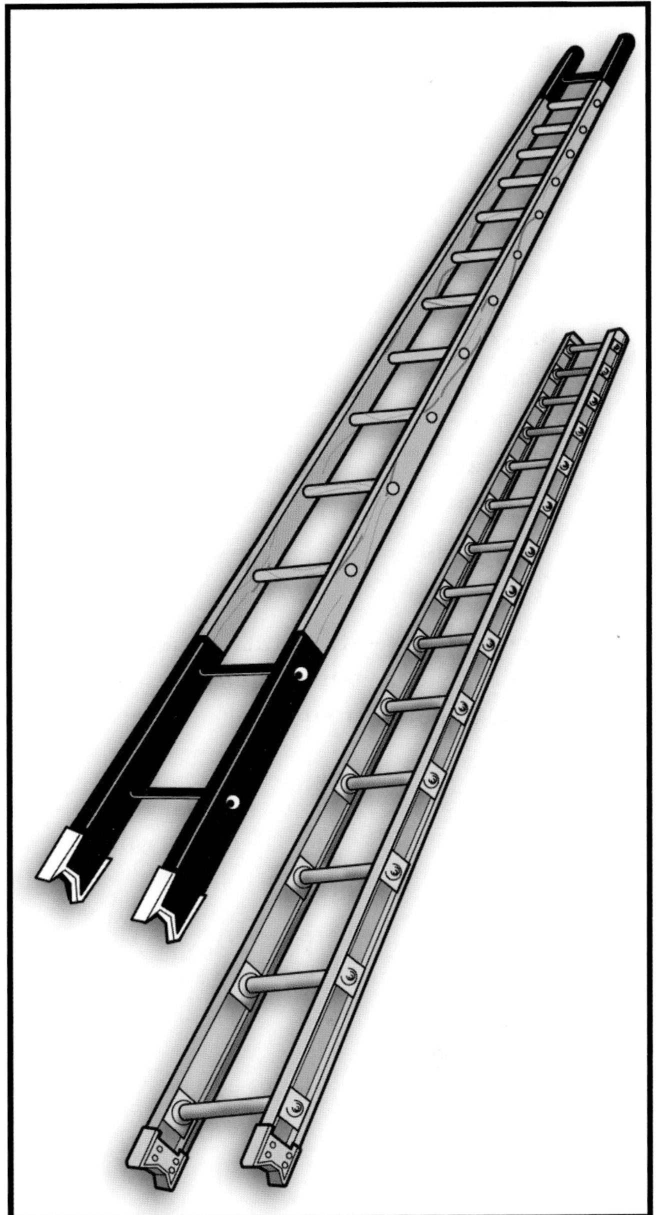

Figure 9.19 Single ladders.

Roof Ladders

Roof ladders are single ladders equipped at the tip with folding hooks that provide a means of anchoring the ladder over the roof ridge or other roof part (Figure 9.20). Roof ladders are generally required to lie flat on the roof surface so that a firefighter may stand on the ladder for roof work. The ladder distributes the firefighter's weight and helps prevent slipping. Roof ladders may also be used as single wall ladders. Their lengths range from 12 to 24 feet (4 m to 8 m).

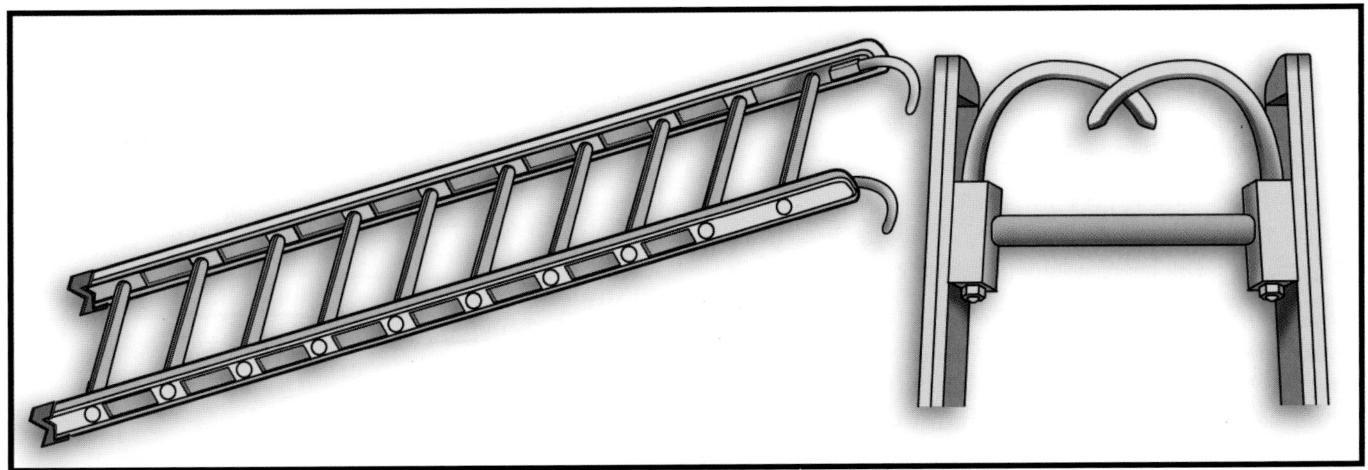

Figure 9.20 Roof ladder: hooks open and hooks nested.

Folding Ladders

Folding ladders are single ladders that have hinged rungs allowing them to be folded so that one beam rests against the other (Figures 9.21 a and b). This allows them to be carried in narrow passageways and used in attic scuttle holes and small rooms or closets. Folding ladders are commonly found in lengths from 8 to 16 feet (2.5 m to 5 m) with the most common being 10 feet (3 m). NFPA 1931 requires folding ladders to have footpads attached to the butt to prevent slipping on floor surfaces.

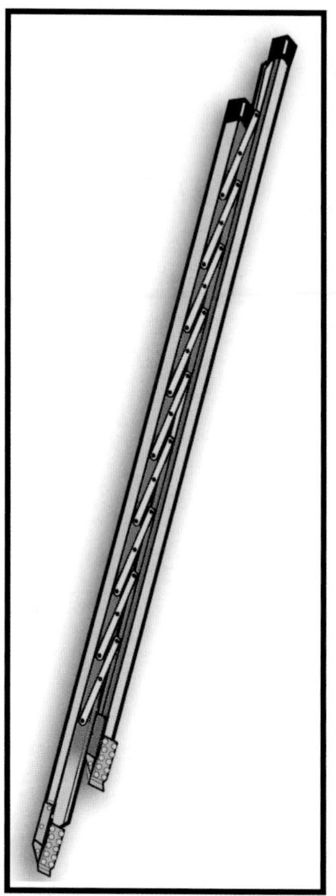

Figure 9.21a In the closed position, a folding ladder is slim and easy to carry.

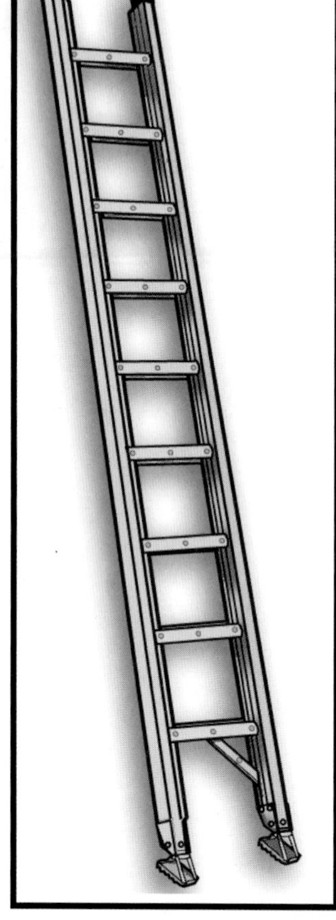

Figure 9.21b The folding ladder is opened when its position is reached.

Extension Ladders

An *extension ladder* is adjustable in length. It consists of a base or bed section and one or more fly sections that travel in guides or brackets to permit length adjustment (Figure 9.22). Its size is designated by the length of the sections and measured along the beams when fully extended. An extension ladder provides access to windows and roofs within the limits of its length. Extension ladders are heavier than single ladders, and more personnel are needed to safely handle them. Extension ladders generally range in length from 12 to 39 feet (4 m to 11.5 m).

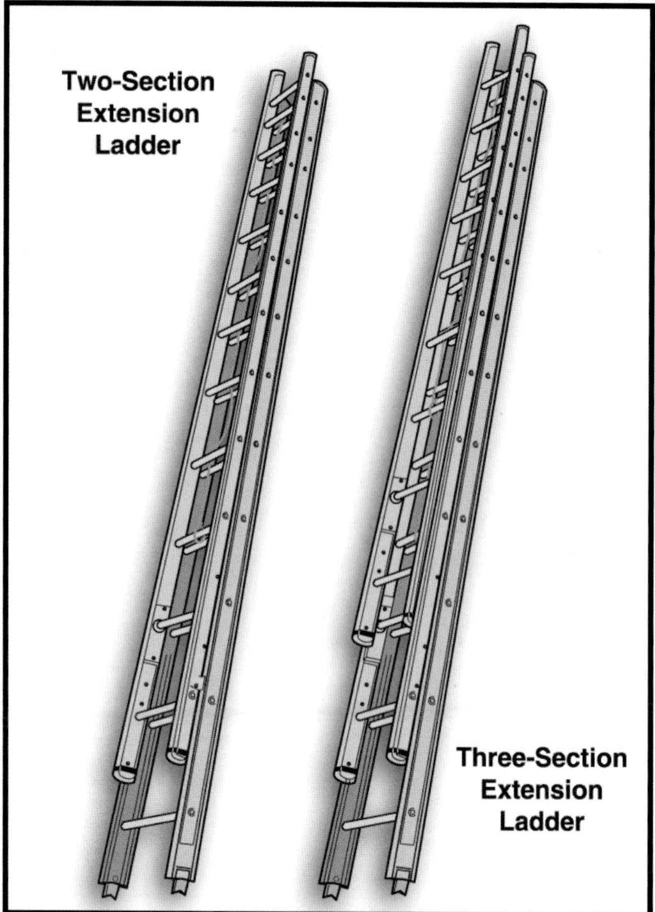

Figure 9.22 Two- and three-section extension ladders.

Pole ladders (Bangor ladders) are extension ladders that have staypoles for added leverage and stability when raising the ladder (Figure 9.23). NFPA 1931 requires all extension ladders that are 40 feet (12 m) or longer to be equipped with staypoles. Pole ladders are manufactured with two to four sections. Most pole ladders do not exceed 50 feet (15 m).

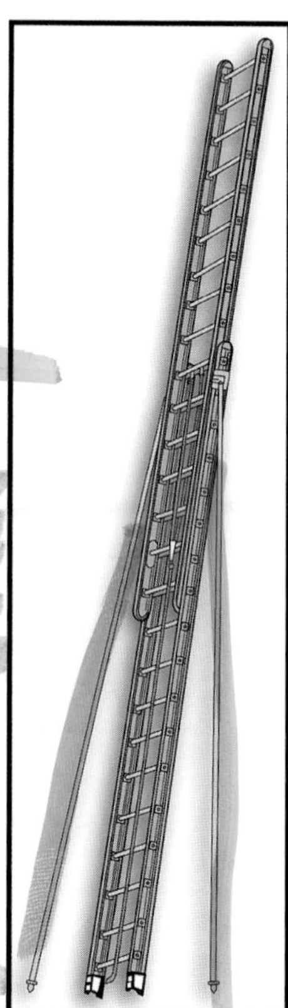

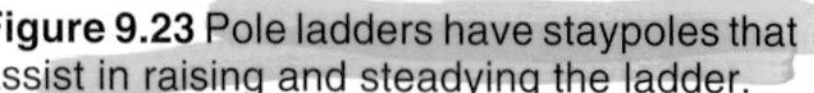

Figure 9.23 Pole ladders have staypoles that assist in raising and steadying the ladder.

Combination Ladders

Combination ladders are designed so that they may be used as a self-supported stepladder (A-frame) and as a single or extension ladder (Figures 9.24 a and b). Lengths range from 8 to 14 feet (2.5 m to 4.3 m) with the most popular being the 10-foot (3 m) model. The ladder must be equipped with positive locking devices to hold the ladder in the open position.

Figure 9.24a A combination extension/A-frame ladder.

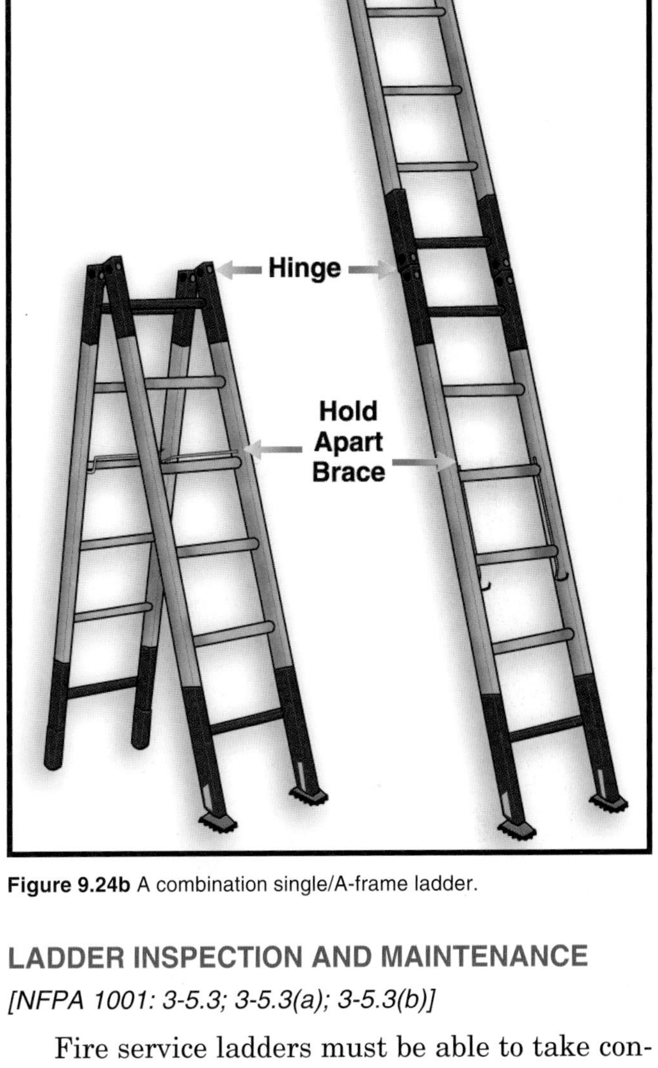

Figure 9.24b A combination single/A-frame ladder.

Pompier Ladders

The *pompier ladder,* sometimes referred to as a *scaling ladder,* is a single-beam ladder with rungs projecting from both sides. It has a large metal "gooseneck" projecting at the top for inserting into windows or other openings (Figure 9.25). It is used to climb from floor to floor, via exterior windows, on a multistory building. Lengths vary from 10 to 16 feet (3 m to 5 m).

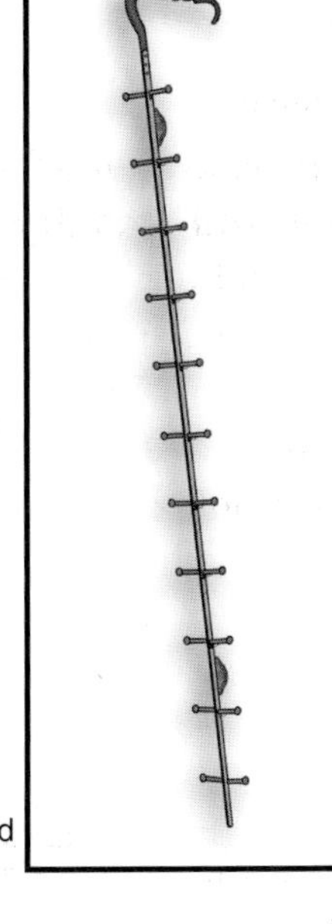

Figure 9.25 Pompier ladders are seldom used in today's fire service.

LADDER INSPECTION AND MAINTENANCE

[NFPA 1001: 3-5.3; 3-5.3(a); 3-5.3(b)]

Fire service ladders must be able to take considerable abuse such as sudden overloading, exposure to temperature extremes, and falling debris. Regardless of what materials or designs are used for ladders, they must conform to NFPA 1931. All ladders meeting NFPA 1931 are required to have a certification label affixed to the ladder by the manufacturer indicating that the ladder meets the standard.

Maintenance

Before discussing ladder maintenance, it is important to understand the difference between maintenance and repair. *Maintenance* means keeping ladders in a state of usefulness or readiness.

Repair means to either restore or replace that which has become inoperable. All firefighters should be capable of performing routine maintenance functions on ground ladders. Any ladders in need of repair require the service of a trained ladder repair technician.

NFPA 1932, *Standard on Use, Maintenance, and Service Testing of Fire Department Ground Ladders,* lists the following general maintenance items that apply to all types of ground ladders:

- Keep ground ladders free of moisture.
- Do not store or rest ladders in a position where they are subjected to exhaust or engine heat (Figure 9.26).
- Do not store ladders in an area where they are exposed to the elements.
- Do not paint ladders except for the top and bottom 12 inches (300 mm) of the beams for purposes of identification or visibility.

Figure 9.26 Never lean the ladder near the apparatus exhaust pipe.

Cleaning Ladders

Regular and proper cleaning of ladders is more than a matter of appearance. Unremoved dirt or debris from a fire may collect and harden to the point where ladder sections are no longer operable. Therefore, it is recommended that ladders be cleaned after every use.

A soft-bristle brush and running water are the most effective tools for cleaning ladders (Figure 9.27). Tar, oil, or greasy residues should be removed with safety solvents. After the ladder is rinsed or anytime a ladder is wet, it should be wiped dry. During each cleaning period, firefighters should look for defects. Any defects should be handled through local fire department procedures. Occasional lubrication where recommended by the manufacturer will maintain smooth operation of the ladder.

Figure 9.27 The ladder should be cleaned of fire-scene and road grime.

Inspecting and Service Testing Ladders

NFPA 1932 requires ladders to be inspected after each use and on a monthly basis. Because fire service ladders are subject to harsh conditions and physical abuse, it is important that they be service tested to ensure they are fit for use. NFPA 1932 should serve as the guideline for ground ladder service testing. This standard recommends that only the tests specified be conducted either by the fire department or an approved testing organization. NFPA 1932 further recommends that caution be used when performing service tests on ground ladders to prevent damage to the ladder or injury to personnel.

When inspecting ground ladders, some of the things that should be checked on all types of ladders include the following:

- Heat sensor labels on metal and fiberglass ladders for a color change indicating heat exposure (**NOTE:** Ladders without a heat sensor label may also show signs of heat exposure such as bubbled or blackened varnish on wood ladders, discoloration of fiberglass ladders, or heavy soot deposits or bubbled paint on the tips of any ladder.)

WARNING

Metal ladders that have been exposed to heat shall be placed out of service until tested. Any metal ladder subjected to direct flame contact or heat high enough to cause water contacting it to sizzle or turn to steam or whose heat sensor label has changed color should be removed from service and tested.

- Rungs for snugness and tightness (Figure 9.28)
- Bolts and rivets for tightness (**NOTE:** Bolts on wood ladders should not be so tight that they crush the wood.)
- Welds for any cracks or apparent defects
- Beams and rungs for cracks, splintering, breaks, gouges, checks, wavy conditions, or deformation

Figure 9.28 Check the rungs to make sure they are tight.

In addition to these general things, there are some other items that need to be checked, depending on the specific type of ladder being inspected. The following sections highlight some of these items.

WOOD LADDERS/LADDERS WITH WOOD COMPONENTS

The following things must be examined on wood ladders or ladders with wood components:

- Look for areas where the varnish finish has been chafed or scraped.
- Check for darkening of the varnish (indicating exposure to heat).
- Check for dark streaks in the wood (indicating deterioration of the wood).

CAUTION: Any indication of deterioration of the wood is cause for the ladder to be removed from service until it can be service tested.

ROOF LADDERS

Make sure that the roof hook assemblies operate with relative ease (Figure 9.29). In addition, the assembly should not show signs of rust, the hooks should not be deformed, and parts should be firmly attached with no sign of looseness. (**NOTE:** Serious problems found should result in removal of the ladder from service pending service testing.)

Figure 9.29 The roof hooks should fully open with ease.

EXTENSION LADDERS

The following must be checked on extension ladders:

- Make sure the pawl assemblies work properly. The hook and finger should move in and out freely.
- Look for fraying or kinking of the halyard (Figure 9.30). If this condition is found, the halyard should be replaced.
- Check the snugness of the halyard cable when the ladder is in the bedded position. This check ensures proper synchronization of the upper sections during operation.
- Make sure the pulleys turn freely.
- Check the condition of the ladder guides and for free movement of the fly sections.

Figure 9.30 Check the halyard for frays or cuts.

- Check for free operation of the pole ladder staypole toggles and check their condition. Detachable staypoles are provided with a latching mechanism at the toggle. This mechanism should be checked to be sure that it is latching properly.

If any of the conditions described are found, the ladder should be removed from service until it can be repaired and tested. Ladders that cannot be safely repaired have to be destroyed or scrapped for parts.

WARNING
Failure to remove a defective ladder from service can result in a catastrophic ladder failure that injures or kills firefighters.

HANDLING LADDERS

[NFPA 1001: 3-3.5; 3-3.5(a); 3-3.5(b); 3-3.10(b); 3-3.11(b)]

NFPA 1901, *Standard for Automotive Fire Apparatus*, sets the minimum lengths and types of ladders to be carried on all pumper or engine companies. Each pumper must carry the following ladders:

- One 10-foot (3 m) folding ladder
- One 14-foot (4.3 m) roof ladder
- One 24-foot (8 m) or larger extension ladder

Ladder Safety

A firefighter's safety and well-being while on a ladder depend on common sense precautions. Firefighters should check important items at every opportunity. Points to ensure safe ladder operation include the following:

- Always wear protective gear, including gloves, when working with ladders.
- Choose the proper ladder for the job.
- Use leg muscles, not back or arm muscles, when lifting ladders below the waist.
- Use the proper number of firefighters for each raise.
- Make sure that ladders are not raised into electrical wires.
- Check the ladder for the proper angle (Figure 9.31).
- Check the pawls to be sure that they are seated over the rungs (Figure 9.32).
- Make sure that the ladder is secure at the top or the bottom (preferably both) before climbing.
- Climb smoothly and rhythmically.
- Do not overload the ladder (Figure 9.33).

Figure 9.31 This label shows in which direction the fly should be facing and helps firefighters assure that the ladder is at a proper climbing angle.

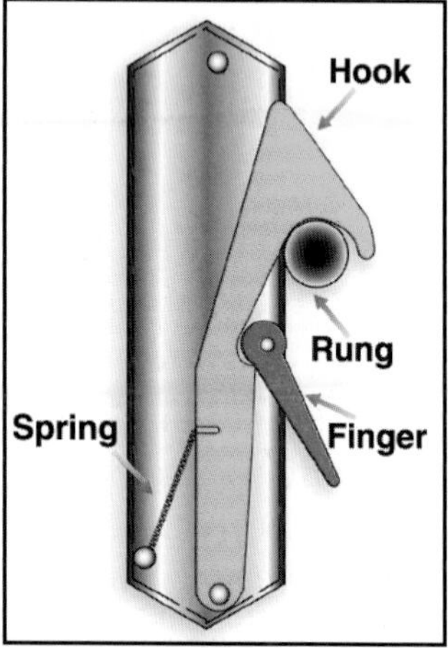

Figure 9.32 An automatic pawl latched onto a rung.

Figure 9.33 Ladders can be damaged by having too many people on them at the same time.

- Always tie in to ground ladders with a leg lock or ladder belt when working from the ladder. (See Working on a Ladder section.)
- Inspect ladders for damage and wear after each use.

WARNING
Extreme caution is necessary whenever metal ladders are used near electrical power sources. Contact with power sources may result in electrocution of anyone in contact with the ladder.

Selecting the Proper Ladder for the Job

Before raising ground ladders, the firefighter(s) must first select the proper ladder for the given job and then carry it to its location for use. It is important that these tasks be accomplished in a safe and efficient manner that will not damage either the ladder or other property. Movements need to be smooth and instinctive because speed is essential in many instances. Because more than one firefighter is frequently required, development of teamwork is another important factor. Therefore, proficiency in handling ladders is realized only with repeated practical training.

Selecting a ladder to do a specific job requires that a firefighter be a good judge of distance. A residential story averages 8 to 10 feet (2.5 m to 3 m), and the distance from the floor to the windowsill averages about 3 feet (1 m). A commercial story averages 12 feet (4 m) from floor to floor, with a 4-foot (1.2 m) distance from the floor to windowsill. Table 9.1 is a general guide that can be used in selecting ladders for specific locations.

Working rules for ladder length include the following:

- The ladder should extend a few feet (meters) (preferably five rungs) beyond the roof edge to provide both a footing and a handhold for persons stepping on or off the ladder (Figure 9.34).
- When used for access from the side of a window or for ventilation, the tip of the ladder should be placed even with the top of the window (Figure 9.35).

TABLE 9.1
Ladder Selection Guide

Working Location of Ladder	Ladder Length
First story roof	16 to 20 feet (4.9 m to 6.0 m)
Second story window	20 to 28 feet (6.0 m to 8.5 m)
Second story roof	28 to 35 feet (8.5 m to 10.7 m)
Third story window or roof	40 to 50 feet (12.2 m to 15.2 m)
Fourth story roof	over 50 feet (15.2 m)

Figure 9.34 A ladder raised to the roof should have at least five rungs above the roof level.

Figure 9.35 A ladder placed for access from the side of a window or for ventilation.

- When rescue from a window opening is to be performed, the tip of the ladder should be placed just below the windowsill (Figure 9.36).

The next step is to determine how far various ladders will reach. Knowledge of the designated length of a ladder can be used to answer this question. Remember that the designated length (normally displayed on the ladder) is a measurement of the maximum extended length

(Figure 9.37). This is *NOT THE LADDER'S REACH,* because ladders are set at angles of approximately 75 degrees. Reach will therefore be *LESS* than the designated length. One more thing needs to be considered: Single, roof, and folding ladders meeting NFPA 1931 are required to have a measured length equal to the designated length. However, the maximum extended length of extension ladders may be as much as 6 inches (150 mm) LESS than the designated length.

Figure 9.36 A ladder used for window rescue should have the tip placed just below the sill.

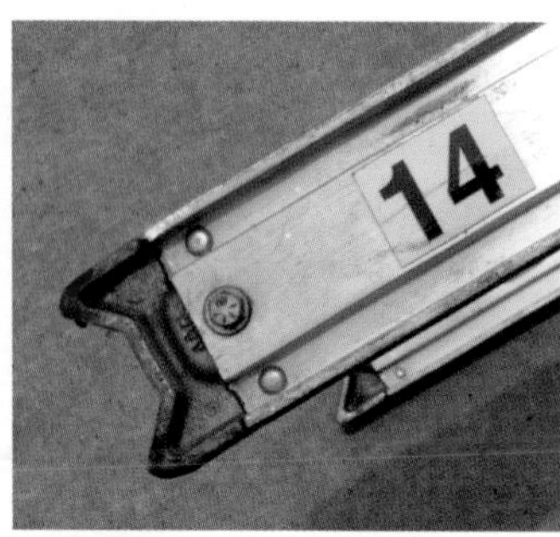

Figure 9.37 The ladder length is marked on the outside of both beams at the butt.

Table 9.2 provides information on the reach of various ground ladders when placed at the proper climbing angle. However, the following should be noted when considering the information contained in Table 9.2.

- For lengths of 35 feet (10.7 m) or less, reach is approximately 1 foot (300 mm) less than the designated length.
- For lengths over 35 feet (10.7 m), reach is approximately 2 feet (600 mm) less than the designated length.

Methods of Mounting Ground Ladders on Apparatus

The method used to mount ground ladders on fire apparatus varies depending on departmental requirements, type of apparatus and body design, type of ladder, type of mounting bracket or racking used, and manufacturer's policies (Figures 9.38 a–c). There are no established standards for the location and mounting of ground ladders on fire apparatus. These differences make it necessary for each fire department to develop and administer its own training procedures for removing and using ground ladders.

TABLE 9.2
Maximum Working Heights for Ladders Set at Proper Climbing Angle

Designated Length of Ladder	Maximum Reach
10 foot (3.0 m)	9 feet (2.7 m)
14 foot (4.3 m)	13 feet (4.0 m)
16 foot (4.9 m)	15 feet (4.6 m)
20 foot (6.1 m)	19 feet (5.8 m)
24 foot (7.3 m)	23 feet (7.0 m)
28 foot (8.5 m)	27 feet (8.2 m)
35 foot (10.7 m)	34 feet (10.4 m)
40 foot (12.2 m)	38 feet (11.6 m)
45 foot (13.7 m)	43 feet (13.1 m)
50 foot (15.2 m)	48 feet (14.6 m)

Figure 9.38a Most pumpers carry their ladders mounted on the right side of the apparatus.

Figure 9.38b These ladders are loaded in a flat position from the rear of the apparatus.

Figure 9.38c Some ladders that are mounted vertically on the side of the apparatus must be loaded/unloaded from the rear.

Removing Ladders From Apparatus

Before firefighters are drilled in removing ground ladders from apparatus, each firefighter should be able to answer the following questions:

- What ladders (types and lengths) are carried and where are they carried on the apparatus?

- Are the ladders racked with the butt toward the front or toward the rear of the apparatus?
- Where ladders are nested together, can one ladder be removed leaving the other(s) securely in place? (In particular, can the roof ladder be removed from the side of the pumper and leave the extension ladder securely in place?)
- In what order do the ladders that nest together rack? (Extension ladder goes on first, roof ladder second, or vice versa?)
- Is the top fly of the extension ladder on the inside or on the outside when the ladder is racked on the side of the apparatus?
- How are the ladders secured?
- Which rungs go in or near the brackets when ladders are mounted vertically on the side of apparatus? (Many departments find it a good practice to mark ladders to indicate when rungs go in or near the brackets as shown in Figure 9.39).

Figure 9.39 Most ladders have marks on them to denote where they should be placed on the mounting brackets.

Proper Lifting and Lowering Methods

Many firefighters are injured from using improper lifting and lowering techniques. Often, these injuries are preventable. The following procedures are recommended:

- Have adequate personnel for the task.
- Bend the knees, keeping back as straight as possible, and lift with the legs, *NOT WITH THE BACK OR ARMS* (Figure 9.40).
- Lifting should be done on the command of a firefighter at the rear who can see the whole operation when two or more firefighters are lifting a ladder (Figure 9.41). If any firefighter is not ready, that person should make it known immediately so that the operation will be halted. Lifting should be done in unison.
- Reverse the procedure for lifting when it is necessary to place a ladder on the ground before raising it. Lower the ladder with the leg muscles. Also, be sure to keep the body and feet parallel to the ladder so that when the ladder is placed it does not injure the toes (Figure 9.42).

Figure 9.40 Always lift with your legs not your back.

Figure 9.41 The firefighter at the butt gives the command to lift.

Figure 9.42 Keep your body and toes parallel to the ladder as it is lowered.

LADDER CARRIES

[NFPA 1001: 3-3.5; 3-3.5(b); 3-3.11(b)]

Once the ladder has been removed from its mounting, there are numerous ways it can be transported to its point of use. The procedures for initiating ladder carries for ladders on the ground differ from those for ladders that are carried on apparatus. Different storage methods require different procedures that must be adapted to the individual situation. Since there are many different types of apparatus and means of mounting ladders, all carries in this section are demonstrated from the ground.

One-Firefighter Low-Shoulder Carry

Single or roof ladders may be safely carried by one firefighter. The low-shoulder carry involves resting the ladder's upper beam on the firefighter's shoulder, while the firefighter's arm goes between two rungs (Figure 9.43). Skill Sheet 9-1 shows the steps for performing the one-firefighter low-shoulder carry from flat on the ground.

Figure 9.43 A completed one-firefighter low-shoulder carry.

WARNING

Carry the forward end of the ladder slightly lowered. Lowering the forward portion provides better balance when carrying, improves visibility by allowing the firefighter to view the way ahead, and if the ladder should strike another person, the butt spurs will make contact with the body area instead of the head.

Two-Firefighter Low-Shoulder Carry

Although the two-firefighter low-shoulder carry may be used with single or roof ladders, it is most commonly used for 24-, 28- and 35-foot (8 m, 8.5 m, and 10.7 m) extension ladders. The two-firefighter low-shoulder carry gives firefighters excellent control of the ladder (Figure 9.44). The forward firefighter places his free hand over the upper butt spur. This is done to prevent injury in case there is a collision with someone while the ladder is being carried. Skill Sheet 9-2 describes the two-firefighter low-shoulder carry from flat on the ground.

Figure 9.44 A completed two-firefighter low-shoulder carry.

Three-Firefighter Flat-Shoulder Carry

The three-firefighter flat-shoulder carry is typically used on extension ladders up to 35 feet (10.7 m). This method has two firefighters, one at each end on one side of the ladder, and one firefighter on the other side in the middle (Figure 9.45). Skill Sheet 9-3 shows the procedure for carrying the ladder from the ground using the three-firefighter flat-shoulder carry.

Figure 9.45 A completed three-firefighter flat-shoulder carry.

Four-Firefighter Flat-Shoulder Carry

The same flat-shoulder method used by three firefighters for carrying ladders is used by four firefighters except that there is a change in the positioning of the firefighters to accommodate the fourth firefighter. When four firefighters use the flat-shoulder carry, two are positioned at each end of the ladder, opposite each other (Figure 9.46).

Figure 9.46 A completed four-firefighter flat-shoulder carry.

Two-Firefighter Arm's Length On-Edge Carry

The two-firefighter arm's length on-edge carry is best performed with lightweight ladders (Figure 9.47). The two-firefighter arm's length on-edge carry described in Skill Sheet 9-4 is based on the fact that the firefighters are positioned on the bed section (widest) side of the ladder when it is in the vertical position.

Figure 9.47 A completed two-firefighter arm's length on-edge carry.

Special Procedures for Carrying Roof Ladders

The procedures previously described are for carrying ladders butt forward. In some cases, a firefighter will carry a roof ladder with the intention of climbing another ground ladder and placing the roof ladder with hooks deployed on a sloped roof. In this situation, the firefighter should use the low-shoulder method and have the tip (hooks) forward (Figure 9.48). Skill Sheet 9-5 describes the procedure for carrying a roof ladder.

Figure 9.48 The roof ladder may be carried with the tip forward.

Normally, the roof ladder is carried with the hooks closed to the foot of the second ladder. A second firefighter opens the hooks while the first firefighter maintains the carry (Figure 9.49). When no second firefighter is present, the firefighter sets the ladder down, moves to the tip, picks up the tip, opens the hooks, lays the tip down, returns to the midpoint, picks up the ladder, and resumes the carry.

There may be occasions when there is no second firefighter to open the hooks, time is critical, and there is no crowd of people through which the ladder must be carried. In this case, the hooks may be opened at the apparatus before the carry is begun; they are turned outward in relation to the firefighter carrying the ladder (Figure 9.50).

Figure 9.49 A second firefighter may open the roof hooks.

Figure 9.50 Carry the ladder with the hooks open facing outward.

POSITIONING (PLACEMENT) OF GROUND LADDERS

[NFPA 1001: 3-3.5; 3-3.5(a); 3-3.5(b); 3-3.8(b); 3-3.11(b)]

Proper positioning, or placement, of ground ladders is important because it affects the safety and efficiency of operations. The following sections contain some of the basic considerations and requirements for ground ladder placement.

Responsibility for Positioning

Normally, an officer designates the general location where the ladder is to be positioned and/or the task is to be performed. However, personnel carrying the ladder frequently decide on the exact spot where the butt is to be placed. The firefighter nearest the butt is the logical person to make this decision because this end is placed on the ground to initiate raising the ladder. When there are two firefighters at the butt, the one on the right side is usually the one responsible for placement (Figure 9.51). However, this designation is an option as far as each department's policy is concerned.

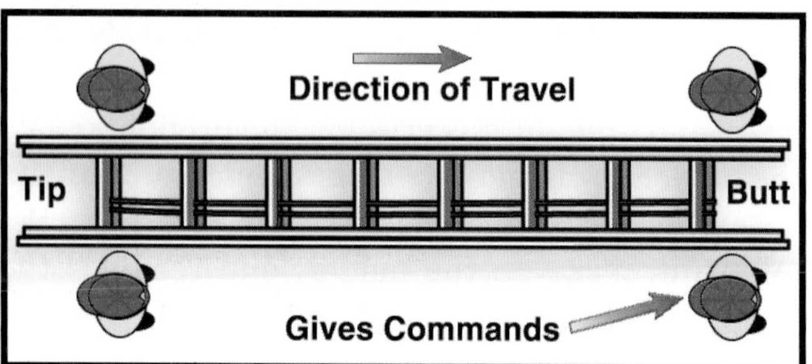

Figure 9.51 The firefighter on the right side of the butt gives the commands.

Figure 9.52 Place the tip adjacent to the top of the window opening.

Figure 9.53 For rescue, place the tip just below the lower sill.

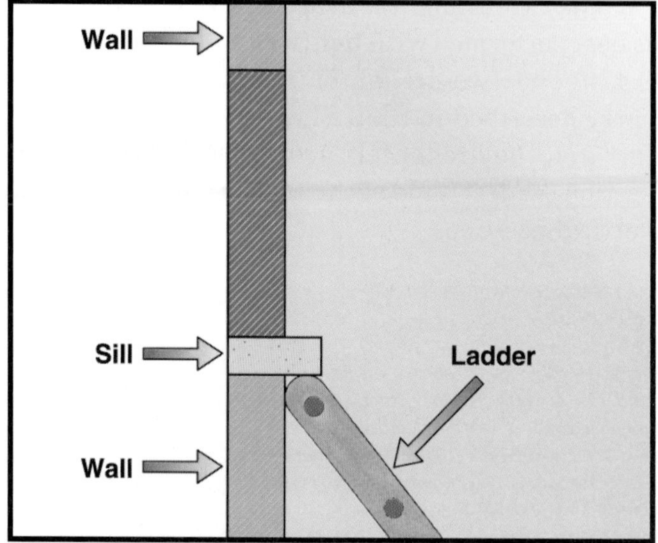

Figure 9.54 Wedging the tip under the sill makes for a more stable placement.

Factors Affecting Ground Ladder Placement

When placing ladders, there are two objectives to be met: first, to place the ladder properly for its intended use and second, to place the butt the proper distance from the building for safe and easy climbing. There are numerous factors that dictate the exact place to position the ladder.

If a ladder is to be used to provide a vantage point from which a firefighter can break a window for ventilation, it should be placed alongside the window to the windward (upwind) side. The tip should be about even with the top of the window (Figure 9.52). The same position can be used when firefighters desire to climb in or out narrow windows.

If a ladder is to be used for entry or rescue from a window, usually the ladder tip is placed slightly below the sill (Figure 9.53). If the sill projects out from the wall, the tip of the ladder can be wedged under the sill for additional stability (Figure 9.54). If the window opening is wide enough to permit the ladder tip to project into it and still allow room beside it to facilitate entry and rescue, the ladder should be placed so that two or three rungs extend above the sill (Figure 9.55).

Figure 9.55 In wide windows the ladder may be extended into one side of the opening.

When a ladder is used as a vantage point from which to direct a hose stream into a window opening and no entry is made, it is raised directly in front of the window with the tip on the wall above the window opening (Figure 9.56). Care must be taken to keep flames from engulfing the tip of the ladder. If this situation cannot be avoided, the ladder is raised just to the sill.

Figure 9.56 Place the ladder directly over the opening so that a hose stream may be discharged into the window.

Other placement guidelines include the following:

- Ladder at least two points on different sides of the building (Figure 9.57).
- Avoid placing ladders over openings such as windows and doors.
- Take advantage of strong points in building construction when placing ladders.
- Raise the ladder directly in front of the window when a ladder is to be used as a support for a smoke fan. Place the ladder tip on the wall above the window opening.
- Avoid placing ladders where they may come into contact with overhead obstructions such as wires, tree limbs, or signs (Figure 9.58).

Figure 9.57 The roof or any portions of the building should be laddered from at least two points. Here the upper story is laddered from two points. *Courtesy of Bill Tompkins.*

- Avoid placing ladders on uneven terrain or on soft spots.
- Avoid placing ladders on main paths of travel that firefighters or evacuees will need to use (Figure 9.59).
- Avoid placing ladders where they may contact either burning surfaces or openings with flames present.
- Avoid placing ladders on top of sidewalk elevator trapdoors or sidewalk deadlights. These areas may give way under the weight of both the ladder and firefighters (Figure 9.60).
- Do not place ladders against unstable walls or surfaces.

Figure 9.58 Watch for trees or other overhead obstructions. *Courtesy of Bill Tompkins.*

Figure 9.59 Do not place ladders in front of doors.

Figure 9.60 Avoid placing ladders on trapdoors, grates, and utility covers.

When the ladder has been raised and lowered into place, the desired angle of inclination is approximately 75 degrees (Figure 9.61). This angle provides good stability and places stresses on the ladder properly. It also provides for easy climbing because it permits the climber to stand perpendicular to the ground, at arm's length from the rungs. The distance of the butt end from the building establishes the angle formed by the ladder and the ground. If the butt is placed too close to the building, its stability is reduced because climbing tends to cause the tip to pull away from the building.

Figure 9.61 Ladders should be placed at a 75-degree angle.

If the butt of the ladder is placed too far away from the building, the load-carrying capacity of the ladder is reduced, and it has more of a tendency to slip. Placement at such a poor angle may sometimes be necessary, so either tie the bottom of the ladder off or heel (steady) it at all times. (See Securing the Ladder section for tying off and heeling instructions.)

An easy way to determine the proper distance between the heel of the ladder and the building is to divide the used length of the ladder by 4. For example, if 20 feet (6 m) of ladder is needed to reach a window, the butt end should be placed 5 feet (1.5 m) from the building (20 feet divided by 4 [6 m divided by 4]) (Figure 9.62). Exact measurements are unnecessary on the fire scene. Firefighters develop the experience to visually judge the proper positioning for the ladder. The proper angle can also be checked by standing on the bottom rung and reaching out for the rung in front. A firefighter should be able to grab the rung while standing straight up, with arms extended straight out (Figure 9.63). Newer ladders are equipped with an inclination marking on the outside of the beam whose lines become perfectly vertical and horizontal when the ladder is properly set (Figure 9.64).

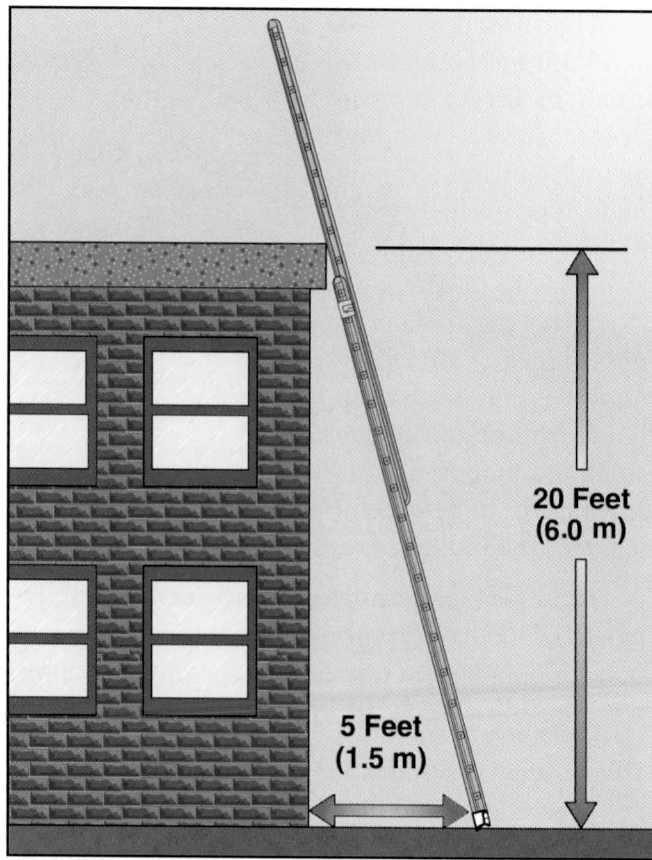

Figure 9.62 A ladder that is raised 20 feet (6 m) should have the base 5 feet (1.5 m) from the building.

Figure 9.63 Check for a proper angle by standing on the bottom rung and reaching for the rung directly at shoulder level.

Figure 9.64 Newer ladders have labels that help firefighters achieve a proper climbing angle.

GENERAL PROCEDURES FOR RAISING AND CLIMBING LADDERS

[NFPA 1001: 3-3.5; 3-3.5(a); 3-3.5(b)]

A well-positioned ladder becomes a means by which important fire fighting operations can be performed. If speed and accuracy are to be developed, teamwork, smoothness, and rhythm are necessary when raising and lowering fire department ladders. However, before learning the technique of raising ladders, firefighters should be aware of certain general procedures that affect the raising of ladders.

Transition From Carry to Raise

The methods and precautions for raising single and extension ladders are much the same. With the exception of pole ladders, it is not necessary to place the ladder flat on the ground prior to raising; only the butt needs to be placed on the ground (Figure 9.65).

Figure 9.65 With the exception of pole ladders, it is not necessary to place the ladder flat on the ground prior to raising.

The transition from carry to raise can and should be smooth and continuous.

This section contains step-by-step information only for raising ladders. In every case, the procedure for lowering the ladder is to reverse the listed steps in the order given. Before raising a ladder, there are a number of things firefighters need to consider and precautions they must take. Some of the more important ones are contained in the sections that follow.

Electrical Hazards

A major concern when raising ladders is possible contact with live electrical wires or equipment, either by the ladder or by the persons who have to climb it. The danger with metal ladders has been stressed previously. However, many firefighters do not realize that *WET* wood or fiberglass ladders present the same hazard. To avoid this hazard, care must be taken *BEFORE BEGINNING A RAISE* (Figure 9.66).

Firefighters need to look overhead for electrical wires or equipment before making the final selection on where to place a ladder or what method to use for raising it. IFSTA recommends that all ladders maintain a distance of at least 10 feet (3 m) from all energized electrical lines or equipment. This distance must be maintained at all times, including during the raise itself. In some cases, the ladder will come to rest a safe distance from the electrical equipment; however, it will come too close to this equipment during the actual raise (Figure 9.67). In these cases, an alternate method for raising the ladder, such as raising parallel to the building as opposed to perpendicular, may be required (Figure 9.68).

Figure 9.66 Always check for electrical hazards before raising a ladder.

Figure 9.67 If this ladder is raised perpendicular to the building, it may come in contact with the power lines.

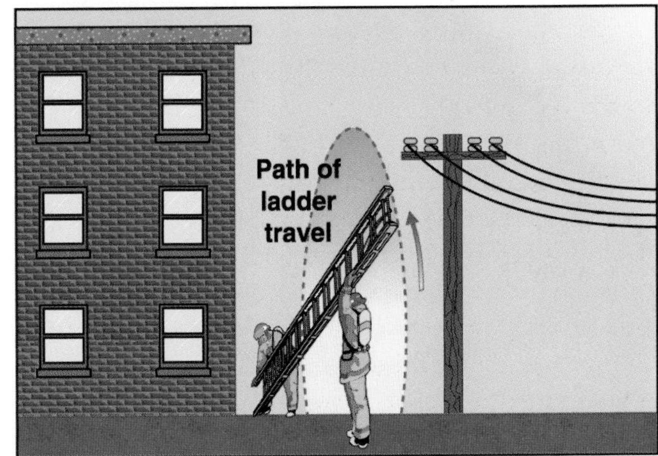

Figure 9.68 By using a parallel raise, the overhead wires are avoided.

Position of the Fly Section on Extension Ladders

The question of whether the fly on an extension ladder should be in (next to the building) or out (away from the building) must be settled before starting the discussion on raises. This question has been a matter of controversy in the fire service for many years.

Each ladder manufacturer specifies whether the ladder should be placed with the fly in or out. This recommendation is based on the design of the ladder and the fly position at which manufacturer's tests show it to be strongest. Failure to follow this recommendation could void the warranty of the ladder should a failure or damage occur.

In general, all modern metal and fiberglass ladders are designed to be used with the *FLY OUT* (away from the building) (Figure 9.69). Wood ladders that are designed with the rungs mounted in the top truss rail (the only type of wood ladder still manufactured today) are intended to be deployed with the *FLY IN* (Figure 9.70). Again, consult department SOPs or the manufacturer of the ladders to find out for certain the correct fly position.

Figure 9.69 Metal extension ladders are deployed with the fly out (away from the building).

Figure 9.70 Wood ladders are deployed with the fly in (toward the building).

Some departments have ladders that are intended to be used with the fly out but prefer that the firefighter extending the halyard be on the outside of the ladder. In this case, firefighters will need to pivot the ladder 180 degrees (discussed later) after it has been extended.

Tying the Halyard

Once an extension ladder is resting against a building and before it is climbed, the excess halyard should be tied to the ladder with a clove hitch (Figure 9.71). This prevents the fly from slipping and prevents anyone from tripping over the rope. The same tie can be used for either a closed- or open-ended halyard. Skill Sheet 9-6 describes the procedure for tying the halyard.

Figure 9.71 An overhand safety knot should be tied to secure the clove hitch.

LADDER RAISES

[NFPA 1001: 3-3.5; 3-3.5(b); 3-3.11(b)]

There are numerous ways to safely raise ground ladders. These methods vary depending on the type and size of the ladder, number of personnel available to help with the raise, and weather and topography considerations. The raises discussed here represent only some of the more commonly used methods; there are many more.

One-Firefighter Raises

One firefighter may safely raise single ladders and small extension ladders. The following procedures should be used to accomplish these raises.

ONE-FIREFIGHTER SINGLE LADDER RAISES

Single and roof ladders of 14 feet (4.3 m) or less are light enough that one firefighter can usually place the butt end at the point where it will be located for climbing without heeling (steadying) it against the building or another object before raising (Figure 9.72). The steps described in Skill Sheet 9-7 should be used to perform the one-firefighter raise for single ladders under 14 feet (4.3 m). On single ladders longer than 14 feet (4.3 m), use the procedures in Skill Sheet 9-8.

Figure 9.72 Place the butt on the ground at an appropriate distance from the building.

ONE-FIREFIGHTER EXTENSION LADDER RAISE

One method of raising extension ladders with one firefighter is from the low-shoulder carry. When using the one-firefighter raise from the low-shoulder carry, the placement of the butt is important. In this instance, a building is used to heel the ladder to prevent the ladder butt from slipping while the ladder is brought to the vertical position (Figure 9.73). Skill Sheet 9-9 describes the procedure for raising the ladder from the low-shoulder carry.

Figure 9.73 One firefighter can raise an extension ladder.

Two-Firefighter Raises

Space permitting, it makes little difference if a ladder is raised parallel with or perpendicular to a building. If raised parallel with the building, the ladder can always be pivoted after it is in the vertical position. Whenever two or more firefighters are involved in raising a ladder, the firefighter at the butt end, the *heeler,* is responsible for placing it at the desired distance from the building and determining whether the ladder will be raised parallel with or perpendicular to the building. There are two basic ways for two firefighters to raise a ladder: the flat raise and the beam raise. Skill Sheet 9-10 describes the procedure for the two-firefighter flat raise. Skill Sheet 9-11 shows the procedure for the two-firefighter beam raise.

Three-Firefighter Flat Raise

As the length of the ladder increases, the weight also increases. This requires more personnel for raising the larger extension ladders (Figure 9.74). Typically, ladders of 35 feet (10.7 m) or longer should be raised by at least three firefighters. Skill Sheet 9-12 describes the procedure for flat-raising ladders with three fire-fighters.

Figure 9.74 More personnel are required for longer extension ladders.

To raise a ladder using the beam method with three firefighters, follow the same procedures for the two-firefighter flat raise. The only difference is that the third firefighter is positioned along the beam (Figure 9.75). Once the ladder has been raised to a vertical position, follow the procedures described for the flat raise.

Figure 9.75 A three-firefighter beam raise.

Four-Firefighter Flat Raise

When personnel are available, four firefighters can be used to better handle the larger and heavier extension ladders (Figure 9.76). A flat raise is normally used, and the procedures for raising the ladder are similar to the three-firefighter raise except for the placement of personnel. A firefighter at the butt is responsible for placing the butt at the desired distance from the building and determining whether the ladder will be raised parallel with or perpendicular to the building. Skill Sheet 9-13 describes the procedure for raising ladders with four firefighters.

Figure 9.76 A four-firefighter flat raise is used for larger extension ladders.

Placing a Roof Ladder

There are a number of ways to get a roof ladder in place on a sloped roof. Once a firefighter has carried the roof ladder to the location, it can be placed by either one or two firefighters. Skill Sheet 9-14 shows the procedure for one firefighter to place a roof ladder in position.

It is, however, much easier to climb another ladder and place the roof ladder using two firefighters (Figure 9.77). There are two methods of accomplishing this task, both named for the way the ladder is carried from the apparatus: hooks-first method and butt-first method. The hooks-first method is described in Skill Sheet 9-15. When a roof ladder has been carried to the scene butt first, there is no need to waste valuable time turning it around. The butt-first method described in Skill Sheet 9-16 can be used.

Figure 9.77 Two firefighters deploying a roof ladder.

SPECIAL PROCEDURES FOR MOVING GROUND LADDERS

[NFPA 1001: 3-3.5; 3-3.5(a)]

Sometimes the basic ladder raising procedures described are not sufficient to get the ladder into its final position for use. In many cases it will be necessary to move the ladder slightly after it has been extended.

Pivoting Ladders With Two Firefighters

Occasionally, a ladder is raised with the fly in the incorrect position for deployment. When this happens, it is necessary to pivot the ladder. Any ladder flat-raised parallel to the building also requires pivoting to align it with the wall upon which it will rest. The beam closest to the building should be used for the pivot. Whenever possible, the ladder should be pivoted before it is extended.

The two-firefighter pivot may be used on any ground ladder that two firefighters can raise (Figure 9.78). The procedure described in Skill Sheet 9-17 is for a ladder that must be turned 180 degrees to get the fly section in the proper position. The same procedure is used for positioning a ladder that was flat-raised parallel to the building. In this case, the beam nearest the building is used to pivot the ladder.

Figure 9.78 Two firefighters pivoting a ladder.

Shifting Raised Ground Ladders

Occasionally, circumstances require that ground ladders be moved while vertical. Shifting a ladder that is in a vertical position should be limited to short distances such as aligning ladders to a building or to an adjacent window.

One firefighter can safely shift a single ladder that is 20 feet (6 m) long or less. The procedure for the one-firefighter shift is described in Skill Sheet 9-18. Because of their weight, extension ladders require two firefighters for the shifting maneuver described in Skill Sheet 9-19.

SECURING THE LADDER

[NFPA 1001: 3-3.5; 3-3.5(a); 3-3.5(b); 3-3.11(b)]

Ground ladders should be secured whenever firefighters are climbing or working from them. Two methods discussed in this section are heeling and tying in.

Heeling

One way of preventing movement of a ladder is to properly heel, or foot, it. There are several methods of properly heeling a ladder. One method is for a firefighter to stand underneath the ladder with feet about shoulder-width apart (or one foot slightly ahead of the other). The firefighter then grasps the ladder beams at about eye level, and pulls backward to press the ladder against the building (Figure 9.79). When using this method, the firefighter must wear head and eye protection and not look up when there is someone climbing the ladder. The firefighter must be sure to grasp the beams and not the rungs.

Figure 9.79 The ladder may be heeled from behind.

Another method of heeling a ladder is for a firefighter to stand on the outside of the ladder and chock the butt end with his feet (Figures 9.80 a and b).

Figure 9.80a The ladder may be held with a foot on the beam at the ground.

Figure 9.80b The ladder may be heeled with one foot on the rung.

With this method, either the firefighter's toes are placed against the butt spur or one foot is placed on the bottom rung. The firefighter grasps the beams, and the ladder is pressed against the building. A firefighter must stay alert for descending firefighters when heeling the ladder in this way.

Tying In

Whenever possible, a ladder should be tied securely to a fixed object. Tying in a ladder is simple, can be done quickly, and is strongly recommended to prevent the ladder from slipping or pulling away from the building. Tying in also frees personnel who would otherwise be holding the ladder in place. A rope hose tool or safety strap can be used between the ladder and a fixed object (Figures 9.81 a and b).

The process of securing a ground ladder may include any or all of the following:

- Make sure the ladder locks are locked (extension ladders only). This should have already been accomplished before the ladder was placed against the structure.
- Tie the halyard (extension ladder only).
- Prevent movement of the ladder away from the building by heeling and/or tying in.

Figure 9.81a The ladder may be tied off near the bottom.

Figure 9.81b . . . or near the top.

CLIMBING LADDERS

[NFPA 1001: 3-3.5; 3-3.11(b)]

Ladder climbing should be done smoothly and rhythmically. The climber should ascend the ladder so that there is the least possible amount of bounce and sway. This smoothness is accomplished if the climber's knee is bent to ease the weight on each rung. Balance on the ladder will come naturally if the ladder is properly spaced from the building because the body will be perpendicular to the ground.

The climb may be started after the climbing angle has been checked and the ladder is properly secured. The climber's eyes should be focused forward, with an occasional glance at the tip of the ladder. The climber's arms should be kept straight during the climb; this action keeps the body away from the ladder and permits free knee movement during the climb (Figure 9.82). When there is no equipment being carried, the hands can be put on the beams or the rungs. When using the rungs, the hands should grasp the rungs with the palms down and the thumbs beneath the rung. Some people find it natural to grasp every rung with alternate hands while climbing; others prefer to grasp alternate rungs (Figure 9.83). An option for hand placement when climbing ground ladders is to climb while sliding both hands up behind the beams to maintain constant contact (Figure 9.84).

Figure 9.82 Proper ladder-climbing techniques include keeping both the back and arms straight as the climb is made.

Figure 9.83 Some firefighters find that there is less bounce to the climb if the foot and hand on the same side are raised together.

If the feet should slip during any of these options, the arms and hands are in a position to stop the fall. All upward progress should be performed by the leg muscles, not the arm muscles. The arms and hands should not reach upward during the climb, because reaching upward will bring the body too close to the ladder.

Practice climbing should be done slowly to develop form rather than speed. Speed develops as the proper technique is mastered. Too much speed results in lack of body control, and quick movements cause the ladder to bounce and sway.

A firefighter is often required to carry equipment up and down a ladder during fire fighting. This procedure interrupts the natural climb either because of the added weight on the shoulder or the necessity of using one hand to hold a tool. If a tool is carried in one hand, it is desirable to slide the free hand under the beam while making the climb (Figure 9.85). This method permits constant contact with the ladder. Whenever possible, a utility rope should be used to hoist tools and equipment rather than carrying them up a ladder.

Figure 9.84 The hands may slide up the underside of the beam.

Figure 9.85 Slide one hand up the underside of the beam and carry the tool in the other hand.

WORKING ON A LADDER

[NFPA 1001: 3-3.5; 3-3.11(b)]

Firefighters must sometimes work while standing on a ground ladder, and both hands must be free. Either a Class I life safety harness (ladder belt) or a leg lock can be used to safely secure the firefighter to a ladder while work is being performed. If a firefighter chooses to apply a leg lock on a ground ladder, the procedure in Skill Sheet 9-20 should be used.

WARNING
Exercise caution to ensure that the rated capacity of the ladder is not exceeded. To avoid overloading the ladder, only one person should be allowed on each section of the ladder at the same time.

A life safety harness must be strapped tightly around the waist during use. The hook may be moved to one side, out of the way, while a firefighter is climbing a ladder. However, after reaching the desired height, the firefighter returns the hook to the center and attaches it to a rung (Figure 9.86). All life safety harnesses should meet the requirements set forth in NFPA 1983, *Standard on Fire Service Life Safety Rope and System Components.*

Figure 9.86 One method of securing a firefighter to a ladder is to use a ladder or safety belt.

ASSISTING A VICTIM DOWN A LADDER

[NFPA 1001: 3-3.5, 3-3.8(a); 3-3.8(b)]

When it is known that a ground ladder will be used for rescue through a window, the ladder tip is raised to just below the sill. This allows the victim easier access to the ladder. The ladder is heeled, and all other loads and activity removed from it during rescue operations. Since even healthy, conscious occupants are probably unaccustomed to climbing down a ladder, care must be exercised to keep them from slipping and possibly hurting themselves. To bring victims down a ground ladder, at least four firefighters are needed: two inside the building, one or two on the ladder, and one heeling the ladder.

Several methods for lowering conscious or unconscious victims are as follows:

- Conscious victims are lowered feet first from the building onto a ladder (Figure 9.87).
- An unconscious victim is held on a ladder in the same way as a conscious victim except that the victim's body rests on the rescuer's supporting knee (Figure 9.88). The victim's feet are placed outside the rails to prevent entanglement.
- A similar way to lower an unconscious victim involves using the same hold by the rescuer described in the previous bulleted paragraph, but the victim is turned around to face the rescuer (Figure 9.89). This position reduces the chances of the victim's limbs catching between the rungs.
- An unconscious victim is supported at the crotch by one of the rescuer's arms and at the chest by the other arm (Figure 9.90). The rescuer may be aided by another firefighter.

Figure 9.87 A conscious victim is lowered onto the ladder feet first.

Figure 9.88 An unconscious victim is supported by the rescuer's knee.

Figure 9.89 The victim faces the rescuer in this method.

Figure 9.90 Another way to rescue an unconscious victim.

- A conscious or unconscious victim is cradled in front of the rescuer, with the victim's legs over the rescuer's shoulders, and the victim's arms draped over the rescuer's arms (Figure 9.91). If the ladder is set at a slightly steeper than normal climbing angle, the unconscious victim's head can be tilted forward to avoid hitting each rung during descent. This method is also very effective with extremely heavy victims, whether they are conscious or not (Figure 9.92).

- Another method of removing extraordinarily heavy victims involves several firefighters. Two ground ladders are placed side by side. One firefighter supports the victim's waist and legs. A second firefighter on the other ladder supports the victim's head and upper torso (Figure 9.93).

- Small children who must be brought down a ladder can be cradled across the rescuer's arms (Figure 9.94).

Figure 9.91 A victim is cradled between the rescuer and the ladder.

Figure 9.92 An effective method for handling very heavy victims.

Figure 9.93 Two ladders and two rescuers are needed for this method.

Figure 9.94 A small child is cradled across the rescuer's arms.

SKILL SHEET 9-1 ONE-FIREFIGHTER LOW-SHOULDER CARRY

From Flat on the Ground

Step 1: Kneel beside the ladder, facing the tip.

Step 2: Grasp the middle rung with your near hand.

Step 3: Lift the ladder.

Step 4: Pivot into the ladder as it rises.

Step 5: Place your free arm between two rungs so that the upper beam comes to rest on the shoulder.

SKILL SHEET 9-2 TWO-FIREFIGHTER LOW-SHOULDER CARRY

From Flat on the Ground

NOTE: Firefighter #1 is located near the butt end of the ladder. Firefighter #2 is located near the tip of the ladder.

Step 1: ***Both Firefighters:*** Kneel on the same side of the ladder facing the tip.

Step 2: ***Both Firefighters:*** Grasp a convenient rung with the near hand, palm forward.

Step 3: ***Both Firefighters:*** Stand the ladder on edge.

Step 4: ***Firefighter #1:*** Give the command to "shoulder the ladder."

Step 5: ***Both Firefighters:*** Stand, using the leg muscles to lift the ladder.

Step 6: ***Both Firefighters:*** Tilt the far beam upward as the ladder and the firefighters rise.

Step 7: ***Both Firefighters:*** Pivot and place the free arm between two rungs.

Step 8: ***Both Firefighters:*** Place the upper beam on the shoulders.

NOTE: Both firefighters should be facing the butt. The lift should be smooth and continuous.

SKILL SHEET 9-3 THREE-FIREFIGHTER FLAT-SHOULDER CARRY

From Flat on the Ground

Step 1: ***Firefighters #1 and #2:*** Kneel on one side of the ladder, one at either end, facing the tip.

Step 2: ***Firefighter #3:*** Kneel on the opposite side at midpoint, also facing the tip end.

NOTE: In each case, the knee closer to the ladder is the one touching the ground.

Step 3: ***All Firefighters:*** Stand and lift the ladder.

Step 4: ***All Firefighters:*** Pivot toward the butt when the ladder is about chest high.

Step 5: ***All Firefighters:*** Place the beam onto the shoulders.

SKILL SHEET 9-4 TWO-FIREFIGHTER ARM'S LENGTH ON-EDGE CARRY

From Flat on the Ground

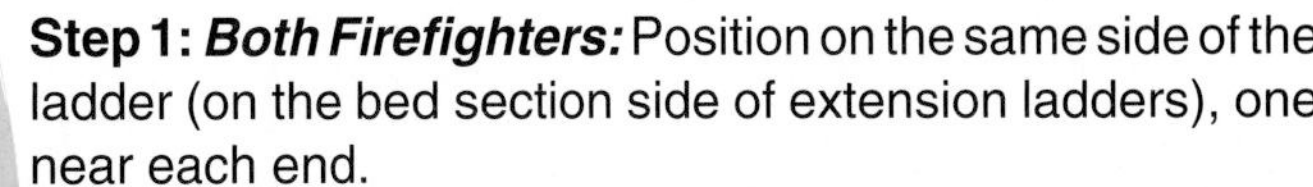

Step 1: ***Both Firefighters:*** Position on the same side of the ladder (on the bed section side of extension ladders), one near each end.

Step 2: ***Both Firefighters:*** Tilt up one beam so that the ladder is resting on the other beam.

Step 3: ***Both Firefighters:*** Squat slightly, facing the butt.

Step 4: ***Both Firefighters:*** Grasp the upper beam with the near hand (the beam of the outermost fly section on an extension ladder).

Step 5: ***Both Firefighters:*** Stand, lifting the ladder until it is at arm's length.

SKILL SHEET 9-5 CARRYING A ROOF LADDER

From Flat on the Ground
One Firefighter

Step 1: Kneel facing the butt end of the ladder.

Step 2: Grasp the middle rung with your near hand, palm facing forward.

Step 3: Stand the ladder on edge.

Step 4: Stand and lift the ladder.

Step 5: Pivot toward the tip of the ladder.

Step 6: Place your free arm between two rungs so that the upper beam comes to rest on the shoulder.

SKILL SHEET 9-6 TYING THE HALYARD

Step 1: Wrap the excess halyard around two convenient rungs.

Step 2: Pull it taut.

Step 3: Hold the halyard between the thumb and forefinger with the palm down.

Step 4: Turn the hand palm up.

Step 5: Push the halyard underneath and back over the top of the rung.

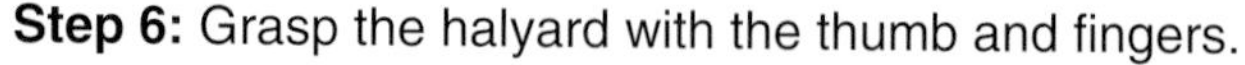

Step 6: Grasp the halyard with the thumb and fingers.

Step 7: Pull it through the loop, making a clove hitch.

Step 8: Finish the tie by making a half hitch or overhand safety on top of the clove hitch.

SKILL SHEET 9-7 ONE-FIREFIGHTER SINGLE LADDER RAISE

Ladders Under 14 Feet (4 m) in Length

Step 1: Lower the butt to the ground at the proper distance from the building for climbing.

Step 2: Raise the ladder simultaneously to a vertical position.

Step 3: Grasp both beams.

Step 4: Heel the butt end of the ladder.

Step 5: Lower the ladder into the objective.

SKILL SHEET 9-8 ONE-FIREFIGHTER SINGLE LADDER RAISE

Ladders Over 14 Feet (4 m) in Length

Step 1: Place the butt end against the building to heel the ladder as it is raised.

Step 2: Raise the ladder to a vertical position.

Step 3: Pull the butt end out away from the building to the proper distance for a good climbing angle.

SKILL SHEET 9-9 ONE-FIREFIGHTER EXTENSION LADDER RAISE

From the Low-Shoulder Carry

Step 1: Place the butt end of the ladder on the ground with the butt spurs against the wall of the building.

Step 2: Grasp a rung in front of your shoulder with your free hand.

Step 3: Remove the opposite arm from between the rungs.

Step 4: Step beneath the ladder.

Step 5: Grasp a convenient rung with the other hand.

NOTE: At this point, the ladder should be flat with both butt spurs against the building.

CAUTION: The area overhead should be visually checked for obstructions before bringing the ladder to a vertical position. The terrain in front of the firefighter should also be visually checked before stepping forward.

Step 6: Advance hand-over-hand down the rungs toward the butt until the ladder is in a vertical position.

Step 7: Extend the ladder by pulling the halyard until the ladder has been raised to the desired level and the pawls are engaged.

NOTE: Care must be taken to pull straight down on the halyard so that the ladder is not pulled over.

Step 8: Position the ladder for climbing by pushing against an upper rung to keep the ladder against the building.

Step 9: Grasp a lower rung with your other hand.

Step 10: Move the ladder butt carefully out from the building to the desired location.

NOTE: If necessary, turn the ladder to bring the fly to the out position.

SKILL SHEET 9-10 TWO-FIREFIGHTER FLAT RAISE

NOTE: **Firefighter #1 is located near the butt end of the ladder. Firefighter #2 is located near the tip end of the ladder.**

Step 1: ***Both Firefighters:*** Carry the ladder to the desired location for the raise.

Step 2: ***Firefighter #1:*** Place the butt end on the ground.

Step 3: ***Firefighter #2:*** Rest the ladder beam on a shoulder.

Step 4: ***Firefighter #1:*** Heel the ladder by standing on the bottom rung.

Step 5: ***Firefighter #1:*** Crouch down to grasp a convenient rung or the beams with both hands.

Step 6: ***Firefighter #1:*** Lean back.

Step 7: ***Firefighter #2:*** Step beneath the ladder.

Step 8: ***Firefighter #2:*** Grasp a convenient rung with both hands.

CAUTION: Visually check the area overhead for obstructions before bringing the ladder to a vertical position. Before stepping forward, visually check the terrain.

Step 9: ***Firefighter #2:*** Advance hand-over-hand down the rungs toward the butt end until the ladder is in a vertical position.

Step 10: ***Firefighter #1:*** Grasp successively higher rungs or higher on the beams as the ladder comes to a vertical position until standing upright.

Step 11: ***Both Firefighters:*** Face each other.

Step 12: ***Both Firefighters:*** Heel the ladder by placing toes against the beams.

NOTE: When raising extension ladders, pivot the ladder to position the fly away from the building (fly in for wooden ladders) if it is not already in that position.

Step 13: ***Firefighter #1:*** Grasp the halyard.

Step 14: ***Firefighter #1:*** Extend the fly section with a hand-over-hand motion until the tip reaches the desired elevation.

Step 15: ***Firefighter #1:*** Check that pawls are engaged.

Step 16: ***Firefighter #2:*** Place one foot against a butt spur or on the bottom rung.

Step 17: ***Firefighter #2:*** Grasp the beams.

Step 18: ***Both Firefighters:*** Lower the ladder gently onto the building.

NOTE: If the ladder has not yet been turned to position the fly in the out position, it can be done at this time.

SKILL SHEET 9-11 TWO-FIREFIGHTER BEAM RAISE

NOTE: Firefighter #1 is located near the butt end of the ladder. Firefighter #2 is located near the tip end of the ladder.

Step 1: *Both Firefighters:* Carry the ladder to the desired location for the raise.

Step 2: *Firefighter #1:* Place the ladder beam on the ground.

Step 3:*Firefighter #2:* Rest the beam on one shoulder.

Step 4: *Firefighter #1:* Place the foot closest to the lower beam on the lower beam at the butt spur.

Step 5: *Firefighter #1:* Grasp the upper beam with hands apart and the other foot extended back to act as a counterbalance.

Step 6: *Firefighter #2:* Advance hand-over-hand down the beam toward the butt until the ladder is in a vertical position.

CAUTION: Visually check the area overhead for obstructions before bringing the ladder to a vertical position. Before stepping forward, visually check the terrain.

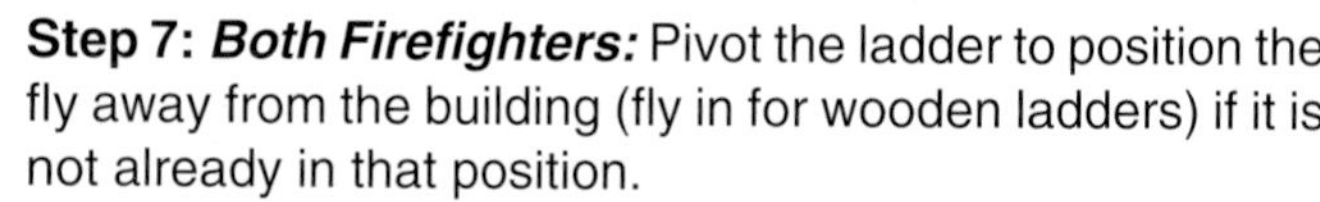

Step 7: ***Both Firefighters:*** Pivot the ladder to position the fly away from the building (fly in for wooden ladders) if it is not already in that position.

Step 8: ***Firefighter #2:*** Grasp the halyard.

Step 9: ***Firefighter #2:*** Extend the fly section with a hand-over-hand motion until the tip reaches the desired elevation.

Step 10: ***Firefighter #2:*** Check that ladder locks are in place.

Step 11: ***Firefighter #1:*** Place one foot against a butt spur or on the bottom rung and grasp the rung or beams.

Step 12: ***Both Firefighters:*** Lower the ladder gently onto the building.

SKILL SHEET 9-12 THREE-FIREFIGHTER FLAT RAISE

NOTE: **Firefighter #1 is located near the butt end of the ladder. Firefighters #2 and #3 are located near the tip end of the ladder.**

Step 1: ***All Firefighters:*** Carry the ladder to the desired location for the raise.

Step 2: ***Firefighter #1:*** Place the ladder butt end on the ground. ***Firefighters #2 and #3:*** Rest the ladder flat on the shoulders.

Step 3: ***Firefighter #1:*** Heel the ladder by standing on the bottom rung (A) or by placing the toes or insteps on the beam (B).

Step 4: ***Firefighter #1:*** Crouch down to grasp a convenient rung with both hands.

Step 5: ***Firefighter #1:*** Lean back.

Step 6: ***Firefighters #2 and #3:*** Advance in unison, with outside hands on the beams and inside hands on the rungs, until the ladder is in a vertical position.

NOTE: If necessary, the firefighters pivot the ladder to position the fly section away from the building. If using a wood ladder, the fly should be in toward the building.

CAUTION: Visually check the area overhead for obstructions before bringing the ladder to a vertical position. Before stepping forward, visually check the terrain.

Step 7: ***Firefighters #2 and #3:*** Place the inside of a foot against the butt spur.

Step 8: ***Firefighters #2 and #3:*** Steady the ladder with both hands on the beam.

Step 9: ***Firefighter #1:*** Grasp the halyard.

Step 10: ***Firefighter #1:*** Place the toe of one foot on the butt spur.

Step 11: ***Firefighter #1:*** Extend the fly section with a hand-over-hand motion until the tip reaches the desired elevation.

Step 12: ***Firefighter #1:*** Check that the ladder locks are in place.

Step 13: ***Firefighters #2 and #3:*** Grasp the beam or a convenient rung.

NOTE: Either method is acceptable as long as both do it the same way.

Step 14: ***Firefighter #1:*** Steady the ladder from the inside position.

Step 15: ***All Firefighters:*** Lower the ladder gently onto the building.

SKILL SHEET 9-13 FOUR-FIREFIGHTER FLAT RAISE

NOTE: **Firefighters #1 and #2 are located near the butt end of the ladder. Firefighters #3 and #4 are located near the tip end of the ladder.**

Step 1: ***All Firefighters:*** Carry the ladder to the desired location for the raise.

Step 2: ***Firefighters #1 and #2:*** Place the ladder butt on the ground.

Step 3: ***Firefighters #3 and #4:*** Rest the ladder flat on the shoulders.

Step 4: ***Firefighters #1 and #2:*** Heel the ladder by placing the inside feet on the bottom rung and the outside feet on the ground outside the beam.

Step 5: ***Firefighters #1 and #2:*** Grasp a convenient rung with the inside hands and the beam with the other hands.

Step 6: ***Firefighters #1 and #2:*** Pull back.

Step 7: ***Firefighters #3 and #4:*** Advance in unison, with the hands on the beams until the ladder is in a vertical position.

NOTE: If necessary, pivot the ladder to position the fly section away from the building. Wood ladders should be positioned with the fly section toward the building.

CAUTION: Visually check the area overhead for obstructions before bringing the ladder to a vertical position. Before stepping forward, visually check the terrain.

Step 8: ***Firefighter #1 or #2:*** Grasp the halyard.

Step 9: ***Firefighter #1 or #2:*** Extend the fly section with a hand-over-hand motion until the tip reaches the desired elevation.

Step 10: ***Firefighters #1 and #2:*** Check that the ladder locks are in place.

Step 11: ***Firefighters #3 and #4:*** Place the inside feet against the butt spur or bottom rung.

Step 12: ***Firefighters #3 and #4:*** Grasp the beams.

Step 13: ***All Firefighters:*** Lower the ladder gently onto the building.

SKILL SHEET 9-14 ROOF LADDER DEPLOYMENT

One-Firefighter Method

Step 1: Carry the roof ladder to the ladder that is to be climbed.

Step 2: Set the roof ladder down.

Step 3: Open the hooks.

Step 4: Face the hooks outward.

Step 5: Tilt the roof ladder up so that it rests against the other ladder.

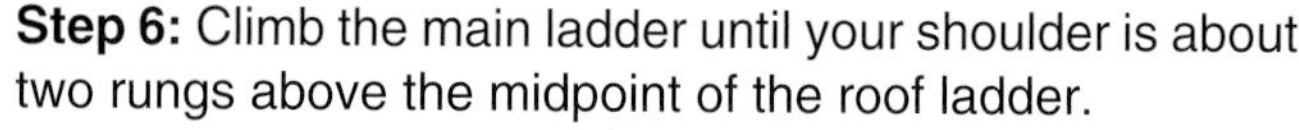

Step 6: Climb the main ladder until your shoulder is about two rungs above the midpoint of the roof ladder.

Step 7: Reach through the rungs of the roof ladder.

Step 8: Hoist the ladder onto the shoulder.

Step 9: Climb to the top of the ladder.

Step 10: Lock into the ladder using a leg lock or life safety harness.

Step 11: Take the roof ladder off the shoulder.

Step 12: Use a hand-over-hand method to push the roof ladder onto the roof.

NOTE: The ladder should be pushed onto the roof so that the hooks are in the down position.

Step 13: Push the roof ladder up the roof until the hooks go over the edge of the peak and catch solidly.

NOTE: Remove the roof ladder by reversing the process.

SKILL SHEET 9-15 ROOF LADDER DEPLOYMENT

Hooks-First Method
Two Firefighters

NOTE: Firefighter #1 is located near the butt of the roof ladder. Firefighter #2 is located near the tip of the roof ladder.

Step 1: ***Both Firefighters:*** Carry the roof ladder to the ladder that has been raised.

NOTE: Use the low-shoulder carry, hooks (tip) first.

Step 2: ***Firefighter #2:*** Open the hooks in such a manner that the hooks face outward.

Step 3: ***Firefighter #2:*** Ascend the raised ladder using a free hand on the beam for support.

NOTE: Both firefighters will complete Step 3 if the height of the roof requires both of them to climb the ladder.

Step 4: ***Firefighter #2:*** Leg lock in or connect a safety belt to the ladder when reaching the roof edge.

Step 5: ***Both Firefighters:*** Remove the roof ladder from the shoulders.

Step 6: ***Both Firefighters:*** Push the ladder on its beam onto the roof.

Step 7: ***Firefighter #2:*** Slide the roof ladder up the roof on its beam until the balance point is reached.

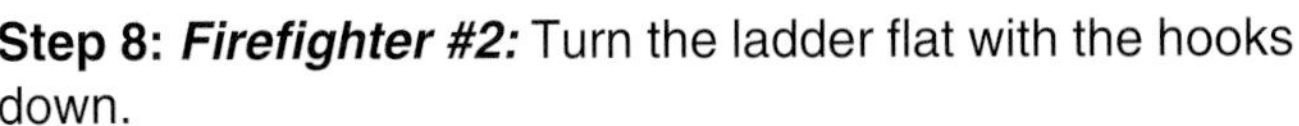

Step 8: ***Firefighter #2:*** Turn the ladder flat with the hooks down.

Step 9: ***Firefighter #2:*** Slide the ladder the remaining distance to the roof peak on the hooks until the hooks drop over the peak.

Step 10: ***Firefighter #2:*** Pull back on the roof ladder to snug it in.

SKILL SHEET 9-16 ROOF LADDER DEPLOYMENT

Butt-First Method
Two Firefighters

NOTE: **Firefighter #1 is located near the butt of the roof ladder. Firefighter #2 is located near the tip of the roof ladder.**

Step 1: ***Firefighter #1:*** Lower the butt of the roof ladder to the ground adjacent to the ladder that has been raised.

Step 2: ***Firefighter #2:*** Maintain the carry position.

Step 3: ***Firefighter #1:*** Assume a heeler position.

Step 4: ***Firefighter #2:*** Shift out of the carry position and raise the ladder to vertical.

Step 5: ***Firefighter #2:*** Stand the roof ladder alongside the other ladder.

Step 6: ***Firefighter #1:*** Steady the roof ladder.

Step 7: ***Firefighter #1:*** Heel the climbing ladder while holding the roof ladder.

NOTE: Step 7 may be omitted if it is possible to either secure the butt of the climbing ladder with a rope or use a third firefighter as the heeler.

Step 8: ***Firefighter #2:*** Climb to a point near the tip.

Step 9: ***Firefighter #2:*** Leg lock in.

Step 10: ***Firefighter #2:*** Open the hooks away from the body.

Step 11: ***Both Firefighters:*** Push the ladder upward.

Step 12: ***Firefighter #2:*** Slide the roof ladder up the roof on its beam until the balance point is reached.

Step 13: ***Firefighter #2:*** Turn the ladder flat, hooks down.

Step 14: ***Firefighter #2:*** Slide the ladder the remaining distance to the roof peak on its hooks until the hooks drop over the peak.

Step 15: ***Firefighter #2:*** Pull back on the roof ladder to snug it in.

SKILL SHEET 9-17 TWO-FIREFIGHTER LADDER PIVOT

Step 1: ***Both Firefighters:*** Face each other through the ladder.

Step 2: ***Both Firefighters:*** Grasp the ladder with both hands.

Step 3: ***Appropriate Firefighter:*** Place a foot against the side of the beam on which the ladder will pivot.

Step 4: ***Both Firefighters:*** Tilt the ladder onto the pivot beam.

Step 5: ***Both Firefighters:*** Pivot the ladder 90 degrees. Simultaneously adjust positions as necessary.

Step 6: ***Both Firefighters:*** Repeat the process until the ladder is turned a full 180 degrees and the fly is in the proper position.

NOTE: **When firefighters become proficient in this maneuver, they may be able to pivot the ladder 180 degrees in one step.**

SKILL SHEET 9-18 ONE-FIREFIGHTER LADDER SHIFT

Step 1: Face the ladder.

Step 2: Heel the ladder.

Step 3: Grasp the beams.

Step 4: Bring the ladder outward to vertical.

Step 5: Shift your grip on the ladder, one hand at a time, so that one hand grasps as low a rung as convenient, palm upward.

Step 6: Grasp a rung as high as convenient with the other hand, palm downward.

Step 7: Turn slightly in the direction of travel.

Step 8: Check visually the terrain and the area overhead.

Step 9: Lift the ladder and proceed forward a short distance.

Step 10: Watch the tip as it is being moved.

WARNING
Do not attempt this procedure close to overhead wires.

Step 11: Set the ladder down at the new position.

Step 12: Switch your grip back to the beams.

Step 13: Heel the ladder.

Step 14: Lower the ladder into position.

SKILL SHEET 9-19 TWO-FIREFIGHTER LADDER SHIFT

Step 1: ***Both Firefighters:*** Position on opposite sides of the ladder.

NOTE: If the ladder is not vertical, it is brought to vertical; if extended, it is fully retracted.

Step 2: ***Both Firefighters:*** Position hands.

NOTE: One hand grasps as low a rung as convenient, palm upward. The other hand grasps a rung as high as convenient, palm downward. The side grasped low by one firefighter is grasped high by the other.

Step 3: ***Both Firefighters:*** Lift the ladder just clear of the ground.

Step 4: ***Both Firefighters:*** Watch the tip while shifting the ladder to the new position.

Step 5: ***Both Firefighters:*** Re-extend the ladder (if necessary).

Step 6: ***Both Firefighters:*** Lower the ladder gently into position.

SKILL SHEET 9-20 APPLYING A LEG LOCK ON A GROUND LADDER

Step 1: Climb to the desired height.

Step 2: Advance one rung higher.

Step 3: Slide the leg on the opposite side from the working side over and behind the rung that you will lock onto.

Step 4: Hook your foot either on the rung (A) or on the beam (B).

Step 5: Rest on your thigh.

Step 6: Step down with the opposite leg.

Chapter 10
Ventilation

Chapter 10
Ventilation

INTRODUCTION

Ventilation is the systematic removal and replacement of heated air, smoke, and gases from a structure with cooler air. The cooler air facilitates entry by firefighters and improves life safety for rescue and other fire fighting operations. The importance of ventilation cannot be overlooked. It increases visibility for quicker location of the seat of the fire. It decreases the danger to trapped occupants by channeling away hot, toxic gases. Ventilation also reduces the chance of flashover or backdraft.

Modern technology requires a greater emphasis on ventilation. As a result of the increased use of plastics and other synthetic materials, the fuel load in all occupancies has increased dramatically. The products of combustion produced during fires are becoming more dangerous and are in larger quantities than ever before. Prompt ventilation for the saving of lives, suppression of fire, and reduction of damage becomes more important every day.

Modern energy conservation practices using increased insulation may be creating additional ventilation problems. In addition, energy-saving glass, insulated steel entry doors, and entire building vapor barriers make heat retention much greater. This means that the heat from a fire is retained better, and flashover may occur much faster than it would in a less-insulated structure.

A *roof covering* is the exposed part of the roof. Its primary purpose is to afford protection against the weather. Roof coverings may be wood shingles, composition shingles, composition roofing paper, tile, slate, synthetic membrane, or a built-up tar and gravel surface. The type of roof covering is important from a fire protection standpoint because it may be subjected to sparks and blazing embers. Insulation installed over roof coverings of fire-rated roof construction effectively retains heat and may reduce the fire rating drastically, causing premature roof failure. Therefore, the need for ventilation is increased, and it must be accomplished much sooner than has been practiced in the past.

Firefighters must be aware of how the roofs in their response areas are constructed. Pre-incident surveys should note roof construction and areas where extra insulation has been added to existing roofs and attic areas. Also, current methods for reducing costs in building construction consist of using less expensive materials in the construction process. One of these methods is the use of lightweight building materials. Lightweight roofs consisting of wooden I beam and truss construction create a collapse hazard when involved in fires. Information gathered during pre-incident planning will alert firefighters to these possible problems while performing ventilation.

When a fire officer determines the need for ventilation, he must consider the precautions necessary to control the fire and assure the safety of firefighters performing the ventilation. Firefighters must wear full protective clothing, including SCBA. A charged hoseline should be available (Figure 10.1). Before, during, and following the ventilation operation, it is important to consider the possibility of fire spreading throughout a building and the danger of exposure fires.

This chapter covers the basics of ventilation operations, the advantages of proper ventilation,

Figure 10.1 A charged hoseline should be present when cutting a vent hole.

and considerations for deciding if and where to ventilate. Also, vertical (roof or topside ventilation), horizontal (using wall openings such as doors or windows), and forced (using fans or fire streams) ventilation procedures are covered. Finally, the effects of building ventilation systems in fire situations are covered.

ADVANTAGES OF VENTILATION

[NFPA 1001: 3-3.10 (a)]

Ventilation during fire fighting aids in meeting fire fighting objectives. There are certain advantages to the overall fire fighting operation that result from proper ventilation. The following sections describe some of the advantages.

Rescue Operations

Proper ventilation simplifies and expedites rescue by removing smoke and gases that endanger trapped or unconscious occupants. The replacement of heat, smoke, and gases with cooler, fresh air helps victims breathe better (Figure 10.2). Proper ventilation also makes conditions safer for firefighters and improves visibility so that unconscious victims may be located more easily.

Fire Attack and Extinguishment

Ventilation must be closely coordinated with fire attack. When a ventilation opening is made in the upper portion of a building, a *chimney* effect (drawing air currents from throughout the building in the direction of the opening) occurs (Figure 10.3). For example, if this opening is made directly over a fire, it tends to localize the fire. If it is made elsewhere, it may contribute to the spread of the fire (Figure 10.4). The channeling effect from a properly placed hole aids in the removal of smoke, gases, and heat from a building, which in turn permits firefighters to more rapidly locate the fire and proceed with extinguishment. It also reduces the chance of firefighters receiving steam burns when the water converts to steam. Proper ventilation reduces obstacles, such as limited visibility and excessive heat, that hinder firefighters while they perform fire extinguishment, salvage, rescue, and overhaul procedures.

Property Conservation

Rapid extinguishment of a fire reduces water, heat, and smoke damage. Proper ventilation assists in making this damage reduction possible. One method of ventilation that may prove advantageous is applying water to the heated area in the form of water fog or spray. The gases and smoke may be dissipated, absorbed, or expelled by the rapid expansion of the water when it is converted to steam. In addition to removing gases, smoke,

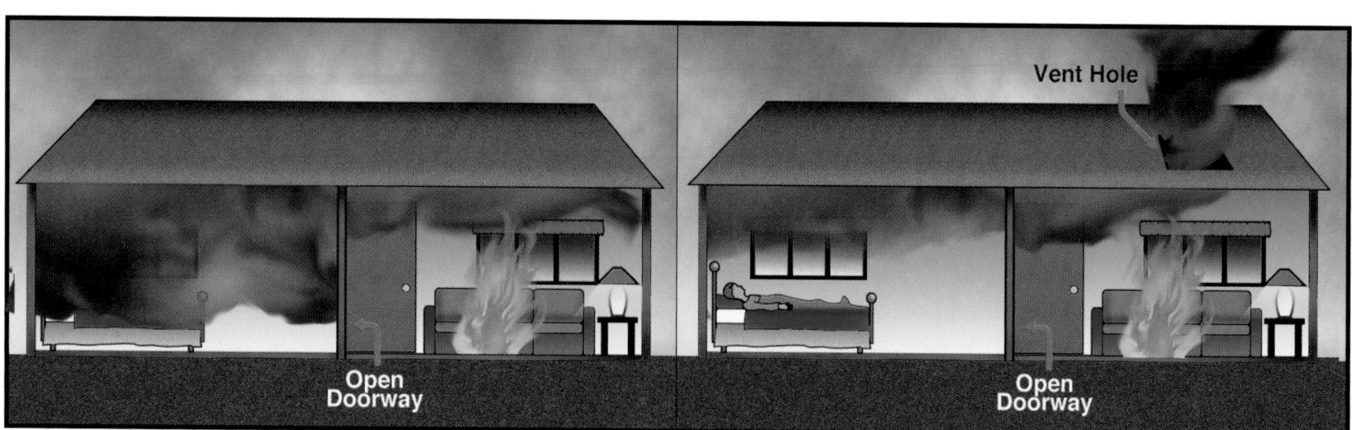

Figure 10.2 Ventilation increases the firefighter's ability to see in the structure and helps lift the smoke and toxic gases from around the victims.

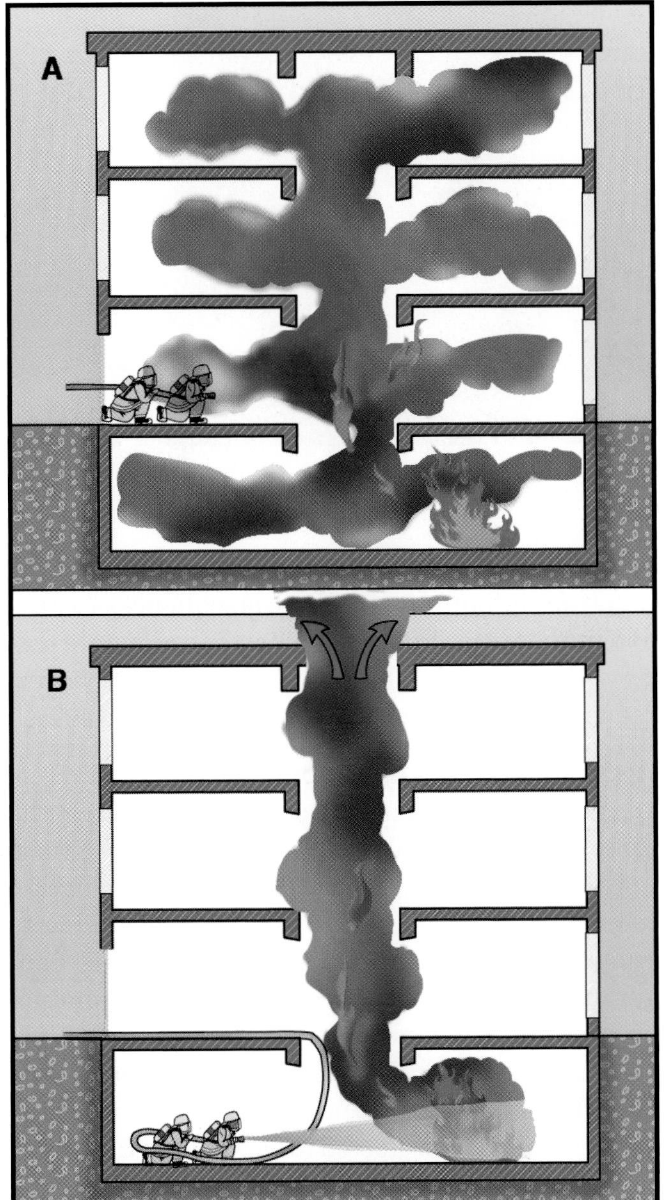

Figure 10.3 This illustrates the chimney effect in the building.

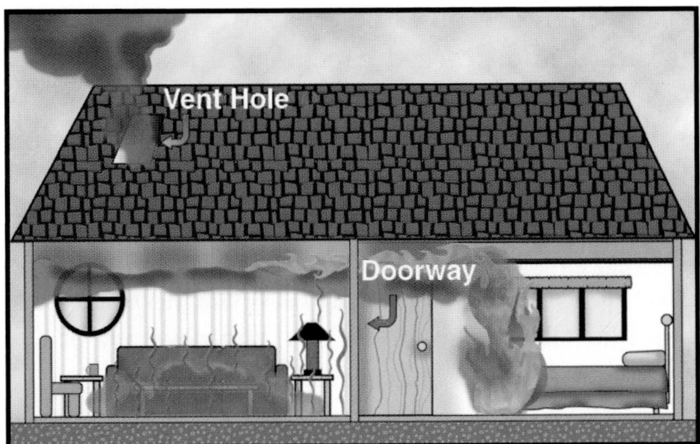

Figure 10.4 If the vent opening is not placed directly over the fire, the fire will be pulled through the structure toward the opening, thus increasing the fire damage to the structure.

and heat, this method also reduces the amount of water that may be required to extinguish the fire.

Smoke may be removed from burning buildings by controlling heat currents, by dissipating smoke through the expansion of water as it turns into steam, or by mechanical processes. Mechanical processes include the use of fans (Figure 10.5). Regardless of the method used, ventilation reduces smoke damage.

When smoke, gases, and heat are removed from a burning building, the fire can be quickly confined to an area. This permits effective salvage operations to be initiated even while fire control is being accomplished.

Figure 10.5 Mechanical fans help speed the ventilation process.

Fire Spread Control

Convection causes heat, smoke, and fire gases to travel upward to the highest point in an area until they are trapped by a roof or ceiling. As the heat, smoke, and fire gases are trapped and begin to accumulate, they bank down and spread laterally to involve other areas of the structure (Figure 10.6). This process is generally termed *mushrooming*.

Proper ventilation of a building during a fire reduces the possibility of mushrooming. This in turn reduces the rate at which fire will spread over an area by providing an escape for the rising

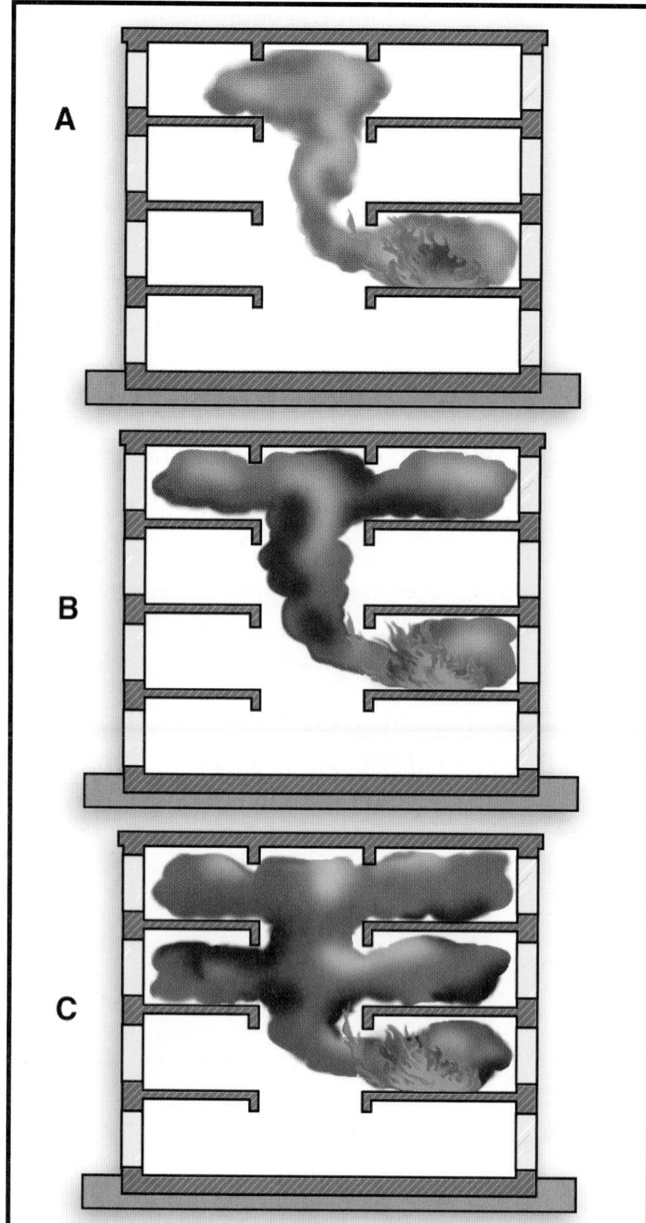

Figure 10.6 Mushrooming will occur in unventilated structures.

heated gases, at least for a short time. However, even with proper ventilation, if the fire is not extinguished soon after ventilation is accomplished, the increased supply of fresh air will feed the fire and eventually allow it to grow. Therefore ventilation should occur when the hoseline crews are ready to move in and attack the fire (Figure 10.7).

Reduction of Flashover Potential

Flashover is the transition between growth and the fully developed fire stages. As the fire continues to burn, combustibles in the room are heated to their ignition temperatures. Once their ignition

Figure 10.7 A hoseline must be ready and in place when ventilation is being performed.

temperatures are reached, the entire room will be involved in flames with dire consequences to anyone in the room. Ventilation helps to alleviate this condition because the heat is removed before it reaches the necessary levels for mass ignition.

Reduction of Backdraft Potential

When sufficient heat is confined in an area, the temperatures of combustible materials rise to their ignition points. These materials will not ignite, however, unless sufficient oxygen is available to support combustion. In this situation, a very dangerous condition exists because the admittance of an air supply (which provides the necessary oxygen) is all that is needed to explosively change the superheated area into an instant inferno. This sudden ignition is referred to as *backdraft*. In order to prevent this critical situation from occurring, top (vertical) ventilation must be provided to release superheated fire gases and smoke.

Firefighters must be aware of this explosion potential and proceed cautiously in areas where excessive amounts of heat have accumulated. As previously explained in Chapter 2, Fire Behavior, firefighters should be observant for the signs of possible backdraft conditions. If any signs of backdraft are present, firefighters should stay away from doors and windows until vertical ventilation has had the chance to reduce the severity of the situation. These signs include the following:

- Smoke-stained windows
- Smoke puffing at intervals from the building (appearance of breathing)

- Pressurized smoke coming from small cracks
- Little visible flame from the exterior of the building
- Black smoke becoming dense gray yellow
- Confinement and excessive heat

CONSIDERATIONS AFFECTING THE DECISION TO VENTILATE

[NFPA 1001: 3-3.10(a); 3-3.11(a)]

The requirements for a plan of attack must be considered before a fire officer directs or orders ventilation to be started. A series of decisions should first be made that pertain to ventilation needs. These decisions, by the nature of fire situations, fall into the following order:

- *Is there a need for ventilation at this time?* The need must be based upon the heat, smoke, and gas conditions within the structure, structural conditions, and the life hazard.
- *Where is ventilation needed?* This involves knowing construction features of the building, contents, exposures, wind direction, extent of the fire, location of the fire, location of top or vertical openings, and location of cross or horizontal openings.
- *What type of ventilation should be used?* Horizontal (natural or mechanical)? Vertical (natural or mechanical)?
- *Do fire and structural conditions allow for safe roof operations?*

To answer these questions, firefighters have to evaluate several pieces of information and take into account numerous factors. These are detailed in the following sections.

Life Safety Hazards

Dealing with the danger to human life is of utmost importance. The first consideration is the safety of firefighters and occupants. Life hazards are generally reduced in an occupied building involved by fire if the occupants are awake. If, however, the occupants were asleep when the fire developed and are still in the building, either of two situations may be expected: First, the occupants may have been overcome by smoke and gases. Second, they might have become lost in the building and are probably panicking. In either case, proper ventilation will be needed in conjunction with rescue operations. Depending on fire conditions, ventilation may need to be performed before rescue operations begin, or if conditions warrant, spreading flames may need to be attacked first; sometimes both must be performed simultaneously.

In addition to the hazards that endanger occupants, there are potential hazards to firefighters and rescue workers. The type of structure involved, whether natural openings are adequate, and the need to cut through roofs, walls, or floors (combined with other factors) add more problems requiring consideration to the decision process. The hazards that can be expected from the accumulation of smoke and gases in a building include the following:

- Obscurity caused by dense smoke
- Presence of poisonous gases
- Lack of oxygen
- Presence of flammable gases
- Backdraft
- Flashover

Visible Smoke Conditions

When first arriving at the scene of a fire, firefighters can make some ventilation decisions, as well as other tactical decisions, based on visible smoke conditions. Smoke accompanies most ordinary forms of combustion, and it differs greatly with the substance of the materials being burned. The density of the smoke is in direct ratio to the amount of suspended particles. Smoke conditions vary according to how burning has progressed. A developing fire must be treated differently than one in the decaying phase. A fire that is just beginning and is consuming wood, cloth, and other ordinary furnishings ordinarily gives off smoke that is of no great density (Figure 10.8). As burning progresses, the density may increase; the smoke may become darker because of the presence of large quantities of carbon particles (Figure 10.9).

The Building Involved

Knowledge of the building involved is a great asset when making decisions concerning ventila-

Figure 10.8 In the early stages of a fire, smoke is of no great density. *Courtesy of Chris Mickal.*

Figure 10.9 During the latter stages of a fire, smoke becomes markedly darker.

Figure 10.10 Note the availability and condition of fire escapes.

tion. Building type and design are the initial factors to consider in determining whether to use horizontal or vertical ventilation. Other determining factors include the following:

- Number and size of wall openings
- Number of stories, staircases, shafts, dumbwaiters, ducts, and roof openings
- Availability and involvement of exterior fire escapes and exposures (Figure 10.10)

Building permits that are issued within the fire department's jurisdiction may enable the department to know when buildings are altered or subdivided. Checking these permits will also often reveal information concerning heating, ventilating, and air-conditioning (HVAC) systems and avenues of escape for smoke, heat, and fire gases. The extent to which a building is connected to adjoining structures also has a bearing on the decision to ventilate. In-service company inspection and pre-incident planning may provide more valuable and detailed information.

HIGH-RISE BUILDINGS

The danger to occupants from heat and smoke is a major consideration in high-rise buildings. High-rise buildings are normally occupied by hospitals, hotels, apartments, and business offices. In any case, a great number of people may be exposed to danger.

Fire and smoke may spread rapidly through pipe shafts, stairways, elevator shafts, air-handling systems, and other vertical openings. These openings contribute to a *stack effect* (natural, vertical heat and smoke movement throughout a building), creating an upward draft and interfering with evacuation and ventilation (Figure 10.11).

The creation of layers of smoke and fire gases on floors below the top floor of unvented multistory buildings is possible. Smoke and fire gases travel through a building until their temperatures are reduced to the temperature of the surrounding air. When this stabilization of temperature occurs, the smoke and fire gases form layers or clouds within the building. The mushrooming effect, which is usually expected on top floors, does not occur in tall buildings until sufficient heat is built up to move the stratified, or layered, smoke and fire-gas clouds that have gathered on lower floors in an upward direction. Pre-incident planning should include tactics and strategy that can cope with the ventilation and life hazard problems inherent in stratified smoke.

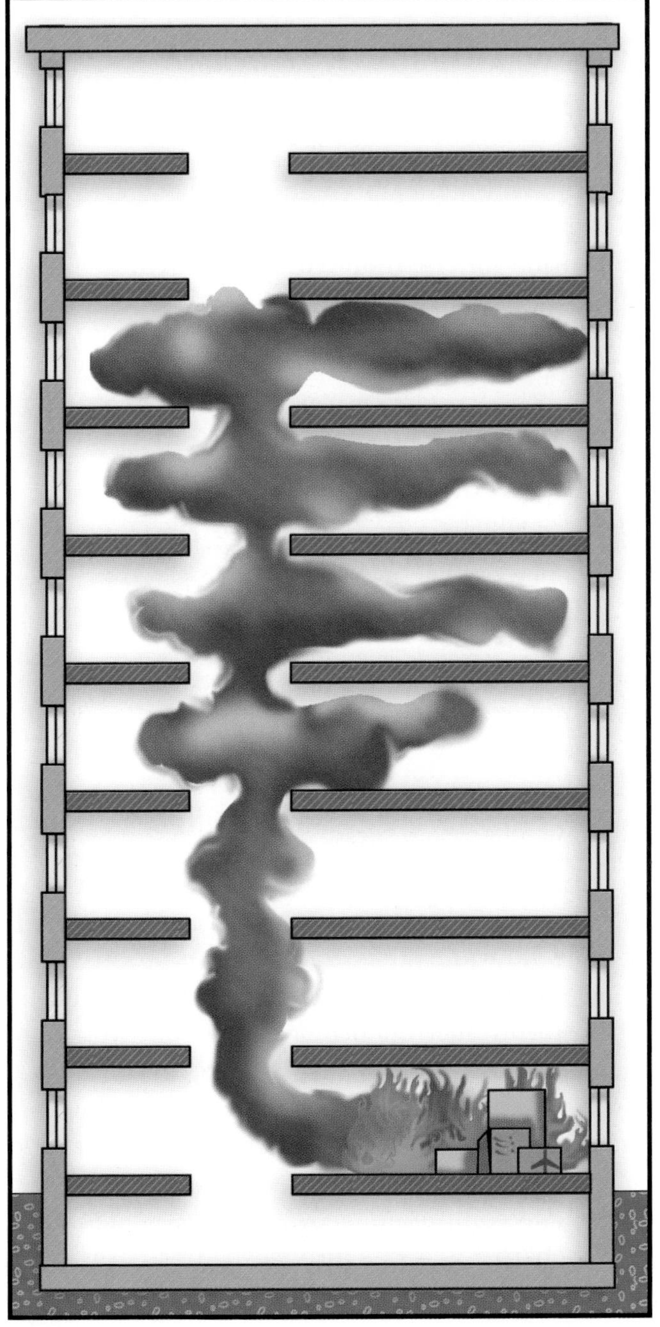

Figure 10.11 The stack effect occurs in high-rise occupancies.

Ventilation in a high-rise building must be carefully coordinated to ensure the effective use of personnel, equipment, and extinguishing agents. The personnel demand for this type of building is approximately four to six times as great as required for an average residential fire. In many instances, ventilation must be accomplished horizontally with the use of mechanical ventilation devices. Protective breathing equipment will be in great demand, and the ability to provide large quantities of fresh SCBA cylinders must be addressed. The problems of communication and coordination among the various attack and ventilating teams become more involved as the number of participants increases.

Top ventilation in high-rise buildings must be considered during pre-incident planning. In many buildings, only one stairwell pierces the roof. This vertical "chimney" must be used to ventilate smoke, heat, and fire gases from various floors. Before the doorways on the fire floors are opened and the stair shaft is ventilated, the door leading to the roof must be blocked open or removed from its hinges (Figure 10.12). Removal of the door at the top of the shaft ensures that it cannot close and allow the

Figure 10.12 Roof doors should be blocked open to facilitate ventilation.

shaft to become filled with superheated gases after ventilation tactics are started. Many elevator shafts penetrate the roof line and may be used for ventilation. Using stairwells or elevator shafts for evacuation and ventilation simultaneously is potentially life-threatening.

BASEMENTS AND WINDOWLESS BUILDINGS

Basement fires are among the most challenging that firefighters will face. Access into the basement is difficult because firefighters have to descend through the worst heat and smoke to get to the seat of the fire. Access to the basement may be via interior or exterior stairs, exterior windows, or hoistways. Many outside entrances to basements may be blocked or secured by iron gratings, steel shutters, wooden doors, or combinations of these for protection against weather and burglars (Figure 10.13). All of these features serve to impede attempts at natural ventilation.

Many buildings, especially in business areas, have windowless wall areas. While windows may not be the most desirable means for escape from burning buildings, they are an important consideration for ventilation. Windowless buildings create an adverse effect on fire fighting and ventilation operations (Figure 10.14). The ventilation of a windowless building may be delayed for a considerable time, allowing the fire to gain headway or to create backdraft conditions.

Figure 10.13 Exterior basement doors are commonly made of steel and are difficult to force open. *Courtesy of Bob Esposito.*

Figure 10.14 Buildings with few or no windows are difficult to ventilate. In this particular building, the windows above the second floor have been bricked in.

There are problems inherent in ventilating this type of building, and the problems vary depending on the size, occupancy, configuration, and type of material from which the building is constructed. Windowless buildings usually require mechanical ventilation for the removal of smoke. Most buildings of this type are automatically cooled and heated through ducts (Figure 10.15). Mechanical ventilation equipment can sometimes effectively clear the area of smoke by itself; however, these systems may also cause the spread of heat and fire.

Figure 10.15 HVAC ducts can spread the products of combustion throughout a structure.

Location and Extent of the Fire

The fire may have traveled some distance throughout the structure by the time fire fighting forces arrive, and consideration must be given to the extent of the fire as well as to its location. Opening for ventilation purposes before the fire is located may spread the fire throughout areas of the building that otherwise would not have been affected. The severity and extent of the fire usually depend upon the type of fuel and the amount of time it has been burning, installed early warning and fire protection devices, and degree of confinement of the fire. The phase to which the fire has progressed is a primary consideration in determining ventilation procedures.

Some of the ways vertical fire extension occurs are as follows:

- Through stairwells, elevators, and shafts by direct flame contact or by convected air currents (Figure 10.16)

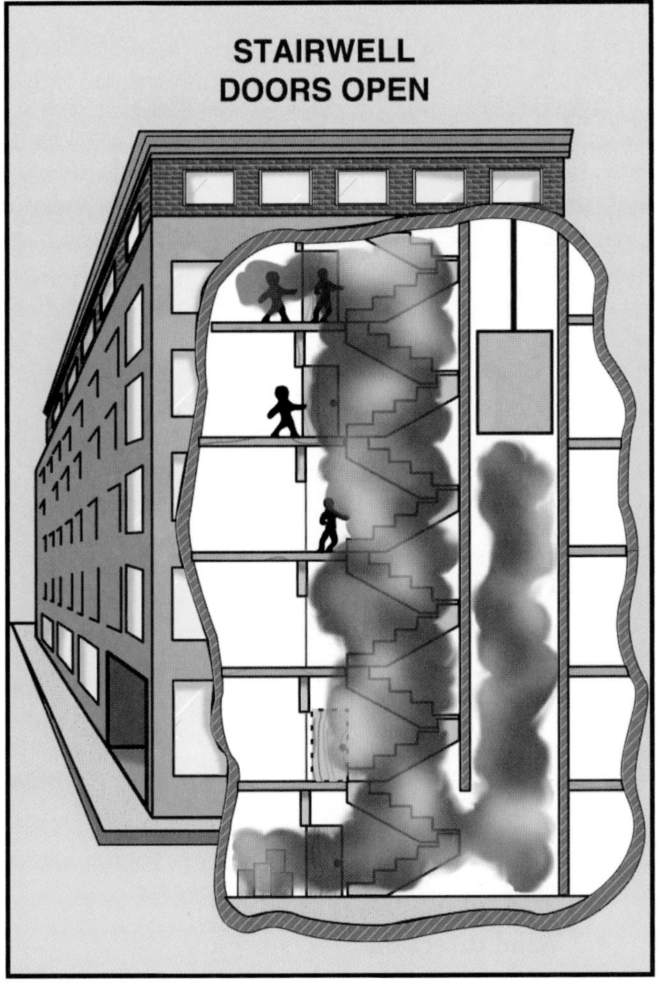

Figure 10.16 Fire and smoke can spread throughout the stairwell system if doors are left open.

- Through partitions and walls and upward between the walls by flame contact and convected air currents (Figure 10.17)
- Through windows or other outside openings where flame extends to other exterior openings and enters upper floors (commonly called *lapping*) (Figure 10.18)
- Through ceilings and floors by conduction of heat through beams, pipes, or other objects that extend from floor to floor (Figure 10.19)

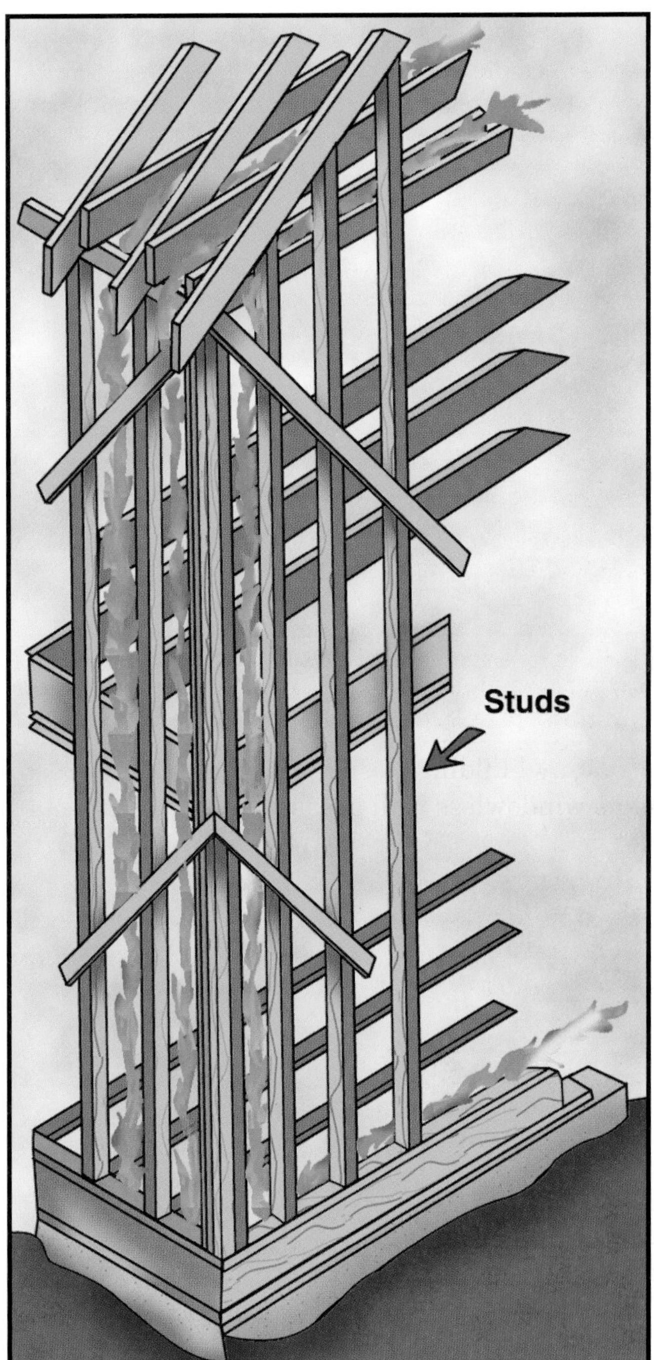

Figure 10.17 Fire will extend through open spaces within the wall assembly.

- Through floor and ceiling openings where sparks and burning material fall through to lower floors
- By the collapse of floors and roofs

Selecting the Place to Ventilate

The ideal situation in selecting a place to ventilate is one in which firefighters have prior knowledge of the building and its contents.

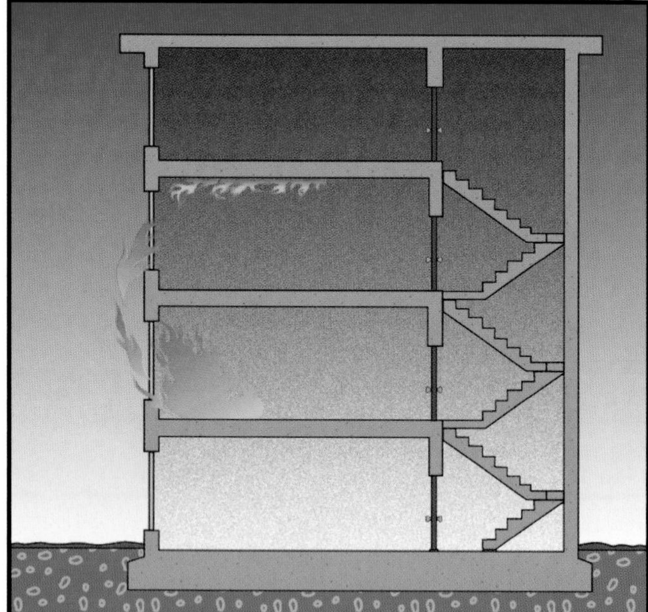

Figure 10.18 Fire can spread vertically by lapping from window to window.

Figure 10.19 Fire can spread to floors above the original fire floor by conduction through building materials or systems.

There is no rule of thumb in selecting the exact point at which to open a roof except "as directly over the fire as possible." Many factors will have a bearing on where to ventilate. Some of these factors include the following:

- Availability of natural openings such as skylights, ventilator shafts, monitors, and hatches (Figure 10.20)

Figure 10.20 Existing roof openings include (from top to bottom) scuttle hatches, skylights or monitors, chimneys and vent pipes, and stairwell openings.

- Location of the fire and the direction in which the incident commander wishes it to be drawn
- Type of building construction
- Wind direction
- Extent of progress of the fire and the condition of the building and its contents
- Bubbles or melting of roof tar
- Indications of lessening structural integrity of the roof
- Effect that ventilation will have on the fire
- Effect that ventilation will have on exposures
- Attack crew's state of readiness
- Ability to protect exposures prior to actually opening the building (Figure 10.21)

Before ventilating a building, adequate personnel and fire control equipment must be ready because the fire may immediately increase in inten-

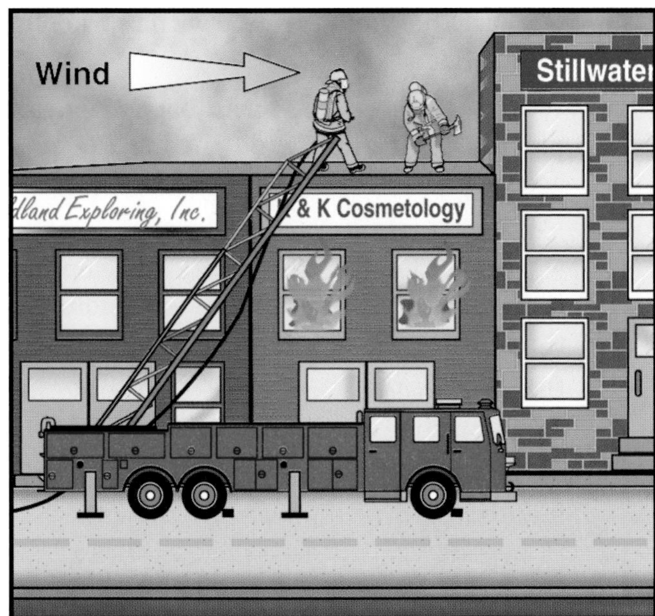

Figure 10.21 Use caution when venting next to an exposure that is taller than the fire building.

sity when the building is opened. These resources should be provided for both the building involved and other exposed buildings. As soon as the building has been opened to permit hot gases and smoke to escape, an effort to reach the seat of the fire for extinguishment should be made at once if conditions permit this to be done safely. If wind direction permits, entrance should be made into the building as near the fire as possible. It is at this opening that charged hoselines should be positioned in case of violent burning or an explosion. Charged lines should also be in place at critical points of exposure to prevent the fire from spreading.

VERTICAL VENTILATION

[NFPA 1001: 3-3.10(a); 3-3.11; 3-3.11(a); 3-3.11(b); 4-3.2; 4-3.2(a); 4-3.2(b)]

Vertical ventilation generally means opening the roof or existing roof openings for the purpose of allowing heated gases and smoke to escape to the atmosphere. In order to properly ventilate a roof, the firefighter must understand the basic types and designs of roofs. Many designs are used, and their names vary with the locality.

A study of local roof types and the manner in which their construction affects opening procedures is necessary to develop effective vertical ventilation policies and procedures. The firefighter is concerned with three prevalent types of roof shapes: *flat, pitched,* and *arched* (Figure 10.22). Buildings may be constructed with a combination of roof designs. Some of the more common styles are the flat, gable, gambrel, shed, hip, mansard, dome, lantern, and butterfly (Figure 10.23).

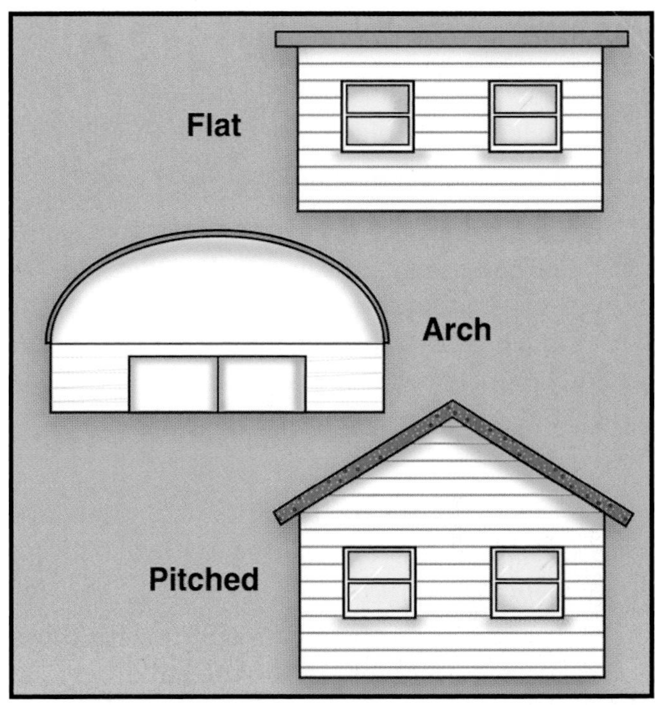

Figure 10.22 The three basic shapes of roofs are flat, arched, and pitched.

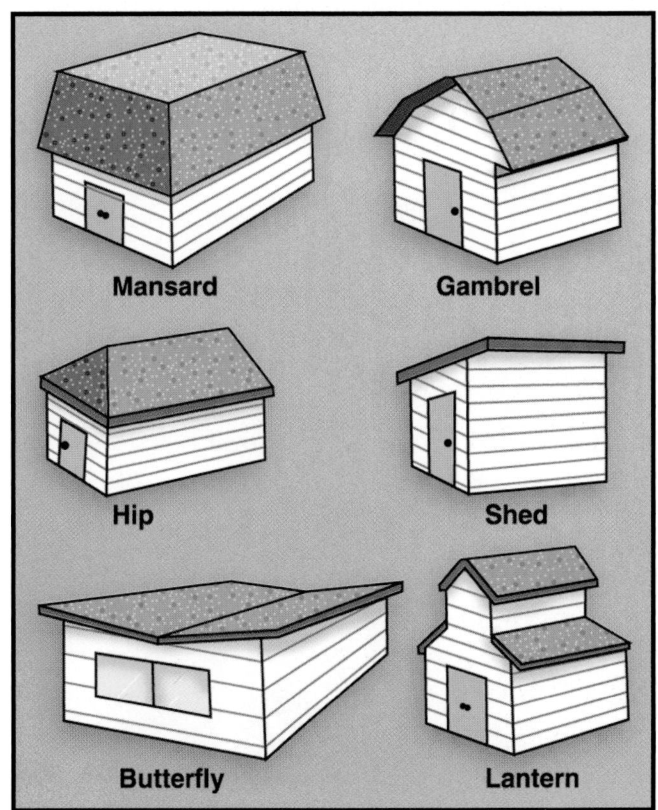

Figure 10.23 Common roof styles.

Vertical ventilation can be undertaken after the fire officer has completed the following:

- Considered the type of building involved
- Considered the location, duration, and extent of the fire
- Observed safety precautions
- Identified escape routes
- Selected the place to ventilate
- Moved personnel and tools to the roof

The roof team should be in constant communication with the incident commander. Portable radios are most adaptable to this type of communication (Figure 10.24). Responsibilities of the leader on the roof include the following:

- Ensuring that only the required openings are made
- Directing efforts to minimize secondary damage (damage caused by fire fighting operations)
- Coordinating the crew's efforts with those of the firefighters inside the building
- Ensuring the safety of all personnel who are assisting in the opening of the building

Safety Precautions

Some of the safety precautions that should be practiced include the following:

- Observe the wind direction with relation to exposures.
- Work with the wind at your back or side to provide protection while cutting the roof opening.
- Note the existence of obstructions or excess weight on the roof. These may make operations more difficult or reduce the amount of time before a roof fails.
- Provide a secondary means of escape for crews on the roof (Figure 10.25).
- Exercise care in making the opening so that main structural supports are not cut.
- Guard the opening to prevent personnel from falling into the building.
- Evacuate the roof promptly when ventilation work is complete.
- Use lifelines, roof ladders, or other means to protect personnel from sliding and falling off the roof.
- Make sure that a roof ladder (if used) is firmly secured over the peak of the roof before operating from it.
- Exercise caution in working around electric wires and guy wires.
- Ensure that all personnel on the roof are wearing full personal protective equipment including SCBA.

Figure 10.24 By using a portable radio, the leader on the roof can have constant communication with the incident commander.

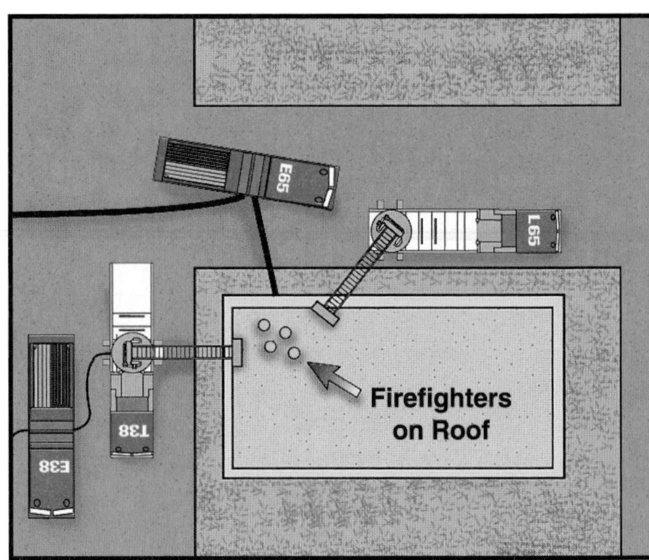

Figure 10.25 Provide two means of escape from the roof. They should be well apart from each other but close to the work area.

- Keep other firefighters out of the range of those handling axes and operating power saws.
- Caution axe users to beware of overhead obstructions within the range of their axe.
- Start power tools on the ground to ensure operation; however, it is important that the tools be shut off before hoisting or carrying them to the roof.
- Make sure that the angle of the cut is not toward the body.
- Extend ladders at least five rungs above the roof line and secure the ladder. When using elevating platforms, the floor of the platform should be even with or slightly above roof level.
- Check the roof for structural integrity before stepping on it; do not jump onto a roof without checking it first (Figure 10.26).
- Use pre-incident planning and surveys to identify buildings that have roofs supported by lightweight or wooden trusses. Realize that these roofs may fail early into a fire and are extremely dangerous to be on or under.
- Be aware of the following warning signs of an unsafe roof condition:
  — melting asphalt
  — “spongy” roof (a normally solid roof that springs back when walked upon)
  — smoke coming from the roof
  — fire coming from the roof

Figure 10.26 Check the roof for structural integrity before stepping on it.

- Work in groups of at least two, with no more people than absolutely necessary to get the job done.

Existing Roof Openings

Existing roof openings, such as scuttle hatches, skylights, monitors, ventilating shafts, and stairway doors, may be found on various types of roofs (Figure 10.27). Almost every roof opening will be locked or secured in some manner. Scuttle hatches are normally square and large enough to permit a person to climb onto the roof (Figure 10.28). A scuttle hatch may be metal or wood, and generally, it does not provide an adequate opening for ventilation purposes. If skylights contain ordinary shatter-type glass, they may be conveniently opened. If they contain wired glass, Plexiglas® acrylic plastic, or Lexan® plastic, they are very difficult to shatter and are more easily opened by removing the frame. The sides of a monitor may contain glass (which is easily removed) or louvers made of wood or metal (Figure 10.29). The sides, which are hinged,

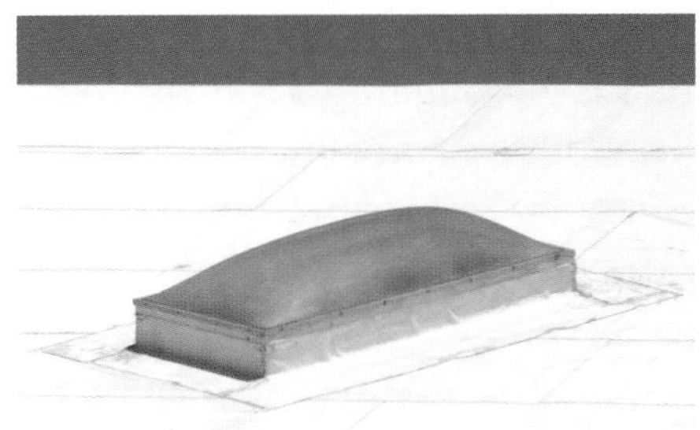

Figure 10.27 A skylight is one type of existing roof opening.

Figure 10.28 A scuttle hatch.

Figure 10.29 Louvers on a roof monitor.

are easily forced at the top. If the top of the monitor is not removable, at least two sides should be opened to create the required draft. Stairway doors may be forced open in the same manner as other doors of the same type.

Existing openings should be used for vertical ventilation purposes whenever possible. Typically, it is quicker to open one of these existing openings than it is to cut a hole in the roof. However, firefighters must realize that these openings are rarely in the best location or large enough for adequate ventilation. Most often they will simply supplement holes that have to be cut.

Roofs

The best way for fire departments to determine the material from which roofs are constructed is through pre-incident planning surveys. When cutting through a roof, the firefighter should make the opening rectangular or square to facilitate repairs to the roof. One large opening, at least 4 x 4 feet (1.2 m by 1.2 m), is much better than several small ones (Figure 10.30).

Power equipment for opening roofs is most useful and often provides a means by which ventilation procedures may be accelerated. Rotary rescue saws, carbide-tipped chain saws, or ventilation saws are excellent for roof-cutting operations. Care should always be taken to ensure that the saw operator has good footing and does not operate the saw in a manner that might allow it to accidentally come in contact with any parts of the body. Always turn off the saw when it is being transported to or from the point of operation.

Fire departments should formulate plans for dealing with types of roof construction specific to their jurisdiction. Other types of openings used in roof operations include kerf cuts, inspection openings, and louvered cuts (Figure 10.31).

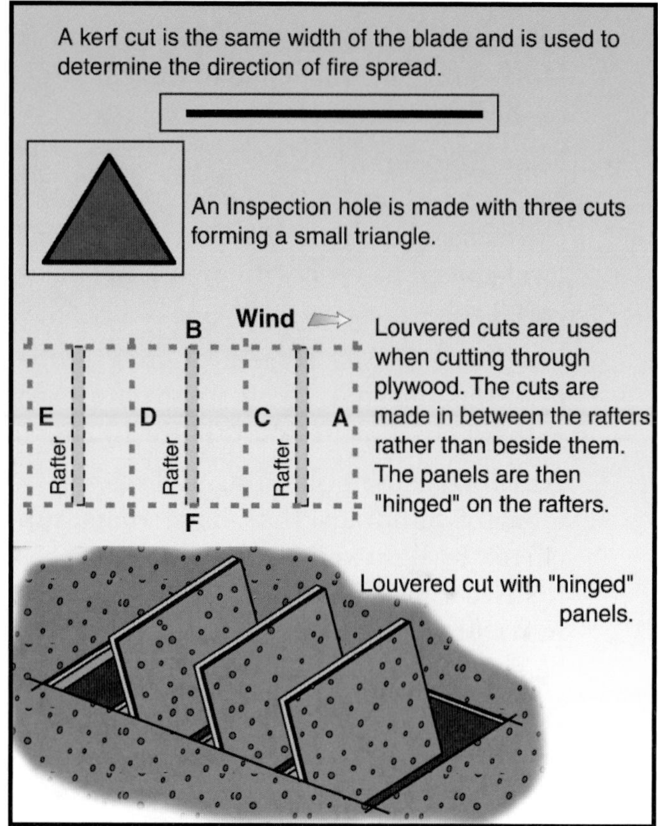

Figure 10.31 Kerf, inspection, and louvered cuts.

Figure 10.30 One large opening is better than several small openings.

FLAT ROOFS

Flat roofs are most commonly found on commercial, industrial, and apartment buildings. This type of roof may or may not have a slight slope to facilitate water drainage. The flat roof is frequently pierced by chimneys, vent pipes, shafts, scuttles, and skylights (Figure 10.32). The roof may be surrounded and/or divided by parapets, and it may support water tanks, air-conditioning equipment, antennas, and other obstructions that may interfere with ventilation operations.

The structural part of a flat roof is generally similar to the construction of a floor that consists of wooden, concrete, or metal joists covered with sheathing. The sheathing is covered with a layer of waterproofing material and an insulating material. Instead of joists and sheathing construction, flat roofs are sometimes poured reinforced concrete or lightweight concrete, precast gypsum, or concrete slabs set within metal joists. The materials used in flat-roof construction determine what equipment will be necessary to open holes in it.

A procedure for opening a flat wood roof with a power saw is suggested in the sequence given in Skill Sheet 10-1.

Figure 10.32 The flat roof is frequently pierced by vent pipes.

PITCHED ROOFS

The pitched roof is elevated in the center and thus forms a pitch to the edges (Figure 10.33). Pitched-roof construction involves rafters or trusses that run from the ridge to a wall plate on top of the outer wall at the eaves level. The rafters or trusses that carry the sloping roof can be made of various materials. Over these rafters, the sheathing material is applied either squarely or diagonally. Sheathing is sometimes applied solidly over the entire roof. Pitched roofs sometimes have a covering of roofing paper applied before shingles are laid. Shingles may be wood, metal, composition, asbestos, slate, or tile.

Figure 10.33 Most single-family dwellings have pitched roofs.

Pitched roofs on barns, churches, supermarkets, and industrial buildings may have roll felt applied over the sheathing. This is then usually mopped with asphalt roofing tar. Instead of wood sheathing, gypsum slabs, approximately 2 inches (50 mm) thick, may be laid between the metal trusses of a pitched roof. These conditions can only be determined by pre-incident surveys conducted by fire department personnel.

Pitched roofs have a more pronounced downward incline than flat roofs. This incline may be gradual or steep. The procedures for opening pitched roofs are quite similar to those for flat roofs except that additional precautions must be taken to prevent slipping. Suggested steps for opening pitched roofs are given in Skill Sheet 10-2.

Other types of pitched roofs may require different opening techniques. For example, some slate and tile roofs may require no cutting. Slate and tile roofs can be opened by using a large sledgehammer to smash the slate or tile and the thin lath strips or the 1- x 4-inch boards that support the tile or slate. Tin roofs can be sliced open and peeled back with tin snips or a large device similar to a can opener.

ARCHED ROOFS

Roofs have many desirable qualities for certain types of buildings. One form of arched roof is constructed using the bowstring truss for supporting members. The lower chord of the truss may be covered with a ceiling to form an enclosed cockloft or roof space (Figure 10.34). Such concealed, unvented spaces create dangerous ventilation problems and contribute to the spread of fire and early failure of the roof.

WARNING

Many firefighters have lost their lives when a trussed roof has failed. A good rule to follow is that when a significant amount of fire exists in the truss area of a roof structure, firefighters should not be on or under a truss roof.

Figure 10.34 Notice the large open areas within the truss roof structure.

Trussless arched roofs are made of relatively short timbers of uniform length. These timbers are beveled and bored at the ends where they are bolted together at an angle to form a network of structural timbers. This network forms an arch of mutually braced and stiffened timbers (Figure 10.35). Being an arch rather than a truss, the roof exerts a horizontal reaction in addition to the vertical reaction on supporting structural components. A hole of considerable size may be cut or burned through the network sheathing and roofing anyplace without causing collapse of the roof structure. The loads are distributed to less damaged timbers around the opening.

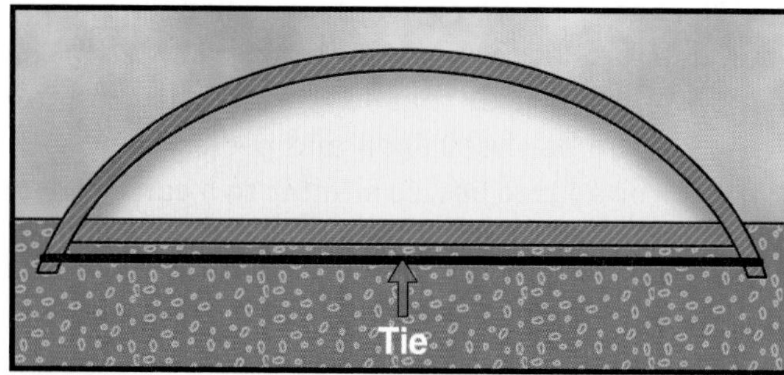

Figure 10.35 A trussless arched roof.

Cutting procedures for opening arched roofs are the same as for flat or pitched roofs except that it is doubtful that a roof ladder can always be used on an arched roof. Regardless of the method used to support the firefighter, the procedure is difficult and dangerous because of the curvature of the roof. Because of the potential for sudden collapse of this type of roof under fire conditions, firefighters should work only from an aerial ladder or platform extended to the roof.

CONCRETE ROOFS

The use of precast concrete is very popular with certain types of construction. Precast roof slabs are available in many shapes, sizes, and designs. These precast slabs are hauled to the construction site, ready for use. Other builders form and pour the concrete on the job. Roofs of either precast or reinforced concrete are extremely difficult to break through, and opening them should be avoided whenever possible (Figure 10.36). Natural roof openings and horizontal openings should be used on buildings with heavy concrete roofs.

A popular lightweight material made of gypsum plaster and portland cement mixed with aggregates, such as perlite, vermiculite, or sand, provides a lightweight floor and roof assembly. This material is sometimes referred to as lightweight concrete. Lightweight precast planks are manufactured from this material, and the slabs

Figure 10.36 Most concrete roofs are of the prefabricated type.

are reinforced with steel mesh or rods. Lightweight concrete roofs are usually finished with roofing felt and a mopping of hot tar to make them watertight.

Lightweight concrete roof decks are also poured in place over permanent form boards, steel roof decking, paper-backed mesh, or metal rib lath. These lightweight concrete slabs are relatively easy to penetrate. Some types of lightweight concrete can be penetrated with a hammerhead pick, power saw with concrete blade, jackhammer, or any other penetrating tool.

METAL ROOFS

Metal roof coverings are made from several different kinds of metal and are constructed in many styles. Light-gauge steel roof decks can either be supported on steel frameworks or they can span wider spaces. Other types of corrugated roofing sheets are made from light-gauge cold-formed steel, galvanized sheet metal, and aluminum. The light-gauge cold-formed steel sheets are used primarily for the roofs of industrial buildings. Corrugated galvanized sheet metal and aluminum are seldom covered with a roof material, and the sheets can usually be pried from their supports (Figure 10.37).

Metal cutting tools or power saws with metal cutting blades must be employed to open metal roofs. Metal roofs on industrial buildings are usually provided with adequate roof openings, skylights, or hatches. Older buildings may have roofs that are made of large, fairly thin sheets of tin laid over lath strips. These can be opened by cutting with a power saw, axe, or a large sheet-metal cutter similar to a can opener.

Figure 10.37 A corrugated sheet-metal roof.

Trench or Strip Ventilation

Trench ventilation (also referred to as *strip ventilation*) is used in a slightly different way than the standard vertical ventilation techniques previously described. Standard vertical ventilation is used simply to remove heated smoke and gases from the structure and is best done directly above the fire. Trench ventilation is used to stop the spread of fire in a long, narrow structure. *Trench ventilation* is performed by cutting a large hole, or trench, that is at least 4 feet (1.2 m) wide and extends from one exterior wall to the opposite exterior wall (Figure 10.38). This hole is usually cut well ahead of the advancing fire for the purpose of setting up a defensive line where the fire's progress will be halted.

Figure 10.38 A trench cut covers the entire width of the roof.

Basement Fires

The importance of ventilation when attacking basement fires cannot be overemphasized. In the absence of built-in vents from the basement, heat and smoke from basement fires will quickly spread upward into the building (Figure 10.39). This is especially true in buildings of balloon-frame construction where the wall studs are continuous from the foundation to the roof. There may be no firestops (wood or other solid material placed within a void to retard or prevent the spread of fire through the void) between the studs. In buildings of this type, the first extension of a basement fire will commonly be into the attic. The likelihood of

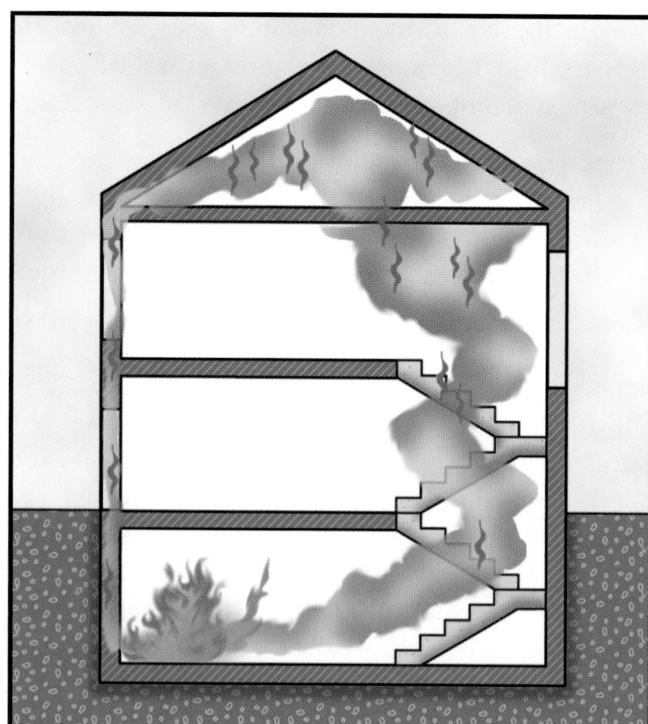
Figure 10.39 Products of combustion from a basement fire will quickly collect in the upper reaches of a structure.

vertical extension of the fire may be reduced by direct ventilation of the basement during fire attack. After the basement fire is confirmed to be extinguished, the attic may be vented to remove residual smoke.

Direct ventilation of a basement can be accomplished several ways. If the basement has ground-level windows or even belowground-level windows in wells, horizontal ventilation can be employed effectively (Figure 10.40). If these windows are not available, interior vertical ventilation must be performed. Natural paths from the basement, such as stairwells and hoistway shafts, can be used to evacuate heat and smoke provided there is a means to expel the heat and smoke to the atmosphere without placing other portions of the building in danger (Figure 10.41). As a last resort, a hole may be cut in the floor near a ground-level door or window, and the heat and smoke can be forced from the hole through the exterior opening using fans (Figure 10.42).

Figure 10.40 Some basement windows will actually be at ground level outside the building.

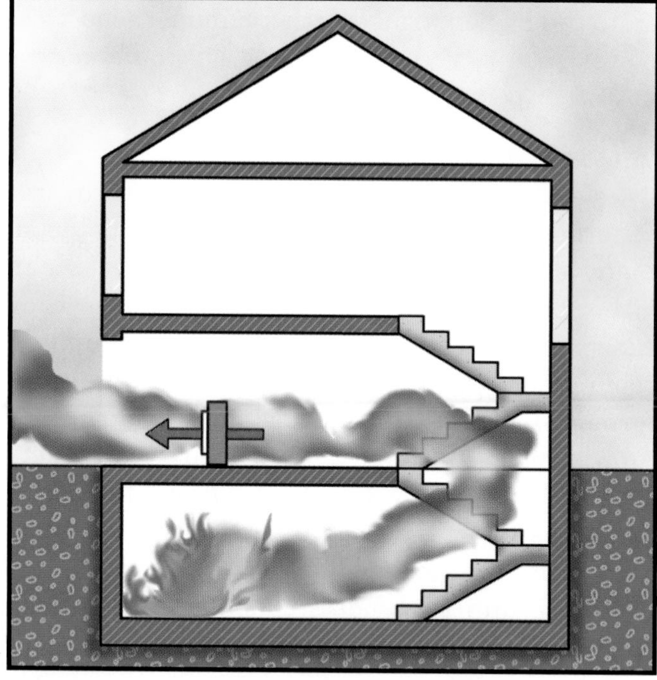
Figure 10.41 Stairways can be used to vent a basement.

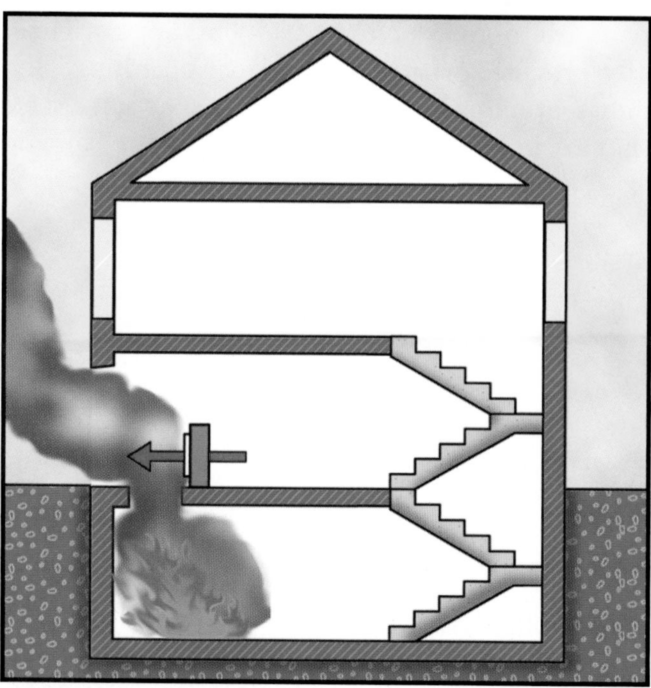
Figure 10.42 A hole may be cut in the floor, if needed, to ventilate a basement.

Precautions Against Upsetting Established Vertical Ventilation

When vertical ventilation is accomplished, the natural convection of the heated gases creates upward currents that draw the fire and heat in the direction of the upper opening. Fire fighting teams take advantage of the improved visibility and less contaminated atmosphere to attack the fire at its lower point.

Elevated streams are frequently used to lessen sparks and flying embers from a burning building or to reduce the thermal column of heat over a building. However, when elevated or handline streams are projected downward through a ventilation opening or are improperly used to reduce the thermal column, the orderly movement of fire gases from the building is either destroyed or upset. This can force superheated air and gases back down on firefighters, causing serious injury or death. At the very least, it will contribute to the spread of fire throughout the structure. Streams that are being operated just above ventilated openings should be projected slightly above the horizontal plane (Figure 10.43). In this position, they help cool the thermal column and extinguish sparks. The stream may even increase the rate of ventilation.

Figure 10.43 A fire stream directed above the vent opening can aid the ventilation process and reduce the chance of secondary fires caused by flying embers.

Ventilation problems can be avoided by well-trained firefighters conducting a well-coordinated attack. Firefighters need to be aware of some common factors that can destroy the effectiveness of vertical ventilation:

- Improper use of forced ventilation
- Excess breakage of glass
- Fire streams directed into ventilation holes
- Breakage of skylights
- Explosions
- Burn-through of the roof, a floor, or a wall
- Additional openings between the attack team and the upper opening

WARNING

Never operate any type of fire stream through a ventilation hole during offensive operations. This stops the ventilation process and places interior crews in serious danger.

Vertical ventilation cannot be the solution to all ventilation problems because there may be many instances where its application would be impractical or impossible. In these cases, other strategies, such as the use of strictly horizontal ventilation, must be employed.

HORIZONTAL VENTILATION

[NFPA 1001: 3-3.10; 3-3.10(a); 4-3.2; 4-3.2(a); 4-3.2(b)]

Horizontal ventilation is the venting of heat, smoke, and gases through wall openings such as windows and doors. Structures that lend themselves to the application of horizontal ventilation include the following:

- Residential-type buildings in which the fire has not involved the attic area
- Involved floors of multistoried structures below the top floor, or the top floor if the attic is uninvolved
- Buildings with large, unsupported open spaces under the roof in which the structure has been weakened by the effects of burning

Many aspects of vertical ventilation also apply to horizontal ventilation. However, a different procedure must be followed in horizontally ventilating a room, floor, cockloft, attic, or basement. The procedure to be followed will be influenced by the location and extent of the fire. Some of the ways by which horizontal extension occurs are as follows:

- Through wall openings by direct flame contact or by convected air
- Through corridors, halls, or passageways by convected air currents, radiation, and flame contact (Figure 10.44)
- Through open space by radiated heat or by convected air currents (Figure 10.45)
- In all directions by explosion or flash burning of fire gases, flammable vapors, or dust
- Through walls and interior partitions by direct flame contact
- Through walls by conduction of heat through beams, pipes, or other objects that extend through walls

Weather Conditions

Weather conditions are always a consideration in determining the proper horizontal ventilation procedure. The wind plays an important role in ventilation. Its direction may be designated as windward or leeward. The side of the building the wind is striking is *windward,* the opposite is *leeward* (Figure 10.46). Under certain conditions, when there is no wind, natural horizontal ventilation is less effective because the force to remove the smoke is absent. In other instances, natural horizontal ventilation cannot be accomplished due to the danger of wind blowing toward an exposure or feeding oxygen to the fire.

Exposures

Because horizontal ventilation does not normally release heat and smoke directly above the fire, some routing is necessary. Firefighters should be aware of internal exposures as well as external exposures. The routes by which the smoke and heated gases would travel to the exit may be the same corridors and passageways that occupants will be using for evacuation. Therefore, the practice of horizontal ventila-

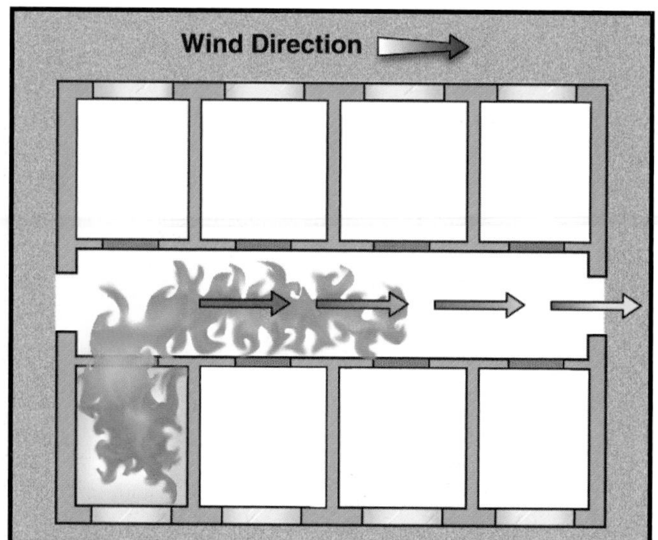

Figure 10.44 Fire can spread down a hallway, especially when the wind is aiding the process.

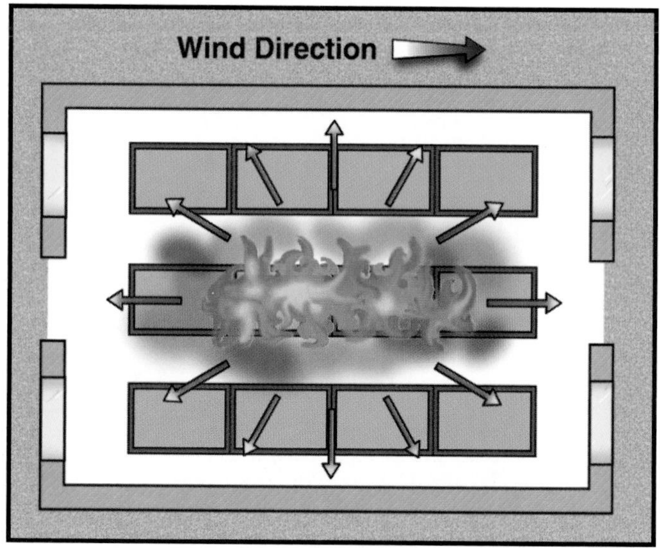

Figure 10.45 Fire can spread rapidly in all directions in a large area.

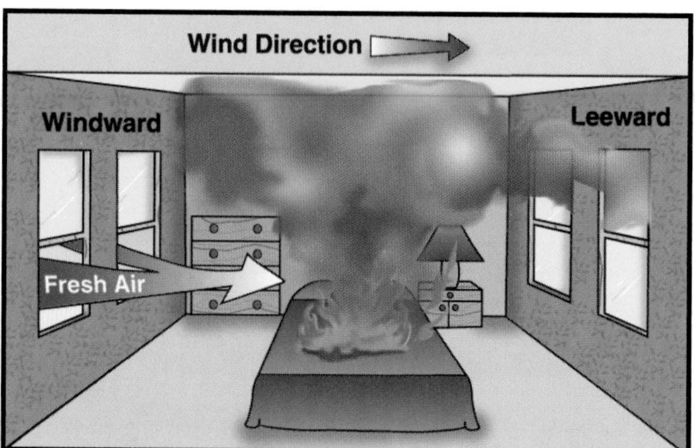

Figure 10.46 Fresh air enters the windward side, and smoke, heat, and gases exit the leeward side.

tion, without first considering occupants and rescue procedures, may block the escape of occupants. The theory of horizontal ventilation is basically the same as that of vertical ventilation inasmuch as the release of the smoke and heat is an aid in fighting the fire and reducing damage.

Because horizontal ventilation is not accomplished at the highest point of a building, there is the constant danger that when the rising heated gases are released, they will ignite higher portions of the fire building. They may ignite eaves of adjacent structures or be drawn into windows above their liberation point (Figure 10.47). Unless for the specific purpose of aiding in rescue, a building should not be opened until charged lines are in place at the attack entrance point, at the intermediate point where fire might be expected to spread, and in positions to protect other exposures (Figure 10.48).

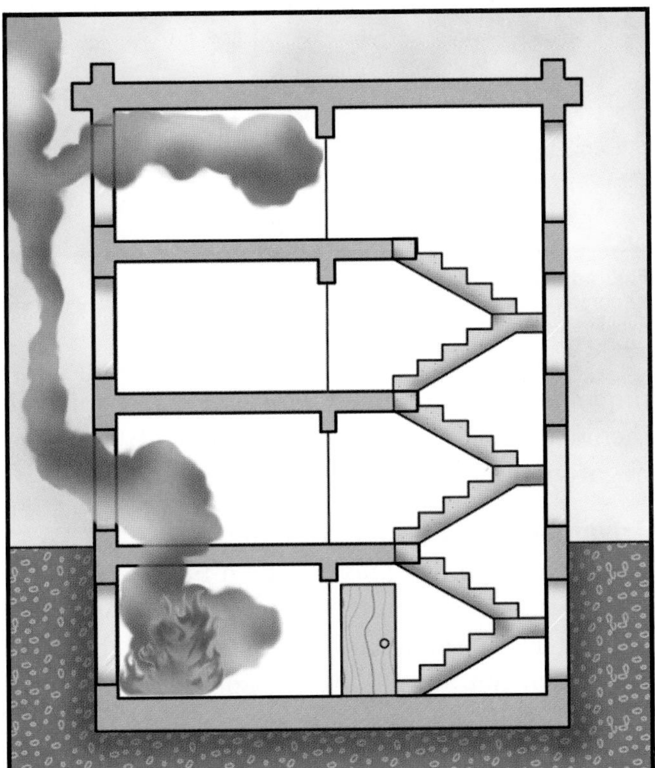

Figure 10.47 If care is not taken when ventilating, smoke removed from lower levels of a structure can be drawn back in at higher parts.

Precautions Against Upsetting Horizontal Ventilation

The opening of a door or window on the windward side of a structure prior to first opening a door on the leeward side may pressurize the building and upset the normal process of thermal layering (Figure 10.49). Opening doors and windows between the advancing fire fighting crews and the established ventilation exit point reduces the intake of fresh air from the opening behind the firefighters. Firefighters following established cross ventilation currents are illustrated in Figure 10.50. The smoke and heat will intensify if the established current is interrupted by a firefighter or other obstruction in the doorway (Figure 10.51).

Figure 10.48 Be ready to enter the structure as soon as ventilation is accomplished.

Figure 10.49 Opening extra doors or windows can disrupt the ventilation process.

Figure 10.50 Proper ventilation allows for more effective fire attacks.

Figure 10.51 A person or object blocking the ventilation opening can disrupt the whole process.

FORCED VENTILATION

[NFPA 1001: 3-3.10; 3-3.10(a); 3-3.10(b); 3-3.11(b); 4-3.2; 4-3.2(a); 4-3.2(b)]

Up to this point, ventilation has been considered from the standpoint of the natural flow of air currents and the currents created by fire. *Forced ventilation* is accomplished mechanically (with fans) or hydraulically (with fog streams). The principle applied is that of moving large quantities of air and smoke. The fact that forced ventilation is effective for heat and smoke removal when other methods are not adequate proves its value and importance.

It is difficult to classify forced ventilation equipment by any particular type. These portable fans are powered by electric motors, gasoline-driven engines, or water pressure from hoselines. Portable fans and several methods of using them to ventilate are shown in Figure 10.52.

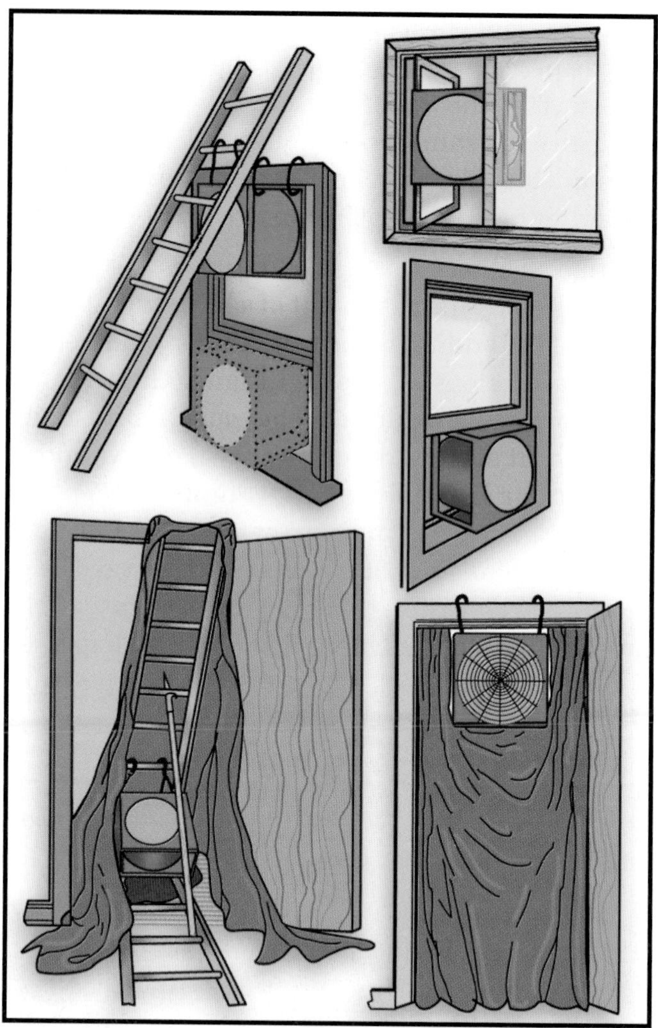

Figure 10.52 There are many ways a portable fan can be positioned in a doorway or window. Be sure to cover any open area around the fan to avoid recirculating air in the opening.

This section discusses the advantages and disadvantages of forced ventilation, the devices necessary to create forced ventilation, and the techniques used in applying forced ventilation. Also included in the discussion of forced ventilation are negative- and positive-pressure ventilation.

Advantages of Forced Ventilation

Even when fire may not be a factor, contaminated atmospheres must be rapidly and thoroughly cleared. Forced ventilation, if not the only means of clearing a contaminated atmosphere, is always a welcome addition to normal ventilation. Some of the reasons for employing forced ventilation include the following:

- Ensures more positive control of the fire
- Supplements natural ventilation
- Speeds the removal of contaminants, facilitating a more rapid rescue under safer conditions

- Reduces smoke damage
- Promotes good public relations

Disadvantages of Forced Ventilation

If forced ventilation is misapplied or improperly controlled, it can cause a great deal of harm. Forced ventilation requires supervision because of the mechanical force behind it. Some of the disadvantages of forced ventilation include the following:

- Introduces air in such great volumes that it can cause the fire to intensify and spread
- Depends upon a power source
- Requires special equipment

Negative-Pressure Ventilation

The term *negative-pressure ventilation* describes the oldest of mechanical forced ventilation techniques: using fans to develop artificial circulation and pull smoke out of a structure. Fans are placed in windows, doors, or roof vent holes, and they pull the smoke, heat, and gases from inside the building and eject them to the exterior (Figure 10.53).

In negative-pressure ventilation the fan should be placed to exhaust in the same direction as the natural wind. This will aid the ejection process by supplying fresh air to replace that which is being drawn from the building. If the natural wind is too light to be effective, fans on one side of the structure can be turned to blow air into the building while the fans on the other side are turned so that they exhaust the smoke and other combustion byproducts from the building.

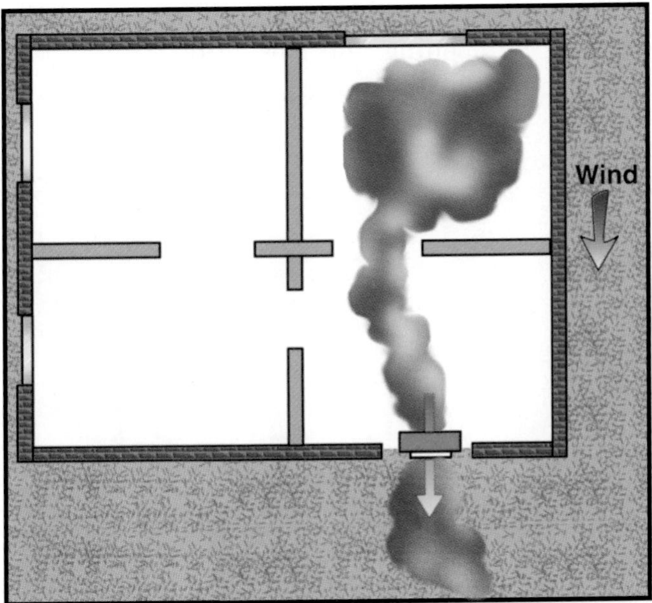

Figure 10.53 Negative-pressure ventilation is accomplished by using a portable fan to draw smoke out of a building.

Recirculating air in the opening around the fan can be a problem. When air is allowed to recirculate around the sides of the fan and in and out of nearby openings, it causes a *churning* action that reduces efficiency (Figure 10.54). If the area surrounding the fan is left open, atmospheric pressure pushes the air through the bottom of the doorway and pulls the smoke back into the room. To prevent churning air, cover the area around the unit with salvage covers or other material.

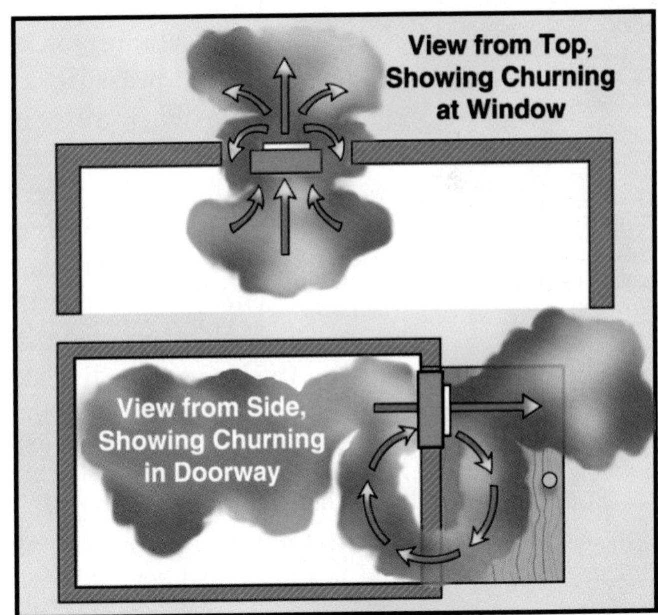

Figure 10.54 Avoid churning situations when using portable fans.

Establish the desired draft path and keep the airflow in as straight a line as possible. Every corner causes turbulence and decreases efficiency. Avoid opening windows or doors near the exhausting fan unless opening them definitely increases circulation. Remove all obstacles to the airflow. Even a window screen will cut effective exhaust by half. Avoid blockage of the intake side of the fan by debris, curtains, drapes, or anything that can decrease the amount of intake air.

Forced-air fans should always be equipped with explosion-proof motors and power cable connections when used in a flammable atmosphere. Forced-air fans should be turned off when they are moved, and they should be moved by the handles provided for this purpose. Before starting forced-air fans, be sure that no one is near the blades and that clothing, curtains, or draperies are not in a position to be drawn into the fan. The discharge stream of air should be avoided because of particles that may be picked up and blown by the venting equipment.

Positive-Pressure Ventilation

Figure 10.55 A positive-pressure ventilation fan.

Positive-pressure ventilation is a forced ventilation technique that uses the principle of creating pressure differentials. By using high-volume fans, a higher pressure is created inside a building than that of the outside environment (Figure 10.55). As long as the pressure is higher inside the building, the smoke within the building will seek an outlet to a lower-pressure zone through openings controlled by firefighters.

The location where positive-pressure ventilation is done, usually an exterior doorway, is called the *point of entry*. The fan is placed several feet (meters) outside the door so that the cone of air from the fan completely covers the door opening (Figure 10.56). The smoke is then ejected from an exhaust opening the same size as the entry opening. It is important that no other exterior openings be opened while the positive-pressure operation is in use except at the point where the smoke is to leave the building.

By closing doors within the structure and pressurizing one room or area at a time, the process of removing the smoke is speeded up because the velocity of the air movement is increased. The process can also be speeded up by placing additional fans at the entry point. If none of the doors inside the structure were opened and closed systematically, the process would still work, but it would take more time.

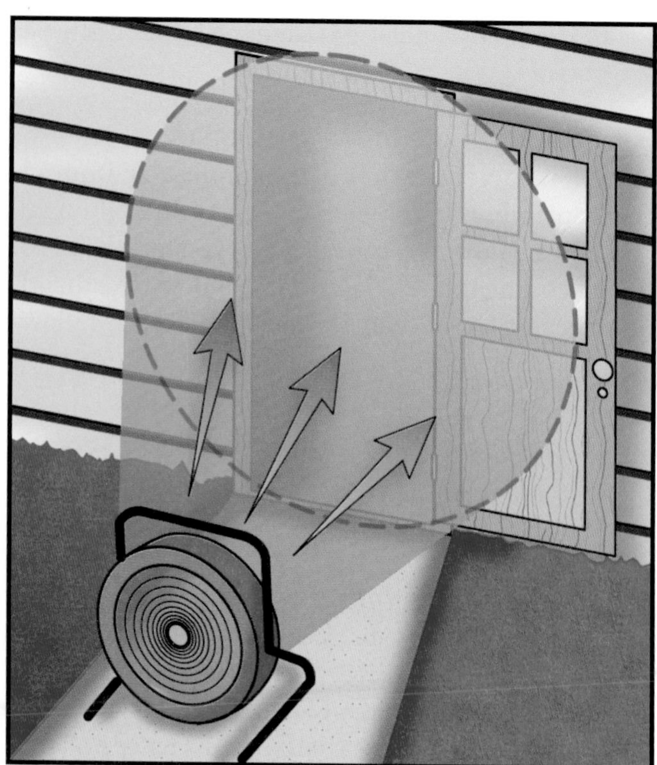

Figure 10.56 A cone of air must cover the entire opening.

When using positive pressure to remove smoke from multiple floors of a building, it is generally best to apply positive pressure at the lowest point (Figure 10.57). Smoke can then be systematically removed one floor at a time starting with the floor most heavily charged with smoke. Note that positive pressure is applied to the building at ground level through the use of one or more fans. The positive pressure is then directed throughout the building by opening and closing doors until the building is totally evacuated of smoke through any

Figure 10.57 Place the fan at a low point of the structure.

opening selected by firefighters (Figure 10.58). This is accomplished by either cross-ventilating fire floors or directing smoke up a stairwell and out the stair shaft rooftop opening.

Positive-pressure ventilation requires good fireground discipline, coordination, and tactics. The main problem in using positive-pressure ventilation in aboveground operations is coordinating the opening and closing of the doors in the stairwell being used to ventilate the building. Curious tenants will often stand with the doors to the stair shaft or the doors to their rooms open and thereby redirect the positive pressure away from the fire floor. To control openings or pressure leaks, place one person in charge of the pressurizing process. It is helpful to use portable radios and to have firefighters patrol the stairwell and hallways.

In order to ensure an effective positive-pressure ventilation operation, the following points should be taken into consideration:

- Take advantage of existing wind conditions.
- Make certain that the cone of air from the fan covers the entire entry opening.
- Reduce the size of the area being pressurized to speed up the process by systematically opening and closing doors or by increasing the number of fans.
- Keep the size of the exit opening in proportion to the entry opening.

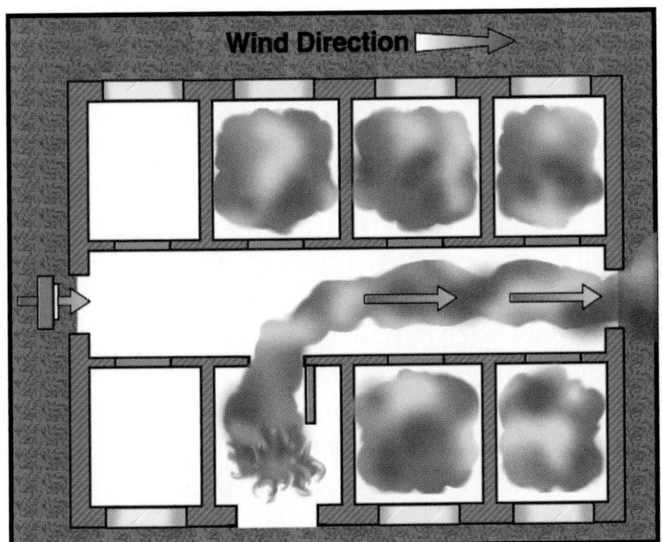

Figure 10.58 Use doors to control the ventilation effort. This could even allow the building to be cleared one room at a time.

The advantages of positive-pressure ventilation compared to negative-pressure ventilation include the following:

- Firefighters can set up forced ventilation procedures without entering the smoke-filled environment.
- Positive-pressure ventilation is equally effective with either horizontal ventilation or vertical ventilation because it merely supplements natural ventilation currents.
- More efficient removal of smoke and heat from a structure or vessel is allowed.
- The velocity of air currents within a building is minimal and has little, if any, effects that disturb the building contents or smoldering debris. Yet, the total exchange of air within the building is faster than using negative-pressure ventilation.
- Fans powered by internal combustion engines operate more efficiently in clean, oxygen-rich atmospheres.
- The placement of fans does not interfere with ingress or egress.
- The cleaning and maintenance of fans used for positive-pressure ventilation is greatly reduced compared to that of those used in negative-pressure ventilation.
- This system is applicable to all types of structures or vessels and is particularly effective at removing smoke from large, high-ceiling areas where negative-pressure ventilation is ineffective.
- Heat and smoke may be directed away from unburned areas or paths of exit.

The disadvantages of positive-pressure ventilation are as follows:

- An intact structure is required.
- Interior carbon monoxide levels may be increased.
- Hidden fires may be extended.

Hydraulic Ventilation

Hydraulic ventilation may be used in situations where other types of forced ventilation are not being used (Figure 10.59). *Hydraulic ventilation* is performed by hose teams making an interior attack on the fire. Typically, this technique is used to clear the room or building of smoke, heat, steam, and gases following the initial knockdown of the fire. This technique takes advantage of the air that is drawn into a fog stream to help push the products of combustion out of the structure.

Figure 10.59 Hydraulic ventilation in progress.

To perform hydraulic ventilation, a fog stream is set on a wide fog pattern that will cover 85 to 90 percent of the window or door opening from which the smoke will be pushed out. The nozzle tip should be at least 2 feet (0.6 m) back from the opening (Figure 10.60). The larger the opening, the faster the ventilation process will go.

There are drawbacks to the use of fog streams in forced ventilation. These drawbacks include the following:

- There may be an increase in the amount of water damage within the structure.
- There will be a drain on the available water supply. This is particularly crucial in rural fire fighting operations where water shuttles are being used.

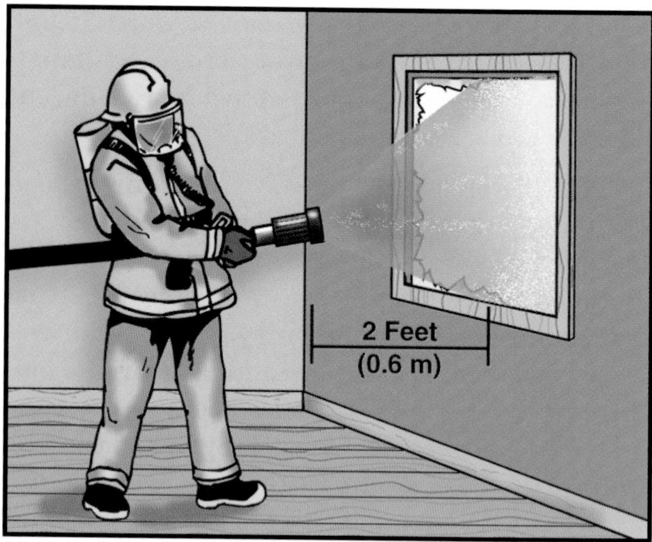

Figure 10.60 The nozzle should be 2 feet (0.6 m) back from the window.

- In climates subject to freezing temperatures, there will be an increase in the problem of ice in the area surrounding the building.
- The firefighters operating the nozzle must remain in the heated, contaminated atmosphere throughout the operation.
- The operation may have to be interrupted when the nozzle team has to leave the area for some reason (reservice SCBA, rest, etc.).

THE EFFECT OF BUILDING VENTILATION SYSTEMS IN FIRE SITUATIONS

[NFPA 1001: 3-3.10(a)]

Most modern buildings have heating, ventilation, and air-conditioning (HVAC) systems. Firefighters should be aware that these systems can significantly contribute to the spread of smoke and fire throughout a structure. Pre-incident planning should include information on the design capabilities of HVAC systems. Also included should be diagrams of the duct system throughout a building and information on fire protection systems (sprinkler, smoke, or heat detection) within the HVAC ductwork. Fire personnel should be familiar with the location and operation of controls that will manually shut down the system when so desired.

Because the system may draw heat and smoke into the duct before it is shut down, firefighters should always check around the ductwork for fire extension during overhaul operations. Also, personnel should be familiar with the best ways to rid the system of smoke before reactivating it.

Smoke control systems are used in buildings involving either a large number of people or a large quantity of combustibles such as high rises, shopping malls, and buildings with open atriums. These systems involve not only the mechanical systems, but the doors, partitions, windows, shafts, ducts, fan dampers, wire controls, and pipes. Smoke control systems should be identified during prefire planning sessions. Because of the variety and complexity of these systems, firefighters should not attempt to operate them under fire conditions. Building engineers should be called to the scene to operate the system under the fire department's direction.

SKILL SHEET 10-1 OPENING A FLAT ROOF

With a Power Saw

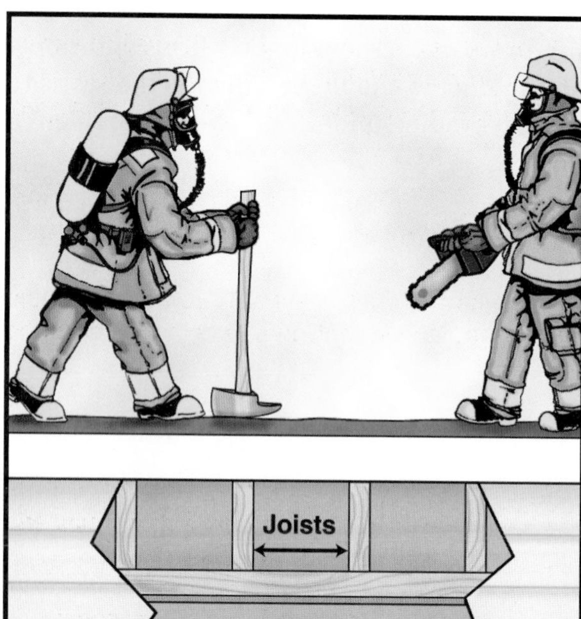

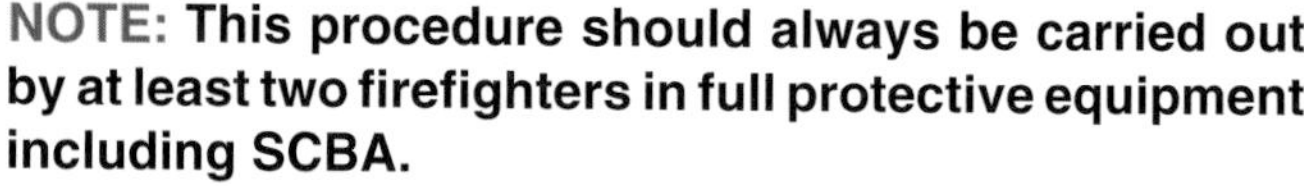

NOTE: This procedure should always be carried out by at least two firefighters in full protective equipment including SCBA.

Step 1: Determine the location for the opening using the following factors:

- — Location of seat of fire
- — Direction of wind
- — Existing exposures
- — Extent of fire
- — Obstructions

Step 2: Locate roof supports by sounding with an axe or other appropriate tool.

NOTE: The roof will sound hollow between the joists, and the axe will bounce. When near or on top of a support, the roof will sound dull and solid.

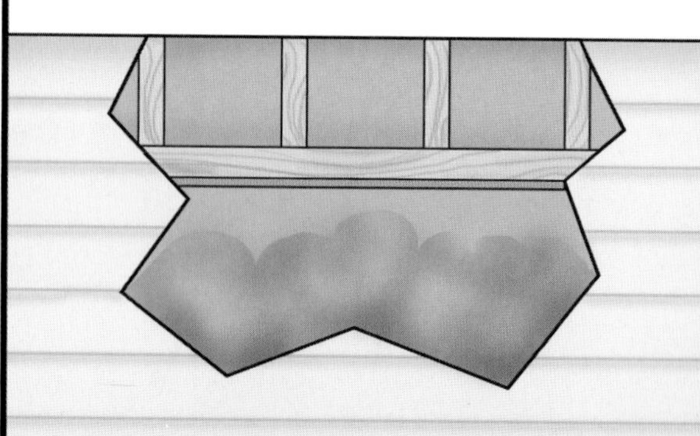

Step 3: Mark the location for the opening by scratching a line on the roof surface with the pick end of an axe.

Step 4: Position yourself on the upwind side of the planned ventilation opening.

Step 5: Prepare to make the first cut into the wood decking. This cut should be made on the side of the planned ventilation opening farthest from the ladder.

NOTE: When making the cuts, work from the area farthest away from the escape route back toward the area of safety. Do not cut a hole between you and the escape route.

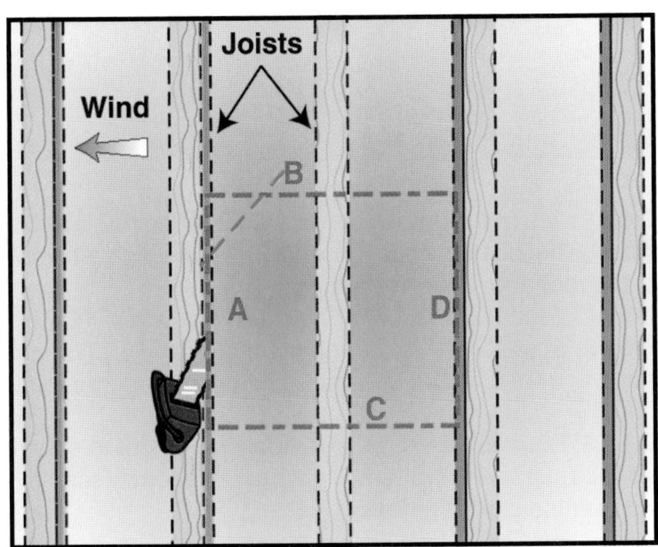

Step 6: Cut the wood decking alongside the joist (A).

NOTE: The joist should never be cut.

Step 7: Cut the decking horizontally (B) across the top of the planned ventilation opening.

NOTE: A connecting cut can be made diagonally between the vertical and horizontal cuts. This can be used as a knockout space so a pike pole or other pushing/pulling tool can be inserted.

Step 8: Cut the decking horizontally (C) across the bottom of the planned ventilation opening.

Step 9: Make the final cut (D) vertically from top to bottom on the side of the planned ventilation opening closest to the ladder.

Step 10: Pry up the sheathing material with the pick end of an axe.

Step 11: Push the blunt end of a pike pole, or some other suitable tool, through the roof opening to open the ceiling below.

NOTE: The procedures described can also be used with an axe instead of a power saw except that before making any cuts into the wood decking, you must remove the built-up roof material or metal by cutting the material and using the pick head to pull the material out of the way.

SKILL SHEET 10-2 OPENING A PITCHED ROOF

With an Axe

Step 1: Locate the position where the opening is to be made. The location should usually be at the highest point of the roof above the fire area.

Step 2: Place a roof ladder on the roof so that personnel working from it will be upwind from the hole.

Step 3: Bounce an axe or other tool on the roof to sound for solid supports or rafters. Mark their location by scratching with the pick of an axe.

Step 4: Strip off the shingles or roofing felt sufficiently to permit the initial cut to be made.

NOTE: In some cases, it is best to first remove all shingles or roofing felt from the entire area where the hole is to be made. The removal of these coverings is most important when the hole is going to be made with an axe.

Step 5: Prepare to make the first cut into the wood decking. This cut should be made on the side of the planned ventilation opening farthest from the ladder.

NOTE: When making the cuts, work from the area farthest away from the escape route back toward the area of safety. Do not cut a hole between you and the route.

Step 6: Cut the wood decking (A) alongside the rafter.

NOTE: The rafter should never be cut.

Step 7: Cut the decking horizontally (B) across the top of the planned ventilation opening.

NOTE: A diagonal connecting cut can be made between the vertical and horizontal cuts. This can be used as a knockout space so a pike pole or other pushing/pulling tool can be inserted.

Step 8: Cut the decking horizontally (C) across the bottom of the planned ventilation opening.

Step 9: Make the final cut (D) vertically from top to bottom on the side of the planned ventilation opening closest to the ladder.

Step 10: Remove sheathing materials with the pick of the axe or some other suitable tool.

Step 11: Push the blunt end of a pike pole or other long-handled tool through the hole to open the ceiling.

Chapter 11
Water Supply

Chapter 11
Water Supply

INTRODUCTION

Technology continues to develop new methods and materials for extinguishing fires. However, water still remains the primary extinguishing agent because of its universal abundance and ability to absorb heat. Two primary advantages of water are that it can be conveyed long distances and it can be easily stored. These are also the fundamental principles of a water supply system. Because water remains the primary extinguishing agent used by firefighters, it is important that they have a good working knowledge of water supply systems.

This chapter covers the principles of municipal water supply systems and the methods of moving water throughout the system. It includes a description of the components of the water distribution system and the types of pressure found within the system. The chapter also explains the components of fire hydrants, how they are located, and how they are maintained. Finally, alternative water supplies such as lakes and ponds are discussed along with the methods of moving the water from the source to the fire by water shuttles and relay pumping.

PRINCIPLES OF MUNICIPAL WATER SUPPLY SYSTEMS

[NFPA 1001: 3-3.14(a); 4-5.1(a); 4-5.4(a)]

Public and/or private water systems provide the methods for supplying water to populated areas. As the population increases in rural areas, rural communities seek to improve water distribution systems from reliable sources.

The water department may be a separate, city-operated utility or a regional or private water authority. Its principal function is to provide potable water. Water department officials should be considered the experts in water supply problems. The fire department must work with the water department in planning fire protection coverage. Water department officials should realize that fire departments are vitally concerned with water supply and work with them on water supply needs and the locations and types of fire hydrants.

The intricate working parts of a water system are many and varied. Basically, the system is composed of the following fundamental components, which are explored in the following subsections (Figure 11.1):

- Source of water supply
- Means of moving water
- Water processing or treatment facilities
- Water distribution system, including storage

Sources of Water Supply

The primary water supply can be obtained from either surface water or groundwater. Although most water systems are supplied from only one source, there are instances where both sources are used. Two examples of surface water supply are rivers and lakes. Groundwater supply can be water wells or water-producing springs (Figures 11.2 a and b).

The amount of water that a community needs can be determined by an engineering estimate. This estimate is the total amount of water needed for domestic and industrial use and for fire fighting use. In cities, the domestic/industrial requirements far exceed that required for fire

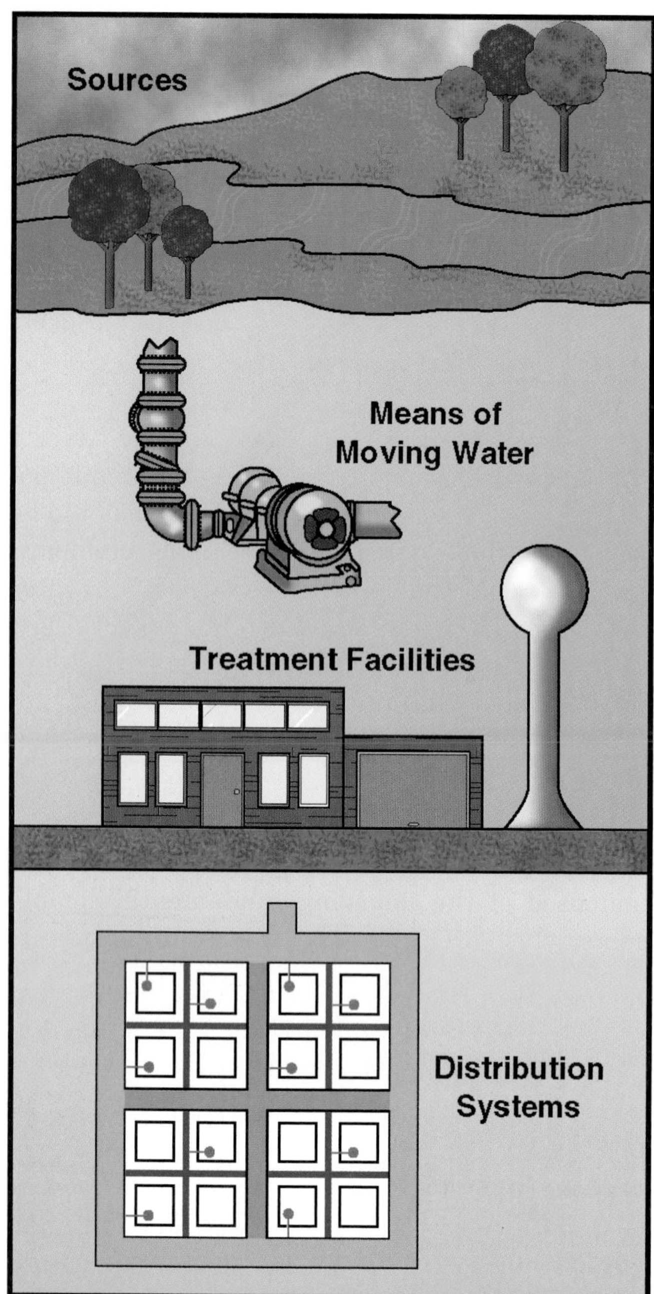

Figure 11.1 There are four components of any municipal water supply system.

Figure 11.2a Open-ground reservoirs are generally located at the water treatment plant and may be used as a water source for fire fighting operations.

Figure 11.2b Ground-level storage tanks can supply large amounts of water for fire fighting operations.

protection. In small towns, the requirements for fire protection exceed other requirements.

Means of Moving Water

There are three methods of moving water in a system:

- Direct pumping system
- Gravity system
- Combination system

DIRECT PUMPING SYSTEM

Direct pumping systems use one or more pumps that take water from the primary source and discharge it through the filtration and treatment processes (Figure 11.3). From there, a series of pumps force the water into the distribution system. If purification of the water is not needed, the water can be pumped directly into the distribution system from the primary source. Failures in supply lines and pumps can usually

Figure 11.3 Municipal water supply pumps.

be overcome by duplicating these units and providing a secondary power source.

GRAVITY SYSTEM

A gravity system uses a primary water source located at a higher elevation than the distribution system. The gravity flow from the higher elevation provides the water pressure (Figure 11.4). This pressure is usually only sufficient when the primary water source is located at least several hundred feet (meters) higher than the highest point in the water distribution system. The most common examples include a mountain reservoir that supplies water to a city below or a system of elevated tanks in a city itself.

COMBINATION SYSTEM

Most communities use a combination of the direct pumping and gravity systems. In most cases, the gravity flow is supplied by elevated storage tanks (Figure 11.5). These tanks serve as emergency storage and provide adequate pressure through the use of gravity. When the system pressure is high during periods of low consumption, automatic valves open and allow the elevated storage tanks to fill. When the pressure drops during periods of heavy consumption, the storage containers provide extra water by feeding it back into the distribution system. Providing a good combination system involves reliable, duplicated equipment and proper-sized storage containers that are strategically located.

Figure 11.5 An elevated storage tank.

The storage of water in elevated reservoirs can also ensure water supply when the system becomes otherwise inoperative. Storage should be sufficient to provide domestic and industrial demands plus the demands expected in fire fighting operations. Such storage should also be sufficient to permit making most repairs, alterations, or additions to the system. Location of the storage and the capacity of the mains leading from this storage are also important factors.

Many industries provide their own private systems, such as elevated storage tanks, that are available to the fire department (Figure 11.6).

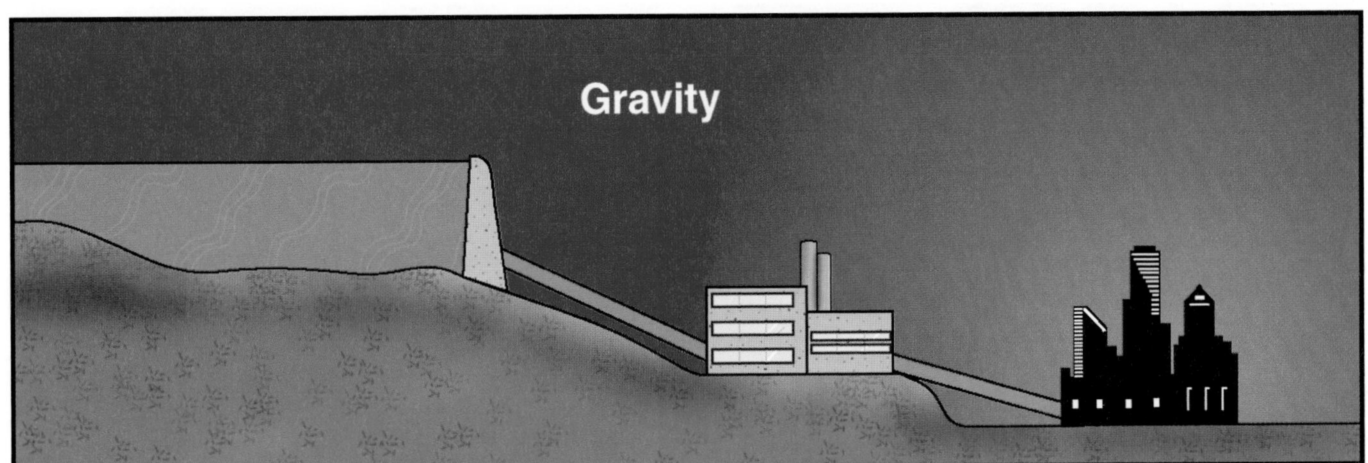

Figure 11.4 Gravity systems rely on natural forces to move water within the system.

Water for fire protection may be available to some communities from storage systems, such as cisterns, that are considered a part of the distribution system. The fire department pumper removes the water from these sources by draft (process of obtaining water from a static source into a pump that is above the source's level) and provides pressure by its pump.

Processing or Treatment Facilities

The treatment of water for the water supply system is a vital process. Water is treated to remove contaminants that may be detrimental to the health of those who use or drink it. Water may be treated by coagulation, sedimentation, filtration, or the addition of chemicals, bacteria, or other organisms. In addition to removing things from the water, some things may be added such as fluoride or oxygen.

The fire department's main concern regarding treatment facilities is that a maintenance error, natural disaster, loss of power supply, or fire could disable the pumping station(s) or severely hamper the purification process. Any of these situations would drastically reduce the volume and pressure of water available for fire fighting operations. Another problem would be the inability of the treatment system to process water fast enough to meet the demand. In either case, fire officials must have a plan to deal with these potential shortfalls.

Figure 11.6 An industry's private water supply.

Distribution System

The distribution system of the overall water supply system is the part that receives the water from the pumping station and delivers it throughout the area served (Figure 11.7). The ability of a water system to deliver an adequate quantity of water relies upon the carrying capacity of the system's network of pipes. When water flows through pipes, its movement causes friction that results in a reduction of pressure. There is much less pressure loss in a water distribution system when fire hydrants are supplied from two or more directions. A fire hydrant that receives water from only one direction is known as a *dead-end hydrant* (Figure 11.8). When a fire hydrant receives water from two or more directions, it is said to have *circulating feed* or a *looped line* (Figure 11.9). A distribution system that provides circulating feed from several mains constitutes a *grid system* (Figure 11.10). A grid system should consist of the following components:

- ***Primary feeders*** — Large pipes (mains), with relatively widespread spacing, that convey large quantities of water to various points of the system for local distribution to the smaller mains

Figure 11.7 A typical distribution system arrangement.

Figure 11.8 Dead-end hydrants receive water from only one direction.

Figure 11.9 A circulating-feed hydrant receives water from more than one direction.

- ***Secondary feeders*** — Network of intermediate-sized pipes that reinforce the grid within the various loops of the primary feeder system and aid the concentration of the required fire flow at any point
- ***Distributors*** — Grid arrangement of smaller mains serving individual fire hydrants and blocks of consumers

To ensure sufficient water, two or more primary feeders should run from the source of supply to the high-risk and industrial districts of the community by separate routes. Similarly, secondary feeders should be arranged in loops as far as possible to give two directions of supply to any point. This practice increases the capacity of the supply at any given point and ensures that a break in a feeder main will not completely cut off the supply.

In residential areas, the recommended size for fire-hydrant supply mains is at least 6 inches (150 mm) in diameter. These should be closely gridded by 8-inch (200 mm) cross-connecting mains at intervals of not more than 600 feet (180 m). In the business and industrial districts, the minimum recommended size is an 8-inch (200 mm) main with cross-connecting mains every 600 feet (180 m). Twelve-inch (300 mm) mains may be used on principal streets and in long mains not cross-connected at frequent intervals.

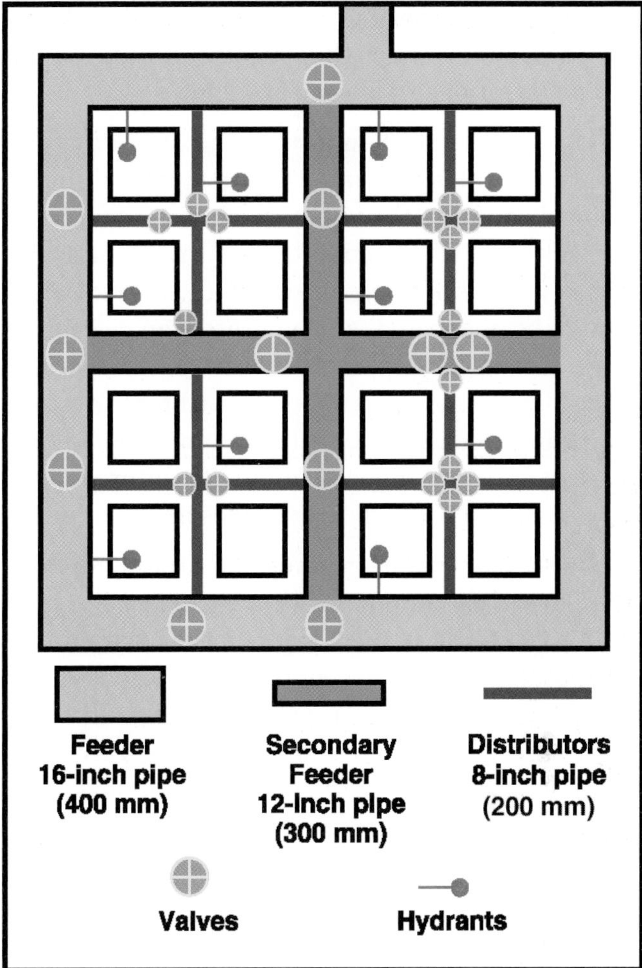

Figure 11.10 Grid system of water mains.

WATER MAIN VALVES

The function of a valve in a water distribution system is to provide a means for controlling the flow of water through the distribution piping. Valves should be located at frequent intervals in the grid system so that only small districts are cut off if it is necessary to stop the flow at specified points (see Figure 11.10). Valves should be operated at least once a year to keep them in good condition. The actual need for valve operation in a water system rarely occurs, sometimes not for many years. Valve spacing should be such that only a minimum length of pipe is out of service at one time.

One of the most important factors in a water supply system is the water department's ability to promptly operate the valves during an emergency or breakdown of equipment. A well-run water utility has records of the locations of all valves. Valves should be inspected and operated on a regular basis. If each fire department company is informed of the locations of valves in the distribution system, their condition and accessibility can be noted during fire-hydrant inspections. The water department is then informed if any valves need attention.

Valves for water systems are broadly divided into *indicating* and *nonindicating* types. An indicating valve visually shows whether the gate or valve seat is open, closed, or partially closed. Valves in private fire protection systems are usually of the indicating type. Two common indicator valves are the *post indicator valve (PIV)* and the *outside screw and yoke (OS&Y) valve.* The post indicator valve is a hollow metal post that is attached to the valve housing. The valve stem inside this post has the words *OPEN* and *SHUT* printed on it so that the position of the valve is shown (Figure 11.11). The OS&Y valve has a yoke on the outside with a threaded stem that controls the gate's opening or closing (Figure 11.12). The threaded portion of the stem is out of the yoke when the valve is open and inside the yoke when the valve is closed.

Nonindicating valves in a water distribution system are normally buried or installed in manholes. If a buried valve is properly installed, the valve can be operated aboveground through a valve box (Figure 11.13). A special socket wrench on the end of a reach rod operates the valve (Figure 11.14).

Figure 11.11 Post indicator valves clearly show whether the water supply is turned on or off.

Figure 11.12 An OS&Y valve.

Figure 11.13 Some valve boxes are belowground.

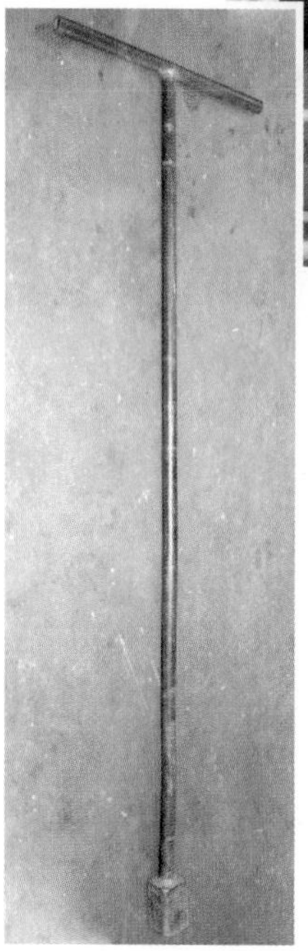

Figure 11.14 A water valve key.

Control valves in water distribution systems may be either gate valves or butterfly valves. Both valves can be of the indicating or nonindicating types. Gate valves are usually the nonrising stem type; as the valve nut is turned by the valve key (wrench), the gate either rises or lowers to control the water flow (Figure 11.15). Gate valves should be marked with a number indicating the number of turns necessary to completely close the valve. If a valve resists turning after fewer than the indicated number of turns, it usually means there is debris or another obstruction in the valve. Butterfly valves are tight closing, and they usually have a rubber or a rubber-composition seat that is bonded to the valve body.

The valve disk rotates 90 degrees from the fully open to the tight-shut position (Figure 11.16). The nonindicating butterfly type also requires a valve key. Its principle of operation provides satisfactory water control after long periods of inactivity.

The advantages of proper valve installation in a distribution system are readily apparent. If valves are installed according to established standards, it normally will be necessary to close off only one or perhaps two fire hydrants from service while a single break is being repaired. The advantage of proper valve installation is, however, reduced if all valves are not properly maintained and kept fully open. High friction loss is caused by valves that are only partially open. When valves are closed or partially closed, the condition may not be noticeable during ordinary domestic flows of water. As a result, the impairment will not be known until a fire occurs or until detailed inspections and fire flow tests are made. A fire department will experience difficulty in obtaining water in areas where there are closed or partially closed valves in the distribution system.

Figure 11.15 A gate valve.

Figure 11.16 A butterfly valve.

WATER PIPES

Water pipe that is used underground is generally made of cast iron, ductile iron, asbestos cement, steel, plastic, or concrete. Whenever pipe is installed, it should be the proper type for the soil conditions and pressures to which it will be subjected. When water mains are installed in unstable or corrosive soils or in difficult access areas, steel or reinforced concrete pipe may be used to give the strength needed. Some locations that may require extra protection include areas beneath railroad tracks and highways, areas close to heavy industrial machinery, areas prone to earthquakes, or areas of rugged terrain.

The internal surface of the pipe, regardless of the material from which it is made, offers resistance to water flow. Some materials, however, have considerably less resistance to water flow than others. Personnel from the engineering division of the water department should determine the type of pipe best suited for the conditions at hand.

The amount of water able to flow through a pipe and the amount of friction loss created can also be affected by other factors. Frequently, friction loss is increased by encrustation of minerals on the interior surfaces of the pipe. Another problem is sedimentation that settles out of the water. Both of these conditions result in a restriction of the pipe size, increased friction loss, and a proportionate reduction in the amount of water that can be drawn from the system.

KINDS OF PRESSURE

[NFPA 1001: 4-5.4(a)]

The term *pressure*, in connection with fluids, has a very broad meaning. Technically, pressure is defined as *force per unit area*. In fire service terms, pressure is most commonly thought of as the velocity of water in a conduit (either pipe or hose) of a certain size. Pressure in the fire service sense is measured in pounds per square inch (psi) or kilopascal (kPa). It is essential to understand the following terms that identify the kinds of pressure with which the fire service is concerned:

- Static pressure
- Normal operating pressure
- Residual pressure
- Flow pressure

Static Pressure

If the water is not moving, the pressure exerted is static (Figure 11.17). *Static pressure* is stored

Figure 11.17 Static pressure on a hydrant.

potential energy that is available to force water through pipe, fittings, fire hose, and adapters. Because true static pressure is rarely found in a water supply system, there is a different use for the term in water supply system application. In these cases, static pressure is defined as the normal pressure existing on a system before a flow hydrant is opened.

Normal Operating Pressure

The flow of water through a distribution system fluctuates during the day and night. *Normal operating pressure* is that pressure found in a water distribution system during periods of normal consumption demand (an average of the total amount of water used each day during a one-year period).

Residual Pressure

The term residual pressure represents the pressure left in a distribution system at a specific location when a quantity of water is flowing (Figure 11.18). *Residual pressure* is that part of the total available pressure that is not used to overcome friction or gravity while forcing water through pipe, fittings, fire hose, and adapters.

Flow Pressure

The forward velocity of a water stream exerts a pressure that can be read on a pitot tube and gauge (Figure 11.19). *Flow pressure* is the forward velocity pressure at a discharge opening, either at a hydrant discharge or a nozzle discharge orifice, while water is flowing.

Figure 11.18 The residual pressure at a hydrant is measured while water is being discharged from another hydrant.

Figure 11.19 A pitot tube in use.

FIRE HYDRANTS

[NFPA 1001: 3-3.14; 3-3.14(a); 4-5.4; 4-5.4(a); 4-5.4(b)]

The two main types of fire hydrants are *dry barrel* and *wet barrel*. The dry-barrel hydrant, used in climates where freezing weather is expected, is usually classified as a compression, gate, or knuckle-joint type that opens either with pressure or against pressure (Figures 11.20 a and b). The actual valve holding back the water is well belowground — below the anticipated frost line for that geographic location. When the hydrant is

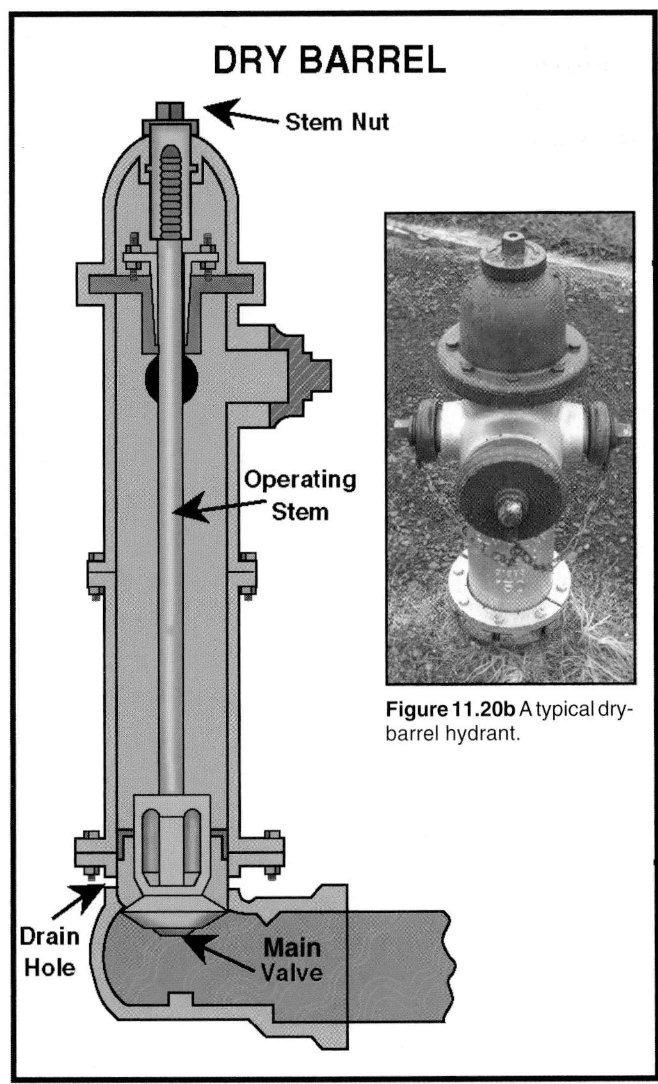

Figure 11.20b A typical dry-barrel hydrant.

Figure 11.20a A dry-barrel hydrant schematic.

closed, the barrel from the top of the hydrant down to the main valve should be empty. Any water that remains in a closed, dry-barrel hydrant empties through a small drain at the bottom of the hydrant near the main valve. This draining feature of a dry-barrel hydrant is very important in determining hydrant usability. The drain on the dry-barrel hydrant is open when the hydrant is not flowing water and is closed when the hydrant is operating. If the hydrant is not completely open, the drain is left partly open. The resulting flow from the hydrant contributes to ground erosion. This explains the old adage that a hydrant must be either completely open or completely shut — there is no halfway point.

The hydrant's ability to drain may be tested in the following manner: After allowing the hydrant to flow some water, close it, and cap all discharges except one. Place a hand over the discharge (Figure 11.21). At this time, a person should feel a slight vacuum pulling the palm toward the discharge. If this vacuum is not felt, notify the waterworks authority and have them inspect the hydrant because it is probably plugged. If this situation occurs in cold climates, the hydrant must be pumped to prevent the water from freezing.

Figure 11.21 Check to see if the hydrant is draining.

Wet-barrel hydrants may only be used in areas that do not have freezing weather. Wet-barrel hydrants usually have a compression-type valve at each outlet, or they may have only one valve in the bonnet that controls the flow of water to all outlets (Figures 11.22 a and b). The entire hydrant is always filled with water to the valves near the discharges.

In general, all hydrant bonnets, barrels, and footpieces are made of cast iron. The important working parts are usually made of bronze, but valve facings may be made of rubber, leather, or composition materials.

The flow of a hydrant varies for several reasons. First, and most obviously, the proximity of feeder mains and the size of the mains to which the hydrant is connected have a major impact on the amount of flow. Sedimentation and deposits within the distribution system may increase the resistance to water flow. These problems may occur over a period of time; therefore, older water systems may experience a decline in the flow available. Firefighters can make better decisions affecting a

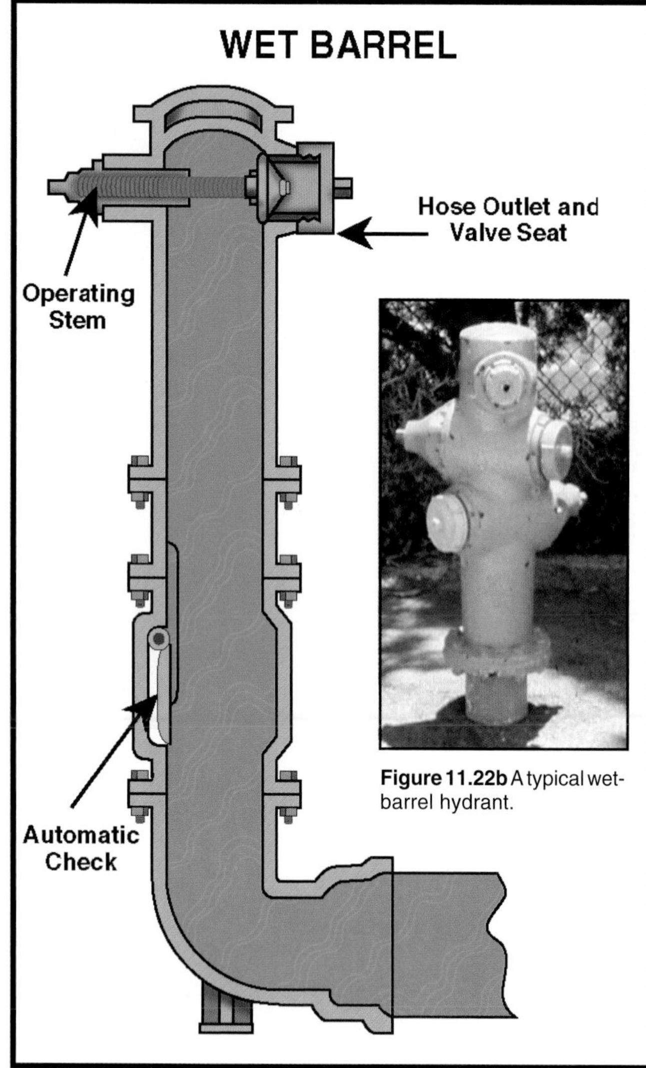

Figure 11.22b A typical wet-barrel hydrant.

Figure 11.22a A wet-barrel hydrant schematic.

fire attack if they at least know the relative available water flow of different hydrants in the vicinity. To aid them, a system of coloring hydrants to indicate a range of water flow was developed by the NFPA. With this system, hydrants are classified as shown in Table 11.1. Local coloring variations may be found, but simplicity is the main intent of any color scheme.

Location of Fire Hydrants

Although the installation of fire hydrants is usually performed by water department personnel, the location, spacing, and distribution of fire hydrants should be the responsibility of the fire chief or fire marshal. In general, fire hydrants should not be spaced more than 300 feet (90 m) apart in high-value districts. A basic rule to follow is to place one hydrant near each street intersection and to place intermediate hydrants where distances between intersections exceed 350 to 400 feet (105 m to 120 m). This basic rule represents a minimum requirement and should be regarded only as a guide for spacing hydrants. Other factors more pertinent to the particular locale include types of construction, types of occupancy, congestion, the sizes of water mains, required fire flows, and pumping capacities.

TABLE 11.1
Hydrant Color Codes

Hydrant Class	Color	Flow
Class AA	Light Blue	1,500 gpm (5 680 L/min) or greater
Class A	Green	1,000–1,499 gpm (3 785 L/min to 5 675 L/min)
Class B	Orange	500–999 gpm (1 900 L/min to 3 780 L/min)
Class C	Red	Less than 500 gpm (1 900 L/min)

Fire Hydrant Inspection and Maintenance

In most cities, repair and maintenance of fire hydrants are the responsibilities of the water department because this department is in a better position to do this work than any other agency. However, in many cases, fire department personnel perform water supply testing and hydrant inspections. Therefore, firefighters should look for the following potential problems when checking fire hydrants:

- Are obstructions, such as sign posts, utility poles, or fences, too near the hydrant to make pumper-to-hydrant connections?
- Do the outlets face the proper direction for pumper-to-hydrant connections, and is there sufficient clearance between the outlets and the ground for hose connections (Figure 11.23)?

Figure 11.23 Check for sufficient clearance between the ground and the hydrant outlet.

- Is the hydrant damaged because of traffic accidents?
- Is the hydrant rusting or corroded?
- Are the hydrant caps stuck in place with paint?
- Is the operating stem easily turned?
- Are there any obstructions (bottles, cans, rocks) inside the hydrant restricting water flow?

Using a Pitot Tube

Firefighters who assist in hydrant testing and inspections will be required to use a pitot tube to measure the flow pressure coming from a hydrant. There are two methods of holding the pitot tube properly. The first is to grasp the pitot tube just behind the blade with the first two fingers and thumb of the left hand while the right hand holds the air chamber. The little finger of the left hand rests upon the hydrant outlet or nozzle to steady the instrument (Figure 11.24). Another method is to have the fingers of the left hand split around the gauge outlet and the left side of the fist placed on the edge of the hydrant orifice or outlet (Figure 11.25). The blade can then be sliced into the stream in a counterclockwise direction. The right hand once again steadies the air chamber. Flow test kits are also available for conducting hydrant tests. Using a "fixed-mount" pitot tube reduces the possibility of human error that may occur when using a handheld pitot tube (Figure 11.26). Skill Sheet 11-1 shows the procedure for using the pitot tube.

Figure 11.24 The little finger is used to help steady the pitot tube.

Figure 11.25 Steady the pitot tube by holding the left side of the fist against the discharge outlet. Then, slice the blade into the stream.

Figure 11.26 A fixed-mount pitot tube.

For more information on testing fire hydrants, see IFSTA's **Water Supplies for Fire Protection** or **Fire Inspection and Code Enforcement** manuals.

ALTERNATIVE WATER SUPPLIES

[NFPA 1001: 3-3.14; 3-3.14(a); 3-3.14(b); 4-5.1(a)]

Fire departments should not limit their study of water supplies to the piped public-distribution system. Areas outside the public system should be studied for available water. Even areas with good water systems should be surveyed for alternative supplies in case the water system fails or a fire occurs that requires more water than the system can supply. For example, a water supply can be supplemented by an industry that has its own private water system. With today's modern pumpers, water can be drawn from many natural sources such as the ocean, lakes, ponds, and rivers (Figure 11.27). Water is also found in farm stock tanks and swimming pools. A good method of providing water for fire protection is to construct storage tanks at strategic locations.

Figure 11.27 A drafting operation is a good way to provide water for fire protection.

The process of raising water from a static source to supply a pumper is known as *drafting*. Almost any static source of water can be used if it is sufficient in quantity and not contaminated to the point of creating a health hazard. The depth for drafting water at a source is an important operational consideration. Silt and debris can render a source useless by clogging strainers, by seizing (stopping) or damaging pumps, and by allowing sand and small rocks to enter attack lines and clogging fog-stream nozzles. All hard suction lines should have strainers on them whenever drafting from a natural source. The suction hose should be located and supported so the strainer does not rest on or near the bottom of the source. A depth of 24 inches (600 mm) of water above and below the hard suction strainer is a good guideline for placing a strainer, although lesser depths have been used successfully (Figure 11.28). Special drafting or floating strainers that can draw water from 1- to 2-inch (25 mm to 50 mm) levels are available to use in shallow sources (Figures 11.29 a and b).

Fire department personnel should make every attempt to identify, mark, and record alternative water supply sources in pre-incident planning. Consideration should be given to the effect that weather has on the amount of water available and the accesses to water sources.

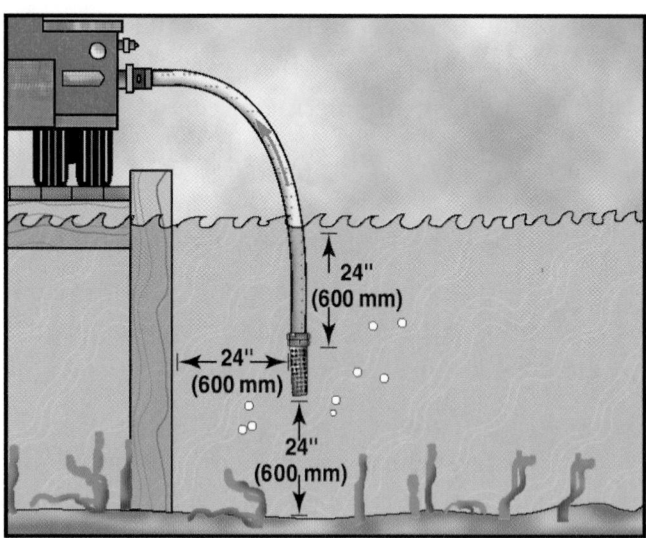

Figure 11.28 A minimum of 24 inches (600 mm) of water should surround all sides of a strainer.

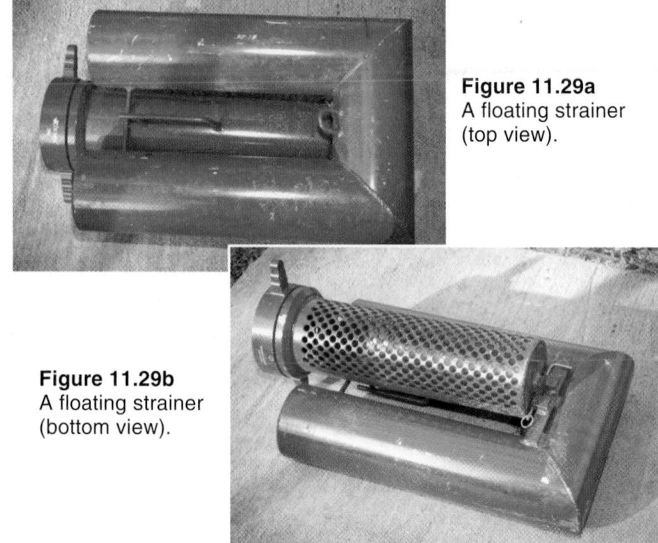

Figure 11.29a A floating strainer (top view).

Figure 11.29b A floating strainer (bottom view).

RURAL WATER SUPPLY OPERATIONS

[NFPA 1001: 3-3.14(a); 3-3.14(b)]

Rural water supply operations consist principally of mobile water supply apparatus (tanker/tender) shuttles and relay pumping. For either type of operation to succeed, pre-incident planning and practice are required. Adequate resources must be dispatched promptly, and an incident management system is necessary for control and coordination. The following subsections briefly highlight each of these operations. For more information on rural water supply operations, see IFSTA **Water Supplies for Fire Protection** manual and NFPA 1231, *Standard for Water Supplies for Suburban and Rural Fire Fighting.*

Water Shuttling/Shuttles

Water shuttling is the hauling of water from a supply source to portable tanks from which water may be drawn to fight a fire (Figure 11.30). Water shuttling is recommended for distances greater than ½ mile (0.8 km) or greater than the fire department's capability of laying supply hoselines. It is critical to have an adequate number of water tankers/tenders for the needed fire flow.

The keys to efficient water shuttles are fast-fill and fast-dump times. Water supply officers should be positioned at both the dump and fill sites. As personnel are available, consideration should be given to assigning people to traffic control, hydrant operations, hookups, and tank venting. If possible, the tanker/tender drivers should remain in their vehicles during filling/dumping operations.

Figure 11.30 Portable tanks are necessary for efficient shuttle operations.

There are three key components to water-shuttle operations:

- Attack apparatus at the fire (dump site)
- Fill apparatus at the fill site
- Mobile water supply apparatus (tankers/tenders) to haul water from the fill site to the dump site (Figure 11.31)

The dump site is generally located near the actual fire or incident. The dump site consists of one or more portable water tanks into which water-hauling apparatus dump water before returning to the fill site. Apparatus attacking the fire may draft directly from the portable tanks, or other apparatus may draft from the tanks and supply the attack apparatus. Low-level intake devices, either commercial or homemade, permit use of most of the water in the portable reservoir (Figure 11.32).

Figure 11.31 Some tankers (tenders) have large pumps and are equipped similar to a standard engine company. *Courtesy of Mike Wieder.*

Figure 11.32 Low-level strainer designed to be used in a portable tank.

When large flows must be maintained, multiple portable tanks are required. Capacities of portable tanks range from 1,000 gallons (4 000 L) upward. When multiple portable tanks are used, a jet siphon maintains the water level in one tank for the pumper, while water tankers/tenders dump into the others. A jet siphon uses a 1½-inch (38 mm) discharge line connected to the siphon. The siphon is then attached to a hard sleeve placed between two tanks (Figures 11.33 a–c). Plain siphons or commercial tank-connecting devices are also sometimes used for this purpose, although they are not generally as efficient as jet siphons.

There are several methods of constructing portable reservoirs. The most common is the collapsible or folding style that uses a square metal frame and a synthetic or canvas duck liner. Another style is a round, synthetic tank with a floating collar that rises as the tank is filled, making it self-supporting. These portable reservoirs should be mounted for easy removal from the apparatus.

Before opening a portable tank, a heavy tarp should be spread on the ground to help protect the liner once water is dumped into it (Figure 11.34). Portable tanks should, if possible, be positioned in a location that allows easy access from multiple directions but does not inhibit access of other apparatus to the fire scene. If more than one water tanker/tender has arrived, empty one completely, and send it for another load before emptying the second one. This procedure sequences the tankers/tenders better at both the dump and fill sites.

There are four basic methods by which tankers/tenders unload water:

- Gravity dumping through large (10- or 12-inch [250 mm or 300 mm]) dump valves (Figure 11.35)
- Jet dumps that create a venturi effect increasing the flow rate (Figure 11.36)
- Apparatus-mounted pumps that off-load the water (Figure 11.37)
- Combination of these methods

NFPA 1901, *Standard for Automotive Fire Apparatus,* requires that apparatus on level ground be capable of dumping or filling at rates of at least

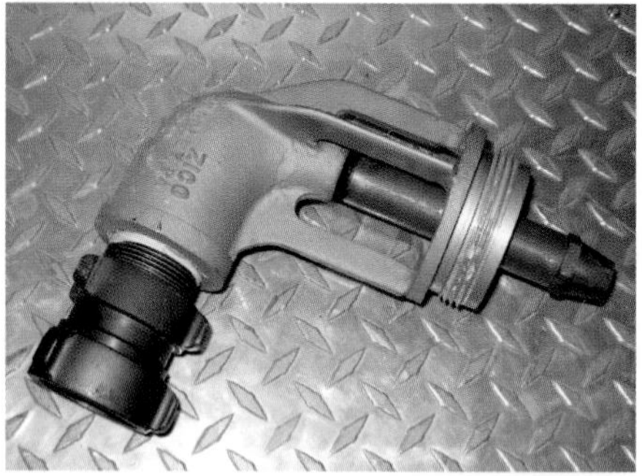

Figure 11.33a A commercially built jet siphon. *Courtesy of Craig L. Hannan.*

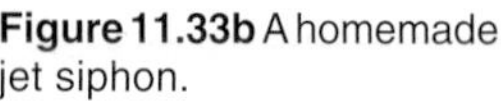

Figure 11.33b A homemade jet siphon.

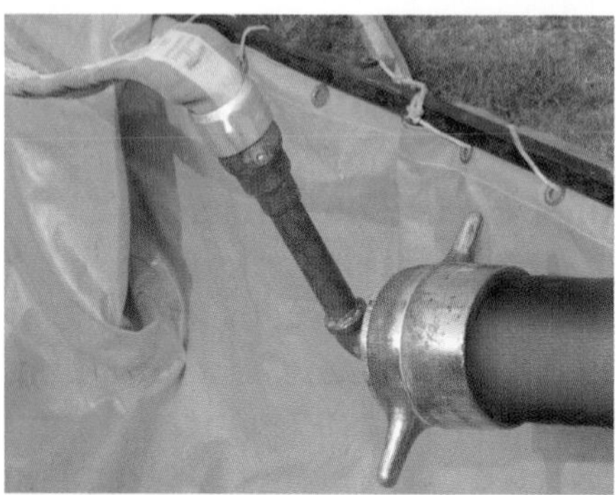

Figure 11.33c A jet siphon attached to a hard sleeve.

Figure 11.34 A tarp placed under the portable tank helps protect the liner.

Figure 11.35 Dump valves allow for maximum unloading capacity. They may be located on any side of the apparatus but are most commonly found on the rear. *Courtesy of Mike Wieder.*

Figure 11.36 Jet dumps increase the flow rate of any dump valve. *Courtesy of Montezuma-Rimrock (AZ) Fire Department.*

Figure 11.37 Stream shapers are used to ensure that all water is directed toward the portable tank. They are most commonly used when water is being pumped off the apparatus.

1,000 gpm (4 000 L/min). This rate necessitates adequate tank venting and openings in tank baffles. Pumping the water from the tanker/tender needs to be done by a trained apparatus driver/operator. Gravity dumps may be activated by a firefighter, which relieves the driver/operator from exiting the cab and saves time in the overall process. Most gravity dumps are activated by a lever near the outlet.

In order to fill water tankers/tenders quickly, use the best fill site or hydrant available, large hoselines, multiple hoselines, and if necessary, a pumper for an adequate flow rate. In some situations multiple portable pumps may be necessary. Both fill sites and dump sites should be arranged so that a minimum of backing (or maneuvering) of apparatus is required.

Relay Pumping

Sometimes a water source is close enough to the fire scene that relay pumping can be used. Some departments use variations of a combination tender shuttle and relay pumping to minimize congestion of apparatus at the fire scene (Figure 11.38). There are two important factors to be considered when contemplating the establishment of a relay operation:

- The water supply must be capable of maintaining the desired volume of water required for the duration of the incident.

Figure 11.38 Some tankers have the capability of pumping their load.

- The relay must be established quickly enough to be worthwhile.

The number of pumpers needed and the distance between pumpers is determined by several factors such as volume of water needed, distance between the water source and the fire scene, hose size available, amount of hose available, and pumper capacities. The apparatus with the greatest pumping capacity should be located at the water source. Large diameter hose or multiple hoselines increase the distance and volume that a relay can supply because of reduced friction loss. A water supply officer must be appointed to determine the distance between pumpers and to coordinate water supply operations.

After considering these factors, a quick calculation must be made by the water supply officer in order to determine the distance between pumpers. It is important to know the friction loss at particular flows for the size hose being used. These figures can be made into a chart and placed on the pumper for quick reference. The best way to prepare for relay operations is to plan them in advance and to practice them during training exercises.

SKILL SHEET 11-1 USING A PITOT TUBE

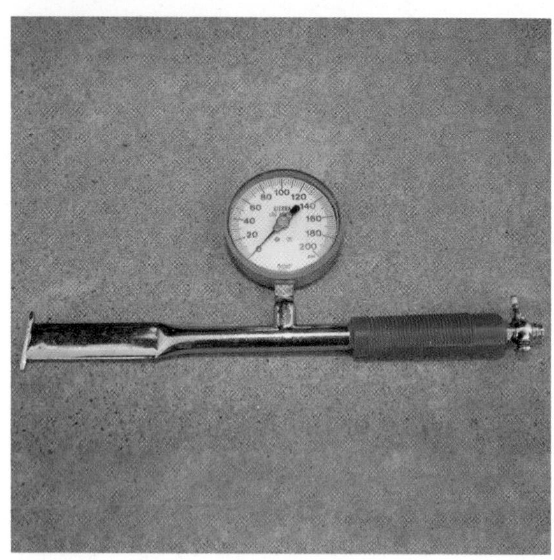

Step 1: Open the petcock on the pitot tube. Make certain the air chamber is drained; then close the petcock.

NOTE: The petcock is located on the bottom of the pitot tube. When properly drained and ready, the gauge needle should read zero.

Step 2: Edge the blade into the stream with the small opening or point centered in the stream and held away from the orifice at a distance approximately half the diameter of the orifice. For a 2½-inch (65 mm) hydrant butt, this distance would be 1¼ inches (32 mm).

Step 3: Keep the air chamber above the horizontal plane passing through the center of the stream. The pitot tube is now parallel to the outlet opening.

NOTE: This position increases the efficiency of the air chamber and helps avoid needle fluctuations.

Step 4: Record the velocity pressure reading from the gauge. If the needle is fluctuating, read and record the value located between the high and low extremes.

Chapter 12
Fire Hose

Chapter 12
Fire Hose

INTRODUCTION

The term *fire hose* identifies a type of flexible tube used by firefighters to carry water under pressure from the source of supply to a point where it is discharged. In order to be reliable, fire hose should be constructed of the best materials, and it should not be used for purposes other than fire fighting. Fire hose is the most used item in the fire service. It must be flexible, be watertight, have a smooth lining, and have a durable covering (also called a hose jacket). Depending on its intended use, fire hose is manufactured in different configurations such as single-jacket, double-jacket, rubber single-jacket, and hard-rubber noncollapsing types (Figure 12.1).

This chapter includes a discussion of fire hose sizes, causes of hose damage and its prevention, and general care and maintenance. The chapter describes the types of hose couplings and their care and use. Also covered are the different types of hose appliances used in water movement and the tools used in hose operations. The procedures for rolling hose, loading supply hose on apparatus, preparing finishes, and loading preconnected attack hoselines are discussed and demonstrated. The chapter reviews hose-lay procedures, hose-handling techniques, and advancing and operating hoselines. Finally the chapter ends with a discussion of the procedures for service testing of fire hose.

FIRE HOSE SIZES

[NFPA 1001: 3-3.7(a); 3-3.9(a)]

Each size of fire hose is designed for a specific purpose. Reference made to the diameter of fire hose refers to the dimensions of the inside diameter of the hose. Fire hose is most commonly cut and coupled into lengths of 50 or 100 feet (15 m or 30 m) for convenience of handling and replacement, but other lengths may be obtained. These lengths are also referred to as *sections,* and they must be coupled together to produce a continuous hoseline.

Intake hose is used to connect a fire department pumper or a portable pump to a nearby water source. There are two groups within this category: *soft sleeve hose* and *hard suction hose.* Soft sleeve hose is used to transfer water from a pressurized water source, such as a fire hydrant, to the pump intake (Figure 12.2). Soft sleeves are available in sizes ranging from 2½ to 6 inches (65 mm to 150 mm). Hard suction hose (also called a *hard sleeve*) is used primarily to draft water from an open water source (Figure 12.3). It is also used to siphon water from one portable tank to another, usually in a tanker shuttle operation. Hard suction hose is constructed of a rubberized, reinforced material designed to withstand the partial vacuum conditions created when drafting. It is also available in sizes ranging from 2½ to 6 inches (65 mm to 150 mm).

NFPA 1961, *Standard on Fire Hose,* lists specifications for fire hose; NFPA 1963, *Standard for Fire Hose Connections,* lists specifications for fire hose couplings and screw threads. NFPA 1901, *Standard for Automotive Fire Apparatus,* requires pumpers to carry 15 feet (4.6 m) of large soft sleeve hose or 20 feet (6 m) of hard suction hose, 1,200 feet (366 m) of 2½-inch (65 mm) or larger supply hose (hose between the water source and the attack pumper to provide large volumes of water), and 400 feet (122 m) of 1½-, 1¾-, or 2-inch (38 mm, 45 mm, or 50 mm) attack hose (hose between the attack

TYPE	HOSE CONSTRUCTION	DESCRIPTION
Booster Hose **¾- or 1-inch** **(20 mm or 25 mm)**	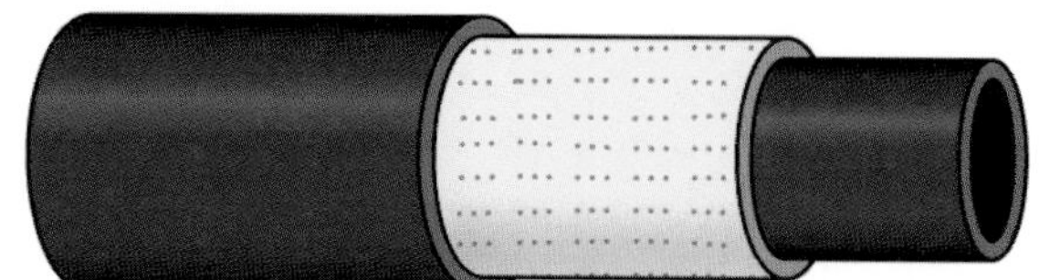	• Rubber Covered • Rubber Lined • Fabric Reinforced
Woven-Jacket Hose 1- to 6-inch (25 mm to 150 mm)	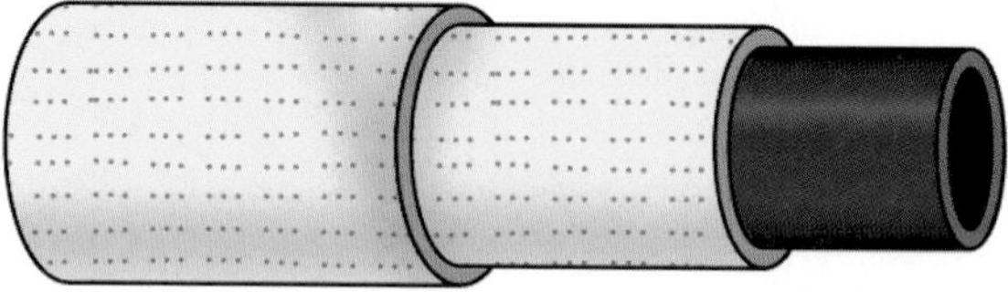	• One or Two Woven-Fabric Jackets • Rubber Lined
Impregnated Single-Jacket Hose 1½- to 5-inch (38 mm to 125 mm)		• Polymer Covered • Polymer Lined
Noncollapsible Intake Hose 2½- to 6-inch (65 mm to 152 mm)	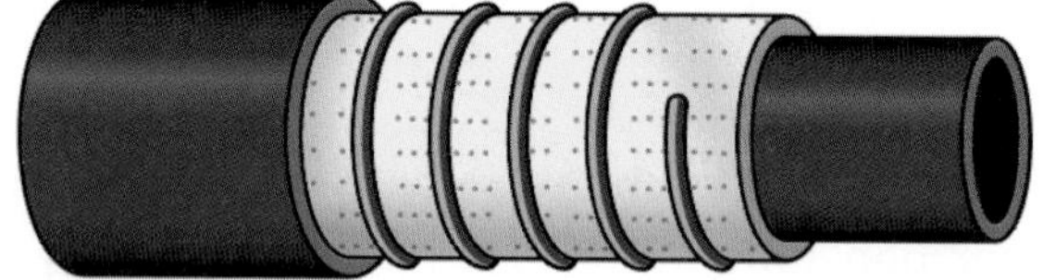	• Rubber Covered • Fabric and Wire (Helix) Reinforced • Rubber Lined
Flexible Noncollapsible Intake Hose 2½- to 6-inch (65 mm to 150 mm)	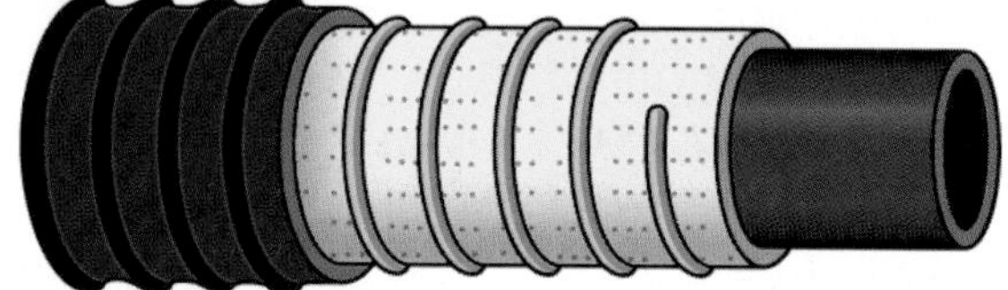	• Rubber Covered • Fabric and Plastic (Helix) Reinforced • Rubber Lined

Figure 12.1 Common types of fire hose.

Figure 12.2 A soft sleeve hose transfers water from the fire hydrant to the pump intake.

Figure 12.3 Hard suction hose is designed to withstand the partial vacuum of drafting.

pumper and the nozzle used to control and extinguish fire). These lengths and sizes may be increased, depending on the needs of the department.

CAUSES AND PREVENTION OF FIRE HOSE DAMAGE

[NFPA 1001: 3-5.4; 3-5.4(a)]

Fire hose is a tool that is subjected to many potential sources of damage during fire fighting. Usually little can be done at fires to provide safe usage and to protect the hose from injury. The most important factor relating to the life of fire hose is the care it gets after fires, in storage, and on the fire apparatus. Fire hose should be selected with caution to ensure its lasting qualities. Even if constructed of quality materials, it cannot endure mechanical injury, heat, mildew and mold, and chemical contacts. The life of fire hose is, however, considerably dependent upon how well the hose is protected against these destructive causes.

Mechanical Damage

Fire hose may be damaged in a variety of ways while being used at fires. Some common mechanical injuries are worn places, rips, and abrasions on the coverings, crushed or damaged couplings, and cracked inner linings (Figure 12.4). To prevent these damages, the following practices are recommended:

Figure 12.4 Typical damage to hose threads.

- Avoid laying or pulling hose over rough, sharp edges or objects.
- Use hose ramps or bridges to protect hose from vehicles running over it (Figure 12.5).
- Open and close nozzles, valves, and hydrants slowly to prevent water hammer (force created by the rapid deceleration of water).
- Change position of bends in hose when reloading hose on apparatus.
- Provide chafing blocks to prevent abrasion to hose when it vibrates near the pumper (Figure 12.6).
- Avoid excessive pump pressure on hoselines.

Figure 12.5 Hose bridges in use.

Figure 12.6 Chafing blocks help prevent hose from being damaged by apparatus vibrations and rubbing on the pavement.

Thermal Damage

The exposure of hose to excessive heat or its contact with fire will char, melt, or weaken the fabric covering and dry the rubber lining. A similar drying effect may occur to inner linings when hose is hung to dry in a drying tower for a longer period of time than is necessary or when it is dried in intense sunlight (Figure 12.7). To prevent thermal damage, firefighters should conform to the following recommended practices:

Figure 12.7 Hose towers are designed for hanging and properly drying hose.

- Protect hose from exposure to excessive heat or fire when possible.
- Do not allow hose to remain in any heated area after it is dry.
- Use moderate temperature for drying. A current of warm air is much better than hot air.
- Keep the outside of woven-jacket fire hose dry.
- Run water through hose that has not been used for some time to prolong its life.
- Avoid laying fire hose on hot pavement to dry.
- Prevent hose from coming in contact with, or being in close proximity to, vehicle exhaust systems.
- Use hose bed covers on apparatus to shield the hose from the sun (Figures 12.8 a and b).

NOTE: Hose can also be damaged by freezing temperatures. Hose, wet or dry, should not be subjected to freezing conditions for prolonged periods of time.

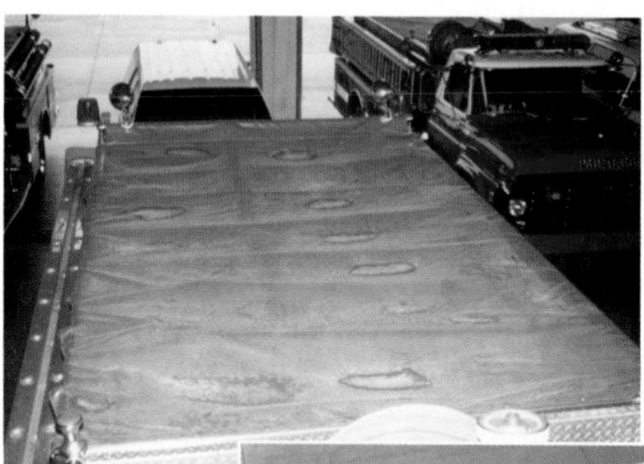

Figure 12.8a Hose can be covered with a tarp to protect it from the sun.

Figure 12.8b Metal hose bed covers have been installed on some apparatus. *Courtesy of Rick Montemorra, Mesa FD.*

Organic Damage

Organic damage such as mildew and mold may occur on woven-jacket hose when moisture remains on the outer surfaces (Figures 12.9). Mildew and mold cause decay and the consequent deterioration of the hose. Rubber-jacket hose is not subject to mold and mildew damage. Some methods of preventing mildew and mold on woven-jacket hose are as follows:

- Remove all wet woven-jacket hose from the apparatus after a fire and replace with dry hose.
- Remove, inspect, sweep, and reload woven-jacket hose if it has not been unloaded from the apparatus during a period of 30 days.
- Exercise woven-jacket hose every 30 days and run water through it every 90 days to prevent drying and cracking of the rubber lining. Some woven-jacket fire hose has been chemically treated to resist mildew and mold but such treatment is not always 100 percent effective.

Figure 12.9 Mold or mildew can weaken the jacket of woven-jacket hose.

Chemical Damage

Chemicals and chemical vapors will damage the rubber lining and often cause the lining and jacket to separate. When hose is exposed to petroleum products, paints, acids, or alkalis, it may be weakened to the point of bursting. Runoff water from a fire may also carry foreign materials that can damage fire hose. After being exposed to chemicals or chemical vapors, hose should be cleaned as soon as practical. Some recommended practices are as follows:

- Scrub hose thoroughly and brush all traces of acid contacts with a solution of baking soda and water. Baking soda neutralizes acids.

- Remove hose periodically from the apparatus, wash it with plain water, and dry it thoroughly.
- Test hose properly if there is the least suspicion of damage (see Service Testing Fire Hose section).
- Avoid laying hose in the gutter or next to the curb where vehicles have been parked because they can drop oil from their mechanical components and acid from batteries (Figure 12.10).
- Dispose of hose properly if it has been exposed to hazardous materials and cannot be decontaminated.

Figure 12.10 Avoid laying hose in the gutter where it would be subjected to debris and runoff.

GENERAL CARE AND MAINTENANCE OF FIRE HOSE

[NFPA 1001: 3-5.4; 3-5.4(a); 3-5.4(b)]

If fire hose is properly cared for, its life span can be extended appreciably. The techniques of washing and drying and the provisions for storage are very important functions in the care of fire hose. The following sections highlight the proper care of fire hose.

Washing Hose

The method used to wash fire hose depends on the type of hose. Hard-rubber booster hose, hard suction hose, and rubber-jacket collapsible hose require little more than rinsing with clear water, although a mild soap may be used if appearance is important (Figure 12.11).

Figure 12.11 Hose can be cleaned by rinsing it with water.

Most woven-jacket fire hose requires a little more care than the previously mentioned ones. After woven-jacket hose is used, the usual accumulation of dust and dirt should be thoroughly brushed from it. If the dirt cannot be removed by brushing, the hose should be washed and scrubbed with clear water.

When fire hose has been exposed to oil, it should be washed with a mild soap or detergent, making sure that the oil is completely removed. The hose should then be rinsed thoroughly. If a commercial hose-washing machine is not available, common scrub brushes or brooms can be used with streams of water from a hoseline and nozzle.

A hose-washing machine is a very important appliance in the care and maintenance of fire hose (Figure 12.12). The most common type washes almost any size of fire hose up to 3 inches (77 mm).

Figure 12.12 A jet-spray washer cleans the hose jacket with a high-pressure water stream that surrounds the hose.

The flow of water into this device can be adjusted as desired, and the movement of the water assists in propelling the hose through the device. The hoseline that supplies the washer with water can be connected to a pumper or used directly from a hydrant. Higher water pressure, obviously, gives better results.

A cabinet-type machine that washes, rinses, and drains fire hose is designed to be used in the station (Figure 12.13). This type of machine can be operated by one person, is self-propelled, and can be used with or without detergents.

Figure 12.13 A commercially produced hose-washing machine. *Courtesy of Thomas Locke and South Union Volunteer Fire Company.*

Drying Hose

The methods used to dry hose depend on the type of hose. Hard-rubber booster hose, hard suction hose, and rubber-jacket collapsible hose may be placed back on the apparatus while wet with no ill effects. Woven-jacket hose requires thorough drying before being reloaded on the apparatus. Hose should be dried in accordance with local procedures and manufacturer's recommendations.

Storing Hose

After fire hose has been adequately brushed, washed, and dried, it should be rolled and stored in suitable racks (see Hose Rolls section). Hose racks should be located in a clean, well-ventilated room in or close to the apparatus room for easy access. Racks can be freestanding on the floor or mounted permanently on the wall (Figure 12.14). Mobile hose racks can be used to both store hose and move hose from storage rooms to the apparatus for loading.

Figure 12.14 Clean, dry hose should be rolled and stored on racks.

FIRE HOSE COUPLINGS

[NFPA 1001: 3-3.9(b); 3-5.4; 3-5.4(a); 3-5.4(b)]

Fire hose couplings are made of durable materials and designed so that it is possible to couple and uncouple them with little effort in a short time. The materials used for fire hose couplings are generally alloys in varied percentages of brass, aluminum, or magnesium. These alloys make the coupling durable and easy to attach to the hose. Much of the efficiency of the fire-hose operation depends upon the condition and maintenance of its couplings. Firefighters should be knowledgeable of the type of couplings with which they work.

Types of Fire Hose Couplings

There are several types of hose couplings used in the fire service. The most commonly used couplings are the *threaded* and *Storz* types (Figures 12.15 a and b). Other types of couplings used with less frequency are the *quarter turn, oilfield rocker lug,* and *snap* (sometimes referred to as the *Jones snap*) (Figure 12.16). Couplings constructed of metals such as brass, aluminum alloy, and aluminum alloy with a hard coating will not rust. Couplings are made by forging, extruding, or casting. Drop-forged couplings are stronger than extruded

Figure 12.15a A threaded coupling.

Figure 12.15b Storz couplings.

Figure 12.16 Snap couplings interlock when the two spring-loaded hooks on the female coupling engage a ring on the shank of the male coupling.

couplings and stand up well to normal use. Even though extruded couplings tend to be somewhat weaker than drop-forged couplings, they are acceptable for fire fighting operations. Cast couplings are the weakest and are rarely used on modern fire hose.

THREADED COUPLINGS

Threaded couplings are either three-piece or five-piece types (Figure 12.17). Five-piece types are reducing couplings that are used when the needed coupling size is smaller than the hose to which it is attached. They are used so that hoses of different sizes can be connected without using adapter fittings (devices used to connect hose couplings with dissimilar threads) (see Hose Appliances section).

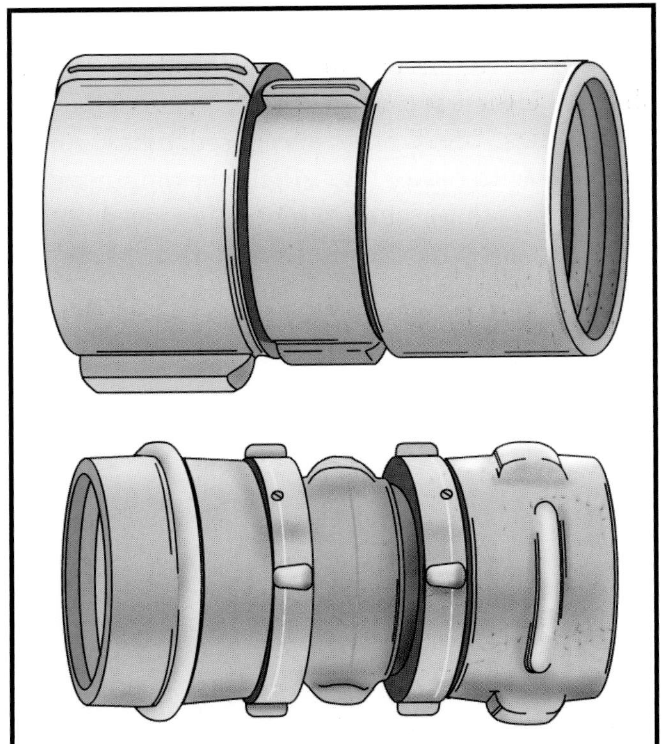

Figure 12.17 Three-piece and five-piece threaded couplings.

Three-piece fire hose couplings are also used for intake hose couplings. Hose couplings for the various sizes of intake hose are equipped with extended lugs that afford convenient handles for attaching intake hose to a hydrant or pump intake (Figure 12.18).

Figure 12.18 An intake hose coupling.

The portion of the coupling that serves as the point of attachment to the hose is the *shank* (also called the *tailpiece, bowl,* or *shell*). The male side of a connected coupling can be distinguished from the female side by noting the lugs. Only male couplings have lugs on the shank. The female coupling has lugs on the swivel (Figure 12.19).

Each threaded coupling is manufactured with lugs to aid in tightening and loosening connections. They also aid in grasping the coupling when making and breaking coupling connections. Connections may be made by hand or with *spanners* (special tools that fit against the lugs) (see Hose Tools section). There are three types of lugs: *pin, rocker,* and *recessed* (Figures 12.20 a–c). Although

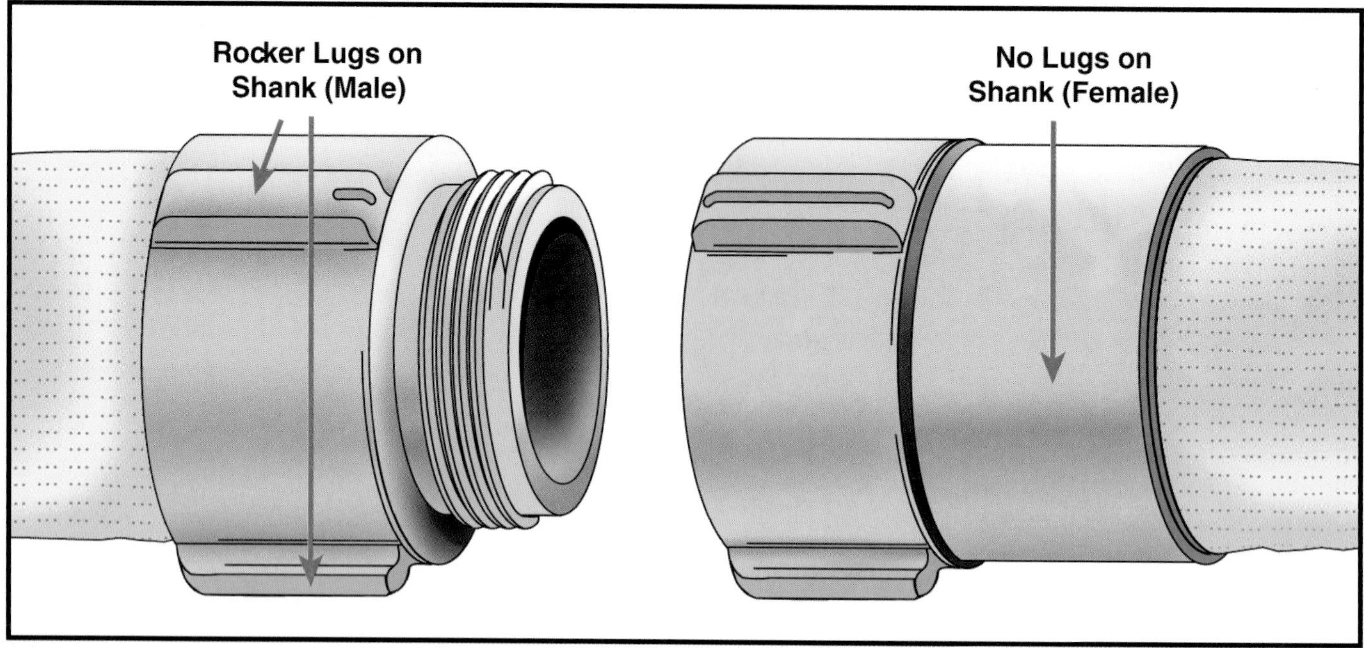

Figure 12.19 The male side of a connected coupling can be distinguished from the female side by noting the rocker lugs on the shank.

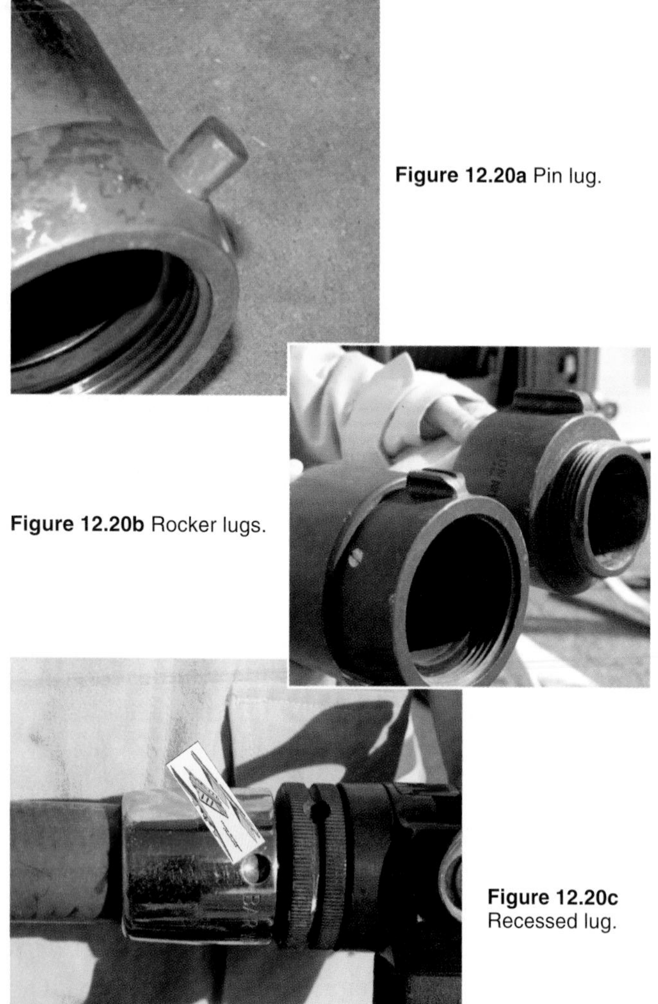

Figure 12.20a Pin lug.

Figure 12.20b Rocker lugs.

Figure 12.20c Recessed lug.

still available, pin-lug couplings are not commonly ordered with new fire hose because of their tendency to snag when hose is dragged over objects. Booster hose normally has couplings with recessed lugs, which are simply shallow holes drilled into the coupling. This lug design prevents abrasion that would occur if the hose had protruding lugs and was wound onto reels. These holes are designed to accept a special spanner wrench that can be used to couple or uncouple the hose (Figure 12.21). Modern threaded couplings have rounded rocker lugs. Most hose purchased today comes equipped with rocker lugs to help the coupling slide over obstructions when the hose is moved on the ground or around objects. Hose couplings may be obtained with either two or three rocker lugs.

An added feature that may be obtained with screw thread couplings is the Higbee cut and indicator. The *Higbee cut* is a special type of thread

Figure 12.21 This special spanner wrench is used to couple or uncouple hose with recessed lug couplings.

design in which the beginning of the thread is "cut" to provide a positive connection between the first threads of opposing couplings, which tends to eliminate cross-threading (Figure 12.22). One of the rocker lugs on the swivel is scalloped with a shallow indention, the *Higbee indicator,* to mark where the Higbee cut begins. This indicator aids in matching the male coupling thread to the female coupling thread, which is not readily visible.

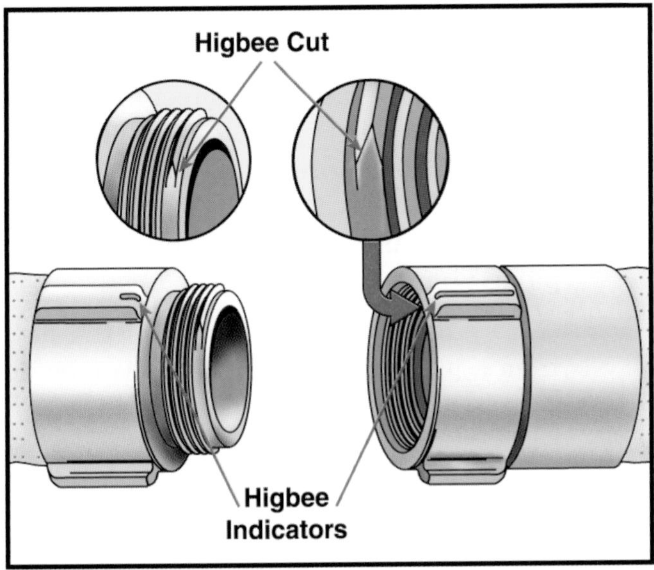

Figure 12.22 This shows the location of the Higbee cuts and indicators.

STORZ-TYPE COUPLINGS

Storz-type couplings are sometimes referred to as *sexless* couplings. This term means that there are no distinct male or female components; both couplings are identical and may be connected to each other. These couplings are designed to be connected and disconnected with only one-third of a turn. The locking components are grooved lugs and inset rings built into the swivels of each coupling (Figure 12.23). When mated, the lugs of each coupling fit into the recesses in the opposing coupling ring and then slide into locking position with a one-third turn.

Figure 12.23 A view of the Storz locking components.

Care of Fire Hose Couplings

All parts of the fire hose coupling are susceptible to damage. On threaded couplings, the male threads are exposed when not connected and are subject to damage. The female threads are not exposed, but the swivel is subject to bending and damage. When either screw-thread couplings or Storz couplings are connected, there is less danger of damage to their parts during common usage; however, they can be bent or crushed if they are run over by vehicles. This is reason enough to prohibit vehicles from running over fire hose. Some simple rules for the care of fire hose couplings are as follows:

- Avoid dropping and/or dragging couplings.
- Do not permit vehicles to run over fire hose.
- Examine couplings when hose is washed and dried.
- Remove the gasket and twist the swivel in warm, soapy water.
- Clean threads to remove tar, dirt, gravel, and oil.
- Inspect gasket, and replace if cracked or creased.

Hose-washing machines will not clean hose couplings sufficiently when the coupling swivel becomes stiff or sluggish from dirt or other foreign matter. The swivel part should be submerged in a container of warm, soapy water and worked forward and backward to thoroughly clean the swivel. The male threads should be cleaned with a suitable brush, and it may be necessary to use a wire brush if threads are clogged by tar, asphalt, or other foreign matter (Figure 12.24).

Figure 12.24 Use a wire brush to clean male threads that are clogged with foreign materials.

Figure 12.25 A view of a swivel gasket in position.

The *swivel gasket* and the *expansion-ring gasket* are two types of gaskets used with fire hose couplings. The swivel gasket is used to make the connection watertight when female and male ends are connected (Figure 12.25). The expansion-ring gasket is used at the end of the hose where it is expanded into the shank of the coupling. These two gaskets are not interchangeable. The difference lies between their thickness and width. Swivel gaskets should occasionally be removed from the coupling and checked for cracks, creases, and general elastic deterioration. The gasket inspection can be made by simply pinching the gasket together between the thumb and index finger. This method usually discloses any defects and demonstrates the inability of the gasket to return to normal shape. Skill Sheet 12-1 shows the procedure for replacing the swivel gasket.

HOSE APPLIANCES AND HOSE TOOLS

[NFPA 1001: 4-3.2(a)]

A complete hose layout for fire fighting purposes includes one end of the hose attached to or submerged in a source of water and the other attached to a nozzle or similar discharge device. There are various devices used with fire hose, other than hose couplings and nozzles, to complete such an arrangement. These devices are usually grouped into two categories: *hose appliances* and *hose tools.* Appliances include valves, valve devices (such as wyes, siameses, water thieves, large diameter hose appliances, and hydrant valves), fittings (which include adapters), and intake devices. Examples of hose tools include hose rollers, spanner wrenches, hose strap and hose rope tools, hose chain tools, hose ramps, hose jackets, blocks, and hose clamps. The following sections highlight some of the more common hose appliances and hose tools.

Hose Appliances

A hose appliance is any piece of hardware used in conjunction with fire hose for the purpose of delivering water. A simple way to remember the difference between hose appliances and hose tools is that appliances have water flowing through them and tools do not.

VALVES

The flow of water is controlled by various valves in hoselines, at hydrants, and at pumpers. These valves include the following types:

- ***Ball valves*** — Used in pumper discharges and gated wyes (Figure 12.26). Ball valves are open when the handle is in line with the hose and closed when it is at a right angle to the hose. Ball valves are also used in fire pump piping systems.
- ***Gate valves*** — Used to control the flow from a hydrant. Gate valves have a baffle that is moved by a handle and screw arrangement (Figure 12.27).
- ***Butterfly valves*** — Used on large pump intakes. A butterfly valve uses a flat baffle operated by a quarter-turn handle. The baffle is in the center of the waterway when the valve is open (Figure 12.28).
- ***Clapper valves*** — Used in siamese appliances (see Valve Devices section) to allow only one intake hose to be connected and charged before the addition of more hoses. The clapper is a flat disk that is hinged on one side and swings in a door-like manner (Figure 12.29).

VALVE DEVICES

Valve devices increase or decrease the number of hoselines operating at the fireground. These devices include wye appliances, siamese appliances, water thief appliances, large diameter hose appliances, and hydrant valves.

Wye appliances. Certain situations make it desirable to divide a line of hose into two or more lines. Various types of wye connections are used for this purpose. The most common wye has a 2½-inch (65 mm) inlet to two 1½-inch (38 mm) outlets, although there are many other combinations commonly found (Figure 12.30). The 2½-inch (65 mm) wye is also used to divide one 2½-inch (65 mm) or larger hoseline into two 2½-inch (65 mm) lines (Figure 12.31). Wye appliances are often gated so that water being fed into the hoselines may be controlled at the gate.

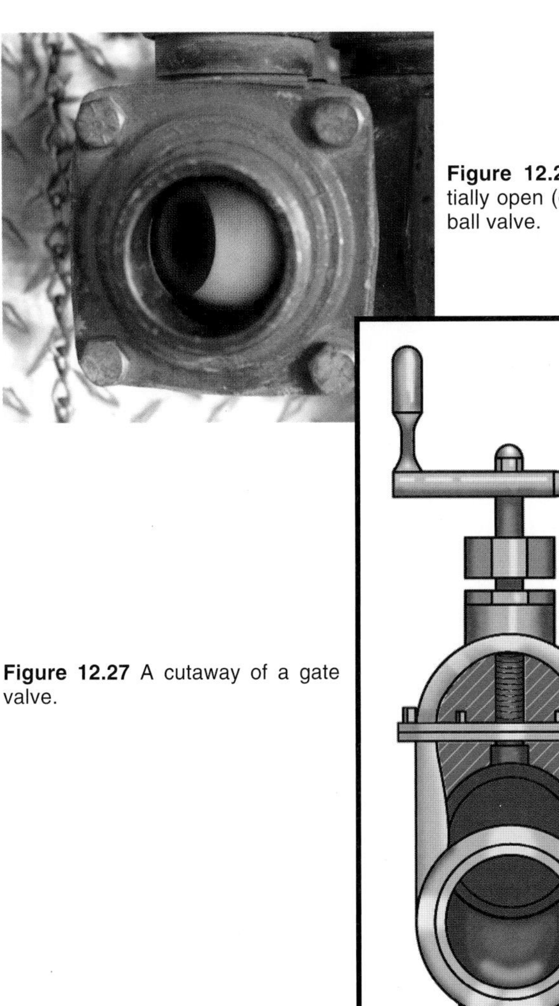
Figure 12.26 A partially open (or closed) ball valve.

Figure 12.27 A cutaway of a gate valve.

Figure 12.30 This common wye has a 2½-inch (65 mm) inlet to two 1½-inch (38 mm) outlets. It is often referred to as a "leader line" wye.

Figure 12.31 A 2½-inch (65 mm) wye is used to divide one 2½-inch (65 mm) or larger hoseline into two 2½-inch (65 mm) lines.

Siamese appliances. The siamese and wye appliances are often confused because of their close resemblance. Siamese fire hose layouts consist of two or more hoselines that are brought into one hoseline or device. The typical siamese has two or three female connections coming into the appliance and one male discharge exiting the appliance (Figure 12.32). Siamese appliances may be equipped with or without clapper valves. The clapper valves allow the siamese to be used with only one incoming supply hoseline attached to it.

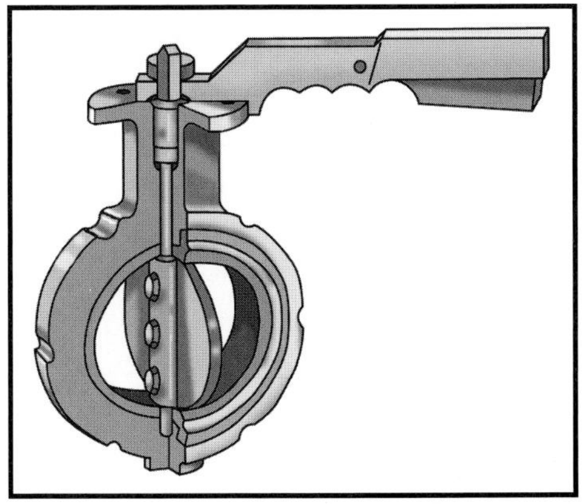
Figure 12.28 A butterfly valve.

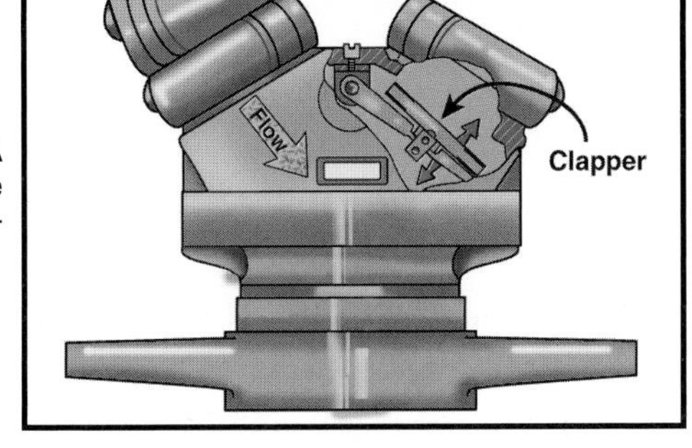

Figure 12.29 A clapper valve inside a siamese appliance.

Figure 12.32 A siamese appliance.

Siamese appliances are commonly used to overcome the problems caused by friction loss in hose lays that carry a large flow or cover a long distance. They are also used quite frequently when supplying ladder pipes that are not equipped with a permanent waterway. Two or three lines are used to supply the one line that is actually going up the ladder. With the increased popularity of large diameter hose (LDH), siamese appliances are being used to feed a large diameter hoseline when multiple smaller hoselines have to be used in the same relay as larger diameter hose.

Water thief appliances. The water thief is a variation of the wye appliance. The most common water thief consists of one 2½-inch (65 mm) inlet with one 2½-inch (65 mm) and two 1½-inch (38 mm) discharge outlets, although other versions, such as a 1½-inch (38 mm) to 1-inch (25 mm) model, are in use (Figures 12.33 a and b). Quarter-turn valves control the outlets. The water thief is intended to be used on a 2½-inch (65 mm) or larger hoseline, usually near the nozzle, so that 2½-inch (65 mm) and 1½-inch (38 mm) hoselines may be used as desired from the same layout.

Figure 12.33a A water thief appliance.

Figure 12.33b This forestry water thief is designed to split a 1-inch (25 mm) hose off the main 1½-inch (38 mm) hose.

Large diameter hose appliances. Large diameter hose operations often necessitate the use of special appliances to distribute the water near the final destination of the hoseline. Depending on the locale and the brand of the appliance, these devices are sometimes called *portable hydrants, manifolds, phantom pumpers,* or *large diameter distributors.* These appliances come in a variety of forms, but in general they have one 4- or 5-inch (100 mm or 125 mm) inlet and two or more smaller outlets (Figure 12.34). Some are similar to water thieves in that they contain one discharge that is the same size as the intake along with several smaller discharges.

Figure 12.34 A manifold distributes water to a number of hoses.

Hydrant valves. A variety of hydrant valves are available for use in supply-line operations (Figures 12.35 a and b). These valves are used when a hose lay is made from the water-supply source to the fire scene (forward or straight lay) (see Supply Hose Lays section). The hydrant valve allows the original supply line to be connected to

Figure 12.35a A four-way hydrant valve.

Figure 12.35b A typical hydrant valve.

the hydrant and charged before the arrival of another pumper at the hydrant. By using the hydrant valve, additional hoselines may be laid to the hydrant, the supply pumper may connect to the hydrant, and pressure may be boosted in the original supply line without having to interrupt the flow of water in the original supply line (Figure 12.36).

(1) The Hydrant Valve is connected to the hydrant with the handle pointing away from the hydrant. Flow is straight through (A to B).

(2) When pressure is required, a hose line is connected from port C to the suction inlet of the pumper. Another line is connected from a discharge on the pumper to port D. The valve handle is then pointed at port D (3 o'clock position).

Flow is then from the valve to suction inlet and pump discharge to valve. There is no interruption in flow. Water continues to flow straight through the valve until pump overcomes hydrant pressure, automatically closing a clapper.

(3) To completely close the hydrant with the Hydrant Valve, change valve handle position to C (9 o'clock). This is accomplished by moving the upper handle toward the B port knob and moving the handle clockwise.

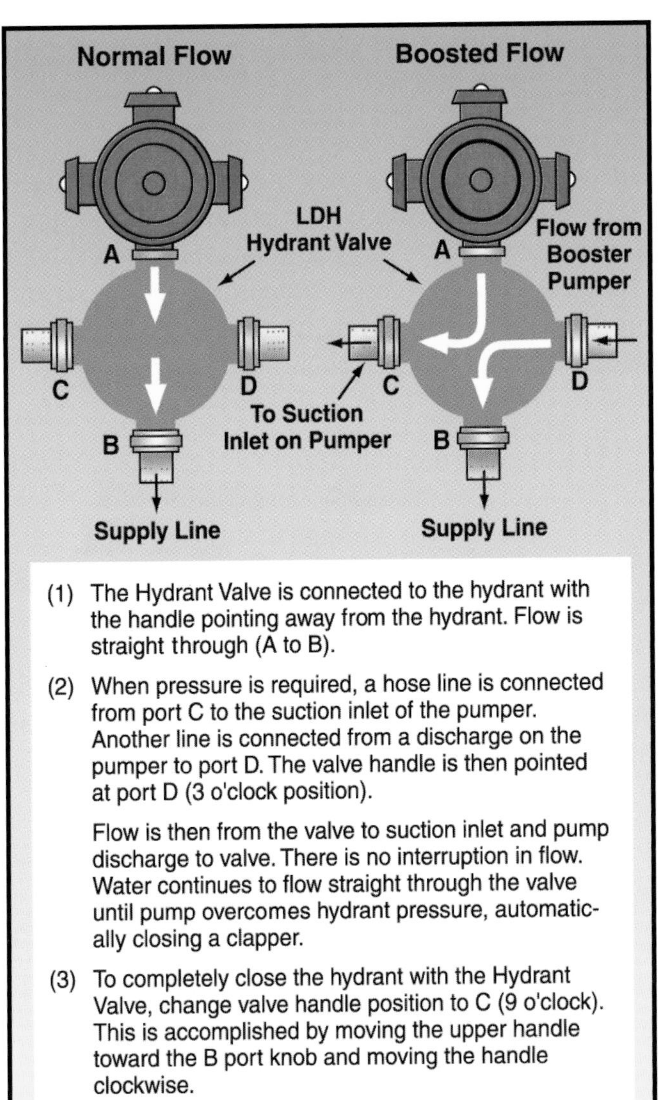

Figure 12.36 Typical operation of a hydrant valve. *Courtesy of Harrington, Inc.*

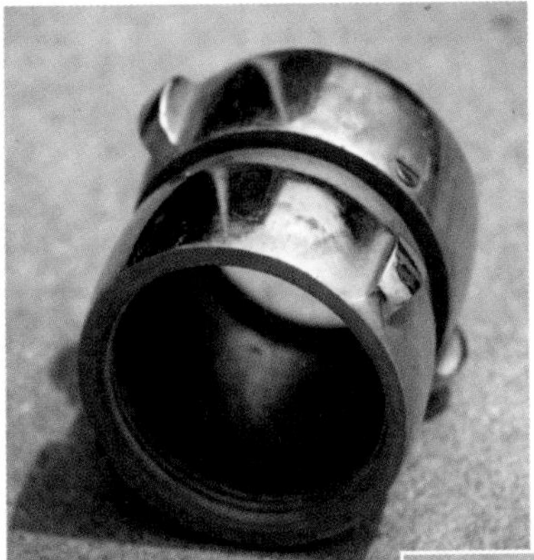

Figure 12.37a A double female adapter.

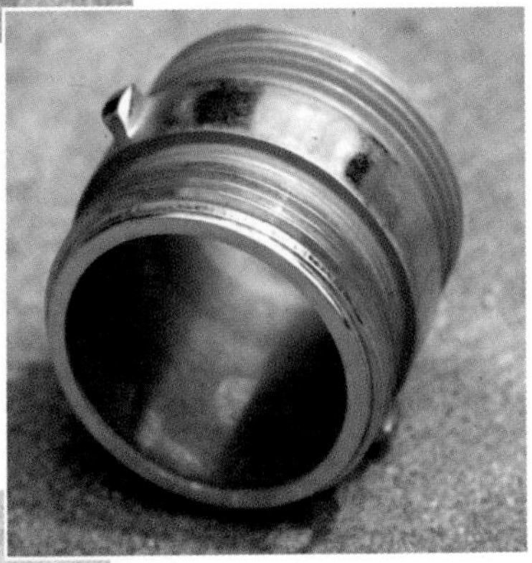

Figure 12.37b A double male adapter.

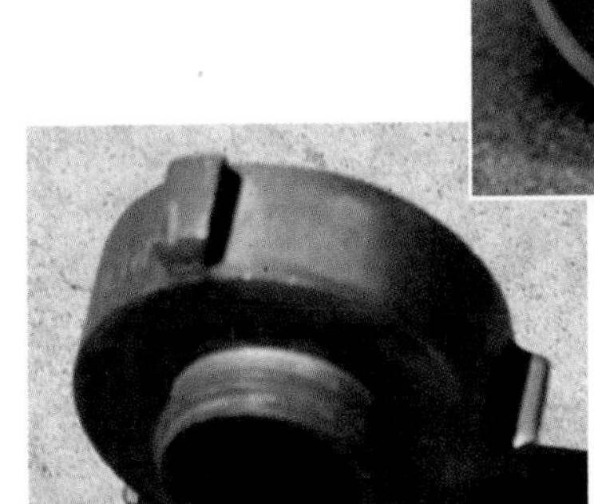

Figure 12.37c A reducer fitting.

FITTINGS

Hardware accessories called *fittings* are available for connecting hoses of different sizes and thread types (Figures 12.37 a–c). An *adapter* is a fitting for connecting hose couplings with dissimilar threads but with the same inside diameter. A variety of special hose appliances are sometimes used in special situations. The double male and double female adapters are probably used more than any other special hose appliance. These appliances allow hoses to be connected when both couplings are of the same sex. This need most frequently occurs when a pumper that is set up for a forward hose lay is used for a reverse lay (laying hose from fire scene back to water supply source) or vice versa (see Supply Hose Lays section).

A reducer, another common hose fitting, is used to extend a larger hoseline by connecting a smaller one to the end. Reducers are also commonly found on pump discharge outlets so that smaller hoselines may be hooked directly to the pump. It should be noted that extending a line with a reducer limits options to just that hoseline whereas using a gated wye at that point allows the option of adding another line if needed.

Other common fittings include *elbows* that change the direction of flow, *hose caps* that close off male couplings, and *hose plugs* that close off female couplings (Figures 12.38 a–c).

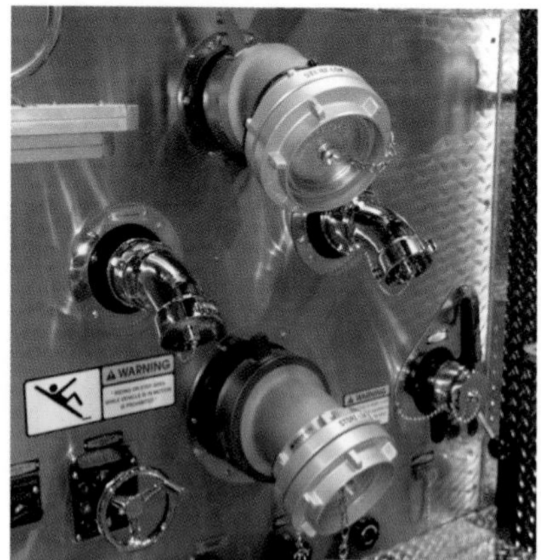

Figure 12.38a An elbow fitting changes the direction of flow.

Figure 12.38b A cap is used to close off male couplings or pump discharges.

Figure 12.38c A plug is used to close a female coupling or a pump intake connection.

INTAKE DEVICES

Suction hose strainers are intake devices attached to the drafting end of a hard suction (sleeve) to keep debris from entering the fire pump. Such debris can pass through the pump and down the line to plug the nozzle. Strainers should not be allowed to rest on the bottom of the water source except when the bottom is clean and hard, such as the bottom of a swimming pool. In order to prevent strainers from resting on the bottom of the water source, some are provided with an eyelet to which a short length of rope can be attached. Some departments keep this rope attached to the strainer as shown in Figure 12.39.

Figure 12.39 A rope attached to a strainer helps keep the strainer from resting on the bottom of the water source.

Hose Tools

There are a variety of tools used in conjunction with hoselines. The following sections highlight some of the more common ones: hose rollers, spanner wrenches, hose strap and hose rope tools, hose chain tools, hose clamps, hose jackets, ramps, and blocks. As stated earlier, hose tools do not have water flowing through them.

HOSE ROLLER (HOIST)

Hose can be damaged when dragged over sharp surfaces such as roof edges and windowsills. A tool for preventing such damage is the *hose roller* (also know as *hose hoist*) (Figure 12.40). The hose roller, consisting of a metal frame with two or more rollers, is placed on the potentially damaging edge and secured with a rope or C-clamp. The hose is then pulled over the rollers. This tool can also be used for handling rope over similar edges.

Figure 12.40 A hose roller prevents hose from being damaged by being dragged over rough or sharp edges.

HOSE JACKET

When a section of hose ruptures, the entire hoseline is unable to transport water effectively. The most practical way to permanently correct the problem is to shut down the line and replace the damaged section of hose. When fire fighting conditions are such that it is not possible to shut down the hoseline and replace the bad section, a *hose jacket* can be installed on the hose at the point of rupture. A hose jacket consists of a two-piece metal cylinder that hinges open and closed (Figure 12.41). Rubber gaskets at each end of the cylinder seal against the hose to prevent leakage. A clamp device locks the cylinder closed when in use. Hose jackets are made in two sizes: 2½ inches and 3 inches (65 mm and 77 mm). The hose jacket encloses the hose so effectively that it can continue to operate at full pressure. A hose jacket can also be used to connect hose with mismatched or damaged screw-thread couplings.

Figure 12.41 A hose jacket in use.

HOSE CLAMP

A hose clamp can be used to stop the flow of water in a hoseline for the following reasons:

- To prevent charging the hose bed during hose-lay operations
- To allow replacement of a burst section of hose without shutting down the water supply (see Replacing Burst Sections section)
- To allow extension of a hoseline without shutting down the water supply (see Extending a Section of Hose section)
- To allow advancement of a charged hoseline up stairs (see Advancing Hose Up a Stairway section)

Based on the method by which they work, there are three types of hose clamps: *screw-down, press-down,* and *hydraulic press* (Figure 12.42). It is important to know that a hose clamp can cause injury to firefighters or damage hose if it is not used correctly. Some general rules that apply to hose clamps are as follows:

- Apply the hose clamp at least 20 feet (6 m) behind the apparatus (Figure 12.43).
- Apply the hose clamp approximately 5 feet (1.5 m) from the coupling on the incoming water side.
- Stand to one side when applying or releasing the press-down type of hose clamp (the operating handle is prone to snapping open suddenly) (Figure 12.44).

CAUTION: Never stand over the handle of a hose clamp when applying or releasing it. The handle may swing upward in a violent motion and injure the firefighter attempting to operate the handle.

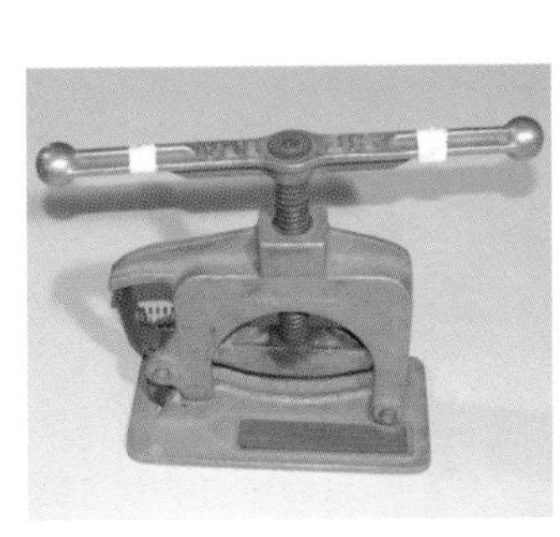

Courtesy of Creston Fire Department, Iowa.

Figure 12.42 Various types of hose clamps.

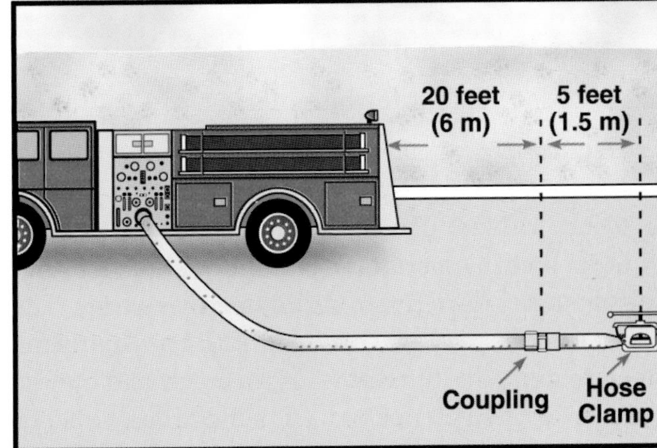

Figure 12.43 Place the hose clamp at least 20 feet (6 m) behind the apparatus and approximately 5 feet (1.5 m) behind the coupling.

- Center the hose evenly in the jaws to avoid pinching the hose.
- Close and open the hose clamp slowly to avoid water hammer.

Figure 12.44 Always stand to one side when applying or releasing the press-down type of hose clamp.

SPANNER, HYDRANT WRENCH, AND RUBBER MALLET

The primary purpose of a *spanner wrench,* or simply *spanner,* is to tighten and loosen hose couplings (Figure 12.45). A number of other features have been built into some spanner wrenches:

- Wedge for prying
- Opening that fits gas utility valves
- Slot for pulling nails
- Flat surface for hammering

Hydrant wrenches are primarily used to remove caps from fire hydrant outlets and to open fire hydrant valves. The hydrant wrench is usually equipped with a pentagon opening in its head that fits most standard fire hydrant operating nuts. The lever handle may be threaded into the operating head to make it adjustable, or the head and handle may be of the ratchet type. The head may also be equipped with a spanner to help make or break coupling connections.

Figure 12.45 Various types of hand tools used in hose operations.

The *rubber mallet* is used to strike the lugs to tighten or loosen intake hose couplings. It is sometimes difficult to get a completely airtight connection with intake hose couplings even though these couplings may be equipped with long operating lugs. Thus, the rubber mallet is used to further tighten the connection.

HOSE BRIDGE OR RAMP

Hose bridges (also called *hose ramps*) help prevent injury to hose when vehicles cross it (Figure 12.46). They should be used wherever a hoseline crosses a street or other area where vehicular traffic cannot be diverted. Some ramps can also be posi-

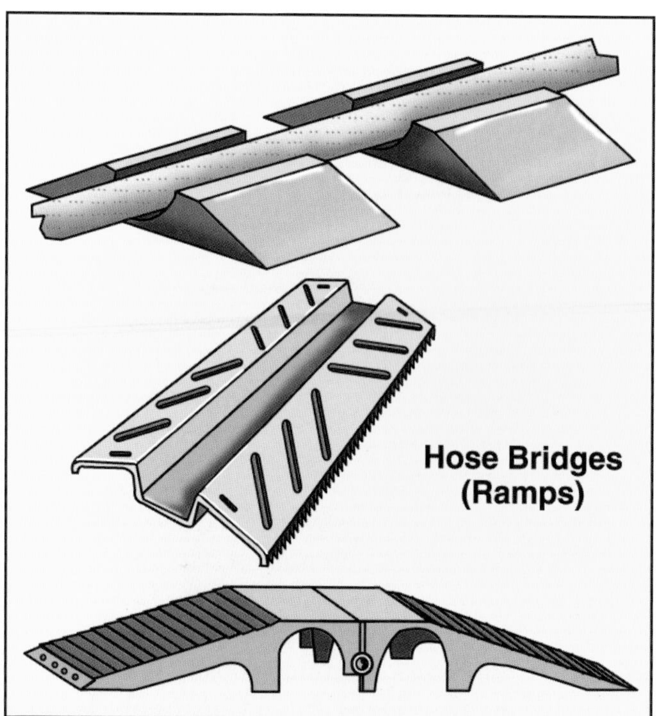

Figure 12.46 Various types of hose bridges (ramps).

tioned over small spills to keep hoselines out of potentially damaging liquids. Hose ramps can also be used as chafing blocks (device used to prevent hose from rubbing against the ground or concrete pavement). See following Chafing Block section.

CHAFING BLOCK

Chafing blocks are devices that are used to protect fire hose where the hose is subjected to rubbing from vibrations (Figure 12.47). Chafing blocks are particularly useful where intake hose comes in contact with pavement or curb steps. At these points, wear on intake hose is most likely because pumper vibrations may be keeping the intake hose in constant motion. Chafing blocks may be made of wood, leather, or sections of old truck tires.

HOSE STRAP, HOSE ROPE, AND HOSE CHAIN

One of the most useful tools to aid in carrying or handling a charged hoseline is a *hose strap.* Similar to the hose strap are the *hose rope* and *hose chain.* These devices can be used to carry and pull fire hose, but their primary value is to provide a more secure means to handle pressurized hose when applying water. Another important use of these tools is to secure hose to ladders and other fixed objects (Figures 12.48 a and b).

HOSE ROLLS

[NFPA 1001: 3-5.4; 3-5.4(a); 3-5.4(b)]

There are a number of different methods for rolling fire hose, depending on its intended use.

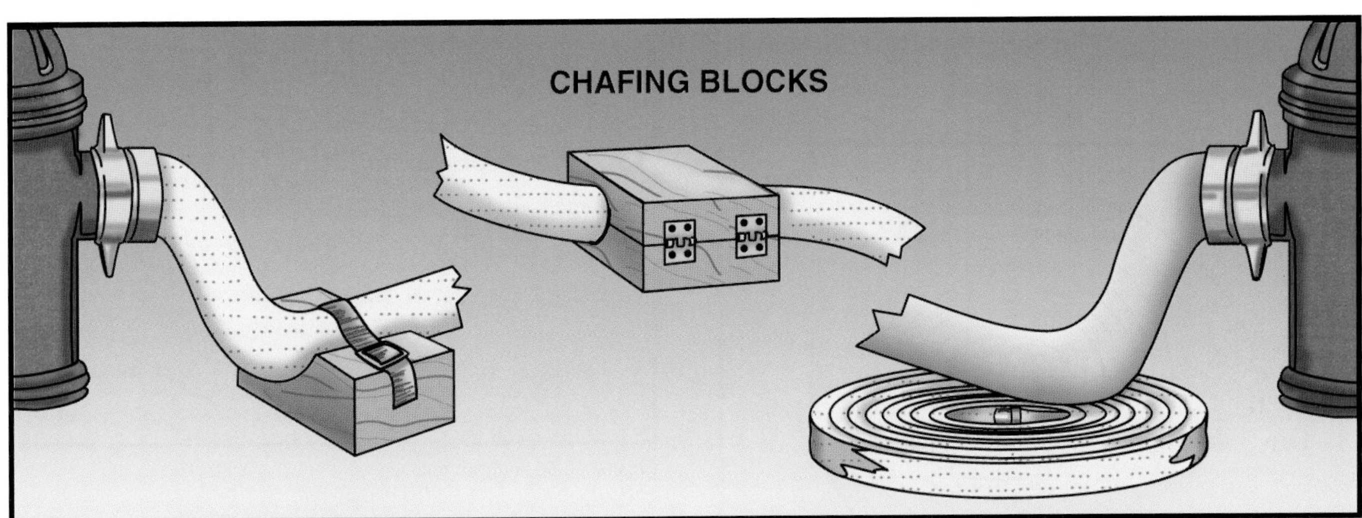

Figure 12.47 Chafing blocks prevent damage to intake hose.

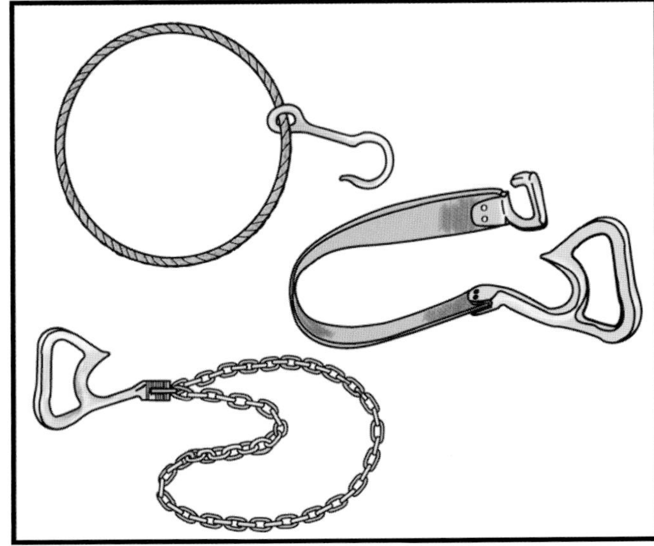

Figure 12.48a Typical hose tools.

Figure 12.48b Attach a hose strap or rope tool so that the nozzle is within easy reach.

In all methods, care must be taken to protect the couplings. Some of the various hose rolls will be discussed in the following sections.

Straight Roll

The straight roll consists of starting at one end, usually at the male coupling, and rolling the hose toward the other end to complete the roll (Figure 12.49). When the roll is completed, the female end is exposed and the male end is protected in the center of the roll. The straight roll is commonly used for hose in the following situations:

- When loaded back on the apparatus at the fire scene
- When returned to quarters for washing
- When placed in storage (especially rack storage)

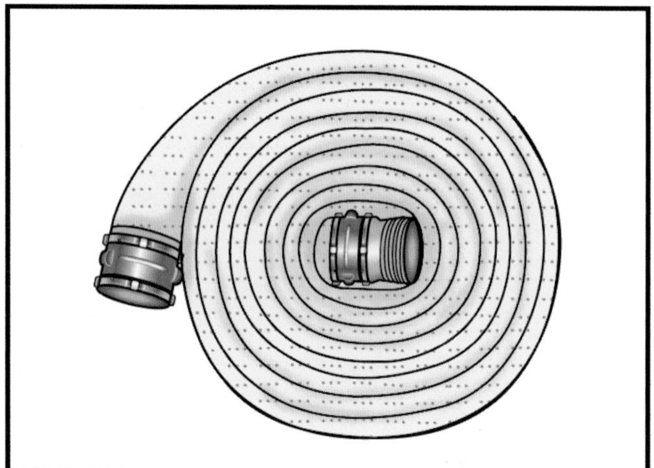

Figure 12.49 A straight roll.

This method is also used for easy loading of the minuteman load (See Preconnected Hose Loads For Attack Lines section).

A variation of the straight roll is to begin the roll at the female coupling so that when the roll is completed, the male coupling is exposed. This method is often done to denote a damaged coupling or piece of hose. A tag is usually attached to the male coupling indicating the type and location of damage. This is also done when the hose is going to be reloaded on the apparatus for a forward (straight) lay. Skill Sheet 12-2 describes the procedure for making the basic straight roll.

Donut Roll

The donut roll is commonly used in situations where hose is going to be deployed for use directly from a roll (Figure 12.50). The donut roll has certain advantages that the straight roll does not possess. Three main advantages are that both ends are available on the outside of the roll, the hose may be quickly unrolled and placed into service, and the hose is less likely to spiral or kink when unrolled. When a section of fire hose needs to be rolled into a donut roll, one or two firefighters may perform the task. Skill Sheets 12-3 and 12-4 describe two methods used to make the donut roll.

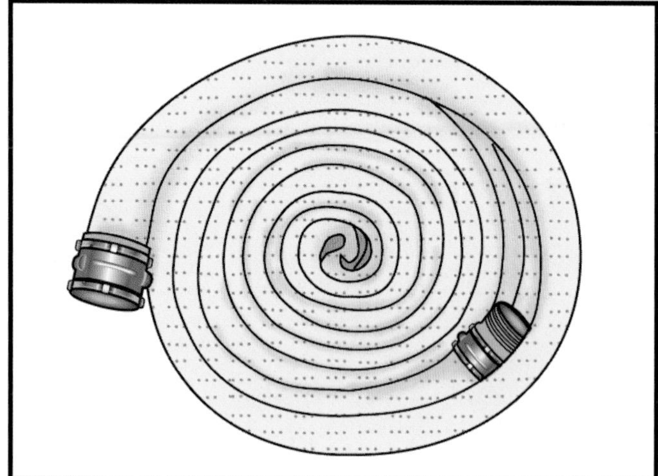

Figure 12.50 A donut roll.

Twin Donut Roll

The twin donut roll is more adaptable to 1½-inch (38 mm) and 1¾-inch (45 mm) hose, although 2-, 2½-, or 3-inch (50 mm, 65 mm, or 77 mm) hose can be used (Figure 12.51). Its purpose is to arrange a compact roll that may be transported and carried for special applications such as high-rise operations. Skill Sheet 12-5 describes how the twin donut roll can be made.

If the couplings are offset by about 1 foot (0.3 m) at the beginning, they can be coupled together after the roll is tied or strapped. This forms a convenient loop that can be slung over one shoulder for carrying while leaving the hands free. By offsetting the couplings at the beginning, they do not dig into the shoulder but are still readily accessible when needed to place the section in service (Figure 12.52).

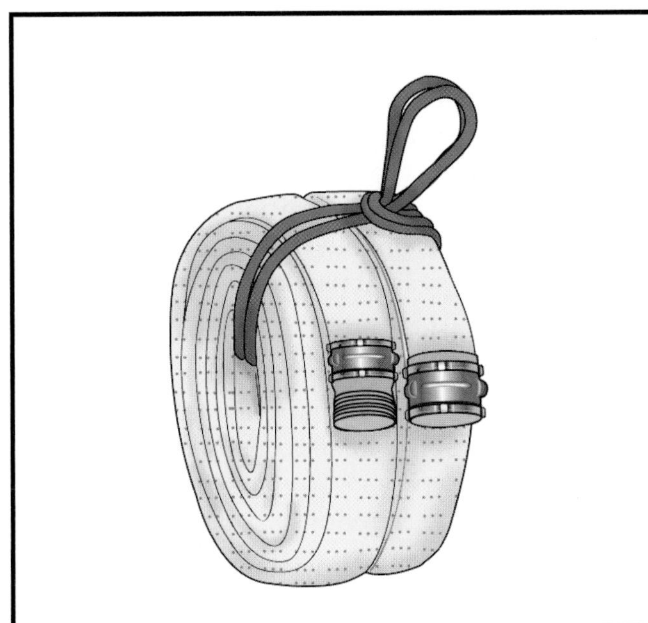

Figure 12.51 A twin donut roll.

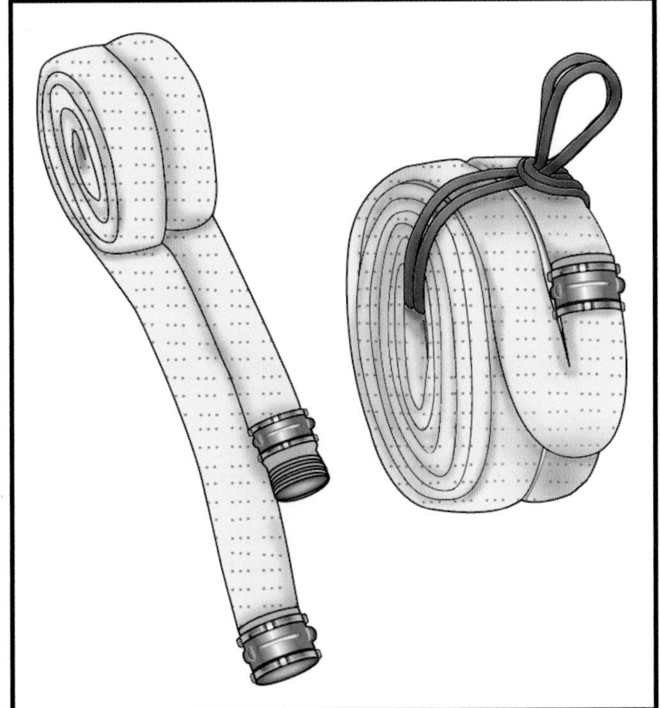

Figure 12.52 If the couplings are offset by about 1 foot (0.3 m) at the beginning, they can be coupled together after the roll is tied or strapped.

Self-Locking Twin Donut Roll

The self-locking twin donut roll is a twin donut roll that has a built-in carrying strap formed from the hose itself (Figure 12.53). This strap locks over the couplings to keep the roll intact for carrying. The length of the carrying strap may be adjusted to accommodate the height of the person carrying the hose. Skill Sheet 12-6 describes how to make the self-locking twin donut roll.

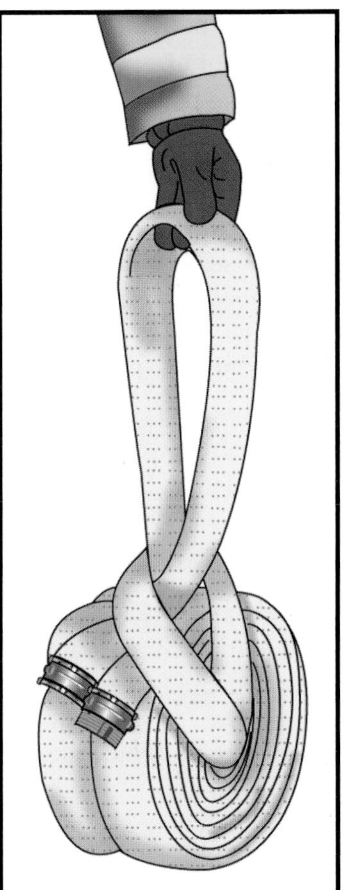

Figure 12.53 A self-locking twin donut roll.

COUPLING AND UNCOUPLING FIRE HOSE

[NFPA 1001: 3-3.9(b)]

The processes of coupling and uncoupling hose are, for the most part, simple procedures for fastening and separating the male and female hose couplings or the sexless couplings in the case of the Storz-type couplings. The need for speed and accuracy under emergency conditions requires that specific techniques for coupling and uncoupling hose be developed. Nozzles may be attached to the hose and separated from the hose by using the same methods as when coupling and uncoupling two sections of hose.

Skill Sheets 12-7 and 12-8 describe two methods of coupling threaded couplings; the same techniques can be applied to Storz (sexless) couplings. It is sometimes necessary to break a tight coupling when spanner wrenches are not available. Skill Sheets 12-9 and 12-10 show two methods by which one or two firefighters may accomplish this task.

BASIC HOSE LOADS AND FINISHES

[NFPA 1001: 3-5.4; 3-5.4(a); 3-5.4(b)]

The most common terminology used to describe a fire hose compartment is *hose bed.* Hose beds vary in size and shape, and they are sometimes built for specific needs. In this manual, the front of the hose bed is designated as that part of the compartment toward the front of the apparatus, and the rear of the hose bed is designated as that part of the compartment toward the rear of the apparatus. Most hose beds have open slats in the bottom that enable air to circulate throughout the hose load. Without this feature, woven-jacket hose could mildew and rot in a very short time.

A hose bed may be divided or separated at some point for the compartment to hold two or more separate loads of hose (split hose bed) (Figure 12.54). The divider (separator) is usually made of sheet metal. A split bed allows the apparatus to have both forward and reverse lays if desired (See Supply Hose Lays section). Hose in a split bed should be stored so that both beds may be connected when a long lay is required.

Another way to arrange hose is to "finish" a hose load with additional hose that can be quickly pulled at the beginning of a forward or reverse lay. *Finishes* are arrangements of hose that are usually placed on top of a hose load and connected to the end of the load.

The following sections provide guidelines for loading hose and highlight the three most common loads for supply hoselines (*accordion, horseshoe,* and *flat*) along with hose load finishes.

Figure 12.54 Hose beds have dividers to separate the various loads.

Hose Loading Guidelines

Although the loading of hose on fire apparatus is not an emergency operation, it is a vital operation that must be done correctly. When fire hose is needed at a fire, the proper hose load permits efficient and effective operations. The following general guidelines should be followed, regardless of the type of hose load used:

- Check gaskets and swivel before connecting any coupling.
- Keep the flat sides of the hose in the same plane when two sections of hose are connected (Figure 12.55). The alignment of the lugs on the couplings is not important.
- Tighten the couplings hand-tight when two sections of hose are connected. Never use wrenches or undue force.
- Remove wrinkles from fire hose when it must be bent to form a loop in the hose bed by pressing with the fingers so that the inside of the bend is smoothly folded.
- Make a short fold or reverse bend (*dutchman*) in the hose during the loading process so that couplings do not have to be turned around to be pulled out of the bed (Figure 12.56).

Figure 12.55 When two sections of hose are connected, keep the flat sides of the hose in the same plane.

Figure 12.56 A short fold or reverse bend in the hose is commonly referred to as a *dutchman*.

- Load large diameter hose (3½-inch [90 mm] or larger) with all couplings placed at the front of the bed. This procedure saves space and allows the hose to lie flat. Couplings should be laid in a manner that does not require them to turn over when the hose pays out of the bed.
- Do not pack hose too tightly. This puts excess pressure on the folds of the hose, and it causes couplings to snag when the hose pays out of the bed. A general rule is that the hose should be loose enough to allow a hand to be easily inserted between the folds (Figure 12.57).

Figure 12.57 If hose is loaded properly, a firefighter should be able to slide a hand between the folds.

Accordion Load

The accordion load derives its name from the manner in which the hose appears after loading (Figure 12.58). The hose is laid progressively on edge in folds that lie adjacent to each other (accordionlike). The first coupling placed in the bed should be located to the rear of the bed. It can be placed on either side if the bed is not split. An advantage of this load is its ease of loading. Its simple design requires only two or three people (although four people are best) to load the hose, and loading can be completed in a matter of minutes. Another advantage is that hose for shoulder carries can easily be taken from the load by simply picking up a number of folds and placing them on the shoulder. Skill Sheet 12-11 shows the procedures for loading an accordion load into a split hose bed for a reverse lay.

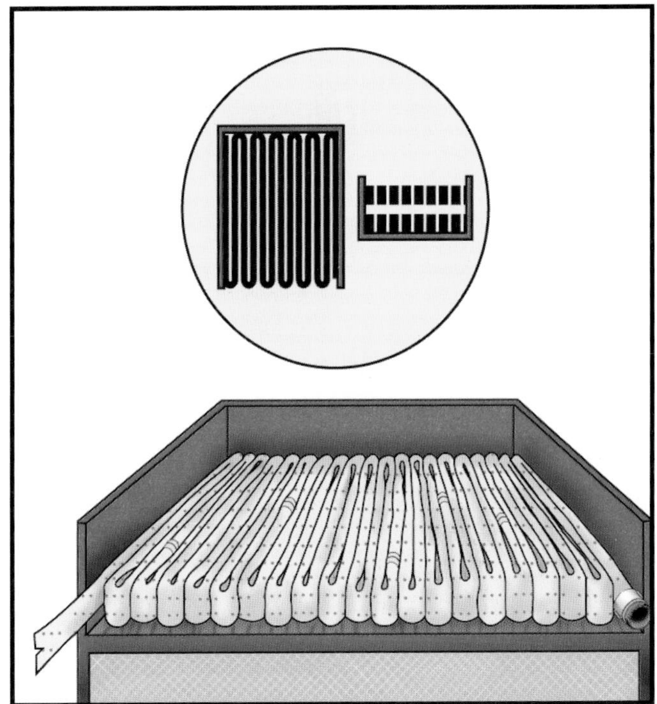

Figure 12.58 The accordion load.

Horseshoe Load

The horseshoe load is also named for the way it appears after loading (Figure 12.59). Like the accordion load, it is loaded on edge, but in this case the hose is laid around the perimeter of the hose bed in a *U*-shaped configuration. Each length is progressively laid from the outside of the bed toward the inside so that the last length is at the center of the horseshoe. The primary advantage of the horseshoe load is that it has fewer sharp bends in the hose than the accordion or flat loads. A disadvantage of the horseshoe load occurs most often in wide hose beds — the hose sometimes comes out in a wavy, or snakelike, lay in the street or on the ground as the hose is pulled alternately from one side of a bed and then the other. Another disadvantage is that folds for a shoulder carry cannot be obtained as easily as with an accordion load. With the horseshoe load, two people are required to make the shoulder folds for the carry. As is the case with the accordion load, the hose is loaded on edge, which can promote wear on hose edges. The horseshoe load does not work for large

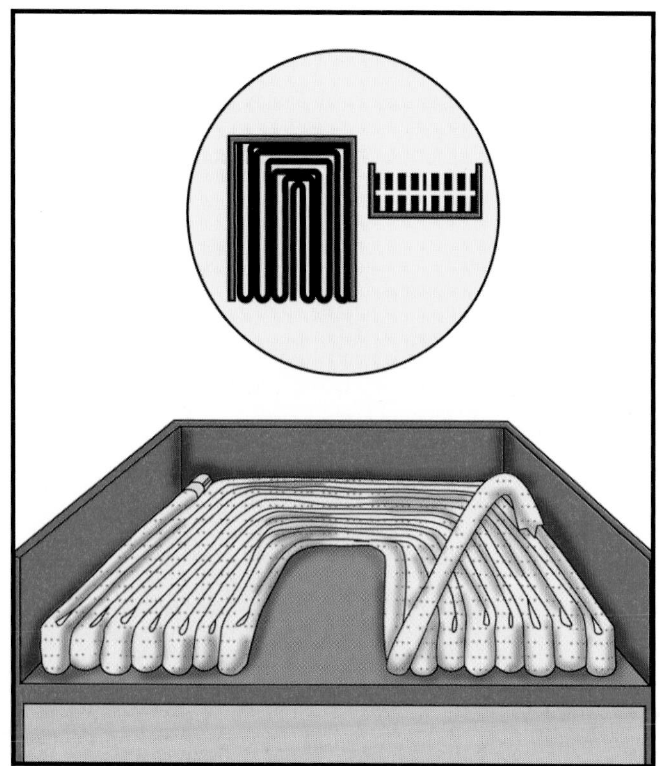

Figure 12.59 The horseshoe load.

diameter hose because the hose remaining in the bed tends to fall over as the hose pays off, which causes the hose to become entangled.

In a single hose bed, the horseshoe load may be started on either side. In a split hose bed, lay the first length against the partition with the coupling hanging an appropriate distance below the hose bed. Determine this distance by estimating the anticipated height of the completed hose load so that the coupling can be connected to the last coupling of the load on the opposite side (crossover) and laid on top of the load. This placement allows easy disconnection of the couplings when the load must be split to lay dual lines. When one side is loaded for a reverse lay and the other is loaded for a forward lay (*combination load*), use an adapter to connect identical couplings. Skill Sheet 12-12 describes the procedures for a single-bed horseshoe load set up for a reverse lay.

Flat Load

Of the three supply hose loads, the flat load is the easiest to load. It is suitable for any size of supply hose and is the best way to load large diameter hose. As the name implies, the hose is laid so that its folds lie flat rather than on edge (Figure 12.60). Hose loaded in this manner is less subject to wear from apparatus vibration during travel. A disadvantage of this load is that the hose folds contain sharp bends at both ends, which requires that the hose be reloaded periodically to relocate bends within each length to prevent damage to the lining.

In a single hose bed, the flat load may be started on either side. In a split hose bed, lay the first length against the partition with the coupling hanging an appropriate distance below the hose bed. Determine this distance by estimating the anticipated height of the hose bed so that the coupling can be connected to the last coupling of the load on the opposite side (crossover) and laid on top of the load. This placement allows easy disconnection of the couplings when the load must be split to lay dual lines. With a combination load, use an adapter to connect identical couplings. Skill Sheet 12-13 shows the procedures for loading a split-bed combination flat load.

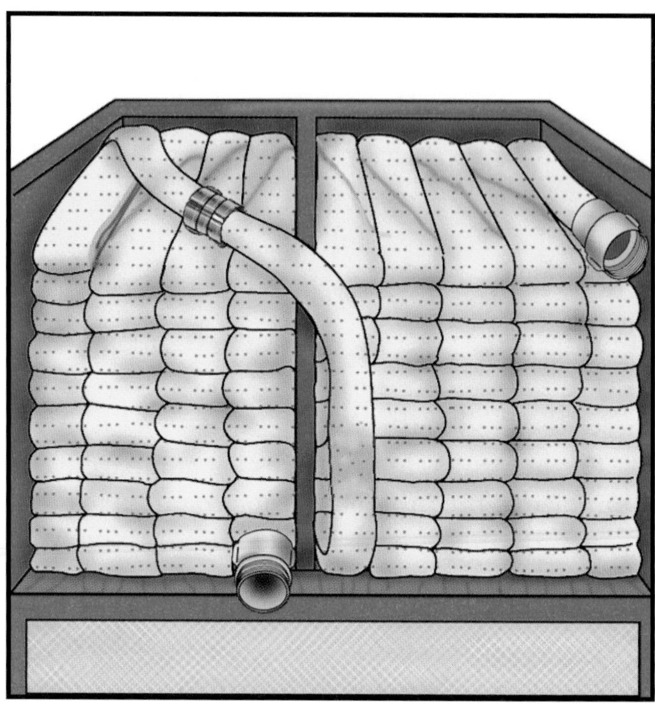

Figure 12.60 The flat load in a split hose bed.

The flat load method described in Skill Sheet 12-13 can be adapted for loading large diameter hose. Large diameter hose can be loaded directly from the street or ground after an incident by straddling the hose with the pumper and driving slowly backward (or according to standard operating procedures) as the hose is progressively loaded into the bed (Figure 12.61). A hose wringer or roller can be used to expel the air and water from the hose as it is placed in the hose bed (Figure 12.62). This procedure creates a neat and space-efficient load of large diameter hose.

Figure 12.61 Large diameter hose can be loaded directly by straddling the hose with the pumper and driving backward slowly as the hose is progressively loaded into the bed.

Figure 12.62 A hose wringer can be used to rid the hose of excess water and air. The reduction of excess water and air makes the hose lie flatter in the bed. *Courtesy of Task Force Tips.*

The hose lay for large diameter hose should be started 12 to 18 inches (300 mm to 450 mm) from the front of the hose bed. This extra space should be reserved for couplings, and all couplings should be laid in a manner that allows them to pay out without turning them over (Figure 12.63). It may be necessary to make a short fold or reverse bend (dutchman) in the hose to do this (Figure 12.64). The dutchman serves two purposes: (1) it changes the direction of a coupling and (2) it changes the location of a coupling.

Figure 12.63 Some departments prefer to load large diameter hose with all the couplings near the front of the hose bed. *Courtesy of Sam Goldwater.*

Figure 12.64 A large diameter hose dutchman.

Hose Load Finishes

Hose load finishes are added to the basic hose load to increase the versatility of the load. Finishes are normally loaded to provide enough hose to make a hydrant connection and to provide a working line at the fire scene.

Finishes fall into two categories: those for forward lays (*straight finish*) and those for reverse lays (*reverse horseshoe finish*). A finish for a reverse lay expedites removing equipment for fire fighting. Finishes for forward lays are usually designed to speed the pulling of hose when making a hydrant connection and are not as elaborate as finishes for reverse lays.

STRAIGHT FINISH

A straight finish consists of the last length or two of hose flaked loosely back and forth across the top of the hose load. This finish is normally associated with forward-lay operation. A hydrant wrench, gate valve, and any necessary adapters should be strapped on the hose at or near the female coupling (Figure 12.65).

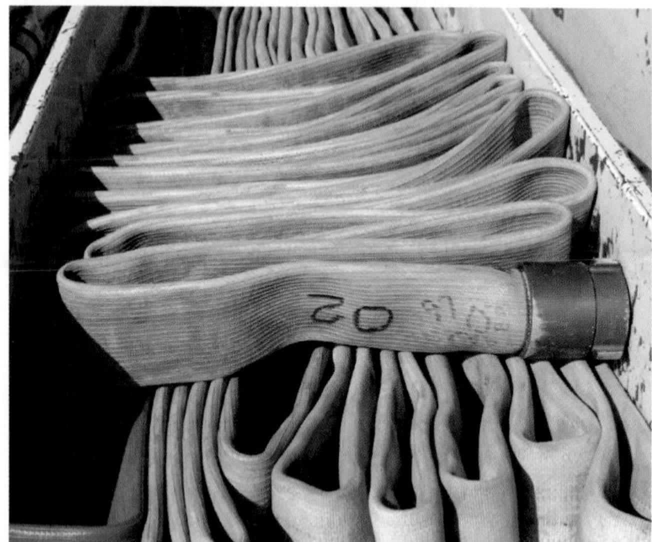

Figure 12.65 A straight finish.

REVERSE HORSESHOE FINISH

This finish is similar to the horseshoe load except that the bottom of the *U* portion of the horseshoe is at the rear of the hose bed (Figure 12.66). It is made of one or two 100-foot (30 m) lengths of hose, each connected to one side of a wye. Any size of attack hose can be used, 1½, 1¾, or 2½ inches (38 mm, 45 mm, or 65 mm). The smaller sizes require a 2½- × 1½-inch (65 mm by 38 mm) gated reducing wye. The 2½-inch (65 mm) hose requires a 2½- × 2½-inch (65 mm by 65 mm) gated wye. Two nozzles of the appropriate size are also needed. Skill Sheet 12-14 outlines the procedures for making a reverse horseshoe finish with 1½-inch (38 mm) hose.

The reverse horseshoe finish can also be used for a preconnected line and can be loaded in two or three layers. With the nozzle extending to the rear, the finish can be placed over a shoulder and the opposite arm extended through the loops of the layers, pulling the hose from the bed for an arm carry. A second preconnected line can be bedded below when there is sufficient depth.

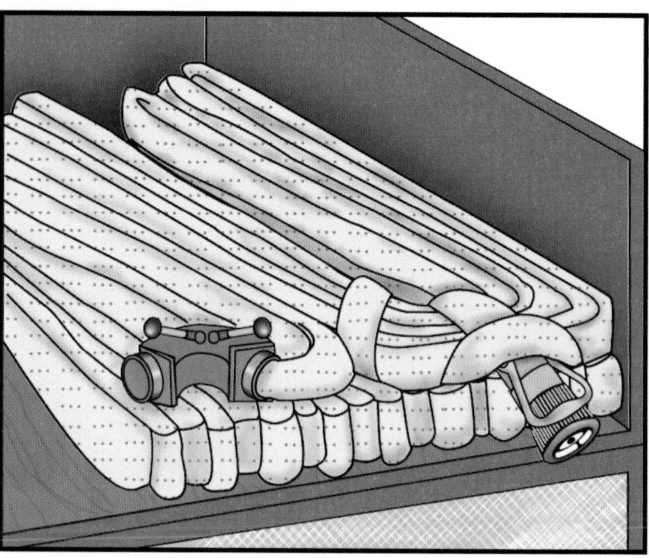

Figure 12.66 A reverse horseshoe finish.

PRECONNECTED HOSE LOADS FOR ATTACK LINES

[NFPA 1001: 3-5.4(b); 3-3.7(a)]

Preconnected hoselines are the primary lines used for fire attack by most fire departments. These hoselines are connected to a discharge valve and placed in an area other than the main hose bed. Preconnected hoselines generally range from 50 to 250 feet (15 m to 75 m) in length. There are several places in which preconnected attack lines can be carried:

- Longitudinal beds (Figure 12.67)
- Raised trays
- Transverse beds (Figure 12.68)
- Tailboard compartments
- Side compartments or bins
- Front bumper wells
- Reels

There are several different loads that can be used for preconnected lines. The following sections detail some of the more common ones. Special loads to meet local requirements may be developed based on individual experiences and apparatus limitations.

Figure 12.67 Some preconnects come off the rear of the apparatus.

Figure 12.68 Mattydale or cross lay preconnects are located above the pump panel.

Preconnected Flat Load

The *preconnected flat load* is adaptable for varying widths of hose beds and is often used in transverse beds (Figure 12.69). This load is similar to the flat load for larger supply hose with two exceptions: (1) It is preconnected and (2) loops are provided to aid in pulling the load from the bed. The pull loops should be placed at regular intervals within the load so that equal portions of the load are pulled from the bed. The number of loops and the intervals at which they are placed are dependent upon the size and total length of the hose. The procedures in Skill Sheet 12-15 can be adapted for any type of hose bed.

Figure 12.69 The preconnected flat load.

Triple Layer Load

The *triple layer load* gets its name because the load begins with hose folded in three layers. The three folds are then laid into the bed in an *S*-shaped fashion (Figure 12.70). The load is designed to be pulled by one person. A disadvantage with the triple layer load is that the three layers, which may be as long as 50 feet (15 m), must be completely removed from the bed before leading in the nozzle end of the hose. This could be a problem if other apparatus are parked directly behind the hose bed.

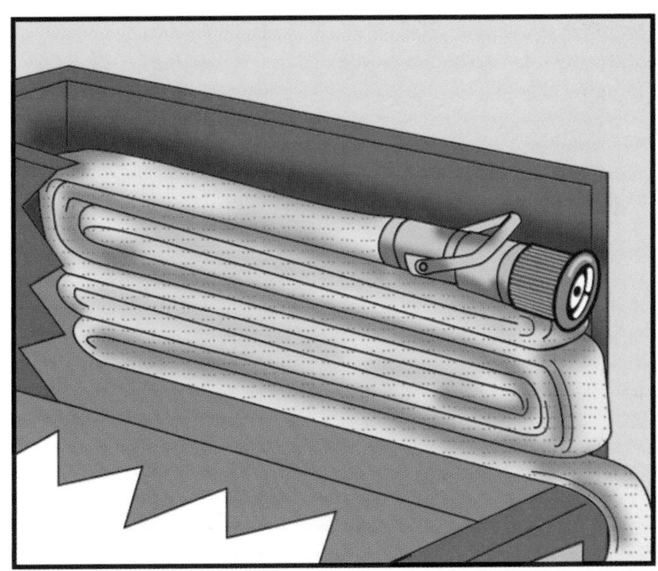

Figure 12.70 The triple layer load.

While this hose load can be used for all sizes of attack lines, it is often preferred for larger (2- and 2½-inch [50 and 65 mm]) lines that may be too cumbersome for shoulder carries. The procedures for making the triple layer load for 200 feet (60 m) of 1½- or 1¾-inch (38 mm or 45 mm) hose are given in Skill Sheet 12-16.

Minuteman Load

The *minuteman load* is designed to be pulled and advanced by one person (Figure 12.71). The primary advantage with this load is that it is carried on the shoulder, completely clear of the ground, so it does not snag on obstacles. The load pays off the shoulder as the firefighter advances toward the fire. The load is also particularly well suited for a narrow bed. A disadvantage with the load is that it can be awkward to carry when wearing an SCBA. If the load is in a single stack, it may also collapse on the shoulder if not held tightly in place. The procedures for making the minuteman load for 150 feet (45 m) of 1½- or 1¾-inch (38 mm or 45 mm) hose loaded in a double stack are described in Skill Sheet 12-17.

Booster Hose Reels

Booster hoselines are preconnected hose that are usually carried coiled upon reels (Figure 12.72). These *booster hose reels* may be mounted several places upon the fire apparatus according to specified needs and the design of the apparatus. Some booster hose reels are mounted above the fire pump and behind the apparatus cab. This arrangement provides booster hose that can be unrolled from either side of the apparatus, but its advancement above ground level is limited to its length. Other booster hose reels are mounted on the front bumper of the apparatus or in rear compartments. Hand- and power-operated reels are available. Noncollapsible hose should be loaded one layer at a time in an even manner. This allows the maximum amount to be loaded and provides for the easiest removal from the reel.

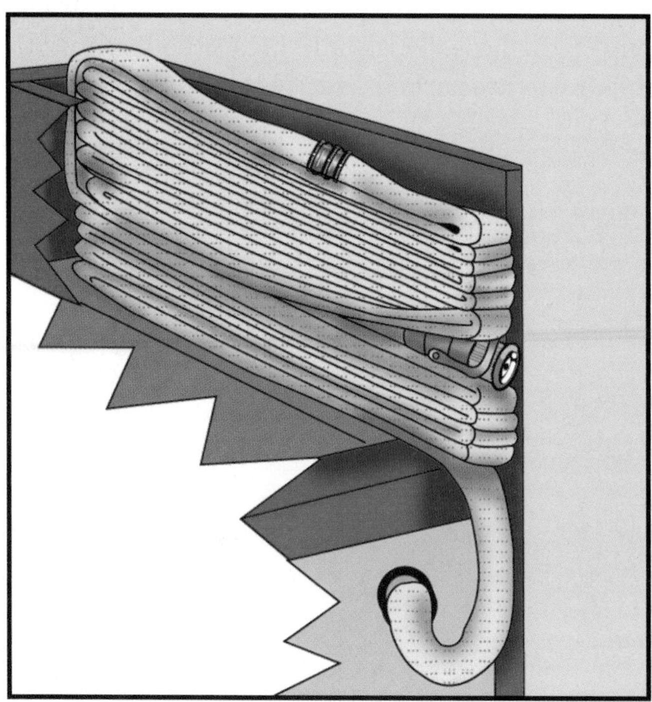

Figure 12.71 The minuteman load.

Figure 12.72 A booster reel located on top of the apparatus.

SUPPLY HOSE LAYS

[NFPA 1001: 3-3.14; 3-3.14(a); 3-3.14(b)]

Threaded-coupling supply hose is usually arranged in the hose bed so that when hose is laid, the end with the female coupling is toward the water source and the end with the male coupling is toward the fire. When hose is arranged in this manner, several hose-lay options are available. At the water source, hose can be connected to the male threads of a pumper discharge valve or to the male threads of a hydrant. At the fire end, hose can be connected to the auxiliary intake valve of a pumper or it can be connected directly to nozzles and appliances, all of which have female threads. There are three basic hose lays for supply hose: *forward lay* (also called *straight lay*), *reverse lay,* and *split lay* (sometimes called *combination lay*).

Hose-lay procedures vary from department to department, but the basic methods of laying hose remain the same. Hose is either laid forward from a water source to the incident scene, reverse from the incident scene to a water source, or split so that the hose can be laid to and from the junction to the water source and the incident scene. These basic methods are presented to provide the foundation for developing hose lays that more specifically suit individual department needs.

Regardless of the method chosen, the following basic guidelines should be followed when laying hose:

- Do not ride in a standing position anytime the apparatus is moving.
- Drive at a speed no greater than that which allows the couplings to clear the tailboard as the hose leaves the bed — generally between 5 and 10 mph (8 kmph and 16 kmph).
- Lay the hose to one side of the roadway so that other apparatus are not forced to drive over it.

Forward Lay

With the forward lay, hose is laid from the water source to the fire. This method is often used when the water source is a hydrant and the pumper must stay at the fire location (Figure 12.73). Hose beds set up for forward lays should be loaded so that the first coupling to come off the hose bed is female (Figure 12.74). The operation consists of stopping the apparatus at the water-supply source and permitting the hydrant person to safely leave the apparatus and secure the hose. Then the apparatus proceeds to the fire laying either single or dual hoselines.

The primary advantage with this lay is that a pumper can remain at the incident scene so that its hose, equipment, and tools can be quickly obtained if needed. The pump operator also has visual contact with the fire fighting crew and can better react

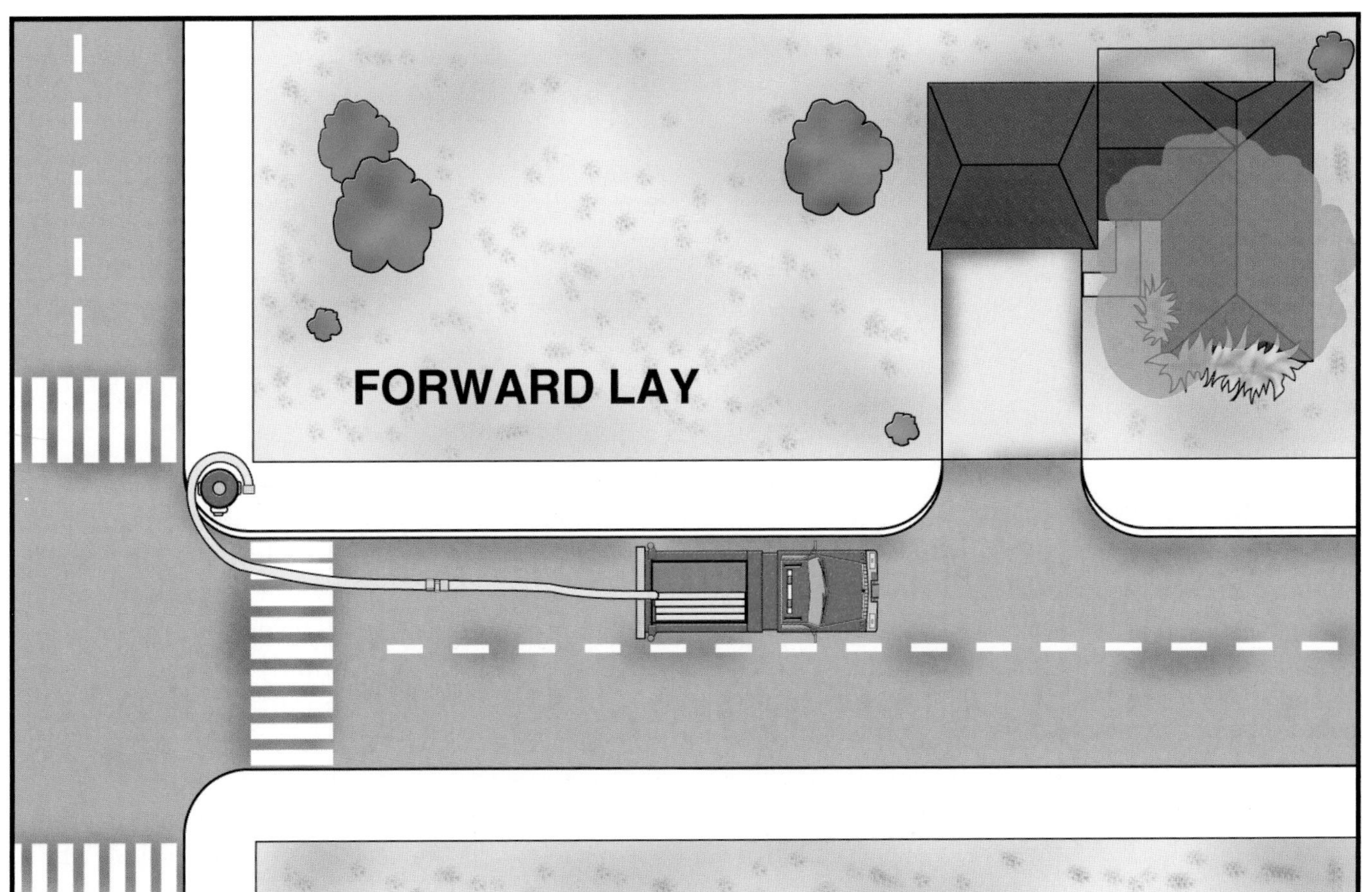

Figure 12.73 The forward lay proceeds from the water source to the scene.

Figure 12.74 The female coupling comes off first on a forward or straight lay.

to changes in the fire operation than if the pumper were at the hydrant. A disadvantage with the lay, however, is that if a long length of medium diameter (2½- or 3-inch [65 mm or 77 mm]) hose is laid, it may be necessary for a second pumper to boost the pressure in the line at the hydrant. This requires the use of a four-way hydrant valve so that the transition from hydrant pressure to pump pressure can be made without interrupting the flow of water in the supply hose (see Using Four-Way Hydrant Valves section). Another disadvantage is that one member of the crew is temporarily unavailable for a fire fighting assignment because that person must stay at the hydrant long enough to make the connection and open the hydrant.

There are two primary skills that the firefighter who is going to make the hydrant connection (also known as *catching the plug* or *making the hydrant*) must know: (1) the proper procedures for wrapping and connecting to the hydrant and (2) the operation of the hydrant valve if one is used.

MAKING THE HYDRANT CONNECTION

The person *catching* the hydrant should have a spanner wrench, hydrant wrench, and a four-way hydrant valve if these are not preconnected to the supply line. Many departments choose to put all of these tools in a jump kit that is kept on the rear step of the apparatus. It is also desirable that the hydrant person have a portable radio so that when the attack engine is ready to receive water, it may be sent immediately when the message is received. However, most departments are not fortunate enough to have a sufficient quantity of radios to do this. In these cases, visual or audible signals are used to tell the firefighter at the hydrant when to start the flow of water. The use of audible warning devices can be a problem when other apparatus are responding to the scene. The hydrant person might mistakenly misunderstand and charge the line before the driver/operator is ready to accept the water. This can result in a charged hose bed, which is useless, or a loose, flowing hose coupling.

The first task to be accomplished when laying a hoseline is to manually remove a small amount of supply hose from the hose bed to start the lay. As a general rule, it is best to start by pulling about 30 feet (9 m) of hose from the apparatus. This amount varies depending on the distance the hydrant or other anchoring object is from the apparatus. When pulling hose from the apparatus, it is important that firefighters also have the necessary tools conveniently located to make the hydrant connection.

Once the appropriate amount of hose is removed and the proper tools are gathered, the firefighter must anchor the hose. It is necessary to anchor the hose at the location from which the lay is being made in order to ensure that the end of the hose remains at the desired location. The best way to do this is to wrap the end of the hose around a stationary object. For the forward lay, this object would be the fire hydrant. However, when making a split lay from a location where there is no hydrant, a service pole, sturdy sign post, mailbox, or parked vehicle can be used as an anchor. The procedures for making a hydrant connection from a forward lay are given in Skill Sheet 12-18.

USING FOUR-WAY HYDRANT VALVES

A four-way hydrant valve allows a forward-laid supply line to be immediately charged and allows a later-arriving pumper to connect to the hydrant (Figure 12.75). The second pumper can then supply additional supply lines and/or boost the pressure to the original line. Typically, the four-way hydrant valve is preconnected to the end of the supply line. This allows the firefighter who is catching the hydrant to hook the valve and the hose to the hydrant in one action. There are several manufacturers providing four-way valves that have the same basic operating prin-

Figure 12.75 A firefighter connecting a four-way hydrant valve to a pumper.

ciples. The steps in Skill Sheet 12-19 describe the typical application of a four-way hydrant valve.

Reverse Lay

With the reverse lay, hose is laid from the fire to the water source. This method is used when a pumper must first go to the fire location so that a size-up can be made before laying a supply line (Figure 12.76). It is also the most expedient way to lay hose if the apparatus that lays the hose must stay at the water source such as when drafting or boosting hydrant pressure to the supply line. Hose beds set up for reverse lays should be loaded so that the first coupling to come off the hose bed is male (Figure 12.77).

Figure 12.77 Set up hose beds for reverse lays so that the first coupling to come off the hose bed is male.

REVERSE LAY

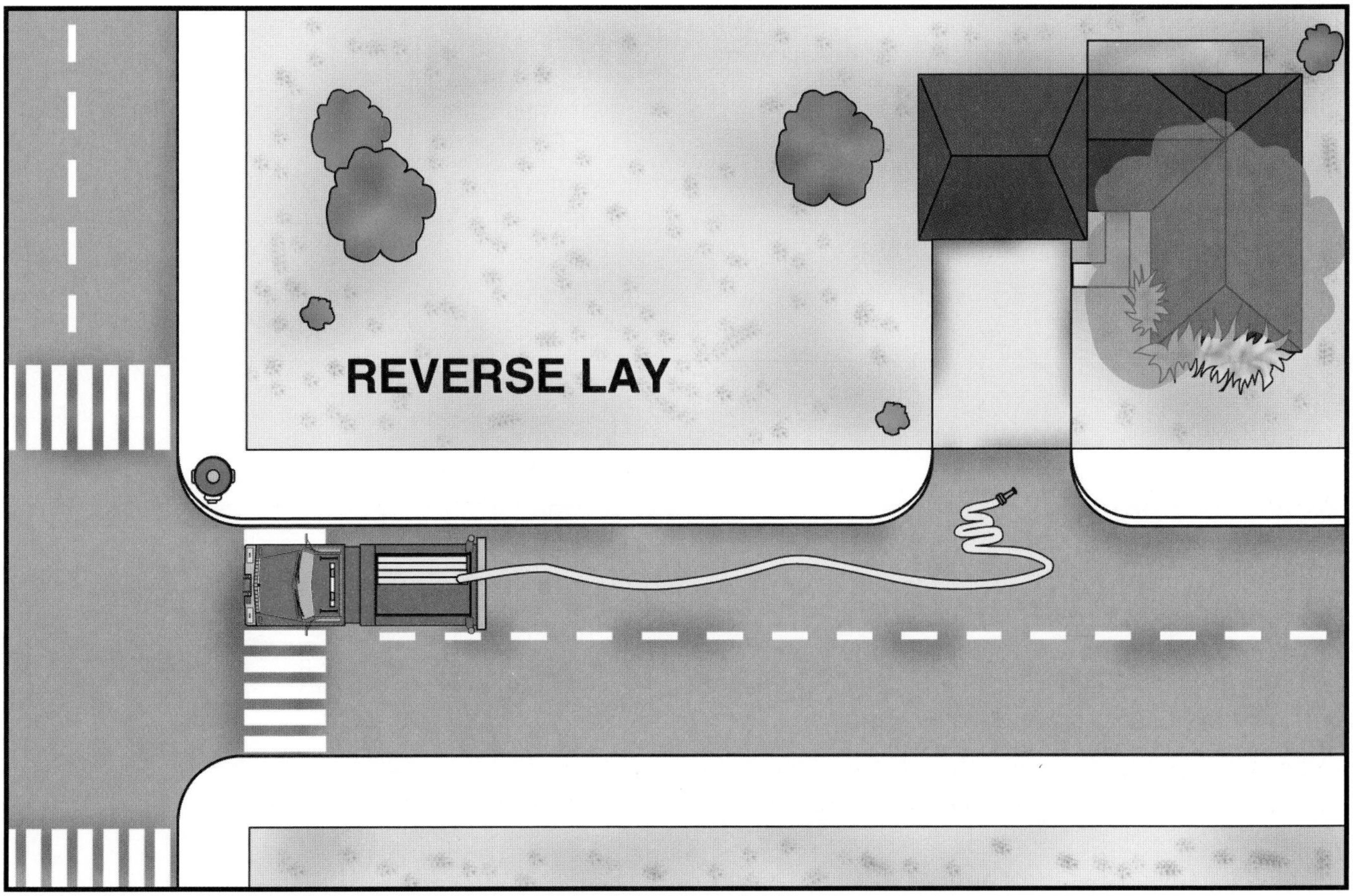

Figure 12.76 The reverse lay proceeds from the scene to the water source.

Laying hose from the incident scene back to the water source has become a standard method for setting up a relay pumping operation when using medium diameter hose as a supply line. With medium diameter hose, it is necessary in most cases to place a pumper at the hydrant to supplement hydrant pressure to the supply hose. It is, of course, always necessary to place a pumper at the water source when drafting. The reverse lay is the most direct way to supplement hydrant pressure and perform drafting operations.

A disadvantage with the reverse lay, however, is that essential fire fighting equipment, including attack hose, must be removed and placed at the fire location before the pumper can proceed to the water source. This causes some delay in the initial attack. The reverse lay also obligates one person, the pump operator, to stay with the pumper at the water source, thus preventing that person from performing other essential fire-location activities.

A common operation involving two pumpers — an attack pumper and a water-supply pumper — calls for the first-arriving pumper to go directly to the scene to start an initial attack on the fire using water from its tank, while the second-arriving pumper lays the supply line from the attack pumper back to the water source. This is a relatively simple operation because the second pumper needs only to connect its just-laid hose to a discharge outlet, connect a suction hose, and begin pumping.

When reverse-laying a supply hose, it is not necessary to use a four-way hydrant valve. One can be used, however, if it is expected that the pumper will later disconnect from the supply hose and leave the hose connected to the hydrant. This situation may be desirable when the demand for water diminishes to the point that the second pumper can be made available for response to other incidents. As with a forward lay, using the four-way valve in a reverse lay provides the means to switch from pump pressure to hydrant pressure without interrupting the flow.

The reverse lay is also used when the first pumper arrives at a fire and must work alone for an extended period of time. In this case, the hose laid in reverse becomes an attack line. It is often connected to a reducing wye so that two smaller hoses can be used to make a two-directional attack on the fire (Figure 12.78). The reverse-lay procedures outlined in Skill Sheet 12-20 describe how the

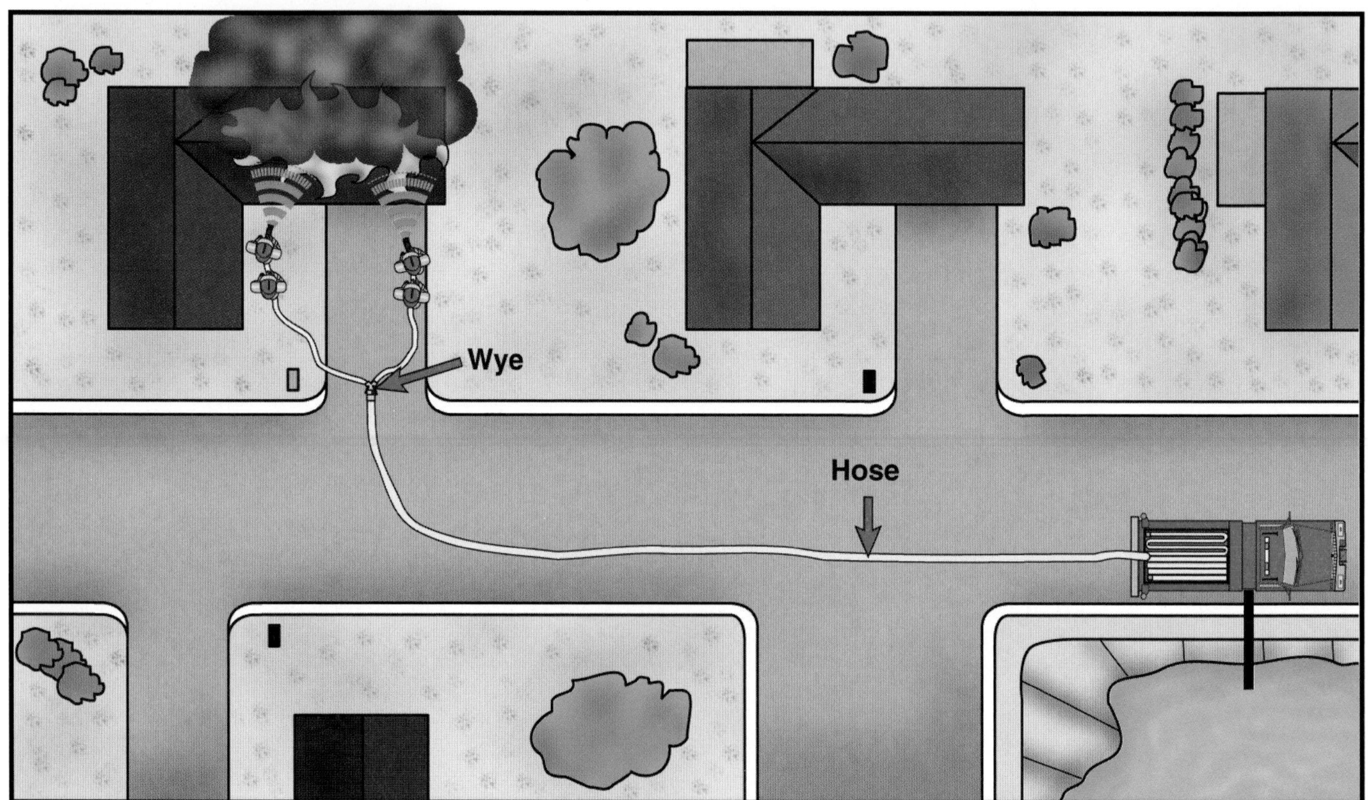

Figure 12.78 A reverse lay using wyed hoselines in operation.

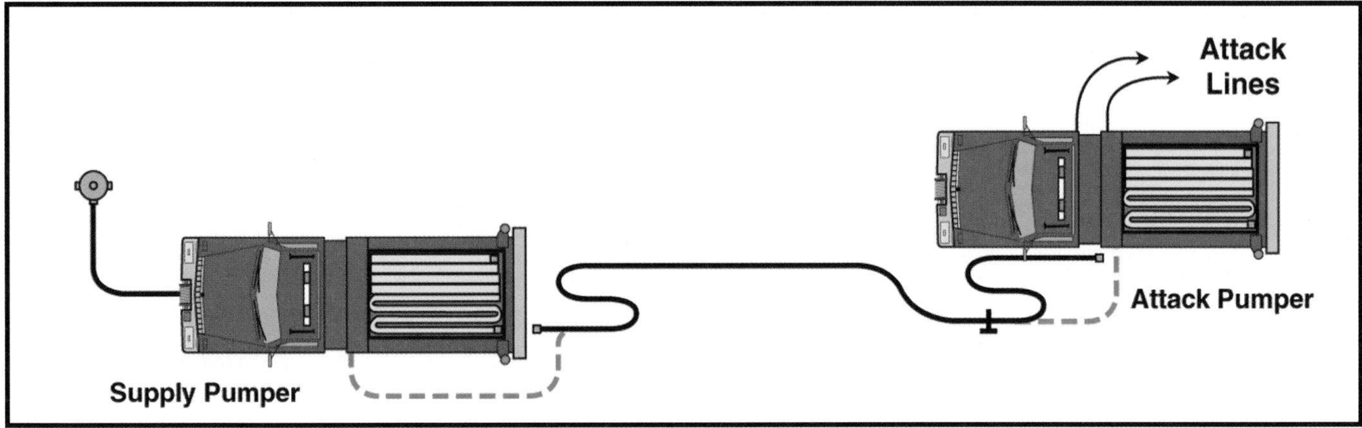

Figure 12.79 The supply pumper lays a reverse supply line from the attack pumper.

second pumper lays a line from an attack pumper to a hydrant (Figure 12.79). They can be modified to accommodate most types of apparatus, hose, and equipment.

MAKING HYDRANT CONNECTIONS WITH SOFT SLEEVE HOSE

Frequently, firefighters will assist pumper driver/operators in making hydrant connections following a reverse lay. Either soft sleeve or hard suction intake hose designed for hydrant operations may be used to connect to hydrants. Hard suction hose must be used when drafting from a static water supply source. Skill Sheet 12-21 illustrates the procedures for making a soft sleeve connection to a hydrant.

Not all hydrants have large steamer discharges capable of accepting direct connections from soft sleeve hoses. Operations on hydrants equipped with two 2½-inch (65 mm) outlets require the use of two 2½- or 3-inch (65 mm or 77 mm) hoselines (Figure 12.80). These smaller intake hoses can be connected to a siamese at the pump. It is more efficient to connect a 4½-inch (115 mm) or larger intake hose to a hydrant with only 2½-inch (65 mm) outlets. Such a connection is made by using a 4½-inch (115 mm) hose, or whatever size intake hose coupling is used, and connecting it to a 2½-inch (65 mm) reducer coupling.

Figure 12.80 Two smaller supply lines may be used to connect the pumper to a hydrant that lacks a large steamer connection.

MAKING HYDRANT CONNECTIONS WITH HARD SUCTION HOSE

Connecting a pumper to a fire hydrant involves coordination and teamwork. More people are needed to connect hard suction hose than are needed to connect soft sleeve hose. Making hydrant connections with a hard suction hose is also considerably more difficult than making connections with a soft sleeve hose. The first aspect that is important is the positioning of the pumper from the hydrant. No definite rule can be given to determine this distance because not all hydrants are the same distance from the curb or road edge, and the hydrant outlet may not directly face the street or road. Another determining factor is that while most apparatus have pump intakes on both sides, others may also have one at the front or rear. It is considered good policy to stop the apparatus with the intake of choice just short of the hydrant outlet. Depending on local preferences, the hard suction hose may be hooked to either the apparatus or the hydrant first when making hydrant connections. Skill Sheet 12-22 describes the procedures for making the hard suction hydrant connection.

NOTE: If the hard suction is marked FOR VACUUM USE ONLY, do not use it for hydrant connections. This type of hard suction is for drafting operations only.

Split Lay

The term *split lay* can refer to any one of a number of ways to lay multiple supply hoses. Dividing a hose bed into two or more separate sections provides the most options for laying multiple lines. Depending upon whether the beds are set up for forward or reverse lays, lines can be laid in the following ways (assume for now that hoses of the same diameter are in two hose beds):

- Two lines laid forward
- Two lines laid reverse
- Forward lay followed by a reverse lay
- Reverse lay followed by a forward lay
- Two lines laid forward followed by one or two lines laid reverse
- Two lines laid reverse followed by one or two lines laid forward

One type of split lay is a hoseline laid in part as a forward lay and in part as a reverse lay. This is accomplished by one pumper making a forward lay from an intersection or driveway entrance toward the fire. A second pumper then makes a reverse lay to the water-supply source from the point where the initial line was laid (Figure 12.81). Care must be taken to avoid making the lay too long for the pump, hose size, and required gallons (liters) per minute delivery.

It must be noted that when using hose equipped with Storz (sexless) couplings, the direction of lay is a moot point. The hose may be laid in either direction with the same result. The only thing that firefighters and driver/operators must be concerned with is making sure that the

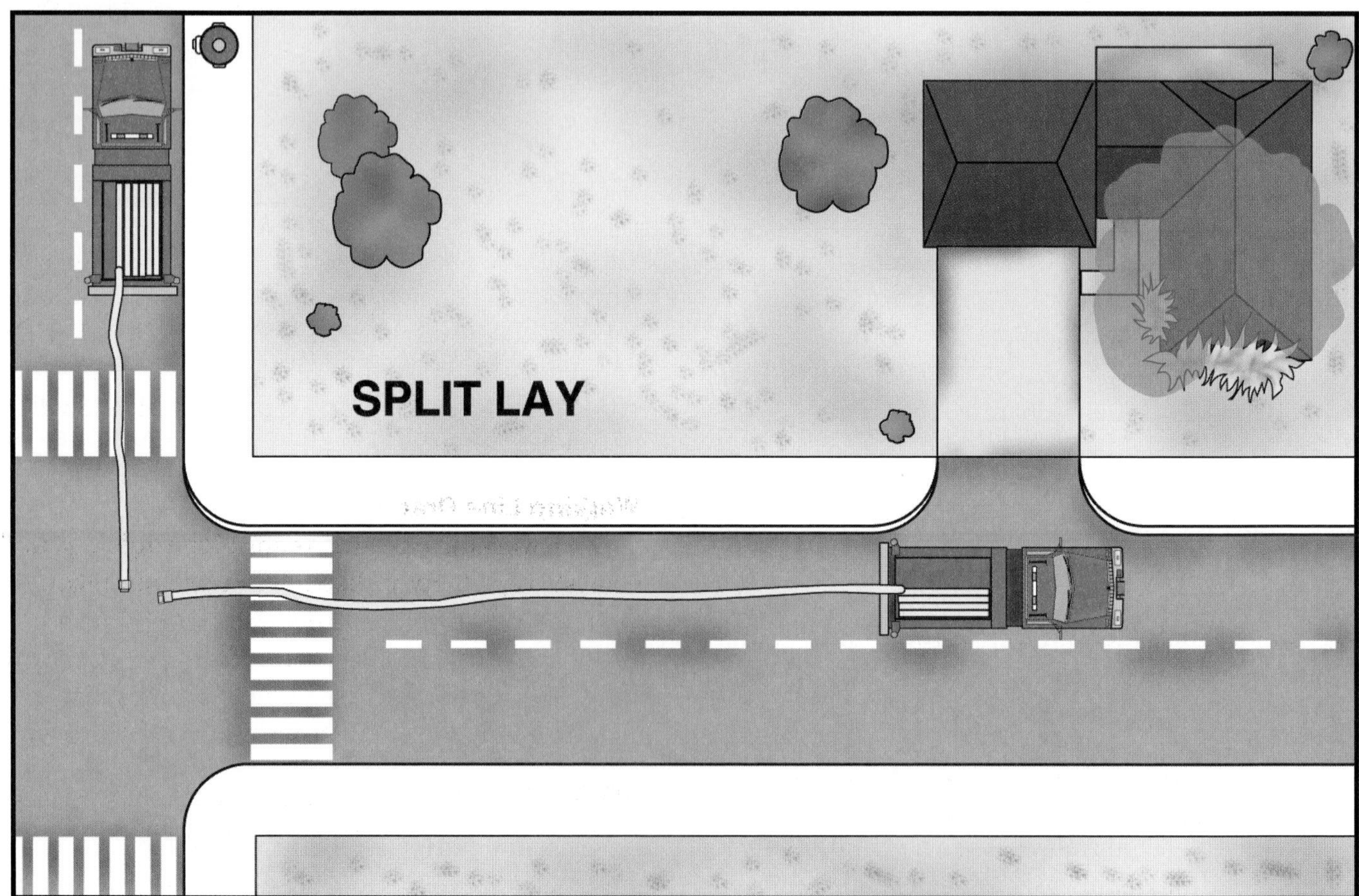

Figure 12.81 A split lay is composed of both a forward and a reverse lay.

proper adapters are present at each end of the lay to make the appropriate connections.

Clearly, there are many other split-lay options when the hose bed is divided. One of the most versatile arrangements is one in which one section of the hose bed contains large diameter hose (LDH) and the other sections contain small diameter hose that can be used for either supply or attack. A pumper set up in this manner can lay LDH when the fire situation requires the pumper to lay its own supply line and work alone (laying it forward so the pumper stays at the incident scene). Firefighters can use small diameter hose as a supply line at fires with less demanding water flow requirements as well as an attack line on large fires. A split hose bed, therefore, gives the fire officer the greatest number of choices when determining the best way to use limited resources.

HANDLING HOSELINES

[NFPA 1001: 3-3.9(a); 3-3.9(b); 3-3.12(b); 3-3.14(b)]

To effectively attack and extinguish a fire, hoselines must be removed from the apparatus and advanced to the location of the fire. The techniques used to advance hoselines depend on how the hose is loaded. Hoselines may be loaded preconnected to a discharge outlet or simply placed in the hose bed unconnected.

Preconnected Hoselines

The method used to pull preconnected hoselines varies with the type of hose load that is used. The following sections describe the methods used to pull and carry preconnected hose from the loads described earlier in this chapter.

PRECONNECTED FLAT LOAD

Advancing the preconnected flat load involves pulling the hose from the compartment and walking toward the fire. This procedure is described in Skill Sheet 12-23.

MINUTEMAN LOAD

The minuteman load is intended to be deployed without dragging any of the hose on the ground. The hose is flaked off the top of the shoulder as the firefighter advances toward the fire. This procedure is described in Skill Sheet 12-24.

TRIPLE LAYER LOAD

Advancing the triple layer load involves placing the nozzle and the fold of the first tier on the shoulder and walking away from the apparatus. This procedure is described in Skill Sheet 12-25.

Other Hoselines

The following procedures are used for handling hose that is not preconnected. This hose is usually 2½ inches (65 mm) or larger.

WYED LINES

The reverse horseshoe finish and other wyed lines are normally used in connection with a reverse layout because the wye connection is fastened to the 2½- or 3-inch (65 mm or 77 mm) hose. The unloading process involves two operations that can be done consecutively by one person. The steps for unloading and advancing wyed lines are contained in Skill Sheet 12-26.

SHOULDER LOADS FROM FLAT OR HORSESHOE LOADS

Because of the way flat and horseshoe loads are arranged in the hose bed, it is necessary to load one section of hose at a time onto the shoulder. Skill Sheet 12-27 describes steps for shoulder loading and advancing hose from either a flat load or a horseshoe load.

SHOULDER LOADS FROM ACCORDION OR FLAT LOADS

Because all of the folds in an accordion load and a flat load are nearly the same length, they can be loaded on the shoulder by taking several folds at a time directly from the hose bed. Skill Sheet 12-28 describes the steps for shoulder loading and advancing hose that is loaded in either an accordion load or a flat load.

Working Line Drag

The working line drag is one of the quickest and easiest ways to move fire hose at ground level. Its use is limited by available personnel, but when adapted to certain situations, it is an acceptable method. Skill Sheet 12-29 contains the procedure for advancing hose using the working line drag.

ADVANCING HOSELINES TO FINAL POSITIONS

[NFPA 1001: 3-3.9(a); 3-3.9(b); 3-3.12(b)]

Once hoselines have been laid out and connected for fire fighting, they must be advanced into final position for applying water on the fire. The methods of deploying hose described to this point work well if the firefighter is simply advancing hose over flat ground with no obstacles. The advancement of hoselines becomes considerably more difficult when other factors come into play. Advancing hose up and down stairways, from standpipes, up ladders, and into buildings are all examples of tasks that require the firefighter to know special techniques. These tasks are more easily accomplished before the hose is charged because water adds considerable weight and makes the lines less maneuverable. However, sometimes it becomes necessary to perform these tasks with charged hoselines, and methods for handling both dry and charged lines, where appropriate, are discussed. Firefighters may also be involved in situations where it is necessary to extend a hoseline by adding additional hose. If a hose bursts, retrieving the loose hoseline and replacing the burst section becomes necessary. These techniques are also discussed.

Advancing Hose Into a Structure

For maximum firefighter safety, it is necessary that firefighters be alert to the potential dangers of backdraft, flashover, and structural collapse, among other things, when advancing hose into a structure. The following are general safety guidelines that should be observed when advancing a hoseline into a burning structure:

- Place the firefighter on the nozzle and the backup firefighter(s) on the same side of the line (Figure 12.82).
- Check the door for heat before entering. This may give an indication of whether there is an extreme amount of heat built up behind the door and alerts firefighters to the possibility of backdraft or flashover conditions.
- Release (bleed) air from the hoseline once it is charged and before entering the building or fire area.
- Stay low and avoid blocking ventilation openings such as doorways or windows.

Advancing Hose Up a Stairway

Hose is difficult to drag in an open space and is exceedingly difficult to drag around the obstructions found in a stairway. When safely possible, hose should be advanced up stairways before it is charged with water. If the line has already been charged, clamp it before advancing up the stairs.

The shoulder carry is adaptable to stairway advancement because the hose is carried into position and fed out as needed. The minuteman load and carry is also excellent for use on stairways. During the advancing process, lay the hose on the stairs against the outside wall to avoid sharp bends and kinks. Excess hose should be flaked up the stairs toward the floor above the fire floor because it will be much easier to advance when the hoseline is carried onto the fire floor. If possible, firefighters should be positioned at every turn or area of resistance to ensure swift, efficient deployment of the hoseline (Figure 12.83).

Figure 12.82 Both firefighters should be on the same side of the hoseline.

Figure 12.83 Lay the hose against the outside wall. *Courtesy of Rick Montemorra, Mesa FD.*

Advancing Hose Down a Stairway

The advancement of an uncharged (dry) hoseline down a flight of stairs is considerably easier than advancing a charged hose. But because advancing a hoseline down a stairway often subjects firefighters to intense heat, the hoseline should be charged in most cases. Advancing an uncharged line downstairs is recommended only when there is no fire present or it is very minor.

Advancing a charged hoseline down a stairway is difficult because of the awkwardness of the fire hose. Increasing heat from the fire floor also makes the surroundings unfavorable. It is necessary to have all available hose at the fire floor because the advance must be made quickly because of these hot conditions. Firefighters must be stationed at critical points — corners and obstructions — to help feed the hose and to keep it on the outside of the staircase.

Advancing Hose From a Standpipe

Fighting fires in tall buildings presents the problem of getting hose to upper floors. While hoselines may be pulled from the apparatus and extended to the fire area, it is not considered good practice. It is more practical to have some hose rolled or folded on the apparatus ready for standpipe use.

The manner in which standpipe hose is arranged is a matter of department standard operating procedures. It may be in the form of folds or bundles that are easily carried on the shoulder or in specially designed hose packs complete with nozzles, fittings, and tools (Figure 12.84).

Hose should be brought to the fire floor by the stairway. Fire crews should stop one floor below the fire floor and make the connection to the standpipe. The standpipe connection is usually in the stairwell or just outside the stairwell door. Firefighters can also use the floor below to get a general idea of the layout of the fire floor.

Upon reaching the standpipe, detach the building hoseline or outlet cap (whichever is present), check the connection for the correct adapters (if needed), check for foreign objects in the discharge, and connect the fire department hose to the standpipe (Figures 12.85 a and b). Be alert for pressure relief devices and follow department standard operating procedures for removal or connection. If 1½-, 1¾-, or 2-inch (38 mm, 45 mm, or 50 mm) hose is used, it is a good practice to place a gated wye either on the standpipe or at the end of a short piece of 2½-inch (65 mm) hose connected to the standpipe. A 2½-inch (65 mm) attack line may also be used depending on the size and nature of the fire. Once the standpipe connection is completed, any extra hose should be flaked up the stairs toward the floor above the fire (Figure 12.86). Charged hoselines as well as dry lines may be advanced in this manner.

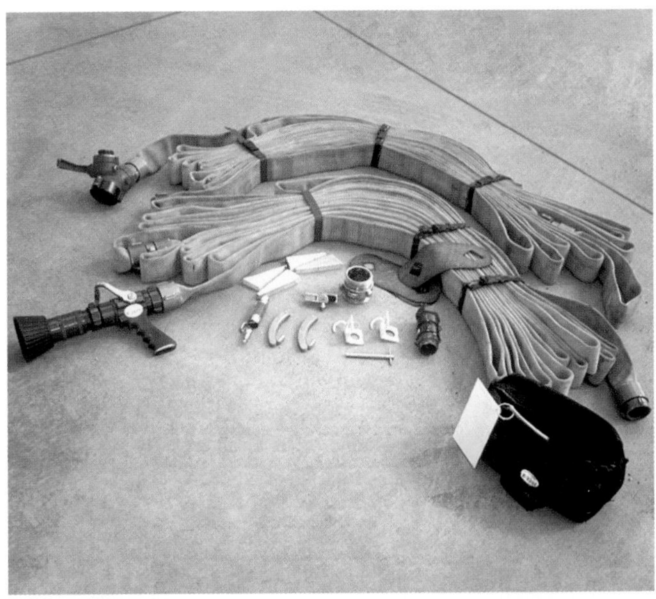

Figure 12.84 The components of a typical hose bundle. *Courtesy of Rick Montemorra, Mesa FD.*

Figure 12.85a Check the discharge for debris. *Courtesy of Rick Montemorra, Mesa FD.*

Figure 12.85b Connect to the discharge. *Courtesy of Rick Montemorra, Mesa FD.*

Figure 12.86 Once the standpipe connection is completed, extra hose should be taken up the stairs toward the floor above the fire. *Courtesy of Rick Montemorra, Mesa FD.*

During pickup operations, the water contained in the hoselines should be carefully drained to prevent unnecessary water damage. This can be accomplished by draining the hose out a window, down a stairway, or down some other suitable drain.

Advancing Hose Up a Ladder

One of the safest ways to get hose to an elevated position is to carry it up the stairs in a bundle and drop the end over a ledge or out a window to connect to a source. Another safe method is to hoist it up to a window or landing using a rope (see Chapter 6, Ropes and Knots). However, sometimes these methods cannot be used, and it is necessary to advance the hose up a ladder. Advancing fire hose up a ladder can be best achieved with a line that is not charged. If the hose is already charged with water, it is safer, quicker, and easier to drain the hose and relieve the pressure before advancement is made.

Whenever possible, it is best to have one firefighter at the base of the ladder to help feed the hose to the carriers and to have one firefighter heel the ladder during the advancement. The best way to advance an *uncharged hoseline* up a ladder is to have the lead firefighter drape the nozzle or end coupling over the shoulder from the front on the side on which the hose is being carried. This firefighter then advances up the ladder until the first fly section is reached and waits until the next firefighter is ready to proceed. At this point, a second firefighter drapes a large loop of hose over the shoulder and starts up the ladder. If the ladder is a three-section ladder, a third firefighter may continue the process once the second firefighter reaches the first fly section (Figure 12.87). To avoid overloading the ladder, only one person should be allowed on each section of the

Figure 12.87 The hose may be carried up the ladder on the firefighters' shoulders.

ladder. Rope hose tools or utility straps can also be used for this advancement (Figure 12.88). The hose can be charged once it has reached its point of intended use.

Figure 12.88 Rope hose tools may be used to carry hose up the ladder.

In those cases where it is absolutely necessary to advance a charged line up a ladder, firefighters should position themselves on the ladder within reach of each other. Each firefighter should be attached to the ladder via a leg lock or ladder belt because both hands are required to move the charged line. The hose is then pushed upward from firefighter to firefighter. The firefighter on the nozzle takes the line into the window, and the other firefighters continue to hoist additional hose as necessary.

WARNING

Caution must be exercised to ensure that the rated capacity of the ladder is not exceeded. If the hose cannot be passed up the ladder without exceeding the load limit, another method of advancing the hoseline, such as by hoisting, should be used.

Sometimes it is necessary to operate the hoseline from the ladder. The hoseline is first passed up the ladder as previously stated. The hose should be secured to the ladder with a hose strap at a point several rungs below the one on which the nozzleperson is standing (Figure 12.89). All firefighters on the ladder must use a leg lock or ladder belt to secure themselves to the ladder. The firefighter on the nozzle projects the nozzle through the ladder and holds it with a rope hose tool or similar aid. When the line and all firefighters on the ladder are properly secured, the nozzle can be opened (Figure 12.90).

Figure 12.89 Secure the hose to the ladder.

Figure 12.90 The firefighter should lock in before operating the fire stream.

Extending a Section of Hose

Occasionally, it becomes necessary to extend the length of a hoseline with hose of the same size or perhaps even smaller hose. Skill Sheet 12-30 describes the procedures that may be used to extend hoselines.

Retrieving a Loose Hoseline

A loose hoseline is one in which water is flowing through a nozzle, an open butt, or a broken line and is not under control by firefighters. This situation is very dangerous because the loose hoseline may whip back and forth and up and down. Firefighters and bystanders may be seriously injured or killed if they are hit by the uncontrolled whipping end.

Closing a valve at the pump or hydrant to turn off the flow of water is the safest way to control a loose line. Another method is to position a hose clamp at a stationary point in the hoseline. It may also be possible to put a kink in the hose at a point away from the break until the appropriate valve is closed (Figure 12.91). To put a kink in the hose, obtain sufficient slack in the line, bend the hose over on itself (this does not apply to LDH because of its size and weight when charged), and apply body weight to the bends in the hose. During this operation, it is helpful to place one knee directly upon the bend and apply pressure at this point.

Figure 12.91 The hose may be kinked to stop the flow of water.

Replacing Burst Sections

A hose clamp or a kink in the hose can also be used to stop the flow of water when replacing a burst section of hose. Two additional sections of hose should be used to replace any one bad section. This is necessary because hoselines stretch to longer lengths when under pressure; thus the couplings in the line are invariably farther apart than the length of a single replacement section.

OPERATING HOSELINES

[NFPA 1001: 3-3.6(b); 3-3.7(b); 3-3.9(a); 3-3.9(b); 3-3.12(b)]

In order to successfully attack a fire, a firefighter must know how to operate and control the hoseline. There are many methods that may be used. The method that any one particular firefighter finds most comfortable varies depending on the size, strength, and personal preference of the firefighter. Some of the more popular techniques are described in the following sections.

Operating Medium-Size Attack Lines

The following methods can be used with medium size attack lines of 1½-, 1¾-, and 2-inch (38 mm, 45 mm, 50 mm) hose.

ONE-FIREFIGHTER METHOD

Whenever one firefighter is required to operate a medium-size hose and nozzle, some means must be provided for bracing and anchoring the hoseline. To accomplish this, the firefighter should hold the nozzle with one hand and hold the hose with the other hand just behind the nozzle (Figure 12.92). The hoseline should be straight for at least 10 feet (3 m) behind the nozzle, and the firefighter should face the direction in which the fire stream is projected. Permit the hose to cradle against the inside of the closest leg, and brace or hold it against the front of the body and hip. Anchor the hose to the ground or floor by placing the foot of the supporting leg upon the hose. If the stream is to be moved or directed at an excessive angle from the centerline, close the nozzle, straighten the hose, and resume the operating position.

TWO-FIREFIGHTER METHOD

The two-firefighter method of handling a nozzle on a medium-size attack line should be used whenever possible because it provides a greater degree of safety than the one-firefighter method. The two-firefighter method is usually necessary when the nozzle is advanced. The person at the nozzle holds the nozzle with one hand and holds the hose just behind the nozzle with the other hand. The hoseline is then rested against the waist and across the hip. The backup firefighter takes a position on the same side of the hose about 3 feet (1 m) behind the nozzleperson. The second firefighter holds the hose with both hands and rests it against the waist and across the hip or braces it with the leg (Figure 12.93). One important function of the backup firefighter is to keep the hose straight behind the person at the nozzle. For extended operation, either one or both firefighters may apply a hose strap or utility strap to reduce the effects of nozzle reaction on the arms.

Figure 12.92 One firefighter can operate small hoselines.

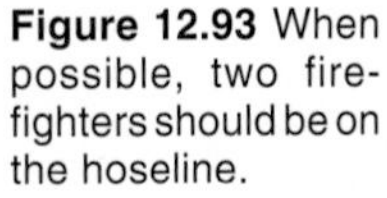

Figure 12.93 When possible, two firefighters should be on the hoseline.

Operating Large-Size Attack Lines

The following methods can be used with large-size attack lines of 2½-, 2¾-, and 3-inch (65 mm, 70 mm, 77 mm) or larger hose.

ONE-FIREFIGHTER METHOD

Whenever a nozzle connected to a large-size attack line is used, a minimum of two firefighters, and preferably three, should be used to operate the line. However, one firefighter may find it necessary to operate a large charged hoseline alone. A reasonably safe way to perform this task is illustrated in Figure 12.94. The firefighter secures slack hose from the line, forms a large loop, and crosses the loop over the line about 2 feet (0.6 m) behind the nozzle. The firefighter then sits where the hose crosses and directs the stream. This method does not permit very much maneuvering of the nozzle, but it can be operated from this point until help is available. If operation continues for a long duration and master stream equipment or personnel are not available, tie the hose at the cross to permit ease of operation and greater safety.

Figure 12.94 When looped, a 2½-inch (65 mm) hoseline can be safely operated by one firefighter.

TWO-FIREFIGHTER METHOD

When only two firefighters are available to handle a nozzle on a large hoseline, some means of anchoring the hose must be provided because of the nozzle reaction. The person at the nozzle holds the nozzle with one hand and holds the hose just behind the nozzle with the other hand. The nozzleperson rests the hoseline against the waist and across the hip. The backup firefighter must serve as an anchor at a position about 3 feet (1 m) behind the nozzleperson. The backup firefighter places the closest knee on the hoseline. In this position, the backup firefighter should be kneeling on one knee with both hands on the hoseline near the other knee. This position prevents the hose from moving back or to either side. Should the hose in front try to move back or up, the backup firefighter is in a position to push it forward.

Another two-firefighter method uses hose rope tools to assist in anchoring the hose. The nozzleperson loops a hose rope tool or utility strap around the hose a short distance from the nozzle and places the large loop across the back and over the outside shoulder. The nozzle is then held with one hand, and the hose just behind the nozzle is held with the other hand. The hoseline is rested against the body. Leaning slightly toward the nozzle helps control the nozzle reaction. The backup firefighter again serves as an anchor about 3 feet (1 m) back. The backup firefighter also has a hose rope tool around the hose and the shoulder and leans forward to absorb some of the nozzle reaction (Figure 12.95).

Figure 12.95 When possible, use a rope hose tool to help control the line.

THREE-FIREFIGHTER METHOD

Handling a nozzle on a large-size hoseline can be more easily accomplished by three firefighters. There are several methods for three firefighters to control large hoselines. In all cases, the positioning of the nozzleperson is the same as previously described for the two-firefighter method. The only differences will be in the position of the second and third firefighters on the hoseline. Some departments prefer the first backup firefighter to stand directly behind the firefighter at the nozzle, with the third firefighter kneeling on the hose behind the second firefighter. Another method is for both firefighters to serve as anchors by kneeling on opposite sides of the hoseline. Another technique is for all firefighters to use hose straps and remain in a standing position, which is the most mobile method (Figure 12.96).

Figure 12.96 Three firefighters make the hoseline more maneuverable.

SERVICE TESTING FIRE HOSE

[NFPA 1001: 3-5.4(a); 3-5.4(b); 4-5.3; 4-5.3(a); 4-5.3(b)]

There are two types of tests for fire hose: *acceptance testing* and *service testing*. At the request of the purchasing agency, coupled hose is acceptance tested by the manufacturer before the hose is shipped. This type of testing is relatively rigorous, and the hose is subjected to extremely high pressures to ensure that it can withstand the most extreme conditions in the field. Acceptance testing should not be attempted by fire department personnel. Service testing, however, is performed periodically by the user to ensure that the hose is being maintained in optimum condition. This testing of in-service hose confirms that it is still able to function under maximum pressure during fire fighting or other operations. Guidelines for both types of tests are in NFPA 1962, *Standard for the Care, Use, and Service Testing of Fire Hose Including Couplings and Nozzles.*

Because fire hose is required to be tested annually, firefighters often assist in the process. Fire department hose should also be tested after being repaired and after being run over by a vehicle.

Before performing a service test, the hose should be examined for jacket defects, coupling damage, and worn or defective gaskets. Any defects should be corrected if possible. If damage is not repairable, the hose should be taken out of service.

Test Site Preparation

Hose should be tested in a place that has adequate room to lay out the hose in straight runs, free of kinks or twists. The site should be isolated from traffic. If testing is done at night, the area should be well lighted. The test area should be smooth and free of dirt and debris. A slight grade to facilitate the draining of water is helpful. A water source sufficient for filling the hose is also necessary.

The following equipment is needed to service test hose:

- Hose testing machine, portable pump, or fire department pumper equipped with gauges certified as accurate within one year before testing (Figure 12.97)
- Hose test gate valve

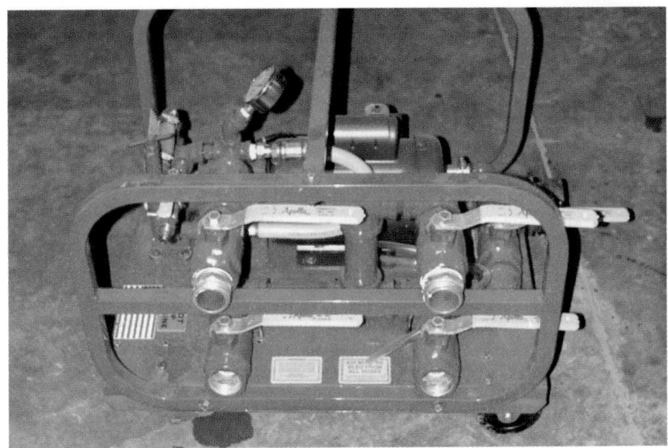

Figure 12.97 A typical hose testing machine. *Courtesy of Rico Hydro Equipment Mfg.*

- Means of recording the hose numbers and test results
- Tags or other means to identify sections that fail
- Nozzles with shutoff valves
- Means of marking each length with the year of the test to easily identify which lengths have been tested and which have not without looking in the hose records

Service Test Procedure

Exercise care when working with hose, especially when it is under pressure. Pressurized hose is potentially dangerous because of its tendency to whip back and forth if a break occurs such as when a coupling pulls loose. To prevent this situation, use a specially designed hose test gate valve (Figure 12.98). This is a valve with a ¼-inch (6 mm) hole in the gate that permits pressurizing the hose but does not allow water to surge through the hose if it fails. Even when using the test gate valve, stand or walk near the pressurized hose only as necessary.

CAUTION: All personnel operating in the area of the pressurized hose should wear at least a helmet as a safety precaution.

When possible, connect the hose to discharges on the side of the apparatus opposite the pump panel. Open and close all valves slowly to prevent water hammer in the hose and pump. Test lengths of hose shall not exceed 300 feet (90 m) in length (longer lengths are more difficult to purge of air).

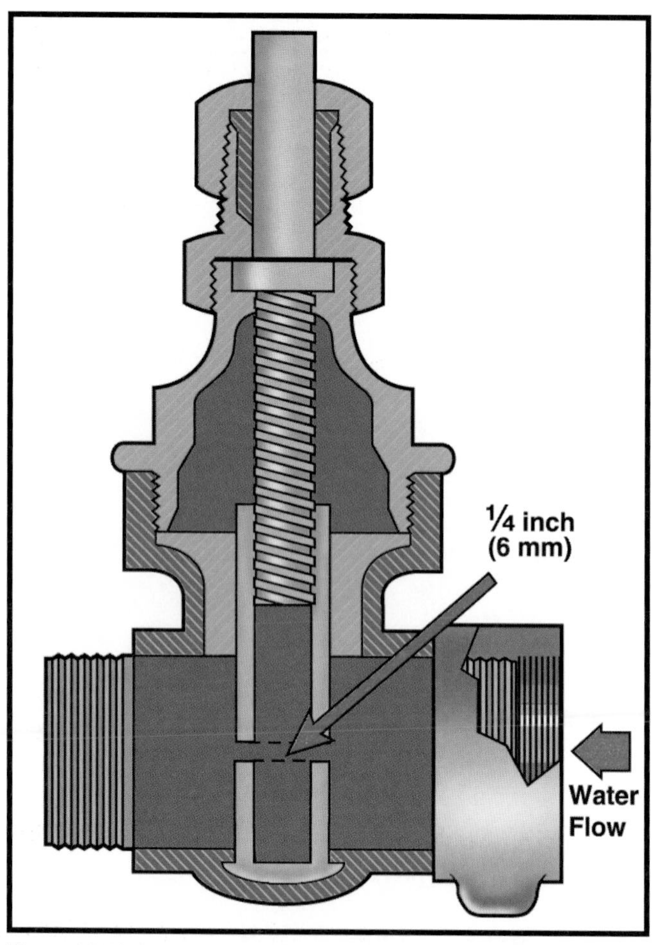

Figure 12.98 A cutaway showing the hole in the valve gate.

Laying large diameter hose flat on the ground before charging helps to prevent unnecessary wear at the edges. Stand away from the discharge valve connection when charging because of the hose's tendency to twist when filled with water and pressurized; this twisting could cause the connection to twist loose.

Keep the hose testing area free of water when filling and discharging air from the hoses. During testing, this air aids in detecting minor leaks around couplings. The procedure for service testing lined fire hose and large diameter hose is described in Skill Sheet 12-31.

SKILL SHEET 12-1 REPLACING THE SWIVEL GASKET

Step 1: Hold the gasket between the middle finger and thumb.

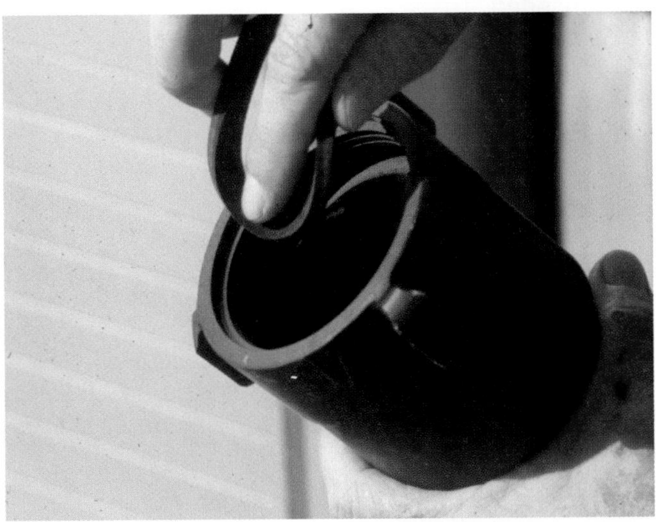

Step 2: Fold the outer rim of the gasket upward by pulling with the index finger.

Step 3. Place the gasket into the swivel by permitting the large loop of the gasket to enter into the coupling swivel at the place provided.

Step 4: Release your grip on the gasket, allowing the small loop to fall into place.

SKILL SHEET 12-2 MAKING THE STRAIGHT ROLL

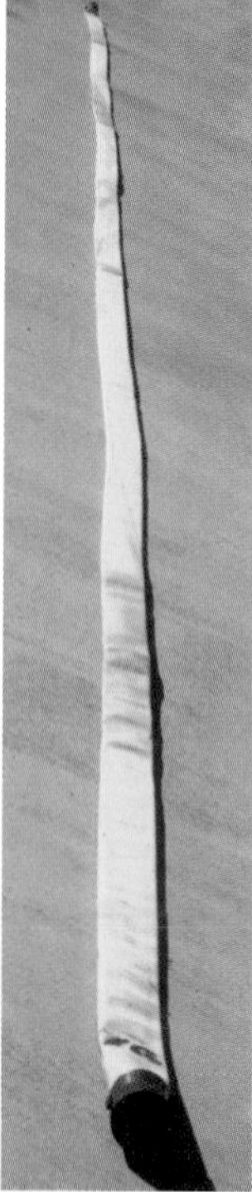

Step 1: Lay out the hose straight and flat on a clean surface.

Step 2: Roll the male coupling over onto the hose to start the roll.

NOTE: Form a coil that is open enough to allow the fingers to be inserted.

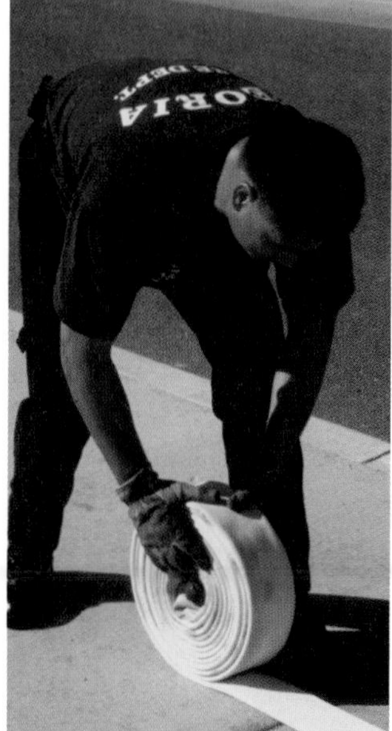

Step 3: Continue rolling the coupling over onto the hose, forming an even roll.

NOTE: Keep the edges of the roll aligned on the remaining hose to make a uniform roll as the roll increases in size.

Step 4: Lay the completed roll on the ground.

Step 5: Tamp any protruding coils down into the roll with a foot.

SKILL SHEET 12-3 MAKING THE DONUT ROLL

Method One

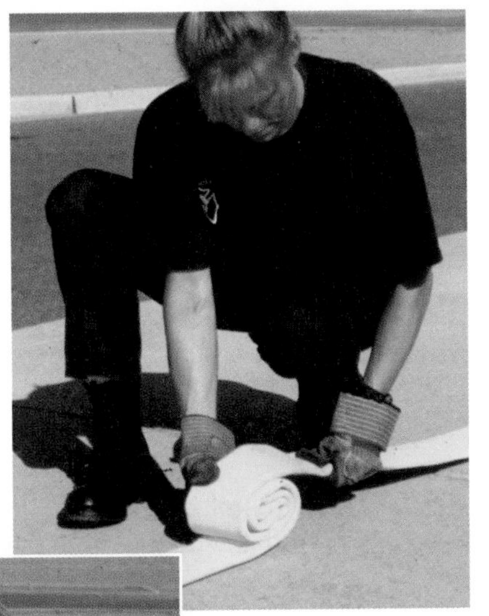

Step 1: Lay the section of hose flat and in a straight line.

Step 2: Start the roll from a point 5 or 6 feet (1.5 m or 1.8 m) off center toward the male coupling.

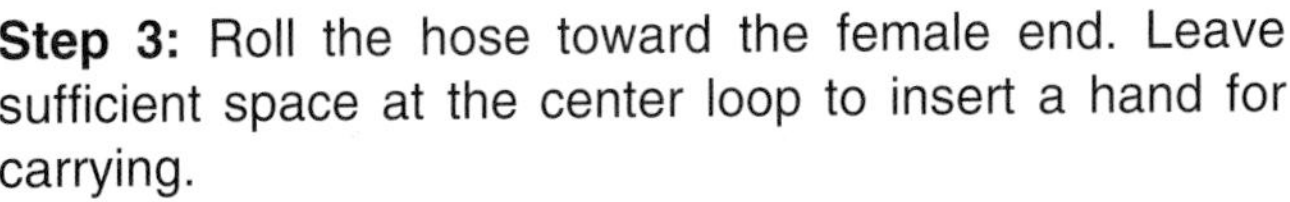

Step 3: Roll the hose toward the female end. Leave sufficient space at the center loop to insert a hand for carrying.

NOTE: When the roll is complete, the male coupling will be inside the roll. The female coupling will be about 3 feet (1 m) ahead of the male coupling.

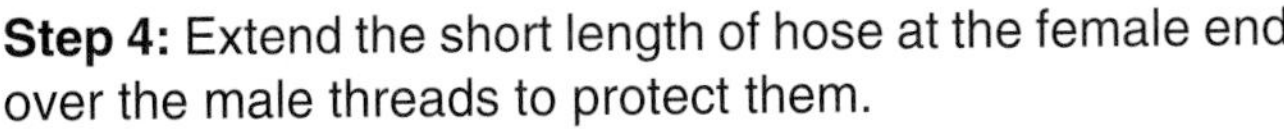

Step 4: Extend the short length of hose at the female end over the male threads to protect them.

SKILL SHEET 12-4 — MAKING THE DONUT ROLL

Method Two

Step 1: Grasp either coupling end, and carry it to the opposite end.

NOTE: The looped section should lie flat, straight, and without twists.

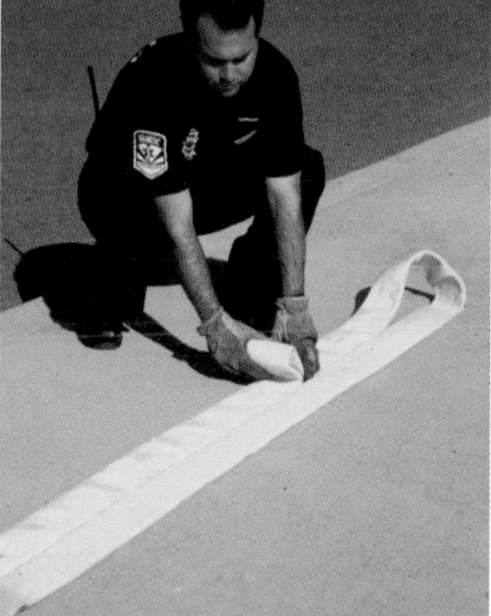

Step 2: Face the coupling ends.

Step 3: Start the roll on the male coupling side about 2½ feet (0.8 m) from the bend (1½ feet [0.5 m] for 1½-inch [38 mm] hose).

Step 4: Roll the hose toward the male coupling.

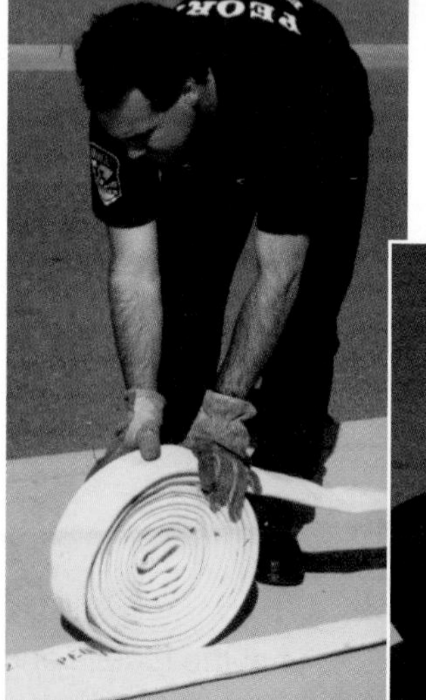

Step 5: Pull the female side back a short distance to relieve the tension if the hose behind the roll becomes tight during the roll.

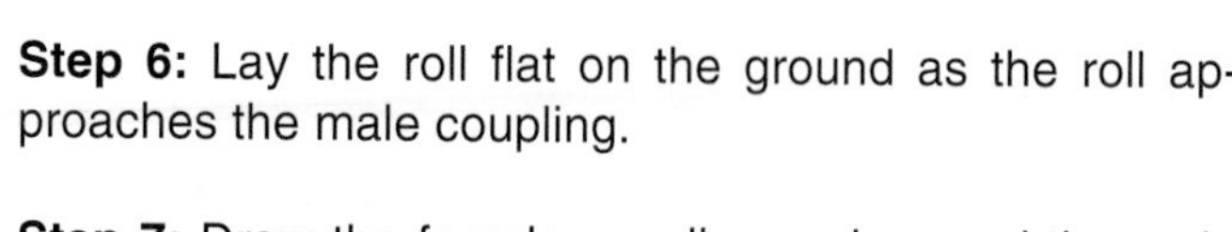

Step 6: Lay the roll flat on the ground as the roll approaches the male coupling.

Step 7: Draw the female coupling end around the male coupling to complete the roll.

SKILL SHEET 12-5 MAKING THE TWIN DONUT ROLL

Step 1: Place the male and female couplings together.

Step 2: Lay the hose flat, without twisting, to form two parallel lines from the loop end to the couplings.

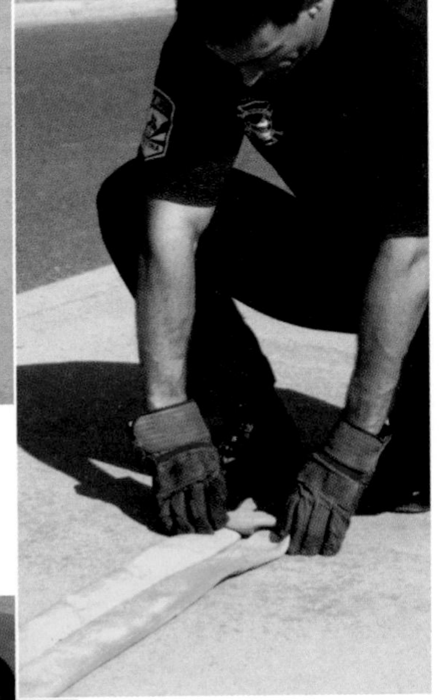

Step 3: Fold the loop end over and upon the two lines to start the roll.

Step 4: Continue to roll both lines simultaneously toward the coupling ends, forming a twin roll with a decreased diameter.

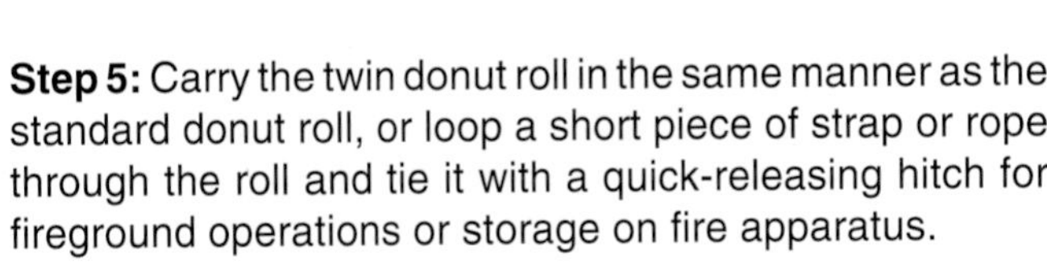

Step 5: Carry the twin donut roll in the same manner as the standard donut roll, or loop a short piece of strap or rope through the roll and tie it with a quick-releasing hitch for fireground operations or storage on fire apparatus.

SKILL SHEET 12-6 MAKING THE SELF-LOCKING TWIN DONUT ROLL

Step 1: Place the male and female couplings together.

Step 2: Lay the hose flat, without twisting, to form two parallel lines from the loop end to the couplings.

Step 3: Move one side of the hose up and over 2½ to 3 feet (0.8 m to 1 m) to the opposite side without turning.

NOTE: This lay-over method prevents a twist in the hose at the big loop. The size of this loop, known as a *butterfly loop,* determines the length of the shoulder loop for carrying.

Step 4: Face the coupling ends, bring the back side of the loop forward toward the couplings, and place it on top of where the hose crosses.

NOTE: This action forms a loop on each side without a twist.

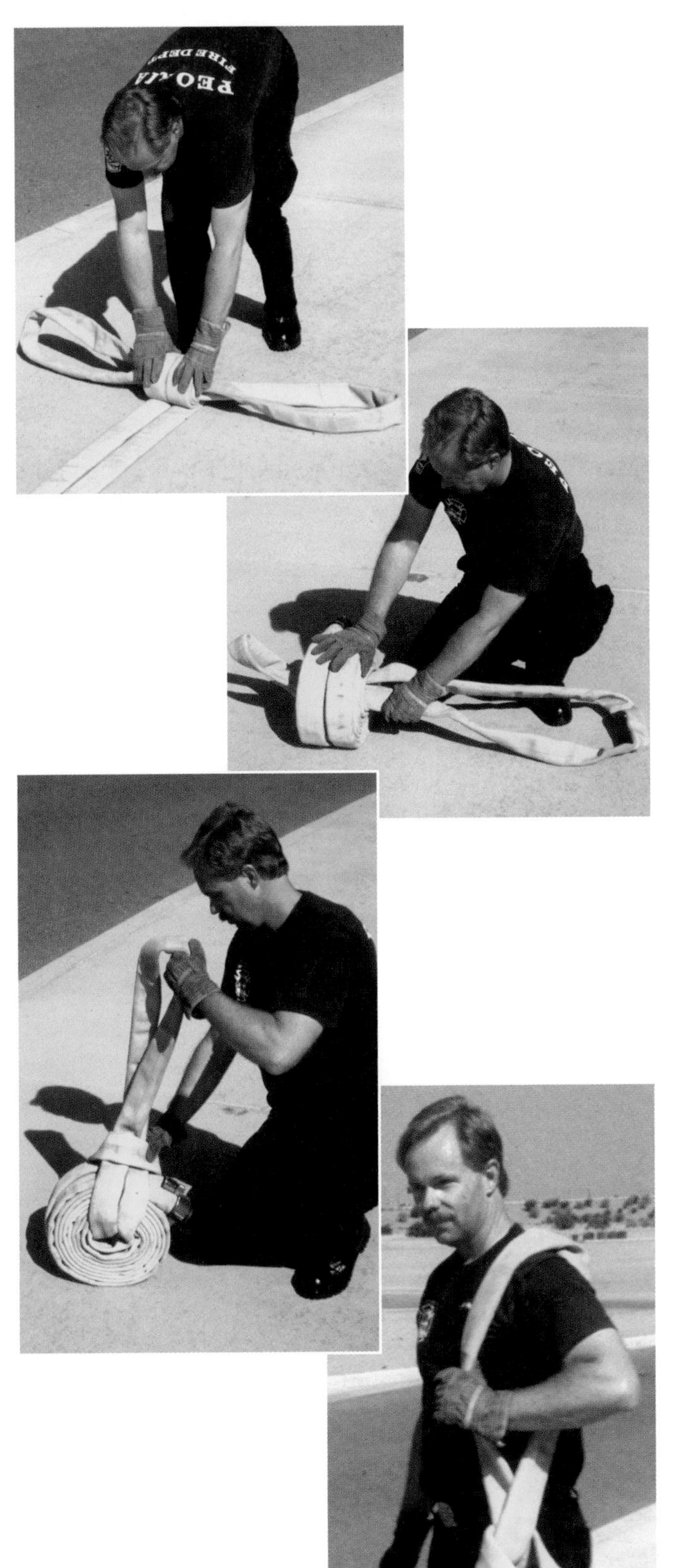

Step 5: Start rolling toward the coupling ends, forming two rolls side by side.

Step 6: Allow the couplings to lie across the top of each roll when the rolls are completed.

Step 7: Adjust the loops, one short and one long, by pulling only one side of the loop through.

Step 8: Place the long loop through the short loop, just behind the couplings, and tighten snugly.

NOTE: The loop forms a shoulder sling.

Step 9: Carry the coupling ends in front or to the rear.

SKILL SHEET 12-7 COUPLING HOSE

One-Firefighter Foot-Tilt Method

Step 1: Stand facing the two couplings so that one foot is near the male end.

Step 2: Place a foot on the hose directly behind the male coupling

Step 3: Apply pressure to tilt it upward.

NOTE: Position the feet well apart for balance.

Step 4: Grasp the female end by placing one hand behind the coupling and the other hand on the coupling swivel.

Step 5: Bring the two couplings together, and turn the swivel clockwise with thumb to make the connection.

SKILL SHEET 12-8 COUPLING HOSE

Two-Firefighter Method

Step 1: ***Firefighter #1:*** Grasp the male coupling with both hands.

Step 2: ***Firefighter #1:*** Bend the hose directly behind the coupling.

Step 3: ***Firefighter #1:*** Hold the coupling and hose tightly against the upper thigh or midsection with the male threads pointed outward.

NOTE: It may help for Firefighter #1 to now look in another direction in order to prevent trying to help align the couplings.

Step 4: ***Firefighter #2:*** Grasp the female coupling with both hands.

Step 5: ***Firefighter #2:*** Bring the two couplings together, and align their positions.

NOTE: The alignment of the hose must be done by the firefighter with the female coupling. The Higbee indicator can be used to align the couplings.

Step 6: ***Firefighter #2:*** Turn the female coupling counter-clockwise until a click is heard. This indicates that the threads are aligned.

Step 7: ***Firefighter #2:*** Turn the female swivel clockwise to complete the connection.

SKILL SHEET 12-9 UNCOUPLING HOSE

One-Firefighter Knee-Press Method

Step 1: Grasp the hose behind the female coupling.

Step 2: Stand the male coupling on end.

Step 3: Set feet well apart for balance.

Step 4: Place one knee upon the hose and shank of the female coupling.

Step 5: Snap the swivel quickly in a counterclockwise direction as body weight is applied to loosen the connection.

SKILL SHEET 12-10 UNCOUPLING HOSE

Two-Firefighter Stiff-Arm Method

Step 1. ***Both Firefighters:*** Take a firm two-handed grip on your respective coupling and press the coupling toward the other firefighter, thereby compressing the gasket in the coupling.

Step 2: ***Both Firefighters:*** Keep arms stiff, and use the weight of both bodies to turn each hose coupling counterclockwise, thus loosening the connection.

SKILL SHEET 12-11 MAKING THE ACCORDION LOAD

Split Bed/Reverse Lay

Step 1: Lay the first length of hose in the bed on edge against the partition.

NOTE: Allow the female coupling to hang below the hose bed so that it can later be placed on top of the hose in the adjacent bed.

Step 2: Fold the hose at the front of the hose bed back on itself.

Step 3: Lay the hose back to the rear next to the first length.

Step 4: Fold the hose at the rear of the hose bed so that the bend is even with the rear edge of the bed.

Step 5: Lay the hose back to the front.

Step 6: Continue laying the hose in folds across the hose bed.

NOTE: Stagger the folds at the rear edge of the bed so that every other bend is approximately 2 inches (50 mm) shorter than the edge of the bed. This stagger may also be done at the front of the bed if desired.

Step 7: Angle the hose upward to start the next tier.

Step 8: Make the first fold of the second tier directly over the last fold of the first tier at the rear of the bed.

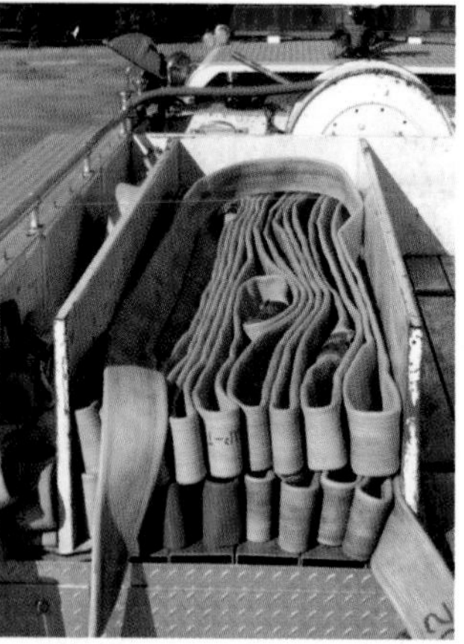

Step 9: Continue with the second tier in the same manner as the first, progressively laying the hose in folds across the hose bed.

NOTE: Stagger the folds as before so that every other bend is approximately 2 inches (50 mm) inside adjacent bends.

Step 10: Make the third and succeeding tiers in the same manner as the first two tiers.

Step 11: Move to the opposite hose bed.

Step 12: Load the hose in the same manner as the first side.

NOTE: Start by placing the first female coupling against the front wall of the hose bed so that it will be pulled straight from the bed when this section of hose is pulled.

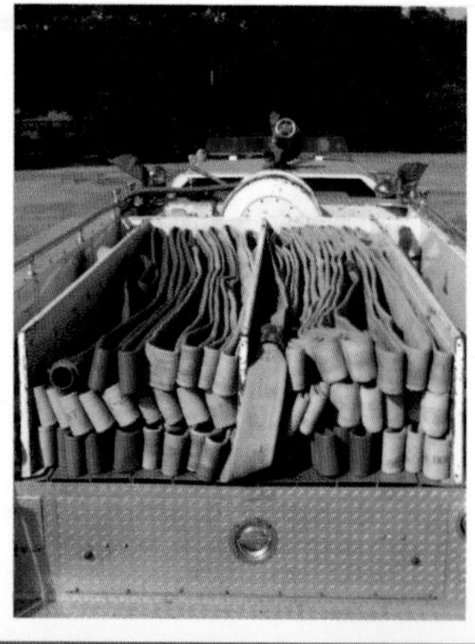

Step 13: Connect the last coupling on top with the female coupling from the first side when the load is completed.

Step 14: Lay the connected couplings on top of the hose load.

Step 15: Pull out the slack so that the crossover loop lies tightly against the hose load.

SKILL SHEET 12-12 MAKING THE HORSESHOE LOAD

Single-Bed/Reverse Lay

Step 1: Place the female coupling in a front corner of the hose bed.

Step 2: Lay the first length of hose on edge against the wall.

Step 3: Make the first fold at the rear even with the edge of the hose bed.

Step 4: Lay the hose to the front and then around the perimeter of the bed so that it comes back to the rear along the opposite side.

Step 5: Make a fold at the rear in the same manner as done before.

Step 6: Lay the hose back around the perimeter of the hose bed inside the first length of hose.

Step 7: Lay succeeding lengths progressively inward toward the center until the entire space is filled.

NOTE: If desired, stagger the folds so that every other bend is approximately 2 inches (50 mm) inside adjacent bends.

Step 8: Start the second tier by extending the hose from the last fold directly over to a front corner of the bed, laying it flat on the hose of the first tier.

Step 9: Make the second and succeeding tiers in the same manner as the first.

NOTE: Lay the crossover length flat on the second tier, but lay it to the opposite corner from that of the first tier. Make crossovers in succeeding tiers to alternate corners.

SKILL SHEET 12-13 MAKING THE FLAT LOAD

Split-Bed Combination Load

NOTE: The right bed is loaded for a reverse lay, and the left bed is loaded for a forward lay (combination load). When both beds are connected, they can be used for a reverse lay.

Step 1: Lay the first length of hose flat in the bed against the partition with the female coupling (which will be connected later to hose in the adjacent bed) hanging below the hose bed.

Step 2: Fold the hose back on itself at the front of the hose bed.

Step 3: Lay the hose back to the rear on top of the previous length.

Step 4: Fold the hose so that the bend is even with the rear edge of the bed.

Step 5: Lay the hose back to the front of the bed, angling it to make the front fold adjacent to the previous fold.

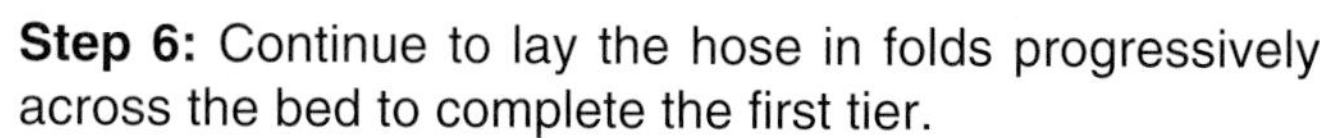

Step 6: Continue to lay the hose in folds progressively across the bed to complete the first tier.

Step 7: Continue with the second tier in the same manner as the first, laying the hose in folds progressively across the hose bed.

NOTE: If desired, make the folds of the second tier approximately 2 inches (50 mm) shorter than the folds of the first tier.

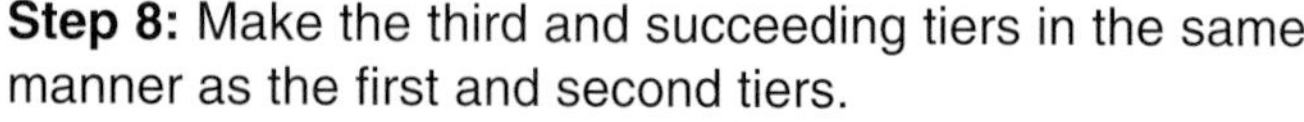

Step 8: Make the third and succeeding tiers in the same manner as the first and second tiers.

NOTE: Align the bends of the third tier even with those of the first tier, the bends of the fourth tier even with those of the second tier, and so on until the load is completed.

Step 9: Move to the opposite hose bed.

Step 10: Load the hose in the same manner as the first side.

NOTE: Start by placing the first male coupling against the front wall of the hose bed so that it will be pulled straight from the bed when this last section of hose is pulled.

Step 11: Connect the last coupling on top when the opposite side is loaded with the female coupling from the first side (use a double male coupling).

Step 12: Lay the connected couplings on top of the hose load.

Step 13: Pull out the slack so that the crossover loop lies tightly against the hose load.

SKILL SHEET 12-14 MAKING THE REVERSE HORSESHOE FINISH

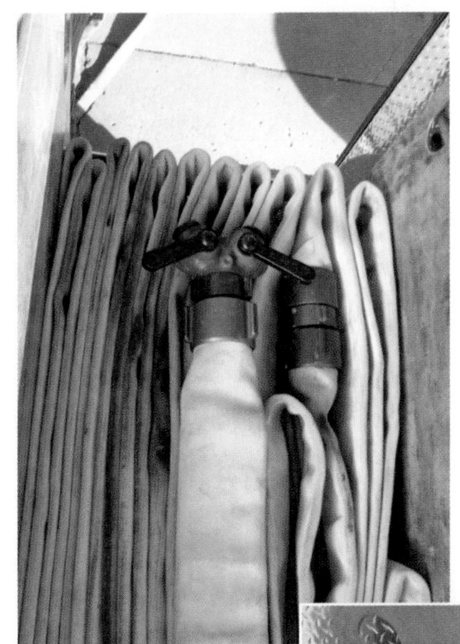

Step 1: Connect the wye to the end (male) coupling of the hose load at the rear of the bed.

Step 2: Place the wye in the center of the hose load with the two male openings toward the rear of the bed.

Step 3: Connect one 1½-inch (38 mm) hose to the wye.

Step 4: Lay the hose on edge to the front of the bed and make a fold.

Step 5: Lay the hose back to the rear alongside the first length.

Step 6: Form a *U* at the edge of the bed.

Step 7: Return the hose to the front and make a fold.

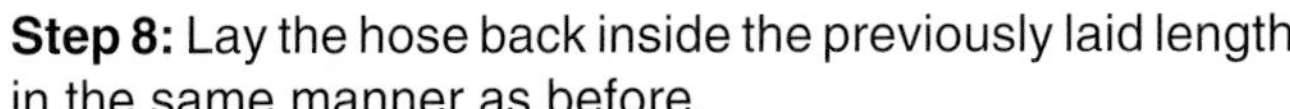

Step 8: Lay the hose back inside the previously laid length in the same manner as before.

Step 9: Continue until the entire length has been loaded.

Step 10: Wrap the male end of the hose once around the horseshoe loops.

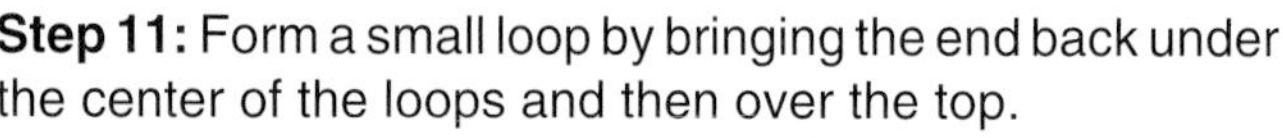

Step 11: Form a small loop by bringing the end back under the center of the loops and then over the top.

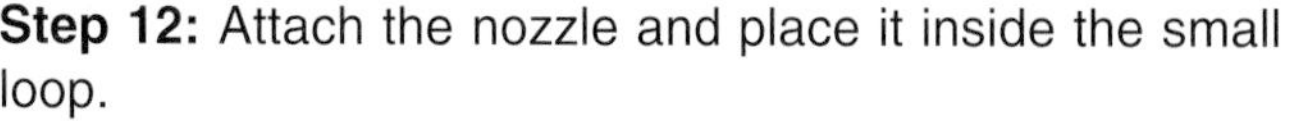

Step 12: Attach the nozzle and place it inside the small loop.

Step 13: Pull the remaining slack hose back into the center of the horseshoe to tighten the loop against the nozzle.

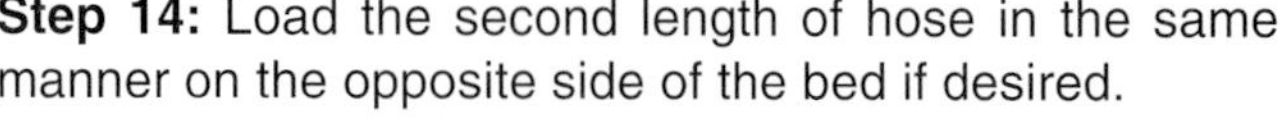

Step 14: Load the second length of hose in the same manner on the opposite side of the bed if desired.

SKILL SHEET 12-15 MAKING THE PRECONNECTED FLAT LOAD

Step 1: Attach the female coupling to the discharge outlet.

Step 2: Lay the first length of hose flat in the bed against the side wall.

Step 3: Angle the hose to lay the next fold adjacent to the first fold.

Step 4: Continue building the first tier in this manner.

Step 5: Make a fold that extends approximately 8 inches (200 mm) beyond the load at a point that is approximately one-third the total length of the load.

NOTE: This loop will later serve as a pull handle.

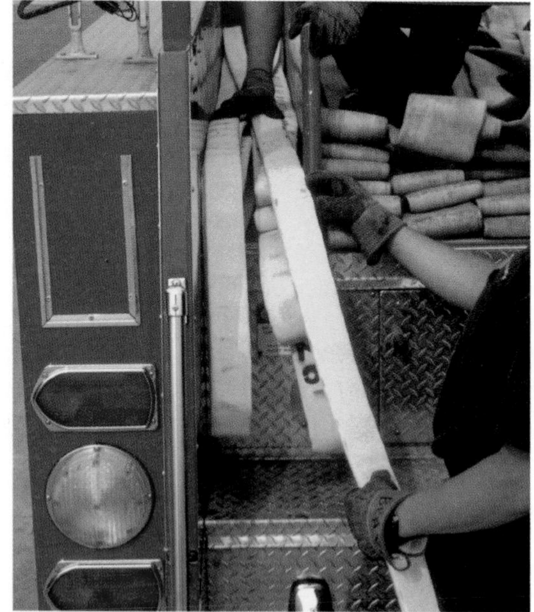

Step 6: Continue laying the hose in the same manner, building each tier with folds laid progressively across the bed.

Step 7: Make a fold that extends approximately 14 inches (350 mm) beyond the load at a point that is approximately two-thirds the total length of the load.

NOTE: This loop will also serve as a pull handle.

Step 8: Complete the load.

Step 9: Attach the nozzle and lay it on top of the load.

SKILL SHEET 12-16 MAKING THE TRIPLE LAYER LOAD

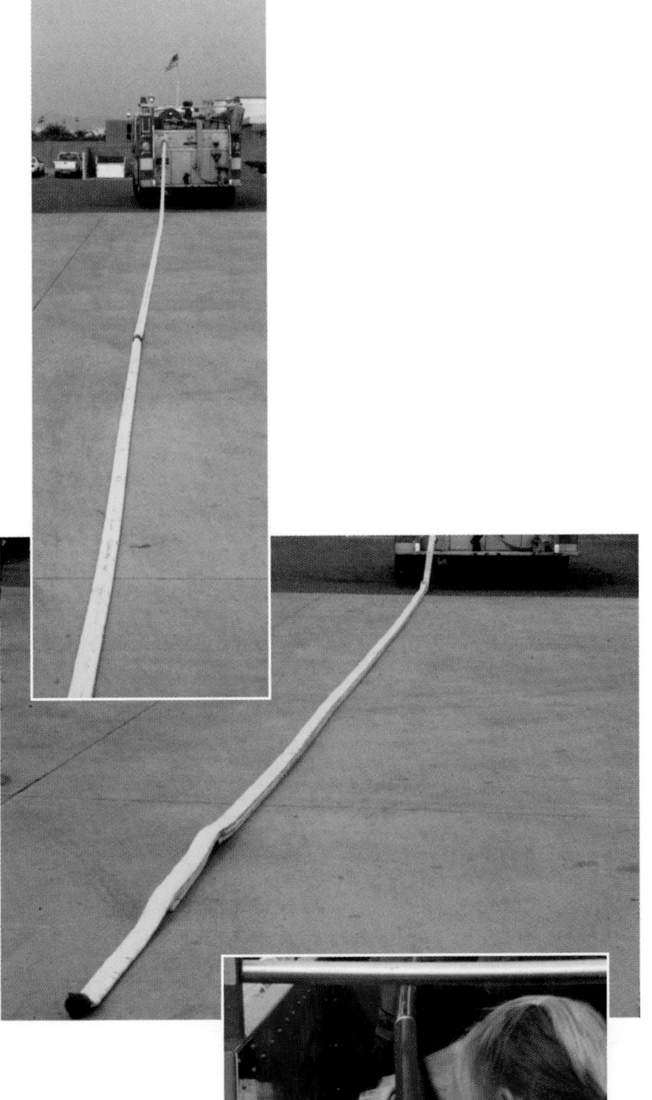

NOTE: Start the load with the sections of hose connected.

Step 1: Connect the female coupling to the discharge outlet.

Step 2: Extend the hose in a straight line to the rear.

Step 3: Pick up the hose at a point two-thirds the distance from the tailboard to the nozzle.

Step 4: Carry this hose to the tailboard.

NOTE: This will form three layers of hose stacked one on the other with a fold at each end.

Step 5: Use several people to pick up the entire length of the three layers.

Step 6: Begin laying the hose into the bed by folding over the three layers into the hose bed.

Step 7: Fold the layers over at the front of the bed.

Step 8: Lay them back to the rear on top of the previously laid hose.

NOTE: If the hose compartment is wider than one hose width, alternate folds on each side of the bed. Make all folds at the rear even with the edge of the hose bed.

Step 9: Continue to lay the hose into the bed in an *S*-shaped configuration until the entire length is loaded.

Step 10: ***Optional:*** Secure the nozzle to the first set of loops using a rope or strap if desired.

NOTE: Some departments like to pull the loop at the end through the nozzle bale. This can be a problem if the line is charged before removing the loop from the bale. Once the line is charged, it may not be possible to pull the loop through the bale.

SKILL SHEET 12-17 MAKING THE MINUTEMAN LOAD

Step 1: Connect the first section of hose to the discharge outlet. Do not connect it to the other lengths of hose.

Step 2: Lay the hose flat in the bed to the front.

Step 3: Lay the remaining hose out the front of the bed to be loaded later.

NOTE: If the discharge outlet is at the front of the bed, lay the hose to the rear of the bed and then back to the front before it is set aside. This provides slack hose for pulling the load clear of the bed.

Step 4: Couple the remaining hose sections together.

Step 5: Attach a nozzle to the male end.

Step 6: Place the nozzle on top of the first length at the rear.

Step 7: Angle the hose to the opposite side of the bed and make a fold.

Step 8: Lay the hose back to the rear.

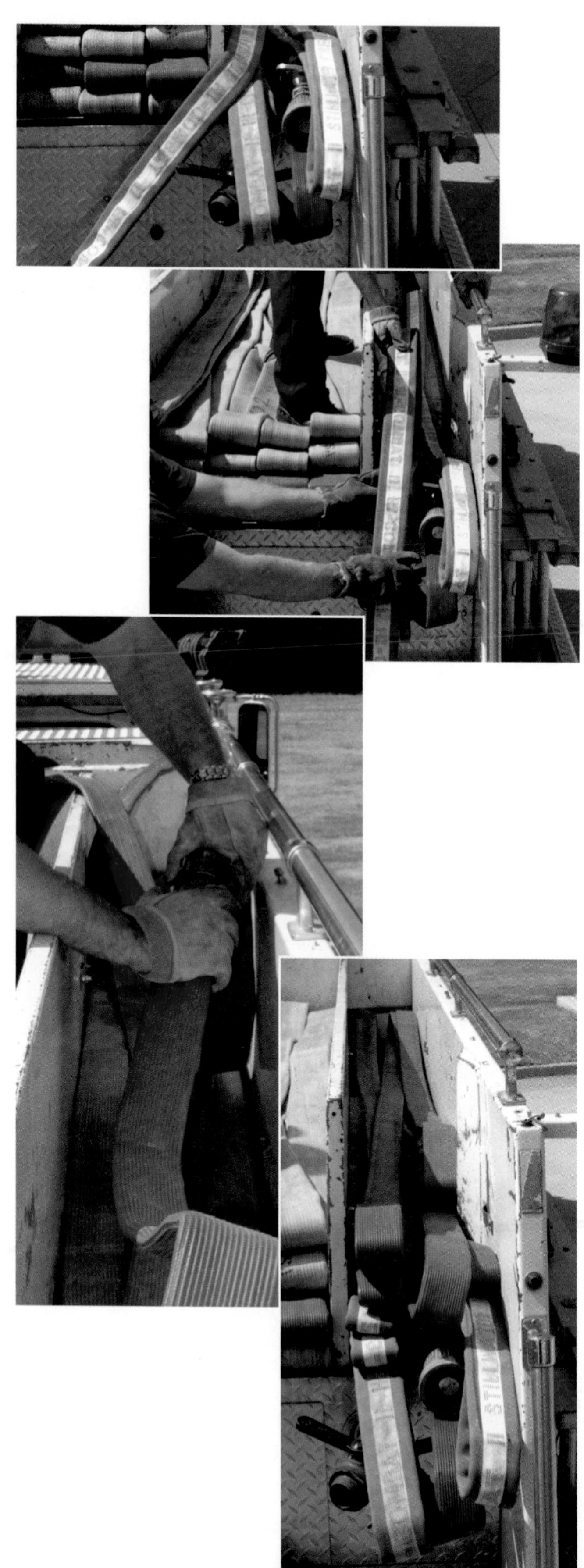

Step 9: Make a fold at the rear of the bed.

Step 10: Angle the hose back to the other side and make a fold at the front.

NOTE: The first fold or two may be longer than the others to facilitate the pulling of the hose from the bed.

Step 11: Continue loading the hose to alternating sides of the bed in the same manner until the complete length is loaded.

Step 12: Connect the male coupling of the first section to the female coupling of the last section.

Step 13: Lay the remainder of the first section in the bed in the same manner.

SKILL SHEET 12-18 MAKING THE HYDRANT CONNECTION

Forward Lay

INSTRUCTIONS: The driver/operator stops the fire apparatus approximately 10 feet (3 m) beyond the hydrant. The hydrant person then performs Steps 1 through 8.

Step 1: Grasp a sufficient amount of hose to reach the hydrant.

Step 2: Step down from the tailboard and face the hydrant with all of the equipment necessary to make the hydrant connection.

Step 3: Approach the hydrant and loop the hydrant in accordance with standard operating procedures.

NOTE: Examples are to wrap the hydrant with the hose and place a foot on the hose or place a rope, which is tied to the hose, around the hydrant.

Step 4: Signal the driver/operator to proceed driving to the fire.

Step 5: Remove the cap from the hydrant.

NOTE: Place a gate valve on the outlet away from the fire if department policy calls for this procedure.

Step 6: Place the hydrant wrench on the valve stem operating nut.

Step 7: Remove the hose loop from the hydrant.

Step 8: Connect the hose to the outlet nearest the fire.

NOTE: When using large diameter hose, a threaded-to-quick-coupling adapter must be placed on the hydrant before the hose can be connected.

INSTRUCTIONS: The driver/operator and other crew members complete Steps 9 through 14.

Step 9: Complete the hose lay to the scene.

Step 10: Apply the hose clamp on the supply line 20 feet (6 m) behind the apparatus. **NOTE:** This allows room to remove fire fighting attack lines.

Step 11: Give the signal to charge the line.

NOTE: Signaling to charge the line can be accomplished by using hand signals, a hand light, a radio, a bell, a siren, or an air horn.

Step 12: Uncouple the hose from the bed (allowing enough hose to reach the pump inlet).

Step 13: Connect the hose to the pump.

Step 14: Release the hose clamp.

INSTRUCTIONS: The hydrant person completes Steps 15 and 16.

Step 15: Open the hydrant fully when the appropriate order or signal is given.

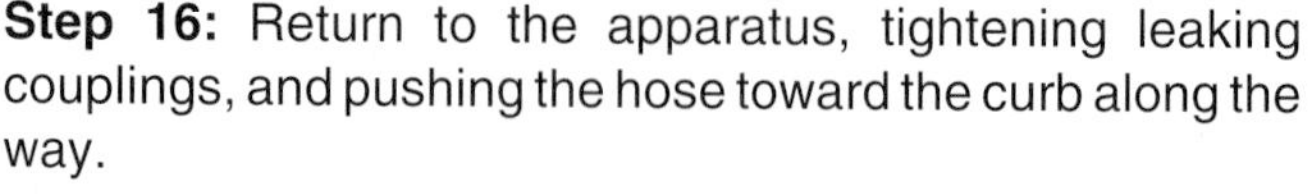

Step 16: Return to the apparatus, tightening leaking couplings, and pushing the hose toward the curb along the way.

NOTE: If multiple lines are laid, follow the same procedures as those given for a single line.

SKILL SHEET 12-19 USING THE FOUR-WAY HYDRANT VALVE

INSTRUCTIONS: The hydrant person on the first pumper completes Steps 1 through 4.

Step 1: Wrap the hydrant as described previously for forward lays.

Step 2: Remove the steamer connection cap.

Step 3: Connect the four-way valve to the hydrant once the hose can be unwrapped from the hydrant.

Step 4: Turn on the hydrant completely when signaled that the pumper at the scene is ready for water.

INSTRUCTIONS: The driver/operator on the second pumper completes Steps 5 through 11.

Step 5: Stop the second pumper at the hydrant.

Step 6: Connect the intake sleeve to the large connection on the four-way hydrant valve.

Step 7: Open the valve to permit water flow into the pump.

Step 8: Connect a discharge line to the four-way valve inlet.

Step 9: Apply proper pressure to the discharge line to support the first pumper through the original supply line.

Step 10: Switch the four-way valve from hydrant to pumper supply where necessary.

Step 11: Charge other supply lines as needed.

Photos courtesy of George Braun, Gainesville (FL) Fire-Rescue

SKILL SHEET 12-20 MAKING THE REVERSE LAY

INSTRUCTIONS: **Both the driver/operator of Pumper #2 and a firefighter perform Steps 1 through 5.**

Step 1: ***Driver/Operator of Pumper #2:*** Stop the second apparatus where its tailboard is slightly beyond the intake valve of the attack pumper.

Step 2: ***Firefighter:*** Pull sufficient hose to reach the intake valve.

Step 3: ***Firefighter:*** Anchor the hose.

NOTE: Anchor the hose to a secure object if possible.

Step 4: ***Driver/Operator of Pumper #2:*** Lay out the hose to the water source when signaled by the firefighter that the hose is anchored.

Step 5: ***Firefighter:*** Apply a hose clamp to the hose at the attack pumper.

INSTRUCTIONS: The driver/operator of Pumper #2 completes Steps 6 through 11.

Step 6: Stop the second apparatus at the hydrant.

Step 7: Make an intake hose connection.

NOTE: Make preparations for drafting if laying to a static source.

Step 8: Pull the remaining length of the last section of hose from the hose bed.

Step 9: Disconnect the couplings, and return the male coupling to the hose bed.

Step 10: Connect the supply hose to a discharge valve.

Step 11: Charge the hose.

SKILL SHEET 12-21 MAKING A SOFT SLEEVE CONNECTION TO A HYDRANT

Step 1: Position the pumper so that the pump intake is either a few feet (meters) ahead or short of the hydrant connection.

NOTE: This allows for a slight bend in the hose, yet avoids kinking.

Step 2: Remove the soft sleeve hose, hydrant wrench, and any adapters necessary from the pumper.

Step 3: Make pumper connection (if the intake hose is not preconnected).

Step 4: Unroll the intake hose.

Step 5: Place the hydrant wrench on the hydrant valve stem operating nut.

NOTE: If it is a dry-barrel hydrant, point the handle away from the outlet.

Step 6: Remove the hydrant cap and add any adapters that may be necessary.

NOTE: If an adapter is needed, it is usually used at the hydrant connection.

Step 7: Place two full twists in the hose to prevent kinking when the hose is charged.

NOTE: Hose with Storz-type couplings should not be twisted.

Step 8: Connect the hose to the hydrant.

Step 9: Open the hydrant slowly.

Step 10: Tighten any leaking connections.

Step 11: Add chafing blocks to the hose where it contacts the ground to prevent rub-induced damage by water and apparatus vibrations.

SKILL SHEET 12-22 MAKING A HARD SUCTION CONNECTION TO A HYDRANT

Step 1: ***Driver/Operator:*** Spot the pumper at a convenient angle to the hydrant and within the limits of the length of the intake hose.

Step 2: ***Driver/Operator:*** Check to see whether the booster tank valve is closed.

Step 3: ***Driver/Operator:*** Remove the pump intake cap.

Step 4: ***Firefighter:*** Remove the hydrant outlet cap.

Step 5: ***Firefighter:*** Place the hydrant wrench on the hydrant valve stem operating nut with the handle pointing away from the outlet.

NOTE: Place an adapter on the hydrant outlet if necessary.

Step 6: ***Driver/Operator:*** Connect the hard suction hose to the large intake.

NOTE: Depending on local preference, the hydrant may be connected first.

Step 7: ***Firefighter:*** Connect the opposite end to the hydrant.

Step 8: ***Driver/Operator:*** Move the apparatus slightly to accomplish this connection if necessary.

NOTE: Put at least a slight bend in the hose.

Step 9: ***Firefighter:*** Open the hydrant.

Step 10: ***Driver/Operator:*** Ready the pump for operation.

SKILL SHEET 12-23 ADVANCING THE PRECONNECTED FLAT LOAD

Step 1: Put one arm through the longer loop.

Step 2: Grasp the shorter pull loop with the same hand.

Step 3: Grasp the nozzle with the opposite hand.

Step 4: Pull the load from the bed using the pull loops.

Step 5: Walk toward the fire.

NOTE: As the hose pulls taut in the hand, release the hand loop.

Step 6: Continue to lead in the hose.

NOTE: As the shoulder loop becomes taut, drop it to the ground.

Step 7: Proceed until the hose is fully extended.

SKILL SHEET 12-24 ADVANCING THE MINUTEMAN LOAD

Step 1: Grasp the nozzle and bottom loops, if provided.

Step 2: Pull the load approximately one-third to one-half of the way out of the hose bed.

Step 3: Face away from the apparatus.

Step 4: Place the hose load on the shoulder with the nozzle against the stomach.

Step 5: Walk away from the apparatus, pulling the hose out of the bed by the bottom loop.

Step 6: Advance toward the fire allowing the load to pay off from the top of the pile.

SKILL SHEET 12-25 ADVANCING THE TRIPLE LAYER LOAD

Step 1: Place the nozzle and fold of the first tier over the shoulder.

Step 2: Face the direction of travel.

Step 3: Walk away from the apparatus.

Step 4: Pull the hose *completely* out of the bed.

Step 5: Drop the folded end from the shoulder when the hose bed has been cleared.

Step 6: Advance the nozzle.

NOTE: If the direction of travel is going to be changed, the firefighter may wish to hold onto the fold and pull all three layers in that direction before dropping the fold and advancing the nozzle.

SKILL SHEET 12-26 UNLOADING AND ADVANCING WYED LINES

Step 1: Grasp the nozzle and small loop of one bundle.

Step 2: Pull the bundle from the bed until it clears the tailboard.

Step 3: Lay the bundle on the ground.

Step 4: Pull the opposite bundle in the same way.

Step 5: Pull the wye and attached hose from the bed.

Step 6: Lay the wye between the bundles near the ties.

Step 7: Pick up the wye when ready to reverse lay to the hydrant.

Step 8: Signal the driver/operator to proceed.

Step 9: Anchor the hose so that it drops from the bed as the apparatus proceeds toward the water source.

Step 10: Place one arm through the horseshoe loops of one bundle after the apparatus completes the lay.

Step 11: Peel off the loops one at a time to lead in the hose.

Step 12: Lay out the second hose in the same manner.

Step 13: Open the wye when ready for water.

SKILL SHEET 12-27 SHOULDER LOADING AND ADVANCING HOSE

Flat or Horseshoe Loads

INSTRUCTIONS: **Firefighter #1 performs Step 1. Firefighter #2 performs Steps 2 through 8.**

Step 1: ***Firefighter #1:*** Attach the nozzle to the end of the hose if desired.

NOTE: Assist other firefighters with loading hose on their shoulders.

Step 2: ***Firefighter #2:*** Position at the tailboard facing the direction of travel.

Step 3: ***Firefighter #2:*** Place the initial fold of hose over the shoulder so that the nozzle can be held at chest height.

Step 4: ***Firefighter #2:*** Bring the hose from behind back over the shoulder so that the rear fold ends at the back of the knee.

Step 5: ***Firefighter #2:*** Make a fold in front that ends at knee height and bring the hose back over the shoulder.

Step 6: ***Firefighter #2:*** Continue to make knee-high folds until an appropriate amount of hose is loaded.

Step 7: ***Firefighter #2:*** Hold the hose to prevent it from slipping off the shoulder.

Step 8: ***Firefighter #2:*** Move forward approximately 15 feet (5 m).

INSTRUCTIONS: Firefighter #3 performs Steps 9 and 10. Firefighter #4 performs Step 11. Firefighter #1 performs Step 12.

Step 9: ***Firefighter #3:*** Position at the tailboard facing the direction of travel.

Step 10: ***Firefighter #3:*** Load hose onto the shoulder in the same manner as Firefighter #2, making knee-high folds until an appropriate amount of hose is loaded.

Step 11: ***Firefighter #4:*** Repeat the loading process.

Step 12: ***Firefighter #1:*** Uncouple the hose from the hose bed, and hand the coupling to the last firefighter.

NOTE: Repeat the process with additional firefighters until the desired length of hose is loaded.

SKILL SHEET 12-28 SHOULDER LOADING AND ADVANCING HOSE

Accordion or Flat Loads

Step 1: Face the hose bed.

Step 2: Grasp the nozzle or coupling.

Step 3: Grasp with both hands the number of folds needed to make up that portion of the shoulder load.

Step 4: Pull the folds about one-third of the way out of the bed.

Step 4a: ***For Accordion Loads Only:*** Twist the folds into an upright position.

Step 5: Turn and pivot into the folds, placing them on top of the shoulder.

NOTE: Make sure that the hose is flat on the shoulder with the nozzle or coupling in front of the body.

Step 6: Grasp the bundle tightly with both hands.

Step 7: Step away from the apparatus, pulling the shoulder load completely out of the bed.

NOTE: **Additional firefighters may remove shoulder loads in the same manner.**

SKILL SHEET 12-29 ADVANCING HOSE

Working Line Drag

Step 1: Stand alongside a single hoseline at a coupling or nozzle.

Step 2: Face the direction of travel.

Step 3: Place the hose over the shoulder with a coupling or nozzle in front, resting it on the chest.

Step 4: Hold the coupling or nozzle in place and pull with the shoulder.

Step 5: Position additional firefighters at each coupling to assist in advancing the hose.

NOTE: About one-third of the hose section should form a loop on the ground between each firefighter.

SKILL SHEET 12-30 EXTENDING A HOSELINE

Step 1: Bring additional sections of hose as needed to the nozzle end of the hoseline.

Step 2: Open the nozzle slightly.

Step 3: Apply a hose clamp approximately 5 feet (1.5 m) behind the nozzle.

NOTE: If the line being extended is equipped with a stacked-tip, solid stream nozzle, or breakaway nozzle, the hoseline may be extended without using a hose clamp. With the nozzle turned off, the tips ahead of the nozzle may be removed to reveal appropriate threads for hose connection. Once the new hose and nozzle are added, the nozzle may be turned on to resume the flow of water.

Step 4: Remove the nozzle.

Step 5: Add the new section of hose.

Step 6: Reattach the nozzle.

Step 7: Release the clamp slowly, allowing water to flow to the nozzle.

SKILL SHEET 12-31 SERVICE TESTING FIRE HOSE

Step 1: Connect a number of hose sections (check the gaskets before connecting) into test lengths of no more than 300 feet (90 m) each.

Step 2: Tighten the connections between the sections with spanners.

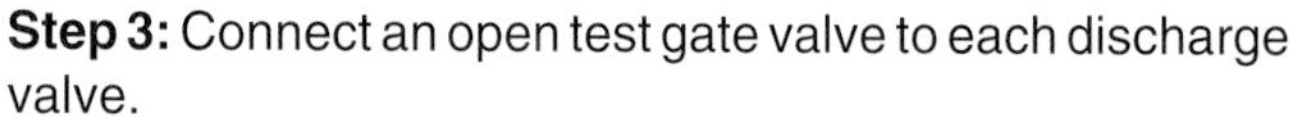

Step 3: Connect an open test gate valve to each discharge valve.

Step 4: Tighten each connection with spanners.

Step 5: Connect a test length to each test gate valve.

Step 6: Tighten each connection with a spanner.

Step 7: Tie a rope, hose rope tool, or hose strap to each test length of hose 10 to 15 inches (250 mm to 375 mm) from the test gate valve connections.

Step 8: Secure the other end to the discharge pipe or other nearby anchor.

Step 9: Attach a shutoff nozzle (or any device that permits water and air to drain from the hose) to the open end of each test length.

Step 10: Fill each hoseline with water with a pump pressure of 50 psi (350 kPa) or to hydrant pressure.

Step 11: Open the nozzles as the hoselines are filling.

Step 12: Hold nozzles above the level of the pump discharge to permit all the air in the hose to discharge.

Step 13: Discharge the water away from the test area.

Step 14: Close the nozzles after all air has been purged from each test length.

Step 15: Make a chalk or pencil mark on the hose jackets against each coupling.

Step 16: Check that all hose is free of kinks and twists and that no couplings are leaking.

NOTE: Any length found to be leaking from *BEHIND* the coupling should be taken out of service and repaired before being tested.

Step 17: Retighten any couplings that are leaking at the connections.

NOTE: If the leak cannot be stopped by tightening the couplings, depressurize, disconnect the couplings, replace the gasket, and start over at Step 10.

Step 18: Close each hose test gate valve.

Step 19: Increase the pump pressure to the required test pressure given in NFPA 1962.

Step 20: Closely monitor the connections for leakage as the pressure increases.

Step 21: Maintain the test pressure for 5 minutes.

Step 22: Inspect all couplings to check for leakage (weeping) at the point of attachment.

Step 23: Slowly reduce the pump pressure after 5 minutes.

Step 24: Close each discharge valve.

Step 25: Disengage the pump.

Step 26: Open each nozzle slowly to bleed off pressure in the test lengths.

Step 27: Break all hose connections and drain water from the test area.

Step 28: Observe marks placed on the hose at the couplings.

NOTES:

- If a coupling has moved during the test, tag the hose section for recoupling. Tag all hose that has leaked or failed in any other way.
- Expect a 1⁄16- to 1⁄8-inch (2 mm to 3 mm) uniform movement of the coupling on newly coupled hose. This slippage is normal during initial testing but should not occur during subsequent tests.

Step 29: Record the test results for each section of hose.

Chapter 13
Fire Streams

Chapter 13
Fire Streams

INTRODUCTION

A *fire stream* can be defined as a stream of water or other extinguishing agent after it leaves a fire hose and nozzle until it reaches the desired point. The perfect fire stream can no longer be sharply defined because individual desires and extinguishing requirements vary. During the time a stream of water or extinguishing agent passes through space, it is influenced by its velocity, by gravity, by wind, and by friction with the air. The condition of the stream when it leaves the nozzle is influenced by operating pressures, nozzle design, nozzle adjustment, and the condition of the nozzle orifice.

Fire streams are intended to reduce high temperatures from a fire and provide protection to firefighters and to exposures through the following methods:

- Applying water or foam directly to burning material to reduce its temperature
- Applying water or foam over an open fire to reduce the temperature so firefighters can advance handlines closer to effect extinguishment
- Reducing high atmospheric temperature
- Dispersing hot smoke and fire gases from a heated area by using a fire stream
- Creating a water curtain to protect firefighters and property from heat
- Creating a barrier between a fuel and a fire by covering with a foam blanket

This chapter focuses on several aspects of water and foam fire streams. The first portion of the chapter includes the elements of what is required for the production of water fire streams, the different types of streams, and the different types of nozzles used to produce them. The second part of the chapter focuses on the basic principles related to fire fighting foams such as how and why foam works, types of foam concentrates, the general characteristics of foam, how foam is mixed (proportioned) with water, application equipment, and foam application techniques.

EXTINGUISHING PROPERTIES OF WATER

[NFPA 1001: 3-3.7(a); 3-3.9(a)]

Water has the ability to extinguish fire in several ways. The primary way is by cooling, which removes the heat from the fire. Another way is by smothering, which includes water's ability to absorb large quantities of heat and also to dilute oxygen. When heated to its boiling point, water absorbs heat by converting into a gas called *water vapor* or *steam*, which cannot be seen (vaporization). When steam starts to cool, however, its visible form is called *condensed steam* (Figure 13.1).

Complete vaporization does not happen the instant water reaches its boiling point because additional heat is required to completely turn the water into steam. When a water fire stream is broken into small particles, it absorbs heat and converts into steam more rapidly than it would in a compact form because more of the water's surface is exposed to the heat. For example, 1 cubic inch (1 638.7 mm^3) of ice dropped into a glass of water takes some time to absorb its capacity of heat. This is because a surface area of only 6 square inches (3 870 mm^2 or 38.7 cm^2) of the ice is exposed to the water. However, if that cube of ice is divided into ⅛-cubic inch (204.8 mm^3) cubes and dropped into the water, a surface area of 48

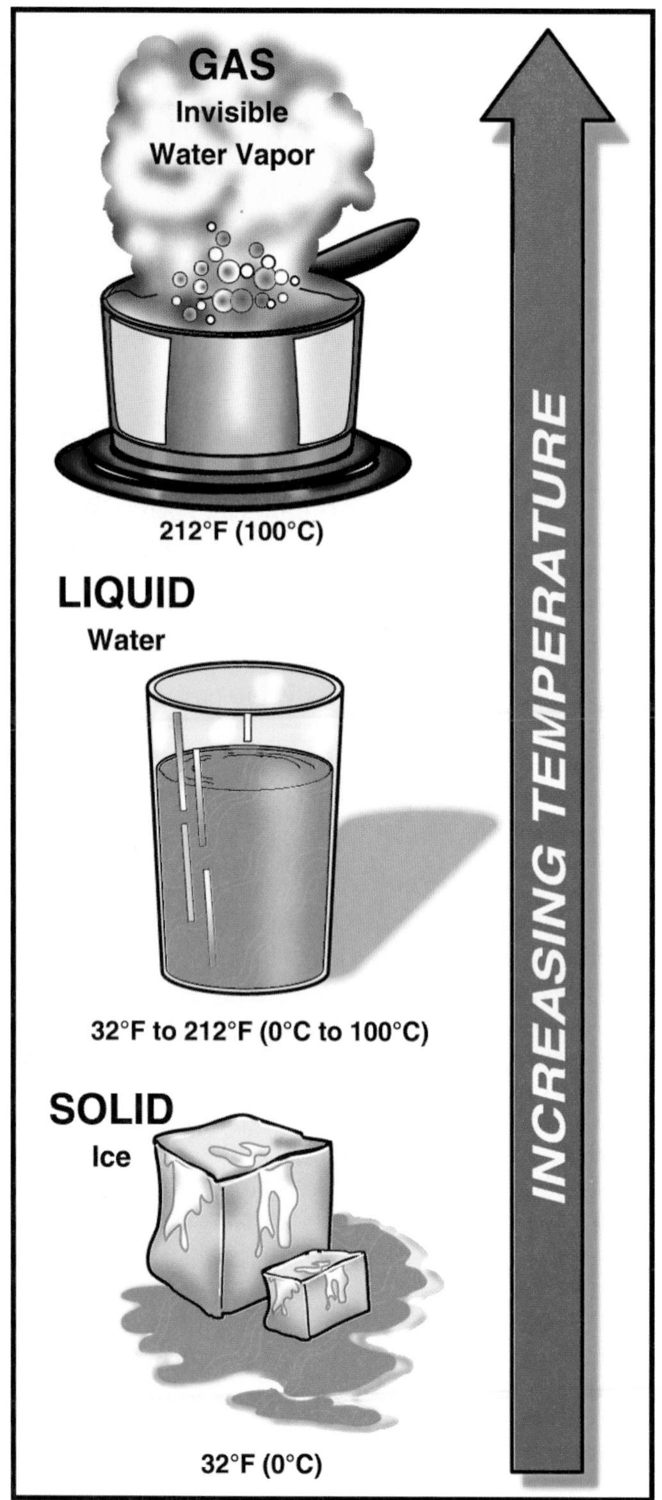

Figure 13.1 Water is found in the solid, liquid, and gaseous states.

square inches (30 967 mm² or 309.7 cm²) of ice is exposed to the water. The finely divided particles of ice absorb heat more rapidly. This same principle applies to water in the liquid state.

Another characteristic of water that is sometimes an aid to fire fighting is its expansion capability when converted into steam. This expansion helps cool the fire area by driving heat and smoke from the area. This steam, however, can cause serious burn injuries to firefighters and occupants. The amount of expansion varies with the temperatures of the fire area. At 212°F (100°C), water expands approximately 1,700 times its original volume (Figure 13.2).

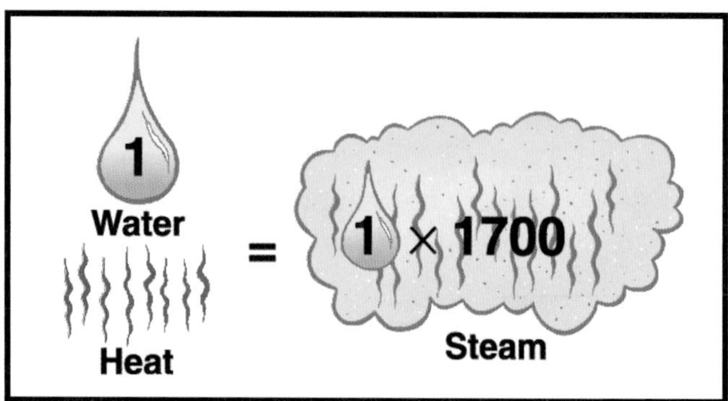

Figure 13.2 Water expands to 1,700 times its original volume when it converts to steam.

To illustrate steam expansion, consider a nozzle discharging 150 gallons (568 L) of water fog every minute into an area heated to approximately 500°F (260°C), causing the water fog to convert into steam. During one minute of operation, 20 cubic feet (0.57 m³) of water is discharged and vaporized. This 20 cubic feet (0.57 m³) of water expands to approximately 48,000 cubic feet (1 359 m³) of steam. This is enough steam to fill a room approximately 10 feet (3 m) high, 50 feet (15 m) wide, and 96 feet (29 m) long (Figure 13.3). In hotter atmospheres, steam expands to even greater volumes.

Steam expansion is not gradual, but rapid. If a room is already full of smoke and gases, the steam that is generated displaces these gases when adequate ventilation openings are provided. As the room cools, the steam condenses and allows the room to refill with cooler air (Figure 13.4). The use of a fog stream in a direct or combination fire attack requires that adequate ventilation be provided ahead of the hoseline (see Chapter 14, Fire Control). Otherwise, there is a high possibility of steam or even fire rolling back over and around the hose team, and the potential for injury is great. There are some observable results of the proper application of a

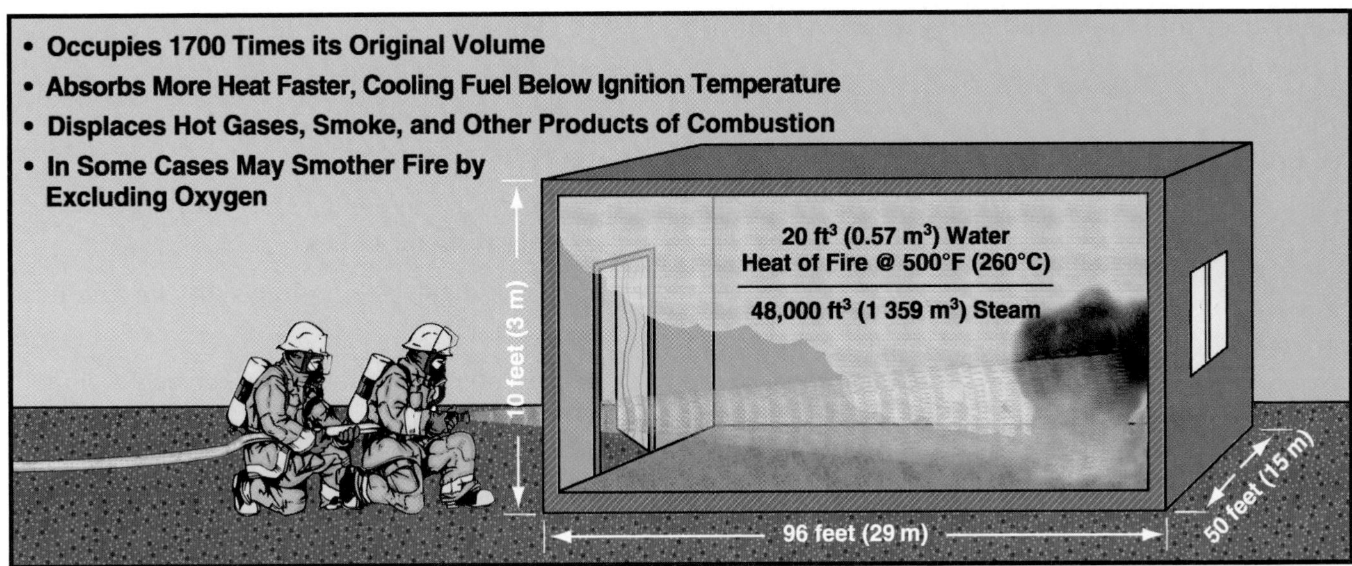

Figure 13.3 Water's expansion rate makes it very effective for fire extinguishment.

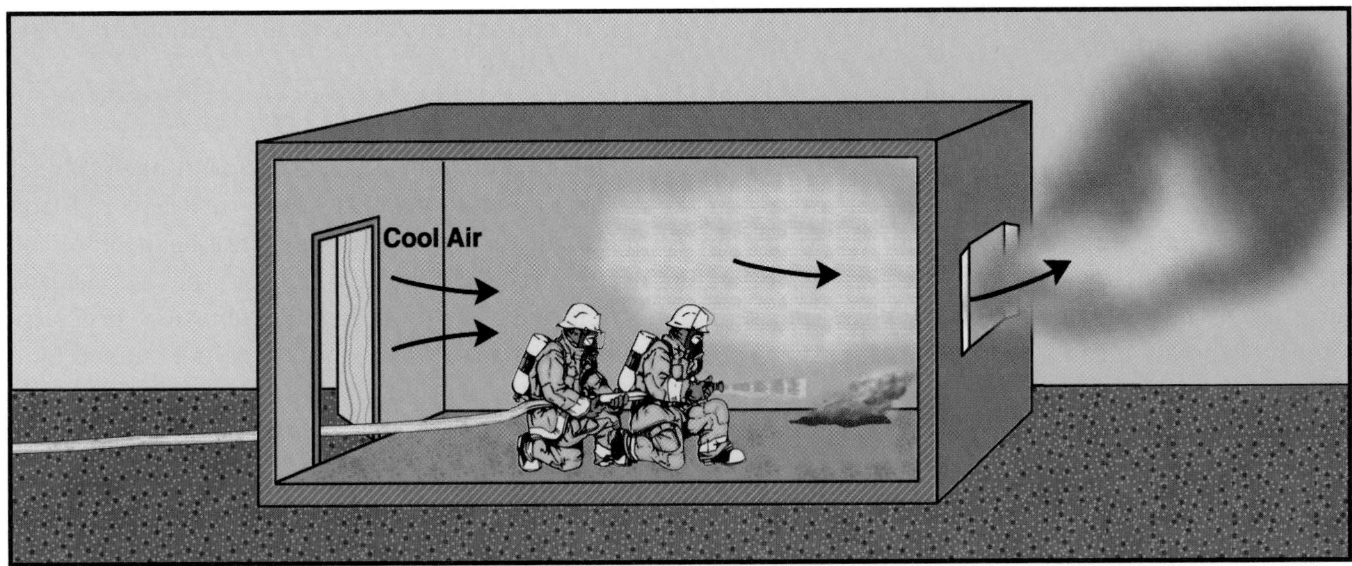

Figure 13.4 Steam will disperse the products of combustion from an enclosed area with adequate ventilation.

water fire stream into a room: Fire is extinguished or reduced in size, visibility may be maintained, and room temperature is reduced.

The steam produced by a fire stream can also be an aid to fire extinguishment by smothering, which is accomplished when the expansion of steam reduces oxygen in a confined space.

Several characteristics of water that are extremely valuable for fire extinguishment are as follows:

- Water is readily available and inexpensive.
- Water has a greater heat-absorbing capacity than other common extinguishing agents.
- Water changing into steam requires a relatively large amount of heat.
- The greater the surface area of the water exposed, the more rapidly heat is absorbed.

PRESSURE LOSS/GAIN

[NFPA 1001: 3-3.9(a); 3-3.9(b)]

To produce effective fire streams, it is necessary to know the effects of factors affecting pressure loss and gain. Two important factors that affect pressure loss and gain in a fire stream are friction loss and elevation. Pressure changes are possible due to friction loss in hose and appliances. A loss or gain in pressure may result due to elevation and the direction of water flow uphill or downhill.

Friction Loss

A definition of friction loss as it relates to water fire streams is as follows: *Friction loss is that part of total pressure that is lost while forcing water through pipes, fittings, fire hose, and adapters.* The difference in pressure on a hoseline between the nozzle and the pumper (excluding pressure lost due to a change in elevation between the two; see Elevation Loss/Gain section) is a good example of friction loss. Friction loss can be measured by inserting in-line gauges at different points in a hoseline (Figure 13.5). The difference in the pressures between gauges when water is flowing through the hose is the friction loss for the length of hose between those gauges for that rate of flow.

One point to consider in applying pressure to water in a hoseline is that there is a limit to the velocity or speed at which the water can travel. If the velocity is increased beyond these limits, the friction becomes so great that the water in the hoseline is agitated by resistance. Certain characteristics of hose layouts such as hose size and length of the lay also affect friction loss.

In order to reduce pressure loss due to friction, consider the following guidelines:

- Check for rough linings in fire hose.
- Replace damaged hose couplings.
- Eliminate sharp bends in hose when possible.
- Use adapters to make hose connections only when necessary.
- Keep nozzles and valves fully open when operating hoselines.
- Use proper size hose gaskets for the hose selected.
- Use short hoselines as much as possible.
- Use larger hose (for example, increase from booster hose to 1¾-inch [45 mm] hose or from 1¾-inch [45 mm] hose to 2½-inch [65 mm] hose) or multiple lines when flow must be increased.
- Reduce the amount of flow (for example, change nozzle tips or reduce flow setting).

Elevation Loss/Gain

Elevation refers to the position of an object above or below ground level. In a fire fighting operation, elevation refers to the position of the nozzle in relation to the pumping apparatus, which is at ground level. Elevation pressure refers to a gain or loss in a hoseline caused by a change in elevation. When a nozzle is *above* the fire pump, there is a *pressure loss* (Figure 13.6). When the nozzle is *below* the pump, there is a *pressure gain*. These losses and gains occur because of gravity.

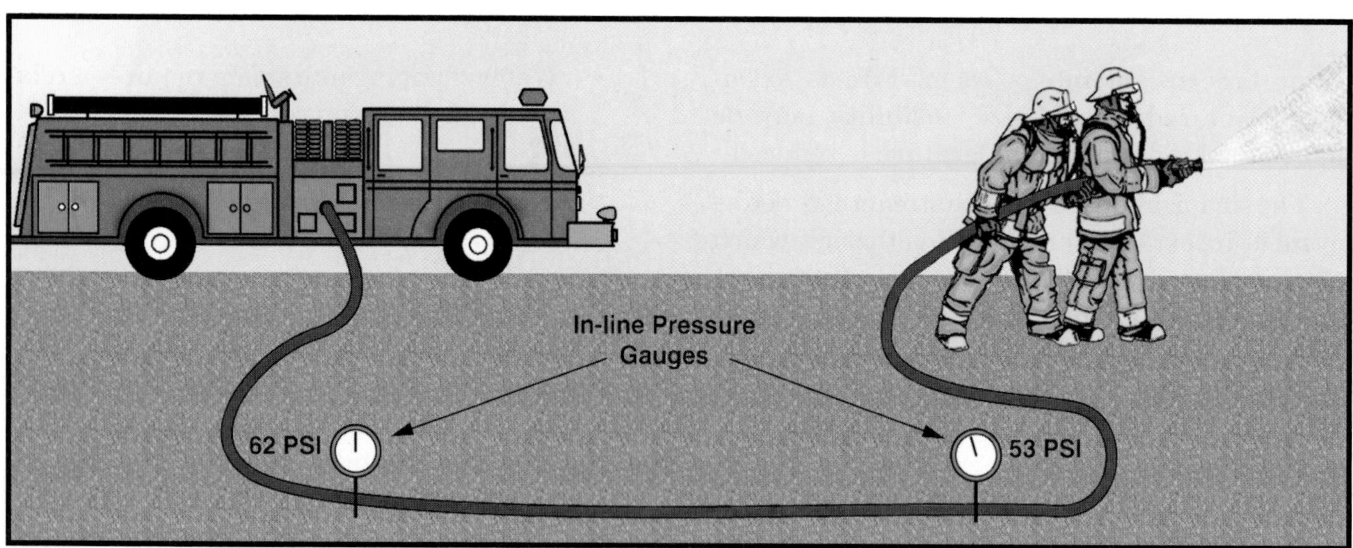

Figure 13.5 In-line gauges can be used to show the friction loss between the gauges.

Figure 13.6 Elevation pressure loss occurs when the nozzle is above the fire pump.

WATER HAMMER

[NFPA 1001: 3-3.9(b)]

When the flow of water through fire hose or pipe is suddenly stopped, the resulting surge is referred to as *water hammer*. Water hammer can often be heard as a distinct sharp clank, very much like a hammer striking a pipe. This sudden stopping results in a change in the direction of energy. This energy creates excessive pressures that can cause considerable damage to water mains, plumbing, fire hose, hydrants, and fire pumps. Operate nozzle controls, hydrants, valves, and hose clamps slowly to prevent water hammer (Figure 13.7).

WATER FIRE STREAM PATTERNS AND NOZZLES

(NFPA 1001: 3-3.6(b); 3-3.9(a); 3-3.9(b); 4-3.2(a)]

A water fire stream is identified by its size and type. The size refers to the volume of water flowing per minute; the type indicates a specific pattern of water. Fire streams are classified into one of three sizes: *low-volume streams, handline streams,* and *master streams.* The rate of discharge of a fire stream is measured in gallons per minute (gpm) or liters per minute (L/min).

- ***Low-volume stream*** — Discharges less than 40 gpm (160 L/min) including those fed by booster hoselines.
- ***Handline stream*** — Supplied by 1½- to 3-inch (38 mm to 77 mm) hose, which flows from 40 to 350 gpm (160 L/min to 1 400 L/

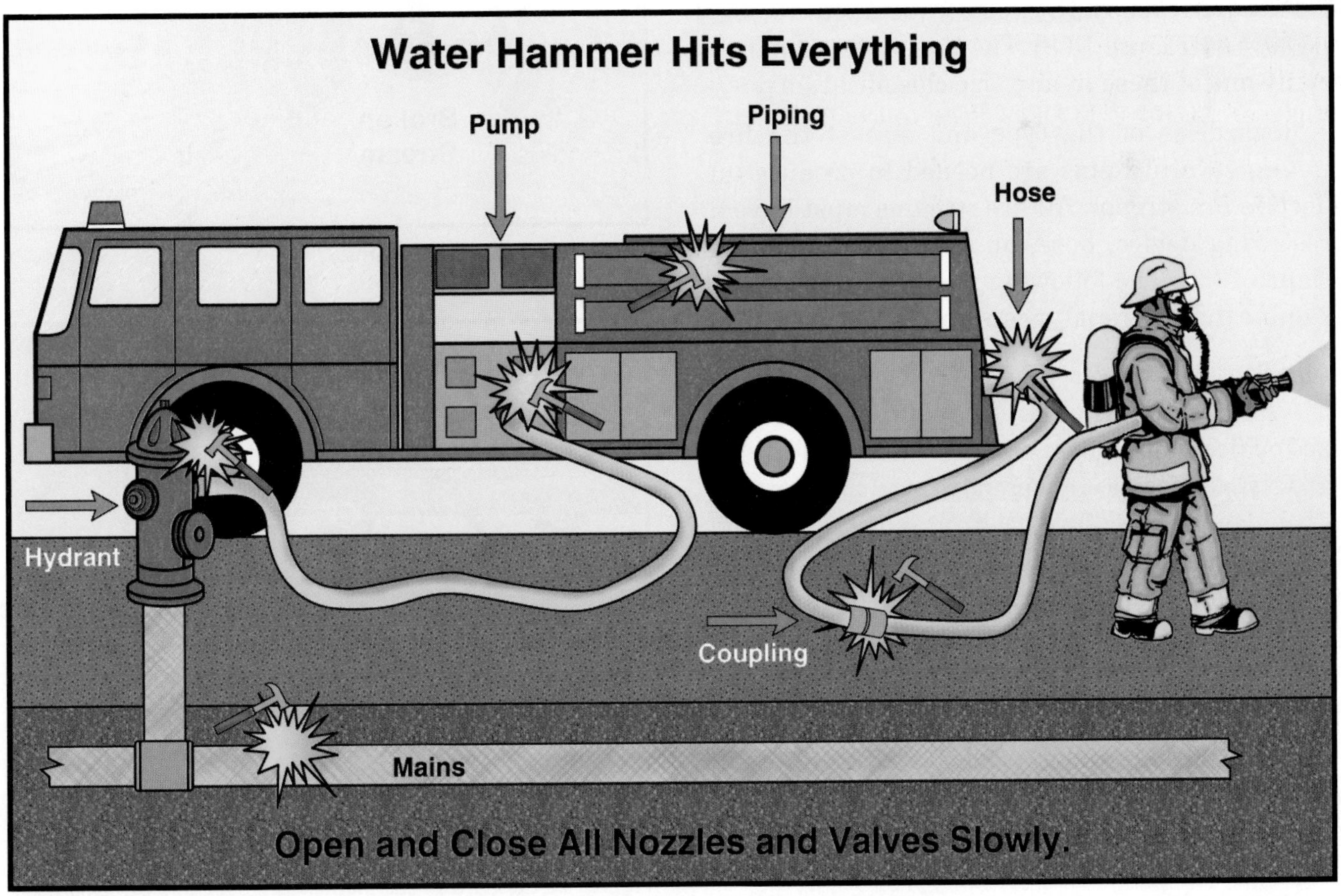

Figure 13.7 Water hammer can cause damage to all parts of the water system and to fire equipment.

min). Nozzles with flows in excess of 350 gpm (1 400 L/min) are not recommended for handlines.

- ***Master stream*** — Discharges more than 350 gpm (1 400 L/min) and is fed by multiple 2½- or 3-inch (65 mm or 77 mm) hoselines or large diameter hoselines connected to a master stream nozzle. Master streams are large-volume fire streams.

The volume of water discharged is determined by the design of the nozzle and the pressure at the nozzle. It is essential for a fire stream to deliver a volume of water sufficient to absorb heat more rapidly than it is generated. Fire stream patterns must have sufficient volume to penetrate the heated area. If a low-volume nozzle producing finely divided particles is used where heat is generated faster than it is absorbed, extinguishment will not be accomplished until the fuel is completely consumed or its supply is turned off.

The type of fire stream indicates a specific pattern of water needed for a specific job. There are three major types of fire stream patterns: *solid, fog,* and *broken* (Figure 13.8). The stream pattern may be any one of these in any size classification.

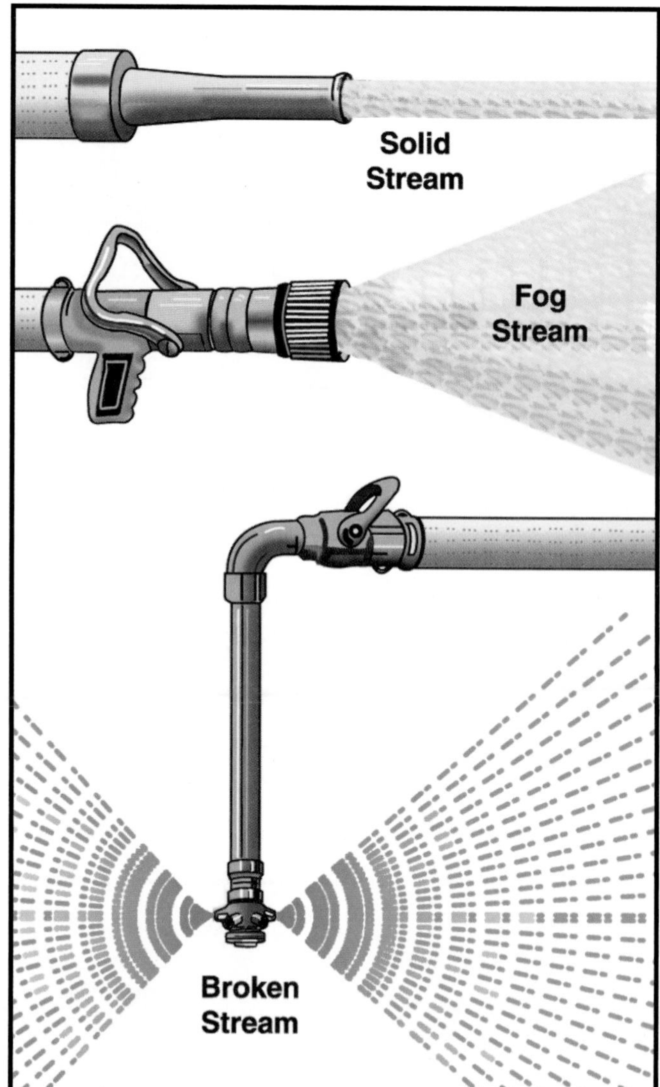

Figure 13.8 Fire stream patterns.

Regardless of the type and size of the fire stream, several items are needed to produce an effective fire stream. All fire streams must have a pressuring device, hose, an agent, and a nozzle (Figure 13.9). The following sections more closely examine the different types of streams and nozzles.

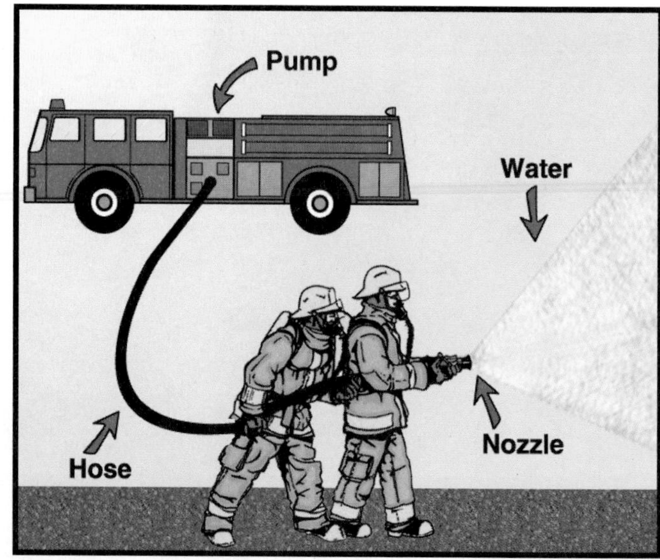

Figure 13.9 The four elements that make a fire stream are a pump, hose, a nozzle, and water.

Solid Stream

A *solid stream* is a fire stream produced from a fixed orifice, smoothbore nozzle (Figure 13.10). The solid stream nozzle is designed to produce a stream as compact as possible with little shower or spray. A solid stream has the ability to reach areas that other streams might not reach and also minimizes the chance of steam burns to firefighters. The reach of a solid stream can be affected by gravity, friction of the air, and wind.

Solid stream nozzles are designed so that the shape of the water in the nozzle is gradually reduced until it reaches a point a short distance from the outlet (Figure 13.11). At this point, the nozzle becomes a cylindrical bore whose length is

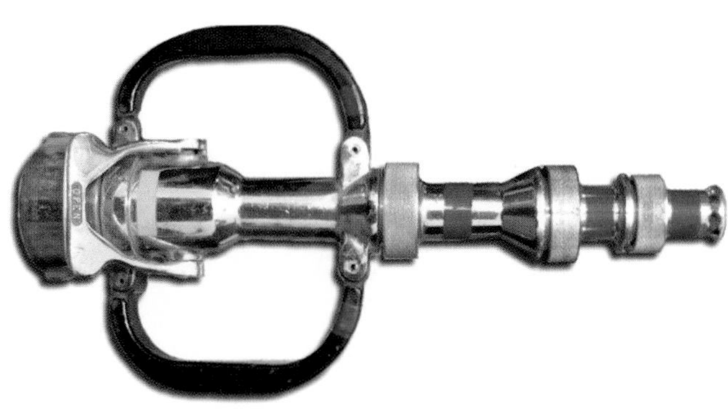

Figure 13.10 A solid stream nozzle.

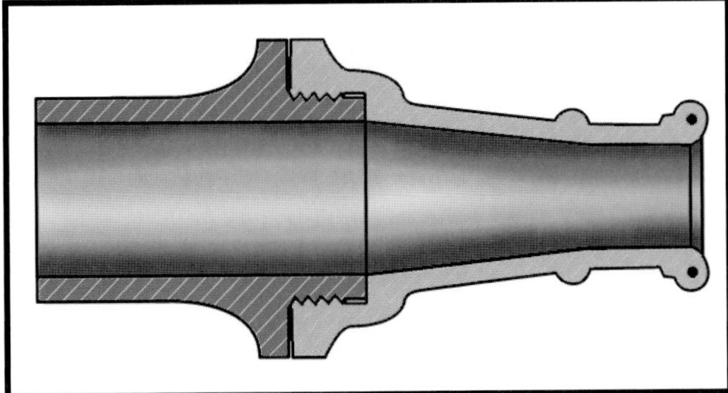

Figure 13.11 The basic design of a solid stream nozzle.

from one to one and one-half times its diameter. The purpose of this short, truly cylindrical bore is to give the water its round shape before discharge. A smooth-finish waterway contributes to both the shape and reach of the stream. Alteration or damage to the nozzle can significantly alter stream shape and performance.

The velocity of the stream (nozzle pressure) and the size of the discharge opening determine the flow from a solid stream nozzle. When solid stream nozzles are used on handlines, they should be operated at 50 psi (350 kPa) nozzle pressure. A solid stream master stream device should be operated at 80 psi (560 kPa).

The extreme limit at which a solid stream of water can be classified as a good stream cannot be sharply defined and is, to a considerable extent, a matter of judgment. It is difficult to say just exactly where the stream ceases to be good. Observations and tests covering the effective range of fire streams classify effective streams as follows:

- A stream that does not lose its continuity until it reaches the point where it loses its forward velocity (*breakover*) and falls into showers of spray that are easily blown away (Figure 13.12)
- A stream that is stiff enough to maintain its original shape and attain the required height even in a light, gentle wind (breeze)

HANDLING SOLID STREAM NOZZLES

When water flows from the nozzle, the reaction is equally strong in the opposite direction, thus a force pushes back on the person handling the hoseline (nozzle reaction). This reaction is caused by the velocity and quantity of the stream, which acts against the nozzle and the curves in the hose, making the nozzle difficult to handle. The greater

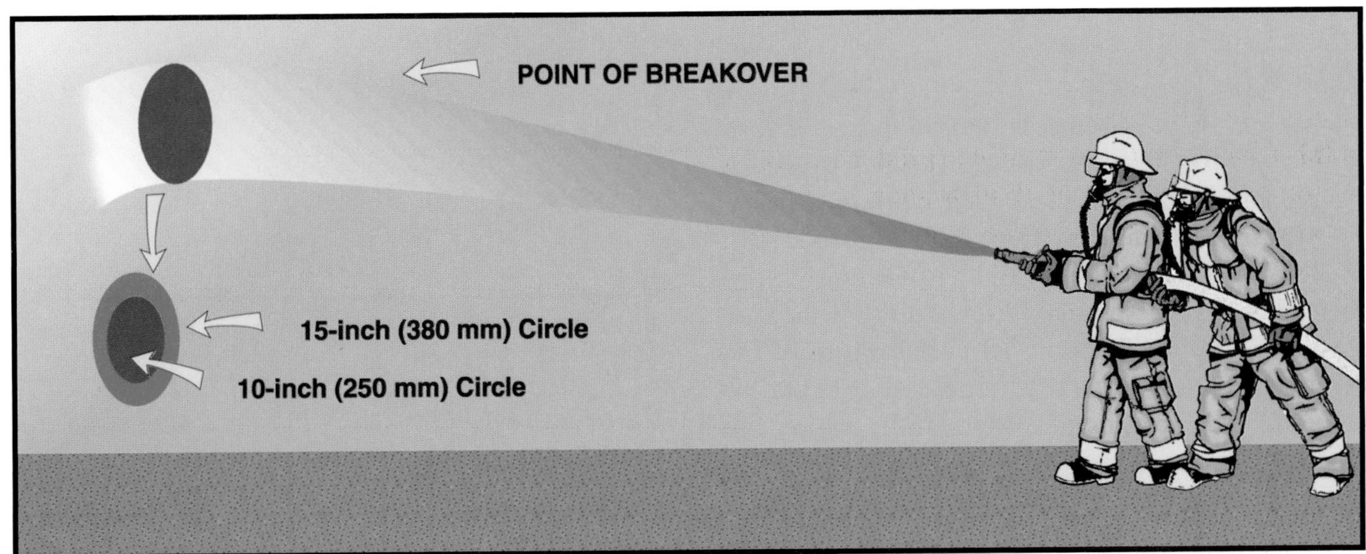

Figure 13.12 The breakover point is that point at which the stream begins to lose its forward velocity.

the nozzle discharge pressure, the greater the resulting nozzle reaction.

ADVANTAGES

- Solid streams maintain better visibility for firefighter than other types of streams.
- Solid streams have greater reach than other types of streams (see Fog Stream section).
- Solid streams operate at reduced nozzle pressures per gallon (liter) than other types of streams thus reducing the nozzle reaction.
- Solid streams have greater penetration power than other types of streams.
- Solid streams are less likely to disturb normal thermal layering of heat and gases during interior structural attacks than other types of streams.

DISADVANTAGES

- Solid streams do not allow for different stream pattern selections.
- Solid streams cannot be used for foam application.
- Solid streams provide less heat absorption per gallon (liter) delivered than other types of streams.

CAUTION: Do not use solid streams on energized electrical equipment. Use fog patterns with at least 100 psi (700 kPa) nozzle pressure. Do not use wand applicators because they can be conductors.

Figure 13.13 Adjustable fog nozzles. *Courtesy of Elkhart Brass Mfg. Company, Inc.*

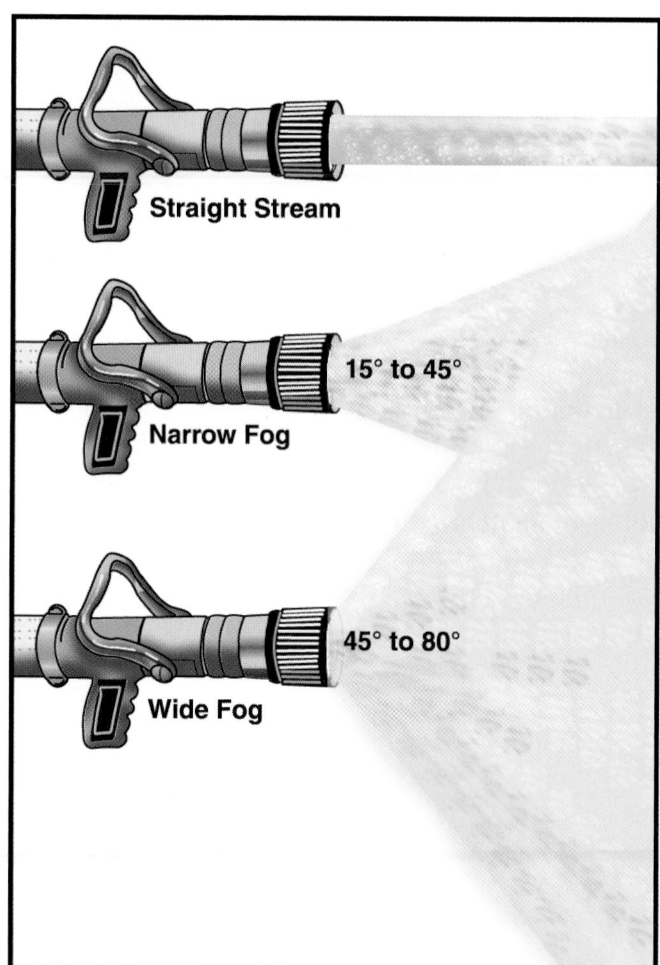

Figure 13.14 Fog nozzles are most commonly set to a straight stream, a narrow fog, or a wide fog.

Fog Stream

A *fog stream* is a fire stream composed of very fine water droplets. The design of most fog nozzles permits adjustment of the fog tip to produce different stream patterns from the nozzle (Figure 13.13). Water droplets, in either a shower or spray, are formed to expose the maximum water surface for heat absorption. The desired performance of fog stream nozzles is judged by the amount of heat that a fog stream absorbs and the rate by which the water is converted into steam or vapor. Fog nozzles permit settings of *straight stream, narrow-angle fog* and *wide-angle fog* (Figure 13.14). It should be understood that a straight stream is a pattern of the adjustable fog nozzle, whereas a solid stream is discharged from a smoothbore nozzle.

A wide-angle fog pattern has less forward velocity and a shorter reach than the other fog settings. A narrow-angle fog pattern has considerable forward velocity, and its reach varies in proportion to the pressure applied (Figure 13.15). Fog nozzles should be operated at *their designed* nozzle pressure. Of course, there is a maximum reach to any fog pattern, which is true with any stream. Once the nozzle pressure has produced a stream with maximum reach, further increases in nozzle pressure have little effect upon the stream except to increase the volume.

There are five factors that affect the reach of a fog stream:

- Gravity
- Water velocity
- Fire stream pattern selection
- Water droplet friction with air
- Wind

The interaction of these factors on a fog stream results in a fire stream with less reach than that of a solid stream. As the list shows, there are more factors that may negatively affect a fog stream than were given earlier for a solid stream. The more negative factors there are, the less the reach of the stream is likely to be. This shorter reach is why fog streams are seldom useful for outside, defensive fire fighting operations (Figure 13.16). The fog stream, however, is useful for fighting enclosed fires.

WATER-FLOW ADJUSTMENT

It is often desirable to control the rate of water flow through a fog nozzle such as when the water supply is limited. Two types of nozzles provide this capability: *manually adjustable* and *automatic (constant pressure).*

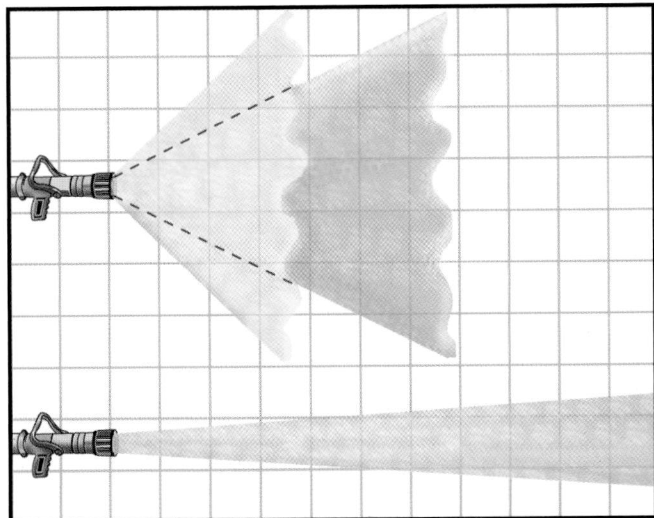

Figure 13.15 A wide-angle fog has less reach and forward velocity. A narrow-angle fog has greater reach and forward velocity.

Figure 13.16 Fine water particles are easily affected by wind and air currents.

Manually adjustable nozzles. Firefighters can change the rate of discharge from a manually adjustable fog nozzle by rotating the selector ring — usually located directly behind the nozzle tip — to a specific gpm (L/min) setting (Figure 13.17). Each setting provides a constant rate of flow as long as the operator maintains the proper nozzle pressure. The firefighter has the choice of making flow-rate adjustments either before opening the nozzle or while water is flowing. Depending upon

Figure 13.17 Fog nozzle with selective flow. *Courtesy of Akron Brass Company.*

the size of the nozzle, the firefighter may adjust flow rates from 10 gpm to 250 gpm (40 L/min to 1 000 L/min) for handlines and from 300 gpm to 2,500 gpm (1 200 L/min to 10 000 L/min) for master streams. Most of these nozzles also have a "flush" setting to rinse debris from the nozzle.

CAUTION: Make adjustments to the rate of flow in increments. Major adjustments can cause an abrupt change in the reaction force of the hoseline that may throw a firefighter off balance.

Automatic (constant-pressure) nozzles. Constant-pressure nozzles automatically vary the rate of flow to maintain an effective nozzle pressure (Figure 13.18). Obviously, a certain minimum nozzle pressure is needed to maintain a good spray pattern. With this type of nozzle, the nozzleperson can change the rate of flow by opening or closing the shutoff valve (see Nozzle Control Valves section). Automatic nozzles allow the nozzleperson to deliver large quantities of water at constant operating pressures or to reduce the flow to allow for mobility while maintaining an efficient discharge pattern.

Figure 13.18 An automatic nozzle.

CAUTION: Water flow adjustments in manual and automatic fog nozzles require close coordination between the nozzleperson, the company officer, and the pump operator.

HANDLING FOG STREAM NOZZLES

Although nozzle designs differ, the water pattern that is produced by the nozzle setting may affect the ease with which a particular nozzle is operated. Fire stream nozzles, in general, are not easy to control. If water travels at angles to the direct line of discharge, the reaction forces may be made to more or less counterbalance each other and reduce the nozzle reaction. This balancing of forces is the reason why a wide-angle fog pattern can be handled more easily than a straight-stream pattern.

ADVANTAGES

- The discharge pattern of fog streams may be adjusted to suit the situation (see Chapter 14, Fire Control).
- Some fog stream nozzles have adjustable settings to control the amount of water being used.
- Fog streams aid ventilation (see Chapter 10, Ventilation).
- Fog streams dissipate heat by exposing the maximum water surface for heat absorption.

DISADVANTAGES

- Fog streams do not have the reach or penetrating power of solid streams.
- Fog streams are more susceptible to wind currents than are solid streams.
- Fog streams may contribute to fire spread, create heat inversion, and cause steam burns to firefighters when improperly used during interior attacks (see Chapter 14, Fire Control).

Broken Stream

A *broken stream* is a stream of water that has been broken into coarsely divided drops (Figure 13.19). While a solid stream may become a broken stream past the point of breakover, a true broken stream takes on that form as it exits the nozzle. The coarse drops of a broken stream absorb more heat

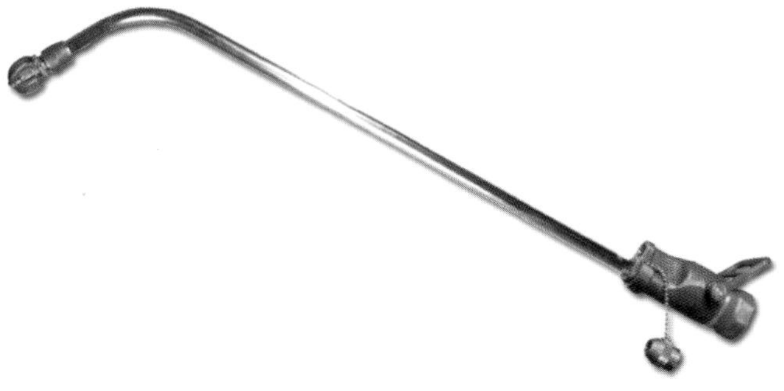

Figure 13.19 A broken stream nozzle with applicator. *Courtesy of Akron Brass Company.*

per gallon (liter) than a solid stream, and a broken stream has greater reach and penetration than a fog stream, so it can be the most effective stream in certain situations. Firefighters use broken streams most often on fires in confined spaces, such as those in belowground areas, attics, and wall spaces. Because a broken stream may have sufficient continuity to conduct electricity, it is not recommended for use on Class C fires.

Nozzle Control Valves

Nozzle control (shutoff) valves enable the operator to start, stop, or reduce the flow of water, thereby maintaining effective control of the handline or master stream appliance. These valves allow nozzles to open slowly so that the operator can adjust as the nozzle reaction increases; they also allow nozzles to close slowly to prevent water hammer (force created by the rapid deceleration of water). There are three principal types of control valves: *ball, slide,* and *rotary control.*

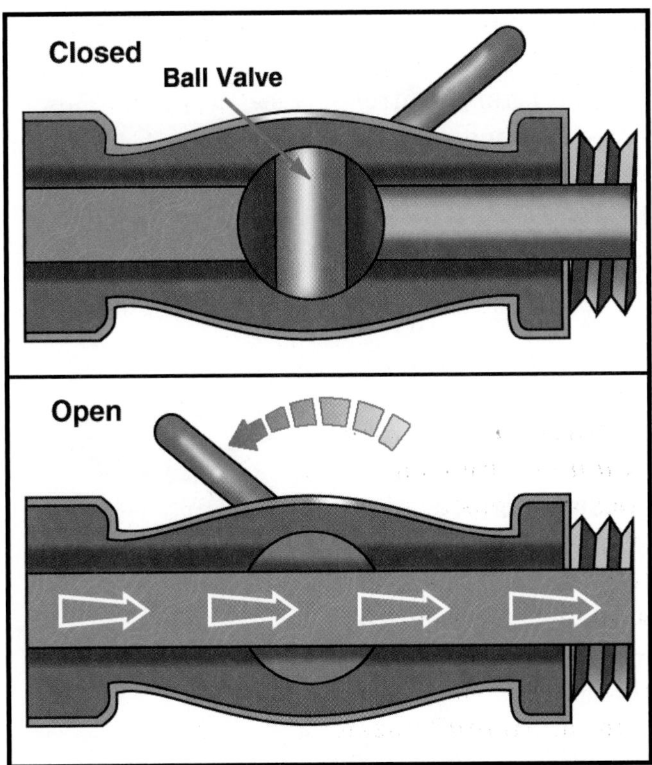

Figure 13.20 The operation of a ball valve.

BALL VALVE

The design and construction of the ball valve in handline nozzles provide effective control during fire fighting with a minimum of effort. The ball, perforated by a smooth waterway, is suspended from both sides of the nozzle body and seals against a seat (Figure 13.20). The ball can be rotated up to 90 degrees by moving the valve handle backward to open it and forward to close it. With the valve in the closed position, the waterway is perpendicular to the nozzle body, effectively blocking the flow of water through the nozzle. With the valve in the open position, the waterway is in line with the axis of the nozzle, allowing water to flow through it. While it will operate in any position between fully closed and fully open, operating the nozzle with the valve in the fully open position gives maximum flow and performance. When a ball valve is used with a solid stream nozzle, turbulence caused by a partially open valve may affect the desired stream or pattern.

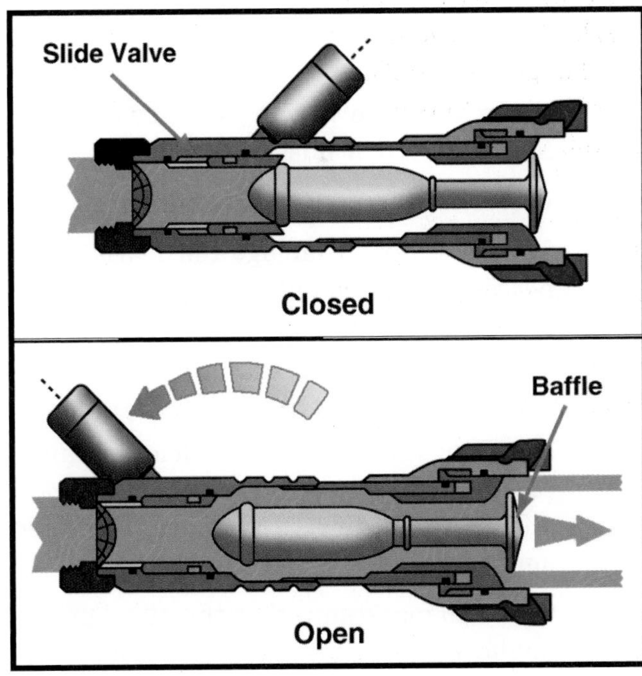

Figure 13.21 The operation of a slide valve.

SLIDE VALVE

The cylindrical slide valve control seats a movable cylinder against a shaped cone to turn off the flow of water (Figure 13.21). Flow increases or decreases as the shutoff handle is moved to change the position of the sliding cylinder relative to the cone. This stainless steel slide valve controls the flow of water through the nozzle without creating turbulence. The pressure control then compensates for the increase or decrease in flow by moving the baffle to develop the proper tip size and pressure.

ROTARY CONTROL VALVE

The rotary control valve is found only on rotary control fog nozzles (Figure 13.22). It consists of an exterior barrel guided by a screw that moves it forward or backward, rotating around an interior barrel. A major difference between rotary control valves and other control valves is that rotary control valves also control the discharge pattern of the stream.

Figure 13.22 A rotary control nozzle. *Courtesy of Elkhart Brass Manufacturing Company.*

Maintenance of Nozzles

Firefighters should inspect nozzles periodically and after each use to make sure that they are in proper working condition. This inspection includes the following checks:

- Check the swivel gasket for damage or wear. Replace worn or missing gaskets.
- Look for external damage to the nozzle.
- Look for internal damage and debris. When necessary, thoroughly clean nozzles with soap and water, using a soft bristle brush (Figure 13.23).
- Check for ease of operation by physically operating the nozzle parts. Clean and lubricate any moving parts that appear to be sticking according to manufacturer's recommendations.
- Check to make sure that the pistol grip (if applicable) is secured to the nozzle.

Figure 13.23 Clean nozzles with soap and water when necessary.

EXTINGUISHING FIRE WITH FIRE FIGHTING FOAM

[NFPA 1001: 3-3.7(a); 4-3.1; 4-3.1(a); 4-3.1(b)]

In general, fire fighting foam works by forming a blanket of foam on the burning fuel. The foam blanket excludes oxygen and stops the burning process. The water in the foam is slowly released as the foam breaks down. This action provides a cooling effect on the fuel. Fire fighting foam extinguishes and/or prevents fire in several ways:

- ***Separating*** — Creates a barrier between the fuel and the fire
- ***Cooling*** — Lowers the temperature of the fuel and adjacent surfaces
- ***Suppressing*** (sometimes referred to as *smothering*) — Prevents the release of flammable vapors and therefore reduces the possibility of ignition or reignition (Figure 13.24)

Water alone is not always effective as an extinguishing agent. Under certain circumstances, foam is needed. Fire fighting foam is especially effective on the two basic categories of flammable liquids: *hydrocarbon fuels* and *polar solvents.* Even fires that can be fought successfully using plain water may be more effectively fought if a fire fighting foam concentrate is added.

Hydrocarbon fuels, such as crude oil, fuel oil, gasoline, benzene, naphtha, jet fuel, and kero-

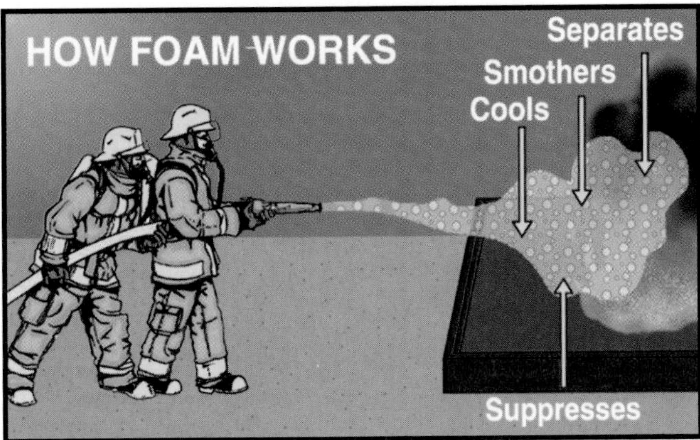

Figure 13.24 Foam cools, smothers, separates, and suppresses vapors.

sene, are petroleum-based and float on water. Fire fighting foam is effective as an extinguishing agent and vapor suppressant on Class B liquids because it can float on the surface of these fuels.

Polar solvents, such as alcohols, acetone, lacquer thinner, ketones, esters, and acids, are flammable liquids that have an attraction for water, much like a positive magnetic pole attracts a negative pole. Fire fighting foam, in special formulations, is effective on these fuels.

Specialized foams are also used for acid spills, pesticide fires, confined- or enclosed-space fires, and deep-seated Class A fires. In addition to regular fire fighting foams, there are special foams designed solely for use on unignited spills of hazardous liquids. These special foams are necessary because unignited chemicals have a tendency to either change the pH of water or remove the water from fire fighting foams, thereby rendering them ineffective.

Before discussing types of foams and the foam-making process, it is important to understand the following terms:

- ***Foam concentrate*** — Raw foam liquid as it rests in its storage container before the introduction of water and air
- ***Foam proportioner*** — Device that introduces foam concentrate into the water stream to make the foam solution
- ***Foam solution*** — Mixture of foam concentrate and water before the introduction of air
- ***Foam (finished foam)*** — Completed product after air is introduced into the foam solution

How Foam Is Generated

Foams in use today are of the mechanical type and must be *proportioned* (mixed with water) and *aerated* (mixed with air) before they can be used. Foam concentrate, water, air, and mechanical aeration are needed to produce quality fire fighting foam (Figure 13.25). These elements must be present and blended in the correct ratios. Removing any element results in either no foam production or poor-quality foam.

Aeration should produce an adequate amount of foam bubbles to form an effective foam blanket. Proper aeration produces uniform-sized bubbles to provide a longer lasting blanket. A good foam blanket is required to maintain an effective cover over either Class A or Class B fuels for the period of time desired.

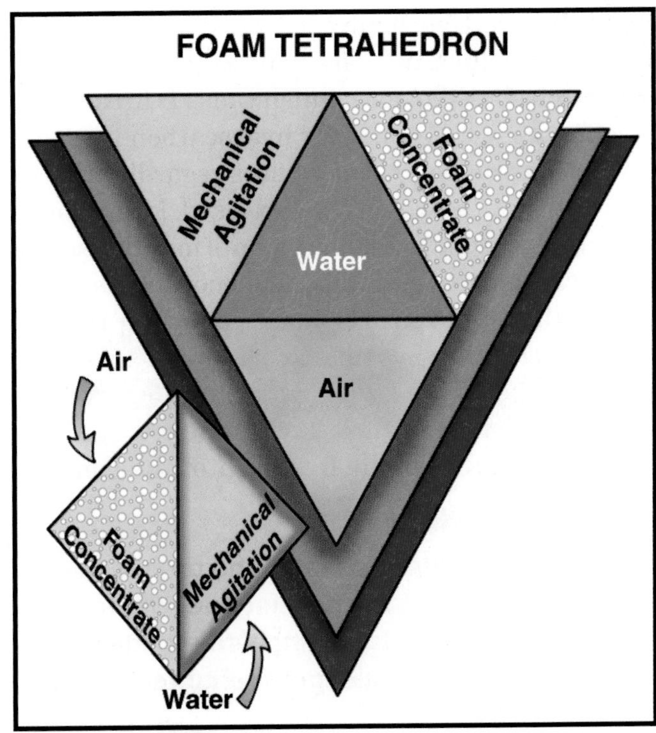

Figure 13.25 The foam tetrahedron.

Foam Expansion

Foam expansion refers to the increase in volume of a foam solution when it is aerated. This is a key characteristic to consider when choosing a foam concentrate for a specific application. The method of aerating a foam solution results in varying degrees of expansion that depend on the following factors:

- Type of foam concentrate used
- Accurate proportioning (mixing) of the foam concentrate in the solution
- Quality of the foam concentrate
- Method of aspiration

Depending on its purpose, foam can be described by three types: *low-expansion, medium-expansion,* and *high-expansion.* NFPA 11, *Standard for Low-Expansion Foam,* states that low-expansion foam has an air/solution ratio up to 20 parts finished foam for every part of foam solution (20:1 ratio). Medium-expansion foam is most commonly used at the rate of 20:1 to 200:1 through hydraulically operated nozzle-style delivery devices. In the high-expansion foams, the rate is 200:1 to 1000:1.

Foam Concentrates

To be effective, foam concentrates must match the fuel to which they are applied. Class A foams are not designed to extinguish Class B fires. Class B foams designed solely for hydrocarbon fires will not extinguish polar solvent fires regardless of the concentration at which they are used. Many foams that are intended for polar solvents may be used on hydrocarbon fires, but this should not be attempted unless the manufacturer of the particular concentrate specifically says this can be done. This incompatibility factor is why it is extremely important to identify the type of fuel involved before applying foam. Table 13.1 highlights each of the common types of foam concentrates.

CAUTION: Failure to match the proper foam concentrate with the burning fuel will result in an unsuccessful extinguishing attempt and could endanger firefighters.

CLASS A FOAM

Foams specifically designed for use on Class A fuels (ordinary combustibles) are becoming increasingly popular for use in both wildland and structural fire fighting. Class A foam is a special formulation of hydrocarbon surfactants. These surfactants reduce the surface tension of water in the foam solution. Reducing surface tension provides for better water penetration, thereby increasing its effectiveness. When aerated, Class A foam coats and insulates fuels, protecting them from ignition.

Class A foam may be used with fog nozzles, aspirating foam nozzles, medium- and high-expansion devices, and compressed air foam systems (CAFS) (Figure 13.26). Class A foam concentrate has supercleaning characteristics and is mildly corrosive. It is important to thoroughly flush equipment after use. For more information on Class A foam, refer to IFSTA's **Principles of Foam Fire Fighting** manual.

Figure 13.26 Class A foam may be discharged through a standard fog nozzle.

CLASS B FOAM

Class B foam is used to extinguish fires involving flammable and combustible liquids (Figure 13.27 on page 504). It is also used to suppress vapors from unignited spills of these liquids. There are several types of Class B foam concentrates; each type has its advantages and disadvantages. Class B foam concentrates are manufactured from either a synthetic or protein base. Protein-based foams are derived from animal protein. Synthetic foam is made from a mixture of fluorosurfactants. Some foam is made from a combination of synthetic and protein bases.

Class B foam may be proportioned into the fire stream via a fixed system, an apparatus-mounted system, or by portable foam proportioning equipment. Proportioning equipment is discussed later in this chapter (see Foam Proportioners section).

TABLE 13.1
Foam Concentrate Characteristics/Application Techniques

Type	Characteristics	Storage Range	Application Rate	Application Techniques	Primary Uses
Protein Foam (3% and 6%)	— Protein based — Low expansion — Good reignition (burnback) resistance — Excellent water retention — High heat resistance and stability — Performance can be affected by freezing and thawing — Concentrate can be freeze protected with antifreeze — Not as mobile or fluid on fuel surface as other low-expansion foams	35–120°F (2°C to 49°C)	0.16 gpm/ft^2 (6.5 L/min/m^2)	— Indirect foam stream so as not to mix fuel with foam **NOTE:** Fuel should not be agitated during application because static spark ignition of volatile hydrocarbons can result from plunging and turbulence from a direct foam water stream. — Alcohol-resistant must be used within seconds of proportioning — Not compatible with dry chemical extinguishing agents	— Class B fires involving hydrocarbons — To protect flammable and combustible liquids where they are stored, transported, and processed
Fluoroprotein Foam (3% and 6%)	— Protein and synthetic based; derived from protein foam — Fuel shedding — Long-term vapor suppression — Good water retention — Excellent, long-lasting heat resistance — Performance not affected by freezing and thawing — Maintains low viscosity at low temperatures — Can be freeze protected with antifreeze — May be used with fresh or salt water — Nontoxic and biodegradable after dilution — Good mobility and fluidity on fuel surface — Premixable for short periods of time	35–120°F (2°C to 49°C)	0.16 gpm/ft^2 (6.5 L/min/m^2)	— Direct plunge technique — Subsurface injection — Compatible with simultaneous application of dry chemical extinguishing agents — Must be delivered through air-aspirating equipment	— Hydrocarbon vapor suppression — Subsurface application to hydrocarbon fuel storage tanks — Extinguishing in-depth crude petroleum or other hydro-carbon fuel fires

continued

Table 13.1 continued

Type	Characteristics	Storage Range	Application Rate	Application Techniques	Primary Uses
Film Forming Fluoroprotein Foam (FFFP) (3% and 6%)	— Protein based; fortified with additional surfactants that reduce the burnback characteristics of other protein-based foams — Fuel shedding — Develops a fast-healing, continuous-floating film on hydrocarbon fuel surfaces — Excellent, long-lasting heat resistance — Good low temperature viscosity — Fast fire knockdown — Affected by freezing and thawing — Can be used with fresh or salt water — Can be stored premixed — Can be freeze protected with antifreeze — Alcohol-resistant FFFP can be used on polar solvents at 6% solution and on hydrocarbon fuels at 3% solution — Nontoxic and biodegradable after dilution	35–120°F (2°C to 49°C)	Ignited Hydrocarbon Fuel: 0.10 gpm/ft^2 (4.1 L/min/m^2) Polar Solvent Fuel: 0.24 gpm/ft^2 (9.8 L/min/m^2)	— Must cover entire fuel surface — May be applied with dry chemical agents — May be applied with spray nozzles — Subsurface injection — Can be plunged into fuel during application	— Suppressing vapors in unignited spills of hazardous liquids — Extinguishing fires in hydrocarbon fuels
Aqueous Film Forming Foam (AFFF) (1%, 3%, and 6%)	— Synthetic based — Good penetrating capabilities — Spreads vapor-sealing film over and floats on hydrocarbon fuels — Can be used through nonaerating nozzles — Performance may be adversely affected by freezing and storing — Has good low-temperature viscosity — Can be freeze protected with antifreeze — Can be used with fresh or salt water — Can be premixed	25–120°F (-4°F to 49°C)	Ignited Hydrocarbon Fuel: 0.10 gpm/ft^2 (4.1 L/min/m^2) Polar Solvent Fuel: 0.24 gpm/ft^2 (9.8 L/min/m^2)	— May be applied directly onto fuel surface — May be applied indirectly by bouncing it off a wall and allowing it to float onto fuel surface — Subsurface injection — May be applied with dry chemical agents	— Controlling and extinguishing Class B fires — Handling land or sea crash rescues involving spills — Extinguishing most transportation-related fires — Wetting and penetrating Class A fuels — Securing unignited hydrocarbon spills

continued

Table 13.1 continued

Type	Characteristics	Storage Range	Application Rate	Application Techniques	Primary Uses
Alcohol-Resistant AFFF (3% and 6%)	— AFFF concentrate to which polymer has been added — Multipurpose: Can be used on both polar solvents and hydrocarbon fuels (used on polar solvents at 6% solution and on hydrocarbon fuels at 3% solution) — Forms a membrane on polar solvent fuels that prevents destruction of the foam blanket — Forms same aqueous film on hydrocarbon fuels as AFFF — Fast flame knockdown — Good burnback resistance on both fuels — Not easily premixed	25–120°F (-4°C to 49°C) May become viscous at temperatures under 50°F (10°C)	Ignited Hydrocarbon Fuel: 0.10 gpm/ft² (4.1 L/min/m²) Polar Solvent Fuel: 0.24 gpm/ft² (9.8 L/min/m²)	— Apply gently directly onto fuel surface — May be applied indirectly by bouncing it off a wall and allowing it to float onto fuel surface — Subsurface injection	Fires or spills of both hydrocarbon and polar solvent fuels
High-Expansion Foam	— Synthetic detergent based — Special-purpose, low water content — High air-to-solution ratios: 200:1 to 1,000:1 — Performance not affected by freezing and thawing — Poor heat resistance — Prolonged contact with galvanized or raw steel may attack these surfaces	27–110°F (-3°C to 43°C)	Sufficient to quickly cover the fuel or fill the space	— Gentle application so as not to mix foam with fuel — Must cover entire fuel surface — Usually fills entire space in confined space incidents	— Extinguishing Class A and some Class B fires — Flooding confined spaces — Volumetrically displacing vapor, heat, and smoke — Reducing vaporization from liquefied natural gas (LNG) spills — Extinguishing pesticide fires — Suppressing fuming acid vapors — Suppressing vapors in coal mines and other subterranean spaces; in concealed spaces in basements — As extinguishing agent in fixed extinguishing systems for industrial uses — Not recommended for outdoor use

continued

Figure 13.27 Foam is needed to combat large-scale Class B fires. *Courtesy of Harvey Eisner.*

Foams (such as aqueous film forming foam [AFFF] and film forming fluoroprotein foam [FFFP]) may be applied either with standard fog nozzles or with air-aspirating foam nozzles (all types) (see Foam Delivery Devices section). The rate of application (minimum amount of foam solution that must be applied) for Class B foam varies depending on any one of several variables:

- Type of foam concentrate used
- Whether or not the fuel is on fire (Figure 13.28)
- Type of fuel (hydrocarbon/polar solvent) involved
- Whether the fuel is spilled or in a tank (**NOTE:** If the fuel is in a tank, the type of tank will have a bearing on the application rate.)
- Whether the foam is applied via either a fixed system or portable equipment

Table 13.1 continued

Type	Characteristics	Storage Range	Application Rate	Application Techniques	Primary Uses
Class A Foam	— Synthetic — Wetting agent that reduces surface tension of water and allows it to soak into combustible materials — Rapid extinguishment with less water use than other foams — Can be used with regular water stream equipment — Can be premixed with water in the booster tank — Mildly corrosive — Requires lower percentage of concentration (0.2 to 1.0) than other foams (1%, 3%, or 6% concentrate) — Outstanding insulating qualities — Good penetrating capabilities	25–120°F (-4°C to 49°C) Concentrate subject to freezing but can be thawed and used if freezing occurs	The same as the minimum critical flow rate for plain water on similar Class A Fuels; flow rates should not be reduced when using Class A foam	— Can be propelled with compressed air systems — Can be applied with all conventional fire department nozzles	— Extinguishing Class A combustibles only

Figure 13.28 Whether or not the fuel is on fire affects the needed application rate.

Unignited spills do not require the same application rates as ignited spills because radiant heat, open flame, and thermal drafts do not attack the finished foam as they would under fire conditions. In case the spill does ignite, however, firefighters should be prepared to flow at least the minimum application rate for a specified amount of time based on fire conditions.

All foam concentrate supplies should be on the fireground at the point of proportioning before application is started. Once application has started, it should continue uninterrupted until extinguishment is complete. Stopping and restarting may allow the fire and fuel to consume whatever foam blanket has been established.

Because polar solvent fuels have differing affinities for water, it is important to know application rates for each type of solvent. These rates also vary with the type and manufacturer of the foam concentrate selected. Foam concentrate manufacturers provide information on the proper application rates as listed by UL (Figure 13.29). For more complete information on application rates, consult NFPA 11, the foam manufacturers' recommendations, and IFSTA's **Principles of Foam Fire Fighting** manual.

SPECIFIC APPLICATION FOAMS

Numerous types of foams are selected for specific applications according to their properties and performance. Some are thick and viscous and form tough, heat-resistant blankets over burning liquid surfaces; others are thinner and spread more rapidly. Some foams produce a vapor-sealing film of surface-active water solution on a liquid surface. Others, such as medium- and high-expansion foams, are used in large volumes to flood surfaces and fill cavities.

Foam Proportioning

The term *proportioning* is used to describe the mixing of water with foam concentrate to form a foam solution (Figure 13.30). Most foam concentrates are intended to be mixed with either fresh or salt water. For maximum effectiveness, foam concentrates must be proportioned at the specific percentage for which they are designed. This percentage rate for the intended fuel is clearly marked on the outside of every foam container (Figure 13.31). Failure to follow this procedure, such as trying to use 6% foam at a 3% concentration, will result in poor-quality foam that may not perform as desired.

Most fire fighting foam concentrates are intended to be mixed with 94 to 99.9 percent water. For example, when using 3% foam concentrate, 97 parts water mixed with 3 parts foam concentrate equals 100 parts foam solution (Figure 13.32). For 6% foam concentrate, 94 parts water mixed with 6 parts foam concentrate equals 100 percent foam solution.

Class A foams are an exception to this percentage rule. The proportioning percentage for Class A foams can be adjusted (within limits recommended by the manufacturer) to achieve specific objectives. To produce a dry (thick) foam suitable for exposure protection and fire breaks, the foam concentrate can be adjusted to a higher percentage. To produce a wet (thin) foam that rapidly sinks into a fuel's surface, the foam concentrate can be adjusted to a lower percentage.

ANSUL®

EXTINGUISHING

ANSULITE® ALCOHOL RESISTANT CONCENTRATE (ARC) 3% AND 6%

DESCRIPTION

ANSULITE ARC 3%/6% Alcohol Resistant Concentrate is formulated from special flourochemical and hydrocarbon surfactants, a high molecular weight polymer, and solvents. It is transported and stored as a concentrate to provide ease of use and considerable savings in weight and volume.

It is intended for use as a 3% or 6% proportioned solution (depending on the type of fuel) in fresh, salt or hard water. (Water hardness should not exceed 500 ppm expressed as calcium and magnesium.) It may also be used and stored as a premixed solution in fresh or potable water for use with the Ansul Model AR-33-D wheeled fire extinguisher.

There are three fire extinguishing mechanisms in effect when using ANSULITE ARC concentrate on either a conventional Class B hydrocarbon fuel such as gasoline, diesel fuel, etc., or a Class B polar solvent (water miscible fuel) such as methyl alcohol, acetone, etc. First, an aqueous film is formed in the case of a conventional hydrocarbon fuel, or a polymeric membrane in the case of a polar solvent fuel. This film or membrane forms a barrier to help prevent the release of fuel vapor. Second, regardless of the fuel type, a foam blanket is formed which excludes oxygen and from which drains the liquids that form the film or polymeric membrane. Third, the water content from the foam produces a cooling effect.

Physicochemical Properties at 77°F (25°C)

Appearance	Light Amber Gel-Like Liquid
Density	1.000 ∓ 0.25 gm/ml
pH	7.0 - 8.5
Refractive Index	1.3600 ∓ .0018
Viscosity	2500 ∓ 300 cps*
Chloride Content	Less than 75 ppm

*Brookfield Viscometer Spindle #4, Speed 30

ANSULITE ARC Alcohol Resistant Concentrate is a non-Newtonian fluid that is both pseudoplastic and thixtropic. Because of these properties, dynamic viscosity will decrease as shear increases.

APPLICATION

ANSULITE ARC 3%/6% AFFF is unique among the ANSULITE AFFF agents in that it can be used on either Conventional Class B fuels or the polar solvent type Class B fuels. Its excellent wetting characteristics make it useful in combating Class A Fires as well. Because of the low energy to make foam, it can be used with both aspirating and non-aspirating discharge devices.

To provide even greater fire protection capability, it can be used with dry chemical extinguishing agents without regard to the order of application to provide even greater fire protection capability. Due to the velocity of the dry chemical discharge, care must be taken **not** to submerge the polymeric membrane below the fuel surface.

APPLICATION RATES

Application Rates using U.L. 162 Standard 50 ft.^{2} Fire Test on representative hydrocarbon and polar solvent fuels are listed below.

UL Type II Application [(1)] - Polar Solvents

Fuel Group	Concentration	U.L. Test Application Rate gpm/ft^2	Lpm/m^2	U.L.[(2)] Recommended Application Rate gpm/ft^2	Lpm/m^2
Alcohol					
Methanol (MeOH)	6%	.10	(4.1)	**.17**	(6.9)
Ethanol (EtOH)	6%	.10	(4.1)	**.17**	(6.9)
Isopropanol (IPA)	6%	.10	(4.1)	**.17**	(6.9)
Ketone					
Acetone	6%	.10	(4.1)	**.17**	(6.9)
Methyl Ethyl Ketone (MEK)	6%	.10	(4.1)	**.17**	(6.9)
Carboxylic Acid					
Acetic Acid Glacial	6%	.10	(4.1)	**.17**	(6.9)
Ether					
Diethyl Ether	6%	.10	(4.1)	**.17**	(6.9)
Aldehyde					
Propionaldehyde	6%	.08	(3.3)	**.13**	(5.3)
Ester					
Ethyl Acetate	6%	.06	(2.4)	**.10**	(4.1)
Butyl Acetate	6%	.06	(2.4)	**.10**	(4.1)
UL Type II Application [(3)] - Hydrocarbons					
Heptane	3%	.04	(1.6)	**.10**	(4.1)
Toluene	3%	.04	(1.6)	**.10**	(4.1)
Gasoline	3%	.04	(1.6)	**.10**	(4.1)
10% Gasohol (EtOH)	3%	.04	(1.6)	**.10**	(4.1)

(1) Type III DISCHARGE OUTLET - A device that delivers foam onto the burning liquid and partially submerges the foam or produces restricted agitation of the surface as described in U.L. 162.
(2) U.L. builds in a 5⁄3 safety factor from its test rate to find its recommended rate of application.
(3) TYPE III DISCHARGE OUTLET - A device that delivers the foam directly onto the burning liquid as described in U.L. 162.

Figure 13.29 A sample manufacturer's foam application rate sheet. *Courtesy of Ansul Fire Protection.*

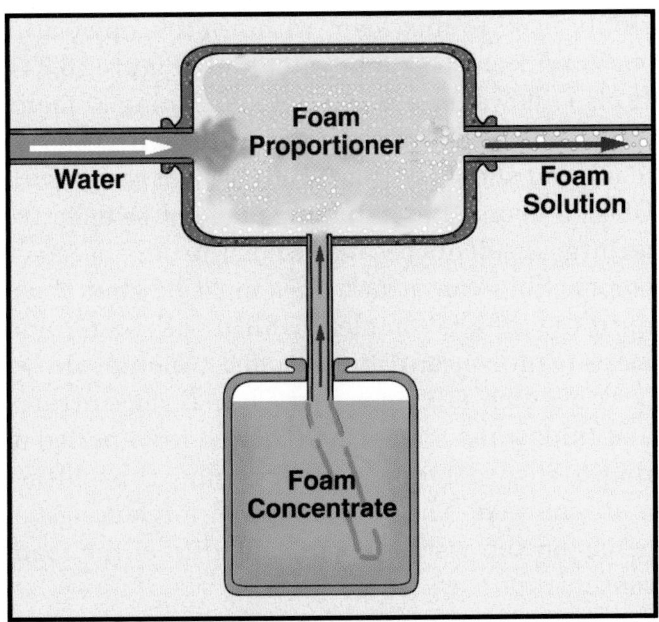

Figure 13.30 Foam solution is comprised of foam concentrate and water.

Figure 13.31 All foam concentrate containers have the proper proportioning percentage clearly marked on the outside of the container.

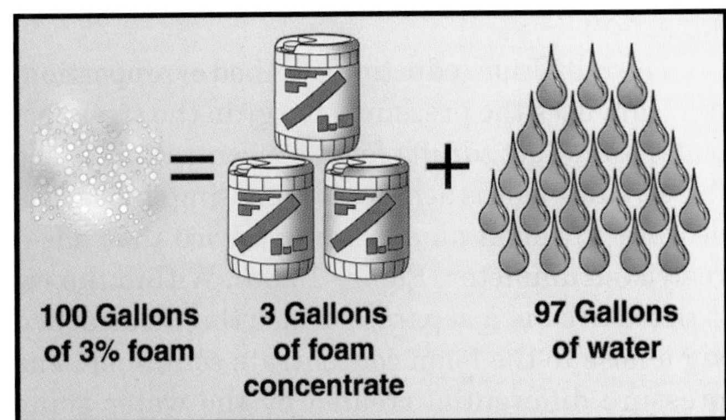

Figure 13.32 Foam generated using a 3% foam concentrate is 3 parts foam concentrate to 97 parts water.

Class B foams are mixed in proportions from 1% to 6%. Some multipurpose Class B foams designed for use on both hydrocarbon and polar solvent fuels can be used at different concentrations, depending on which of the two fuels is burning. These concentrates are normally used at a 3% rate for hydrocarbons and 6% for polar solvents. Newer multipurpose foams may be used at 3% concentrations regardless of the type of fuel. Medium-expansion Class B foams are typically used at either 1½%, 2%, or 3% concentrations. Follow the manufacturer's recommendations for proportioning.

A variety of equipment is used to proportion foam. Some types are designed for mobile apparatus and others are designed for fixed fire protection systems. The selection of a proportioner depends on the foam solution flow requirements, available water pressure, cost, intended use (truck, fixed, or portable), and the agent to be used. Proportioners and delivery devices (foam nozzle, foam maker, etc.) are engineered to work together. Using a foam proportioner that is not compatible with the delivery device (even if the two are made by the same manufacturer) can result in unsatisfactory foam or no foam at all (see Foam Proportioners and Foam Delivery Devices/Generating Systems section).

There are four basic methods by which foam may be proportioned:

- Induction
- Injection
- Batch-mixing
- Premixing

INDUCTION

The induction (eduction) method of proportioning foam uses the pressure energy in the stream of water to induct (draft) foam concentrate into the fire stream. This is achieved by passing the stream of water through an *eductor,* a device that has a restricted diameter (Figure 13.33). Within the restricted area is a separate orifice that is attached via a hose to the foam concentrate container. The pressure differential created by the water going through the restricted area and over the orifice creates a suction that draws the foam concentrate into the fire stream. In-line eductors and foam nozzle eductors are examples of foam proportioners that work by this method.

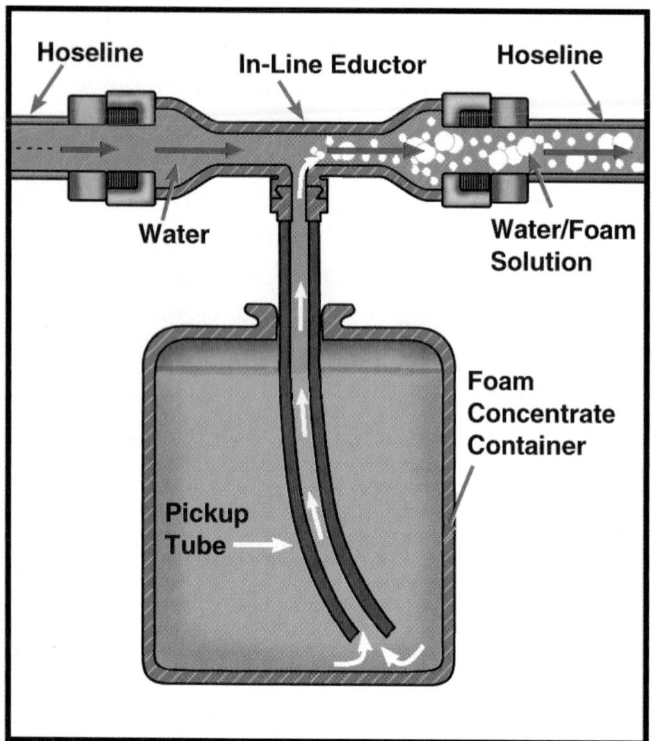

Figure 13.33 The operating principles of an in-line eductor.

INJECTION

The injection method of proportioning foam uses an external pump or head pressure to force foam concentrate into the fire stream at the correct ratio in comparison to the flow. These systems are commonly employed in apparatus-mounted or fixed fire protection system applications.

BATCH-MIXING

Batch-mixing is the most simple method of mixing foam concentrate and water. It is commonly used to mix foam within a fire apparatus water tank or a portable water tank (Figure 13.34). It also allows for accurate proportioning of foam. Batch-mixing is commonly practiced with Class A foams but should only be used as a last resort with Class B foams. Batch-mixing may not be effective on large incidents because when the tank becomes empty, the foam attack lines must be shut down until the tank is completely filled with water and more foam concentrate is added. Another drawback of batch-mixing is that Class B concentrates and tank water must be circulated for a period of time to ensure thorough mixing before the solution is discharged. The time required for mixing depends on the viscosity and solubility of the foam concentrate.

Figure 13.34 Batch-mixing can be accomplished by pouring foam concentrate into the apparatus water tank.

PREMIXING

Premixing is one of the more commonly used methods of proportioning. With this method, premeasured portions of water and foam concentrate are mixed in a container. Typically, the premix method is used with portable extinguishers, wheeled extinguishers, skid-mounted twin-agent units, and vehicle-mounted tank systems (Figures 13.35 a–c).

In most cases, premixed solutions are discharged from a pressure-rated tank using either a compressed inert gas or air. An alternative method of discharge uses a pump and a nonpressure-rated atmospheric storage tank. The pump discharges the foam solution through piping or hose to the delivery devices. Premix systems are limited to a one-time application. When used, the tank must be completely emptied and then refilled before it can be used again.

Figure 13.35a A modern AFFF portable fire extinguisher. *Courtesy of Amerex Corp.*

Figure 13.35b A wheeled AFFF fire extinguisher. *Courtesy of Conoco Oil Co.*

Figure 13.35c Twin-agent units may be mounted on hand-pulled carts or in the back of a truck. *Courtesy of Conoco Oil Co.*

FOAM PROPORTIONERS AND FOAM DELIVERY DEVICES/GENERATING SYSTEMS

[NFPA 1001: 4-3.1; 4-3.1(a); 4-3.1(b)]

In addition to a pump to supply water and fire hose to transport it, there are two basic pieces of equipment needed to produce a foam fire stream: a *foam proportioner* and a *foam delivery device (nozzle or generating system)*. It is important that the proportioner and delivery device/system are compatible in order to produce usable foam (Figure 13.36). Foam proportioning simply introduces the appropriate amount of foam concentrate into the water to form a foam solution. A foam generating system/nozzle adds the air into foam solutions to produce finished fire fighting foam. The following sections detail the various types of foam proportioning devices commonly found in portable and apparatus-mounted applications and various foam delivery devices (nozzles/generating systems).

Figure 13.36 It is important that the proportioner and delivery device match each other to produce usable foam.

Foam Proportioners

Foam proportioners may be portable or apparatus-mounted. In general, foam proportioning devices operate by one of two basic principles:

- The pressure of the water stream flowing through an orifice creates a venturi action that inducts (drafts) foam concentrate into the water stream (see In-line foam eductors section).
- Pressurized proportioning devices inject foam concentrate into the water stream at a desired ratio and at a higher pressure than that of the water.

PORTABLE FOAM PROPORTIONERS

Portable foam proportioners are the simplest and most common foam proportioning devices in use today. Two types of portable foam proportioners are *in-line foam eductors* and *foam nozzle eductors.*

In-line foam eductors. The in-line eductor is the most common type of foam proportioner used in the fire service (Figure 13.37). This eductor is designed to be either directly attached to the pump panel discharge outlet or connected at some point in the hose lay. When using an in-line eductor, it is very important to follow the manufacturer's instructions about inlet pressure and the maximum hose lay between the eductor and the appropriate nozzle.

In-line eductors use the Venturi Principle to draft foam concentrate into the water stream. As water at high pressure passes over a reduced opening, it creates a low-pressure area near the outlet side of the eductor (See Figure 13.33). This low-pressure area creates a suction effect (Venturi Principle). The eductor pickup tube is connected to the eductor at this low-pressure point. A pickup tube submerged in the foam concentrate draws concentrate into the water stream, creating a foam water solution (Figure 13.38). The foam concentrate inlet to the eductor should not be more than 6 feet (1.8 m) above the liquid surface of the foam concentrate. If the inlet is too high, the foam concentration will be very lean, or foam may not be inducted at all (Figure 13.39).

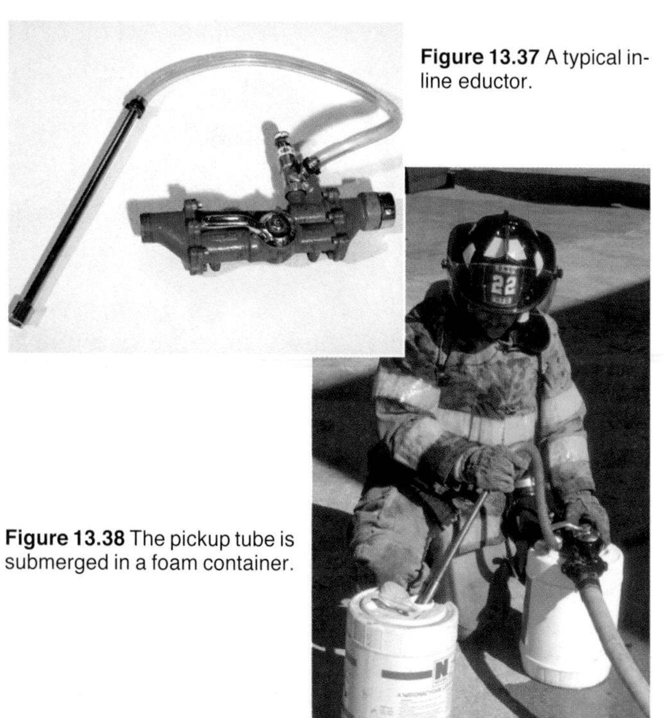

Figure 13.37 A typical in-line eductor.

Figure 13.38 The pickup tube is submerged in a foam container.

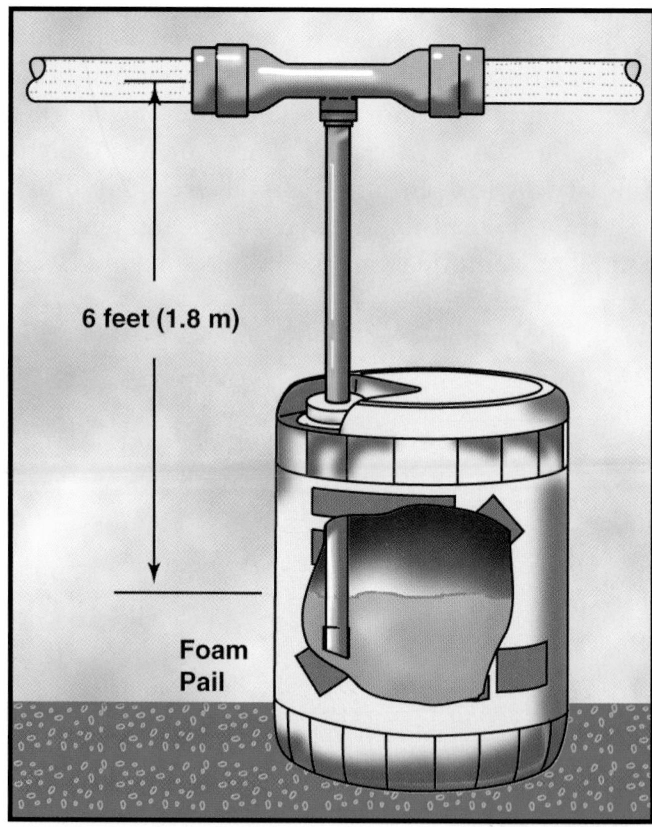

Figure 13.39 The eductor may be no more than 6 feet (1.8 m) above the foam concentrate liquid level.

Foam nozzle eductors. The foam nozzle eductor operates on the same basic principle as the in-line eductor. However, this eductor is built into the nozzle rather than into the hoseline (Figure 13.40). As a result, its use requires the foam concentrate to be available where the nozzle is operated. If the foam nozzle is moved, the foam concentrate also is moved. The logistical problems of relocation are magnified by the gallons (liters) of concentrate required. Use of a foam nozzle eductor compromises firefighter safety. Firefighters cannot always move quickly, and they may have to leave their concentrate supplies behind if they are required to retreat for any reason.

APPARATUS-MOUNTED PROPORTIONERS

Foam proportioning systems are commonly mounted on structural, industrial, wildland, and

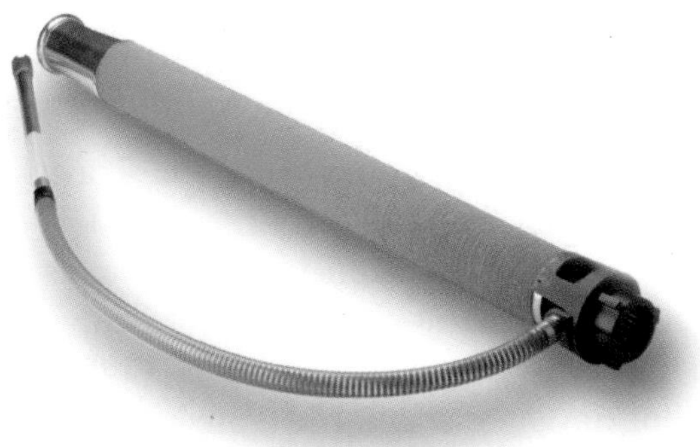

Figure 13.40 A typical handline foam nozzle eductor. *Courtesy of Akron Brass Company.*

aircraft rescue and fire fighting (ARFF) apparatus, as well as fire boats (Figures 13.41 a–c). Three types of the various apparatus-mounted foam proportioning systems are *installed in-line eductors, around-the-pump proportioners,* and *balanced pressure proportioners.* For more information on apparatus-mounted proportioners, refer to IFSTA's **Principles of Foam Fire Fighting** manual.

Figure 13.41a Some structural fire apparatus are equipped with foam systems. *Courtesy of Joel Woods.*

Figure 13.41b Many industrial fire apparatus are equipped with foam systems. *Courtesy of Ron Jeffers.*

Figure 13.41c Virtually all ARFF apparatus are equipped with large foam systems.

Foam Delivery Devices (Nozzles/Generating Systems)

Once the foam concentrate and water have been mixed together to form a foam solution, the foam solution must then be mixed with air (aerated) and delivered to the surface of the fuel. Nozzles/generating systems (foam delivery devices) designed to discharge foam are sometimes called foam makers. There are many types of devices that can be used, including standard water stream nozzles. The following paragraphs highlight some of the more common foam application devices.

NOTE: Foam nozzle eductors are considered portable foam nozzles, but they have been omitted from this section because they were covered earlier in the Portable Foam Proportioners section.

HANDLINE NOZZLES

IFSTA defines a handline nozzle as "*any nozzle that one to three firefighters can safely handle and that flows less than 350 gpm (1 400 L/min).*" Most handline foam nozzles flow considerably less than that figure. The following sections detail the handline nozzles commonly used for foam application.

Solid bore nozzles. The use of solid bore nozzles is limited to certain types of Class A applications. In these applications, the solid bore nozzle provides an effective fire stream that has maximum reach capabilities (Figures 13.42 a and b).

Fog nozzles. Either fixed-flow or automatic fog nozzles can be used with foam solutions to produce a low-expansion, short-lasting foam (Figure 13.43).

Figure 13.42a A solid bore nozzle that is used for a CAFS stream. *Courtesy of Mount Shasta (CA) Fire Protection District.*

Figure 13.42b CAFS fire streams may be discharged through solid bore nozzles. *Courtesy of Mount Shasta (CA) Fire Protection District.*

Figure 13.43 Film forming foams may be discharged through standard fog nozzles.

This nozzle breaks the foam solution into tiny droplets and uses the agitation of water droplets moving through air to achieve its foaming action. Its best application is when it is used with regular AFFF and Class A foams. These nozzles cannot be used with protein and fluoroprotein foams. These nozzles may be used with alcohol-resistant AFFF foams on hydrocarbon fires but should not be used on polar solvent fires. This is because insufficient aspiration occurs to handle the polar solvent fires. Some nozzle manufacturers have foam aeration attachments that can be added to the end of the nozzle to increase aspiration of the foam solution (Figure 13.44). See IFSTA's **Principles of Foam Fire Fighting** manual for more information.

Air-aspirating foam nozzles. The most effective appliance for the generation of low-expansion foam is the air-aspirating foam nozzle. The air-aspirating foam nozzle inducts air into the foam solution by a venturi action (Figure 13.45). This nozzle is especially designed to provide the aeration required to make the highest quality foam possible. These nozzles must be used with protein and fluoroprotein concentrates. They may also be used with Class A foams in wildland applications. These nozzles provide maximum expansion of the agent. The reach of the stream is less than that of a standard fog nozzle.

Figure 13.44 Some fog nozzle manufacturers have aeration attachments for their nozzles that can be used during foam operations. *Courtesy of KK Productions.*

Figure 13.45 A typical air-aspirating foam nozzle.

MEDIUM- AND HIGH-EXPANSION FOAM GENERATING DEVICES

Medium- and high-expansion foam generators produce a high-air-content, semistable foam. For medium-expansion foam, the air content ranges from 20 parts air to 1 part foam solution (20:1) to 200 parts air to 1 part foam solution (200:1). For high-expansion foam, the ratio is 200:1 to 1000:1. There are two basic types of medium- and high-expansion foam generators: the *water-aspirating type nozzle* and the *mechanical blower.*

Water-aspirating type nozzle. The water-aspirating type nozzle is very similar to the other foam-producing nozzles except it is much larger and longer (Figure 13.46). The back of the nozzle is open to allow airflow. The foam solution is pumped through the nozzle in a fine spray that mixes with air to form a moderate-expansion foam. The end of the nozzle has a screen or series of screens that further breaks up the foam and mixes it with air. These nozzles typically produce a lower-air-volume foam than do mechanical blower generators.

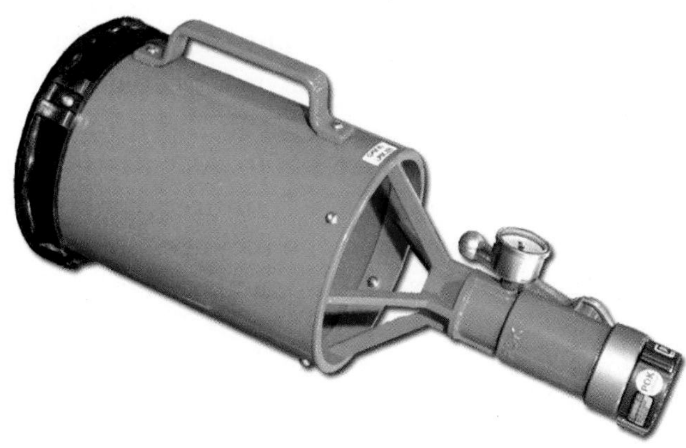

Figure 13.46 A high-expansion foam tube.

Figure 13.47 Mechanical blowers generate massive amounts of high-expansion foam. *Courtesy of Frank Bateman.*

Mechanical blower generator. A mechanical blower generator is similar in appearance to a smoke ejector (Figure 13.47). It operates on the same principle as the water-aspirating nozzle except the air is forced through the foam spray by a powered fan instead of being pulled through by water movement. This device produces a foam with a high air content and is typically associated with total-flooding applications. Its use is limited to high-expansion foam.

ASSEMBLING A FOAM FIRE STREAM SYSTEM

[NFPA 1001: 4-3.1; 4-3.1(a); 4-3.1(b)]

To provide a foam fire stream, a firefighter or apparatus driver/operator must be able to correctly assemble the components of the system and to locate problem areas and make adjustments. Skill Sheet 13-1 describes the steps for placing a foam line in service using an in-line eductor proportioner:

There are a number of reasons for failure to generate foam or for generating poor-quality foam. The most common reasons for failure are as follows:

- Eductor and nozzle flow ratings do not match so foam concentrate cannot induct into the fire stream.
- Air leaks at fittings cause a loss of suction.
- Improper cleaning of proportioning equipment causes clogged foam passages (Figure 13.48).
- Nozzle is not fully open restricting water flow.
- Hose lay on the discharge side of the eductor is too long creating excess back pressure causing reduced foam pickup at the eductor.
- Hose is kinked and stops flow.
- Nozzle is too far above the eductor, which causes excessive elevation pressure.
- Mixing different types of foam concentrate in the same tank results in a mixture too viscous to pass through the eductor.

Figure 13.48 The foam eductor is flushed clean by inserting the pickup tube into a container of clear water for a few moments.

Figure 13.49 Roll-on foam application.

FOAM APPLICATION TECHNIQUES

[NFPA 1001: 4-3.1; 4-3.1(a); 4-3.1(b)]

It is important to use the correct techniques when manually applying foam from handline or master stream nozzles. If incorrect techniques are used, such as plunging the foam into a liquid fuel, the effectiveness of the foam is reduced. The techniques for applying foam to a liquid fuel fire or spill include the *roll-on method, bank-down method,* and *rain-down method.*

Roll-On Method

The roll-on method directs the foam stream on the ground near the front edge of a burning liquid pool (Figure 13.49). The foam then rolls across the surface of the fuel. A firefighter continues to apply foam until it spreads across the entire surface of the fuel and the fire is extinguished. It may be necessary to move the stream to different positions along the edge of a liquid spill to cover the entire pool. This method is used only on a pool of liquid fuel (either ignited or unignited) on the open ground.

Bank-Down Method

The bank-down method may be employed when an elevated object is near or within the area of a burning pool of liquid or an unignited liquid spill. The object may be a wall, tank shell, or similar structure. The foam stream is directed off the object, allowing the foam to run down onto the surface of the fuel (Figure 13.50). As with the roll-on method, it may be necessary to direct the stream off various points around the fuel area to achieve total coverage and extinguishment of the fuel. This method is used primarily in dike fires and fires involving spills around damaged or overturned transport vehicles.

Rain-Down Method

The rain-down method is used when the other two methods are not feasible because of either the size of the spill area (either ignited or unignited) or the lack of an object from which to bank the foam. It is also the primary manual application technique used on aboveground storage tank fires. This method directs the stream into the air above the fire or spill and allows the foam to float gently down onto the surface of the fuel (Figure 13.51). On small fires, a firefighter sweeps the stream back and forth over the entire surface of the fuel until the fuel is completely covered and the fire is extinguished. On large fires, it may be more effective for the firefighter to direct the stream at one location to allow the foam to take effect there and then work its way out from that point.

Figure 13.50 Bank-down method.

Figure 13.51 Rain-down foam application.

FOAM HAZARDS

[NFPA 1001: 4-3.1; 4-3.1(a)]

Foam concentrates, either at full strengths or in diluted forms, pose minimal health risks to firefighters. In both forms, foam concentrates may be mildly irritating to the skin and eyes. Affected areas should be flushed with water. Some concentrates and their vapors may be harmful if ingested or inhaled. Consult the various manufacturers' material safety data sheets (MSDSs) for information on specific foam concentrates.

Most Class A and Class B foam concentrates are mildly corrosive. Although foam concentrate is used in small percentages and in diluted solutions, follow proper flushing procedures to prevent damage to equipment.

When discussing environmental impact, the primary concern is the impact of the finished foam after it has been applied to a fire or liquid fuel spill. The biodegradability of a foam is determined by the rate at which environmental bacteria cause it to decompose. This decomposition process results in the consumption of oxygen by the bacteria "eating" the foam. The subsequent reduction in oxygen from surrounding water can cause damage in waterways by killing fish and other water-inhabiting creatures. The less oxygen required to degrade a particular foam the better or the more environmentally friendly the foam is when it enters a body of water.

The environmental impact of foam concentrates vary. Each foam concentrate manufacturer can provide information on its specific products. In the United States, Class A foams should be approved by the USDA Forest Service for environmental suitability. The chemical properties of Class B foams and their environmental impact vary depending on the type of concentrate and the manufacturer. Generally, protein-based foams are safer for the environment. Consult the various manufacturers' data sheets for environmental impact information.

SKILL SHEET 13-1 PLACING A FOAM LINE IN SERVICE

In-Line Eductor

Step 1: Select the proper foam concentrate for the burning fuel involved.

Step 2: Place the foam concentrate at the eductor.

Step 3: Open enough buckets of foam concentrate to handle the task.

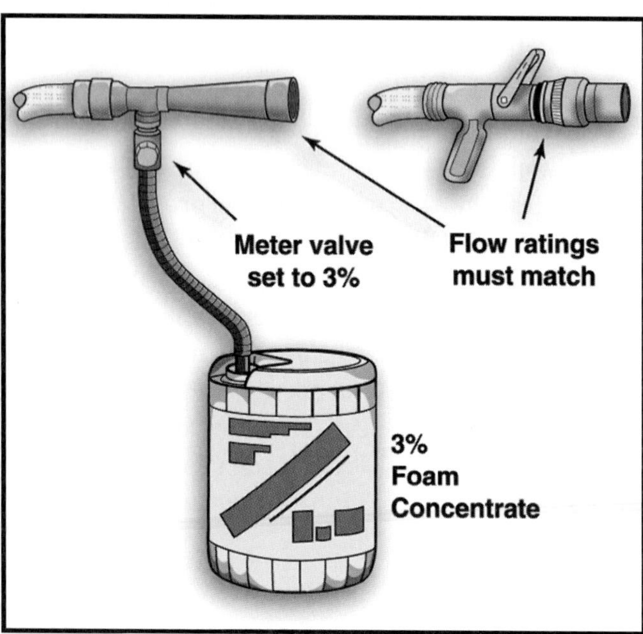

Step 4: Check the eductor and nozzle for hydraulic compatibility (rated for the same flow).

Step 5: Adjust the eductor metering valve to the same percentage rating as that listed on the foam concentrate container.

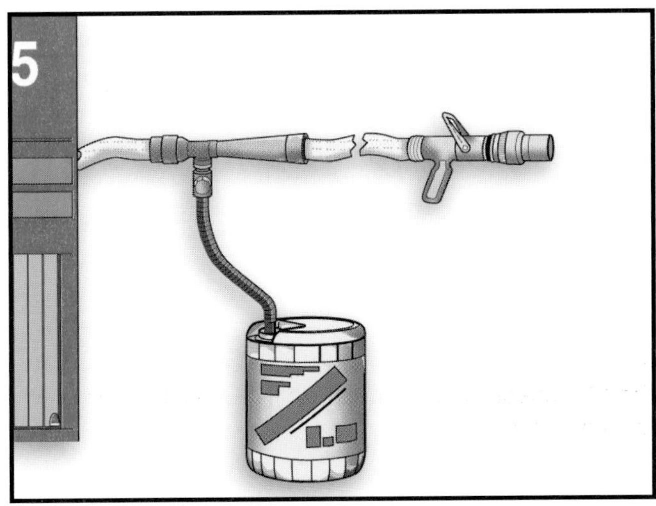

Step 6: Attach the eductor to a hose capable of efficiently flowing the rated capacity of the eductor and the nozzle.

NOTE: If the eductor is attached directly to a pump discharge outlet, make sure that the ball valve gates are completely open. In addition, avoid connections to discharge elbows. This is important because any condition that causes water turbulence will adversely affect the operation of the eductor.

Step 7: Attach the attack hoseline and desired nozzle to the discharge end of the eductor. Avoid kinks in the hose.

NOTE: The length of the hose should not exceed the manufacturer's recommendations.

Step 8: Place the eductor suction hose into the foam concentrate.

Step 9: Open nozzle fully.

Step 10: Increase the water-supply pressure to that required for the eductor. Be sure to consult the manufacturer's recommendations for the specific eductor.

NOTE: Foam should now be flowing.

Chapter 14
Fire Control

Chapter 14
Fire Control

INTRODUCTION

The success or failure of a fire fighting team often depends upon the skill and knowledge of the personnel involved in initial-attack operations. A well-trained team of firefighters with an attack plan and an adequate amount of water, properly applied, can contain most fires in their early stages. Failure to make a well-coordinated attack on a fire can permit or allow the fire to gain headway and burn out of control. Loss of control of the fire can result in increased damage, as well as further endangerment to firefighters and civilians alike (Figure 14.1).

It is important that all personnel be thoroughly trained in the tactics used by their company and in the use of all equipment they will be expected to operate. The quick and efficient use of tools is further enhanced when the companies that frequently work together train together (Figure 14.2).

The need to follow safe procedures and to wear protective clothing during fire control operations cannot be overemphasized. While helmets, gloves, turnout gear, boots, breathing apparatus, and a PASS device protect a firefighter from injury, they also permit the firefighter to apply fire streams from positions close to the fire (Figure 14.3).

Figure 14.2 Companies that frequently work together should also train together.

Figure 14.1 Large fires require firefighters to draw on much of their previous training and experience. *Courtesy of FEMA.*

Figure 14.3 The use of protective equipment allows firefighters to make a close fire attack.

Firefighters should work in pairs or teams whenever they are operating in a hazardous or potentially hazardous location on an emergency scene. Firefighters working alone may overexert themselves or become unable to help themselves if trapped. All team members must watch for a number of potentially hazardous conditions such as the following:

- Imminent building collapse
- Fire that is behind, below, or above the attack team
- Kinks or obstructions to the hoseline
- Holes, weak stairs, or other fall hazards
- Suspended loads on fire-weakened supports (Figure 14.4)
- Hazardous or highly flammable commodities likely to spill
- Backdraft or flashover conditions
- Electrical shock hazards
- Overexertion, confusion, or panic by team members
- Victims

Figure 14.4 Team members must watch for potentially hazardous conditions such as heavy mechanical equipment on the roof.

This chapter looks at some of the common techniques for fighting different types of fires that firefighters face. Hazards peculiar to certain situations are discussed. Finally, basic tactics for commonly encountered fire scenarios are covered.

SUPPRESSING CLASS A (STRUCTURAL) FIRES

[NFPA 1001: 3-3.7; 3-3.7(a); 3-3.7(b); 3-3.9; 3-3.9(a); 3-3.9(b); 4-3.2; 4-3.2(a); 4-3.2(b)]

A fire attack on a structural (Class A) fire must be coordinated to be successful. Firefighters must perform the desired evolutions at the time that the fire officer wants them performed. Depending on the conditions at the fire scene, the fire officer may choose to perform immediate rescue or to protect exposures rather than attack the fire.

Coordination between crews performing different functions is crucial. For example, ventilating a fire before attack lines are in place may result in the unwanted spread of fire due to the increase in air movement through the structure. When properly performed, the ventilation effort substantially aids the entry and attack of hoseline teams. When the attack is coordinated with ventilation, visibility should improve, and entry can be made for rescue, assessment of fire conditions, and suppression.

Teams advancing hoselines should carry equipment that may be needed to force entry/exit or perform other tasks in addition to operating a fire stream (Figure 14.5). This equipment should include at least a portable light, an axe, and a prying tool of some type. Before entering the fire area, the person at the nozzle should bleed the air from the line by opening the nozzle slightly. Opening the handle slightly while waiting for water to arrive hastens this process. The operation of the nozzle should also be checked by testing the range of stream patterns or by setting a proper pattern for the attack based on the conditions found.

Figure 14.5 The backup firefighter on the hoseline is carrying an axe and a Halligan tool into the structure.

Firefighters should wait at the structure's entrance, staying low and out of the doorway, until the fire officer gives the order to advance (Figure 14.6). Any burning fascia and soffit, boxed cornices, or other doorway overhangs should be extinguished before entry. When the fire area is reached, firefighters can attack the fire. The fire should be approached and attacked from the unburned side to keep it from spreading throughout the structure. Sometimes a fire is found in a mattress and it may seem reasonable to take it outside, but removing a burning mattress from a building to extinguish it is a dangerous technique. Mattresses may smolder and when moved, more air may circulate around the fire causing it to erupt in larger flames. In addition, if the mattress gets stuck in a hallway or doorway when this happens, someone could be trapped.

Figure 14.6 Until the officer gives the order to advance, firefighters should stay low and out of the doorway.

Once a fire has been contained, it may be necessary to relieve the initial-attack crew. Breathing apparatus must still be worn during mop-up and overhaul phases of the operation due to the presence of fire gases (Figure 14.7). Special attention should be directed toward walls, partitions, or overhead loads that may be dislodged by fire fighting activities. Valuables found should be taken immediately to a supervisor.

Depending on the size of the fire, the type of nozzle being used, ventilation conditions, and other factors, firefighters may choose to use a direct, indirect, or combination method of attacking the fire. Stream selection and hoseline selection are also made when fire attack is conducted.

Figure 14.7 Firefighters must wear SCBA during overhaul operations.

Stream Selection

When adequate ventilation openings ahead of the nozzles can be provided, it may be possible to use a narrow fog pattern (Figure 14.8). This gives the smoke, heat, and steam a place to go without them rolling back over the nozzle and causing injury to the firefighters. It also helps to maintain the normal thermal layering (the movement of hot gases toward the ceiling) in the area.

If ventilation holes cannot be made large enough for effective ventilation or if ventilation is delayed, then it is important to keep the nozzle on straight stream. By using a straight stream at the base of the fire, the fire may be controlled without upsetting the thermal layering (Figure 14.9). In an unventilated setting, a straight stream does not upset the thermal layering as much as a fog stream does because a straight stream does not push as much air in front of it as a fog stream does. Some disturbance of the thermal layering may occur when straight or solid streams are used because of steam production; however, the disturbance will not be as severe as that created by a fog stream.

Figure 14.8 Adequate ventilation must be provided ahead of fog streams that are used for interior attacks.

Figure 14.9 Using a straight stream to attack the seat of the fire reduces the chance of upsetting the thermal layering in the room.

If a door to the fire area must be opened, position all members of the hose team to one side of the entrance (Figure 14.10). Remember to stay low before entering a fire area in order to allow fire, smoke, and/or heated gases to remain or exit above. Unless a protective stream of water is needed, do not open the nozzle until fire is encountered. Discharging water at smoke decreases visibility and increases water damage. If a fire is localized, direct the stream at the base of the fire. If the area is well involved in fire and sufficiently ventilated, sweeping the ceiling in a side-to-side motion will break up the stream into smaller droplets that rain down on the base of the fire, giving more extinguishing efforts. Rotating the nozzle clockwise will also give this effect. This nozzle action puts water on the fire and in the upper levels of the room.

The safety of the hose crew is imperative. If firefighters are required to back out of an area before full extinguishment, they keep the stream operating until all personnel are in a safe area. There are exceptions to this procedure such as an imminent building collapse where everyone must exit immediately.

Figure 14.10 If a door to the fire area must be opened, all members of the hose team should be positioned to one side of the entrance.

WARNING

Use sirens/air horns to notify firefighters inside a structure to get out immediately.

Hoseline Selection

Water application is only successful if the amount of water applied is sufficient to cool the fuels that are burning. The use of a booster line may not only delay extinguishment but may be of insufficient volume to protect firefighters from advancing flame fronts. Booster lines should only be used for small, exterior fires such as small brush fires or those in Dumpster® trash containers. In large fires or well-involved building fires, 1½-inch (38 mm) hoselines will be insufficient to safely and effectively make an attack.

Hoseline selection should be dependent upon fire conditions and other factors such as the following:

- Fire load and material involved
- Volume of water needed for extinguishment
- Reach needed
- Number of persons available to handle hoseline
- Mobility requirements
- Tactical requirements
- Speed of deployment
- Potential fire spread

Table 14.1 gives a simple analysis of hose stream characteristics and is not meant to replace the judgment of fire personnel in selecting hoselines.

Direct Attack

The most efficient use of water on free-burning fires is made by a direct attack on the base of the fire with a solid stream or straight stream. The water should be applied in short bursts directly on the burning fuels until the fire "darkens down" (Figure 14.11).

Water should not be applied for too long a time otherwise thermal layering will be upset; the steam produced will begin to condense causing the smoke to drop rapidly to the floor and move sluggishly thereafter.

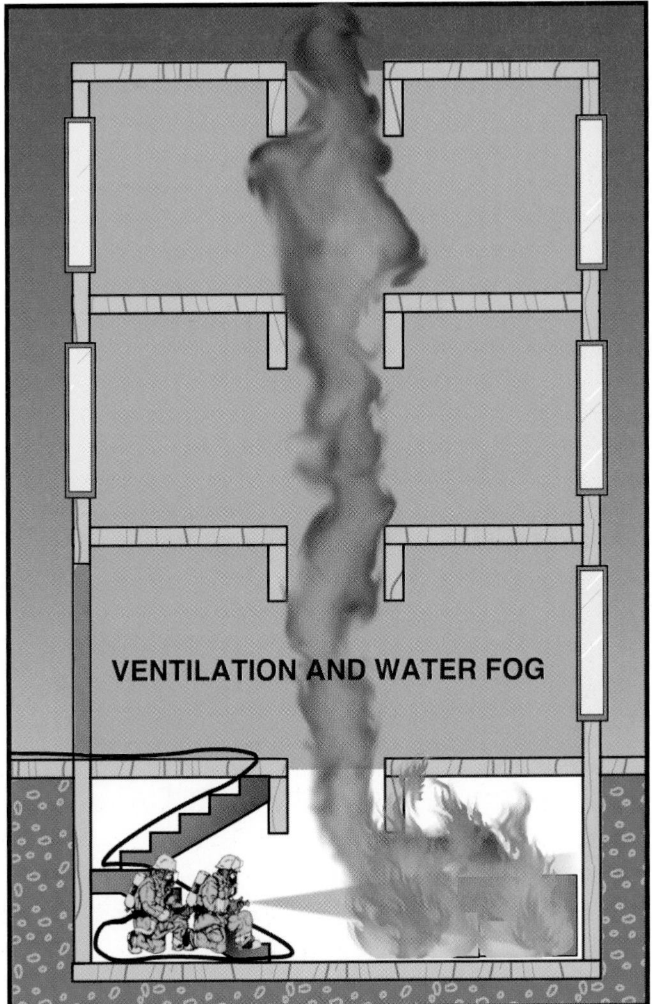

Figure 14.11 A direct attack involves applying water directly to the burning material.

TABLE 14.1
Hose Stream Characteristics

Size in (mm)	GPM (L/min)	Reach (Maximum) ft (m)	No. of Persons on Nozzle	Mobility	Control or Damage	Control of Direction	When used	Estimated Effective Area
1½ inches (38 mm)	40–125 gpm (160 L/min to 500 L/min)	25–50 feet (8 m to 15 m)	1 or 2	Good	Good	Excellent	• Developing fire—still small enough or sufficiently confined to be stopped with relatively limited quantity of water. • For quick attack. • For rapid relocation of streams. • When personnel are limited. • When ratio of fuel load to area is relatively light. • For exposure protection.	One to Three Rooms
1¾ inches (45 mm)	40–175 gpm (160 L/min to 700 L/min)	25–50 feet (8 m to 15 m)	2	Good to Fair	Good	Good		
2 inches (50 mm)	100–250 gpm (400 L/min to 1 000 L/min)	40–70 feet (12 m to 21 m)	2 or 3	Fair	Fair	Good	• When size and intensity of fire are beyond reach, flow, or penetration of 1½-inch (38 mm) line. • When both water and personnel are ample. • When safety of crew dictates. • When larger volumes or greater reach are required for exposure protection.	One Floor or More—Fully Involved
2½ inches (65 mm)	125–350 gpm (500 L/min to 1400 L/min)	50–100 feet (15 m to 30 m)	2 to 4	Fair to Poor	Fair	Good		
Master Stream	350–2,000 gpm (1 400 L/min to 8 000 L/min)	100–200 feet (30 m to 60 m)	1	Poor to None (Aerial master streams can be good)	Poor	Good	• When size and intensity of fire are beyond reach, flow, or penetration of handlines. • When water is ample, but personnel are limited. • When safety of personnel dictates. • When larger volumes or greater reach are required for exposure protection. • When sufficient pumping capability is available. • When massive runoff water can be tolerated. • When interior attack can no longer be maintained.	Large Structures—Fully Involved

Adapted from Joe Batchler, Maryland Fire & Rescue Institute.

Indirect Attack

When firefighters are unable to enter the structure or fire area because of intense fire conditions, an indirect attack can be made from outside the area through a doorway or window (Figure 14.12). This attack is not desirable where victims may yet be trapped or where the spread of fire to uninvolved areas cannot be contained. The fire stream, which could be a solid, straight, or narrow fog pattern, should be directed at the ceiling and played back and forth in the superheated gases at the ceiling level. Directing the stream into the superheated atmosphere near the ceiling results in the production of large quantities of steam, but the stream should be shut down before it disturbs the thermal layering. Once the fire has been darkened down and the space has been ventilated, the hoseline can be advanced to extinguish any remaining hot spots with a direct attack.

Combination Attack

The combination method uses the steam-generating technique of ceiling-level attack combined with a direct attack on materials burning near the floor level. The nozzle may be moved in a *T, Z,* or *O* pattern starting with a solid, straight, or penetrating fog stream directed into the heated gases at the ceiling level and then dropped down to attack the combustibles burning near the floor level (Figure 14.13). The *O* pattern of the combination attack is

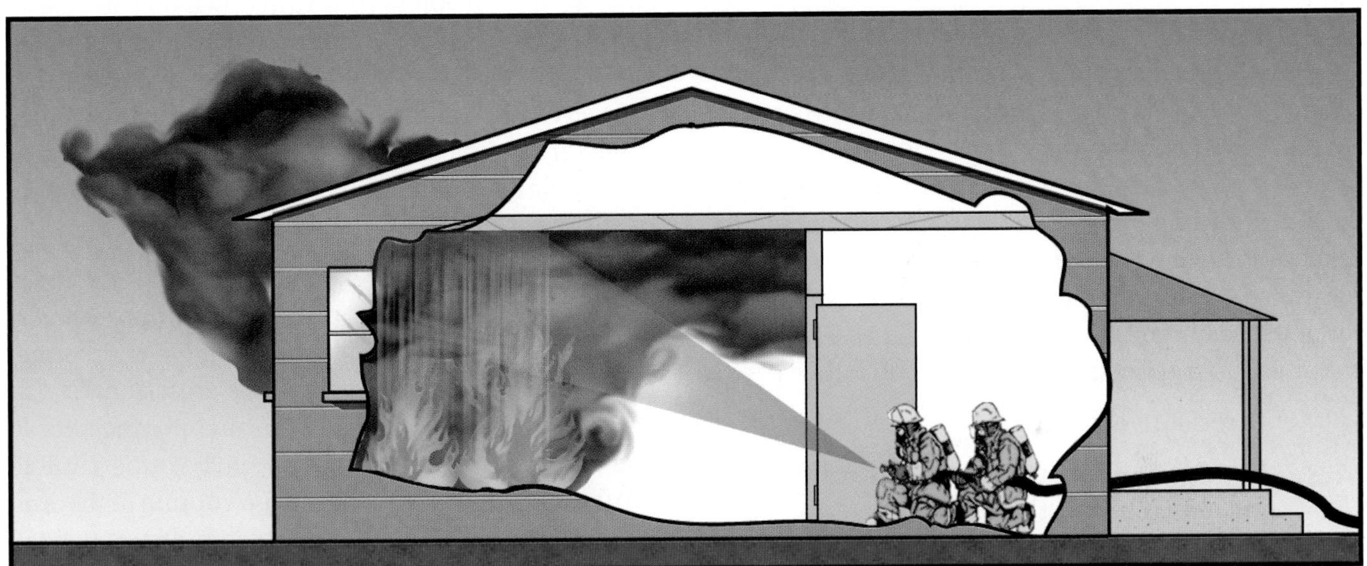

Figure 14.12 An indirect attack involves aiming the stream at the ceiling above the fire and letting the water rain down on the material. Steam conversion by heated air also helps to extinguish the fire.

Figure 14.13 T, Z, or O patterns may be used for a combination attack.

probably the most familiar method of attack. When performing the *O* pattern, the stream should be directed at the ceiling and rotated with the stream edge reaching the ceiling, wall, floor, and opposite wall. Firefighters should keep in mind that applying water to smoke does not extinguish the fire and only causes unnecessary water damage and disturbance of the thermal layering.

Firefighters assisting the person at the nozzle should not crowd behind the nozzle because this makes manipulation of the nozzle difficult. The assisting team members need to advance the hose, as it is needed, to the person at the nozzle.

DEPLOYING MASTER STREAM DEVICES

[NFPA 1001: 3-3.7(a); 3-3.7(b); 3-3.9(a)]

Master stream devices see much less use than other types of nozzles. However, when the need arises for their use, they are generally the only hope left of containing and controlling a large fire. Master streams are deployed in situations where the fire is beyond the control of handlines or there is a need for fire streams in a location that is no longer safe for personnel (Figure 14.14). The three main uses for a master stream are as follows:

- Direct fire attack
- Backup handlines that are already attacking the fire from the exterior
- Exposure protection

The master stream device must be properly positioned to provide an effective stream on a fire because once the line is in operation, it must be shut down to be moved, which can be a time-consuming process. When the master stream is directed into a building, place the device close enough to a window or door so that it can hit the base of the fire. This is particularly important when using a fog nozzle because the fog stream does not have the penetration power of a solid stream.

Figure 14.14 Master streams are used on large fires where mobility is not crucial. *Courtesy of Joseph J. Marino.*

The second aspect of master stream placement is the angle at which the stream enters the structure. The stream should be aimed so that it enters the structure at an upward angle, allowing it to deflect off the ceiling or other overhead objects (Figure 14.15). This makes the stream break up into smaller droplets that rain down on the base of the fire, providing maximum extinguishing effectiveness. Streams that enter the opening at a perfectly horizontal or even less than horizontal angle are not as effective. Operating the stream at too low of an angle might result in a loss of control of the master stream device and hoseline.

It is also desirable to place the master stream device in a location that provides maximum coverage on the face of a building. This gives personnel the opportunity to change the direction of the stream and aim it through another opening should the need arise. This is particularly important in situations where there is a large volume of fire and a limited number of master stream devices.

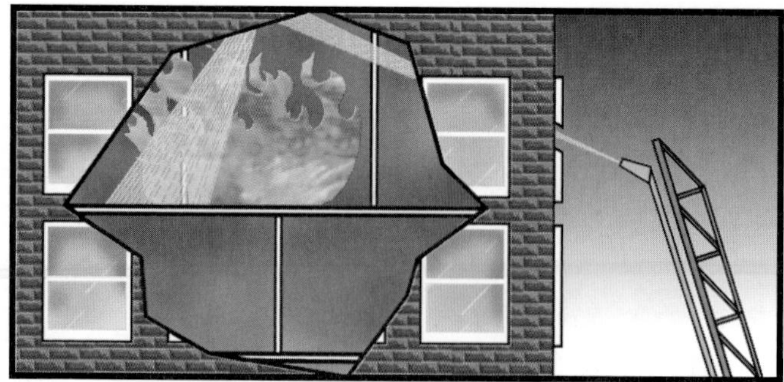

Figure 14.15 Deflect the stream off the ceiling.

Supplying the Master Stream

Master stream devices operate at high flow rates, which in turn means large amounts of friction loss. Typically, a master stream device is expected to flow a minimum of 350 gpm (1 400 L/min). Therefore, it is not practical to supply the master stream appliance with anything less than two 2½-inch (65 mm) hoselines (Figure 14.16). Larger flows will require a third 2½-inch (65 mm) or larger diameter hoseline. Some master stream devices are equipped to handle one large diameter (4-inch [100 mm] or larger) supply line. When possible, it is desirable to supply the device with a maximum of 100 feet (30 m) of hose in order to reduce the amount of pressure lost due to friction.

Hose for the master stream device is generally taken from the main supply in the bed of the engine that supplies the device. If the bed is not set up to provide the deployment of two lines at one time, it will be necessary to pull one line to the device, break the connection, and then pull additional lines as required. In some cases, the portable master stream appliance will be preconnected. This allows for faster deployment of the stream.

Figure 14.16 A typical master stream device with two intake connections.

Staffing the Master Stream Device

It takes a minimum of two firefighters to deploy the master stream device and supply water to it. It would be desirable to have more people available. Once the stream is in place, it can be operated by one firefighter (Figure 14.17). When water is flowing, one firefighter should be stationed at the master stream device at all times. This allows the firefighter to change the direction of the stream when required and prevent the device from *crawling away* (moving). This movement in master stream devices is caused by the pressure in the hoselines, but it is easily controlled by one firefighter. An exception is when the device is being used in hazardous positions such as close to a fire-weakened structure or liquefied petroleum gas (LPG) storage tanks or other objects. These situations may be too dangerous to have personnel that close, so the master stream device should be securely anchored, the desired stream put in place, and then personnel should back away. If the device starts to move, the pressure should be decreased at the supply source to curb the movement.

Figure 14.17 One person can operate the master stream device.

SUPPRESSING CLASS B FIRES

[NFPA 1001: 4-3.3; 4-3.3(a); 4-3.3(b)]

Class B fires are those that involve flammable and combustible liquids and gases (Figure 14.18). *Flammable liquids* are those that have flash points less than 100°F (38°C); examples are gasoline and acetone. *Combustible liquids* are those that have flash points higher than 100°F (38°C); examples are kerosene and vegetable oil. Flammable and combustible liquids can be further divided into hydrocarbons (those that do not mix with water) and polar solvents (those that mix with water).

Figure 14.18 Aircraft emergencies often involve large Class B fires. *Courtesy of Joel Woods, University of Maryland Fire and Rescue Institute.*

Firefighters must exercise caution when attacking fires involving flammable and combustible liquids (Figure 14.19). The first precaution is to avoid standing in pools of fuel or water runoff containing fuel. Protective clothing can absorb fuel in a "wicking" action, which can lead to contact burns of the skin and flaming clothes if an ignition source is present. Even if the wicking action does not occur, extreme danger exists in the event that the pool of liquid ignites.

Unless the leaking product can be turned off, fires burning around relief valves or piping should not be extinguished (Figure 14.20). Simply try to contain the pooling liquid, if any, until the flow can be stopped. Unburned vapors are usually heavier than air and form pools or pockets of gas in low spots where they may ignite. Firefighters must always control all ignition sources in a leak area. Vehicles, smoking materials, electrical fixtures, and sparks from steel tools can all provide an ignition source sufficient to ignite leaking flammable vapors. An increase in the intensity of sound or fire issuing from a relief valve may indicate that rupture of the vessel is imminent. Firefighters should not assume that relief valves are sufficient to safely relieve excess pressures under severe fire conditions. Firefighters have been killed by the rupture of both large and small flammable liquid vessels that have been subjected to flame impingement.

Figure 14.19 Firefighters must exercise great care when attacking large Class B fires.

Figure 14.20 A coordinated effort is required to effectively attack pressurized fuel fires.

In vessels containing flammable liquids, the sudden release and consequent vaporization of the liquids can result in a rupture of the tank — a boiling liquid expanding vapor explosion (BLEVE). A BLEVE results in the explosive release of vessel pressure, pieces of tank, and a characteristic fireball with radiant heat. BLEVEs most commonly occur when flames contact a tank shell above the liquid level or when insufficient water is applied to keep a tank shell cool. When attacking these fires, water should be applied to the upper portions of the tank, preferably from unattended master stream devices (Figure 14.21).

The use of foam is the preferred method to control flammable liquid fires (see Chapter 13, Fire Streams). Water can be used in several forms (cooling agent, mechanical tool, substitute medium, and protective cover) to control Class B fires. Accidents involving vehicles transporting flammable fuels and gas utilities also require Class B fire fighting techniques.

Figure 14.21 Unattended master stream devices should be used to cool tanks.

Using Water to Control Class B Fires

Experience has shown that water in various ways is effective in extinguishing or controlling many Class B fires. Control of these fires can be accomplished safely if proper techniques are used. These techniques require a basic understanding of Class B fuel properties and the effects water has on them. The important thing for firefighters to remember is that hydrocarbon liquids (gasoline, kerosene, and other petroleum products) do not mix with water, and polar solvents (alcohols, lacquers, etc.) mix with water. These facts regarding hydrocarbons and polar solvents affect how fires in each may be extinguished.

COOLING AGENT

Water can be used as a cooling agent to extinguish Class B fires and to protect exposures. Water without foam additives is not particularly effective on lighter petroleum distillates (such as gasoline or kerosene) or alcohols. However, by applying water in droplet form in sufficient quantities to absorb the heat produced, fires in the heavier oils (such as raw crude) can be extinguished.

Water will be most useful as a cooling agent for protecting exposures. To be effective, water streams need to be applied so that they form a protective water film on the exposed surfaces. This applies to materials that might weaken or collapse such as metal tanks or support beams. Water applied to burning storage tanks should be directed above the level of the contained liquid to achieve the maximum efficient use of the water.

MECHANICAL TOOL

Water from hoselines can be used to move Class B fuels (burning or not) to areas where they can safely burn or where ignition sources are more easily controlled. Fuels must never be flushed down drains or sewers. Firefighters should use appropriate fog patterns for protection from radiant heat and to prevent "plunging" the stream into the liquid. Plunging a stream into burning flammable liquids causes increased production of flammable vapors and greatly increases fire intensity. The stream should be slowly played from side to side and the fuel or fire "swept" to the desired location. Care must be taken to keep the leading edge of the fog pattern in contact with the fuel surface or else the fire may run underneath the stream and flashback around the attack crew. When small leaks occur, a solid stream may be applied directly to the opening to keep the escaping liquid back. The pressure of the stream must exceed that of the leaking material in order for this procedure to work properly. Care must be taken to not overflow the container.

Through the use of fog streams, water may also be used to dissipate flammable vapors. Fog streams aid in dilution and dispersion, and these streams control, to a small degree, the movement of the vapors to a desired location.

SUBSTITUTE MEDIUM

Water can be used to displace fuel from pipes or tanks that are leaking. Fires fed by leaks may be extinguished by pumping water back into the leaking pipe or by filling the tank with water to a point above the level of the leak. This displacement floats the volatile product on top of the water as long as the water application rate equals the leak rate. Due to the large water-to-product ratio required, water is seldom used to dilute flammable liquids for fire control. However, this technique may be useful for small fires when the runoff can be contained.

PROTECTIVE COVER

Hoselines can be used as a protective cover for teams advancing to shut off liquid or gaseous fuel valves (Figure 14.22). Coordination and slow, deliberate movements provide relative safety from

Figure 14.22 Fog streams provide protective cover for firefighters when they advance on a fire.

flames and heat. While one hoseline can be used as a protective cover, two lines with a backup line are preferred for fire control and safety.

WARNING

Using hoselines as a protective cover must be practiced through training before being attempted during an emergency.

When containers or tanks of flammable liquids or gases are exposed to flame impingement, solid streams should be applied from their maximum effective reach until relief valves close. This can best be achieved by lobbing a stream along the top of a tank so that water runs down both sides. This film of water cools the vapor space of the tank. Steel supports under tanks should also be cooled to prevent their collapse.

Hose streams can then be advanced under progressively widened protective fog patterns to make temporary repairs or shut off the fuel source. A backup line supplied by a separate pump and water source should be provided to protect firefighters in the event other lines fail or additional tank cooling is needed. Approaches to storage vessels exposed to fire should be made at right angles to the tanks, never from the ends of the vessels.

WARNING

Never approach a horizontal vessel exposed to fire from the ends because of the danger of ruptures; vessels frequently split and become projectiles.

Bulk Transport Vehicle Fires

Pre-incident plans for transportation emergencies should be followed to reduce life loss, property damage, and environmental pollution. The techniques of extinguishment for fires in vehicles transporting flammable fuels are similar in many ways to fires in flammable fuel storage facilities. The difficulties posed by the amount of fuel available to burn, the possibility of vessel failure, and danger to exposures are similar with both. The major differences include the following:

- Increased life safety risks to firefighters from traffic
- Increased life safety risk to passing motorists
- Reduced water supply
- Difficulty in determining the products involved
- Difficulty in containing spills and runoff
- Weakened or damaged tanks and piping caused by the force of collisions
- Instability of vehicles
- Additional concerns posed by the location of the incident (residential neighborhood, schools, etc.)

While a serious accident may bring traffic to a halt, many incidents are handled with traffic passing the scene at near-normal speeds. A lane of traffic in addition to the incident lane should be closed during initial emergency operations (Figure 14.23). The use of open-flame flares should be avoided because of the possibility of igniting leaking fuels. When traffic is passing by closely, firefighters should be careful not to allow tool handles to extend into the traffic lane where they may be struck. When law enforcement personnel are unavailable, a firefighter should be assigned the role of traffic-control officer.

Fire apparatus should be positioned to take advantage of topography (land surface

Figure 14.23 Close at least one additional lane of traffic in each direction.

configuration) and weather conditions (uphill and upwind) and to protect firefighters from traffic. Firefighters should exit the apparatus and work as much as possible from the curbside away from traffic. In addition, firefighters should avoid working where the apparatus could be pushed into them if it were struck by another vehicle.

The techniques of approaching and controlling leaks or fires involving vehicles are the same as for storage vessels. Additionally, firefighters should be aware of the failure of vehicle tires that may cause a flammable load to shift suddenly. Crews need to know the status and limitations of their water supply. It may also be necessary to protect trapped victims with hoselines until they can be rescued.

Firefighters must determine the exact nature of cargos as soon as possible from bills of lading, manifests, placards, or the drivers of the transport vehicles (Figure 14.24). Unfortunately, cases exist where these items could not be found, placards were either wrong or obscured, and drivers were unable to identify their cargos. In these instances, contact should be made with the shippers or manufacturers responsible for the vehicles.

Figure 14.24 Firefighters can determine from the vehicle's placard what type of cargo it is carrying.

Control of Gas Utilities

A working knowledge of the hazards and correct procedures for handling incidents related to natural gas and liquefied petroleum gas (LPG) is important to every firefighter. Many houses, mobile homes, and businesses use natural gas or LPG for cooking, heating, or industrial processes. The firefighter familiar with gas distribution and usage will be able to prevent or reduce damage caused by incidents involving these gases.

Natural gas is mostly methane with small quantities of ethane, propane, butane, and pentane added. The gas is lighter than air so it tends to rise and diffuse in the open. Natural gas is nontoxic, but it is classified as an asphyxiant because it may displace normal breathing air and lead to asphyxiation. The gas has no odor of its own, but a very distinctive odor (mercaptan) is added by the utility. It is distributed from gas wells to its point of usage by a nationwide network of surface and subsurface pipes (Figure 14.25). The pressure in these pipes ranges from ¼ to 1,000 psi (2 kPa to 7 000 kPa). However, the pressure is usually below 50 psi (350 kPa) at the local distribution level. Natural gas is explosive in concentrations between 5 and 15 percent. Natural gas may also be compressed, stored, and shipped in cylinders marked as compressed natural gas (CNG). In this compressed state, it is subject to a BLEVE.

The local utility should be contacted when any emergency involving natural gas occurs in its service area. The local utility will provide an emergency response crew equipped with special (nonsparking) tools, maps of the distribution system, and the training and experience needed to help control the flow of gas. The response time of these crews is usually less than an hour, but the time may be extended in rural areas or in times of great demand. Good relations between the fire department and the utility company are encouraged.

Figure 14.25 Firefighters should be familiar with gas distribution stations and equipment in their area.

LPG, or *bottled gas* as it is sometimes known, is a fuel gas stored in a liquid state under pressure. It is used primarily as a fuel gas in campers, mobile homes, agriculture applications, and rural homes. There is an increased use of LPG as a fuel for motor vehicles. The gas is composed mainly of propane with small quantities of butane, ethane, ethylene, propylene, isobutane, or butylene added. LPG has no natural odor of its own, but a very distinctive odor is added. The gas is nontoxic, but it is classified as an asphyxiant because it may displace normal breathing air and lead to asphyxiation.

LPG is about one and one-half times as heavy as air, and it will generally seek the lowest point possible. The gas is explosive in concentrations between 1.5 and 10 percent. The gas is shipped from its distribution point to its point of usage in cylinders and in tanks on cargo trucks. It is stored in cylinders and tanks near its point of usage, and then the tank or cylinder is connected by underground piping and copper tubing to the appliances the gas serves (Figure 14.26). The supply of gas into a structure may be stopped by shutting a valve on the pipe leading to the building (Figure 14.27). Unburned gas may be dissipated by a fog stream of at least 100 gpm (400 L/min). All LPG containers are subject to BLEVEs when exposed to intense heat or open flame. If there are any problems with a cylinder or tank, the company responsible for it should be contacted.

Incidents involving both CNG and LPG distribution systems are often caused by excavation around underground pipes causing a break. When these breaks occur, the utility company should be contacted immediately. Even if the gas has not yet ignited, apparatus should approach from and stage on the upwind side. Firefighters should be prepared in the event of an explosion and any accompanying fire. The firefighter's first concerns should be the evacuation of the area immediately around the break, evacuation of the area downwind, and elimination of ignition sources. The broken main may have damaged service connections near the break, so surrounding buildings should be checked for gas buildup. Firefighters should not attempt to operate main valves because incorrect action may worsen the situation or cause unnecessary loss of service to areas unaffected by the break. If gas is burning, the flame *should not be extinguished.* Exposures can be protected by hose streams if necessary.

Figure 14.26 Many occupancies use stored fuel gases that are stored in cylinders on their property.

Figure 14.27 Compressed fuel gas cylinders have a shutoff valve that is clearly marked for the direction of operation.

WARNING

If gas is burning from a broken pipe, do not extinguish the fire. Provide protection for exposures.

The most common situation firefighters face in gas utility structure fire incidents involves locating the service meter and turning off the gas. The meter is usually located outside buildings and is normally visible from the street; however, it may also be located inside the building (Figure 14.28).

Firefighters involved in turning off the gas at a meter should advance a hoseline on a fog pattern to protect themselves. Even when the fire is extinguished, vapors could build up and reignite with disastrous results. The flow of gas into the building may be stopped by turning the cutoff valve to the closed position, which is at a right angle to the pipe (Figure 14.29). Any action to stop or reduce the flow of gas should be in accordance with fire department standard operating procedures.

Figure 14.28 A natural gas meter.

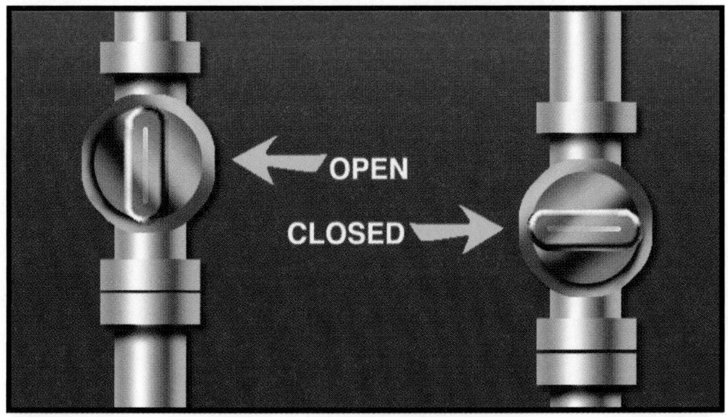

Figure 14.29 This shows the open and closed positions for gas line petcocks.

SUPPRESSING CLASS C FIRES

[NFPA 1001: 3-3.17; 3-3.17(a); 3-3.17(b)]

Fires in electrical equipment (Class C fires) occur quite frequently and, once the equipment is de-energized, they can be handled with relative ease. Other unusual electrical hazards can be found in railroad locomotives, telephone relay switching stations, and electrical substations. Procedures for fighting fires in these occupancies should be established in pre-incident plans.

The primary danger of electrical fires is the failure of emergency personnel to recognize the safety hazard. Although safety is every firefighter's responsibility, it is the responsibility of the fire officer to ensure that appropriate power breakers are opened to control power flow into structures. In certain commercial and/or high-rise buildings, power is necessary to operate elevators and/or air handling equipment; thus the entire building should not be unilaterally de-energized. Similarly, a crew member should be assigned to control the power (disconnect the battery) at vehicle fires and other emergencies (see Controlling Electrical Power section). Once the power has been turned off, these fires may self-extinguish or they will fall into either Class A or Class B fires if they continue to burn.

WARNING

Stop the flow of electricity to the object involved before initiating fire-suppression activities.

When handling fires in delicate electronic or computer equipment, clean extinguishing agents such as carbon dioxide or halon should be used to prevent further damage to equipment (Figure 14.30). Multipurpose dry chemical agents present a considerable cleanup problem in addition to being chemically active with some electrical components. The use of water on energized equipment is discouraged because of the inherent shock hazard. If water must be applied, it should be applied from a distance in the form of a fog stream.

Class C fire suppression techniques are also needed for fires involving transmission lines and equipment, underground lines, and commercial

Figure 14.30 Halon is often used on Class C fires.

high-voltage installations. In addition, the responsibilities for controlling electrical power, the dangers of electric shock, and the guidelines for electrical emergencies should be known by every firefighter.

Transmission Lines and Equipment

A common electrical emergency firefighters face involves fires in bulk electrical transmission lines and equipment (Figure 14.31). When fires occur as a result of transmission lines breaking, an area equal to a span between poles should be cleared on either side of the break. To reduce the risk to life and property in these incidents, consultation and cooperation with power company officials is vital. For maximum safety on the fireground, live wires should not be cut except by experienced power company personnel with proper equipment.

Figure 14.31 Fires at electrical substations are common electrical emergencies firefighters face.

Fires in transformers can present a serious health and environmental risk because of coolant liquids that contain PCBs (polychlorinated biphenyls). These liquids are flammable because of their oil base, and they are carcinogenic (cancer causing). Transformers at ground level should be extinguished carefully with a dry chemical extinguisher. Transformers aboveground should be permitted to burn until qualified utility personnel can extinguish the fire with a dry chemical extinguisher from an aerial device. Placing a ground ladder against a pole holding a burning transformer places personnel under risk from both the power source and from the liquid. Applying hose streams to these fires can result in spreading the material onto the ground.

WARNING

Consider all power line wires live until confirmed otherwise by the power company.

Underground Lines

Underground transmission systems consist of cableways and vaults beneath the surface. The most frequent hazards that these systems present are explosions that may blow utility covers a considerable distance (Figure 14.32). A spark from fuses blowing or short-circuit arcing can ignite an accumulation of gases, causing an explosion. This is a danger to the public as well as firefighters. If these situations are suspected, firefighters must keep the public clear of the area and make sure that apparatus is not parked over a utility cover.

Figure 14.32 Apparatus should never be parked over a utility cover.

Firefighters should not enter a utility vault except to attempt a rescue. Fire fighting can be accomplished from outside. Firefighters should simply discharge carbon dioxide or dry chemical into the utility vault and replace the cover. A wet blanket or salvage cover should be placed over the utility cover to exclude oxygen and assist in extinguishing the fire. Water is not suggested for extinguishment because of the close proximity of electrical equipment. The runoff of water would also create puddles that could become dangerous conductors of electricity.

WARNING

When circumstances dictate that a firefighter must enter a utility vault, it should ONLY be done by personnel properly trained and equipped for confined space entry.

Commercial High-Voltage Installations

Many industries, large buildings, and apartment complexes have electrical equipment that use current in excess of 600 volts. The obvious clue to this condition is *high-voltage* signs on the doors of vaults or fire-resistive rooms housing equipment such as transformers or large electric motors (Figure 14.33). Some transformers use flammable oils as coolants that present a hazard in themselves. Water should not be used in this situation, even in the form of fog, because the hazard of shock is great, and extensive damage may occur to electrical equipment not involved in the fire.

Because of toxic chemicals used in plastic insulation and coolants, smoke becomes an additional hazard. Enter only when rescue operations require it. Wear self-contained breathing apparatus, and use a safety line monitored by someone outside the enclosure. Search with a clenched fist or the back of the hand to prevent reflex actions of grabbing live equipment that may be touched accidentally (Figures 14.34 a and b). If it is believed that toxic materials are involved in a fire, appropriate decontamination procedures should be followed afterward.

Figure 14.33 Exercise caution when entering rooms that contain high-voltage equipment.

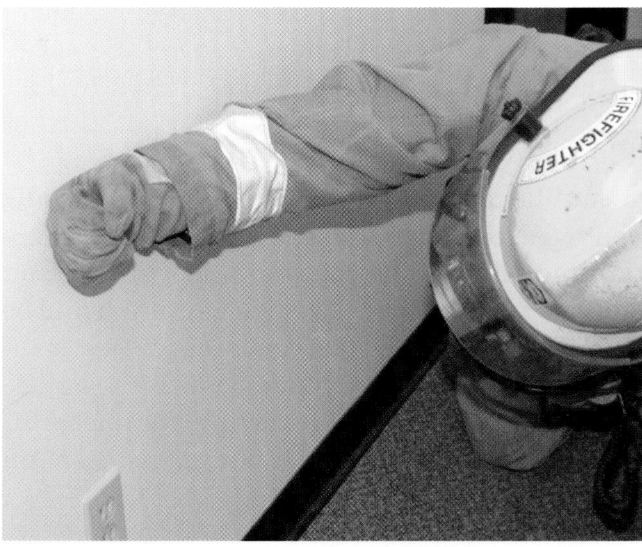

Figure 14.34a Perform search with a clenched fist.

Figure 14.34b Another method used to perform a search is with the back of the hand.

Controlling Electrical Power

From a safety standpoint during structural fire fighting operations, power should remain on as long as possible to provide lighting, run ventilation equipment, or to run special pumps. Firefighters must be able to control the flow of electricity into structures where emergency operations are being performed. Especially when a fire involves only one area, it would be pointless to shut down the entire building. When the building becomes damaged to the point that service is interrupted or an electrical hazard exists, however, power should be turned off by a power company employee if possible. When the fire department must do this, only trained personnel who are aware of the effects should be assigned the task.

It is no longer recommended to pull the electrical meter to turn off the electricity in residential fires. A firefighter should control the power at the panel box by opening the main switch or removing the fuses (Figure 14.35). If further control of the electricity becomes necessary, it should be done only by utility personnel using approved, tested equipment. With some residential and commercial meters, removal does not stop the flow of electricity. Firefighters should be alert for installations with emergency power capabilities such as emergency generators (Figure 14.36). In such cases, removing the meter or turning off the master switch does not turn off the power entirely.

Figure 14.35 The firefighter should control the power at the panel by opening the main switch.

Figure 14.36 Many occupancies have backup electrical generators that become active when the main power supply is interrupted.

Electrical Hazards

In order to avoid injury and to protect electrical equipment, firefighters should be familiar with electrical transmission and its hazards. While high-voltage equipment is usually associated with severe shocks, conventional residential current is sufficiently powerful to deliver fatal shocks (Figure 14.37). In addition to reducing the risk of injury or fatal shock, controlling electrical flow reduces the danger of igniting combustibles or accidentally turning on equipment.

The consequences of electrical shock can include the following:

- Cardiac arrest
- Ventricular fibrillation

Figure 14.37 High-voltage equipment poses severe hazards to firefighters.

- Respiratory arrest
- Involuntary muscle contractions
- Paralysis
- Surface or internal burns
- Damage to joints
- Ultraviolet arc burns to the eyes

Factors most affecting the seriousness of electrical shock include the following:

- Path of electricity through the body
- Degree of skin resistance — wet (low) or dry (high)
- Length of exposure
- Available current — amperage flow
- Available voltage — electromotive force
- Frequency — alternating current (AC) or direct current (DC)

Guidelines For Electrical Emergencies

The following list contains some tips to help deal with electrical emergencies. The list is not totally inclusive but gives principles that should be considered to maintain a safe working environment for personnel.

- Establish a danger zone of one span in either direction for safety when downed power line wires are encountered (Figure 14.38). This is because other wires may have been weakened by a short circuit and may fall at a later time.
- Guard against not only electrical shock and burns but also eye injuries from electrical arcs. Never look directly at arcing electrical lines.
- Treat all wires as energized and high-voltage lines.
- Do not cut any power lines but wait and let trained utility workers do any necessary cutting. This rule is excepted only in the most extreme circumstances and only with personnel who have proper equipment and training.
- Use electrical lockout/tagout devices when working on energized electrical equipment (Figure 14.39). Lockout devices ensure that electrical power will not inadvertently restore after being turned off. Such lockouts are used with padlocks once electricity has been shut off at the control box. The control box should also be tagged (tagout) to indicate that it is out of service. In accordance with all appropriate regulations and fire department guidelines, lockouts and/or

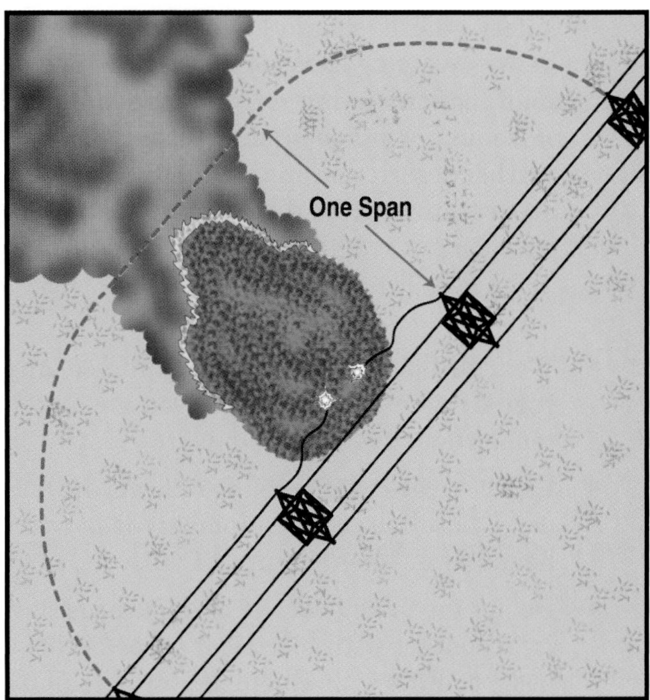

Figure 14.38 A one-span danger zone should be established around a downed wire.

Figure 14.39 Use lockout/tagout devices on electrical equipment when it is de-energized. The use of these devices reduces the possibility of someone turning the power on while equipment is being repaired. *Courtesy of Jim Hanson.*

Figure 14.40 Exercise caution when using ladders in the vicinity of power lines.

Figure 14.41 If it is necessary to leave an apparatus that may be charged, the firefighter should jump clear of the apparatus without touching either the apparatus or the ground at the same time.

safety guards should be used during incidents involving elevator rescue, compactor malfunctions, industrial process equipment mishaps, or other similar situations. For further description and information on lockout/tagout procedures, refer to the department's SOPs and OSHA standard 1910.147, *The control of hazardous energy (lockout / tagout)*.

- Wear full protective clothing and use only regularly tested and approved insulated tools when an electrical hazard exists.
- Exercise care when raising or lowering ladders, hoselines, or equipment near overhead power lines (Figure 14.40).
- Do not touch any vehicle or apparatus that is in contact with electrical wires because body contact will complete the circuit to ground resulting in electrical shock. If it is necessary to leave an apparatus that may be charged, the firefighter should jump clear of the apparatus, touching neither the apparatus nor the ground at the same time (Figure 14.41).
- Consider all downed electrical wires equally dangerous even when one is arcing and others are not.
- Do not use solid and straight streams around energized electrical equipment. Fog patterns are recommended with at least 100 psi (700 kPa) nozzle pressure. Appliances should not be used close to the area.
- Give special considerations for fences. Once an energized electrical line contacts a fence or metal guardrail, the entire length, as long as it is continuous, becomes charged. The length of this fence presents a difficult hazard to protect.
- Proceed carefully in an area where wires are down, and heed any tingling sensation felt in the feet. Because of the carbon in boots, a slight charge is transmitted indicating that the ground may be charged.
- Avoid a ground gradient hazard by maintaining an extra large safety distance between downed electrical wires and operating positions. *Ground gradient* is the tendency of an energized electrical conductor to pass its current along the path of least resistance (from highest to lowest potential) to ground. It is common for downed conductors to discharge their electrical current through surface objects several feet (meters) from their point of contact with the ground. The higher the voltage, the greater the possible travel distance. Dragging a hoseline, ladder, pike pole, or other object in the area of a downed wire, also risks entering a ground gradient condition. If a difference in electrical potential exists between a firefighter's feet and the object contacting the ground, current will pass through the firefighter and return to the ground through the dragged object.

SUPPRESSING CLASS D FIRES

Combustible metals present the dual problem of burning at extremely high temperatures (Class D fires) and being reactive to water. Water is only effective when it can be applied in large enough quantities to cool the metal below its ignition temperature. The usual method of control is to protect exposures and permit the metal to burn itself out. Special extinguishing agents may be manually shoveled or sprayed from special extinguishers in quantities large enough to completely cover the burning metal. Directing hose streams at burning metal can result in the violent decomposition of the water and subsequent release of flammable hydrogen gas. Small metal chips or metal dust are more reactive to water than are ingots or finished products.

Combustible metal fires can be recognized by a characteristic brilliant white light that is given off until an ash layer covers the burning material. Once this layer has formed, it may appear that the fire is out. Firefighters should not assume that these fires are extinguished just because flames are not visible. It may be an extended period of time before the area or substance cools to safe levels. Combustible metal fires are very hot — greater than 2,000°F (1 093°C) — even if they appear suppressed.

FIRE COMPANY TACTICS

[NFPA 1001: 3-3.6; 3-3.6(a); 3-3.6(b); 3-3.9; 3-3.9(a); 3-3.9(b); 3-3.18; 3-3.18(a); 3-3.18(b); 4-3.2; 4-3.2(a); 4-3.2(b)]

As stated earlier, the need to save lives in danger is always the first consideration. Once all possible victims have been rescued, attention is turned to stabilizing the incident. Last, firefighters should make all possible efforts to minimize damage to property. This can be accomplished through proper fire fighting tactics and good loss control techniques.

Fires in Structures

Fires in structures are the most common fire scenarios firefighters face. The rescue, exposure protection, ventilation, confinement, and extinguishment functions must be performed in a coordinated way for the operation to be successful. The following information highlights a typical response to a fire in a residential structure and details the responsibilities of each unit involved.

FIRST-DUE ENGINE COMPANY

The first engine company to arrive at the scene usually initiates incident command and the fire attack, taking into consideration the present and expected behavior of the fire (Figure 14.42). Depending on conditions, the first engine may also need to perform search and rescue functions or exposure protection. The first engine company makes a radio report to the dispatch/telecommunications center regarding the exact location, exposures, conditions found at the incident, and, if necessary, additional resources needed.

If smoke or fire is visible as the engine company approaches the scene, firefighters should stop and lay a supply line from a hydrant or from the end of the driveway into the scene (Figure 14.43). If a hydrant valve is used on the supply line, the line may be charged as soon as a hose clamp is applied at the scene.

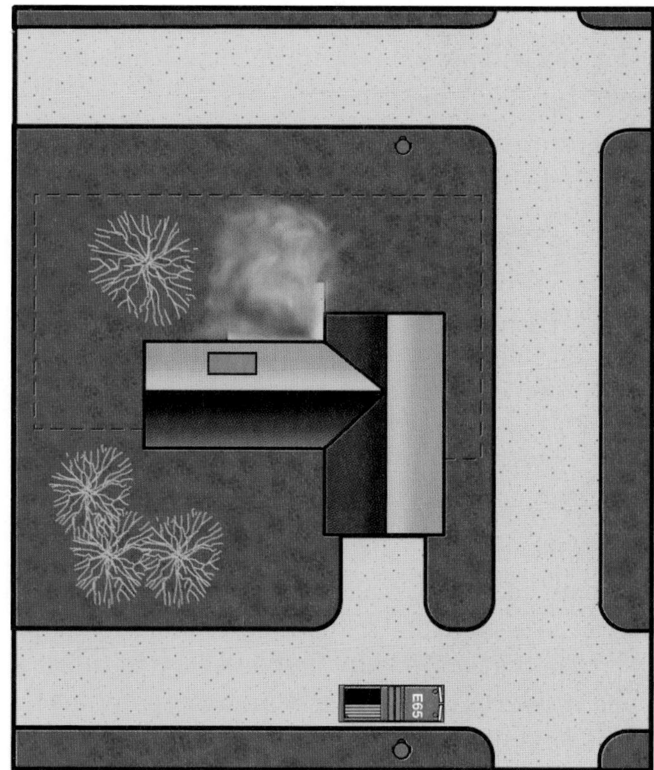

Figure 14.42 More times than not, an engine company will be the first to arrive on scene.

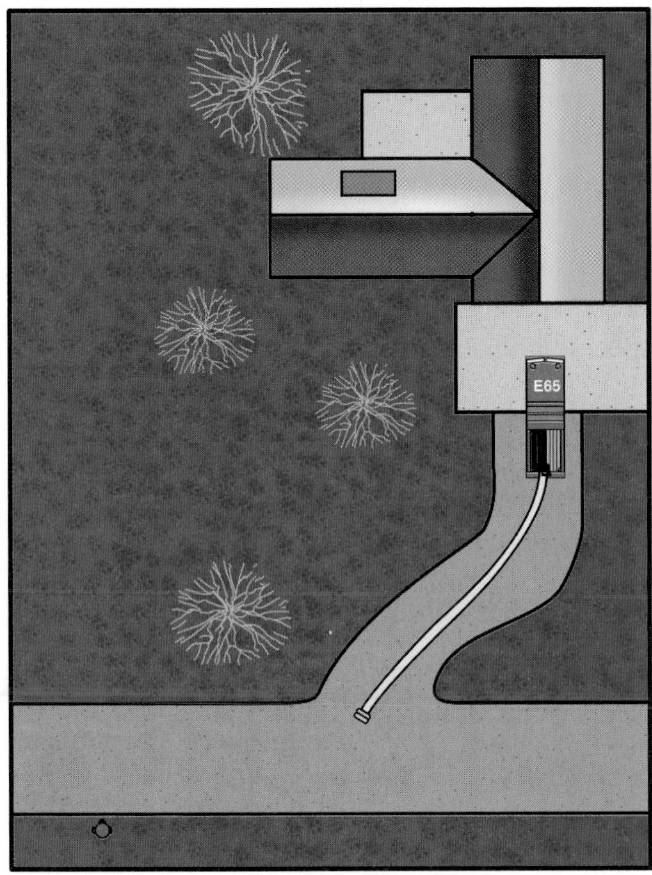

Figure 14.43 If the fire building is at the end of a long driveway, the first engine company should lay a supply line up the driveway.

Once the location of the fire is known, the first company will position the initial attack hoseline to cover the following priorities:

- Intervene between trapped occupants and the fire or protect rescuers.
- Protect primary means of egress.
- Protect interior exposures (other rooms).
- Protect exterior exposures (other buildings).
- Initiate extinguishment from the unburned side.
- Operate master streams.

SECOND-DUE ENGINE COMPANY

Unless otherwise assigned, the second engine company must first make sure that adequate water supply is established to the fireground. Depending on the situation, it may be necessary to finish a hose lay started by the first engine company, lay an additional line, or connect to a hydrant to support the original or additional lines that have been laid (Figures 14.44 a and b). The need to pump the lines from the hydrant depends on local factors that include the size of the hose being used, the distance from the hydrant to the scene, and the residual water pressure available.

Figure 14. 44a Firefighters must understand the operation of hydrant valves used by their department.

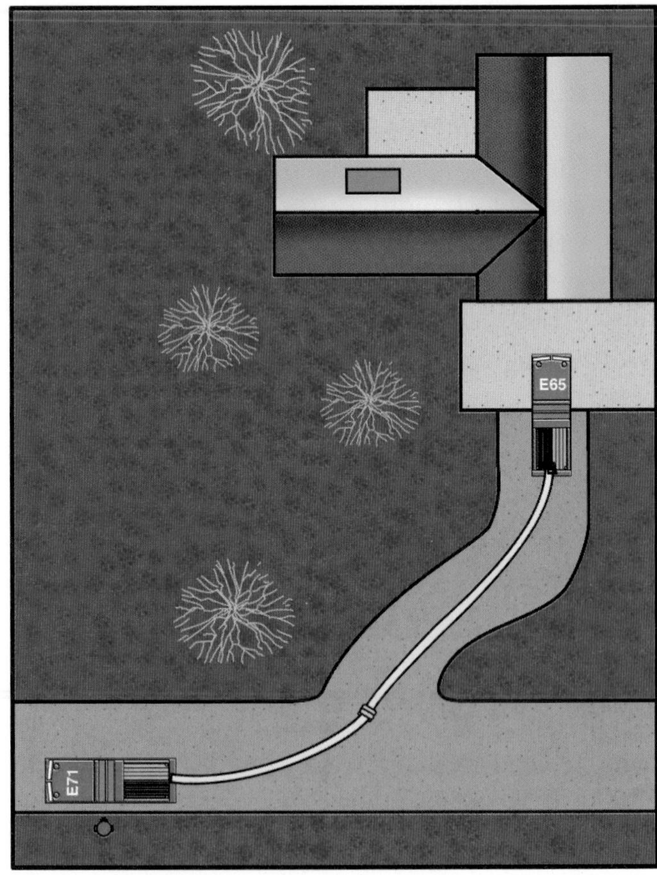

Figure 14. 44b A second engine lays a line from the end of the driveway to the water supply source.

Once the water supply has been established, the second company proceeds according to the following priorities:

- Back up the initial attack line.
- Protect secondary means of egress.
- Prevent fire extension (confinement).
- Protect the most severe exposure.
- Assist in extinguishment.
- Assist with truck/rescue company operations.

TRUCK/RESCUE COMPANY

The truck company most often arrives with or after the first engine company and, among other things, is responsible (in no particular order) for the following:

- Make entry.
- Conduct search and rescue.
- Conduct salvage (property conservation).
- Place ladders.
- Perform ventilation.
- Check for fire extension.
- Set up lighting equipment.
- Control utilities.
- Place elevated fire streams.
- Conduct overhaul.

These functions may be performed by engine or squad personnel when truck companies are not available. Initially, the truck company observes the outside of the building for signs of victims needing immediate rescue. The truck company can then begin to search for victims either using interior or exterior routes of entry. The truck company will also ventilate as it advances (unless positive-pressure ventilation is being done), staying on the alert for signs of fire spread above the fire floor. The truck company must always be prepared to enter the building upon arrival. This means that SCBA should be donned en route to the incident if safely possible (Figure 14.45). A team equipped with tools enters to begin the interior search. Simultaneously, another team raises the necessary ladders to enter or ventilate the building from the outside. For instance, a ground ladder may be used to effect a second floor window rescue while the elevating device is raised to the roof for the purpose of ventilation.

Figure 14.45 A firefighter donning his SCBA en route to the scene.

Depending upon the situation, both interior and exterior teams should search first in areas that are most likely to be inhabited. As previously described, searches must be conducted systematically in accordance with department SOPs to avoid missing areas.

In addition to search and ventilation procedures, it is often necessary for the truck company to assist engine companies in making the fire attack. This can be done by the placement of ground ladders as requested by the fire officer. The truck company can frequently be used to knock down large fires above the first floor with an aerial master stream device (Figure 14.46) This action must be coordinated with other operations to both prevent injury to personnel inside and to avoid spreading the fire to uninvolved parts of the building. Interior attack teams have frequently been

Figure 14.46 A blitz attack may be accomplished with an elevated master stream device.

injured or forced to retreat by poorly directed outside master streams. Elevating devices, ladders, platforms, and ladder towers can also be used as substitute standpipes from which engine company fire fighting teams can advance hoselines.

RAPID INTERVENTION CREW (RIC)

Fire department members and the incident commander continually evaluate the incident scene to determine any possible safety concerns that may develop. However, it is not always possible to predict when an emergency situation (trapped firefighters) or equipment failure (SCBA failure) will occur. When these situations arise, the incident commander must be prepared to immediately deploy rescue personnel, or rapid intervention crews (RICs), to assist other firefighters. With these situations in mind, NFPA 1500, *Standard on Fire Department Occupational Safety and Health Program,* requires fire departments to *". . . provide personnel for the rescue of members operating at emergency incidents . . . ."*

The exact number of rapid intervention crews is determined by the incident commander during the initial phases of the incident. Then crews are added as necessary as the incident escalates or the number of operations increases. This allows flexibility in RIC composition based on the type of incident and numbers of personnel on scene. The rapid intervention crew consists of at least two members wearing appropriate personal protective clothing, equipment, and any special rescue tools and equipment that may be necessary to effect rescue of other emergency personnel. The RIC may be assigned other emergency scene duties; however, they must be prepared to deploy immediately if needed.

CHIEF OFFICER/INCIDENT COMMANDER

Upon arriving at the scene, the chief officer may choose to assume command from the person who is in command at that time and become the new incident commander. If the chief officer takes command, he coordinates the overall activities at the scene (Figure 14.47). The situation must be constantly evaluated to be sure that the resources on the scene are being properly allocated. The need for additional resources should be constantly considered. If additional companies are called for, the chief officer should assign them according to the action plan as they arrive. The chief officer may also have to coordinate between other entities such as mutual aid units, EMS personnel, utility crews, and members of the media. Depending on the number of fire personnel at hand and the scope of the incident, the chief officer may assign personnel to act as liaisons to these other organizations operating on the fire scene.

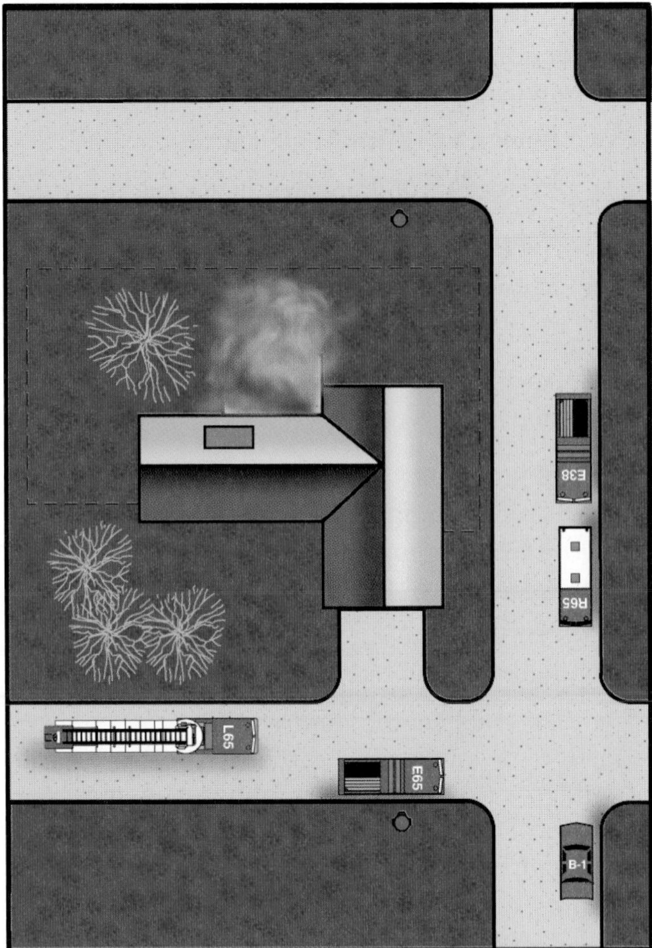

Figure 14.47 The chief officer should be in a position to view the scene but not close enough to get in the way of operations.

Fires in Upper Levels of Structures

Fires in upper levels of structures, such as high-rise buildings, can be very challenging to fire fighting personnel. In order for these fires to be successfully handled, they usually require a response greater than the two engines and one truck used in the single-family dwelling scenario. This is because extensive personnel may be required to first transport fire fighting equipment to the level of the building where it is needed. This can tire personnel before they get a chance to begin the actual attack on the fire. In most cases, equipment must be hand carried up the stairs. Elevators should not be relied upon to provide transportation to the fire level because they can be unreliable and can accidentally deposit firefighters directly into the fire area. If the building has low-rise elevators that end before the level of the fire, they may be used to their top level, and then the equipment can be hand carried the rest of the way.

Typically, the fire attack will be initiated from the floor below the fire floor (Figure 14.48). Firefighters may wish to look at the floor below to get a general idea of the layout of the fire floor. Standpipe pack hoselines should be connected to the standpipe connections on the floor below the fire floor. Extra hose should be flaked up the stairs above the fire floor so that it will feed more easily into the fire floor as the line is advanced. In addition to attacking the fire directly, crews should be checking floors above the main fire floor for fire extension and any victims who may have been unable to escape. The staging of extra equipment and personnel is usually established two floors below the fire floor.

Caution must be exercised around the outside perimeter of a high-rise building that is on fire. Falling glass and other debris from many stories above the street can severely damage equipment, cut hoselines, and injure or kill firefighters. The area should be cordoned off and safe methods of entry to the building identified. Conditions dictate how large an area has to be cordoned off.

Figure 14.48 Organize the fire attack from the floor below the fire.

Fires Belowground in Structures

Structural fires that are belowground expose firefighters to extremely punishing conditions. In order to enter a burning basement, firefighters usually have to descend the stairs to that level (Figure 14.49). These stairways are in effect chimneys for the superheated air and fire gases being given off by the fire. Thus, it is imperative that firefighters descend the stairs as quickly as possible, preferably after proper ventilation has been effected. Once firefighters reach the basement, the heat conditions are similar to those in a standard structural fire attack.

Good ventilation techniques are extremely important when fighting fires that are belowground. The ground-level floor should be vented to remove as much smoke and heat from the basement as possible. The ventilation point will ideally be at a point far away from the stairs that are going to be descended. Firefighters descending the stairs may put the nozzle on a wide fog pattern to make the trip down the stairs more bearable (Figure 14.50). This is only possible if proper ventilation has been performed away from the stairs.

When fighting fires belowground, it is desirable to have other engine/truck companies locate heavy objects above the fire fighting teams. Firefighters should be aware that unprotected steel supports elongate when exposed to temperatures of 1,000°F (538°C) or more. These have been know to push over walls during a fire. The longer steel supports are subjected to fire, the more likely they are to fail, regardless of their composition. In addition, contents such as piled stock will become more unstable as fire damage progresses and water is absorbed. The application of hose streams must be performed with prudent care because of the difficulty of ventilating generated steam.

Fires in belowground areas may be attacked indirectly with piercing nozzles, cellar nozzles, or distributors. Cellar nozzles can be put through

Figure 14.49 Descending the stairs to a belowground fire can be very difficult and dangerous.

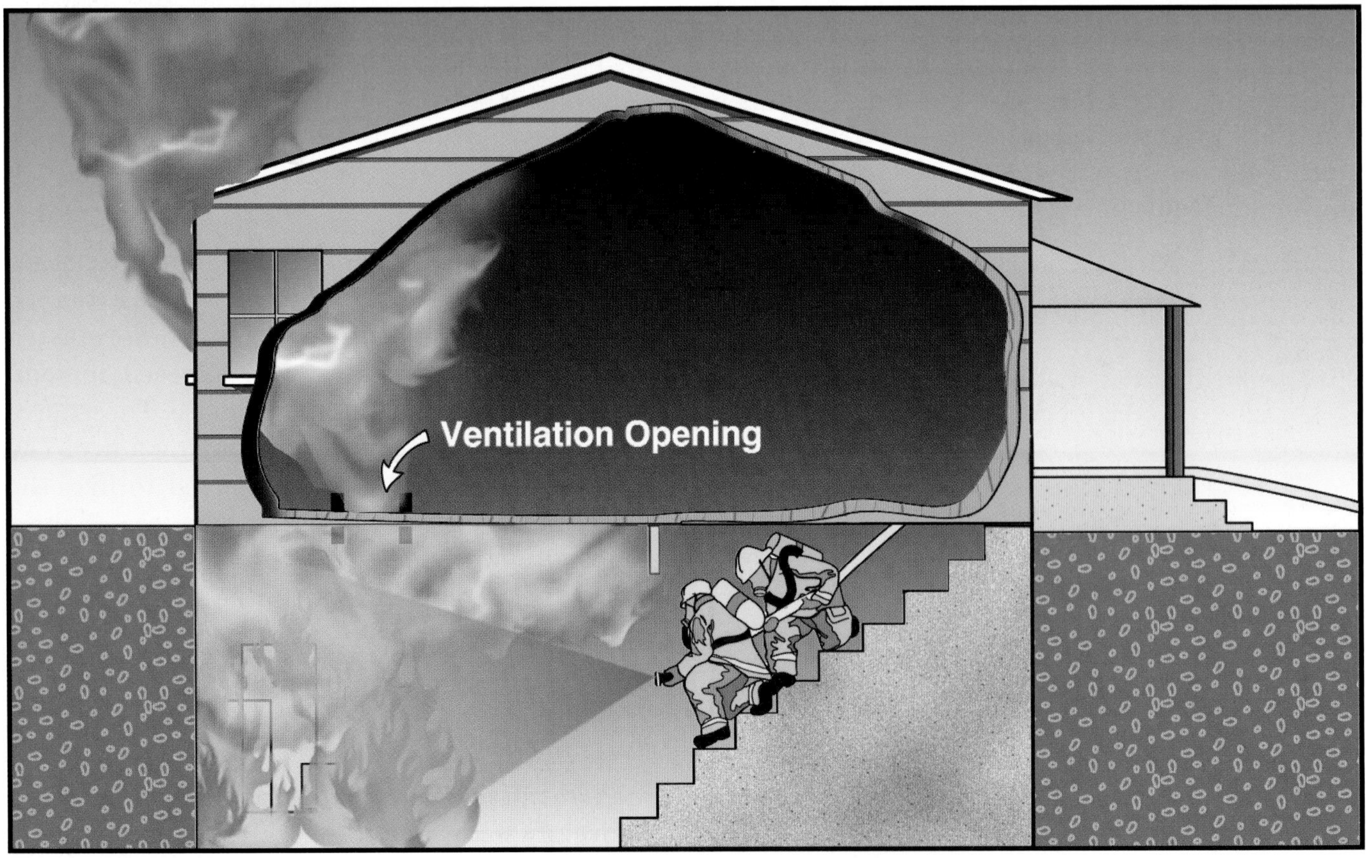

Figure 14.50 Proper ventilation ahead of attack crews can make entering the basement considerably easier.

holes in the floor above the fire to provide water to fires in inaccessible areas (Figure 14.51). In cases where a basement is inaccessible, high-expansion foam may be used to flood the basement and extinguish the fire. This must be carefully coordinated with ventilation activities to keep from pushing all the heat and fire gases to other areas and causing major fire extension or a flashover.

While the fire is being attacked, special attention should be paid to vertical means of fire spread. It is important not to overlook ventilating the attic or highest floor of the building. Heat and fire gases accumulate in these areas and could eventually flashover if not vented.

Figure 14.51 Cellar nozzles can be used to knock down fires in areas that are otherwise inaccessible.

WARNING

During any belowground fire, the firefighter must be aware of the hazards of a weakened floor above and the danger of poor visibility, even with proper ventilation.

Fires at Properties Protected by Fixed Fire Extinguishing Systems

Fire personnel should be familiar with the fixed fire extinguishing systems in buildings protected by their department (Figure 14.52). Fire department operating procedures at these occupancies must take into account the necessity of supporting these systems. The fire department must establish the support of fixed fire suppression systems as a high priority. These systems include the following:

- Sprinkler systems
- Carbon dioxide systems
- Standpipe systems
- Halogenated systems
- Dry chemical hood systems
- Foam systems

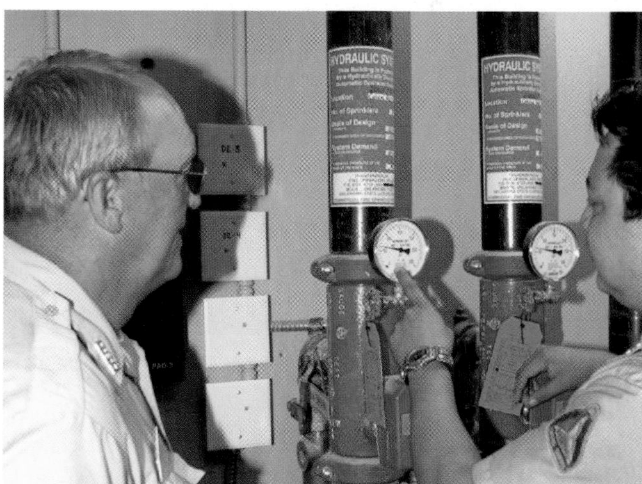

Figure 14.52 Fire protection personnel inspecting a fixed fire extinguishing system.

Some of the dangers involved when dealing with fires in a fixed fire extinguishing system occupancy include the following:

- Oxygen depletion during carbon dioxide activation
- Poor visibility
- Energized electrical equipment
- Toxic environmental exposures

Standard operating procedures used at these occupancies are most likely to be incorporated as a part of the pre-incident plan. This plan includes a detailed account of the construction features, contents, protection systems, and surrounding properties. The pre-incident plan also outlines the procedures to be used by each engine/truck company according to the conditions they find. A building map showing water supplies, protection system connections, and engine/truck company placement should be an integral part of the plan and must be updated to reflect changes affecting fire department operations.

Vehicle Fires

Fires in single-passenger vehicles are among the most common types of fires that firefighters encounter (Figure 14.53). These fires should be treated with the same basic care that structural fires are treated. Firefighters should be in full personal protective clothing, including SCBA.

Figure 14.53 Motor vehicle fires pose many hazards to firefighters. *Courtesy of Bob Esposito.*

The attack line should be at least a 1½-inch (38 mm) hoseline. Booster lines do not provide the protection or rapid cooling needed to effectively and safely fight a vehicle fire. The fire should be attacked from the upwind and uphill side when possible. A backup line should be deployed as soon as possible. For small engine compartment fires, such as carburetor fires, portable extinguishers may be used.

BASIC PROCEDURES

The basic procedure for attacking a vehicle fire is to extinguish any ground fire around or under the vehicle first and then attack the remaining fire in the vehicle. The hoseline should be placed between the vehicle fire and any exposures. If the vehicle has combustible metal components that become involved, large amounts of water or Class D extinguishing agents will be needed. Firefighters should use extra caution when water is first applied to these burning parts because fire intensity will be greatly increased initially. If a large amount of burning fuel has been spilled, foam may be required for successful extinguishment and continued flame suppression.

In most engine compartment fires, the fire must be knocked down before the hood can be opened. One method is to use a piercing nozzle through areas such as the hood, fenders, or wheel wells. Another method requires firefighters to first make an opening for a hoseline stream to enter. The firefighters can use a Halligan tool to make an opening between the hood and the fender. A straight stream or narrow fog can then be directed into the opening.

Another type of vehicle fire occurs in the vehicle's passenger compartment. When firefighters attack this type of fire, they should approach the vehicle from the front or rear corner with a wide fog stream. The firefighters should then attempt to open the door (the driver may have the key if it is locked). If normal entry is not possible, break a window. Once entry has been made, attack the fire with a medium fog pattern in a circular motion. If normal access to the trunk is not possible, forcible entry may be necessary. The spike from a Halligan tool or a pick-head axe should be used to knock the lock barrel out of place. A screwdriver or similar object can then be inserted into the remaining locking mechanism and turned to release the lock.

There are three methods that can be used for vehicle undercarriage fires. If there is a hazard in getting close to the vehicle, a straight stream can be used from a distance to reach under the vehicle. If close proximity to the vehicle is possible, a straight stream can be deflected off the pavement to hit the undercarriage. The third alternative is to open the hood and direct the stream through the engine compartment.

Once the fire has been controlled, overhaul should be conducted as soon as possible to check for extension and hidden fires. Other overhaul considerations include disconnecting the battery, securing air bags (Supplemental Restraint System [SRS] or Side-Impact Protection System [SIPS]), and cooling fuel tanks and any intact sealed components.

HAZARDS

As with any other fire situation, there are hazards associated with vehicle fires. Catalytic converters, used to clean vehicle emissions, can act as an ignition source to grass or other fuels underneath the vehicle. The external temperature of a catalytic converter on a properly operating engine is about 1,300°F (704°C); for a poorly running engine, temperatures can reach as high as 2,500°F (1 371°C). The majority of a vehicle's interior components are plastic, which burn rapidly at high temperatures and give off toxic gases. A hazard with modern vehicles is the air bag (SRS or SIPS) that could possibly deploy from the steering wheel, dashboard, or door of the vehicle.

Modern vehicles have many sealed components. As these components are heated, the gases inside expand, pressurizing the component. When the sealed component fails, projectiles such as shock-absorber-type bumpers, hollow driveshafts, and hatchback supports can be shot from the vehicle with great force. Tires may also blow out as a result of pressure buildup. Firefighters standing around any of these components when they fail may be seriously injured or killed.

Firefighters should not assume that any vehicle is without extraordinary hazards such as saddle fuel tanks, LPG or CNG tanks, alternative fuel tanks, explosives, or hazardous materials (Figure 14.54). Vans and other small passenger-type vehicles are often used to transport small amounts of radioactive materials for hospital use. Also, large money losses can occur from fires in messenger or courier vehicles. Certainly, firefighters should view any military vehicle as a hazard. They could be carrying munitions or other hazardous cargos.

Figure 14.54 Some vehicles are powered by compressed gas rather than liquids. *Courtesy of Conoco, Inc.*

Trash and Dumpster® Container Fires

The possibility of exposure to dangerous byproducts of combustion should not be overlooked when dealing with trash and Dumpster® container fires. In many cases, the contents of the trash pile or Dumpster® may include hazardous materials or plastics that give off highly toxic smoke and gases. Aerosol cans and batteries, which may explode, may also be present. For this reason, full personal protective equipment and SCBA should be worn when attacking all rubbish fires.

The size hoseline to use depends on the size of the fire and its proximity to exposures. Small piles of trash, garbage cans, and small Dumpsters® can be handled with a booster line. Larger piles, large Dumpsters®, and fires that pose exposure problems should be attacked with at least a 1½-inch (38 mm) line. Firefighters should make sure that the fire has not extended into any exposures in the proximity of the fire. Standard overhaul techniques can be used to accomplish this objective.

Fires and Emergencies in Confined Enclosures

Fire fighting and rescue operations must often be carried out in locations that are belowground or otherwise cut off from either natural or forced ventilation. Belowground in structures, caves, sewers, storage tanks, and trenches are just a few examples of these types of areas (Figure 14.55). The single most important factor in safely operating at these emergencies is recognition of the inherent hazards of confined enclosures. The atmospheric conditions that may be expected include the following:

- Oxygen deficiencies
- Flammable gases and vapors
- Toxic gases
- Elevated temperatures
- Explosive dusts

Figure 14.55 Most response areas have a variety of potential confined-space rescue sites.

In addition, physical hazards may also be present such as the following:

- Limited means of entrance and egress
- Cave-ins or unstable support members
- Deep standing water or other liquids
- Utility hazards — electricity, gas, sewage

Communications with plant or building supervisors or other knowledgeable people at the scene are also important because they may be able to give valuable information on hazards that are present and the number of victims and their probable location. Likewise, pre-incident plans of existing enclosed spaces in the fire department's jurisdiction reduce guesswork and should be referred to during operations in these locations. The firefighter should be ready to implement prearranged methods of extinguishment or rescue without delay. These plans should include provisions for victim and rescuer protection, control of utilities and other physical hazards, communications, ventilation, and lighting. Power equipment that is used during nonfire rescue operations should be rated for use in explosive atmospheres; this includes flashlights, portable fans, and radios.

Because the entrances to these incidents are generally restricted, the establishment of a command post and a staging area is vital to a successful operation. The staging area should be near, but not obstructing, the entrance and supplied with the personnel and equipment to be used (Figure 14.56). Firefighters should not enter these enclosures until the incident commander has decided upon a course of action and issued specific orders. A safety officer should be stationed at the entrance to keep track of personnel and equipment entering and leaving the enclosure.

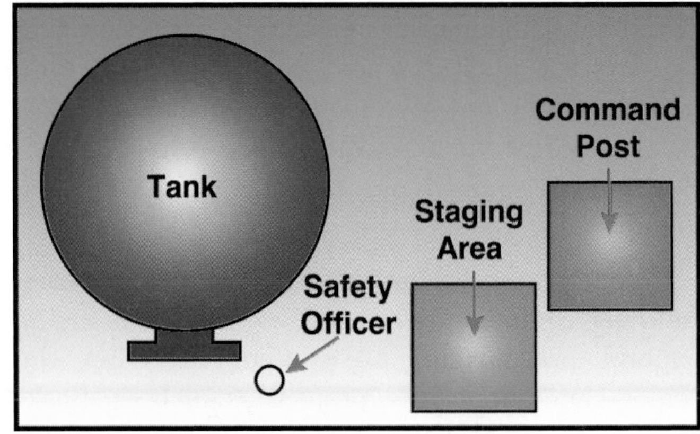

Figure 14.56 The command post and staging areas should be fairly close to the working area. A safety officer should control the entry and exit of rescuers from the work hazard area.

The importance of wearing personal protective equipment (PPE) including SCBA, using air monitoring, and using accountability systems cannot be overemphasized. When a rescuer must enter a confined space separately from the cylinder and harness of the SCBA, extreme caution must be exercised to not pull the mask off the rescuer. Air-supply masks are also available with long supply hoses that eliminate the need for bulky tanks (Figure 14.57).

A lifeline should be tied to each rescuer before entry (Figure 14.58). This line must be constantly monitored, and a properly outfitted standby crew equal in number to the rescuers working inside must be available. A system of communication between inside and outside team members should be prearranged because portable radios may prove unreliable. One method of signaling is called the *O-A-T-H* method. The letter *O* stands for *OK*, *A* stands for *Advance*, *T* for *Take-up*, and *H* for *Help*. One tug means everything is OK, two tugs Advance, three tugs Take-up, and four tugs Help. This system may be used by the firefighter inside the space to communicate to the safety person outside or for the safety person to communicate

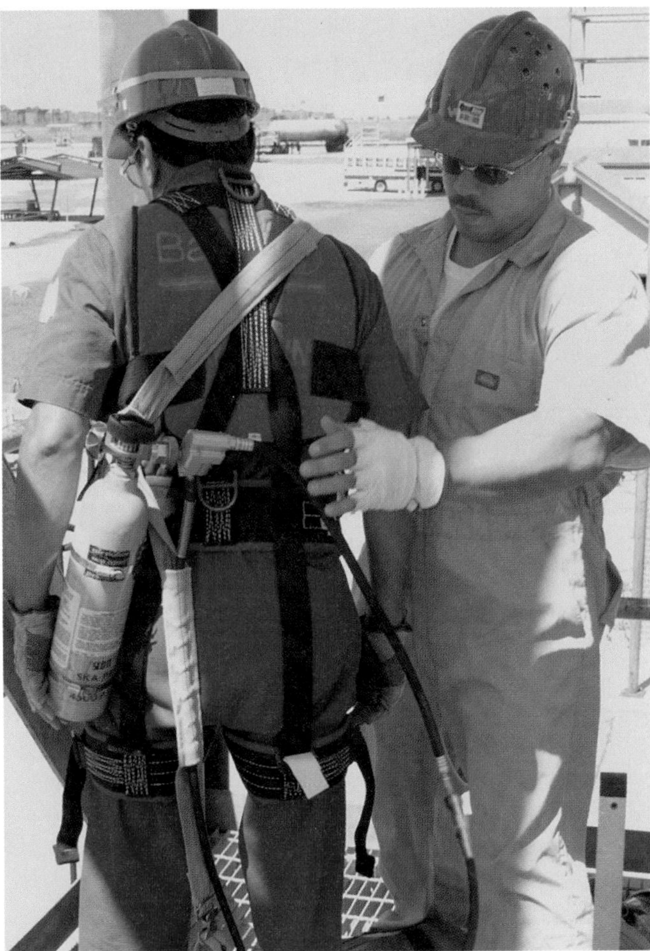

Figure 14.57 The rescuer's partner inspects the rescuer's airline unit to ensure that it has been donned properly.

Figure 14.58 Rescuers entering a confined space must have a lifeline attached to them.

with the firefighter. Any signal given, such as a sharp pull on the line, should be acknowledged by the other party on the line.

Another method of safe communication is the use of sound-powered phones that do not require a power source. The rescuer must be able to use the selected communication system without removing the SCBA mask.

AIR MONITORING

Atmospheric monitoring devices are available in a wide variety of models and configurations to match the particular settings for their intended use. The devices used must be capable of indicating the level of the gas or gases detected; most do so in percentages or parts per million (ppm) or both. The atmosphere is first checked for oxygen because most combustible-gas meters do not give an accurate reading in an oxygen-deficient atmosphere. The results of these tests are used to determine whether and/or when it is safe for rescuers to enter, what type and level PPE is required, and the likelihood of finding viable victims inside. As long as a space continues to be occupied by rescuers or victims, the atmosphere within the space should be monitored. The monitor should be removed from time to time and recalibrated in clear air.

ACCOUNTABILITY SYSTEM

The purpose of the accountability system is to ensure that only those who are authorized and properly equipped to enter a hazardous area are allowed to do so and that both their location and their status are known as long as they remain inside the controlled zone. The safety officer should check and record each entering member's mission, tank pressure, name, and estimated safe working time. This procedure allows for the accounting of all team members and reduces the possibility of a member being unaccounted for after the safe working time limit has passed.

FIRE ATTACK

Fires in confined areas may also be attacked indirectly with piercing nozzles, cellar nozzles, distributors, or high-expansion foam (Figure 14.59). Due to the confinement of heat, firefighters may find that they tire more quickly and use their SCBA air supply faster. Firefighters should call for relief

High-Expansion Foam in Basement Fire

Figure 14.59 High-expansion foam may be used to control basement fires.

before they are exhausted and must pay rigid attention to air-conservation techniques and pressure gauges. Firefighters must not advance into confined spaces farther than their air supplies will allow them safe margins for retreating.

Wildland Fire Techniques

Wildland fires include fires in weeds, grass, field crops, brush, forests, and similar vegetation (Figure 14.60). Wildland fires have characteristics of their own that are not comparable to other forms of fire fighting. Local topography, fuel type, water availability, and weather present different challenges. The local experiences of fire suppression forces determine the methods and techniques used to control wildland fires.

Once a wildland fire starts, burning is generally rapid and continuous. There are many factors that affect wildland fire behavior, but the three most important factors are fuel, weather, and topography. Any one factor may be dominant in influencing what an individual fire does, but usually the combined strength of all three factors dictates a fire's behavior.

Figure 14.60 Large brush fires are difficult to extinguish and are taxing on firefighters. *Courtesy of National Interagency Fire Center (NIFC).*

WARNING

Fighting wildland fires is very dangerous. Many firefighters have lost their lives or have been seriously injured while trying to control these fires. Thoroughly think out the situation. Remember the safety of personnel always comes first.

FUEL

Fuels are generally classified by grouping fuels with similar burning characteristics together. This method classifies wildland fuels as subsurface, surface, and aerial fuels (Figures 14.61 a–c).

- ***Subsurface fuels*** — Roots, peat, duff, and other partially decomposed organic matter that lie under the surface of the ground
- ***Surface fuels*** — Needles, twigs, grass, field crops, brush up to 6 feet (2 m) in height, downed limbs, logging slash, and small trees on or immediately adjacent to the surface of the ground
- ***Aerial fuels*** — Suspended and upright fuels (brush over 6 feet [2 m], leaves and needles on tree limbs, branches, hanging moss, etc.) physically separated from the ground's surface (and sometimes from each other) to the extent that air can circulate freely between them and the ground

Figure 14.61a Typical subsurface fuels.

Figure 14.61b Surface fuels are those on the ground.

Figure 14.61c Aerial fuels are those at the tops of the brush and trees.

Several factors affect the burning characteristics of fuels such as the following:

- ***Fuel size*** — Small or light fuels burn faster.
- ***Compactness*** — Tightly compacted fuels, such as the subsurface or surface types, burn slower than the aerial types.
- ***Continuity*** — When fuels are close together, the fire spreads faster because of the effects of heat transfer. In patchy fuels (those growing in clumps), the rate of spread is less predictable than in continuous fuels.
- ***Volume*** — The amount of fuel present in a given area (its volume) influences the fire's intensity and the amount of water needed to perform extinguishment.
- ***Fuel moisture content*** — As fuels become dry, they ignite easier and burn with greater intensity (amount of heat produced) than those with a higher moisture content.

WEATHER

All aspects of the weather have some effect upon the behavior of a wildland fire. Some weather factors that influence wildland fire behavior are the following:

- ***Wind*** — Fans the flames into greater intensity and supplies fresh air that speeds combustion; medium- and large-sized fires may create their own winds
- ***Temperature*** — Has effects on wind and is closely related to relative humidity; primarily affects the fuels as a result of long-term drying

- ***Relative humidity*** — Impacts greatly on dead fuels that no longer have any moisture content of their own
- ***Precipitation*** — Largely determines the moisture content of live fuels; dead flashy fuels (those easily ignited) may dry quickly, large dead fuels retain this moisture longer and burn slower

TOPOGRAPHY

Topography refers to the features of the earth's surface, and it has a decided effect upon fire behavior. The steepness of a slope affects both the rate and direction of a wildland fire's spread. Fires will usually move faster uphill than downhill, and the steeper the slope, the faster the fire moves (Figure 14.62). Other topographical factors influencing wildland fire behavior are the following:

- ***Aspect*** — The compass direction a slope faces (aspect) determines the effects of solar heating. Full southern exposures (north of the equator) receive more of the sun's direct rays and therefore receive more heat. Wildland fires typically burn faster on southern exposures.

Figure 14.62 Slope can greatly increase the rate of fire spread. *Courtesy of Tony Bacon.*

- ***Local terrain features*** — Obstructions such as ridges, trees, and even large rock outcroppings may alter airflow and cause turbulence or eddies, resulting in erratic fire behavior.
- ***Drainages (or other areas with wind-flow restrictions)*** — These terrain features create turbulent updrafts causing a chimney effect. Wind movement can be critical in chutes (steep V-shaped drainage) and saddles (depression between two adjacent hilltops). Fires in these drainages can spread at an extremely fast rate, even in the absence of winds, and are very dangerous.

PARTS OF A WILDLAND FIRE

The typical parts of a wildland fire are shown in Figure 14.63. Any wildland fire will contain at least two of these parts if not all of them.

Origin. The area where the fire started, and the point from which it spreads, is the *origin.* The origin is often next to a trail, road, or highway, but it also may be in very inaccessible areas (lightning strikes or campfires).

Head. The *head* is the part of a wildland fire that travels or spreads most rapidly. The head is usually found on the side of the fire opposite the direction from which the wind is blowing. The head burns intensely and usually does the most damage. Usually, the key to controlling the fire is to control the head and prevent the formation of a new head.

Finger. *Fingers* are long narrow strips of fire extending from the main fire. They usually occur when the fire burns into an area that has both light fuel and patches of heavy fuel. Light fuel burns faster than the heavy fuel, which gives the finger effect. When not controlled, these fingers can form new heads.

Perimeter. The *perimeter* is the outer boundary, or the distance around the outside edge, of the burning or burned area — also commonly called the *fire edge.* Obviously, the perimeter continues to grow until the fire is suppressed.

Heel. The *heel,* or *rear,* of a wildland fire is the side opposite the head. The heel usually burns slowly and quietly and is easier to control than the head. In most cases, the heel will be found burning downhill or against the wind.

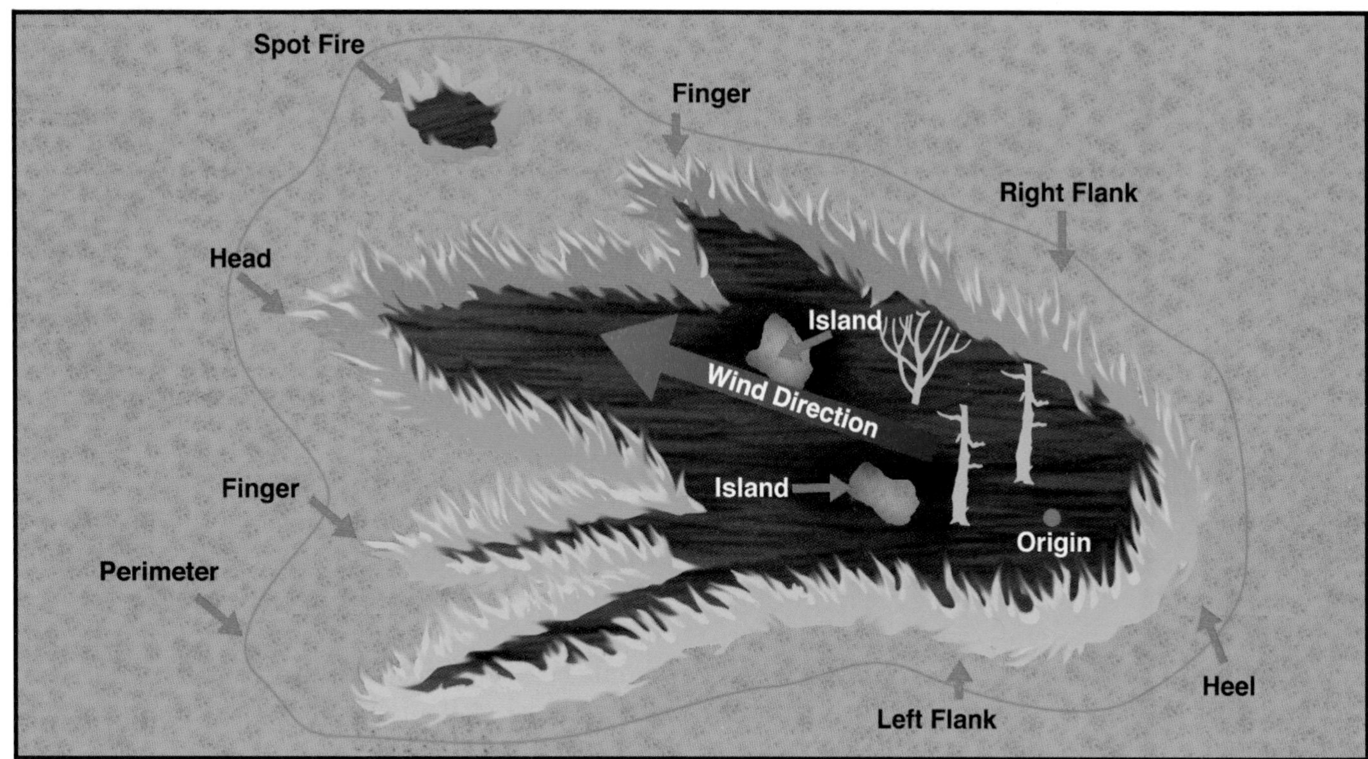

Figure 14.63 The parts of a wildland fire.

Flanks. The *flanks* are the sides of a wildland fire, roughly parallel to the main direction of fire spread. The right and left flanks separate the head from the heel. It is from these flanks that fingers can form. A shift in wind direction can change a flank into a head.

Islands. Unburned areas inside the fire perimeter are called *islands.* Because they are unburned potential fuels for more fire, they must be patrolled frequently and checked for spot fires (see following paragraph).

Spot Fire. *Spot fires* are caused by flying sparks or embers landing outside the main fire. Spot fires present a hazard to personnel (and equipment) working on the main fire because they could become trapped between the two fires. Spot fires must be extinguished quickly or they will form a new head and continue to grow in size.

Green. The area of unburned fuels next to the involved area is called the *green.* While the term refers to the color of some of the fuels in the area, the "green" may not be green at all. The green does not indicate a safe area. It is simply the opposite of the burned area (the black) (see following paragraph) (Figure 14.64).

Black. The opposite of the green — the *black* — is the area in which the fire has consumed or "blackened" the fuels. The black is a relatively safe area during a fire but can be a very hot and smoky environment (Figure 14.65).

Figure 14.64 The green.

Figure 14.65 The black.

ATTACKING THE FIRE

The methods used to attack wildland fires revolve around perimeter control. The control line may be established at the burning edge of the fire, next to it, or at a considerable distance away. The objective is to establish a control line that completely encircles the fire with all the fuel inside rendered harmless.

The direct and indirect approaches are the two basic attack methods for attacking wildland fires. A *direct attack* is action taken directly against the flames at its edge or closely parallel to it (Figure 14.66). The *indirect attack* is used at varying distances from the advancing fire to halt its progress. A line is constructed some distance from the fire's edge and the unburned intervening fuel is burned out. This method is generally used against fires that are either *too hot, too fast,* or *too big* (Figure 14.67). Because a wildland fire is constantly changing, it is quite possible to begin with one attack method and end with another. Size-up must be continued during the fire so that these adjustments can be made when required.

Figure 14.66 Firefighters in a direct attack. *Courtesy of Monterey County Training Officers.*

Figure 14.67 Firefighters burn out from a road in an indirect attack.

TEN STANDARD FIRE FIGHTING ORDERS

The fire-behavior characteristics listed earlier are not the only fire conditions that are dangerous to fire personnel, only the most common ones. Other situations can put firefighters at risk. Studies of firefighter deaths led to the development of the Ten Standard Fire Fighting Orders. In every case where a firefighter was killed while fighting a wildland fire, it was shown that one or more of the Ten Standard Orders had been ignored. Violating one or more of these orders may result in firefighter deaths.

The orders are guidelines that help firefighters identify and avoid high-risk situations. Every firefighter should know and follow them. Being able to recite them is commendable, but putting them into practice is more important. Each order should be considered separately so firefighters will recognize when it applies during a fire and respond correctly.

FIRE ORDERS

- **Fight fire aggressively but provide for safety first.**
- **Initiate all action based on current and expected fire behavior.**
- **Recognize current weather conditions and obtain forecasts.**
- **Ensure instructions are given and understood.**
- **Obtain current information on fire status.**
- **Remain in communication with crew members, your supervisor, and adjoining forces.**
- **Determine safety zones and escape routes.**
- **Establish lookouts in potentially hazardous situations.**
- **Retain control at all times.**
- **Stay alert, keep calm, think clearly, act decisively.**

Chapter 15
Fire Detection, Alarm, and Suppression Systems

Chapter 15
Fire Detection, Alarm, and Suppression Systems

INTRODUCTION

There are a number of reasons for installing fire detection and alarm systems. Each system is designed to fulfill specific needs. The following are recognized functions:

- To notify occupants of a facility to take necessary evasive action to escape the dangers of a hostile fire
- To summon organized assistance to initiate or to assist in fire control activities
- To initiate automatic fire control and suppression systems and to sound an alarm
- To supervise fire control and suppression systems to assure that operational status is maintained
- To initiate a wide variety of auxiliary functions involving environmental, utility, and process controls

Fire detection and alarm systems may incorporate one or all of these features. Such systems may include components that operate mechanically, hydraulically, pneumatically, or electrically, but most state-of-the-art systems operate electronically.

Despite advances in other forms of fixed fire protection suppression systems, automatic sprinkler systems remain the most reliable form for commercial, industrial, institutional, residential, and other occupancies. It is proven that fires controlled by sprinklers result in less business interruption and water damage than those that have to be extinguished by traditional fire department methods. In fact, data compiled by Factory Mutual Research Corporation indicates that about 70 percent of all fires are controlled by the activation of five or fewer sprinklers.

The first part of this chapter discusses the most common types of fire detection and alarm systems and devices in use in North America. The second part of the chapter describes automatic sprinkler systems. Factors to consider during fires at protected properties are also given.

TYPES OF ALARM SYSTEMS

[NFPA 1001: 4-5.1; 4-5.1(a); 4-5.1(b)]

The most basic alarm system is designed to only be initiated manually. This is a local warning system similar to the type installed in schools or theaters, and the signal alerts occupants of the need to evacuate the premises. While alarm standards have traditionally called this type a *local system,* contemporary terminology uses *protected premises fire alarm systems* (Figure 15.1).

A wide variety of optional features are available to expand the capabilities of an alarm system.

Figure 15.1 The two components of a local alarm system.

Automatic fire detection devices may be added, allowing the system to sense the presence of a fire and to initiate a signal. These are discussed in the following sections.

Four basic types of automatic alarm-initiating devices are designed to detect heat, smoke, fire gases, and flame. The following sections describe the most common types of devices in use.

Heat Detectors

Several different types of heat detection devices, such as fixed-temperature devices and rate-of-rise detectors, are discussed in the following sections.

FIXED-TEMPERATURE HEAT DETECTORS

Systems using fixed-temperature heat-detection devices are among the oldest types of fire detection systems in use. They are relatively inexpensive compared to other types of systems, and they are least prone to false activations. But heat detectors are also the slowest to activate of all the various types of alarm-initiating devices.

Because heat rises, heat detectors must be placed in high portions of a room, usually on the ceiling (Figure 15.2). Detectors should have an activation temperature rating slightly above the highest ceiling temperatures normally expected in that space.

The various types of fixed-temperature devices discussed detect heat by one or more of three primary principles of physics:

- Expansion of heated material
- Melting of heated material
- Changes in resistance of heated material

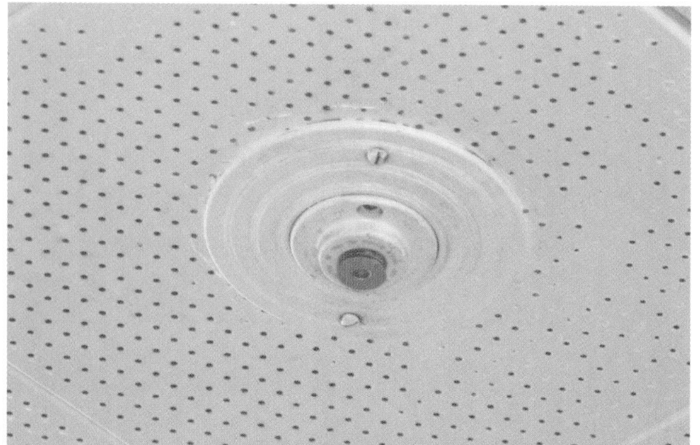

Figure 15.2 A fixed-temperature heat detector.

Fusible devices/frangible bulbs. While these two devices are more commonly associated with automatic sprinklers, they are also used in fire detection and signaling systems. The operating principles of these devices are identical to the fusible links and frangible bulbs used with automatic sprinklers; only their applications differ (see Sprinklers section under Automatic Sprinkler Systems).

A *fusible device* is normally held in place by a solder with a known melting (fusing) temperature. Under normal conditions, the device holds a spring-operated contact inside the detector in the open position (Figure 15.3). When a fire raises the ambient temperature to the fusing temperature of the device, the solder melts, allowing the spring to move the contact point. This action completes the alarm circuit, which initiates an alarm signal. Some of these detectors may be restored by replacing the fusible device; others require the entire heat detector to be replaced.

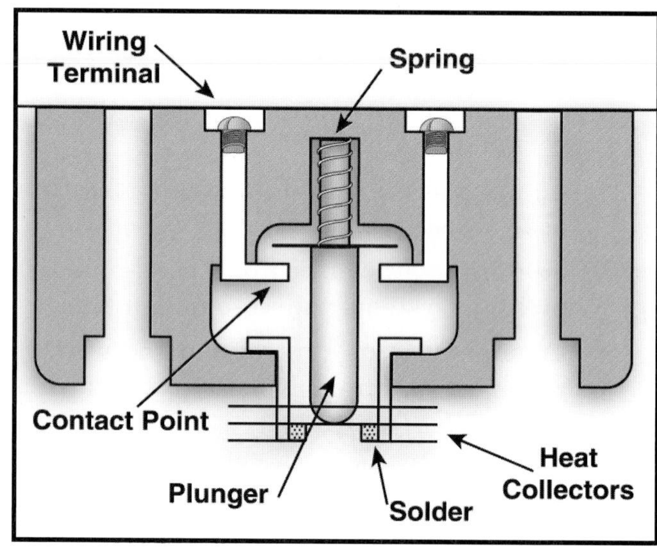

Figure 15.3 Cutaway of a fusible fixed-temperature heat detector.

A *frangible bulb* in a detection device holds electrical contacts apart, much in the way that a fusible device does. The little glass vial (frangible bulb) contains a liquid with a small air bubble. The bulb is designed to break when the liquid is heated to a predetermined temperature. When the rated temperature is reached, the liquid expands and absorbs the air bubble, the bulb fractures and falls out, and the contacts complete the circuit to initiate an alarm. In order to restore the system, the entire detector must be replaced. While detectors of this type are still in service, their manufacture has been discontinued.

Continuous line detector. Most of the detectors described in this chapter are of the *spot* type; that is, they are designed to detect heat only in a relatively small area surrounding the specific spot where they are located. However, *continuous line detection devices* can detect heat over a linear area parallel to the detector.

One such device consists of a cable with a conductive metal inner core sheathed with stainless steel tubing (Figure 15.4). The inner core and the sheath are separated by an electrically insulating semiconductor material that keeps them from touching but allows a small amount of current to flow between the two. This insulation loses some of its electrical resistance capabilities at a predetermined temperature anywhere along the line. When this condition happens, the current flow between the two components increases, initiating an alarm signal through the system control unit. This type of detection device restores itself when the level of heat is reduced.

Another type uses two insulated wires within an outer covering. When the rated temperature is reached, the insulation melts and allows the two wires to touch. This action completes the circuit and initiates an alarm signal through the system control unit (Figure 15.5). To restore this type of line detector, the fused portion of the wires must be cut out and replaced with new wire.

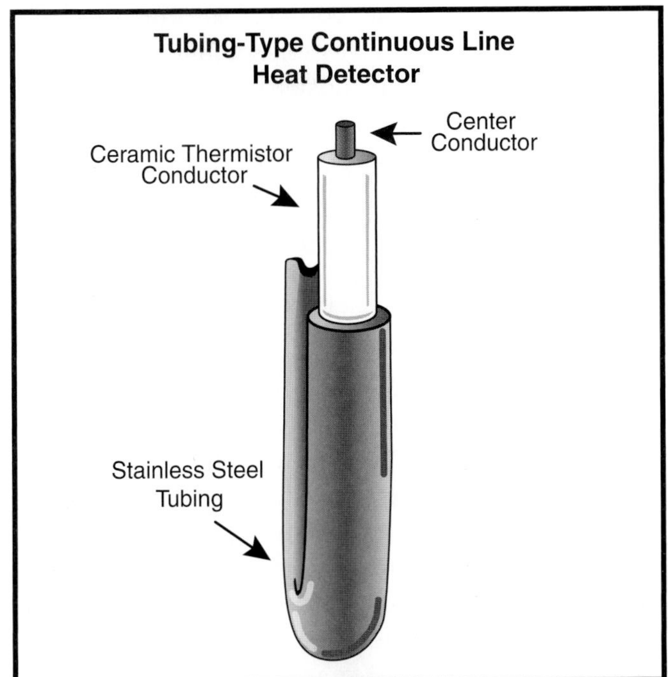

Figure 15.4 Tubing-type continuous line heat detector.

Bimetallic detector. One type of bimetallic detector uses two metals that have different thermal expansion characteristics. Thin strips of the metals are bonded together, and one or both ends of the strips are attached to the alarm circuit. When heated, one metal expands faster than the

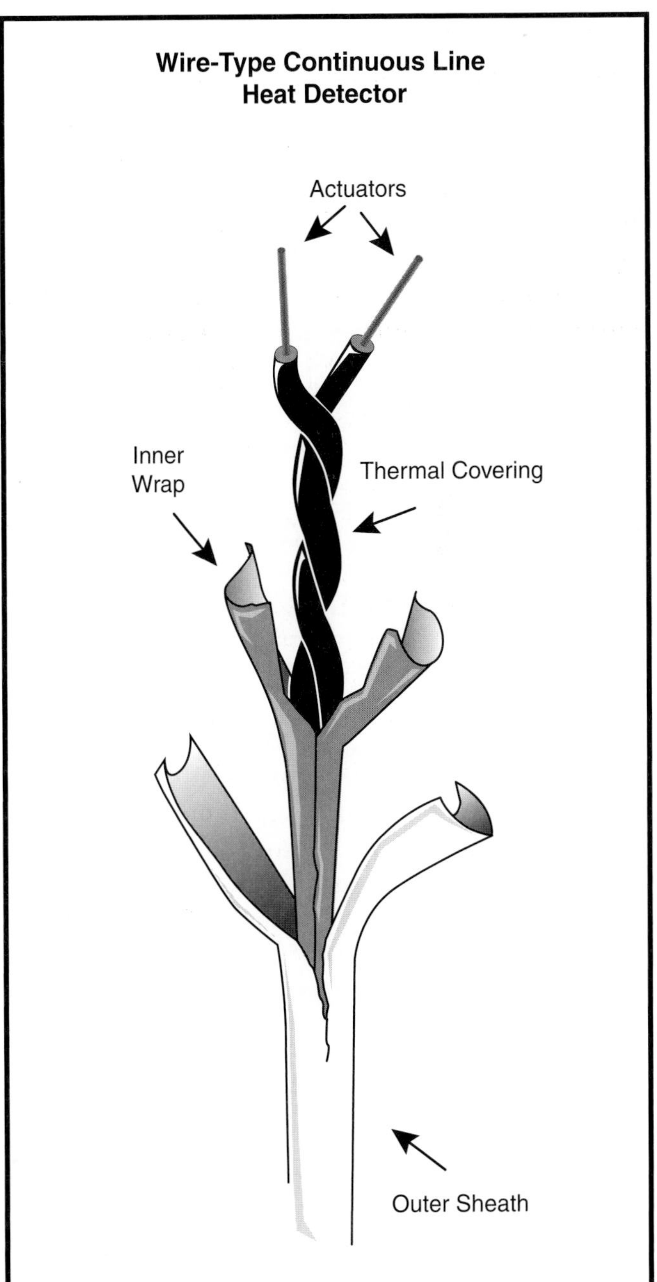

Figure 15.5 Wire-type continuous line heat detector.

other, causing the strip to arch or bend. The deflection of the strip either makes or breaks contact in the alarm circuit, initiating an alarm signal through the system control unit. Another type of bimetallic detector utilizes a snap disk and microswitch (Figure 15.6). Most bimetallic detectors will reset automatically when cooled. After a fire, however, they do need to be checked to ensure that they were not damaged.

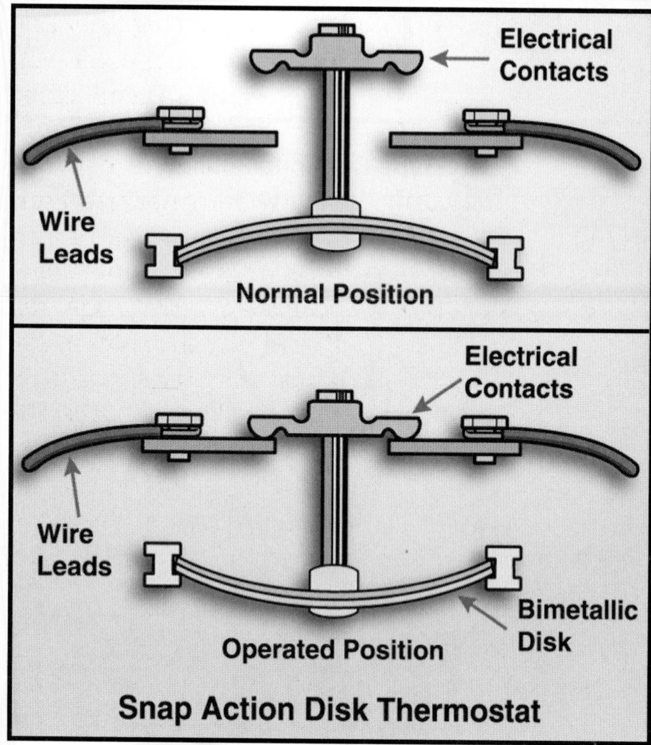

Figure 15.6 Cutaway of a bimetallic detector.

RATE-OF-RISE HEAT DETECTORS

A *rate-of-rise heat detector* operates on the principle that the temperature in a room will increase faster from fire than from atmospheric temperature. Typically, rate-of-rise heat detectors are designed to initiate a signal when the rise in temperature exceeds 12 to 15°F (7°C to 8°C) per minute. Because the alarm is initiated by a sudden rise in temperature, regardless of the initial temperature, an alarm can be initiated at a temperature far below that required for a fixed-temperature device.

Most rate-of-rise heat detectors are reliable and not subject to false activations. However, if not properly installed, they can be activated under nonfire conditions. For example, a rate-of-rise detector is installed just inside an exterior door in an air-conditioned building. If the door is opened on a hot day, the influx of heated air can rapidly increase the temperature around the detector and cause it to actuate. Relocating the detector farther from the doorway should alleviate the problem.

There are several different types of rate-of-rise heat detectors in use; all automatically reset if undamaged. The different types are discussed in more detail in the following paragraphs.

Pneumatic rate-of-rise spot detector. A pneumatic spot detector is the most common type of rate-of-rise detector in use (Figure 15.7). It consists of a small dome-shaped air chamber with a flexible metal diaphragm in the base. A small metering hole allows air to enter and exit the chamber during the normal rise and fall of atmospheric temperature and barometric pressure. In the heat of a fire, however, the air within the chamber expands faster than it can escape through the metering hole. This expansion causes the pressure within the chamber to increase, forcing the metal diaphragm against contact points in the alarm circuit. An alarm signal to the system control unit results. This type of detector is most often combined in one unit that also has fixed-temperature capability.

Figure 15.7 A rate-of-rise spot heat detector.

Pneumatic rate-of-rise line detector. The spot detector monitors a small area surrounding its location; however, a line detector can monitor large areas. A *line detector* consists of a system of tubing arranged over a wide area of coverage (Figure 15.8). The space inside the tubing acts as the air chamber described in the preceding paragraph on spot detectors. The line detector also contains a diaphragm and is vented. When any portion of the tubing is subjected to a rapid increase in temperature, the detector functions in the same manner as the spot detector. The tubing in this system must be limited to about 1,000 feet (300 m) in length. The tubing should be arranged in rows that are not more than 30 feet (9 m) apart and 15 feet (5 m) from walls.

Figure 15.8 A pneumatic rate-of-rise heat detector.

Rate-compensated detector. This detector is designed for use in areas that are normally subject to regular temperature changes that are slower than those under fire conditions. The detector consists of an outer metallic sleeve that encases two bowed struts that have a slower expansion rate than the sleeve (Figure 15.9). The bowed struts have electrical contacts on them. In the normal position, these contacts do not come together. When the detector is heated rapidly, the outer sleeve expands in length. This expansion reduces the tension on the inner strips and allows the contacts to come together, thus initiating an alarm signal through the system control unit.

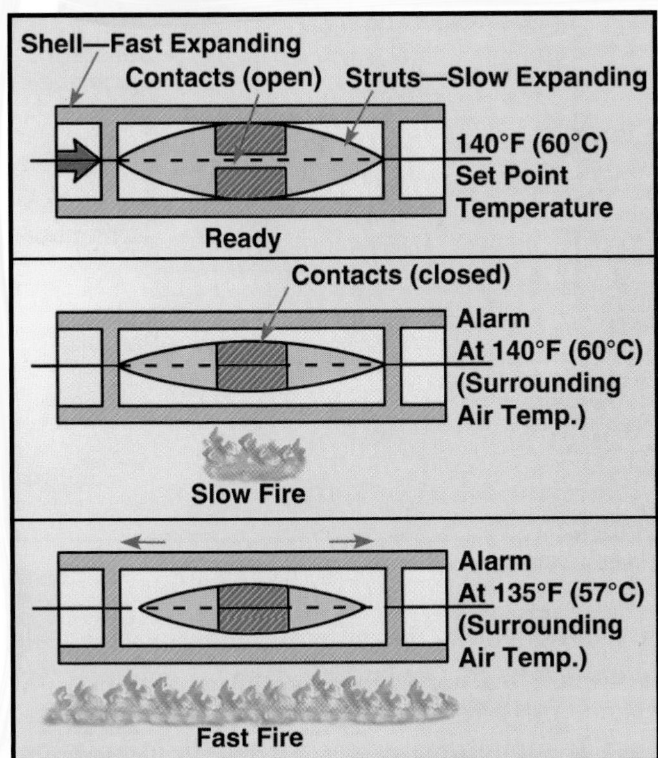

Figure 15.9 Cutaway of a rate-compensated heat detector.

If the rate of temperature rise is fairly slow, such as 5°F (2°C to 3°C) per minute, the sleeve expands slowly enough to maintain tension on the inner strips. This tension prevents unnecessary system activations. However, regardless of the rate of temperature increase, when the surrounding temperature reaches a predetermined point, an alarm signal will be initiated.

Thermoelectric detector. This rate-of-rise detector operates on the principle that when two wires of dissimilar metals are twisted together and heated at one end, an electrical current is generated at the other end (Figure 15.10). The rate at which the wires are heated determines the amount of current that is generated. These detectors are designed to "bleed off" or dissipate small amounts of current. This reduces the chance of a small temperature change activating an alarm unnecessarily. Rapid changes in temperature result in larger amounts of current flowing and activation of the alarm system.

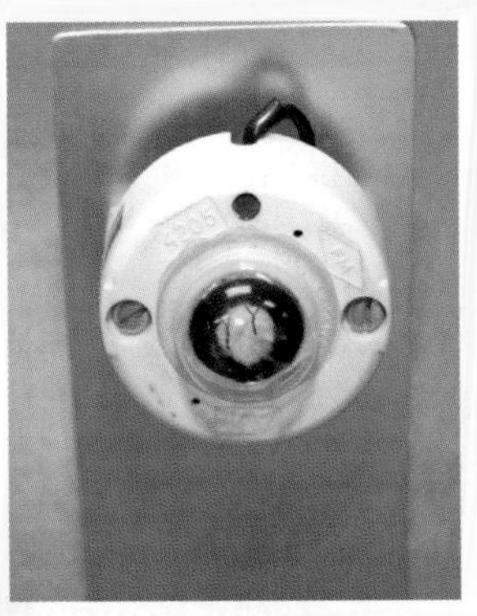

Figure 15.10 A typical thermoelectric heat detector.

Smoke Detectors

Because a smoke detector can respond to smoke produced very early in a fire's development and does not have to wait for heat to be generated, it can initiate an alarm of fire much more quickly than a heat detector. For this reason, the smoke detector is the preferred detector in many types of occupancies. The two basic types, photoelectric and ionization, are described in the following sections, along with a discussion of power sources for smoke detectors.

PHOTOELECTRIC SMOKE DETECTOR

A *photoelectric detector,* sometimes called a *visible products-of-combustion detector*, uses a photoelectric cell coupled with a specific light source. The photoelectric cell functions in two ways to detect smoke: beam application and refractory application.

The *beam application* type uses a beam of light focused across the area being monitored and onto a photoelectric cell. The cell constantly converts the beam into current, which keeps a switch open.

When smoke obscures the path of the light beam, the required amount of current is no longer produced, the switch closes, and an alarm signal is initiated (Figure 15.11).

The *refractory photocell* uses a light beam that passes through a small chamber at a point away from the light source. Normally, the light does not strike the photocell, and no current is produced. This allows the switch to remain open. When smoke enters the chamber, it causes the light beam to be refracted (scattered) in all directions. A portion of the scattered light strikes the photocell, causing current to flow. This current causes the switch to close and initiates the alarm signal (Figure 15.12).

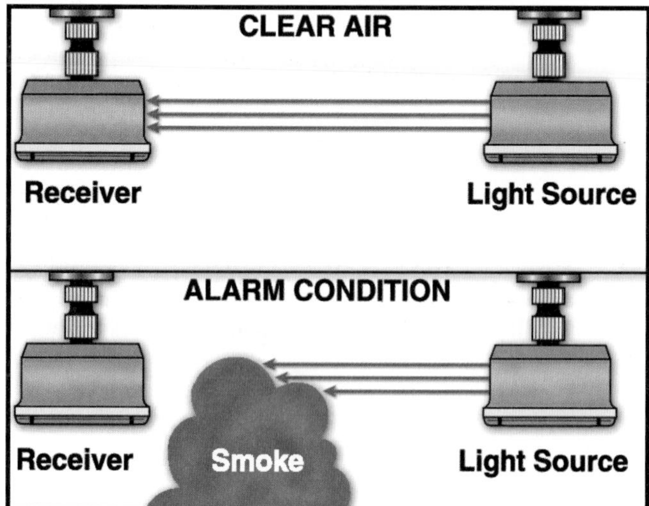

Figure 15.11 Principle of a beam-application photoelectric smoke detector.

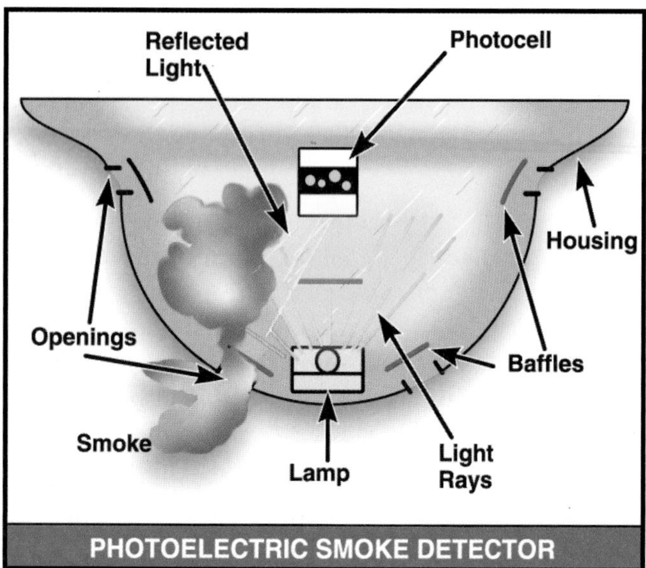

Figure 15.12 Principle of a refractory photoelectric smoke detector.

A photoelectric detector works satisfactorily on all types of fires and automatically resets when the atmosphere is clear. Photoelectric detectors are generally more sensitive to smoldering fires than are ionization detectors.

IONIZATION SMOKE DETECTOR

During combustion, minute particles and aerosols too small to be seen by the naked eye are produced. These invisible products of combustion can be detected by devices that use a tiny amount of radioactive material (usually americium) to ionize air molecules as they enter a chamber within the detector. These ionized particles allow an electrical current to flow between negative and positive plates within the chamber. When the particulate products of combustion (smoke) enter the chamber, they attach themselves to electrically charged molecules of air (ions), making the air within the chamber less conductive. The decrease in current flowing between the plates initiates an alarm signal (Figure 15.13).

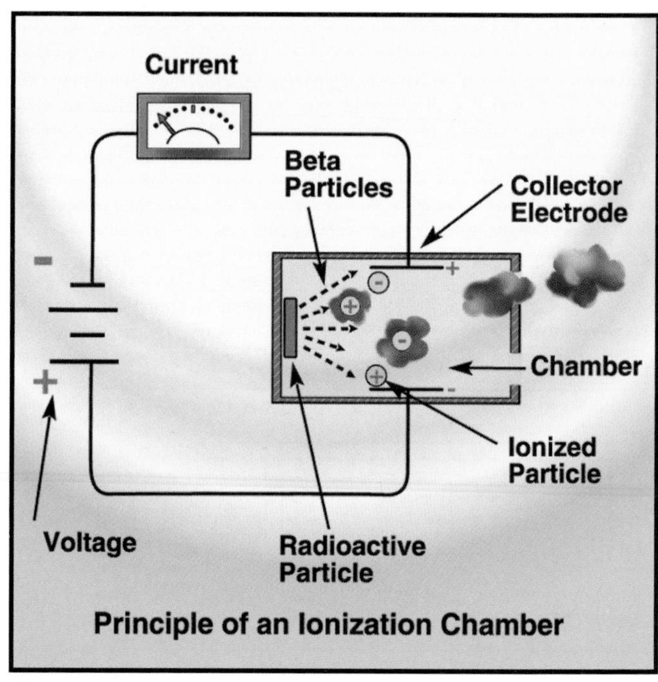

Figure 15.13 Principle of an ionization smoke detector.

An ionization detector responds satisfactorily to most fires; however, this detector generally responds faster to flaming fires than to smoldering ones. It automatically resets when the atmosphere has cleared.

POWER SOURCES

Either batteries or household current can power residential smoke detectors. Battery-operated detectors offer the advantage of easy installation — a screwdriver and a few minutes are all that are needed (Figure 15.14). Battery models are also independent of house power circuits and operate during power failures.

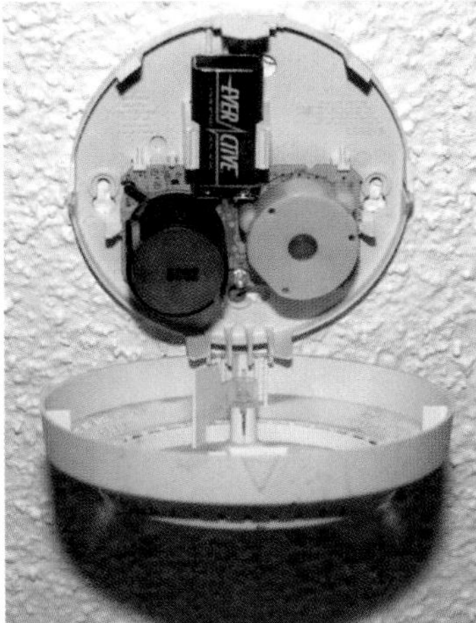

Figure 15.14 The most common type of residential smoke detector is battery operated.

Firefighters should be aware of any state/province or local laws that deal with smoke detectors. Such legislation, in addition to spelling out minimum installation requirements for given occupancies (including homes), may designate the power source to be used. Laws requiring hard-wired units were precipitated by statistics showing a growing lack of maintenance in battery-operated detectors (worn-out batteries not being replaced). Consequently, many codes requiring detectors in newly constructed homes specify 110-volt, hard-wired units. The detector powered by household current is usually the most reliable mechanism. However, in some rural areas and areas with high thunderstorm occurrence, power failures may be more frequent, and battery-operated units may be more appropriate.

It is critical that the specific battery type recommended by the detector's manufacturer be used for replacement. The batteries should be changed at least twice a year or more often if necessary. One way firefighters can get citizens to remember when to change smoke-detector batteries is by suggesting the change be made in the spring and fall at the same time clocks are reset for daylight savings or returned to standard time.

Flame Detectors

There are three basic types of flame detectors (sometimes called *light detectors*).

- Those that detect light in the ultraviolet wave spectrum (UV detectors) (Figure 15.15)
- Those that detect light in the infrared wave spectrum (IR detectors) (Figure 15.16)
- Those that detect both types of light

While these types of detectors are among the most sensitive in detecting fires, they are also easily activated by nonfire conditions such as welding, sunlight, and other sources of bright light. They should only be located in areas where these conditions are unlikely. They must be positioned so that they have an unobstructed view of the protected area. If their line of sight is blocked, they will not activate.

Figure 15.15 A UV detector.

Figure 15.16 An IR detector. *Courtesy of Detector Electronics Corp.*

Because some single-band IR detectors are sensitive to sunlight, they are usually installed in fully enclosed areas. To reduce the likelihood of false alarms, most IR detectors are designed to require the flickering motion of a flame to initiate an alarm.

Ultraviolet detectors are virtually insensitive to sunlight, so they can be used in areas not suitable for IR detectors. However, they are not suitable for areas where arc welding is done or where intense mercury-vapor lamps are used.

Fire-Gas Detectors

When a fire burns in a confined space, it changes the makeup of the atmosphere within the space. Depending on the fuel, some of the gases released by a fire may include the following:

- Water vapor (H_2O)
- Carbon dioxide (CO_2)
- Carbon monoxide (CO)
- Hydrogen chloride (HCl)
- Hydrogen cyanide (HCN)
- Hydrogen fluoride (HF)
- Hydrogen sulfide (H_2S)

Only water vapor, carbon dioxide, and carbon monoxide are released from all fires. Other gases released vary with the specific chemical makeup of the fuel. Therefore, it is only practical to monitor the levels of carbon dioxide and carbon monoxide for general fire-detection purposes (Figure 15.17). This type of detector will initiate an alarm signal somewhat faster than a heat detector but not as quickly as a smoke detector.

Of more importance than the speed of response is the fact that a fire-gas detector can be more discriminating than other types of detectors. A fire-gas detector can be designed to be sensitive only to the gases produced by specific types of hostile fires and to ignore those produced by friendly fires. This detector uses either semiconductors or catalytic elements to sense the gas and trigger the alarm. Compared to the number of other types of detectors, few fire-gas detectors are in use.

Figure 15.17 A fire-gas detector. *Courtesy of RKI Instruments, Inc.*

Combination Detectors

Depending on the design of the system, various combinations of the previously described means of detection may be used in a single device. These combinations include fixed temperature/rate-of-rise heat detectors, combination heat/smoke detectors, and combination smoke/fire-gas detectors (Figure 15.18). The different combinations make these detectors more versatile and more responsive to fire conditions.

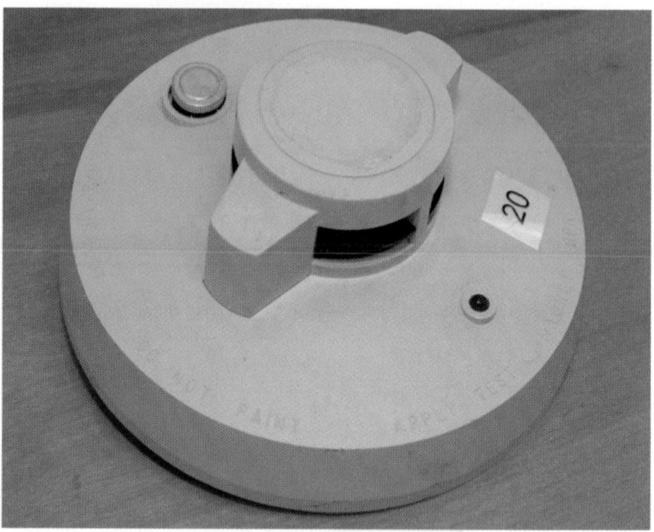

Figure 15.18 A combination heat/smoke detector.

Indicating Devices

A large assortment of audible and visible alarm-indicating devices have also been developed. Some are loud to attract attention in high-noise areas; some generate an electronic tone that is audible in almost any type of environment. Some systems employ bells, horns, or chimes; others use speakers that broadcast prerecorded evacuation instructions (Figure 15.19).

To accommodate special circumstances or populations, such as people who must wear personal noise-attenuation devices because of very high noise levels in their work areas, visual alarm indicators that employ an extremely high intensity clear strobe may be used. These indicators may be used singularly or in combination with other alarm devices (Figure 15.20). Appropriate strobe devices may also be used to meet the requirements of the Americans with Disabilities Act (ADA) in areas where there may be people with hearing impairments.

Figure 15.19 Typical evacuation alarm devices.

Figure 15.20 A strobe evacuation alarm device.

Automatic Alarm Systems

Under some circumstances, insurance carriers may require occupancies to have a system that will transmit a signal to an off-site location for the purpose of summoning organized assistance in fighting a fire. This signal produces an automatic response upon activation of the local alarm at the protected premises. Various brands of alarm systems do this signaling with dedicated wire pairs, leased telephone lines, fiber-optic cable, or wireless communication links.

AUXILIARY SYSTEM

There are three basic types of auxiliary systems: the local energy system, the shunt system, and the parallel telephone system.

Local energy system. A local energy system is used only in those communities that are served by a municipal fire-alarm-box system. A *local energy system* is an auxiliary alarm system within an occupancy that is attached directly to a hard-wired or radio-type municipal fire alarm box (Figure 15.21). When an alarm activates in the protected occupancy, it trips the alarm box to which it is attached and transmits an alarm to the fire alarm center. An alarm can be initiated by manual pull stations, automatic fire detection devices, or water flow devices. Each community has its own requirements for local energy systems, and some do not allow them at all.

Figure 15.21 A municipal master box.

Shunt system. A *shunt system* is one in which the municipal alarm circuit extends (is "shunted") into the protected property. When an alarm is initiated on the premises, whether manually or automatically, the alarm is instantly transmitted to the alarm center over the municipal system.

Parallel telephone system. A *parallel telephone system* does not interconnect with a municipal alarm circuit. It transmits an alarm from the protected property directly to the alarm center over a municipally controlled telephone circuit that serves no other purpose.

REMOTE STATION SYSTEM

A *remote station system* is similar to an auxiliary system but is connected to the fire department telecommunication center directly or through an answering service by some means other than the municipal fire-alarm-box system (Figure 15.22). This connection can be done by leased telephone line or, where permitted, by a radio signal on a dedicated frequency. A remote station is common in localities that are not served by central station systems described later.

A remote station system may transmit either a coded or a noncoded signal. A noncoded system is only allowable where a single occupancy is protected by the system. Up to five facilities may be protected by one coded system. These five facilities usually have a common connection to the remote station. A remote station system must have the ability to transmit a trouble signal to the fire alarm center when the system becomes impaired (Figure 15.23). This type of system may not have local alarm capabilities if evacuation is not the desired action in the event of a fire.

Depending on local policy, the fire department may allow another entity, such as the police department, to monitor the remote station (Figure 15.24). This situation is particularly common in communities that have volunteer fire departments whose stations are not constantly staffed. In these cases, it is important that police telecommunicators are appropriately trained in the actions that need to be taken immediately upon receiving a fire-alarm signal.

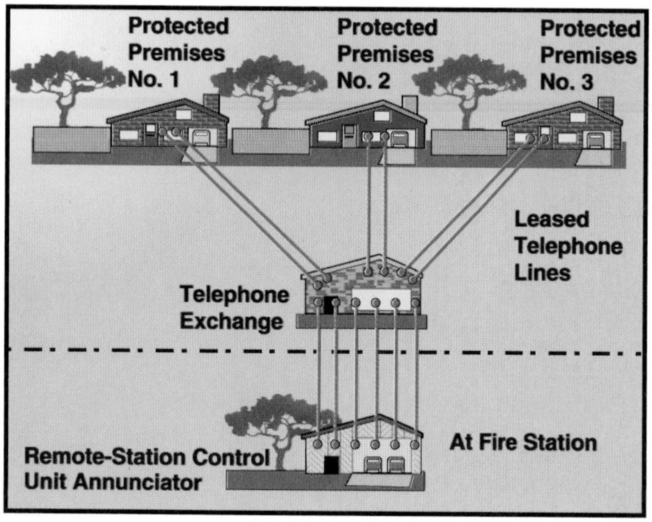

Figure 15.22 A remote system connects directly to the fire department telecommunication center.

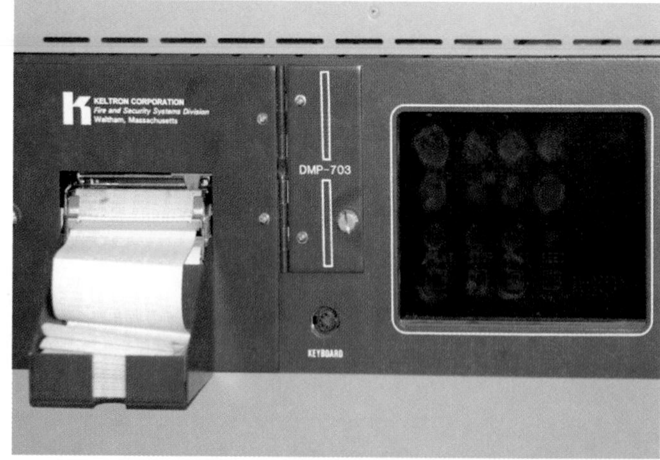

Figure 15.23 A remote system must be capable of transmitting a trouble signal if the system is impaired.

Figure 15.24 Police telecommunications personnel monitor many remote stations.

PROPRIETARY SYSTEM

A *proprietary system* is used to protect large commercial and industrial buildings, high rises, and groups of commonly owned buildings in a single location such as a college campus or industrial complex (Figure 15.25). Each building or area has its own system that is wired into a common receiving point somewhere on the facility. The receiving point must be in a separate structure or in a part of the structure that is remote from any hazardous operations. The receiving station is constantly staffed by representatives of the occupant who are trained in system operation and the actions to take when an alarm is received (Figure 15.26). The operator should be able to summon the fire department either through the system or by telephone.

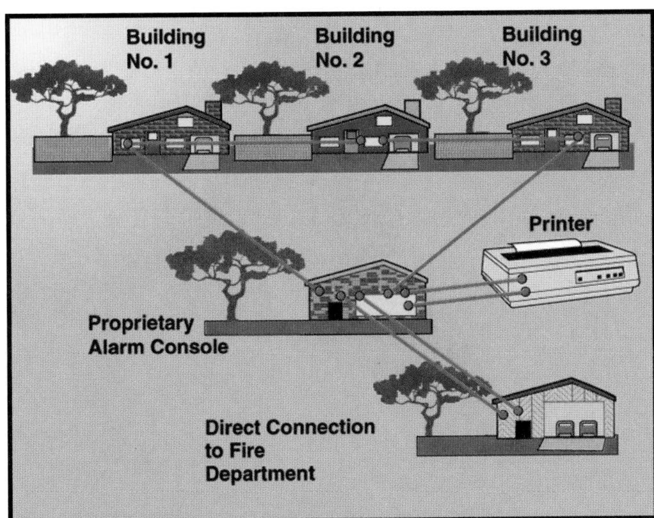

Figure 15.25 Components of a proprietary system.

Figure 15.26 Trained personnel staff the receiving station.

Modern proprietary systems can be very complex with a wide range of capabilities. Some of these capabilities include the following:

- Transmitting coded-alarm and trouble signals
- Monitoring building-utility controls
- Monitoring elevator status
- Monitoring fire and smoke dampers
- Performing security functions

CENTRAL STATION SYSTEM

A central station system is very similar to a proprietary system. The primary difference is that instead of having the alarm-receiving point monitored by the occupant's representative on the protected premises, the receiving point is at an off-site, contracted service point called a *central station* (Figure 15.27). Typically, the central station is an alarm company that contracts with individual customers. When an alarm is initiated at a contracting occupancy, central station employees take that information and initiate an appropriate emergency response. This response usually includes calling the fire department and representatives of the protected occupancy. The alarm system at the

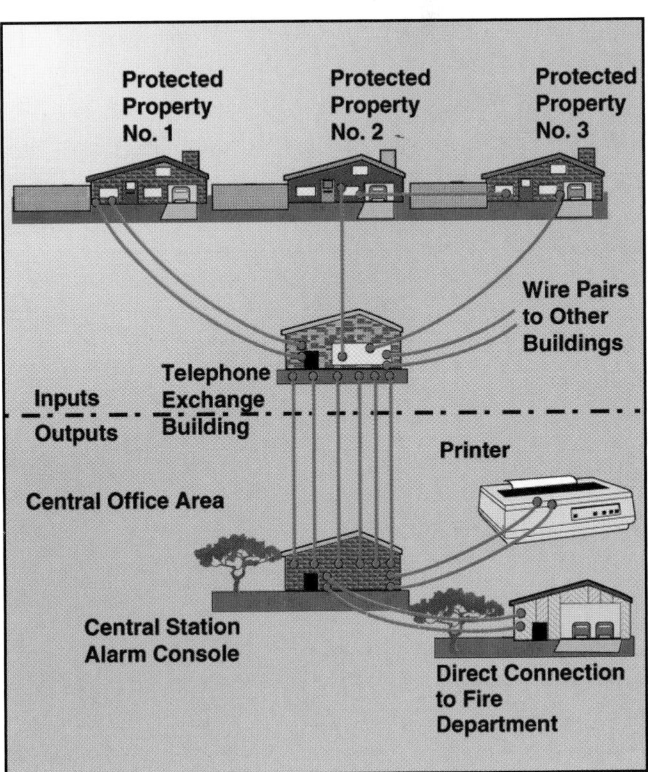

Figure 15.27 Components of a central station system.

protected property and the central station are most commonly connected by supervised telephone lines. All central station systems should comply with the requirements of NFPA 72, *National Fire Alarm Code*.

Supervising Fire Alarm Systems

Unlike ordinary electrical distribution systems that provide power, fire alarm systems are designed to be self-supervising. This means that anytime the system is not operating normally, a distinct trouble signal is generated to attract attention to the system problem (Figure 15.28). This may happen when the system switches to battery power because of a utility power outage or when there is a break in a detector or notification circuit.

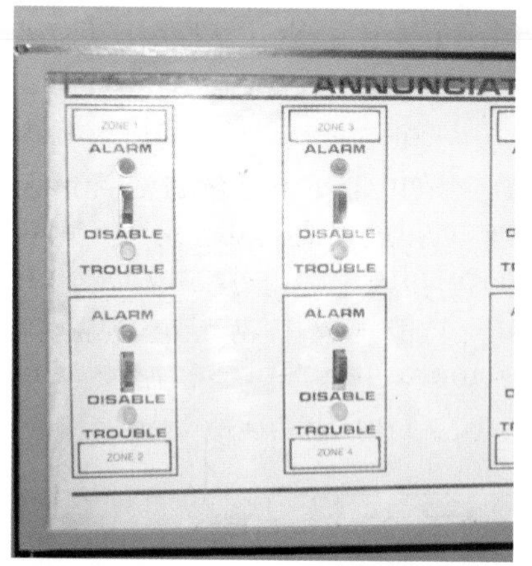

Figure 15.28 An alarm panel with a "trouble" light.

Most older systems operated with looped, supervised circuits in which a tiny current was constantly flowing. A detector would initiate a signal by closing contact points to create a short circuit. In the same way, if a break occurred in one of the detector circuits, a trouble signal would be initiated because the supervisory current would be interrupted.

Many newer systems incorporate some of the latest technological developments of the telecommunication and computer industries. Some of these systems have built-in microprocessors programmed to initiate an internal diagnostic test at specified intervals. The results are recorded on a printer or displayed on a computer screen (Figure 15.29). The same printer or screen is also used to present alarm information.

The sounds of the alarm and trouble signals may be different with each brand of system in use. Firefighters should familiarize themselves with both signals from each brand of system in their area.

Figure 15.29 Modern systems provide a printout to indicate trouble with the system.

Many fixed fire suppression systems depend on a signal from a manual pull station or from an automatic fire detection device to trigger the suppression system (Figure 15.30). The control panel must be specifically listed by a testing laboratory for this purpose, and all control circuits must be supervised.

Figure 15.30 A manual station that initiates an alarm and actuates the suppression system.

Depending on the application, a number of specific features may be incorporated. NFPA 12, *Standard on Carbon Dioxide Extinguishing Systems*, requires a predischarge alarm. Some systems may be programmed to provide an extended discharge period to ensure that a fire-suppressing concentration of CO_2 is achieved. To prevent people from entering the oxygen-deficient environment, automatic door-locking devices may be provided. Older CO_2 systems may be equipped with a pneumatic means of activation. These systems represent another generation of technology that continues to provide reliable protection as long as the systems are tested regularly and maintained in accordance with the manufacturers' instructions.

Fire alarm and supervisory systems may be installed to complement either wet-pipe or dry-pipe sprinkler systems (see Applications of Sprinkler Systems section). Devices capable of sensing a sudden increase or decrease in pressure can detect movement or flow of water; others are actuated by the actual movement of water within a pipe (Figure 15.31). This movement would indicate that either a sprinkler had opened in response to a fire or that water was leaking from a broken pipe. To minimize false alarms but maintain appropriate sensitivity, the detection device is set to respond to flow equal to that of a single sprinkler.

Figure 15.31 A water flow alarm device.

Auxiliary Services

While the primary objective of a fire detection and alarm system is to prevent loss of life and to conserve property in the event of a fire, technological improvements have enhanced the capabilities of contemporary emergency signals systems. Systems that integrate process and environmental controls, security, and personnel-access controls are now common. The following are some of the other auxiliary services available:

- Shutting down or altering airflow in heating, ventilating, and air-conditioning (HVAC) systems for smoke control
- Closing smoke or fire-rated doors and dampers
- Facilitating evacuation by increasing air pressure in stairwells to exclude smoke
- Overriding elevator controls
- Monitoring operation of burner management systems
- Monitoring refrigeration systems and cold-storage areas
- Controlling personnel access to hazardous process or storage areas
- Detecting combustible or toxic gases

AUTOMATIC SPRINKLER SYSTEMS

[NFPA 1001: 3-3.13(a); 3-3.13(b);4-5.1; 4-5.1(a); 4-5.1(b)]

Automatic sprinkler protection consists of a series of sprinklers (sometimes called *sprinkler heads*) arranged so that the system will automatically distribute sufficient quantities of water directly to a fire to either extinguish it or hold it in check until firefighters arrive (Figure 15.32). Water is supplied to the sprinklers through a system of piping. The sprinklers can either extend from exposed pipes or protrude through the ceiling or walls from hidden pipes.

There are two general types of sprinkler coverage: complete sprinkler coverage and partial sprinkler coverage. A *complete sprinkler system* protects the entire building. A *partial sprinkler system* protects only certain areas such as high-hazard areas, exit routes, or places designated by code or by the authority having jurisdiction.

Standards such as NFPA 13, *Standard for the Installation of Sprinkler Systems,* or NFPA 13D, *Standard for the Installation of Sprinkler Systems in One- and Two-Family Dwellings and Manufactured Homes,* are the primary guides used for establishing sprinkler protection in occupancies. These standards have requirements on the spacing of sprinklers in a building, the size of pipe to be

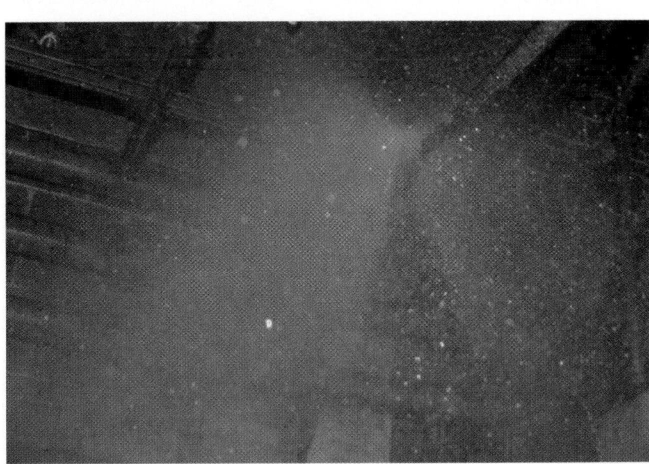

Figure 15.32 A sprinkler in operation.

used, the proper method of hanging the pipe, and all other details concerning the installation of a sprinkler system. These standards specify the minimum design area that should be used to calculate the system. This area is the maximum number of sprinklers that might be expected to activate (Figure 15.33). This is done because most public water supply systems could not be expected to adequately supply 500 or 1,000 operating sprinklers. Thus, the design of the system is based upon the assumption that only a portion of the sprinklers will operate during a fire.

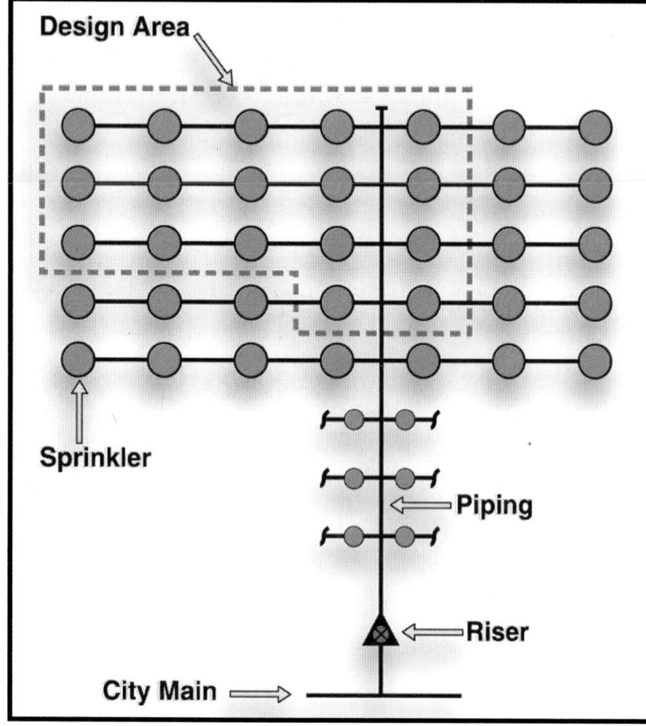

Figure 15.33 Sprinkler systems are designed under the assumption that only a portion of the total number of sprinklers will actually operate during a fire.

The automatic sprinkler and all component parts of the system should be listed by a nationally recognized testing laboratory such as Underwriters Laboratories Inc. or Factory Mutual. Automatic sprinkler systems are now recognized as the most reliable of all fire protection devices, and it is essential for the firefighter to understand the basic system and the operation of pipes and valves. The various applications of sprinkler systems should also be known by firefighters along with sprinkler system effects on life safety.

In general, reports reveal that only in rare instances do automatic sprinkler systems fail to operate. When failures are reported, the reason is rarely because of failure of the actual sprinklers. A sprinkler system may not perform properly because of the following:

- Partially or completely closed main water control valve
- Interruption to the municipal water supply
- Damaged or painted-over sprinklers
- Frozen or broken pipes
- Excess debris or sediment in the pipes
- Failure of a secondary water supply
- Tampering and vandalism

Sprinkler System Effects on Life Safety

The life safety of building occupants is enhanced by the presence of a sprinkler system because it discharges water directly on a fire while it is relatively small. Because the fire is extinguished or controlled in the early growth stage, combustion products are limited. Sprinklers are also effective in the following situations:

- Preventing fire spread upwards in multiple story buildings
- Protecting the lives of occupants in other parts of the building

There are also times, however, when sprinklers alone are not as effective, such as in the following situations:

- Fires are too small to activate the sprinkler system.
- Smoke generation reaches occupants before the sprinkler system activates.
- Sleeping, intoxicated, or handicapped persons occupy the fire building.

Sprinkler System Fundamentals

The principal parts of an automatic sprinkler system are illustrated in Figure 15.34. The system starts with a water main and continues into the control valve. The *riser* is the vertical piping to which the sprinkler valve, one-way check valve, fire department connection (FDC), alarm valve, main drain, and other components are attached. The *feed main* is the pipe connecting the riser to the cross mains. The *cross mains* directly service a number of branch lines on which the sprinklers are installed. Cross mains extend past the last branch lines and are capped to facilitate flushing. System

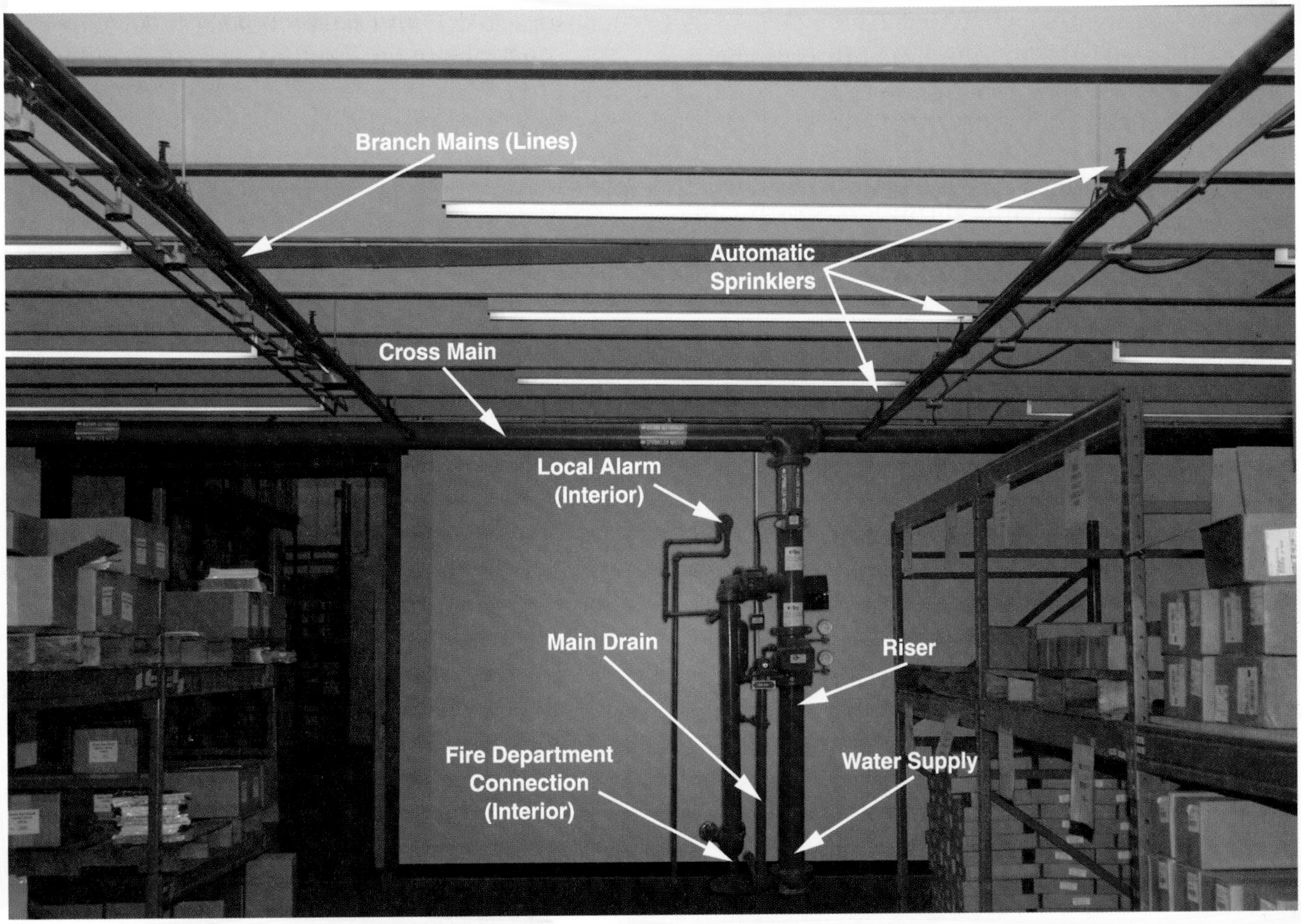

Figure 15.34 Components of a complete sprinkler system.

piping decreases in size from the riser outward. The entire system is supported by hangers and clamps.

Along with discussions of the various types of sprinklers, control valves, operating valves, and water flow alarms, an explanation of how sprinkler systems are supplied with water is also given in the following sections. The various applications of sprinkler systems (dry-pipe, wet-pipe, preaction, deluge, and residential) are also described.

SPRINKLERS

Sprinklers discharge water after the release of a cap or plug that is activated by some heat-responsive element (Figure 15.35). This sprinkler may be thought of as a fixed-spray nozzle that is operated individually by a thermal detector. There are numerous types and designs of sprinklers.

Sprinklers are commonly identified by the temperature at which they are designed to operate. Usually this temperature is identified either

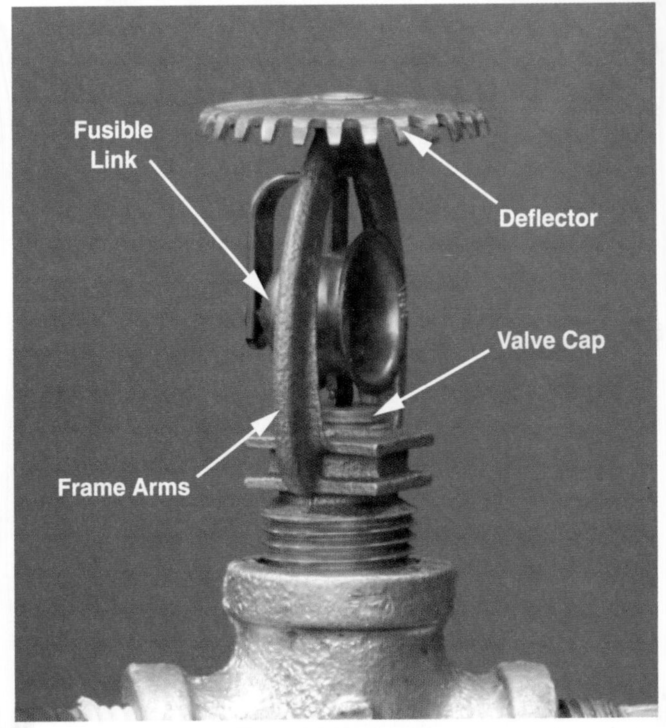

Figure 15.35 Components of an upright fusible-link sprinkler.

by color-coding the sprinkler frame arms, by using different colored liquid in bulb-type sprinklers, or by stamping the temperature into the sprinkler itself (see Table 15.1).

Three of the most commonly used release mechanisms to activate sprinklers are fusible links, frangible bulbs, and chemical pellets. All of these sprinkler mechanisms fuse or open in response to heat (Figures 15.36 a–c).

Fusible link. The design of a sprinkler that uses a fusible link involves a frame that is screwed into the sprinkler piping. Two levers press against the frame, and a cap over the orifice in the frame holds back the water. The fusible link holds the levers together until the link is melted during a fire, after which the water pushes the levers and cap out of the way and strikes the deflector on the end of the frame. The deflector converts the standard ½-inch (13 mm) stream into water spray for more efficient extinguishment.

A quick-response mechanism was developed for life safety purposes. This specially designed fusible link offers increased surface area to collect the heat generated by a fire faster than a standard fusible-link sprinkler. This results in a faster opening of the sprinkler and quicker extinguishment of the fire.

TABLE 15.1
Temperature Ratings, Classifications, and Color Codings

Maximum Ceiling Temperature °F	Maximum Ceiling Temperature °C	Temperature Rating °F	Temperature Rating °C	Temperature Classification	Color Code	Glass Bulb Colors
100	38	135 to 170	57 to 77	Ordinary	Uncolored or Black	Orange or Red
150	66	175 to 225	79 to 107	Intermediate	White	Yellow or Green
225	107	250 to 300	121 to 149	High	Blue	Blue
300	149	325 to 375	163 to 191	Extra High	Red	Purple
375	191	400 to 475	204 to 246	Very Extra High	Green	Black
475	246	500 to 575	260 to 302	Ultra High	Orange	Black
625	329	650	343	Ultra High	Orange	Black

Reprinted with permission from NFPA 13-1996, *Installation of Sprinkler Systems,* Copyright © 1996, National Fire Protection Association, Quincy, MA 02269. This reprinted material is not the complete and official position of the National Fire Protection Association on the referenced subject, which is represented only by the standard in its entirety.

Figure 15.36a A fusible-link sprinkler.

Figure 15.36b A frangible-bulb sprinkler.

Figure 15.36c A chemical-pellet sprinkler.

Frangible bulb. Some sprinklers use a small bulb filled with liquid and an air bubble to hold the orifice shut. Heat expands the liquid until the bubble is absorbed into the liquid. This increases the internal pressure until the bulb shatters at the proper temperature. The breaking temperature is regulated by the amount of liquid and the size of the bubble in the bulb. The liquid is color-coded to designate the designed breaking temperature (refer to Table 15.1). When the bulb shatters, the valve cap is released. The quantity of liquid in the bulb determines when it will shatter.

Chemical pellet. A pellet of solder, under compression, within a small cylinder melts at a predetermined temperature, allowing a plunger to move down and release the valve cap parts.

SPRINKLER POSITION

There are three basic positions for sprinklers: pendant, upright, and sidewall. They cannot be interchanged because they are not designed to provide the proper spray pattern and coverage in any other position. There are also special-purpose sprinklers used in other applications.

Pendant. The *pendant* sprinkler, the most common type in use, extends down from the underside of the piping. This sprinkler sprays a stream of water downward into a deflector that breaks the stream into a hemispherical pattern (Figure 15.37).

Upright. The *upright* sprinkler sits on top of the piping and sprays water into a solid deflector that breaks it into a hemispherical pattern that is redirected toward the floor. The upright standard sprinkler cannot be inverted for use in the hanging or pendant position because the spray of water would be directed toward the ceiling in the inverted position (Figure 15.38).

Sidewall. The *sidewall* sprinkler extends from the side of a pipe and is used in small rooms where the branch line runs along a wall. It has a special deflector that creates a fan-shaped pattern of water (Figure 15.39).

Special-purpose. *Special-purpose* sprinklers are those that are used in specific applications because of their unique characteristics. There are several special-purpose sprinklers for special areas and uses. Special-purpose sprinklers with corrosive-resistant coatings are designed to be installed in areas with corrosive atmospheres. Other special-purpose sprinklers are designed for certain specific applications such as being recessed in the ceiling to blend in with the room decor.

SPRINKLER STORAGE

A storage cabinet for housing extra sprinklers and a sprinkler wrench should be installed in the area protected by the sprinkler system. Normally, these cabinets hold a minimum of six sprinklers and a sprinkler wrench in accordance with NFPA

Figure 15.37 The pendant sprinkler is the most common type of sprinkler in use.

Figure 15.38 An upright sprinkler.

Figure 15.39 A sidewall sprinkler installed.

13 and 13D (Figure 15.40). Typically, the function of changing sprinklers is performed by representatives of the building's occupants who are qualified to perform work on sprinkler systems.

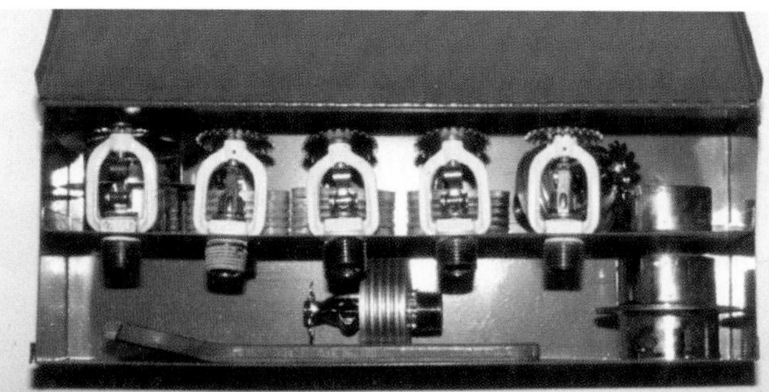

Figure 15.40 A sprinkler storage cabinet.

CONTROL VALVES

Every sprinkler system is equipped with a main water control valve. *Control valves* are used to turn off the water supply to the system so that sprinklers can be replaced, maintenance can be performed, or operations can be interrupted. These valves are located between the source of water supply and the sprinkler system (Figure 15.41). The control valve is usually located immediately under the sprinkler alarm valve, the dry-pipe or deluge valve (see Dry-Pipe System and Deluge System sections), or outside the building near the sprinkler system that it controls. The main control

Figure 15.41 In this case, the OS&Y valve serves as the main control valve.

valve should always be returned to the open position after maintenance is complete. The valves should be secured in the open position or at least supervised to make sure that they are not inadvertently closed.

Main water control valves are indicating and manually operated. An *indicating control valve* is one that shows at a glance whether it is open or closed. There are four common types of indicator control valves used in sprinkler systems: outside screw and yoke (OS&Y), post indicator, wall post indicator, and post indicator valve assembly (PIVA).

Outside screw and yoke (OS&Y) valve. This valve has a yoke on the outside with a threaded stem that controls the opening and closing of the gate. The threaded portion of the stem is out of the yoke when the valve is open and inside the yoke when the valve is closed (Figure 15.42).

Figure 15.42 A small OS&Y valve.

Post indicator valve (PIV). The PIV is a hollow metal post that is attached to the valve housing. The valve stem is inside this post; a movable target is on the stem with the words *OPEN* and *SHUT* at the opening. A PIV with the operating handle in the stored and locked position is shown in Figure 15.43.

Wall post indicator valve (WPIV). A WPIV is similar to a PIV except that it extends through the wall with the target and valve operating nut on the outside of the building (Figure 15.44).

Post indicator valve assembly (PIVA). The PIVA does not use a target with words *OPEN* and *SHUT*, but it has a sight area that is open when the valve is open and closed when the valve is closed (Figure 15.45).

Figure 15.43 A post indicator valve (PIV) showing that it is open. Note the operating handle locked in place.

Figure 15.44 Wall post indicator valves (WPIV).

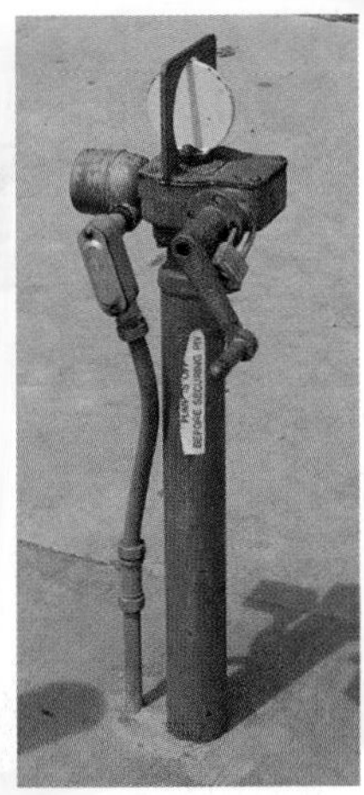

Figure 15.45 A post indicator valve assembly (PIVA) uses a butterfly valve instead of a gate valve.

OPERATING VALVES

Sprinkler systems employ various valves such as alarm test valve, inspector's test valve, and a main drain valve. The alarm test valve is located on a pipe that connects the supply side of the alarm check valve to the retard chamber (place that catches excess water from momentary water pressure surges) (Figure 15.46). This valve is provided to simulate the actuation of the system by allowing

Figure 15.46 A quarter-turn alarm test valve.

water to flow into the retard chamber and operate the water flow alarm devices.

An inspector's test valve is located in a remote part of the sprinkler system (Figure 15.47). The inspector's test valve is equipped with the same size orifice as one sprinkler and is used to simulate the activation of one sprinkler. The water from the inspector's test valve should drain to the outside of the building.

Figure 15.47 The inspector's test valve is equipped with the same size orifice as one sprinkler and is used to simulate the activation of one sprinkler.

Every sprinkler system riser has a main drain valve. The primary purpose of the main drain is to simply drain water from the system for maintenance purposes. However, because a large volume of water flows when the main drain valve is opened, it can also be used to check the system water supply.

WATER FLOW ALARMS

Actuation of fire alarms by sprinkler systems is accomplished when water flows through the system. Sprinkler water flow alarms are normally operated either hydraulically or electrically. The hydraulic alarm is a local alarm used to alert the personnel in a sprinklered building or a passerby that water is flowing in the system (Figure 15.48). This type of alarm uses the water in the system to branch off to a water motor that drives a local alarm gong. The electric water flow alarm is also employed to alert building occupants, and in addition, it can be arranged to notify the fire department.

Figure 15.48 A water flow alarm (also known as a *water gong*) warns that water is moving through the system.

Water Supply

Every sprinkler system should have a water supply of adequate volume, pressure, and reliability. A minimum water supply must be able to deliver the required volume of water to the highest sprinkler in a building at a residual pressure of 15 psi (100 kPa). The minimum flow depends on the hazard to be protected, its occupancy, and the building contents. A connection to a public water system that has adequate volume, pressure, and reliability is a good source of water for automatic sprinklers. This type of connection is often the only water supply available.

The water supply for sprinkler systems is designed to supply only a fraction of the sprinklers actually installed on the system. If a large fire should occur or if a pipe breaks, the sprinkler system will need an outside source of water and pressure to do its job effectively. This additional water and pressure can be provided by a pumper that is connected to the sprinkler fire department connection (Figure 15.49). Fire department connections for sprinklers usually consist of a siamese with at least two 2½-inch (65 mm) female connections with a clapper valve (Figure 15.50) or one large-diameter connection that is connected to a clappered inlet.

Figure 15.49 A supply hose connected to a fire department connection (FDC).

Figure 15.50 The clapper valve is clearly visible inside this FDC.

Sprinkler fire department connections should be supplied with water from pumpers that have a capacity of at least 1,000 gpm (4 000 L/min) or greater. A minimum of two 2½-inch (65 mm) or larger hoses should be attached to the FDC. Whenever possible, fire department pumpers supplying attack lines should operate from hydrants connected to mains other than the main supplying the sprinkler system (Figure 15.51).

After water flows through the fire department connection into the system, it passes through a check valve. This valve prevents water flow from the sprinkler system back into the fire department connection; however, it does allow water from the fire department connection to flow into the sprinkler system (Figure 15.52). The proper direction of water flow through a check valve may be denoted by arrows on the valve or by observing the appearance of the valve casting. A ball drip valve may also be installed at the check valve and FDC. This will keep the valve and connection dry and operating properly during freezing conditions.

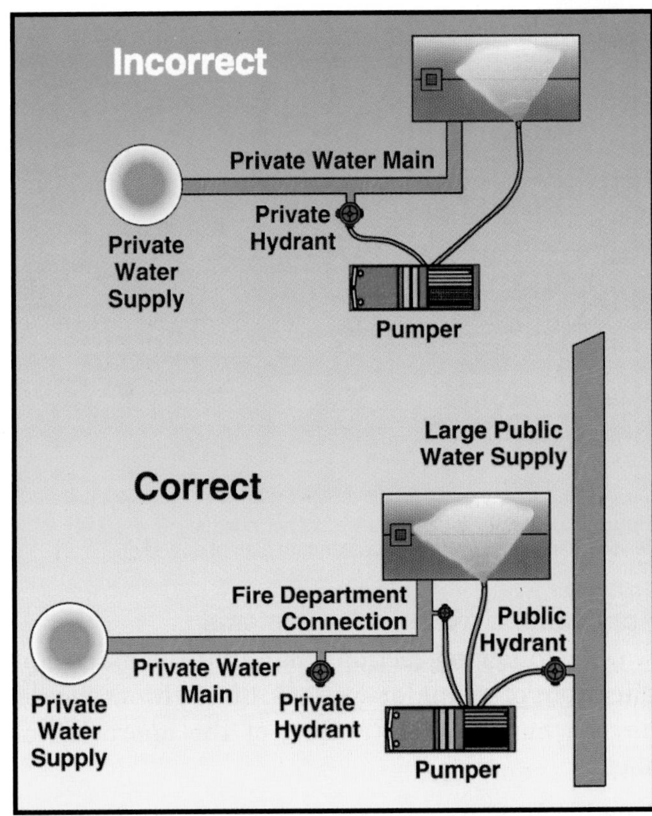

Figure 15.51 The correct and incorrect methods for boosting the water supply to a sprinkler system.

Figure 15.52 A check valve between the FDC and the riser.

A fire department's standard operating procedure should identify the pressure at which a sprinkler system should be supported as well as that needed for special circumstances. Such a plan cannot be established until fire department personnel become familiar with sprinklered properties under their jurisdiction. A standard plan of operation should cover the buildings in the department's jurisdiction, including type of occupancy, type of system, and extent of the system. Therefore, a pre-incident survey is a prerequisite for a plan of operation (Figure 15.53). A thorough knowledge of the public water system is important, including knowing the vol ume and pressure available.

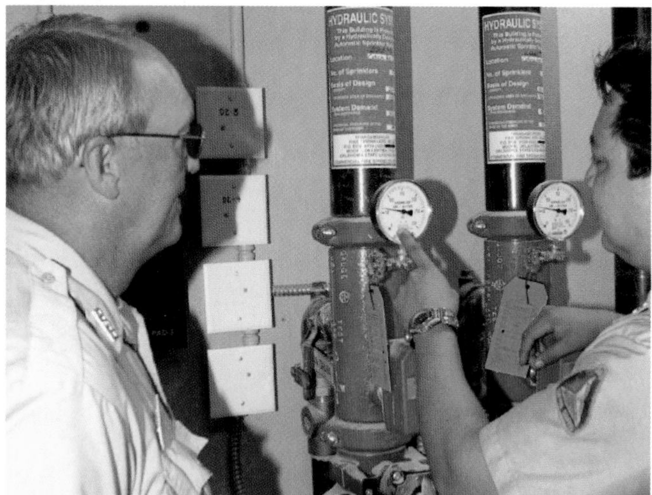

Figure 15.53 Fire department personnel conducting a survey of a building's sprinkler system for a standard plan of operation.

Figure 15.54a A wet-pipe system with an alarm check valve.

Figure 15.54b A "straight stick" wet-pipe sprinkler system. Note the lack of an alarm check valve on the riser.

Applications of Sprinkler Systems

The following sections highlight the major applications of sprinkler systems. Firefighters should have a basic understanding of the operation of each.

- Wet-pipe
- Dry-pipe
- Preaction
- Deluge
- Residential

WET-PIPE SYSTEM

A *wet-pipe sprinkler system* is used in locations that will not be subjected to temperatures below 40°F (4°C). It is the simplest type of automatic fire sprinkler system and generally requires little maintenance. This system contains water under pressure at all times. It is connected to the public water supply so that a fused sprinkler will immediately discharge a water spray in the area and actuate an alarm. This type of system is usually equipped with an alarm check valve that is installed in the main riser adjacent to where the feed main enters the building (Figure 15.54 a). Newer wet-pipe sprinkler systems may not have an alarm check valve. Instead they may have a backflow prevention check valve and an electronic flow alarm. These are sometimes referred to as *straight stick* systems (Figure 15.54 b). To shut down the system, turn off the main water control valve and open the main drain. A pressure gauge on the riser should indicate system pressure.

A wet-pipe sprinkler system may be equipped with a retarding device, commonly called a *retard chamber,* as part of the alarm check valve. This chamber catches excess water that may be sent through the alarm valve during momentary water pressure surges. This reduces the chance of a false alarm activation.

DRY-PIPE SYSTEM

A *dry-pipe sprinkler system* is used in locations where the piping may be subjected to temperatures below 40°F (4°C). All pipes in dry-pipe systems are pitched to help drain the water in the system back toward the main drain. In this system, air under pressure replaces water in the sprinkler piping above the *dry-pipe valve* (device that keeps water out of the sprinkler piping until a fire actuates a sprinkler). When a sprinkler fuses, the pressurized air escapes first, and then the dry-pipe valve automatically opens to permit water into the piping system (Figure 15.55).

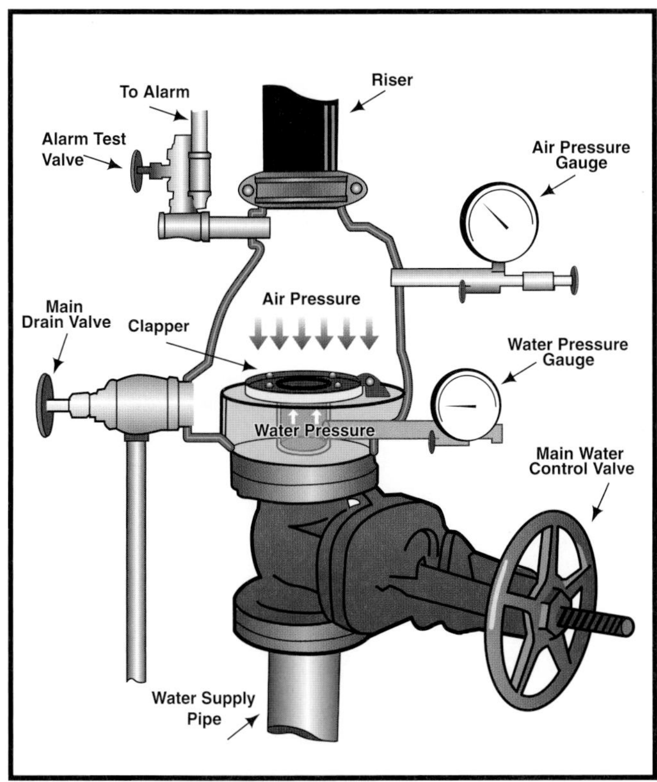

Figure 15.55 Components of a dry-pipe valve.

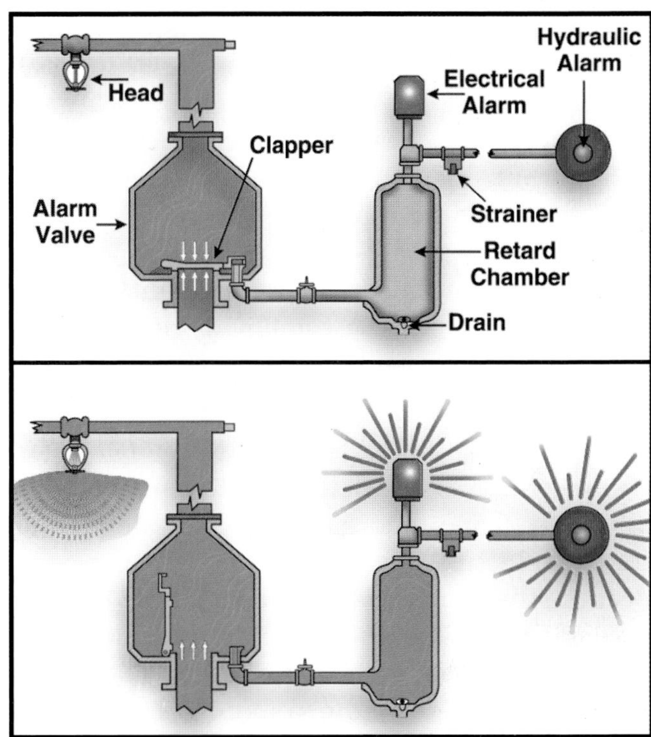

Figure 15.56 This illustrates the dry-pipe valve in both the standby and fire positions.

A dry-pipe valve is designed so that a small amount of air pressure above the dry-pipe valve will hold back a much greater water pressure on the water supply side of the dry-pipe valve. This is accomplished by having a larger surface area on the air side of the clapper valve than on the water side of the valve. The valve is equipped with an air-pressure gauge above the clapper and a water-pressure gauge below the clapper. The required air pressure for dry-pipe systems should be 20 psi (140 kPa) above the trip pressure. Under normal circumstances, the air pressure gauge will read a pressure that is substantially lower than the water-pressure gauge. If the gauges read the same, the system has been tripped, and water has been allowed to enter the pipes. Figure 15.56 illustrates the dry-pipe valve in both the standby and fire positions. Dry-pipe systems are equipped with either electric or hydraulic alarm-signaling equipment.

In a large dry-pipe system, several minutes could be lost while the air is being expelled from the system. Standards require that a quick-opening device be installed in systems that have a water capacity of over 500 gallons (2 000 L). An accelerator is one type of quick-opening device. The basic purpose of this device is to accelerate the opening of the dry-pipe valve. This allows water to be admitted into the sprinkler system quicker, resulting in quicker sprinkler discharge.

PREACTION SYSTEM

A *preaction sprinkler system* is a dry system that employs a deluge-type valve (see Deluge System section), fire detection devices, and closed sprinklers. This type of system is used when it is especially important to prevent water damage, even if pipes are broken. The system will not discharge water into the sprinkler piping except in response to either smoke- or heat-detection system actuation.

Fire detection devices operate a release located in the system actuation unit. This release opens the deluge valve and permits water to enter the distribution system so that water is ready when the sprinklers fuse. When water enters the system, an alarm sounds to give a warning before the opening of the sprinklers.

DELUGE SYSTEM

The purpose of a *deluge sprinkler system* is to wet down the area where a fire originates by discharging water from all open heads in the system. Along with open heads, this system is ordinarily equipped with a deluge valve. When the deluge valve is activated, water is discharged from every open head that is connected to the system controlled by that specific deluge valve. This system is normally used to protect extra-hazardous occupancies. Many modern aircraft hangars are equipped with an automatic deluge system, which may be combined with either a wet- or dry-pipe sprinkler system (Figure 15.57). A system using partly open and partly closed sprinklers is a variation of the deluge system.

Activation of the deluge system may be controlled by flame- and heat-detecting devices or smoke-detecting devices plus a manual device. Because the deluge system is designed to operate automatically and the sprinklers do not have heat-responsive elements, it is necessary to provide a separate detection system. This detection system is connected to a tripping device that is responsible for activating the system. Because there are several different modes of detection, there are also many different methods of operating the deluge valve. Deluge valves may be operated electrically, pneumatically, or hydraulically.

RESIDENTIAL SYSTEMS

Residential sprinkler systems are installed in one- and two-family dwellings (Figure 15.58). This type of sprinkler system is designed to prevent total fire involvement in the room of origin and to give occupants of the dwelling a chance to escape. These systems are covered by NFPA 13D and may be either wet- or dry-pipe systems.

Figure 15.58 Residential sprinkler systems are designed to provide life safety in the occupied portions of one- and two-family dwellings.

Residential sprinkler systems employ quick-response sprinklers. This type of sprinkler is available in both conventional and decorative models. There are several types of piping systems that can be used in this type of system (steel, copper, plastic). Residential sprinkler systems must have a

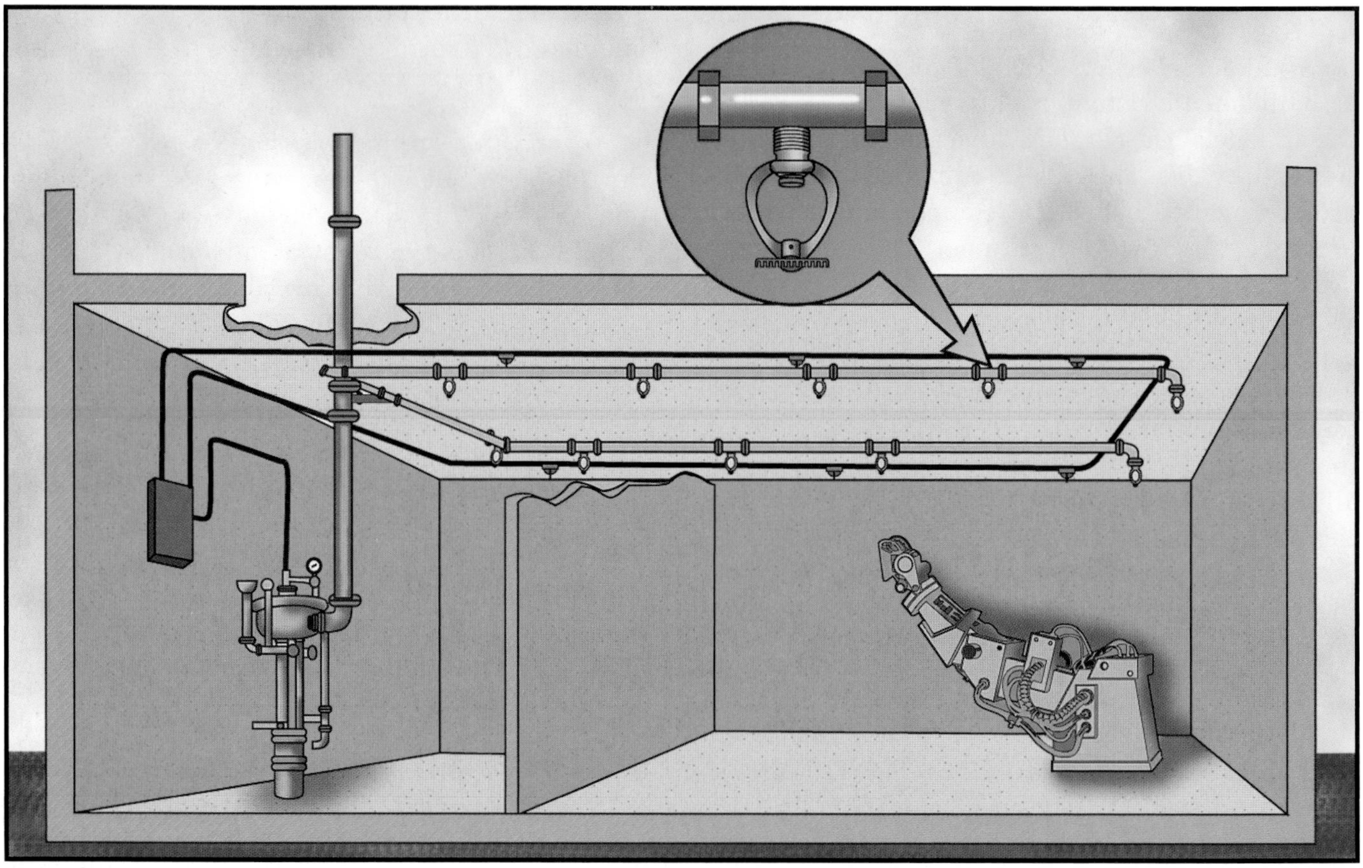

Figure 15.57 A typical deluge system arrangement.

pressure gauge (to check air pressure on dry-pipe systems and water pressure on wet-pipe systems), a flow detector, and a means for draining and testing the system (Figure 15.59). These systems can either be connected directly to the public water supply or tied to the dwelling's domestic water system. A control valve is required to turn off the water to the sprinkler system and to the domestic water system if they are connected. If the sprinkler system is supplied separately from the domestic water system, the sprinkler control valve must be supervised in the open position.

Residential sprinkler systems operate in the same manner as other wet-pipe or dry-pipe systems. (Figure 15.60). Some residential systems may be equipped with a fire department connection (usually a 1½-inch [38 mm] connection), while others have no FDC.

Figure 15.59 A residential sprinkler system drain.

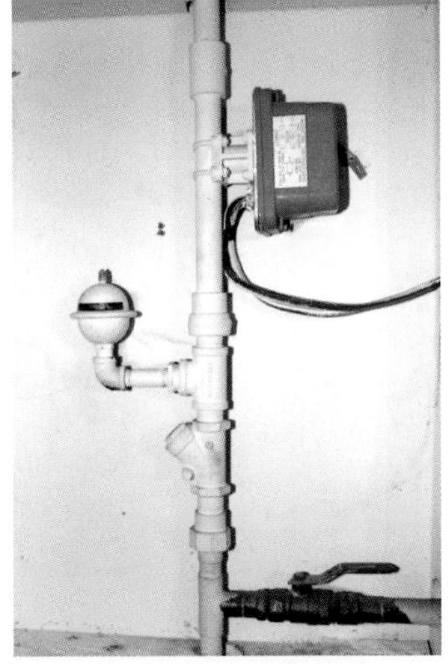

Figure 15.60 Residential sprinkler systems are shut down by closing the sprinkler control valve.

FACTORS TO CONSIDER DURING FIRES AT PROTECTED PROPERTIES

[NFPA 1001: 3-3.13(a); 3-3.13(b)]

Several important considerations must be made when fighting fires in occupancies that have activated sprinkler systems.

- In addition to normal fire fighting operations, an early arriving pumper should connect to the FDC in accordance with the pre-incident plan (Figure 15.61). A maximum effort should be made to supply adequate water to a sprinkler system. The water supply should be conserved for this purpose by limiting the use of direct hoselines from the water supply system serving the sprinkler system. Establish a second water supply for hoselines if necessary. The discharge from sprinklers can be improved by increasing the pressure on the system.

Figure 15.61 A firefighter is connecting the pumper to the fire department connection.

- Sprinkler system control valves must be open for proper operation; check to see that they are. Observe the discharge of sprinklers in the area of the fire and maintain pressure at the pumper to adequately serve the needs of the sprinkler system.
- Sprinkler control valves should not be closed until fire officers are convinced that further operations will simply waste water, produce heavy water damage, or hamper the progress of final extinguishment by fire fighting personnel. Premature closure of

the control valve could lead to a dramatic increase in the intensity of a fire. When a sprinkler control valve is closed, a firefighter with a portable radio should be stationed at the valve in case it needs to be reopened should the fire rekindle. Pumpers should not be disconnected until after extinguishment has been determined by a thorough overhaul.

- Sprinkler equipment should be restored to service before leaving the premises. All sprinkler system maintenance should be performed by representatives of the occupant who are qualified to perform work on sprinkler systems. For liability purposes, it is not recommended that fire department personnel service system components.

Firefighters may be required to stop the flow of water from a single sprinkler that has been activated. This may even be necessary after the main water control valve has been closed because residual water in the system will continue to drain through the open sprinkler until the system is drained below that level. To stop the flow of water, sprinkler wedges may be inserted between the discharge orifice and the deflector and tapped together by hand until the flow is stopped (Figure 15.62). Commercially made stoppers are also available that can be inserted to plug the orifice (Figure 15.63).

Figure 15.62 To stop the water, insert small wooden wedges between the discharge orifice and the detector, and tap together by hand until the flow stops.

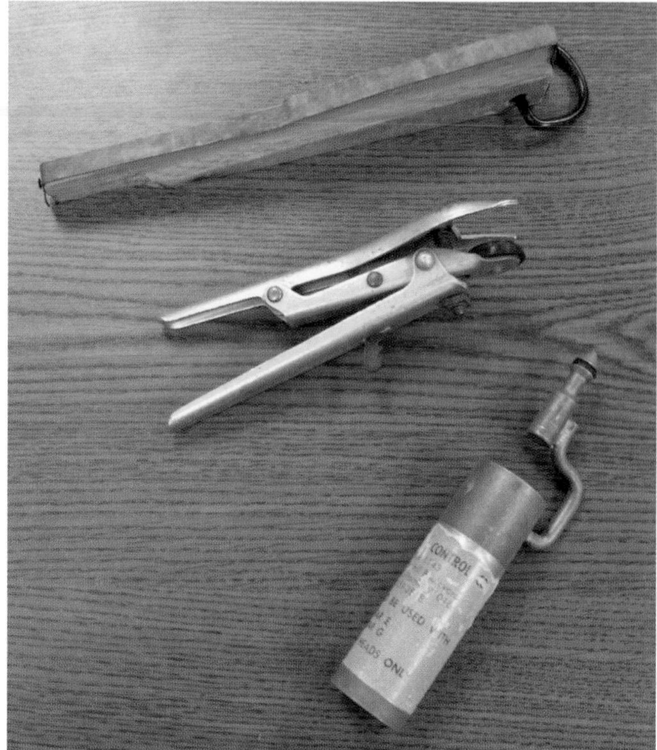

Figure 15.63 Commercially made stoppers are available.

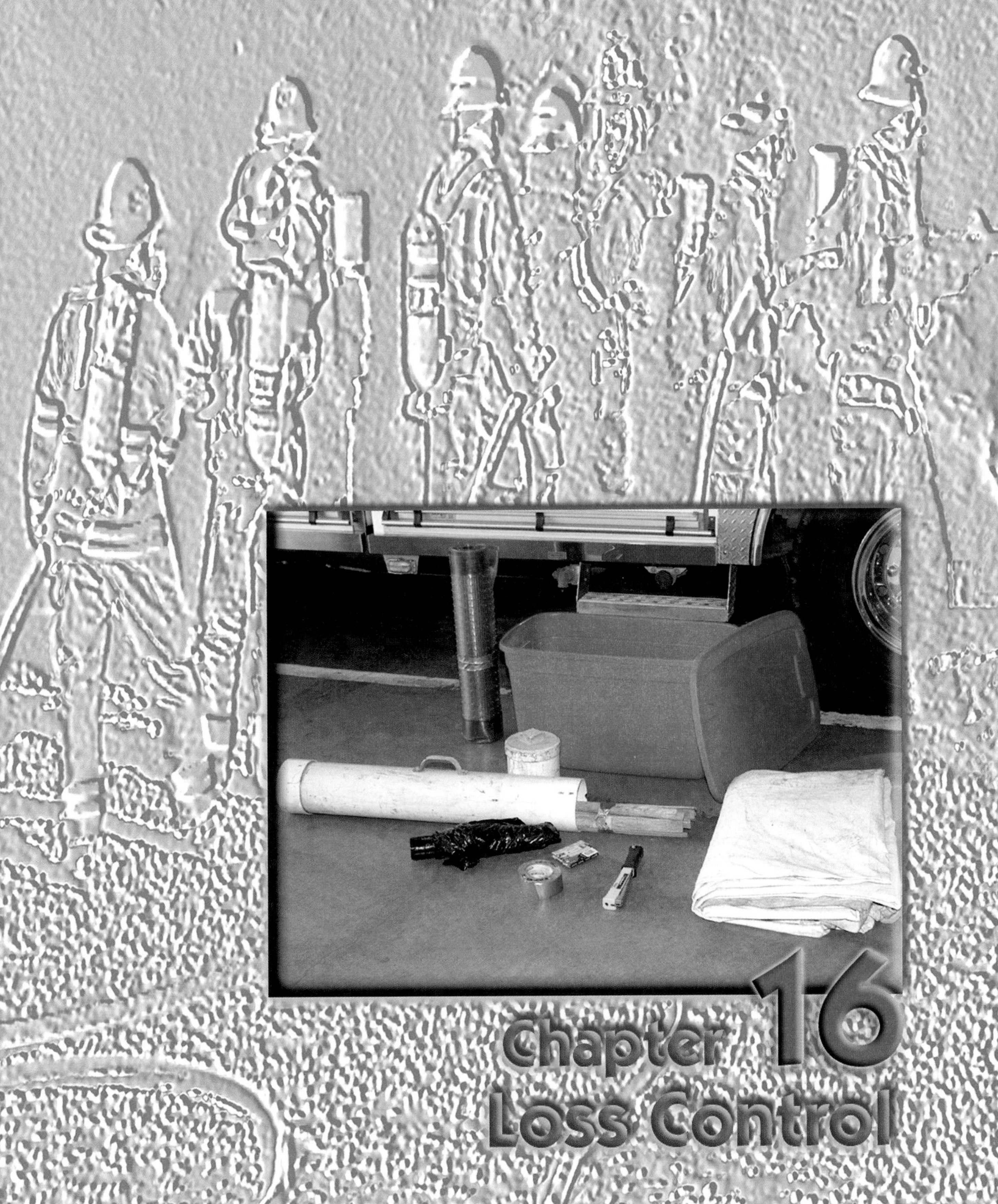

Chapter 16
Loss Control

Chapter 16
Loss Control

INTRODUCTION

Loss control is a component of service delivery that adds value to the only product the fire service has to offer: service. Loss control is the practice that promotes fire fighting as a craft. Most fire events abound with loss control opportunities. It is common for fire officials to notice better morale and efficiency among firefighters who have significantly contributed to reducing fire loss by successfully applying loss control principles.

This chapter explains the philosophy of loss control and gives details of two of the most effective means of loss control: performing proper salvage and overhaul. Planning, procedures, and equipment are discussed.

PHILOSOPHY OF LOSS CONTROL

[NFPA 1001: 3-3.13; 3-3.13(a)]

The philosophy of *loss control* is to minimize damage and provide customer service through effective mitigation and recovery efforts before, during, and after an incident. Loss control is a major part of customer service, which builds goodwill within the community. A fire department often receives words of appreciation and praise by the news media and letters of thanks from citizens for saving their property and cherished possessions. This praise gives firefighters a feeling of accomplishment — particularly when the appreciation comes from those people who have had their belongings saved.

Performing proper salvage and overhaul — two aspects of loss control — are of significant importance to both firefighters and property owners/occupants because they are the most effective means of loss control. Properly applied suppression techniques plus prompt and effective use of good salvage and overhaul procedures will minimize the total losses. Effective salvage procedures coordinated with a thorough and systematic overhaul will also facilitate prompt restoration of the property to full use.

Salvage operations consist of those methods and operating procedures associated with fire fighting that aid in reducing primary and secondary damage during fire fighting operations. Primary damage is caused by the fire; secondary damage is caused by the suppression activities. Both can be minimized through salvage efforts. Some of these damages cannot be avoided because of the need to do forcible entry, apply water, vent the building, and search for fires throughout a structure. Salvage starts as soon as adequate personnel are available and may be done simultaneously with fire attack.

Overhaul operations consist of searching for and extinguishing hidden or remaining fires. Protecting the scene after the fire and preserving evidence of the fire's origin and cause are components of overhaul. Overhaul operations are not normally started until the fire is under control.

SALVAGE

[NFPA 1001: 3-3.13(a); 3-3.13(b); 3-5.3(a); 3-5.3(b)]

Proper salvage operations involve early planning, knowing the procedures necessary to do the job, and being familiar with the various covers and equipment used. Some improvising can be done when equipment is limited. A final part of salvage is protecting the property from the weather.

Planning

Efficient salvage operations require planning and training for fire officers and firefighters (Figure 16.1). Standard operating procedures (SOPs) should be developed to address early and well-coordinated salvage operations. Special preplans may be needed for buildings with high-value contents that are especially susceptible to heat and smoke damage.

In commercial occupancies, an awareness of the value associated with contents vital to business survival is important. The value of contents, particularly in commercial occupancies, can easily exceed the replacement cost of structural materials. Equipment such as computers and file cabinets filled with records, computer disks, etc., are critical to a business's restoration.

Figure 16.1 Firefighters must train regularly in salvage and overhaul techniques.

Fire organizations can facilitate salvage efforts before a fire incident by working with the loss control representatives of various local businesses. Identifying critical records and components needed for business continuation and suggesting implementation of continuous loss control measures, such as stock protection, benefits both the business firm and the fire department.

Procedures

Salvage operations can often be started at the same time as fire attack. For instance, the contents of the room(s) immediately below the fire floor can be covered while fire is being combated on the floor above. Contents of a room can be gathered and covered quickly before a ceiling is pulled. Catching debris with a cover also saves time and effort in cleanup as well as leaving a more professional appearance.

When possible, building contents should be arranged into close piles that can be covered with a minimum of salvage covers. This allows more contents to be protected than if the contents were covered in their original position. When arranging household furnishings, group the furniture in an area of the room away from the wall (Figure 16.2). If a reasonable degree of care is taken, one average-sized cover will usually protect the contents of one residential room. If the floor covering is a removable rug, slip the rug out from under furniture as each piece is moved, and roll the rug to make it easier to move.

A dresser, chest, or high object may be placed at the end of a bed. If there is a rolled rug, place it on top to serve as a ridge pole, allowing water to run off both sides of the covered furniture. Other furniture can be grouped close by; pictures, curtains, lamps, and clothing can be placed on the bed. It may sometimes be necessary to place a salvage cover into position before some articles are placed on the bed. In this event, the bed and furniture will be protected until other items are placed under the cover.

Furniture sitting on wet carpet will absorb water like a sponge, quickly ruining the items even though they may be well covered. To prevent this damage, raise the furniture off the wet floor with

Figure 16.2 Gather furniture into the center of the room so that it can be covered.

water-resistant materials. Precut plastic foam blocks are ideal but canned goods from the kitchen can also be used (Figure 16.3).

Commercial occupancies provide challenges for firefighters who are trying to perform salvage functions (Figure 16.4). The actual arranging of contents to be covered may be limited when large stocks and display features are involved. Display shelves are frequently built to the ceiling and directly against the wall. This construction feature makes contents difficult to cover. When water flows down a wall, it naturally comes into contact with shelving and wets the contents. Contents

Figure 16.3 Precut plastic foam blocks can be used to raise furniture off a wet floor.

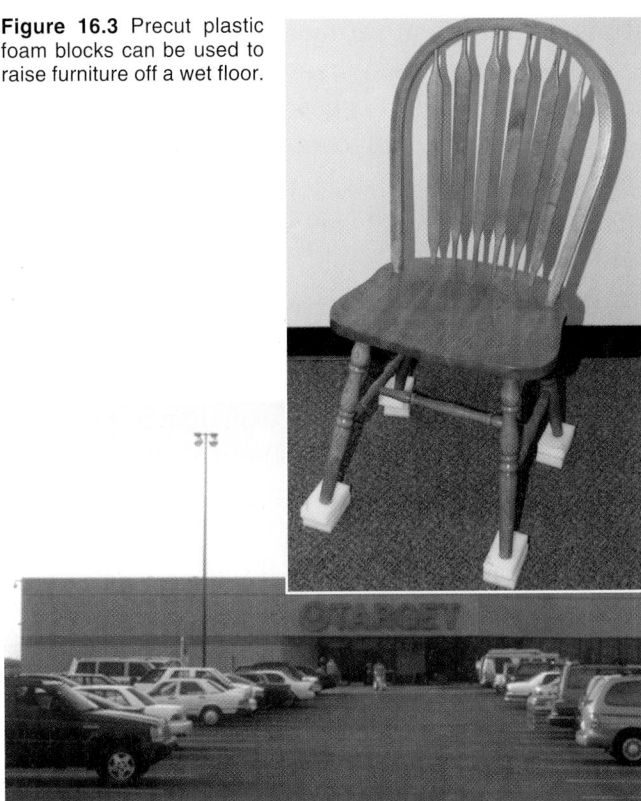

Figure 16.4 Some commercial occupancies, such as this department store, have large quantities of stock to protect.

stacked too close to the ceiling also present a salvage problem. Ideally, there should be enough space between the stock and ceiling to allow firefighters to easily apply salvage covers.

Stock should be stored on pallets, and a lack of skids or pallets under stock susceptible to water damage is a common obstacle to efficient salvage operations (Figure 16.5). Some examples of contents that have perishable characteristics are food

Figure 16.5 Placing boxes on skids provides protection from small amounts of water on the floor.

stuffs, materials in cardboard boxes, feed, paper, and other dry goods. When the number of salvage covers are limited, it is good practice to use available covers for water chutes and catchalls even though the water must be routed to the floor and cleaned up afterward (see Improvising With Salvage Covers section).

Firefighters must be extremely cautious of high-piled stock such as boxed materials or rolled paper that has become wet at the bottom. The wetness often causes the material to expand and push out interior or exterior walls. Wetness also reduces the strength of the material and may cause the piles to collapse (Figure 16.6). Some rolls of paper can weigh a ton (1.02 t) or more. If these rolls of paper were to fall on firefighters, it could seriously injure or kill them.

Large quantities of water may be removed by locating and cleaning clogged drains, removing toilet fixtures, creating scuppers, making use of existing sanitary piping systems, or affixing chutes made of salvage covers, plastic, or other available materials. Water left on cabinets and other horizontal surfaces may ruin finishes over a period of hours. Cabinets and table tops can be wiped off quickly and easily with disposable paper towels. This simple service can save the building owner/occupant a great deal of potential loss.

Figure 16.6 Rolled paper storage can be hazardous to firefighters. If the bottom rolls become wet, the entire stack is likely to collapse. The rolls in this picture each weigh in excess of 2,500 pounds (1 130 kg).

Salvage Covers and Equipment

Salvage covers are made of waterproof canvas materials or vinyl and are found in various sizes. These covers have reinforced corners and edge hems into which grommets are placed for hanging or draping. Synthetic covers are lightweight, easy to handle, economical, and practical for both indoor and outdoor use (Figure 16.7). Many departments use disposable plastic covers. Suitable plastic material is available on rolls, and covers can be cut from the rolls in different sizes to serve various cover needs. Firefighters must be familiar with the salvage covers used in their departments.

NOTE: It is very difficult to adapt plastic covers to traditional salvage cover folds (see Methods of Folding/Rolling and Spreading Salvage Covers section).

A variety of equipment is used along with salvage covers. Every firefighter needs to know how to use these pieces of equipment and how to properly maintain salvage covers.

Figure 16.7 Synthetic salvage covers require less care than natural fiber covers.

SALVAGE COVER MAINTENANCE

The proper cleaning, drying, and repairing of reusable salvage covers increases their span of service. Ordinarily, the only cleaning that is required for canvas salvage covers is wetting or rinsing with a hose stream and scrubbing with a broom. Covers that are extremely dirty and stained may be scrubbed with a detergent solution and then thoroughly rinsed (Figure 16.8). Permitting canvas salvage covers to dry when dirty is not a good practice. After carbon and ash stains have dried, a chemical reaction takes place that rots the canvas. Foreign materials are difficult to remove when dry, even with a detergent. Canvas salvage covers should be completely dry before they are folded and placed in service. This practice is essential to prevent mildew and rot. There is no particular objection to outdoor drying of salvage covers except that the wind tends to blow and whip them about.

Synthetic salvage covers do not require as much maintenance as canvas ones. These covers may be folded wet, but it is usually better to let them dry first so they will not get a mildew smell (Figure 16.9).

After salvage covers are dry, they should be examined for damage. Holes can be located by first placing three or four firefighters side-by-side along one end. Then have them pick up the end and pass it back over their heads while walking toward the other end, looking up at the underside of the cover (Figure 16.10). Light will show through even the smallest holes. Holes should be marked with chalk (Figure 16.11). The holes can be repaired by placing duct or mastic tape over them or by patching with iron-on or sew-on patches, depending upon the material.

Figure 16.8 Extremely dirty salvage covers can be scrubbed with a detergent solution.

Figure 16.9 Synthetic salvage covers can be rinsed clean.

Figure 16.10 Firefighters are examining a salvage cover for damage.

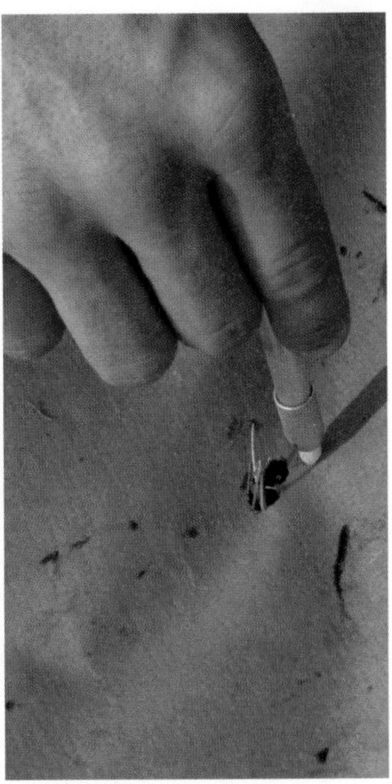

Figure 16.11 Any holes in the canvas should be marked with chalk.

SALVAGE EQUIPMENT

For conducting salvage operations at a fire, it is suggested that salvage equipment be located in a readily accessible area on the apparatus. Individual fire department SOPs dictate on which apparatus the salvage equipment is carried and who performs the primary salvage operations on the fire scene. This avoids delay in beginning salvage operations, although salvage and loss control are everyone's responsibilities.

Smaller tools and equipment should be kept in a specially designated salvage toolbox or other container to make them easier to carry. Loss control materials and supplies may be kept in a plastic tub and brought into the structure early in the fire event (Figure 16.12). The materials and supplies

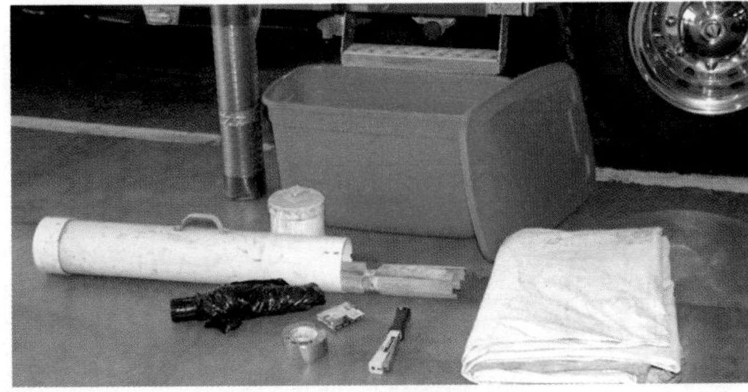

Figure 16.12 Salvage equipment stored in a plastic tub is easily carried.

are then readily available for loss control activities. The tub itself provides a useful water-resistant container to protect items such as computers, pictures, and other water-sensitive materials.

The following is a list of typical salvage equipment that should be carried on the apparatus. The use of this equipment, however, is not limited to salvage operations. Some of these items are discussed in the paragraphs that follow.

- Electrician's pliers
- Sidecutters
- Various chisels
- Tin snips
- Tin roof cutter (can opener)
- Adjustable wrenches
- Pipe wrenches
- Hammer(s)
- Sledgehammer
- Hacksaw
- Crosscut handsaw
- Heavy-duty stapler and staples
- Linoleum knife
- Wrecking bar
- Padlock and hasp
- Hinges
- Screwdriver(s)
- Battery-operated power tools
- Hydraulic jack
- Assortment of nails
- Assortment of screws
- Roofing or tar paper and roofing nails
- Plastic sheeting
- Wood lath
- Mops
- Squeegees
- Water scoops
- Scoop shovels
- Brooms
- Mop wringers with buckets
- Automatic sprinkler kit
- Water vacuum
- Submersible pump and discharge hose
- Sponges
- Chamois
- Paper towels
- Assortment of rags
- 100-foot (30 m) length of electrical cable with locking-type connectors, 14-3 gauge or heavier
- Pigtail ground adapters, 2 wire to 3 wire, 14-3 gauge or heavier with 12-inch (300 mm) minimum length
- An approved ground fault interruption device
- Salvage covers
- Floor runners
- Duct tape
- Plastic bags
- Cardboard boxes with tape dispenser
- Styrofoam blocks

Automatic sprinkler kit. The tools found in a sprinkler kit are needed when fighting fires in buildings protected by automatic sprinkler systems. These tools are used to stop the flow of water from an open sprinkler before it has had time to completely drain after shutdown. A flow of water from an open sprinkler can do considerable damage to merchandise on lower floors after a fire has been controlled in a commercial building. Sprinkler tongs or a stopper and a sprinkler wedge are suggested tools for a sprinkler kit. The SOPs of some departments may require additional equipment necessary to restore a system to service. Use caution when replacing sprinklers; some types require use of a special wrench to prevent damage to the sprinkler.

Carryalls (debris bags). *Carryalls,* sometimes referred to as *debris bags* or *buckets,* are used to carry debris, catch falling debris, and provide a water basin for immersing small burning objects (Figure 16.13).

Floor runners. Costly floor coverings can be ruined by mud and grime tracked in by firefighters. These floor coverings may be protected by using floor runners. Floor runners can be unrolled from an entrance to almost any part of a building.

Figure 16.13 A carryall can be used to haul debris from a building.

Commercially prepared vinyl-laminated nylon floor runners are lightweight, flexible, tough, heat and water resistant, and easy to maintain (Figure 16.14).

Dewatering devices. Dewatering devices are used to remove water from basements, elevator shafts, and sumps (Figure 16.15). Fire department pumpers should not be used for this purpose because they are intricate and expensive machines and are not intended to pump the dirty, gritty water found in such places. Trash-type pumps are best suited for salvage operations. A jet siphon device may be used for the removal of excess water. These devices can be moved to any point where a line of hose can be placed and an outlet for water can be provided.

Water vacuum. One of the easiest and fastest ways to remove water is the use of a water vacuum device. Dirt and small debris may also be removed from carpet, tile, and other types of floor coverings with this equipment. The water vacuum appliance consists of a tank (worn on the back or placed on wheels) and a nozzle. Backpack-type tanks normally have a capacity of 4 to 5 gallons (15 L to 20 L) and can be emptied by simply pulling a lanyard that empties the water through the nozzle or through a separate drain hose (Figure 16.16). Floor models on rollers may have capacities up to 20 gallons (80 L) (Figure 16.17).

Figure 16.14 Floor runners are used to protect carpets and other floor coverings from foot traffic.

Figure 16.15 Portable pumps may be used to remove water from a structure.

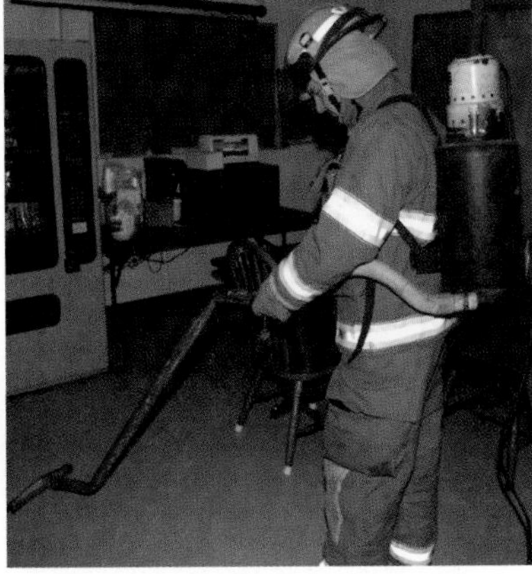

Figure 16.16 Some water vacuums are designed to be worn on the back.

Figure 16.17 A floor model water vacuum.

Methods of Folding/Rolling and Spreading Salvage Covers

One of the key factors in successful salvage operations is the proper handling and deployment of salvage covers. The following sections highlight basic principles of salvage cover storage and deployment by one or two firefighters.

ONE-FIREFIGHTER SPREAD WITH A ROLLED SALVAGE COVER

The principal advantage of the one-firefighter salvage cover roll is that one person can quickly unroll a cover across the top of an object and unfold it (Figure 16.18). Skill Sheet 16-1 describes the procedure for preparing a salvage cover roll for a one-firefighter spread. Folding the cover requires two firefighters. A salvage cover rolled for a one-firefighter spread may be carried on the shoulder or under the arm. Use the steps described in Skill Sheet 16-2 for a one-firefighter spread with a rolled salvage cover.

Figure 16.18 A firefighter performing a one-firefighter salvage cover spread.

ONE-FIREFIGHTER SPREAD WITH A FOLDED SALVAGE COVER

Some departments prefer to carry salvage covers folded, as opposed to rolled (Figure 16.19). The procedures in Skill Sheet 16-3 highlight folding a salvage cover for one-firefighter deployment. Two firefighters are needed to make this fold, and they will be performing the same functions simultaneously. This folded salvage cover may be carried in any manner, however, carrying it on the shoulder is convenient. The steps described in Skill Sheet 16-4 can be used when one firefighter spreads a folded salvage cover.

Figure 16.19 A folded salvage cover.

TWO-FIREFIGHTER SPREAD WITH A FOLDED SALVAGE COVER

Large salvage covers cannot be easily handled by a single firefighter. Therefore, they must be folded for two-firefighter deployment. The procedure in Skill Sheet 16-5 can be used to make the two-firefighter fold. The most convenient way to carry this fold is on the shoulder with the open edges next to the neck. It makes little difference which end of the folded cover is placed in front of the carrier because two open-end folds will be exposed. The cover should be carried so that the carrier can grab the lower pair of corners, and the second firefighter can grab the uppermost pair (Figure 16.20).

The balloon throw is the most common method for two firefighters to deploy a large salvage cover. The balloon throw gives better results when sufficient air is pocketed under the cover. This

Figure 16.20 Proper method for carrying a folded cover for a two-firefighter spread.

pocketed air gives the cover a parachute effect that floats it in place over the article to be covered (Figure 16.21). The steps described in Skill Sheet 16-6 are for making the balloon throw.

Figure 16.21 Firefighters performing a balloon throw.

Improvising With Salvage Covers

In addition to simply covering building contents, salvage covers may also be used to catch and route water from fire fighting operations. The following section details some of these special situations.

REMOVING WATER WITH CHUTES

Using a chute is one of the most practical methods of removing water that comes through the ceiling from upper floors. Water chutes may be constructed on the floor below fire fighting operations to drain runoff through windows or doors (Figure 16.22). Some fire departments carry prepared chutes, approximately 10 feet (3 m) long, as regular equipment, but it may be more practical to make chutes using one or more covers. Effective water diversion is limited only to the imagination of the fire personnel. Plastic sheeting, a hammer-type stapler, and duct tape allow quick and easy construction of water diversion chutes. Skill Sheets 16-7 and 16-8 describe the procedures for constructing water chutes.

CONSTRUCTING A CATCHALL

A catchall is constructed from a salvage cover that has been placed on the floor to hold small amounts of water (Figure 16.23). The catchall may also be used as a temporary means to control large amounts of water until chutes can be constructed to route the water to the outside. Properly constructed catchalls will hold several hundred gallons (liters) of water and often save considerable time during salvage operations. The cover should be placed into position as soon as possible, even before the sides of the cover are rolled. Two people are usually required to prepare a catchall to make more uniform rolls on all sides. The steps required to make a catchall are shown in Skill Sheet 16-9.

Figure 16.22 Ladders and tarps can be used to create water chutes that will channel excess water from the building.

Figure 16.23 A completed catchall.

Covering Openings

One of the final parts of salvage operations is the covering of openings to prevent further damage to the property by weather. Cover any doors or windows that have been broken or removed with plywood, heavy plastic, or some similar materials to keep out rain (Figure 16.24). Plywood, hinges, a hasp, and a padlock can be used to fashion a temporary door. Cover openings in roofs with plywood, roofing paper, heavy plastic sheeting, or tar paper. Use appropriate roofing nails if roofing, tar paper, or plastic is used. Tack down the edges with laths between the nails and the material.

Figure 16.24 Window openings may be covered with plastic to temporarily keep out the elements.

OVERHAUL

[NFPA 1001: 3-3.7(b); 3-3.12; 3-3.12(a); 3-3.12(b); 4-3.2(a)]

Overhaul is the practice of searching a fire scene to detect hidden fires or sparks that may rekindle and to identify the possible point of origin and cause of fire (Figure 16.25). Afterwards, the building, its contents, and the fire area are placed in as safe and habitable a condition as possible and protected from the elements. Any possible incendiary evidence of arson is preserved. Salvage operations performed during fire fighting will directly affect any overhaul work that may be needed later. Many of the tools and equipment used for overhaul are the same as those used for other fire fighting operations. Some of the tools and equipment used specifically for overhaul, along with their uses, may include the following:

Figure 16.25 Proper overhaul reduces the chance of a rekindled fire.

- Pike poles and plaster hooks — Open ceilings to check on fire extension.
- Axes — Open walls and floors.
- Battery-powered saws, drills, and screwdrivers — Make small, neat openings and square up larger holes made with larger tools. These tools are also useful for building "packaging" tasks such as constructing temporary doors and window coverings.
- Carryall, buckets, and tubs — Carry debris or provide a basin for immersing smoldering material.
- Shovels, bale hooks, and pitchforks — Move baled or loose materials.

It is essential for firefighters to wear proper protective clothing including self-contained breathing apparatus (SCBA) while performing overhaul and extinguishing hidden fires unless the atmosphere has been proven safe for a lower level of protection by reliable testing methods. Overhaul operations should be visually directed by a supervisor/officer not directly engaged in overhaul tasks. If a fire investigator is on the scene, he should be involved in planning and supervising the overhaul activities.

Charged hoselines should always be available for the extinguishment of hidden fires; however, the same size lines used to bring the fire under control are not always necessary. Fire department pumpers can often be disconnected from hydrants; however, local policies may dictate that at least one supply line be left in place as a precaution. Typically, 1½-inch (38 mm) or 1¾-inch (45 mm) attack lines are used for overhaul. During minor overhaul operations, water fire extinguishers or booster hoses may be used to extinguish small fires. However, at least one attack line should still be available in the event it is needed. Regardless of the type of hose being used, the nozzle should be placed so that it will not cause additional water damage. Leaking couplings should be tightened or repaired. Don't allow water damage from leaking hoselines. Using a 100-foot (30 m) length of hose as the first section on attack lines would greatly lower the probability that any couplings other than those at the nozzle would even be inside a building.

Before starting a search for hidden fires, it is important to evaluate the condition of the area to be searched. The intensity of the fire and the amount of water used for its control are two important factors that affect the condition of the building. The intensity of the fire determines the extent to which structural members have been weakened, and the amount of water used determines the additional weight placed on floors and walls due to the absorbent qualities of the building contents. Consideration should be given to these factors for the protection of personnel during overhaul.

The firefighters should also be aware of other dangerous building conditions such as the following:

- Weakened floors due to floor joists being burned away
- Concrete that has spalled due to heat
- Weakened steel roof members (tensile strength is affected at about 500°F (260°C)
- Walls offset because of elongation of steel roof supports
- Weakened roof trusses due to burn-through of key members
- Mortar in wall joints opened due to excessive heat
- Wall ties holding veneer walls melted from heat

The firefighter can often detect hidden fires by sight, touch, sound, or electronic sensors (Figure 16.26). The following are some of the indicators for each:

- Sight
  — Discoloration of materials
  — Peeling paint
  — Smoke emissions from cracks
  — Cracked plaster
  — Rippled wallpaper
  — Burned areas

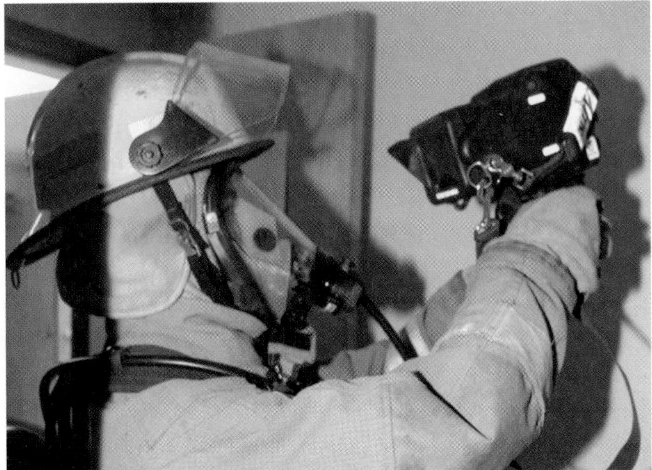

Figure 16.26 A firefighter checking for hidden fire with an infrared heat detector.

- Touch — Feel walls and floors for heat with the back of the hand (Figure 16.27).
- Sound
  — Popping or cracking of fire burning
  — Hissing of steam
- Electronic Sensors
  — Thermal (heat) signature detection
  — Infrared signature detection

Typically, overhaul begins in the area of actual fire involvement. The process of looking for extension should begin as soon as possible after the fire has been declared under control. The overhaul plan can then be systematically carried out. If it is found that the fire extended to other areas,

Figure 16.27 A firefighter using the back of his hand to feel for heat.

firefighters must determine through what medium it traveled. When floor beams have burned at their ends where they enter a party wall, fire personnel overhaul the ends by flushing the voids in the wall with water. The far side of the wall should also be checked to see whether the fire or water has come through. Insulation materials should be thoroughly checked because they can harbor hidden fires for a prolonged period. Usually, it is necessary to remove the material in order to properly check it or extinguish fire in it.

An understanding of basic building construction will assist the firefighter in searching for hidden fires. When the fire has burned around windows or doors, there is a possibility that there is fire remaining within the frames or casings. These areas should be opened to ensure complete extinguishment. These areas can be opened by simply pulling off the molding to expose the inner parts of the frame or casing (Figure 16.28). When fire has burned around a combustible roof or cornice, it is advisable to open the cornice and inspect for hidden fires.

When concealed spaces below floors, above ceilings, or within walls and partitions must be opened during the search for hidden fires, the furnishings of the room should be moved to locations where they will not be damaged. Only enough wall, ceiling, or floor covering should be removed to ensure complete extinguishment. Weight-bearing members should not be disturbed.

Figure 16.28 A firefighter pulling molding to look for hidden fire.

When opening concealed spaces, consideration should also be given to the restoration of the area. During fire fighting, openings should be made in construction to check for extension and allow extinguishment. However, when conditions allow, firefighters should make neat, planned openings to assure extinguishment, demonstrating workmanlike professionalism for future restoration.

If it is not appropriate to use small inspection openings due to obvious fire involvement, ceilings are opened from below using an appropriate overhaul tool. To open a plaster ceiling, a firefighter must first break the plaster and then pull off the lath. Metal or composition ceilings may be pulled from the joist in a like manner. When pulling, the firefighter should not stand under the space to be opened. The firefighter should always be positioned between the area being pulled and the doorway to keep the exit route from being blocked with falling debris (Figure 16.29). The pull should be down and away to prevent the ceiling from dropping on the firefighter's head. No firefighter should attempt to pull a ceiling without wearing full protective clothing, including eye and respiratory protection.

Figure 16.29 Pull ceilings to check for fire extension.

Quite frequently, small burning objects are uncovered during overhaul. Because of their size and condition, it is better to submerge entire objects in containers of water than to try drenching them with streams of water. Bathtubs, sinks, lavatories, and wash tubs are all useful for this purpose. Larger furnishings, such as mattresses, stuffed furniture, and bed linens, should be removed to the outside where they can be easily and thoroughly extinguished. Firefighters should remember that all scorched or partially burned articles may prove helpful to an investigator in preparing an inventory or determining the cause of the fire.

The use of wetting agents is of considerable value when extinguishing hidden fires. The penetrating qualities of wetting agents usually permit complete extinguishment in cotton, upholstery, and baled goods. The only way to assure fires in bales, rags, cotton, hay, alfalfa, etc., is to break them apart. Special care should be taken to eliminate indiscriminate use and direction of hose streams.

SKILL SHEET 16-1 PREPARING A SALVAGE COVER ROLL FOR A ONE-FIREFIGHTER SPREAD

NOTE: Two firefighters must make initial folds to reduce the width of the cover to form this roll. Steps 1 through 8 are performed simultaneously by both firefighters on opposite sides of the cover. Steps 9 through 12 may be performed by both firefighters who are stationed at the same end of the roll.

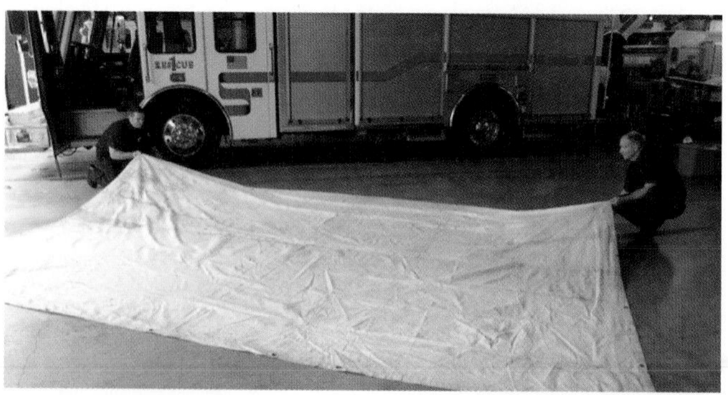

Step 1: Grasp the cover with the outside hand midway between the center and the edge to be folded.

Step 2: Place the other hand on the cover as a pivot midway between the outside hand and the center.

Step 3: Bring the fold over to the center of the cover. This creates an inside fold (center) and an outside fold.

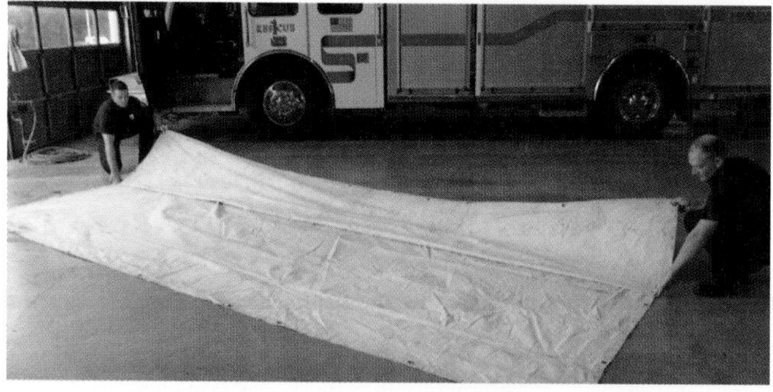

Step 4: Grasp the cover corner with the outside hand.

Step 5: Place the other hand as a pivot on the cover over the outside fold.

Step 6: Bring this outside edge over to the center, and place it on top of and in line with the previously placed first fold.

Step 7: Fold the other half of the cover in the same manner by using Steps 1 through 6.

Step 8: Straighten the folds if they are not straight.
Step 9: Fold over about 12 inches (300 mm) at each end of the cover to make clean, even ends for the completed roll.

Step 10: Start the roll by rolling and compressing one end into a tight compact roll; roll toward the opposite end.

Step 11: Tuck in any wrinkles that form ahead of the roll as the roll progresses.

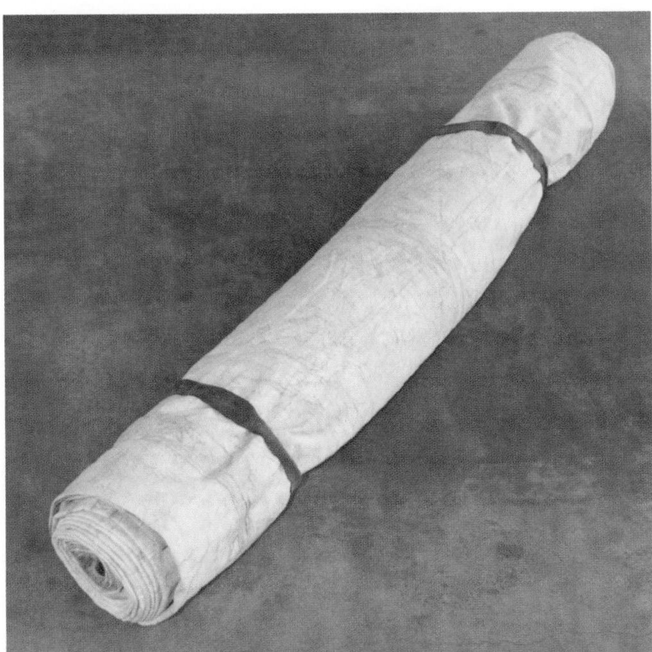

Step 12: Secure the completed roll with inner tube bands or Velcro® straps or tie with cords.

SKILL SHEET 16-2 ONE-FIREFIGHTER SPREAD WITH A ROLLED SALVAGE COVER

Step 1: Start at one end of the object to be covered.

Step 2: Unroll a sufficient amount to cover the end.

Step 3: Continue to unroll toward the opposite end.

Step 4: Let the rest of the roll fall into place at the other end.

Step 5: Stand at one end.

Step 6: Grasp the open edges where convenient, one edge in each hand.

Step 7: Open the sides of the cover over the object by snapping both hands up and out.

Step 8: Open the other end of the cover over the object in the same manner.

Step 9: Tuck in all loose edges at the bottom.

NOTE: **In addition to covering stacks of objects, the rolled cover may be used as a floor runner. Just unroll the cover and spread it out as wide as necessary.**

SKILL SHEET 16-3 PREPARING A FOLDED SALVAGE COVER FOR A ONE-FIREFIGHTER SPREAD

NOTE: Two firefighters must make initial folds to reduce the width of the cover. Steps 1 through 8 are performed simultaneously by both firefighters on opposite sides of the cover. Steps 9 through 13 may be performed by both firefighters who are stationed at the same end of the fold.

Step 1: Grasp the cover with the outside hand midway between the center and the edge to be folded.

Step 2: Place the other hand on the cover as a pivot midway between the outside hand and the center.

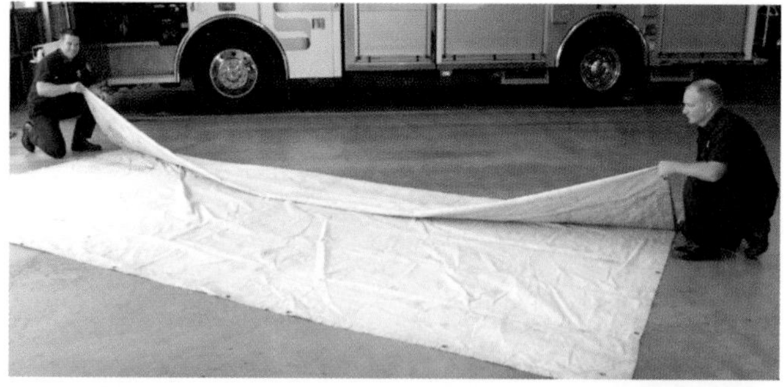

Step 3: Bring the fold over to the center of the cover. This will create an inside fold (center) and an outside fold.

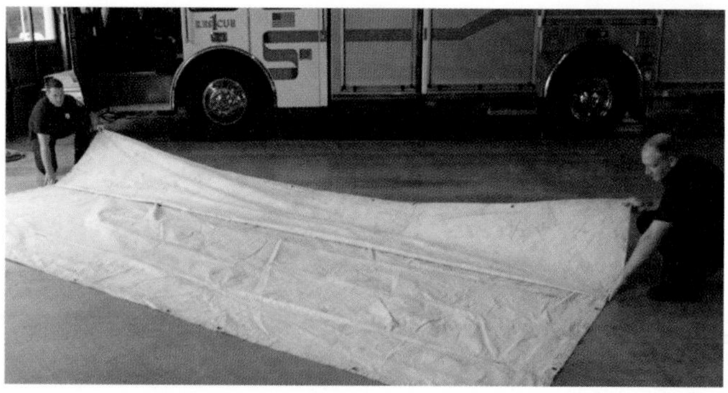

Step 4: Grasp the cover corner with the outside hand.

Step 5: Place the other hand as a pivot on the cover over the outside fold.

Step 6: Bring this outside edge over to the center, and place it on top of and in line with the previously placed first fold.

Step 7: Fold the other half of the cover in the same manner by using Steps 1 through 6.

Step 8: Straighten the folds if they are not straight.

Step 9: Grasp the same end of the cover, with the cover folded to reduce width.

Step 10: Bring this end to a point just short of the center.

Step 11: Use one hand as a pivot, and bring the folded end over and place on top of the first fold.

Step 12: Fold the other end of the cover toward the center, leaving about 4 inches (100 mm) between the two folds.

Step 13: Place one fold on top of the other for the completed fold; the space between the folds now serves as a hinge.

SKILL SHEET 16-4 ONE-FIREFIGHTER SPREAD WITH A FOLDED SALVAGE COVER

Step 1: Lay the folded cover on top of and near the center of the object(s) to be covered.

Step 2: Separate the cover at the first fold.

Step 3: Select either end, and continue to unfold the salvage cover by separating the next fold.

Step 4: Continue to unfold this same end toward the end of the object(s) to be covered.

Step 5: Grasp the end of the cover near the center with both hands to prevent the corners from falling outward.

Step 6: Bring the end of the cover into position over the end of the object(s) being covered.

Step 7: Unfold the other end of the cover in the same manner over the object(s).

Step 8: Stand at one end.

Step 9: Grasp the open edges where convenient, one edge in each hand.

Step 10: Open the sides of the cover over the object(s) by snapping both hands up and out.

Step 11: Open the other end of the cover over the object(s) in the same manner.

Step 12: Tuck in all loose edges at the bottom.

SKILL SHEET 16-5 PREPARING A FOLDED SALVAGE COVER FOR A TWO-FIREFIGHTER SPREAD

NOTE: **Two firefighters must make initial folds to reduce the width of the cover. Steps 1 through 11 are performed simultaneously by both firefighters. Steps 12 through 19 are performed by the respective firefighters. Steps 20 through 24 are performed simultaneously by both firefighters.**

Step 1: Grasp opposite ends of the cover at the center grommet with the cover stretched lengthwise.

Step 2: Pull the cover tightly between each firefighter.

Step 3: Raise this center fold high above the ground.

Step 4: Shake out the wrinkles to form the first half-fold.

Step 5: Spread the half-fold upon the ground.

Step 6: Smooth the half-fold flat to remove the wrinkles.

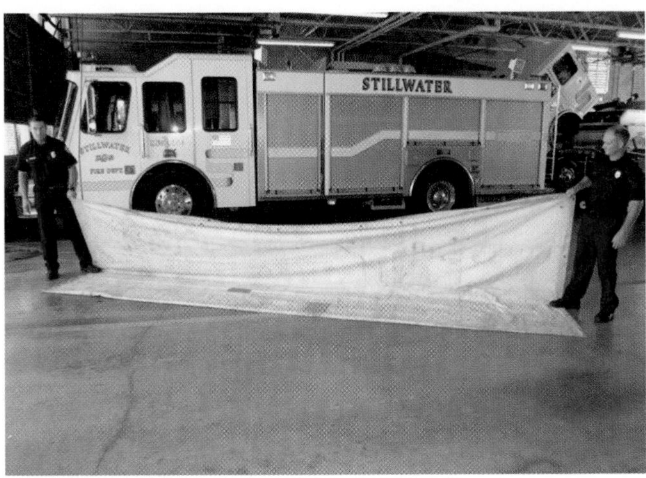

Step 7: Stand at each end of the half-fold and face the cover.

Step 8: Grasp the open-edge corners with the hand nearest to these corners.

Step 9: Place the corresponding foot at the center of the half-fold, thus making a pivot for the next fold.

Step 10: Stretch that part of the cover being folded tightly between each firefighter.

Step 11: Make the quarter-fold by folding the open edges over the folded edge.

Step 12: ***Firefighter #1:*** Stand on one end of the quarter-fold.

Step 13: ***Firefighter #2:*** Grasp the opposite end and shake out all the wrinkles.

Step 14: ***Firefighter #2:*** Carry this end to the opposite end, maintaining alignment of outside edges.

Step 15: ***Both Firefighters:*** Place the carried end on the opposite end, aligning all edges.

Step 16: ***Both Firefighters:*** Position at opposite ends.

Step 17: ***Firefighter #2:*** Stand on the folded end of the cover.

Step 18: ***Firefighter #1:*** Shake out all wrinkles.

Step 19: ***Firefighter #1:*** Align all edges.

Step 20: Grasp the open ends, and use the inside foot as a pivot for the next fold.

Step 21: Bring these open ends over, and place them just short of the folded center fold.

Step 22: Continue this folding process by bringing the open ends over and just short of the folded end.

NOTE: During this fold, the free hand may be used as a pivot to hold the cover straight.

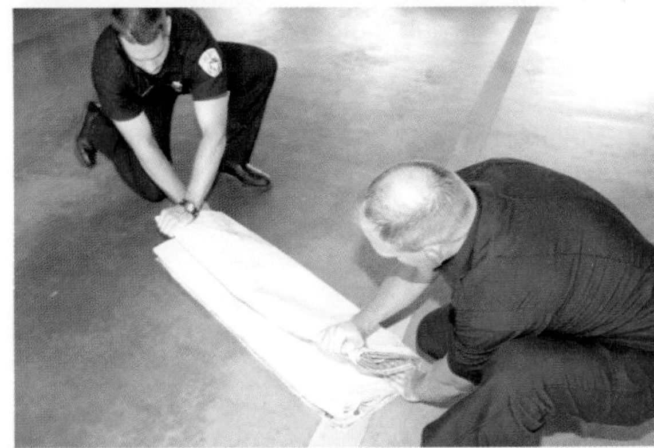

Step 23: Complete the operation by making one more fold in the same manner.

Step 24: Bring the open ends over and to the folded end using the free hand as a pivot during the fold.

SKILL SHEET 16-6 TWO-FIREFIGHTER SPREAD WITH A FOLDED SALVAGE COVER

Balloon Throw

Step 1: Stretch the cover along one side of the object(s) to be covered.

Step 2: Separate the last half-fold by grasping each side of the cover near the ends.

Step 3: Lay the side of the cover closest to the object(s) on the ground.

Step 4: Make several accordion folds in the inside hand.

Step 5: Place the outside hand about midway down the end hem.

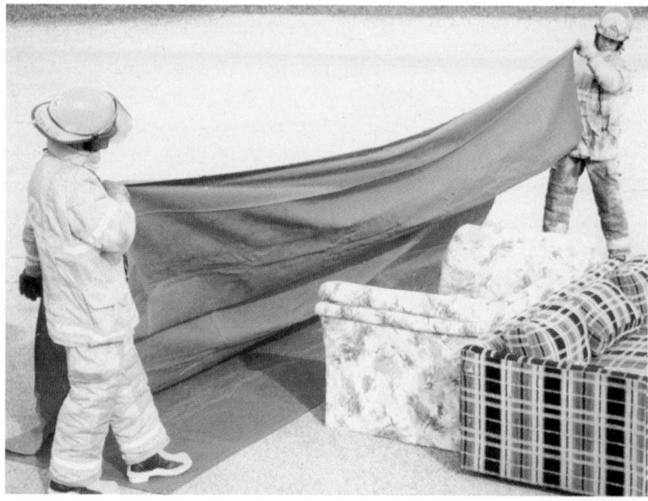

Step 6: Place the inside foot on the corner of the cover to hold it in place.

Step 7: Pull the cover tightly between each firefighter.

Step 8: Swing the folded part down, up, and out in one sweeping movement in order to pocket as much air as possible.

Step 9: Pitch or carry the accordion folds across the object(s) when the cover is as high as each firefighter can reach. This action causes the cover to float over the object(s).

Step 10: Guide the cover into position as it floats over the object(s).

Step 11: Straighten the sides for better water runoff.

SKILL SHEET 16-7 CONSTRUCTING A WATER CHUTE

Without Pike Poles

Step 1: Open the salvage cover.

Step 2: Lay the cover flat at the desired location.

Step 3: Roll the opposite edges of the salvage cover toward the middle until there is a 3-foot (1 m) width between the rolls.

Step 4: Turn the cover over.

Step 5: Adjust the chute to collect and channel water by elevating one end.

Step 6: Extend the other end out a door or window.

SKILL SHEET 16-8 CONSTRUCTING A WATER CHUTE

With Pike Poles

Step 1: Open the salvage cover.

Step 2: Lay the cover flat at the desired location.

Step 3: Place pike poles at opposite edges of the salvage cover with the pike extending off the end of the cover.

Step 4: Roll the edges over the pike poles toward the middle until there is a 3-foot (1 m) width between the rolls.

Step 5: Turn the cover over, keeping the folds in place.

Step 6: Place the chute to collect and channel water.

NOTE: This can be done by hooking the pike poles over a ladder rung or similar object.

Step 7: Extend the other end out a door or window.

NOTE: A completed water chute is shown.

SKILL SHEET 16-9 CONSTRUCTING A CATCHALL

Step 1: Open the salvage cover.

Step 2: Lay the cover flat at the desired location.

Step 3: Roll the sides inward approximately 3 feet (1 m).

Step 4: Lay the ends of the side rolls over at a 90-degree angle to form the corners of the basin.

Step 5: Roll one end into a tight roll on top of the side roll and form a projected flap.

Step 6: Lift the edge roll.

Step 7: Tuck the end roll to lock the corners.

Step 8: Roll the other end in a like manner.

Step 9: Lock the corners. A completed catchall is shown.

Photo courtesy of Sheldon Levi, IFPA.

Chapter 17
Protecting Evidence for Fire Cause Determination

Chapter 17
Protecting Evidence for Fire Cause Determination

INTRODUCTION

[NFPA 1001: 4-3.4; 4-3.4(a); 4-3.4(b)]

Fire departments should investigate all fires to determine the cause of the fire (Figure 17.1). The cause of a fire is a combination of three factors:

- Fuel that ignited
- Form and source of the heat of ignition
- Act or omission that helped to bring these two factors together

Figure 17.1 As a result of thorough fire investigations, investigators can identify and explain the origin and cause of the fire and who or what was responsible. *Courtesy of Scott L. Davidson.*

In order to properly analyze fire cause, it is necessary for firefighters to protect evidence at the scene. A fire officer, fire investigator, or firefighter trained in collecting and preserving evidence collects and analyzes the evidence to determine the exact cause.

Knowing the cause of fires helps prevent fires in the future. Reduced fire loss means that the public is getting the protection that it expects and that the fire department is fulfilling its obligation to provide that protection. As fire incidents decrease, so do loss of life and adverse economic impact.

The firefighter on the scene has the best opportunities to observe evidence of cause and to assist in the fire cause determination effort. The firefighter is an important link in the chain for determining how a fire started and why it spread as it did.

It is the responsibility of the fire department to respond and extinguish a fire as quickly as possible. However, the results of the fire fighting operation may impair an investigator in conducting a fire origin and cause determination investigation. The firefighters' actions may move evidence from its original location or completely sweep it away (Figure 17.2). It is extremely important that firefighters take precautions in protecting evidence while fighting a fire.

Information gathered at the scene is of critical importance to the fire investigator. Investigators are seldom present while firefighters fight a fire, perform overhaul, and interview occupants and witnesses to obtain information. Legal proceedings concerning a particular fire may also become necessary. For these reasons, firefighters must be responsible for noting everything that could point to the origin and cause of a fire.

This chapter contains information on the responsibilities of the firefighter and the fire investigator. Observations that the firefighter can make en route, upon arrival, and during and after the fire which could assist in a subsequent fire

Figure 17.2 Overhauled materials are often piled outside the structure. *Courtesy of Sheldon Levi, IFPA.*

investigation are also covered. The chapter also discusses steps for securing the fire scene and protecting evidence. Finally, the chapter covers the firefighter's conduct at the scene and legal considerations. For more information, refer to IFSTA's **Introduction to Fire Origin and Cause** manual.

RESPONSIBILITIES OF THE FIREFIGHTER

[NFPA 1001: 4-3.4(a)]

The fire chief has the legal responsibility within most jurisdictions for determining the cause of a fire. The fire chief relies on the fire officers and firefighters at the scene to make sure that the true and specific cause of the fire can be determined. Proper training enables firefighters to recognize and collect important information by observing the fire and its behavior during the response, upon arrival, when entering the structure, and while locating and extinguishing the fire. More than anyone else, the firefighter is aware of unusual conditions that may indicate an incendiary (arson) fire.

The first-arriving firefighters may be able to best answer some important questions such as the following:

- Are the contents of the rooms as they normally would be? Are the rooms either ransacked or unusually bare?
- Are the doors and windows locked or open? Is there evidence of forced entry prior to the arrival of firefighters?
- Are there indications of unusual fire behavior or more than one area of origin?
- Are vehicles or people present in the area?

Firefighters must be aware during fireground operations that what they do and how they do it can affect the determination of the origin and cause of the fire. Having an alert and open mind combined with performing judicious and careful overhaul might also uncover important evidence that would otherwise be lost.

ROLE OF THE INVESTIGATOR

[NFPA 1001: 4-3.4(a)]

Fire marshals, fire inspectors, or other members of a fire prevention bureau are usually responsible for carrying fire cause investigations beyond the level of the fire company (Figure 17.3). Firefighters may be questioned by an investigator or asked to assist in some aspect of an investigation.

Some fire departments have special fire investigation or arson squads. In other departments, fire department and law enforcement personnel work together. There are also localities where the police department has sole responsibility for handling an investigation. In other areas, the responsibility for cause determination and investigation lies with the state fire marshal or some other state agency rather than with local agencies. Private companies may con-

Figure 17.3 A fire investigator is responsible for conducting, coordinating, and completing a fire investigation. *Courtesy of Ron Jeffers.*

duct separate investigations when a fire involves their property, or the investigation may be conducted by an insurance company.

OBSERVATIONS OF THE EMERGENCY RESPONDERS

[NFPA 1001: 3-3.7; 3-3.7(a); 3-3.7(b); 3-3.12(a); 3-3.12(b); 4-3.4(a); 4-3.4(b)]

Some of the observations that firefighters make and some of the actions that firefighters perform may be done at different times throughout an incident. For example, firefighters may not find evidence of unusual fire behavior until performing overhaul. A thorough search for containers and signs of forcible entry may not be feasible until the fire is extinguished. The important point is not when the firefighter notices something that can lead to the cause but that the firefighter takes the proper steps afterwards.

Observations En Route

The firefighters' responsibility for gathering information begins when the alarm is received. The firefighter should gather information on the following factors:

- ***Time of day*** — Are people and circumstances at the scene as they normally would be this time of day? For example, if a fire is in a dwelling at 3 a.m., the building occupants would probably be wearing night clothes, not work clothes. If a fire is in an office building after working hours, the owner or employees should have a valid reason for being present at that hour.
- ***Weather and natural hazards*** — Is it hot, cold, or stormy? Is there heavy snow, ice, high water, or fog? If the outside temperature is high, the furnace in the structure would not be operating. If the outside temperature is low, the windows normally should not be wide open. Arsonists sometimes set fires during inclement weather because the fire department's response time may be longer.
- ***Man-made barriers*** — Are there any barriers such as barricades, fallen trees, cables, trash containers, or vehicles blocking access to hydrants, sprinkler and standpipe connections, streets, and driveways? These situations could indicate an attempt on someone's part to delay fire fighting efforts (Figure 17.4).

Figure 17.4 Man-made barriers are sometimes used to delay the fire department's response to an emergency.

- ***People leaving the scene*** — Are people leaving the scene? Most people are intrigued by a fire and stay to watch (Figure 17.5). On the other hand, people leaving the scene by vehicle or on foot may be suspicious. Therefore, when a person leaves the scene by vehicle, make note of the color of the vehicle, its approximate year, its model, the

Figure 17.5 Observe bystanders at the scene. Note if the same person seems to be present at several fires.

body style and condition, and the license plate number. Notice if any occupants are in the vehicle. If a person leaves the scene on foot, note the person's attire, general physical appearance, and any peculiarities such as someone trying to leave undetected, walking briskly, or looking over his shoulder.

Observations Upon Arrival

Additional information that firefighters should gather upon arrival at the scene may include the following:

- ***Time of arrival and extent of fire*** — Ask the person who reported the fire or other witnesses about the extent of the fire at the time it was discovered and reported. The person who reported the fire can be questioned thoroughly at a later time. Note the locations of smoke columns and flames and determine whether flashover or self-ventilation occurred. If the fire self-ventilated, was it vertical or horizontal?
- ***Wind direction and velocity*** — Note wind direction and velocity. These factors may have a great effect on the natural path of fire spread.
- ***Doors or windows locked or unlocked*** — Note the position and condition of doors and windows upon arrival. Before opening doors and windows, determine whether they are locked, are unlocked, or show any signs of forcible entry such as broken glass or split frames (Figure 17.6). Sometimes doors and windows are covered with blankets, paint, and paper to delay discovery of the fire.
- ***Location of the fire*** — Determine the location of the fire. This information helps to identify the area of origin. Also note whether there were separate, seemingly unconnected fires. If so, the fire might have been set in several locations or spread by trailers (combustible material used to spread fire from one area to another).
- ***Containers or cans*** — Note metal cans or plastic containers found inside or outside the structure. They may have been used to transport accelerants.
- ***Burglary tools*** — Note tools such as pry bars and screwdrivers found in unusual areas. They may have been used by a person to enter the facility to set the fire (Figure 17.7).

Figure 17.6 Check for signs that forcible entry was made before fire department personnel arrival.

Figure 17.7 Look for forcible entry tools that may have been used to enter the facility.

- ***Familiar faces*** — Look for familiar faces in the crowd of bystanders. They may be fire buffs, or they may be habitual firesetters.

Observations During Fire Fighting

Firefighters should continue to observe the following conditions that may lead to the determination of the fire cause:

- ***Unusual odors*** — Note unusual odors. Firefighters should always wear self-contained breathing apparatus (SCBA) during fire suppression and overhaul operations. However, unusual odors may sometimes be detected at the fire scene.
- ***Abnormal behavior of fire when water is applied*** — Observe fire behavior when applying water on a fire. Flashbacks, reignition, several rekindles in the same area, and an increase in the intensity of the fire indicate possible accelerant use. Water applied to a burning liquid accelerant may cause it to splatter, allowing flame intensity to increase and the fire to spread in several directions. Water applied to fires involving ordinary combustibles usually reduces flame propagation.
- ***Obstacles hindering fire fighting*** — Note whether doors are tied shut or furniture is placed in doorways and hallways to hinder fire fighting efforts (Figure 17.8). Holes may be cut in the floors that not only hinder fire suppression activities but also spread the fire.
- ***Incendiary devices*** — Note pieces of glass, fragments of bottles or containers, and metal parts of electrical or mechanical devices. Most incendiary devices (any device designed and used to start a fire) leave evidence of their existence (Figure 17.9). More than one device may be found, and sometimes a faulty functioning device can be found during a thorough search.
- ***Trailer*** — Note combustible materials such as rolled rags, blankets, newspapers, or ignitable liquid (trailer) that could be used to spread fire from one point to another. Trailers usually leave char or burn patterns and may be used with incendiary ignition devices (Figure 17.10).

Figure 17.9 Most incendiary devices leave evidence of their existence.

Figure 17.8 Furniture may be placed in front of doors to block the firefighter's entry.

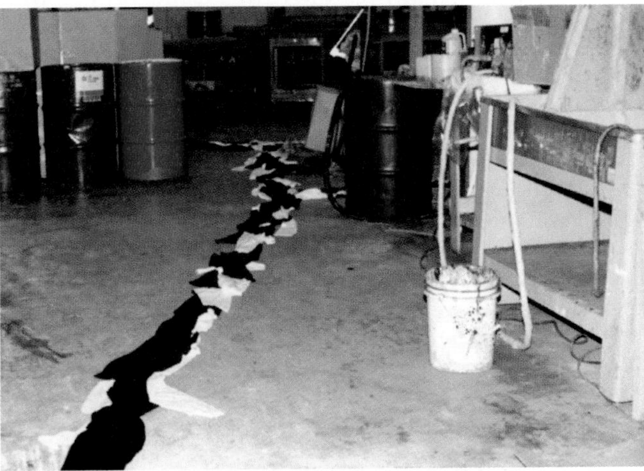

Figure 17.10 A trailer is used to spread fire from one point to another. *Courtesy of Elk Grove Village (IL) Fire Department.*

- ***Structural alterations*** — Observe alterations to the structure: removal of plaster or drywall to expose wood; holes made in ceilings, walls, and floors; and fire doors secured in an open position (Figure 17.11). All of these methods are designed to abnormally spread or move fire quickly through the structure.
- ***Fire patterns*** — Note the fire's movement and intensity patterns. These can trace how the fire spread, identify the original ignition source, and determine the fuel(s) involved. Carefully note areas of irregular burning or locally heavy charring in areas of little fuel.
- ***Heat intensity*** — Look for evidence of high heat intensity, especially in relation to other areas of the same room. This may indicate the use of accelerants. However, other factors may contribute to variations in heat intensity. One of these factors is synthetic materials, such as polyurethane, that may produce areas of normal high heat intensity and may be confused with the use of accelerants.
- ***Availability of documents*** — Be alert to the sudden production of insurance papers, inventory lists, deeds, or other legal documents that may indicate that the fire was premeditated.
- ***Fire detection and protection systems*** — Check for evidence of tampering or intentional damage if fire detection and protection systems and devices are inoperable (Figure 17.12).
- ***Intrusion alarms*** — Check intrusion alarms to see whether they have been tampered with or intentionally disabled.
- ***Location of fire*** — Note ignition sources or possible ignition sources in the area of the fire. Fires in areas remote from normal ignition sources may indicate suspicious activity. Some examples are fires in closets, bathtubs, file drawers, or in the center of the floor.
- ***Personal possessions*** — Look for the following indicators that preparations were made for a fire: absence or shortage of clothing, furnishings, appliances, food, and dishes; absence of personal possessions such as diplomas, financial papers, and toys; absence of items of sentimental value such as photo albums, special collections, wedding pictures, and heirlooms; absence of pets that would ordinarily be in the structure (Figure 17.13). (**NOTE:** Do not read too much into a lack of material possessions. A person's economic status may dictate his lifestyle, and some people just do not have as much as others.)
- ***Household items*** — Note whether major household items appear to be removed or replaced with junk. Other items may also

Figure 17.11 Fire spreads more quickly throughout a structure when a hole has been chopped in the floor.

Figure 17.12 A smoke detector may be disabled to delay the detection of the fire.

Figure 17.13 Missing personal possessions, such as clothes, may indicate that preparations were made for a fire.

appear to be replaced with items of inferior quality. Check to see whether major appliances were disconnected or unplugged and determine why they were in this condition.

- ***Equipment or inventory*** — Look for obsolete equipment or inventory, fixtures, display cases, equipment, and raw materials.
- ***Business records*** — Determine if important business records are out of their normal places and left where they would be endangered by fire. Check safes, fire-resistant files, etc., to determine whether they are open and exposing the contents.

Responsibilities After the Fire

Firefighters should report all facts concerning the fire to the officer in charge as soon as possible. Each firefighter should write a chronological account of important circumstances personally observed. A written account will be valuable if the firefighter must testify in court later. Cases often come to trial several years after an incident, and a person should not rely on memory alone. Report hearsay to the investigator for validation. Example: *"The neighbor told me that he saw the lights flickering for a few days before the fire."* This is hearsay, but it may be very helpful to the investigator.

Salvage and overhaul are probably the pivotal operations in determining fire cause. Some departments take great pride in their salvage and overhaul work and boast that they leave a building neater, cleaner, and more orderly than it was before the fire. This thoroughness in salvage and overhaul is admirable, but in many cases it destroys evidence of how a fire started. Delay thorough salvage and overhaul work until the area of origin and cause have been determined.

Fire personnel should perform salvage and overhaul carefully. They should not move more debris than is necessary, especially in the area of origin, because it may hamper the investigation. Neither should firefighters throw debris outside into a pile — evidence is buried this way and may be permanently lost (Figure 17.14).

Figure 17.14 Evidence may be found in debris that has been thrown outside.

CONDUCT AND STATEMENTS AT THE SCENE

[NFPA 4-3.4(a)]

Although firefighters and the fire officer should obtain all information possible pertaining to a fire, they should not attempt to interrogate a potential arson suspect. The moment one suspects a particular person of arson, he should call a trained investigator to conduct an interview. It is the trained investigator's job to interview an arson suspect. Allow the owners or occupants of the property to talk freely if they are inclined to do so, and give them a sympathetic ear. Some valuable information is often gathered this way.

Fire personnel should never make statements of accusation, personal opinion, or probable cause to anyone. These opinions easily could be overheard by the property owner, news media, or other bystanders who could consider such statements fact. Careless joking and unauthorized or premature remarks that are published or broadcast can be very embarrassing to the fire department. Many times these remarks impede the efforts of an investigator to prove malicious intent as the fire cause. A sufficient reply to any question concerning cause is *"The fire is under investigation."*

After the investigator arrives, personnel should make their statements only to this individual. Any public statement regarding the fire cause should be made only after the investigator and ranking fire officer have agreed to its accuracy and validity and have given permission for it to be released.

SECURING THE FIRE SCENE

[NFPA 4-3.4(a); 4-3.4(b); 3-3.13(b)]

The most efficient and complete efforts to determine the cause of a fire are wasted unless the building and premises are properly secured and guarded until an investigator has finished evaluating the evidence exactly as it appears at the scene. Firefighters should take care not to contaminate the scene while operating power tools, hoselines, or other equipment.

If an investigator is not immediately available, the premises should be guarded and kept under the control of the fire department until all evidence has been collected (Figure 17.15). All evidence should be marked, tagged, and photographed at this time because in many instances a search warrant or written consent to search will be needed for further visits to the premises. This duty might be given to law enforcement personnel, depending on local policies and personnel availability, but whenever possible it should be carried out by fire department personnel trained in evidence collection and preservation.

Figure 17.15 Cordon off the scene with fire line tape, and secure the area with law enforcement personnel.

The fire department has the authority to bar access to any building during fire fighting operations and for a reasonable length of time after fire suppression is terminated. Fire department authority ends when the last fire department representative leaves the scene. Further visits to the scene require either the owner's written permission or a search warrant. Fire personnel should be aware of any local laws pertaining to the right of access by owners or occupants.

Fire personnel should not allow anyone to enter a fire scene without the investigator's permission, and an authorized individual should escort the subject. During fireground operations and the investigation, make a recorded log of any such entry. The log should show the person's name, the time of entry, the time of departure, and a description of any items the person took from the scene.

The premises can be secured and protected in several ways with the use of few personnel. Areas that are fenced can be monitored by one person at a locked gate. At large fire scenes, a full-time guard force is often employed to handle the situation. In some extreme instances, all doors, windows, or other entrances could be completely closed with plywood or similar material. Cordoning off the area also can help provide a safe and secure fire scene. With the area cordoned, bystanders are kept at a safe distance from the incident and out of the way of emergency personnel. There are no specific boundaries for the cordon.

Cordoning can be accomplished with rope or specially designed fire and police line tape. It may be attached to signs, utility poles, parking meters, vehicles, or any other objects readily available. Once in place, law enforcement personnel should monitor the line to make sure people do not cross it. Be aware of seemingly innocent persons (including curious people and the press) attempting to cross a line. Escort from the area anyone in the cordoned area who is not a part of the operation. Record any information obtained from them for future reference.

LEGAL CONSIDERATIONS

[NFPA: 4-3.4(a)]

As previously stated, firefighters may remain on the location as long as necessary, but once they leave they may be required to get a search warrant to reenter the scene. This is based on the case of Michigan vs. Tyler (436 U.S. 499, 56 L.Ed. 2d 486 [1978]). The U.S. Supreme Court held in that case that *"once in a building [to extinguish a fire], firefighters may seize [without a warrant] evidence of arson that is in plain view . . . [and] officials need no warrant to remain in a building for a reasonable time to investigate the cause of a blaze after it has been extinguished."*

The Court agreed, with modification, with the Michigan State Supreme Court's statement that *"[if] there has been a fire, the blaze extinguished and the firefighters have left the premises, a warrant is required to re-enter and search the premises, unless there is consent . . . ."*

The impact of these decisions seems to be that if there is incendiary evidence, the fire department should leave at least one person on the premises until an investigator arrives. To leave the premises, return later without a search warrant, and make a search might be enough to make prosecution impossible or for an appellate court to overturn a conviction (Figure 17.16).

Each department should learn the legal opinions that affect its jurisdiction in this regard. These opinions or interpretations can be obtained from such persons as the district attorney or state attorney general. The fire department should write a standard operating procedure around these opinions.

Figure 17.16 Fire department personnel and the investigator should know the jurisdiction's legal requirements concerning reentry of a fire scene. *Courtesy of Sheldon Levi, IFPA.*

PROTECTING AND PRESERVING EVIDENCE

[NFPA 1001: 3-3.13(b); 4-3.4(a); 4-3.4(b)]

Firefighters should protect evidence, untouched and undisturbed if at all possible, when it is found and provide security for the area until an investigator arrives. They should not gather or handle evidence unless it is absolutely necessary in order to preserve it. If a firefighter handles or procures evidence, he then becomes a link in the chain of custody for that evidence. The firefighter should accurately document all actions as soon as possible. It may be necessary for this individual to subsequently appear in court. Because the amount of time involved in a court trial can be extensive, most departments do not want operational personnel to gather evidence.

No changes of any kind should be permitted in the evidence other than those absolutely necessary in the extinguishment of the fire. Firefighters should avoid trampling over possible evidence and obliterating it. The same precaution applied to the excessive use of water may help avoid similar unsatisfactory results. Human footprints and tire marks must be protected. Boxes placed over prints prevent dust from blowing over otherwise clear prints and keep them in good condition for either photographs or plaster casts at a later time (Figure 17.17). Completely or partially burned papers found in a furnace, stove, or fireplace should be protected by immediately closing dampers and other openings. Leave charred documents found in containers such as wastebaskets, small file cabinets, and binders that can be moved easily. Keep these items away from drafts.

Figure 17.17 Footprints may be covered with a box until they are photographed or plaster casts made.

After evidence has been properly collected by an investigator, debris may be removed. Charred materials should be removed to prevent the possibility of rekindle and to help reduce smoke damage. Any unburned materials should be separated from the debris and cleaned. Debris may be shoveled into large containers, such as buckets or tubs, to reduce the number of trips back and forth to the fire area. It causes poor public relations to dump debris onto streets and sidewalks or to damage costly shrubbery. Rather, dump the debris in a backyard or alley that is not as visible.

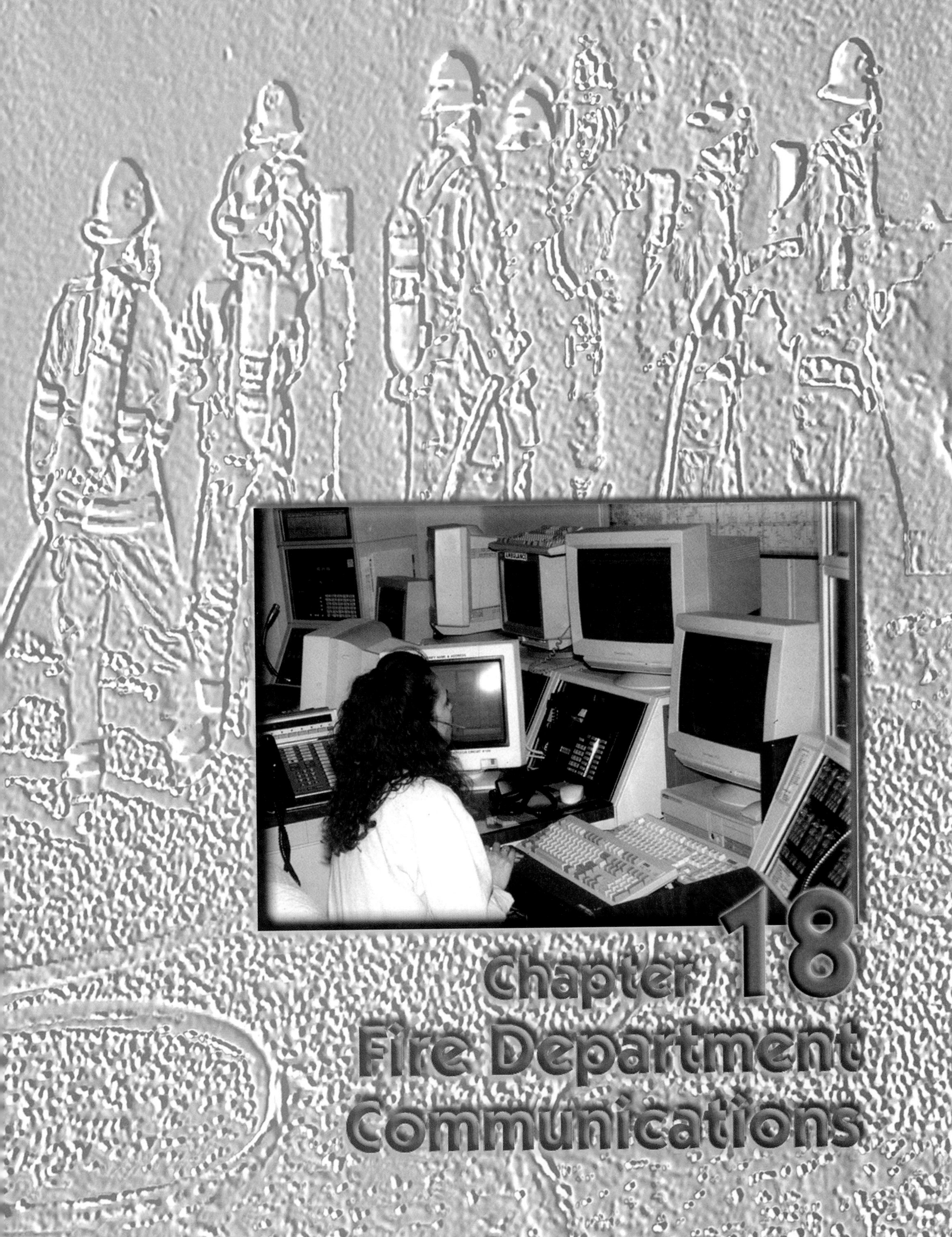

Chapter 18
Fire Department Communications

Chapter 18
Fire Department Communications

INTRODUCTION

The expedient and accurate handling of fire alarms or calls for help are significant factors in the successful outcome of any incident. History has proven time and time again that failure to quickly communicate the need for help can result in large and tragic losses. Fire department communications play a critical role in the successful outcome of an incident.

Fire department communications include the methods by which the public can notify the telecommunications center (also known as *communications center*) of an emergency, the methods by which the center can notify the proper fire fighting forces, and the methods by which information is exchanged at the scene. Firefighters must also know how to handle routine communications, including nonemergency calls for business purposes or public inquiries made directly to the station.

This chapter provides information on the basics of fire department communications. It describes the role of the telecommunicator and provides an overview of the telecommunications center and the basic equipment found in such a center (Figure 18.1). The chapter also describes the procedures for receiving nonemergency telephone calls, receiving reports of emergencies, and alerting fire department personnel. The last section of the chapter describes the use of the incident report.

TELECOMMUNICATIONS CENTER PERSONNEL

[NFPA 1001: 3-2.1(a); 3-2.1(b); 3-2.2(a); 3-2.2(b); 3-2.3(a); 3-2.3(b)]

Most of the people who contact a telecommunicator are not having a good day! The person calling is usually experiencing some kind of difficulty or problem that is upsetting enough that they want assistance. Because of this, the telecommunicator has a very important role. To fulfill this role, a telecommunicator needs to be skilled in customer service, and personal communications.

Figure 18.1 The telecommunications center is the nerve center of emergency response.

Role of the Telecommunicator

A telecommunicator has a role different from but as important as other emergency personnel. In calls for emergency service, time is of the essence. Given the generally accepted time period of one minute to effect dispatch, a telecommunicator must determine required actions very quickly. Time lost in the dispatch function cannot be "made up" by the responders.

A telecommunicator must process calls from unknown and unseen individuals, usually calling under stressful conditions (Figure 18.2). A telecommunicator must be able to obtain complete, reliable information from the caller and prioritize requests for assistance. These decisions and the

Figure 18.2 A telecommunicator receives a fire emergency call.

ability to swiftly and accurately carry out the total dispatch function are often a matter of life and death to citizens.

Once the necessary information is gathered, a telecommunicator must dispatch the emergency responders needed to stabilize the incident. In order to provide timely response, a telecommunicator must know where emergency resources are in relation to the reported incident as well as their availability status (Figure 18.3). It is critical that the appropriate unit closest to the incident be dispatched on emergency responses. A telecommunicator needs to know not only who to assign but also how to alert them. During an incident, the telecommunicator must stay in contact with the incident commander (IC) to receive requests for information and/or additional resources. However, a telecommunicator's job does not stop when the incident is terminated. Records must be kept of each request for assistance and how each one was handled.

Figure 18.3 The telecommunicator must know the status of all resources.

Customer Service

The consumer of emergency services is the general public. Members of the general public (customers) expect and are entitled to professional service. A telecommunicator is likely to have more contact with the public than any other member of the emergency services organization. On a daily basis, a telecommunicator receives calls from any number of people in the community seeking assistance or information. Some examples might include requests for various social services such as homeless shelters, emergency financial assistance, and counseling services. These calls may come from victims of crimes, fires, or other disasters. They may simply come from people who do not know any other way to get assistance such as in the case of a power outage. As the customer's first contact with emergency services, the telecommunicator must project a sense of competence to the caller.

With each request for assistance, even if the question is not directly answered, a telecommunicator makes the decision to refer the caller to an appropriate person or agency. If a nonemergency call comes in over 9-1-1 or another locally used emergency line, the customer may be transferred or referred to another number to have his service request processed. Once on a nonemergency line, the telecommunicator should provide necessary information to the customer about agencies in the area that can help him (Figure 18.4). The worthiness of individuals for

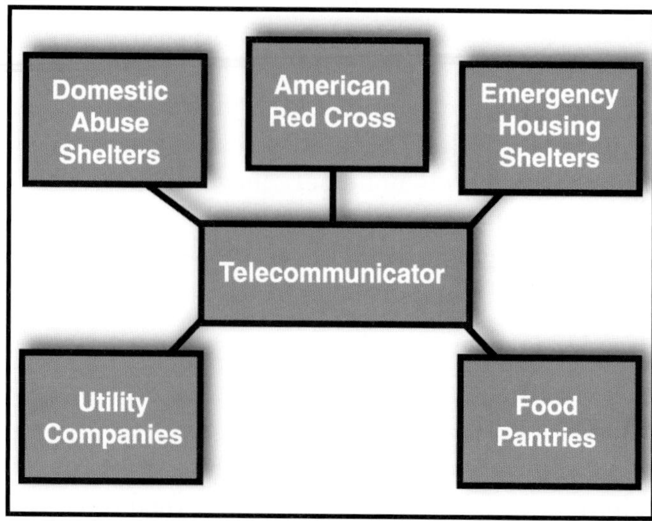

Figure 18.4 Some of the agencies to whom the caller may be referred.

assistance should not be evaluated by the telecommunicator. All requests should be referred to the agency that is best suited to do that evaluation.

Telecommunicator Skills

NFPA 1061, *Standard for Professional Qualifications for Public Safety Telecommunicator*, contains the minimum job performance requirements for public safety telecommunicators. The appendix to NFPA 1061 states that firefighters used as telecommunicators should meet the requirements of that standard. This chapter provides basic information to assist the firefighter in receiving and processing information received at the station.

An important skill of a telecommunicator is to be able to maintain a positive attitude throughout the communication process. It is also vitally important for a telecommunicator to be able to effectively work with and interact with other members of a team. This includes not only other telecommunicators but also other members of the emergency services organization. The appendix material in NFPA 1061 suggests several other traits or personal characteristics that a telecommunicator should possess. They include the following:

- Adjust to various levels of activity.
- Handle multitasking.
- Make decisions and judgments based on common sense and values.
- Maintain composure.
- Form conclusions from disassociated facts.
- Handle criticism.
- Remember and recall information.
- Deal with verbal abuse.
- Function under stress.
- Maintain confidentiality.

Other skills a telecommunicator should possess include communications and map-reading skills.

COMMUNICATION SKILLS

The communication skills required by a telecommunicator include the following:

- ***Basic reading skills*** — Sufficient ability to read and understand so that basic policies, instructions, and direction can be given in writing and be understood.
- ***Basic writing skills*** — Keyboarding and computer literacy are vital when a written description of an event or a problem is required. A telecommunicator must be able to create readable reports, memos, and letters. The reports generated may be used by the media, the courts, or the general public.
- ***Ability to speak clearly*** — This means not only annunciation but with proper grammar and sentence construction. Telecommunicators should know how to control voice tone and speed. The telecommunicator is the voice of the agency. It is oftentimes how people form their initial judgment of an agency.
- ***Ability to follow written and verbal instructions*** — Telecommunicators must know local procedures and local, state/province, and federal radio regulations. They are responsible for taking directions from a number of different sources through their job functions. It is important that they have the ability to read or listen to instructions and then execute those directions without further instruction.

MAP READING

In an age of computers and electronic displays, it is still vital for a telecommunicator to be able to look at a map and locate specific points (Figure 18.5). In fact, many of the newest Computer-Aided Dispatch (CAD) systems contain sophisticated

Figure 18.5 The telecommunicator must be able to use a map.

mapping displays. These maps can be flashed on a screen to help a telecommunicator advise the best route for a responding unit or to select the closest unit to a call.

Use of an Automatic Vehicle Locating (AVL) system adds even more need for the ability to read and use maps. In this technology, the location of a fire department unit is displayed on a map as the vehicle moves up and down the streets. Instead of reducing the need for using maps, the opposite is the case with this technology.

Added to this is the current growth in wireless communications devices such as cellular phones. Future standards will require that a 9-1-1 call placed by a wireless phone provide X and Y coordinates for the location of the caller. This information will be translated by computers and either given as an address or placed on a map display for the telecommunicator to see. This requirement may also include a Z coordinate (the altitude of the caller) to help identify if they are in a high-rise building or on a mountainside. All of this will be displayed to the telecommunicator in some form of graphical representation that is similar to a map.

TELECOMMUNICATIONS CENTER

[NFPA 1001: 3-2.1(a); 3-2.1(b); 3-2.2(a); 3-2.2(b); 3-2.3(a); 3-2.3(b); 4-2.2(a); 4-2.2(b)]

A telecommunication center is the nerve center of emergency response. It is the point through which nearly all information flows, is processed, and then acted upon. The telecommunications center houses the personnel and equipment to receive alarms and dispatch resources. Depending on the size and communications requirements of the department, the telecommunications center may be located in a fire station or in a separate building. In some jurisdictions, the fire telecommunications center will be part of a larger, joint telecommunications center for all emergency services (Figure 18.6).

Figure 18.6 A fire telecommunicator may be a part of a much larger joint operation.

Figure 18.7 This tone-alert equipment is part of the telecommunicator's console.

A telecommunications center may be equipped with a variety of communications equipment, depending on local capabilities. Some of the more common pieces of equipment include the following:

- Two-way radio for communicating with personnel at the emergency scene.
- Tone-alert equipment for dispatching resources (Figure 18.7)
- Telephones for handling both routine and emergency phone calls
- Direct-line phones for communications with hospitals, utilities, and other response agencies
- Computers for dispatch information and communications

- Tape recorders to record phone calls and radio traffic
- Alarm-receiving equipment for municipal alarm box systems and private fire alarm systems

Communications Equipment

A telecommunicator must be able to operate fire department communications equipment. The following sections describe some of the basic communications equipment used in a telecommunications center.

ALARM RECEIVING EQUIPMENT

Fire alarms may be received from the public in different ways: public alerting systems (See Public Alerting Systems section) and private alarm systems. Fire detection and alarm systems were covered in Chapter 15, Fire Detection, Alarm, and Suppression Systems. Receiving alarms from the public is covered later in this chapter.

TELEPHONES

It is difficult to imagine life today without the telephone and all of its related services. The telephone is used to transmit voice messages, computer information, and documents (Figure 18.8). Telephones have grown to resemble computers, and today computers provide many of the same functions that the telephone has historically provided.

The public telephone system is the most widely used method for transmitting fire alarms. In many areas, such as outlying suburbs or rural settings, it is the only method of rapid communication. A major advantage of telephones is that the telecommunicator can ask the caller about the nature of the emergency, and obtain the address or the callback number.

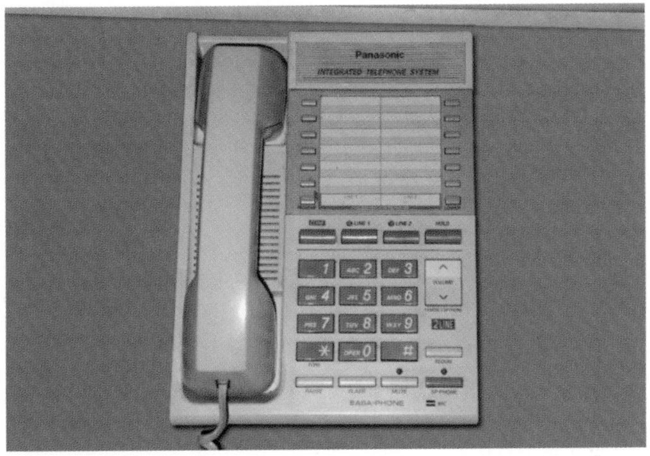

Figure 18.8 The telephone must be convenient for the telecommunicator to use. *Courtesy of Tom Jenkins.*

Commercial phone systems. Commercial phone systems access the public switch network. This means that when the phone is taken off the hook or a button is depressed, the caller hears a dial tone. While many people visualize phone service as basically a residential single-line service, there are in fact vast numbers of commercial phone systems that offer access to multiple phone lines and provide features such as hold, conference calling, speaker phones, and other features.

Direct lines. Direct lines differ from the normal phone lines in that they do not have access to the public switch network and do not have a dial tone. The line is directly connected between point A and point B. When one party picks up the phone, it immediately starts ringing at the other end. No numbers are dialed, and there are no choices for the caller to make. Common applications for these lines would be between the telecommunications center and a fire station or a hospital to request an ambulance or a helicopter. These may be just a button on the phone that a telecommunicator depresses to activate the circuit. Oftentimes, these types of circuits connect emergency communications centers with power plants, airport control towers, or weather services. Another common application for direct lines is to support signals from alarm systems and radio systems.

TDD/TTY/Text phones. A special communications device has been designed to allow the hearing- or speech-impaired community to communicate over the telephone system (Figure 18.9). *Telecommunications Device for the Deaf (TDD), Teletype (TTY),* and *Text phone* are phones that can visually display text. These names are interchangeable and denote a device that permits communications with the fire

Figure 18.9 A telephone designed to facilitate communications with the hearing impaired using a keyboard and visual text readout.

department by the hearing- and speech-impaired people in the local jurisdiction. The current term used most often is Text phone, since it is the most descriptive of the actual functioning of the device.

Wireless. Not only have telephone consumers learned to depend on telephones in their homes, but they have also taken their phones on the road with them. Wireless phones are basically phone devices that are sophisticated two-way radios. The wireless phone devices use radio frequencies to communicate with a base site, which may be several miles (kilometers) from the location of the phone.

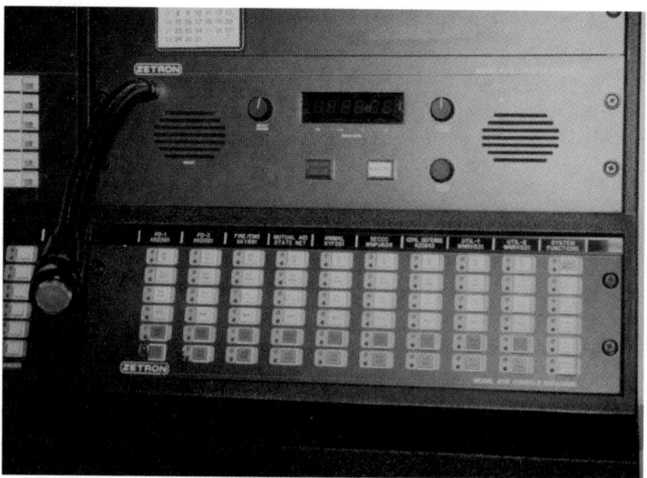

Figure 18.10 The radio is the primary means of communication in the fire service.

FAX MACHINES

At its basics, a fax machine takes a picture, writing, or a diagram and converts the image into digital signals. These digital signals are sent over a communications medium, most commonly a telephone line, although radio is another possible medium. At the receiving end, the other fax machine takes the digital signals and converts them back into the picture or text. While many of these machines are stand-alone machines, which often double as telephone devices, they can be built into a computer. A computer-generated document does not need to be converted to hard copy (paper)—the file is digitized and sent over the transmission medium. At the other end is either a stand-alone fax machine or another computer with a fax modem hooked to it to convert the document.

RADIOS

The purpose of fire department radio communications is to tie all elements of the organization together so that each element can perform its task in an efficient, informed manner (Figure 18.10). Radio equipment is designed to provide a method of transmitting and receiving critical or pertinent tactical information to or from other field units, the telecommunications center, or the incident commander. This information can be task-related (*"Command, this is Engine 7; we need an additional supply line to support Truck 37's ladder pipe."*) or the information can be a direct order based upon the decision of the incident commander (*"Alarm, this is Command; transmit a third alarm. All companies report to staging."*).

Individuals who operate radio equipment should realize that all radio transmissions can be monitored by the news media and the public (Figure 18.11). Any communications that are transmitted via radio could very well be repeated on the front page of tomorrow's newspaper. Radio operators should always be aware of what they are saying and never transmit a message that may bring liability or embarrassment to the department.

Figure 18.11 Anyone can hear radio transmissions on a scanner.

Computer-Aided Dispatch (CAD)

In some jurisdictions, computers (not people) perform many dispatch functions. The term *Computer-Aided Dispatch (CAD)* is also known as *Computer-Assisted Dispatch.* Both imply that the telecommunicator is assisted or aided in the performance of his duties by a computer system. Many departments have found that computer-aided dispatch can significantly shorten response time or enable dispatchers to handle a greater volume of calls. CAD can also reduce the amount of voice communications between telecommunicators and responding units.

CAD systems come in various designs to meet the needs of the telecommunications center and the departments it supports. A CAD system can be as simple as one that retrieves run card information or as complex as one that selects and dispatches units, determines the quickest route to the scene of an emergency, monitors the status of units, and transmits additional information via mobile data terminals. All of these are functions that can appropriately be handled by the computer to assist the telecommunicator.

Small organizations with only a couple of pieces of apparatus may not need a CAD system, or they may not require a complex system capable of tracking multiple stations and dozens of units. On the other hand, it is difficult to imagine a major operation involving dozens of units and many calls a day functioning without some type of computer assistance.

Recording Information

Recording information communicated during emergency operations is very important. Two methods of recording information for future review by telecommunicators or other authorized personnel are voice recorders and radio logs.

VOICE RECORDERS

Recorders document telephone calls on emergency lines, radio traffic, and dispatching information and provide an accurate account of operations (Figure 18.12). They protect the department and its members when questions are raised about communications and operations. They also document such evidence as dispatch time and company arrival on the scene in case of litigation.

Telephone lines connected to a recording device in the alarm center offer several benefits to the telecommunicator. If the caller hangs up or is disconnected, the information received can be played back. The recording device also is important when callers are so excited that they cannot be understood or when they speak a foreign language.

Figure 18.12 All radio transmissions and telephone calls should be recorded.

The recording devices run either continuously or intermittently. The continuous type operates even when no transmissions are taking place; the intermittent units run only when traffic is on the air. Because they run all the time, continuous units use more tape and are more expensive to operate than intermittent types. Intermittent units can miss the beginning of a transmission because they are actuated when traffic is broadcasted, and it takes a little time for the recording to begin. If an operator speaks before recording begins, the recorder misses the first part of the message. Operators can overcome the problem by using proper procedures: Pause after keying the microphone and before speaking. Recorders should be capable of instant playback. Equipment should also automatically record the time of the call.

RADIO LOGS

Radio logs are used to record the incident and location of each activity being performed by a public safety unit. This is basically a manual system entered onto paper. It is usually a chronological recording of each and every activity that has been reported or dispatched over the radio. In addition to the time of the incident, there is generally an entry as to the location and the nature of the incident, along with a notation of which unit(s) responded to this call. By reviewing the current entries, a telecommunicator can determine which units are currently on assignments and which ones are not. A typical series of entries might read as follows:

- 1827 hours: Alarm box 263, Engine 12, Engine 9, Ladder 6, Battalion 2 assigned to 3723 E. Main, Sue's Flower Shop
- 1829 hours: Engine 12 on-scene, light smoke visible
- 1830 hours: Dispatch call letters
- 1831 hours: Battalion 2, Engine 9, Ladder 6 on-scene
- 1844 hours: Battalion 2 transmitted control of fire, placed Engine 9 in-service

- 1857 hours: Battalion 2, Engine 12, Ladder 6 in-service, returning to quarters
- 1901 hours: Battalion 2, Engine 12, Ladder 6 in-quarters
- 1902 hours: Engine 9 in-quarters

RECEIVING NONEMERGENCY TELEPHONE CALLS FROM THE PUBLIC

[NFPA 1001: 3-2.2(a)]

It is vital that any call to a telecommunications center be treated as a possible emergency until it is determined that it is not. A telecommunicator must be able to differentiate between those requests that are emergencies and those that are not. Nonemergency service calls can be directly handled by the telecommunicator, referred to the fire department, or referred to other agencies.

Many business calls come in on the public telephones. Each department will have its own procedures and greeting for answering business calls. For this reason, it is important to know the correct procedure for processing these calls in your department. The following list describes the basic procedures for answering business calls:

- Answer calls promptly.
- Be pleasant and identify the department or company and yourself. For example, *"Good morning, Station 61, Firefighter Krill speaking."*
- Be prepared to take accurate messages by including date, time, name of caller, caller's number, message, and your name.
- Never leave the line open or someone on hold for an extended period of time.
- Post the message or deliver the message promptly to the person to whom it is intended.
- Terminate calls courteously. Always hang up last.

RECEIVING REPORTS OF EMERGENCIES FROM THE PUBLIC

[NFPA 1001: 3-2.1(a); 3-2.1(b); 3-2.3(a); 3-2.3(b)]

One of the most critical periods for telecommunicators is when an alarm is received. A telecommunicator should be well trained to get the right information quickly to start units on their way. Skill is especially important when the public alerts the department by telephone (Figure 18.13). When a citizen calls in an alarm, the telecommunicator should proceed in the following manner:

- Identify the agency (for example, *"Metro 9-1-1"*).
- Ask if there is an emergency and, if so, ask about the problem.
- Have questions organized to control the conversation to get the information needed. Ask questions in an assertive voice. Follow the department's SOPs.

Figure 18.13 Callers may have difficulty describing their locations.

- Get the kind of information that pictures what type of emergency really exists:
  - — Incident location
  - — Type of incident/situation
  - — When the incident occurred
- Make sure to get the exact location of the alarm. Ask about cross streets and other identifying landmarks if necessary.

- Get information about the caller by asking the following:
  - — Name
  - — Location if different from the incident location
  - — Callback phone number
  - — Address
- Do not let the caller off the phone until all necessary information has been obtained to dispatch responding units or until it is certain there is no emergency.

The answers should be recorded on some type of emergency alarm report. Then the telecommunicator can dispatch the necessary units. Telecommunicators must realize that dispatching delays can increase response time.

Figure 18.14 Every phone should be equipped with an emergency sticker.

Public Alerting Systems

Public alerting systems are those systems that may be used by anyone to report an emergency. These systems include telephones, two-way radios, walk-in reports, wired telegraph circuit boxes, telephone fire alarm boxes, and radio alarm boxes.

TELEPHONE

Depending on the phone system capabilities, the fire department emergency number may be 9-1-1, a 7-digit number, or "0" for the operator. Emergency telephone number stickers placed directly on the telephone help customers reduce time delays when calling the fire department (Figure 18.14). The telecommunicator must be prepared to provide the caller with helpful directions or advice. There are generally two types of 9-1-1 service: basic and enhanced.

Basic 9-1-1 service can be as fundamental as dialing 9-1-1 and the phone rings at the telecommunications center. Basic 9-1-1 can also have additional features, the most common of which are called party hold, forced disconnect, and ringback.

- *Called party hold* is a feature that allows a telecommunicator to maintain access to a caller's phone line. By the telecommunicator not hanging up or disconnecting, this call will, in essence, seize the calling person's phone line. If the original calling person hangs up and then picks up the phone to place another call, they will find that they are still connected to the telecommunicator.
- *Forced disconnect* is, in a way, the reverse feature for telecommunicators. When the called party (telecommunicator) hangs up after someone places a call to 9-1-1, the calling party can keep the line active or tied up for a short time period. The forced disconnect feature drops the call out of the system and opens the 9-1-1 line for the next caller.
- *Ringback* is a feature that allows the telecommunicator to call back a calling party's phone, after he has hung up.

Some basic systems will offer one of the features of an enhanced system: Automatic Number Identification (ANI). This features displays the calling party's phone number on a display screen at the telecommunicator's position.

Some states and municipalities are equipped with enhanced 9-1-1 (E-9-1-1) systems. E-9-1-1 systems combine telephone and computer equipment to provide the telecommunicator with instant information such as the caller's location and phone number, directions to the location, and other information about the address. As soon as the telecommunicator picks up the phone, the computer shows the location from where the call is being made through Automatic Location Identification (ALI). Business extensions may not allow the caller's exact location to be displayed on the computer. This system allows help to be sent even if the person on the other end of the line is incapable of giving proper information. Wireless telephones will not activate the E-9-1-1 system ALI.

RADIO

On occasion, the report of an emergency may be received via radio. This type of report will most likely come from fire department personnel who are already on the streets for some reason and happen upon an emergency. The firefighter or telecommunicator in the station monitoring the radio should get the same kind of information that would be taken from a telephone caller. Once all the appropriate information is received, additional resources should be dispatched if required.

Some fire departments also monitor citizens band (CB) radio frequencies for reports of emergencies. The universal frequency for reporting emergencies, and the one most commonly monitored by emergency providers, is CB channel 9. Reports taken via CB radio should be handled in the same way as those taken by telephone; however, in place of the callback number, the caller's radio "handle" or designation should be recorded.

WALK-INS

Occasionally, a citizen will walk into a fire station and report an emergency that has just occurred in the vicinity of the station (Figure 18.15). Firefighters in the station should get the location and type of incident from the person as well as the person's name and address if possible.

Figure 18.15 Firefighters must know what information to get from a civilian who walks into the fire station to report an emergency.

Once the information is obtained, local policy dictates the next step. Some departments require the person taking the complaint to first notify the telecommunications center by phone before taking further action. Other jurisdictions allow personnel in the station to immediately start their response and radio the telecommunicator with information on the incident while they are responding.

WIRED TELEGRAPH CIRCUIT BOX

Wired telegraph circuit boxes were commonly used, particularly in metropolitan and heavy industrial areas, to provide a means for people on the street to summon the fire department (Figure 18.16). This alarm system is operated by pressing a lever in the alarm box that starts a wound-spring mechanism. The rotating mechanism transmits a code by opening and closing a circuit. Each box transmits a different code to specify its location. The system is limited in that the only information transmitted is the location, which presents the problem of malicious false alarms. For this reason, these systems have been eliminated in many localities.

Figure 18.16 Old fire alarm systems use telegraph boxes.

TELEPHONE FIRE ALARM BOX

A telephone is installed in the fire alarm box for direct voice contact, allowing for exchange of more information on the type of response needed. Some municipalities have used the combination telegraph and telephone-type circuits. This gives the best of both systems. The pull-down hook is used to send the coded signal, and a telephone is included for additional use (Figure 18.17).

Figure 18.17 Some localities have emergency phones.

RADIO FIRE ALARM BOX

A radio alarm box contains an independent radio transmitter with a battery power supply (Figure 18.18). Solar recharging is available for some systems. Others feature a wound-spring alternator to provide power when the operating handle is pulled.

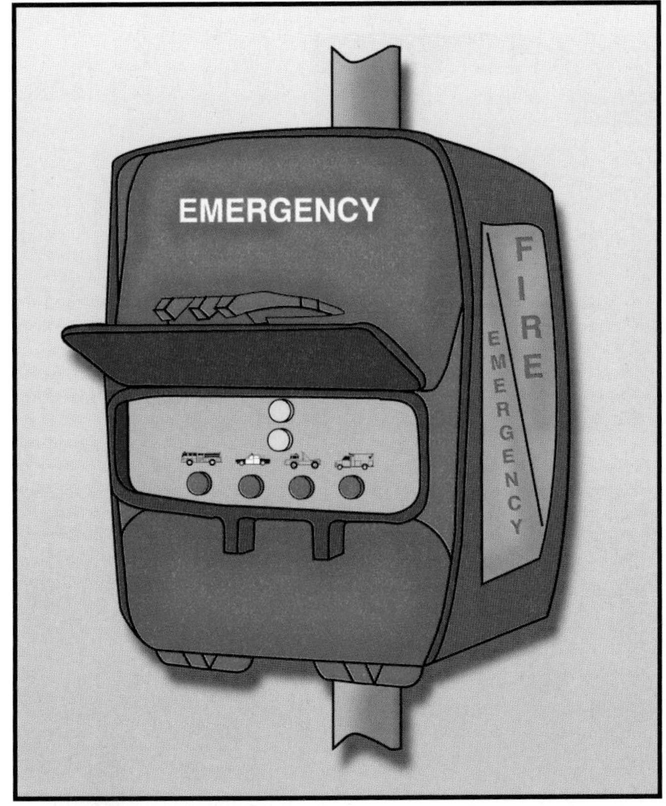

Figure 18.18 A radio alarm box.

There are different types of radio boxes. Activating the alarm in radio boxes alerts the telecommunicator by an audible signal, visual light indicator, and a printed record indicating the location. Some models have, besides a red alarm-light indicator, a different colored light indicator that shows a test or tamper signal. By using a time clock within the box, it can test itself every 24 hours. If the box pole is struck or tampered with, the tamper light comes on and gives the box location. Some boxes are numbered, and this number also appears on the telecommunications center display panel, informing the telecommunicator of the box involved and its location.

When activated by the incoming radio signal, some printing systems print the date, the time of day in 24-hour time, the message sent by the box, the box number, and a coded signal that indicates the strength of the battery within the box. Some radio alarm boxes are designed so that a person can select fire, police, or ambulance service. Some radio alarm boxes located at intervals along roads, along highways, and in rural areas have two-way communications capabilities. Telecommunicators answering these radio alarm box reports should get the same information as they would by telephone.

Procedures for Reporting a Fire/Emergency

A department's fire and life safety education program should include information on how to report an emergency correctly. The public should be trained to report emergencies using the methods given in the following sections.

BY TELEPHONE

- Dial the appropriate number:
  — 9-1-1
  — Fire department 7-digit number
  — "0" for the operator
- Give address, with cross streets or landmarks if possible.
- State your name and location.
- Give the telephone number from which you are calling.
- State the nature of the emergency.
- Stay on the line if requested to do so by the telecommunicator.

FROM A FIRE ALARM TELEGRAPH BOX

- Send signal as directed on the box (Figure 18.19).
- Stay at the box until fire personnel arrive so that you can provide the exact location of the emergency to them.

FROM A LOCAL ALARM BOX

- Send signal as directed on the box (Figure 18.20).
- Notify the fire department by telephone using the guidelines given earlier.

Figure 18.19 A municipal fire alarm box.

Figure 18.20 A local fire alarm station that is commonly found in many occupancies.

ALERTING FIRE DEPARTMENT PERSONNEL

[NFPA 1001: 3-2.1(a); 3-2.1(b)]

Different departments have different ways of alerting units about an alarm location and notifying firefighters of an emergency. Some use a system of bells or other sounding devices, others use radio-voice communications, and still others have an automatic, computer-operated system. The particular way that fire stations and personnel are alerted depends on whether or not the station is staffed.

Some fire departments try to give information from prefire plans about the location of the alarm to fire companies as they respond. Some departments transmit the information by radio; others have the information available in individual fire vehicles on transparencies or microfiche. Some departments "pre-alert" their stations by transmitting the address of a call while researching the dispatch information. Units respond when the address is in their area. The emergency information is then transmitted while the emergency vehicles are en route. This system reduces the amount of time it takes for units to leave the station, reducing the overall response time.

Staffed Stations

Technological advances have brought about new, modern alerting systems to accompany the more traditional types. These types of alerting systems include the following:

- Computerized line printer or terminal screen with alarm (Figure 18.21)
- Vocal alarm
- Teletype
- House bell or gong
- House light (Figure 18.22)
- Telephone from telecommunicator on secure phone line
- Telegraph register
- Radio with tone alert

Alerting methods employed should be effective but not startling to personnel. Extremely loud audible devices or bright lights that come on in the middle of the night can be somewhat shocking to a

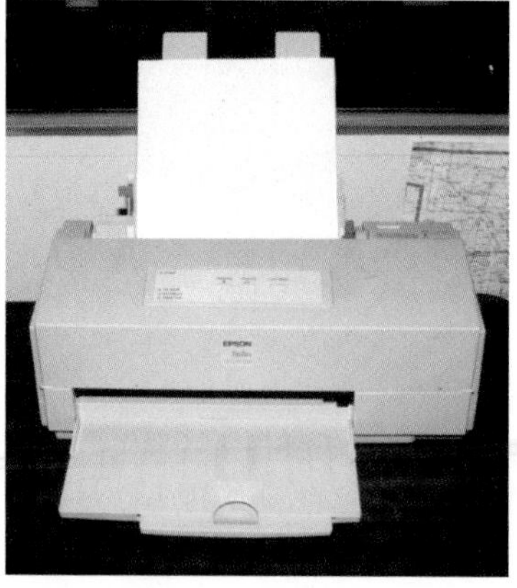

Figure 18.21 Computerized fire departments may have printers in the fire station that give a hard copy of the address of the emergency as well as other pertinent information.

Figure 18.22 A sign board displays which unit is being alerted to respond. *Courtesy of Foster Cryer.*

firefighter and can raise stress and anxiety levels. Dull lights and reasonable audible devices are more effective.

Unstaffed Stations

In order to facilitate as quick a response as possible from unstaffed stations, some method of simultaneously notifying all personnel must be used. These systems include the following:

- Pagers (Figure 18.23)
- Home electronic monitors
- Telephones
- Sirens (Figure 18.24)
- Whistles or air horns

Figure 18.23 Many volunteer firefighters rely on pagers to alert them to respond to an emergency.

Figure 18.24 Fire sirens are used in many municipalities.

Pagers and home electronic monitors are activated by tone signals that are sent over radio waves. The advantage of pagers is that firefighters can carry them wherever they go. Home monitors and telephones require the firefighter to be at home to be aware of the need to respond. Home monitors are quickly being phased out of the fire service with the accessibility of pagers.

A paging system is a transmitter on a given frequency that will speak to a specific pager or specific group of pagers, which are really just miniature receivers. A pager, or receiver, is set on a specific radio frequency and given an address of some specific tones, codes, or frequency. Individual tones, such as for a chaplain or a fire investigator, can also be used. When the pager receives its codes, it turns on and alerts the wearer by tone, light, or vibration. The pager will then either open the speaker for a voice message or display the alpha numeric message sent to it. Most pagers have an alert setting that only activates if that pager's tone is transmitted. Usually when a number of different departments or public safety agencies operate from one dispatch frequency, it is desirable to set pagers to the alert setting to avoid hearing unwanted radio traffic.

Sirens, whistles, and air horns are employed most commonly in small communities. These devices make a loud noise that all members of the community can hear. This noise is both an advantage and a disadvantage. Civilians will be aware that emergency traffic may be on the streets; however, many may also be inclined to follow the apparatus and congest the emergency scene.

RADIO COMMUNICATIONS

[NFPA 1001: 3-2.1(a); 3-2.1(b); 3-2.3(a); 3-2.3(b); 4-2.2(a); 4-2.2(b)]

All radio communication in the United States is under authorization from the Federal Communications Commission (FCC). Fire departments that operate radio equipment must hold radio licenses from the FCC. Depending on the radio system in a particular locality, one license may cover several departments that operate a joint system. Local department rules should specify who is authorized to transmit on the radio. It is a federal offense to send personal or nonemergency service messages over a designated fire department radio channel.

Radio Procedures

Departmental operating procedures such as department codes, test procedures, and time limits on radios need to be established. Ten-codes were popular in the early days of radio equipment because of poor transmission and reception. Advances in radio technology have reduced the need for ten-codes, and many departments have converted to simple English (*clear text*).

A telecommunicator can create an impression of a call by vocal inflection; the most routine call can sound like a major emergency. Likewise, a major call can sound extremely routine. Most agencies strive for a professional approach, which is a constant steady voice level — never too excited or too routine. There are a number of keys to this approach and some guidelines are as follows:

- Use a moderate rate of speaking — not too slow or too fast — designed for easy understanding. This includes not using pauses such as "*ah*" or "*uhm*" during the dispatch.
- Use a moderate expression in speech — not a monotone and not overemphasized — with carefully placed emphasis. Avoid anger or shouting over the radio and be careful to articulate properly. Strive for the correct pronunciation of words.
- Use a vocal quality that is not too strong or weak. Finish every comment, and avoid a voice that trails off towards the end of the transmission. Keep the pitch in a midrange — not too high or too low. Avoid dialects or regionalisms in transmissions, and strive for a good voice quality.
- Keep things such as gum and candy out of the mouth. Be confident in what you say, and position the microphone appropriately to make the best use of the system.
- Be concise and to the point — don't talk around the issue — and give the information required in a logical and complete manner that best addresses the service requested.

Telecommunicators are judged by their effectiveness on the radio. In the real world of reluctant, scared, upset, and lost victims/witnesses, extracting information is oftentimes difficult and time-consuming. While time is very critical, so is accuracy. It does very little good to say that help is needed on Main street if Main Street is 30 blocks long. Nor is it beneficial to quickly say that the help is needed on North Main street, when in fact it turns out to be South Main street. It is necessary to provide responders with the most accurate location where help is needed. The guiding principle is to transmit accurate information as quickly as possible.

Another difficult area for telecommunicators is using enough words but not too many to give a clear picture of the event or service request. Most people find it extremely frustrating to have a telecommunicator use 50 words to describe a request for service when 20 well-chosen words would have been adequate. The reverse can also be a problem: being sent to a report of smell of smoke and finding a building fully involved in fire.

An additional but often forgotten factor when transmitting information and orders via radio is that *only essential information* should be transmitted. For example, consider the difference in the following two radio communications:

Radio Communication 1: *"Communications, this is Lieutenant Thompson on Engine 57 portable. I need another truck company at this location, Box 1333, for personnel."*

Radio Communication 2: *"Communications, this is Engine 57 portable — dispatch 1 truck company to Box 1333."*

In Radio Communication 1, it serves absolutely no purpose to identify oneself by name. *"Engine 57 portable"* communicates to the telecommunicator that the company officer is requesting the truck company. In addition, the use of phrases such as *"I need another"* and *"at this location"* waste radio air time, which is critical during most incidents. Furthermore, communicating *"for personnel"* is not necessary. The company officer does not provide the reason or justification for requests over the air. He simply transmits the request.

Note that Radio Communication 2 required only 13 words. Although brevity is of the utmost importance, this is only part of the advantage of this transmission. Note that it communicates the request in a concise, clear manner. The telecommunicator relays the request to the appropriate source. Chances of error or miscommunication are significantly reduced.

Everyone on the fireground should follow two basic rules to control communications. First, units must identify themselves in every transmission as outlined in the standard operating procedures. Second, the receiver must acknowledge every message by repeating the essence of the message to the sender.

Example:

Engine 4: *"Communications, this is Engine 4. We are on the scene and have a Dumpster® fire. We will handle. Cancel all other units."*

Communications: *"Engine 4, this is Communications. Understand this is a Dumpster® fire, and you will handle. Other dispatched units can be canceled."*

Requiring the receiver to acknowledge every message ensures that everyone understands the same terms. This feedback either tells the sender the message was understood as transmitted or it notifies the sender that the message was not understood and further clarification is necessary.

Other important considerations that should be remembered include the following:

- Do not transmit until the airwaves are clear.
- Think about what is going to be said before transmitting.
- Remember that any unit working at a fire or rescue scene has priority over any other transmission.
- Do not use profane or obscene language on the air.
- Hold the radio/microphone 1 to 2 inches (25 mm to 50 mm) from the mouth at a 45-degree angle (Figure 18.25).

Firefighters on the emergency scene have other considerations as well:

- Avoid laying the microphone on the seat of the vehicle because the switch may become pressed and cause interference.
- Do not touch the antenna when transmitting. Radio frequency burns might result.

Dispatching for Emergency Medical Services (EMS)

Dispatching for EMS can differ from fire response. The telecommunicator must not only obtain the address of the emergency but may also need to determine the type and severity of the medical emergency. Although many departments respond to all calls, the greater number of EMS calls (two to three times as many as fire calls) and often a scarce number of ambulances have led many localities to institute a system of call screening and priority dispatching. Department protocols must be strictly followed and the rule is: When in doubt, send an ambulance.

Arrival and Progress Reports

First-arriving companies should use the radio to provide a description of the conditions found at the scene. All firefighters should know how to provide an accurate report of the conditions they see as they arrive on the scene (Figure 18.26). This process is often referred to as *size-up*. Each department should establish its own format for size-up reports. A good size-up report establishes a time of arrival and allows other responding units to anticipate what actions might be taken upon their arrival. The following is a typical size-up report:

Base, Engine 611 on location at Knik and Railroad Avenue of a two-story, wood frame residential structure; smoke showing from second-story windows; there is an all-clear on the structure. We are stretching 2 handlines; Central 2-1 establishing Railroad Avenue Command at this location.

Figure 18.25 Hold the radio 1 to 2 inches (25 mm to 50 mm) from the mouth at a 45-degree angle.

Figure 18.26 The first arriving unit should provide a size-up report for other responding units.

Or like this:

Base, Central 6-5 on location at Bogard at Postishek Place, stand by for a size-up.

Base, Central 6-5, we have a two-story, wood frame, residential structure. Heavy fire in the first floor rear. No apparent exposure hazards; all occupants are out of the structure. Have Engine 611 prepare for interior attack and Engine 651 start ventilation. Dispatch Support 62 and Central 2. Central 6-5 is establishing Bogard Command at this location.

When giving a report of conditions upon arrival, the following information should be included:

- Address, particularly if other than the one initially reported
- Building and occupancy description
- Nature and extent of fire
- Attack mode selected
- Rescue and exposure problems
- Instructions to other responding units
- Location of incident command position
- Establishing command

Once fire fighting operations have begun, it is important that the telecommunications center is continually advised of the actions taken at the emergency scene. Such progress reports should indicate the following as applicable:

- Transfer of command (Figure 18.27)

Figure 18.27 Any transfer of command should be included in a progress report. *Courtesy of Bill Tompkins.*

- Change in command location
- Progress (or lack of) in situation control
- Direction of fire spread
- Exposures by direction, height, occupancy, and distance
- Any problems or needs
- Anticipated actions

Tactical Channels

The radio communications system should reflect the size and complexity of the incident. Routine, day-to-day incidents can usually be handled on a single channel, but larger incidents may require using several channels to allow for clear and timely exchanges of information. Separate channels may be needed for command, tactical, and support functions.

Often, when one radio channel is primarily used for dispatching, it is necessary to operate on a different channel for an incident. This allows the IC to have an open communication channel between the telecommunications center and an open channel to the fireground officers without having units "talking over each other" (interruptions by other transmissions). Tactical channels are most often used for larger incidents such as structure fires. Smaller routine incidents such as fire alarms and vehicle fires usually do not require the use of a tactical channel.

Units are initially dispatched on the primary dispatch channel. Upon arrival on the scene, they switch to the tactical channel to communicate with the IC. Some of the roles the telecommunicator plays, depending on the severity or extent of the operation, are as follows:

- Assign an operational frequency for the management of the operation or the incident (Figure 18.28).
- Ensure that the current response of additional units to the incident is acknowledged.
- Notify other agencies and services of the incident and the need for them to respond or take other appropriate action.
- Provide updated information that affects the incident.

Figure 18.28 A telecommunicator announces the tactical channel to be used at the scene.

Calls for Additional Response

At some fires, it may be necessary to call for additional units. Normally, only the incident commander may order multiple alarms or additional responses. Depending on who arrives first, the incident commander may be a company officer, chief officer, or a firefighter.

All firefighters need to know the local procedure for requesting additional alarms (Figure 18.29 on next page). They must also be familiar with alarm signals (multiple or special alarms) and know what to do when they are received. Personnel should know the number and types of units that respond to these alarms.

When multiple alarms are given for a single fire, maintaining communications with each unit becomes more difficult as radio traffic increases. To reduce the load on the telecommunications center, a mobile, radio-equipped, command vehicle can be used at large fires (Figure 18.30).

When firefighters function as part of a team, they must be able to communicate the need for team assistance through the fire department's communications equipment. The designated supervisor must be in constant communications with the team and follow an incident management system with local SOPs. Some of these communications might be requests for additional personnel or special equipment or to notify others on the fireground of any apparent hazards.

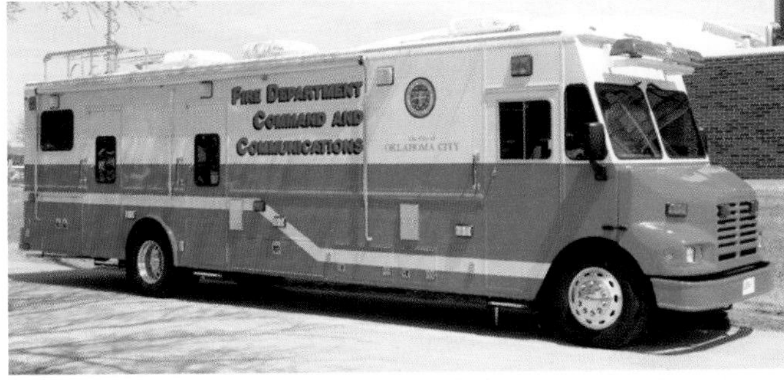

Figure 18.30 This mobile command post and communications vehicle can be used at large incidents.

Emergency Radio Traffic

At times, it may be necessary to broadcast emergency traffic (urgent message) over the radio. Telecommunicators at telecommunications centers are better equipped than on scene personnel to hear weak signals from portable and mobile radios. When firefighters radio that they are in distress, the telecommunicator can make a significant difference in firefighter survival.

Emergency traffic is also necessary when additional resources are needed or detailed instructions must be relayed through the telecommunicator. When the need occurs to transmit emergency traffic, the person transmitting the message should make the urgency clear to the telecommunicator. For example: *"Dispatch, this is Central 1, Emergency Traffic!"* At that point, the telecommunicator should give an attention tone (if used in that system), advise all other units to stand by, and then advise the caller to proceed with the emergency message. After the emergency communication is complete, the telecommunicator should notify all units to resume normal or routine radio traffic.

Evacuation Signals

Evacuation signals are used when command personnel decide that all firefighters should be pulled from a burning building or other hazardous area because conditions have deteriorated beyond the point of reasonable safety. All firefighters should be familiar with their department's method of sounding an evacuation signal. There are several ways this communication may be done. The two most common are to broadcast a radio message ordering them to evacuate and to sound the audible warning devices on the apparatus at the fire scene for an extended period of time.

The radio broadcast of an evacuation signal should be handled in a manner similar to that

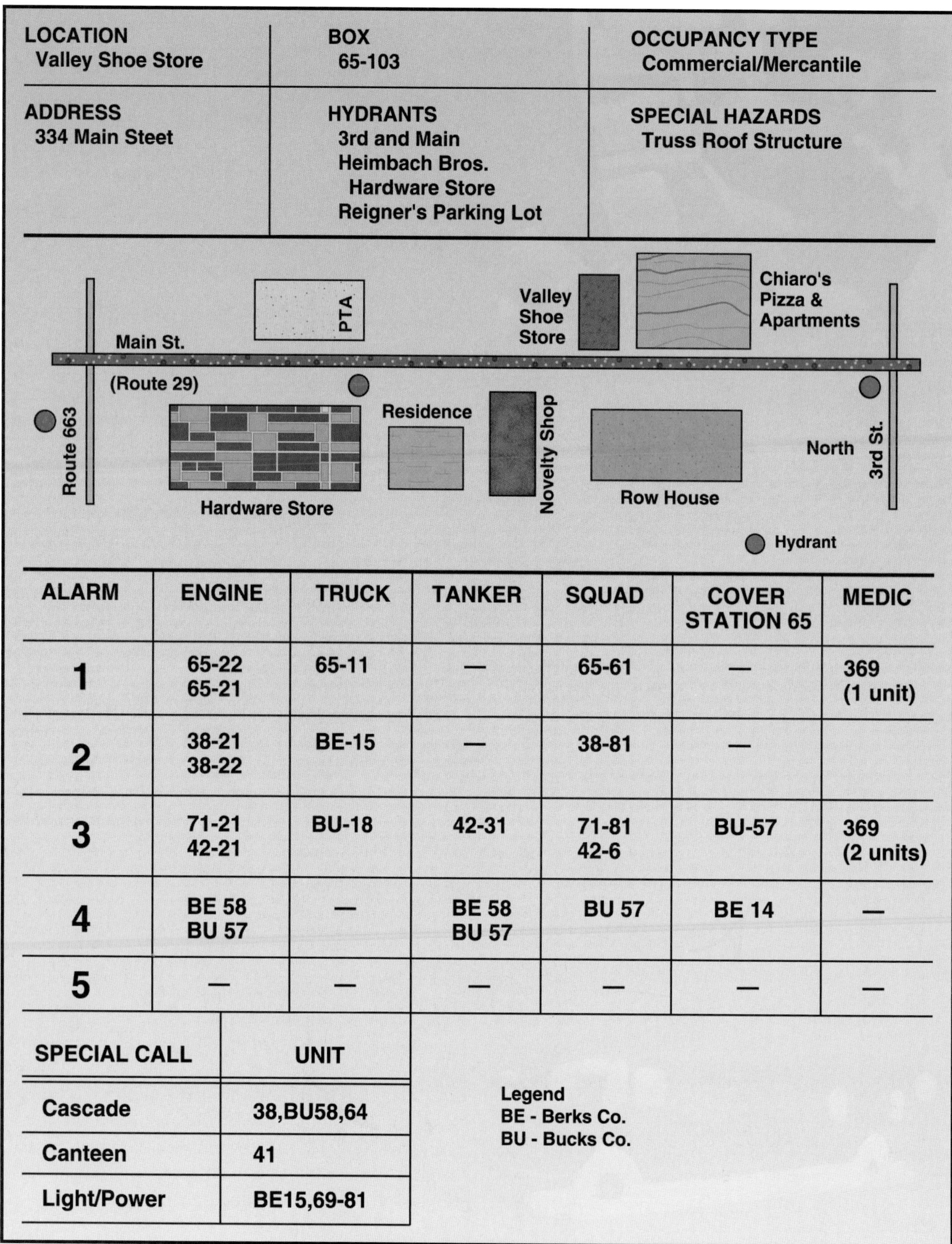

LOCATION	BOX	OCCUPANCY TYPE
Valley Shoe Store	65-103	Commercial/Mercantile
ADDRESS	**HYDRANTS**	**SPECIAL HAZARDS**
334 Main Steet	3rd and Main Heimbach Bros. Hardware Store Reigner's Parking Lot	Truss Roof Structure

ALARM	ENGINE	TRUCK	TANKER	SQUAD	COVER STATION 65	MEDIC
1	65-22 65-21	65-11	—	65-61	—	369 (1 unit)
2	38-21 38-22	BE-15	—	38-81	—	
3	71-21 42-21	BU-18	42-31	71-81 42-6	BU-57	369 (2 units)
4	BE 58 BU 57	—	BE 58 BU 57	BU 57	BE 14	—
5	—	—	—	—	—	—

SPECIAL CALL	UNIT
Cascade	38,BU58,64
Canteen	41
Light/Power	BE15,69-81

Legend
BE - Berks Co.
BU - Bucks Co.

Figure 18.29 This information sheet lists the unit assignments for specific alarm requests.

Chapter 19
Fire Prevention and Public Fire Education

Chapter 19
Fire Prevention and Public Fire Education

INTRODUCTION

Without a doubt, "live" media coverage of events has elevated public concern regarding potential hazards, dangerous practices, and safe environments. The increased demands now placed on firefighters and other emergency responders necessitate a clear understanding of and active participation in all aspects of public fire prevention and education.

Every public action taken by a fire service organization can project a lasting consequence on both its firefighters and the citizens it serves. Firefighters respond to fires and other life-threatening incidents that could have been prevented had the individual causing the situation clearly understood the consequence of the unsafe act or the danger posed by the hazards encountered. The foundation of every successful fire prevention endeavor is grounded in a clear understanding of past problems and current needs (Figure 19.1). Before anyone (emergency services provider or citizen) can institute corrective action, they must be able to recognize and properly interpret the potential risk, need, or condition being encountered.

Firefighters must direct their efforts at reducing known hazardous conditions or preventing dangerous acts before tragedy strikes. This may be accomplished in many innovative ways such as conducting presentations, distributing safety brochures, providing news articles, writing public safety announcements (PSAs), or establishing meaningful displays in well-visited areas (Figure 19.2). Proper utilization of the news media during or just after a preventable accident has occurred can many times pay public awareness dividends.

Figure 19.1 Firefighters must determine where fire problems may be encountered.

PUBLIC SERVICE ANNOUNCEMENTS

SUBJECT: Fire Prevention
FROM: Your Fire Department
PROGRAM: Christmas Holiday Season

(20-Second Announcement)

Decided what to give the children this Christmas? Whatever toys you select, make sure they're safe. Look for the UL label of fire and shock safety on all electrical and heat producing toys. Choose chemistry sets carefully. Supervise play with toys involving fuels and chemicals. Protect your youngsters from fire!

****

(30-Second Announcement)

Is anyone in your family taking part in a Christmas pageant or choir recital during this holiday season? Costumes and choir robes are often made of loosely woven material which is highly combustible. No one wearing such a garment should ever get close to a flame or source of heat. If candles are to be carried, they should be electric candles. Don't let fire turn Christmas joy into tragedy!

****

Figure 19.2 Typical public service announcements (PSAs).

On such occasions, normally the department's public information officer or other chief officer works with media contacts (Figure 19.3).

This chapter provides information to aid the firefighter in fire prevention and public fire education. It begins with an overview of fire prevention and the various types of fire hazards. The next section discusses the two types of fire safety surveys: the pre-incident survey and the residential fire safety survey. The last part of the chapter discusses public fire and life safety education. It provides information on fire safety topics a firefighter may be asked to provide to the public during presentations or as part of a fire station tour.

Figure 19.3 The public information officer acts as a liaison between the incident commander and the media.

FIRE PREVENTION

[NFPA 1001: 3-5.1]

Fire safety surveys in public, commercial, and residential occupancies can have an important effect on community fire prevention and pre-incident planning. *Fire safety surveys* involve those activities that have been planned or legislated to ensure that citizens have a safe physical environment in which to live, work, study, worship, or play. The survey process requires firefighters to become familiar with community structures and to recognize safety hazards quickly. Observed problems can then be resolved diplomatically.

Fire incident records, which contain a wealth of knowledge that represent the documented fire history of a community, can also further fire prevention efforts. Many informational benefits can be accomplished by studying previous incidents, reviewing data obtained from various fire reports and comparing statistical data. Such a review helps identify major fire causes and raises questions about possible solutions.

Another important fire prevention activity is the fire safety inspection. Fire inspections are usually conducted by fire inspectors specially trained in these procedures (Figure 19.4). Their findings not only make firefighters aware of potential hazards but permit inspection personnel to communicate unsafe conditions to building occupants and owners. Any person expected to take a more authoritative role in public safety inspections should be trained to meet the objectives found in NFPA 1031, *Standard for Professional Qualifications for Fire Inspector*. Additional guidance regarding inspection practices can be found in IFSTA's **Fire Inspection and Code Enforcement** manual.

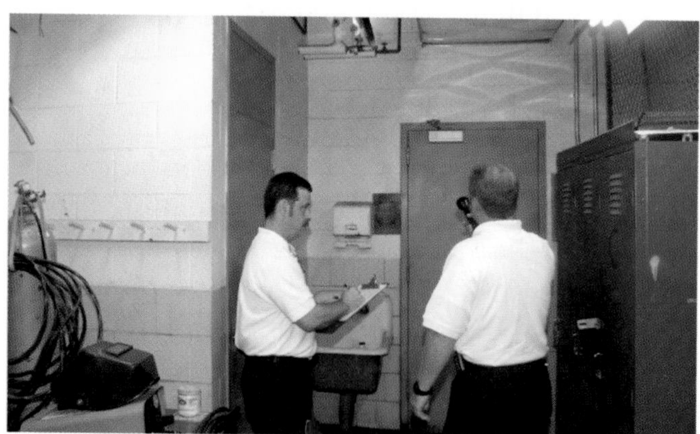

Figure 19.4 Fire inspections are usually conducted by fire inspectors trained in these procedures.

FIRE HAZARDS

[NFPA 1001: 3-5.1; 3-5.1(a); 3-5.1(b); 4-5.1; 4-5.1(b)]

A *fire hazard* is a condition that encourages a fire to start or increases the extent or severity of a fire (Figure 19.5). Basic fire chemistry suggests fire cannot survive without a fuel supply, sufficient heat source, oxygen supply, and a self-sustained chemical reaction (fire tetrahedron). Therefore, hazardous fire conditions can be prevented by eliminating one or all of these elements.

Control of the oxygen-supply hazard is only practical under special cases because 21 percent oxygen is normally present in air. The self-sus-

Figure 19.5 Spray booths are examples of hazards that firefighters should especially notice.

Figure 19.6 Look for obvious hazards such as this flammable liquid container stored next to a heater.

tained chemical chain reaction is required after a fire starts. Control of the hazards associated with fuel supply and heat sources are the most manageable. Any heat source may be dangerous. If heat sources are kept separated from fuel supplies, the condition remains safe. Not all fuel supplies can be ignited easily, but misuse of any fuel under extreme heat conditions can lead to a fire. Some common fuel and heat-source hazards include the following:

Fuel Hazards

- Ordinary combustibles such as wood, cloth or paper
- Flammable and combustible gases such as natural gas, liquefied petroleum gas (LPG), and compressed natural gas (CNG)
- Flammable and combustible liquids such as gasoline, oils, lacquers, or alcohol (Figure 19.6)
- Chemicals such as nitrates, oxides, or chlorates
- Dusts such as grain, wood, metal, or coal
- Metals such as magnesium, sodium, or potassium
- Plastics, resins, and cellulose

Heat Source Hazards (See Chapter 2, Fire Behavior)

- ***Chemical heat energy*** — Materials being improperly stored can result in chemical heat energy. Materials may come in contact with each other and react (oxidizer and reducing agent), or they may decompose and generate heat.
- ***Electrical heat energy*** — Poorly maintained electrical appliances, exposed wiring, and lighting are sources of electrical heat sources.
- ***Mechanical heat energy*** — Moving parts on machines, such as belts and bearings, are a source of mechanical heating (Figure 19.7).
- ***Nuclear heat energy*** — Nuclear heat is created by fission; however, this is not commonly encountered by most firefighters.

Figure 19.7 Many types of machines, even automobiles, have belts and bearings that can overheat and start a fire if not operating properly.

Common Fire Hazards

The term *common* could be misleading to some individuals. It refers to the probable frequency of a hazard being found, not to the severity of the hazard. A *common fire hazard* is a condition that is prevalent in almost all occupancies and encourages a fire to start. Firefighters need to be alert to the dangers posed by the following common hazards:

- Poor housekeeping and improper storage of packing materials and other combustibles
- Defective or improperly used heating, lighting, or power equipment
- Improper disposal of floor cleaning compounds
- Misuse of fumigation substances and flammable or combustible liquids

Poor housekeeping can make maneuvering through an area difficult. Poor housekeeping also increases the fire load in an area and increases the chance that a flammable or combustible material may come in contact with an ignition source. It also hides fire hazards in the clutter.

Improperly functioning heating, lighting, or other electrical equipment can provide an ignition source for nearby combustibles. Floor cleaning compounds, fumigating substances, and other flammable and combustible liquids that are improperly used and stored can provide a volatile fuel source should an ignition source be present.

Common fire hazards also have a personal component. The term *personal* takes into consideration the individual traits, habits, and personalities of the people who work, live, or visit the occupancy, structure, or property in question. *Personal fire hazards* refer to those common hazards caused by the unsafe acts of individuals. Personal hazards, often considered intangible, are always present. A comprehensive program geared toward public awareness, fire and life safety education, and good safety practices can reduce the hazards caused by unsafe personal acts.

Special Fire Hazards

A *special fire hazard* is one that arises as a result of the processes or operations that are characteristics of the individual occupancy. Commercial, manufacturing, and public-assembly occupancies each have their own particular hazards.

COMMERCIAL OCCUPANCIES

- Lack of automatic sprinklers or other relevant fixed fire protection system
- Display or storage of large quantities of products
- Mixed varieties of contents
- Difficulties in entering occupancies during closed periods
- Existence of party walls, common attics, cocklofts and other open voids in multiple occupancies (Figure 19.8)

Figure 19.8 Strip shopping centers often have common cocklofts that promote the spread of fire throughout the building.

MANUFACTURING

- High-hazard processes using volatile substances, oxidizers, or extreme temperatures
- Flammable liquids in dip tanks, ovens, and dryers; plus those used in mixing, coating, spraying, and degreasing processes
- High-piled storage of combustible materials (Figure 19.9)
- Operation of vehicles, fork trucks, and other trucks inside buildings (use, storage, and reservice hazards)
- Large, open areas
- Large-scale use of flammable and combustible gases
- Lack of automatic sprinklers or other fixed fire protection systems

Figure 19.9 High-piled storage produces extreme fire loads.

PUBLIC ASSEMBLY

- Lack of automatic sprinklers, detection systems, or fire notification systems
- Large numbers of people present, exceeding posted occupant limits
- Insufficient, blocked, or locked exits
- Storage of materials in paths of egress
- Highly combustible interior finishes

Target Hazard Properties

A *target hazard* is viewed as a facility in which there is a great potential likelihood of life or property loss from a fire (Figure 19.10). The occupancies should receive special attention during surveys. Some examples include the following:

- Lumberyards
- Bulk oil storage facilities
- Area shopping centers
- Hospitals

Figure 19.10 Pay particular attention to target hazards such as this hospital.

- Theaters
- Nursing homes
- Rows of frame tenements
- Schools

FIRE SAFETY SURVEYS

[NFPA 1001: 3-5.1; 3-5.1(a); 4-5.1; 4-5.1(a)]

Fire safety surveys include *pre-incident surveys* for public and commercial occupancies and *residential fire safety surveys* (Figure 19.11). All of these surveys are used to make citizens aware of hazards or dangerous conditions. Of course, these citizens must then understand the urgency of the situation and take appropriate action to resolve the problems identified, otherwise the fire department's effort may be of little consequence.

The pre-incident survey in public and commercial occupancies gives "up-front" information

Figure 19.11 Firefighters conducting a pre-incident survey of a commercial occupancy.

required to better assess conditions during any emergency situation that might occur in these occupancies. Such insight is a must if firefighters expect to safely, efficiently, and effectively control fire incidents.

Residential fire safety surveys may be accomplished as part of a house-to-house fire prevention program, or they can be done when requested on an individual basis. Fire departments that provide residential fire safety survey services usually do so as part of a public awareness and education program.

Firefighters need a wide range of personal and technical skills to conduct fire safety surveys properly. Development of good interpersonal skills may include those needed for communication, mitigation, facilitation, negotiation, or mediation. Technical knowledge and skills require firefighters to understand building construction, fire and life safety requirements, common and special hazards, building utilities, energy systems, and various fire protection appliances and systems.

Personal Requirements

In the public's eyes, the firefighter's uniform and badge indicate that the wearer is professionally qualified to discuss important aspects of fire prevention and give reliable advice on how fire safety hazards can be corrected. When performing any public fire prevention activity, the firefighter should project a well-groomed, neat appearance. The uniform should be clean and in good condition. A neat appearance gains the respect of the citizens being dealt with and bolsters the fire department's public image (Figure 19.12).

Figure 19.12 A firefighter should present a neat, professional image when performing public fire prevention activities.

Level I and Level II firefighters who meet the fire prevention and public fire and life safety education objectives found in NFPA 1001 will possess a basic understanding of fire prevention principles and can approach their assignments with confidence. Firefighters who perform fire safety surveys are expected to recognize basic hazards and report them through the appropriate channels established within their organization. They can aggressively tackle assigned situations they are trained to handle and may offer corrective advice, especially during voluntary residential surveys. However, firefighters are by no means fully qualified inspectors or public fire and life safety educators; therefore, they must be aware of their limitations in the field.

Each firefighter engaged in public prevention efforts such as fire safety surveys must be capable of meeting with the public and clearly communicating solutions to problems encountered. Firefighters who exhibit technical confidence convey a favorable public impression that benefits both the firefighter and the organization they represent.

A firefighter's ability to judge conditions will improve with study, experience, and on-the-job training. Skills required to transpose visual information into written reports and sketches will also improve with time and practice. The company officer, fire prevention officer, and other individuals from a wide range of technical backgrounds can provide needed insight on those occasions when answers to unexplainable situations are required. Firefighters should have the confidence to ask for the help of the company officer or other available experts.

Survey Equipment

The equipment needed by firefighters to adequately perform a fire safety survey may be divided into equipment needed at the place of survey and equipment used at the fire station (Figure 19.13).

AT THE SURVEY SITE

- Coveralls for crawling into attics and concealed spaces
- Safety glasses, hard hat, gloves

Figure 19.13 The equipment pictured may be needed at the fire safety survey site.

- Approved respirator when airborne particles exist
- Clipboard, survey forms, grid or engineering paper, and standard plan symbols
- Pen or pencil
- 50-foot (15 m) tape measure
- Flashlight
- Camera equipped with flash attachment
- Pitot tube and gauges when water flow tests are required

AT THE FIRE STATION

- Reference books
- Survey reports and forms
- Survey file, preferably on a computer database
- Code and inspection manuals
- Adequate records
- Drawing board
- Drawing scales, rulers, and materials

Scheduling the Fire Safety Survey

A primary management endeavor of every fire department administrator is the balancing of the multiple demands competing for the firefighter's time (Figure 19.14). Fire organizations cannot choose the moment when fires or other emergencies require their investment of time, but they have the prerogative of selecting the time and place to perform fire prevention activities. Because of this choice, the fire department administration should set a schedule for survey activities.

Figure 19.14 A fire officer uses a computer to manage schedules.

The company officer should contact the building owner or occupant ahead of time to arrange for the survey (Figure 19.15). The company officer will inform the owner of the purpose of the fire safety survey and find out what day and time would be most suitable. This will ensure that the survey schedule coincides with the availability of the building owner or occupants. This procedure enables fire safety surveys to be scheduled at a time that will not inflict a hardship on either the occupants or the fire company. A fire safety survey should never be attempted without proper permission. Commercial surveys are usually made during normal business hours, but night surveys are sometimes necessary because of operating schedules.

Figure 19.15 Call ahead so that the visit will not be a surprise to the occupant.

PRE-INCIDENT SURVEY

[NFPA 1001: 4-5.1; 4-5.1(a); 4-5.1(b)]

Pre-incident surveys provide knowledge of building construction, hazardous materials storage, building layout, special processes, fire notification and suppression features, and occupancy concerns. This knowledge greatly improves fire department operations and substantially improves both firefighter and citizen safety when suppression efforts are required. Firefighters use maps, sketches, photographs, and written notes to complete pre-incident surveys. The information learned helps firefighters achieve the following:

- Become familiar with area structures, their uses, and their associated hazards (Figure 19.16).
- Visualize how existing strategies apply to the occupancies.
- Recognize hazards.
- Aid citizens with fire prevention and life safety endeavors.
- Gain valuable on-site information for pre-incident planning.

Figure 19.16 Hazardous occupancies command extra attention during pre-incident surveys.

The fire department administration must specify how assigned firefighters are to conduct fire safety surveys in various public, commercial, or industrial complexes. Firefighters must be familiar with the department's policy and clearly understand the survey process.

An earnest effort by firefighters to create a favorable impression upon the owner helps to establish a courteous and cooperative relationship (Figure 19.17). Firefighters should enter the premises at the main entrance and contact the individual with whom the survey was scheduled. If necessary, firefighters may have to wait to see the proper individual because this person may be busy with other important matters. Reporting to the person in authority (exit interview) after the survey also shows the owner the importance of the survey.

The company officer should introduce the crew and state their business. If the owner has been informed of the purpose of the survey in advance, this introduction will be much easier. A representative of the occupant should accompany firefighters during the entire survey. Such a guide will help obtain ready access to all areas of the building and provide answers to questions.

Figure 19.17 The firefighter must make a positive impression on the occupant.

Making the Survey

After the initial meeting with the owner (or delegate), the survey team should return to the outside of the building to survey the exterior to make certain observations, preliminary notes, and photographs. This procedure makes the survey of the interior easier and provides the necessary information for drawing the exterior walls on a sketch of the floor plan (layout of each floor of a building).

Firefighters should note the location of fire hydrants, standpipe or sprinkler connections, and existing fire alarm boxes (Figure 19.18). The type of building construction and the height and occu-

Figure 19.18 Firefighters should note the location of fire hydrants, sprinkler connections, and water control valves during a survey.

pancy of adjacent exposures is included. Also worth noting is the general housekeeping in the area surrounding the occupancy, the accessibility to all sides of the property, and the condition of the streets. Such factors become extremely important when considering fire apparatus response. Firefighters should check and include the following in the preliminary notes:

- Are address numbers sufficiently visible?
- Are all sides of the building accessible?
- Is natural cover encroachment a threat?
- Are there forcible entry problems posed by barred windows or high-security doors?
- Are there overhead obstructions or other deterrents (Figure 19.19) that would restrict emergency operations?

When the survey of the exterior is completed, the survey team should go directly to the roof or basement and proceed with a systematic survey. From a survey point of view, it does not matter whether to start on the roof and work downward or start in the basement and work upward. From a practical standpoint, however, many firefighters find it helpful and less confusing to start on the roof. No matter what procedure is used, the route should be planned so that the firefighters can systematically look at each floor in succession.

Figure 19.19 Look for large overhangs and trees that might pose an obstacle to aerial apparatus.

To make a thorough survey, firefighters must take sufficient time to make notes and take photographs. Sketches of the interior layout's functional areas, egress routes, and important features should be drawn (or upgraded on existing sketch) or photographed. The sketching function is particularly important when survey information is used for pre-incident planning purposes. A complete set of notes, photographs, and well-prepared sketches of the building provide dependable information from which a complete preplan can be developed (see Maps and Sketch Making and Photographs sections) (Figure 19.20).

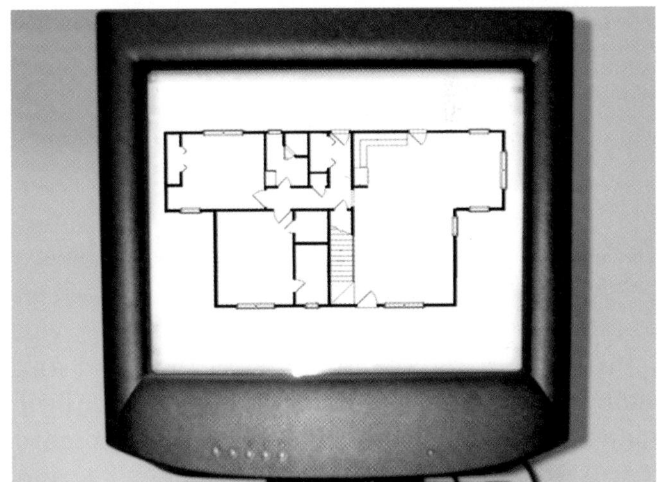

Figure 19.20 Some occupancies may have their building layout on computer. A hard copy of this information would save firefighters from having to sketch the layout.

Firefighters should ask that all locked rooms or closets be opened and explain tactfully why it is necessary to see these areas. For example, if the guide says, *"There is nothing in this locked room,"* a firefighter might say, *"Yes, we understand, but a knowledge of the size, shape, and construction features of the room are necessary."* If admission to an area or room is refused because of a confidential or secure process, the firefighter should suggest that the process be covered or screened to permit the survey to continue. Secret areas from which the survey crew are barred should be reported to the fire marshal, fire prevention officer, or chief administrative official so that appropriate action can be taken.

Firefighters should be particularly observant of hazardous materials (haz mat) commonly used in their response areas. Much of the firefighter's haz mat identification training can be performed at local commercial and industrial facilities. Surveys at these sites allow the firefighter to document the locations of hazardous materials and the physical layout of the plants.

Cleanliness, maintenance, and good housekeeping in hazardous materials areas are important precautions against fire. It should be recommended that a marking system such as that outlined in NFPA 704, *Standard System for the Identification of the Hazards of Material for Emergency Response,* be affixed to the outside of such structures.

In large or complicated buildings, it may be necessary to make more than one visit to complete the survey. If the property includes several buildings, each should be surveyed separately. It is a good idea to start on the roof of the highest building from which the firefighter can get a general view. A sketch of each floor should be completed before proceeding to the next floor. If a floor plan used on a previous survey is available, the survey can proceed more rapidly. Make sure to record any changes that have been made, and update the floor plan sketch accordingly. Allowing adequate time to discuss the survey results as well as any fire and life safety concerns with the owner or occupant usually benefits all concerned.

Maps and Sketch Making

Maps that convey information relative to construction, fire protection, occupancy, fire loading, special hazards, and other details of building complexes are an asset to fire suppression personnel. Large occupancies or complexes may already have maps that were prepared by insurance companies. These maps normally use some form of common map symbols (Figure 19.21).

For buildings where existing maps are not accurate or available, fire department personnel should include some sort of sketch with their survey notes to show the general arrangement of the property with respect to streets, other buildings, and any other important features that will help determine fire fighting procedures. This sketch is commonly called a *plot plan* of the area. A firefighter's sketch of an area frequently constitutes the most informative part of a survey and should be made with neatness and accuracy.

A sketch that is made during a survey may be done with the aid of a clipboard and a rule. Data should be recorded by using standardized plan symbols as much as possible. Engineering or graph paper can make the process somewhat simpler. The use of computerized Graphic Information System (GIS) mapping programs saves hours and should be strongly considered where available.

By using standard symbols on a floor plan, a firefighter can show the type of construction, thickness of walls, partitions, openings, roof types, parapets, and other important features. In addition to these features, fire protection devices, water mains, automatic sprinkler control valves, and other miscellaneous features of fire protection can be included (Figure 19.22).

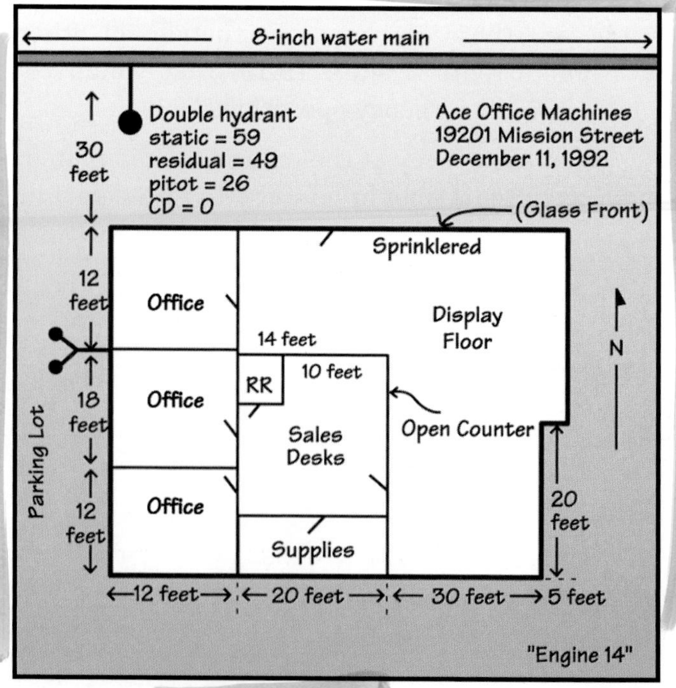

Figure 19.22 A basic floor plan sketch.

Standard Map Symbols

FIRE PROTECTION

Fire Department Connection

AS THRU-OUT — Automatic Sprinklers throughout contiguous sections of single risk

AS — Automatic Sprinklers all floors of building

AS 1st ONLY — Automatic Sprinklers in part of building only (note under symbol indicates protected portion of building

NS — Not Sprinklered

ACS — Automatic Chemical Sprinklers

ACS — Chemical Sprinklers in part of building only (note under symbol indicates protected portion of building)

V.P. HYD. — Vertical Pipe or StandPipe

AFA — Automatic Fire Alarm

WT — Water Tank

F.E. — Fire Escape

FA — Fire Alarm Box

Single Hydrant

D.H. — Double Hydrant

T.H. — Triple Hydrant

Q.H. H.P.F.S. — Quadruple Hydrant of the High Pressure Fire Service

20" W.P. (H.P.F.S.) — Water Pipes of the High Pressure Service

+ 12" + — Water Pipes of the High Pressure Service as shown on Key Map

6" W.P. / 4" W.P. — Public Water Service

6" W.P. (PRIV.) — Private Water Service

Fire Detection System - label type

Alarm gong, with hood

4" — Sprinkler riser (size indicated)

VERTICAL OPENINGS

Skylight lighting top story only

3 — Skylight lighting 3 stories

WG — Skylight with wired glass in metal sash

E — Open elevator

FE — Frame enclosed elevator

ET — Frame enclosed elevator with traps

ESC — Frame enclosed elevator with self-closing traps

CBET — Concrete block enclosed elevator with traps

TESC — Tile enclosed elevator with self-closing traps

BE — Brick enclosed elevator with wired glass door

H — Open hoist

HT — Hoist with traps

H B. To 1 — Open hoist basement 1st

STAIRS — Stairs

MISCELLANEOUS

MANSARD ROOF 48" — Number of stories
Height in feet
Composition roof covering

Parapet 6 inches above roof
Frame cornice
Parapet 12 inches above roof

W. HO 48" — Parapet 24 inches above roof
Occupied by warehouse
Metal, slate, tile or asbestos
Shingle roof covering
Parapet 48 inches above roof

S. 2B 2-D A. in B. BR. 1st R — 2 stories and basement
1st floor occupied by store
2 residential units above 1st
Auto in basement
Drive or passageway
Wood shingle roof

IR. CH. — Iron chimney

IR. CH. S.A. — Iron chimney (with spark arrestor)

UP. B. — Vertical steam boiler

Horizontal steam boiler

CURB LINE — Width of street between block lines, not curb lines

50' (15) — Ground elevation

CURB LINE 56 41 6 2 0 D — House numbers nearest to buildings are official or actually up on buildings. Old house numbers are farthest from buildings

Brick chimney

GT — Gasoline tank

Fire pump

COLOR CODE FOR CONSTRUCTION

Materials for Walls

Brown- Fire-resistive protected steel
Red-Brick, hollow tile
Yellow-Frame—wood, stucco
Blue-Concrete, stone or hollow concrete block
Gray-Noncombustible unprotected steel

Figure 19.21 Standard map symbols that firefighters may choose to use.

A sectional elevation sketch of a structure, consisting of a cross section or cutaway view of a particular portion of a building along a selected imaginary line, may be needed to show elevation changes, mezzanines, balconies, or other structural features (Figure 19.23). The easiest sectional view to portray is to establish the imaginary line along an exterior wall. This location theoretically removes the exterior wall and exposes such features as roof construction, floor construction,

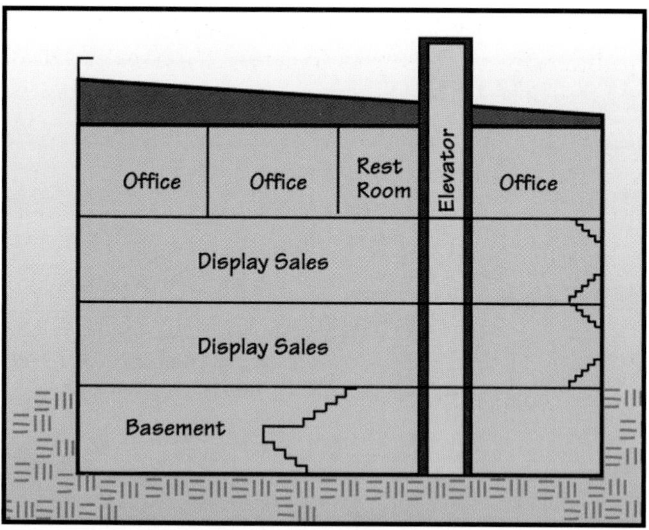

Figure 19.23 Sectional views show a cross section of the occupancy from top to bottom.

parapets, basements, attics, and other items that are difficult to show on a floor plan. Establishing the imaginary line along an exterior wall may not always show the section of the building that is desired. In this case, it may be better to divide the building near the center or along a line where a separate wing is attached to the main structure. From the firefighter's sketch and notes, a permanent drawing can be made to file for future reference and classroom study. The permanent drawing should be drawn to scale.

Photographs

Photographs show worthwhile detail for pre-incident plans, especially if they can be taken from more than one angle. A view that is especially good from a fire fighting standpoint is one from an elevated position (Figure 19.24). An adjoining building or elevated tower can be used for this purpose. Interior and close-up photographs are very effective aids in making a complete pre-incident plan.

Figure 19.24 When possible, get an elevated view of the building.

The Exit Interview

Reporting to the person with authority of the building being surveyed can do much to maintain a cooperative attitude of the owner (Figure 19.25). To leave the premises without consulting that person might give the impression that the survey was unimportant. During this interview, a firefighter or the company officer should comment on the conditions that were found. An exit interview also gives firefighters an opportunity to express thanks for the courtesies extended to the fire department and opens the way to explain how firefighters will study these reports from the standpoint of fire fighting procedures. In the final portion of the exit interview, firefighters should answer any questions they can and refer the owner/occupant to the fire marshal's office for further assistance.

Figure 19.25 Following the pre-incident survey, review the results with the occupant.

RESIDENTIAL FIRE SAFETY SURVEYS

[NFPA 1001: 3-5.1; 3-5.1(a)]

National statistics annually suggest that over 70 percent of all fires and the vast majority of civilian casualties occur in residences. All fire organizations should make a concerted effort to improve safety in the home setting (Figure 19.26). Fire safety surveys in existing residential occupancies (particularly one- and two-family dwellings) can only be accomplished on a voluntary basis. Codes require inspections for structures that house three or more families, but surveys of other than common areas in these structures may still be at the discretion of the occupants.

Figure 19.26 For the firefighter, residential fire safety surveys are educational tools for preventing fires and accidents that could take the lives of occupants and destroy homes.

When residential fire safety surveys are conducted as part of an organized public awareness and education program, a great deal of advanced planning and publicity is necessary to gain full acceptance by the community. It must be made clear that the program is a *fire prevention activity* and not a *code enforcement activity*. In other words, the firefighter is coming to make family members aware of safety hazards, not point out code violations. When firefighters enter the home to conduct a residential fire safety survey, their main objectives should include the following:

- Preventing accidental fires
- Improving life safety conditions
- Helping the owner or occupant to understand and improve existing conditions

Because residential safety surveys are voluntary, the fire department reaps many peripheral benefits in addition to the primary one of reducing loss of life and property. Citizens get to know and trust their firefighters. Safety surveys give residents the feeling that they are getting a complete service, not just an emergency service. This tends to increase citizen support for other fire department programs. During important periods, such as budget time, citizens can remember they get "more for their money." When residents get to know their firefighters, they gain an appreciation of the many duties and services their firefighters provide.

In addition to generating positive public relations and community support, safety surveys in the home increase fire awareness and the interest in public education efforts. The best time to distribute fire prevention literature, promote *exit drills in the home* (EDITH), check emergency telephone stickers, discuss smoke detector and residential sprinkler options, and provide other relevant safety information about such subjects as treatment of burns or CO detectors is during a safety survey (Figure 19.27). Firefighters can explain each item in the literature and possibly tie in a "local angle" of a fire experience.

Figure 19.27 Firefighters can provide homeowners with printed fire prevention materials during the survey.

Some fire departments also print special cards or slips to compliment the homeowner when the dwelling is found in a fire safe condition. Other cards saying *"We're sorry we missed you"* are used to notify absent households that firefighters were in the neighborhood conducting safety surveys.

Firefighters also gain valuable information when performing residential safety surveys. They become more acquainted with home construction, occupancy conditions, local development trends, streets, hydrants, and water supply locations. Notes on these items and other useful information should be made and discussed during training sessions. While these fringe benefits are helpful, the primary reason for conducting surveys is to reduce hazards associated with loss of life and property.

Firefighter Responsibilities

Residential fire safety survey campaigns can reduce the number of fire incidents and deaths occurring in homes. Therefore, it is the firefighter's responsibility to professionally represent his organization. The public has every right to expect firefighters to be qualified on matters pertaining to home fire and life safety. For a successful residential safety survey program, a firefighter should use the following guidelines:

- Provide proper identification.
- Introduce yourself and the purpose of your visit (Figure 19.28).
- Maintain a courteous attitude at all times.
- Request permission to conduct the survey.
- Remember that the primary interest is to prevent a fire.
- Compliment when favorable conditions are found.
- Do not order that corrections must be made when hazards are found.
- Make constructive comments regarding the elimination of hazardous conditions.

Figure 19.28 An occupant may be apprehensive of an unexpected uniformed firefighter conducting a door-to-door program such as a smoke detector survey.

- Survey the basement, attic, utility room, storage rooms, kitchen, and garage.
- Keep the survey confidential; the results are not provided to any outside entity.
- Thank the owners or occupants for the invitation into their home.

If no one is at home, leave appropriate materials between the doors or partially beneath the doormat; do not use the mail box.

Conducting the Residential Fire Safety Survey

There are several items firefighters should look for when conducting residential safety surveys. Fill out an established survey form for each residence and provide a copy to the occupant. The form can serve as a guide for firefighters, and it can be used to make summaries of the survey results.

Firefighters should be alert for the following signs of the most common causes of fires:

- Heating appliances
- Cooking procedures
- Smoking materials
- Electrical distribution
- Electrical appliances (Figure 19.29)
- Combustible or flammable liquids

Firefighters must know the common causes of home fires in order to conduct meaningful residential surveys and make citizens aware of dangerous conditions. For the homeowner or tenant, the residential fire safety survey provides a valuable life safety service. There is no better way for a firefighter to effectively carry out the responsibility of protecting lives and property. Firefighters should note the concerns in the following sections during the residential fire safety survey.

Figure 19.29 The firefighter should be alert for unsafe conditions when conducting residential fire safety surveys.

INTERIOR SURVEY CONCERNS

- ***Combustible materials*** — Are clothing, unused furniture, cardboard boxes, papers, and other materials stored properly? Are combustibles stored in close proximity to registers or heating appliances?
- ***Appliances*** — Inquire about proper operations, maintenance, and conditions, including electrical cords.
- ***Electrical wiring and equipment*** — Check for old, frayed, or exposed wiring and improperly installed electrical conductors. Check for unprotected light bulbs or improperly maintained equipment such as exhaust fans encrusted with dust and dirt.
- ***Portable heating units*** — Note if equipment is listed with Underwriters Laboratories (UL), Factory Mutual (FM), or some other laboratory and is adequately separated from combustible furniture or other materials.
- ***Woodstoves or fireplaces*** — Note whether the unit is properly installed and clear of combustibles and that the vent pipe is in good condition. Inquire as to the frequency of chimney cleaning and maintenance.
- ***Heating fuel*** — Inquire as to where wood or other fuel is stored. What is the procedure for ash disposal?
- ***General housekeeping practices*** — Does the occupant use ash trays for smoking materials? Are matches and lighters kept out of the reach of children? Are open flame items such as candles used safely? Are exhaust vents and dryer vents cleaned regularly of lint?
- ***Smoke detectors*** — Check for and encourage the installation of smoke detectors and testing on a regular basis.
- ***Electrical distribution panels*** — Check for proper circuit protection and clearance.
- ***Gas appliances*** — Check for improper clearance to combustible materials, the existence of automatic gas control safety devices, for manual supply line shutoff, for corroded piping, the condition of vents, and for possible gas leaks.
- ***Oil burning installations*** — Check for the existence of annual service records and the condition of oil burners, chimney pipes, supply tanks, and piping.
- ***Furnaces, hot water heaters, and vent pipes*** — Is the unit properly installed and clear of combustibles? Is the vent pipe in good condition? Inquire about hot water temperature settings to protect against scalds and burns (Figure 19.30).
- ***Shop or work rooms*** — Encourage good housekeeping in work areas and the safe, orderly storage of materials.
- ***Accumulated waste*** — Note stacks of paper, discarded furniture, old rags, and improperly stored items.
- ***Flammable liquids*** — Are flammable sprays, chemicals and other dangerous solutions properly stored and out of the reach of children?

Figure 19.30 Check the home heating unit for obvious problems.

OUTSIDE SURVEY CONCERNS

- ***Roof*** — Check condition of roof. Does it have composition roofing instead of wood shingles or shakes?
- ***Chimneys and spark arrestors*** — Check condition of chimneys and spark arrestors (Figures 19.31 a and b).
- ***Yard and porch areas*** — Is there unkempt vegetation? Are items stored under porches (Figure 19.32)?
- ***Barbecues and fuel*** — How are barbecues used? Is fuel stored properly?

Figure 19.31a Masonry chimneys are found on many homes with fireplaces.

Figure 19.31b Many newer chimneys are of the metal, prefabricated type.

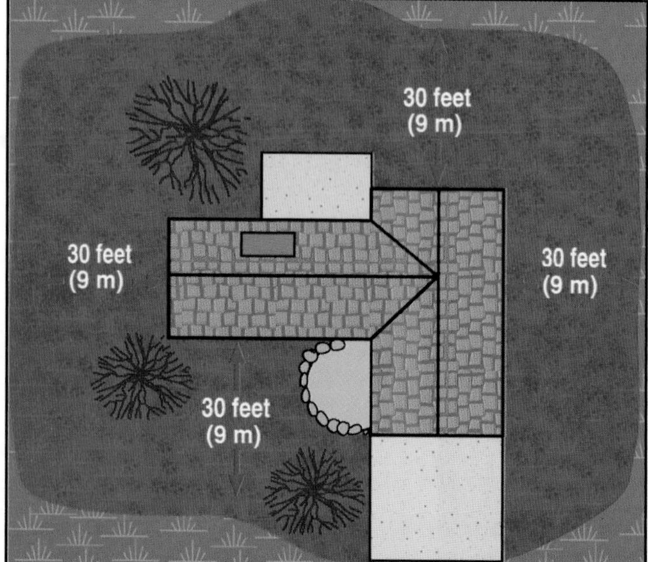

Figure 19.32 Keep tall grass at least 30 feet (9 m) from all structures.

- ***Outside waste burners*** — Discourage use of outside waste burners. Check for conformity to local restrictions.
- ***Garages, sheds, barns, and outbuildings*** — Note storage of dangerous chemicals or other substances (swimming pool chlorine, propane cylinders, charcoal, lighter fluid, gasoline, ammonium nitrate fertilizer, and pesticides). Are old paint cans, turpentine, and similar liquids stored properly?
- ***Flammable liquids and gases*** — Recommend that flammable liquids be kept in a safety-type can and stored only in an outside storage area (Figure 19.33). Gasoline, propane, and other similar flammable liquids and gases should never be brought into a dwelling. Remember flammable liquids should never be used, inside or out, for cleaning or other purposes that would expose the user and\or structure to explosive vapors.
- ***Lightning protection*** — Recommend that the system components of fixed lightning protection systems on structures be tested periodically.
- ***Security devices*** — Note security devices and pets that may hamper ingress or egress (window bars, security fences, dogs).

Figure 19.33 Flammable and combustible liquids should be stored in tool sheds.

HOME SAFETY ISSUES

In addition to performing a residential fire safety survey of the premises, firefighters should also provide occupants with fire and life safety awareness information. Firefighters must skillfully sell occupants on the value of making safe practices a way of life.

PUBLIC FIRE AND LIFE SAFETY EDUCATION

[NFPA 1001: 3-5.2; 3-5.2(a); 3-5.2(b)]

Educating citizens, at all ages, to recognize potential hazards and take appropriate corrective action is a fire department function. The teaching of fire survival techniques such as *stop, drop and roll* or *crawl low under smoke* can favorably alter behavior and impact life safety (Figure 19.34).

Figure 19.34 By playing the "blanket game" with the children, a firefighter teaches a desired behavior: Crawl low under smoke.

Presenting Fire and Life Safety Information

Although this section is not designed to make a firefighter an accomplished speaker or instructor, it presents some basic information that will assist a firefighter in presenting fire and life safety information to a small group of individuals. When making a fire and life safety presentation, a firefighter can take certain steps to make sure that all the information is presented and that the audience can perform basic fire and life safety skills such as calling the fire department or testing a smoke detector.

The first step in making a presentation is to prepare the audience to learn. Preparation involves gaining the attention of participants and letting them know why the material is important to them. An example of motivating a parent during a presentation on smoke detectors might be to appeal to their desire to protect their children. Arousing curiosity, developing interest, and developing a sense of personal involvement on the part of the participants are all parts of preparation.

The second step in making a presentation is for the firefighter to actually transfer facts and ideas (make the subject come alive) to the audience. Presentation involves explaining information, using visual aids (smoke detector, telephone for dialing 9-1-1, fire alarm pull station), and demonstrating techniques (stop, drop, and roll; crawl low in smoke; alert others of an emergency).

In the third step — perhaps the most important one — the participants use or apply the information they have been taught. This step provides the audience with the opportunity to practice using new ideas, information, techniques, and skills. Whenever possible, each person should apply new knowledge by performing the task. For example, the participants could demonstrate how to report a fire, perform the stop, drop, and roll technique, or test a smoke detector. The firefighter should supervise the application step closely, checking key points and correcting errors (Figure 19.35).

Figure 19.35 A firefighter watches as a participant demonstrates the ability to change a smoke-detector battery.

Fire and Life Safety Presentation Topics

Because fire stations are such busy places, a firefighter may be asked to assist in or teach a fire and life safety class from time to time. Some of the topics a firefighter may be asked to present during a fire safety presentation include the following:

- Stop, drop, and roll technique
- Home safety practices
- Placing, testing, and maintaining smoke detectors

STOP, DROP, AND ROLL

Firefighters should do more than simply inform people of what action to take if their clothing catches on fire. Both adults and children need to be effectively educated with firefighters first demonstrating and then soliciting individuals to perform the action. Demonstrate that if their clothes catch on fire, they must immediately STOP moving, DROP to the ground (covering their faces with both hands as they drop), and ROLL over and over until the flames are smothered (Figure 19.36).

Figure 19.36 Young children learn fire and life safety skills by practicing the skills they have been taught.

Point out that if someone's clothes catch on fire, an observer may need to assist the person in dropping to the ground and smothering the flames. Coats, rugs, blankets or other heavy cloth items in close proximity to the victim can be used to help smother the flames. Once the fire is out, cool the area with cold water (if available), and remove burned clothing that is not adhered to the victim's skin if possible. Summon emergency medical assistance immediately.

HOME SAFETY

The discussion of home safety can be presented as part of a group presentation or as part of the residential fire safety survey. As stated earlier, firefighters must skillfully sell the audience or occupants on the value of making safe practices a way of life. Firefighters should promote and favorably comment on escape plans, EDITH, and other practiced safety efforts conducted by home members (Figure 19.37). The proven fact is that citizens can safely escape during home fire emergencies with proper preparation and practice. Communicate the following fire and life safety rules to occupants:

- Keep doors to bedrooms closed during sleeping hours.
- Have two (or more) escape exits from every room.
- Ensure that windows can be easily opened by anyone to indicate his location to someone outside, to get fresh air, or for purposes of secondary escape in case of a fire emergency (exact action must be a predetermined activity).
- Train children properly if they are expected to use fire escape ladders (particularly in two- to three-story dwellings).
- Alert other family members of possible danger if awakened by the smell of smoke (for example, by blowing a whistle kept by every bed).

Figure 19.37 During a home visit, a firefighter demonstrates how to draw a home escape plan.

- Roll out of bed onto the floor (if awakened by a smoke detector sounding an alarm) (Figure 19.38).
- Stay low because dangerously heated gases may be at the top of the room (Figure 19.39).
- Crawl to the door. Feel the door; if it is warm, use the window for escape (Figure 19.40).
- Establish a meeting place outside the home so that all members can be accounted for after escaping. Never go back inside the house once outside.
- Call the fire department from a cellular telephone or a neighbor's house (Figure 19.41).

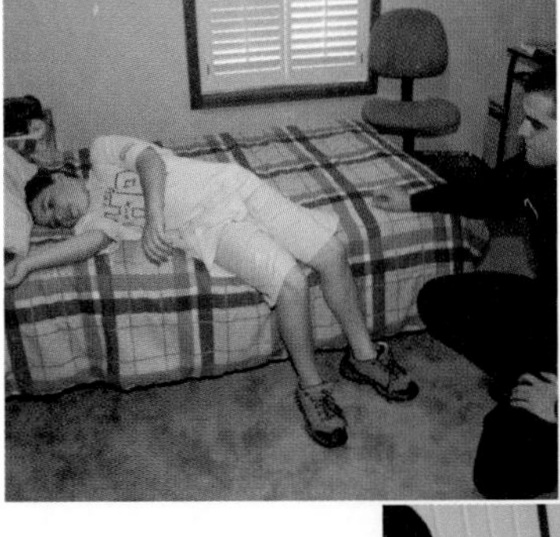

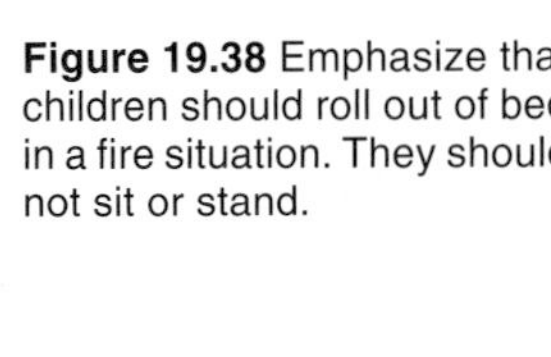

Figure 19.38 Emphasize that children should roll out of bed in a fire situation. They should not sit or stand.

Figure 19.39 To teach the "crawl low under smoke" behavior, the firefighter demonstrates the appropriate action.

Figure 19.40 Children should be instructed to feel the door for heat before opening it.

Figure 19.41 Children should demonstrate that they know how to call the fire department.

SMOKE DETECTORS

As was stated earlier in this chapter, an important part of conducting home safety surveys is to communicate to occupants the importance of smoke detectors (Figure 19.42). Therefore, it is essential that firefighters have a good working knowledge of various residential smoke detectors (ionization and photoelectric types).

Figure 19.42 All homes need smoke detectors.

Smoke detectors provide early warning and facilitate egress for responsive occupants faced with a fire emergency, especially during sleeping hours. This factor has been the key to survival of residents during fire situations.

Fire fighting has been listed as one of the most hazardous occupations in the United States (search and rescue services significantly increase the possibility of firefighter casualties). While an early warning is often credited with saving the lives of a home's occupants, do not forget that the smoke detector's warning may have also protected a firefighter from having to enter the burning structure. When smoke detectors are not in the home, a firefighter may be forced to enter the home to rescue the occupants.

Smoke detector location. A smoke detector in every room would provide the fastest detection times. However, this may not be economically feasible for many residents. What should be strongly recommended is the placement of a smoke detector in every bedroom and at every level of the living unit (Figure 19.43). When providing this "every-level" detection, the user should consider locations such as hallways, stairways, and normal exit routes.

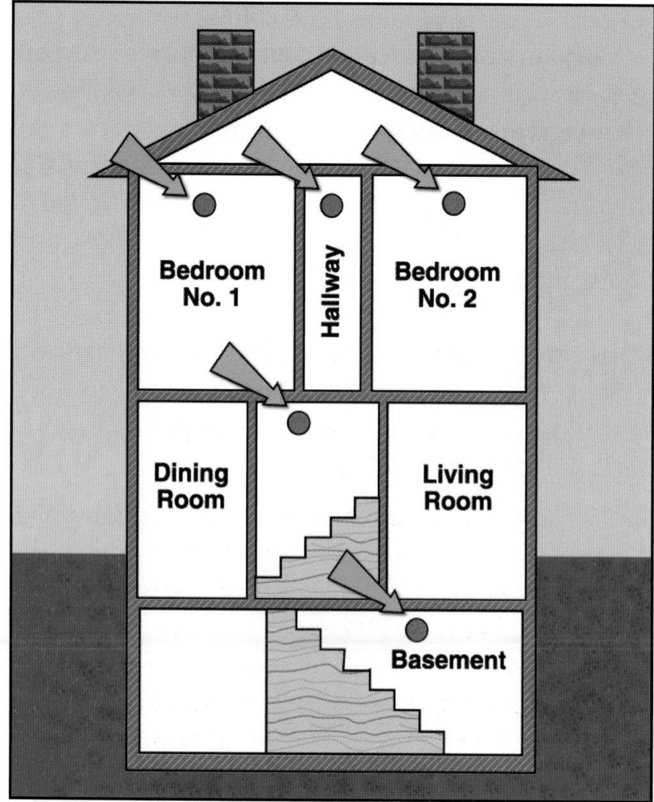

Figure 19.43 Smoke detectors should be located on each level of the structure.

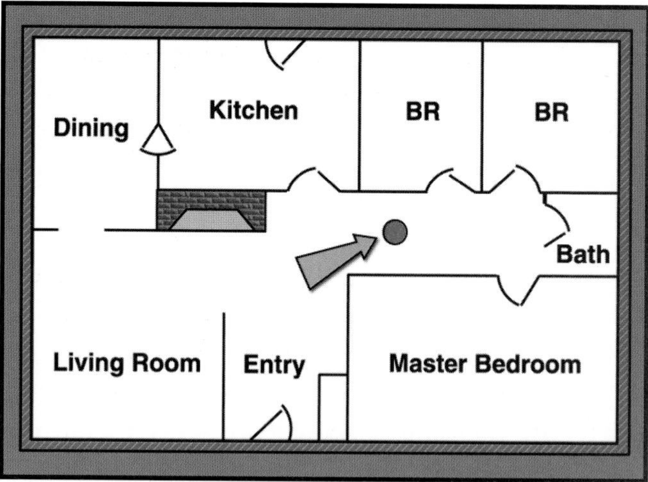

Figure 19.44 Locate smoke detectors outside the sleeping area of the home.

As a minimum, users should install a smoke detector in the hallway outside each sleeping area and between the sleeping area and other rooms in the house (Figure 19.44). The detectors should be close enough to the bedrooms so that the alarm can be heard when the bedroom door is closed. It is most desirable to mount the detectors on the ceiling. However, if the ceiling mount is not possible, position them as high on the walls as possible, but not within a dead air space (Figure 19.45).

Maintenance and testing. The smoke detector should be maintained and tested in accordance with the manufacturer's instructions. Maintenance is usually just a matter of keeping the detector always in operating condition and clean and free of dust by occasional vacuuming. Never disable detectors because of nuisance alarms.

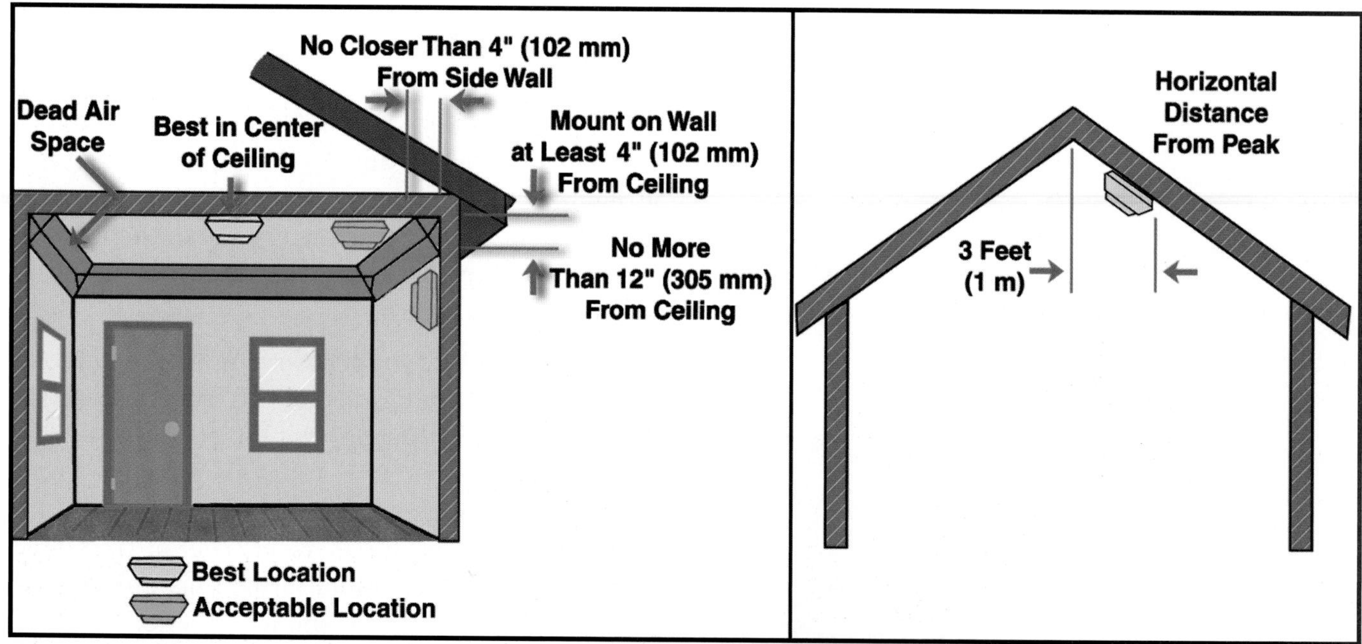

Figure 19.45 Smoke detectors should be placed to avoid dead air space such as corners between the wall and the ceiling.

The test buttons on some detectors may only check the device's horn circuit so being cognizant of the manufacturer's smoke-test procedure is vital to maintaining a functional detector. Only when the detector incorporates a test button that simulates smoke or checks the detector's sensitivity can the "smoke test" be eliminated (Figure 19.46). Smoke detectors with test buttons that simulate smoke or check sensitivity are recommended for those locations where the physically impaired or elderly live and smoke testing cannot realistically be conducted.

When testing with smoke is necessary, only safe ways of producing smoke should be used. Smoke from burning incense or a small piece of smoldering cotton rope or string in an ashtray can be used, especially when testing photoelectric detectors. Ionizing detectors can also be tested by blowing over the top of the flame on a wood or paper match, directing the invisible or visible smoke particles toward the direction of the smoke detector.

Aerosol spray (canned smoke) is also available for smoke-detector testing. Some aerosol sprays can contaminate the ionization chamber, resulting in nuisance alarms. These types of sprays should only be used if they are acceptable to the smoke detector manufacturer.

Figure 19.46 Test the smoke detector to make sure that it functions properly.

FIRE STATION TOURS

[NFPA 1001: 3-5.2(a)]

Firefighters are frequently required to give tours of the fire station to civilians. These may be either spur-of-the-moment visits from people who walk in off the street or organized citizens' groups. Fire Prevention Week tours for groups of children are common (Figure 19.47).

Firefighters should consider such tours more than just an opportunity to impress the public. It is important to fortify such visits with a strong safety message and relevant awareness materials. Such an approach not only helps support fire safety efforts but fosters a good image for the fire department as well. During the time that citizens are in the station, firefighters should be dressed appropriately. What citizens witness while at the station remain in their memories for a very long time, so activities should be productive.

Firefighters should answer all questions courteously and to the best of their ability. Fire safety and prevention information can also be passed on to visitors at this time. When firefighters allow visitors to climb on apparatus or don equipment items, they should be confident that no form of injury will result.

Figure 19.47 Firefighters often provide tours of the fire station to groups of children.

Never allow visitors, especially children, to roam around the fire station unescorted (Figure 19.48). Visiting groups should be met by an assigned firefighter or officer who carefully explains what steps citizens must take in case of an alarm. Special care should be taken to protect curious children or other individuals around shop areas or sliding poles. All groups should be kept together and, if necessary, rearranged into smaller groups with a firefighter assigned to each group.

Figure 19.48 Safety is a primary concern when conducting station tours.

Equipment and apparatus should be demonstrated with considerable caution to ensure that no one gets into a dangerous position. Place a firefighter at each corner of an apparatus to prevent young visitors from getting near the apparatus during demonstrations. Taking visitors on elevating platforms or aerial ladders should be prohibited. Firefighters should refrain from blowing sirens in the presence of children because the decibels produced can be detrimental to their hearing.

Station mascots (dogs, cats, etc.) can be potential safety and liability hazards. Excited animals have been known to strike out and bite visitors; therefore, many organizations restrict the presence of animals. If animals are kept in the fire station, they should be cared for by a veterinarian and receive all the necessary inoculations to assure good health.

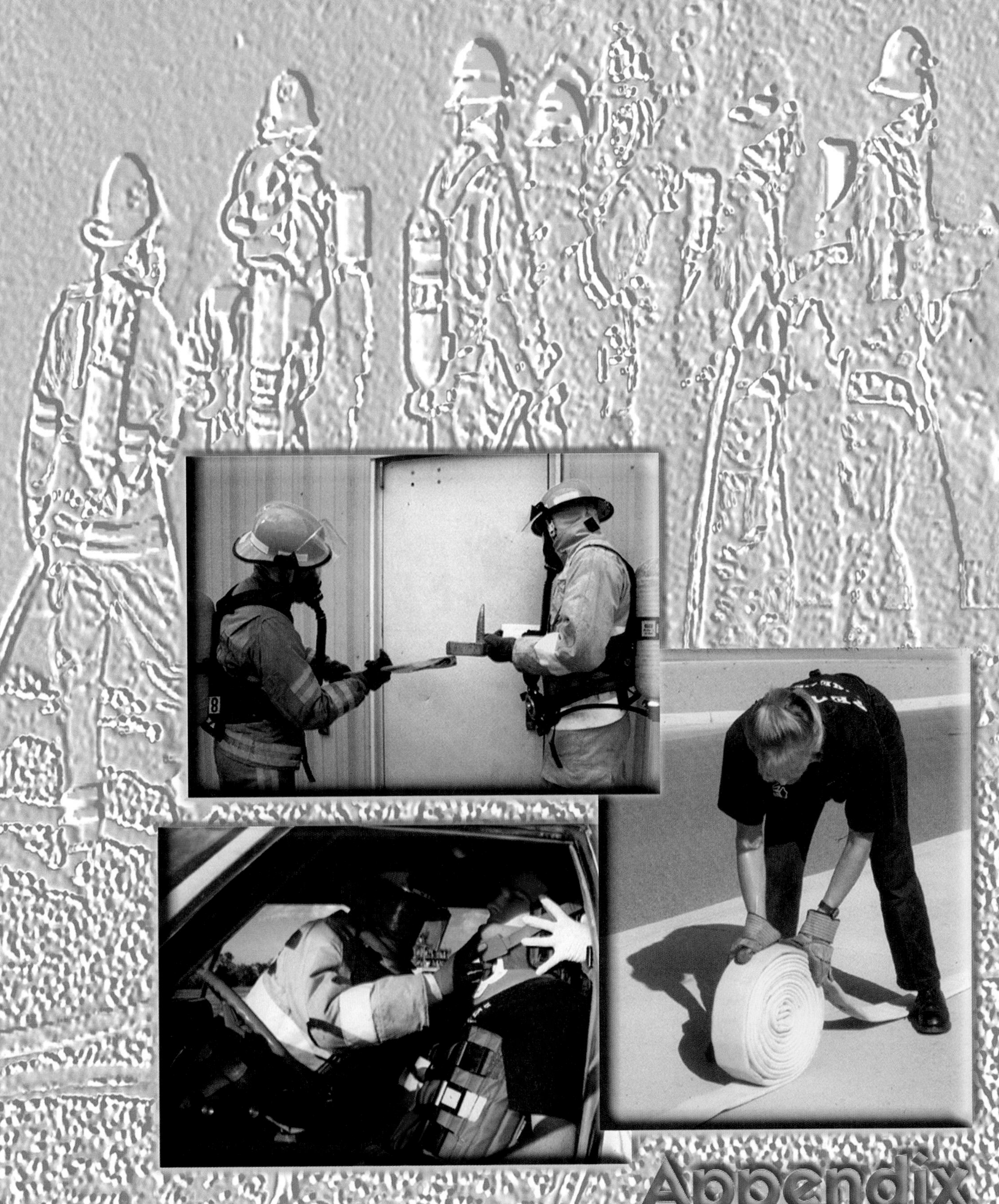

Appendix and Index

Appendix

NFPA 1001 Standard for Fire Fighter Professional Qualifications 1997 Edition

CHAPTER 3 FIREFIGHTER I

3-1 General

3-1.1 For certification at Level I, the fire fighter shall meet the job performance requirements defined in Sections 3-2 through 3-5 of this standard and the requirements defined in Chapter 2, Competencies for the First Responder at the Awareness Level, of NFPA 472, *Standard for Professional Competence of Responders to Hazardous Materials Incidents.*

3-1.1.1 General Knowledge Requirements. The organization of the fire department; the role of the Fire Fighter I in the organization; the mission of the fire service; the fire department's standard operating procedures and rules and regulations as they apply to the Fire Fighter I; the role of other agencies as they relate to the fire department; aspects of the fire department's member assistance program; the critical aspects of NFPA 1500, *Standard on Fire Department Occupational Safety and Health Program,* as they apply to the Fire Fighter I; knot types and usage; the difference between life safety and utility rope; reasons for placing rope out of service; the types of knots to use for given tools, ropes, or situations; hoisting methods for tools and equipment; and using rope to support response activities.

3-1.1.2 General Skill Requirements. The ability to don personal protective clothing within one minute; doff personal protective clothing and prepare for reuse; hoist tools and equipment using ropes and the correct knot; tie a bowline, clove hitch, figure eight on a bight, half hitch, becket or sheet bend, and safety knots; and locate information in departmental documents and standard or code materials.

3-2 Fire Department Communications. This duty involves initiating responses, receiving telephone calls and using fire department communications equipment to effectively relay verbal or written information, according to the following job performance requirements.

3-2.1* Initiate the response to a reported emergency, given the report of an emergency, fire department standard operating procedures, and communications equipment, so that all necessary information is obtained, communications equipment is operated properly, and the information is promptly and accurately relayed to the dispatch center.

(a) *Prerequisite Knowledge:* Procedures for reporting an emergency, departmental standard operating procedures for taking and receiving alarms, radio codes or procedures, and information needs of dispatch center.

(b) *Prerequisite Skills:* The ability to operate fire department communications equipment, relay information, and record information.

3-2.2 Receive a business or personal telephone call, given a fire department business phone, so that proper procedures for answering the phone are used and the caller's information is relayed.

(a) *Prerequisite Knowledge:* Fire department procedures for answering nonemergency telephone calls.

(b) *Prerequisite Skills:* The ability to operate fire station telephone and intercom equipment.

3-2.3 Transmit and receive messages via the fire department radio, given a fire department radio and operating procedures, so that the information is promptly relayed and is accurate, complete, and clear.

(a) *Prerequisite Knowledge:* Departmental radio procedures and etiquette for routine traffic, emergency traffic, and emergency evacuation signals.

(b) *Prerequisite Skills:* The ability to operate radio equipment and discriminate between routine and emergency traffic.

3-3 Fireground Operations. This duty involves performing activities necessary to ensure life safety, fire control, and property conservation, according to the following job performance requirements.

3-3.1* Use SCBA during emergency operations, given SCBA and other personal protective equipment, so that the SCBA is properly donned and activated within one minute, the SCBA is correctly worn, controlled breathing techniques are used, emergency procedures are enacted if the SCBA fails, all low-air warnings are recognized, respiratory protection is not intentionally compromised, and hazardous areas are exited prior to air depletion.

(a) *Prerequisite Knowledge:* Conditions that require respiratory protection, uses and limitations of SCBA, components of SCBA, donning procedures, breathing techniques, indications for and emergency procedures used with SCBA, and physical requirements of the SCBA wearer.

(b) *Prerequisite Skills:* The ability to control breathing, replace SCBA air cylinders, use SCBA to exit through restricted passages, initiate and complete emergency procedures in the event of SCBA failure or air depletion, and complete donning procedures.

3-3.2* Respond on apparatus to an emergency scene, given personal protective clothing and other necessary personal protective equipment, so that the apparatus is safely mounted and dismounted, seat belts are used while the vehicle is in motion, and other personal protective equipment is correctly used.

(a) *Prerequisite Knowledge:* Mounting and dismounting procedures for riding fire apparatus; hazards and ways to avoid hazards associated with riding apparatus; prohibited practices; types of department personal protective equipment and the means for usage.

(b) *Prerequisite Skills:* The ability to use each piece of provided safety equipment.

3-3.3* Force entry into a structure, given personal protective equipment, tools and an assignment, so that the tools are used properly, the barrier is removed, and the opening is in a safe condition and ready for entry.

(a) *Prerequisite Knowledge:* Basic construction of typical doors, windows, and walls within the department's community or service area; operation of doors, windows, and locks; and the dangers associated with forcing entry through doors, windows, and walls.

(b) *Prerequisite Skills:* The ability to transport and operate hand and power tools and to force entry through doors, windows, and walls using assorted methods and tools.

3-3.4* Exit a hazardous area as a team, given vision-obscured conditions, so that a safe haven is found before exhausting the air supply, others are not endangered, and the team integrity is maintained.

(a) *Prerequisite Knowledge:* Personnel accountability systems, communication procedures, emergency evacuation methods, what constitutes a safe haven, elements that create or indicate a hazard, and emergency procedures for loss of air supply.

(b) *Prerequisite Skills:* The ability to operate as a team member in vision-obscured conditions, locate and follow a guideline, conserve air supply, and evaluate areas for hazards and identify a safe haven.

3-3.5* Set-up ground ladders, given single and extension ladders, an assignment, and team members as appropriate, so that hazards are assessed, the ladder is stable, the angle is proper for climbing, extension ladders are extended to the proper height with the fly locked, the top is placed against a reliable structural component, and the assignment is accomplished.

(a) *Prerequisite Knowledge:* Parts of a ladder, hazards associated with setting up ladders, what constitutes a stable foundation for ladder placement, different angles for various tasks, safety limits to the degree of angulation, and what constitutes a reliable structural component for top placement.

(b) *Prerequisite Skills:* The ability to carry ladders, raise ladders, extend ladders and lock flies, determine that a wall and roof will support the ladder, judge extension ladder height requirements, and place the ladder to avoid obvious hazards.

3-3.6* Attack a passenger vehicle fire operating as a member of a team, given personal protective equipment, attack line, and hand tools, so that hazards are avoided, leaking flammable liquids are identified and controlled, protection from flash fires is maintained, all vehicle compartments are overhauled, and the fire is extinguished.

(a) *Prerequisite Knowledge:* Principles of fire streams as they relate to fighting automobile fires; precautions to be followed when advancing hose lines toward an automobile; observable results that a fire stream has been properly applied; identifying alternative fuels and the hazards associated with them; dangerous conditions created during an automobile fire; common types of accidents or injuries related to fighting automobile fires and how to avoid them; how to access locked passenger, trunk, and engine compartments; and methods for overhauling an automobile.

(b) *Prerequisite Skills:* The ability to identify automobile fuel type; assess and control fuel leaks; open, close, and adjust the flow and pattern on nozzles; apply water for maximum effectiveness while maintaining flash fire protection; advance 1½-in. (38-mm) or larger diameter attack lines; and expose hidden fires by opening all automobile compartments.

3-3.7* Extinguish fires in exterior Class A materials, given fires in stacked or piled and small unattached structures or storage containers that can be fought from the exterior, attack lines, hand tools and master stream devices, and an assignment, so that exposures are protected, the spread of fire is stopped, collapse hazards are avoided, water application is effective, the fire is extinguished, and signs of the origin area(s) and arson are preserved.

(a) *Prerequisite Knowledge:* Types of attack lines and water streams appropriate for attacking stacked, piled materials, and outdoor fires; dangers — such as collapse — associated with stacked and piled materials; various extinguishing agents and their effect on different material configurations; tools and methods to use in breaking up various types of materials; the difficulties related to complete extinguishment of stacked and piled materials; water application methods for exposure protection and fire extinguishment; dangers such as exposure to toxic or hazardous materials associated with storage building and container fires; obvious signs of origin and cause; and techniques for the preservation of fire cause evidence.

(b) *Prerequisite Skills:* The ability to recognize inherent hazards related to the material's configuration, operate handlines or master streams, break up materials using hand tools and water streams, evaluate for complete extinguishment, operate hose lines and other water application devices, evaluate and modify water application for maximum penetration, search for and expose hidden fires, assess patterns for origin determination, and evaluate for complete extinguishment.

3-3.8 Conduct a search and rescue in a structure operating as a member of a team, given an assignment, obscured vision conditions, personal protective equipment, a flashlight, forcible entry tools, hose lines and ladders when necessary, so that ladders are correctly placed when used, all assigned areas are searched, all victims are located and removed, team integrity is maintained, and team members' safety — including respiratory protection — is not compromised.

(a) *Prerequisite Knowledge:* Use of forcible entry tools during rescue operations, ladder operations for rescue, psychological effects of operating in obscured conditions and ways to manage them, methods to determine if an area is tenable, primary and secondary search techniques, team members' roles and goals, methods to use and indicators of finding victims, victim removal methods (including various carries), and considerations related to respiratory protection.

(b) *Prerequisite Skills:* The ability to use SCBA to exit through restricted passages, set up and use different types of ladders for various types of rescue operations, rescue a fire fighter with functioning respiratory protection, rescue a fire fighter whose respiratory protection is not functioning, rescue a person who has no respiratory protection, and assess areas to determine tenability.

3-3.9* Attack an interior structure fire operating as a member of a team, given an attack line, ladders when needed, personal protective equipment, tools, and an assignment, so that team integrity is maintained, the attack line is properly deployed for advancement, ladders are correctly placed when used, access is gained into the fire area, effective water application practices are used, the fire is approached safely, attack techniques facilitate suppression given the level of the fire, hidden fires are located and controlled, the correct body posture is maintained, hazards are avoided or managed, and the fire is brought under control.

(a) *Prerequisite Knowledge:* Principles of fire streams; types, design, operation, nozzle pressure effects, and flow capabilities of nozzles; precautions to be followed when advancing hose lines to a fire; observable results that a fire stream has been properly applied; dangerous building conditions created by fire; principles of exposure protection; potential long-term consequences of exposure to products of combustion; physical states of matter in which fuel are found; common types of accidents or injuries and their causes; and the application of each size and type of attack line, the role of the backup team in fire attack situations, attack and control techniques for grade level and above and below grade levels, and exposing hidden fires.

(b) *Prerequisite Skills:* The ability to prevent water hammers when shutting down nozzles; open, close, and adjust nozzle flow and patterns; apply water using direct, indirect, and combination attacks; advance charged and uncharged 1½-in. (38-mm) diameter or larger hose lines up ladders and up and down interior and exterior stairways; extend hose lines; replace burst hose sections; operate charged hose lines of 1½-in. (38-mm) diameter or larger while secured to a ground ladder; couple and uncouple various handline connections; carry hose; attack fires at grade level and above and below grade levels; and locate and suppress interior wall and subfloor fires.

3-3.10 Perform horizontal ventilation on a structure operating as part of a team, given an assignment, personal protective equipment, ventilation tools, equipment, and ladders, so that the ventilation openings are free of obstructions, tools are safely used, ladders are properly used, ventilation devices are properly placed, and the structure is cleared of smoke.

(a) *Prerequisite Knowledge:* The principles, advantages, limitations, and effects of horizontal, mechanical, and hydraulic ventilation; safety considerations when venting a structure; fire behavior in a structure; the products of combustion found in a structure fire; the signs, causes, effects, and prevention of backdrafts; and the relationship of oxygen concentration to life safety and fire growth.

(b) *Prerequisite Skills:* The ability to transport and operate ventilation tools and equipment and ladders and to use safe procedures for breaking window and door glass and removing obstructions.

3-3.11 Perform vertical ventilation on a structure operating as part of a team, given an assignment, personal protective equipment, ground and roof ladders, and tools, so that ladders are properly positioned for ventilation, a sufficient opening is created, all ventilation barriers are removed, structural integrity is not compromised, products of combustion are released from the structure, and the team retreats from the area when ventilation is accomplished.

(a) *Prerequisite Knowledge:* The methods of heat transfer; the principles of thermal layering within a structure on fire; the techniques and safety precautions for venting flat roofs, pitched roofs, and basements; basic indicators of potential collapse or roof failure; the effects of construction type and elapsed time under fire conditions on structural integrity; and the advantages and disadvantages of vertical and trench/strip ventilation.

(b) *Prerequisite Skills:* The ability to transport and operate ventilation tools and equipment; hoist ventilation tools to a roof; cut roofing and flooring materials to vent flat roofs, pitched roofs, and basements; sound a roof for integrity; clear an opening with hand tools; select, carry, deploy, and secure ground ladders for ventilation activities; deploy roof ladders on pitched roofs while secured to a ground ladder; and carry ventilation-related tools and equipment while ascending and descending ladders.

3-3.12 Overhaul a fire scene, given personal protective equipment, attack line, hand tools, a flashlight, and an assignment, so that structural integrity is not compromised, all hidden fires are discovered, fire cause evidence is preserved, and the fire is extinguished.

(a) *Prerequisite Knowledge:* Types of fire attack lines and water application devices most effective for overhaul, water application methods for extinguishment that limit water damage, types of tools and methods used to expose hidden fire, dangers associated with overhaul, obvious signs of area of origin or signs of arson, and reasons for protection of fire scene.

(b) *Prerequisite Skills:* The ability to deploy and operate an attack line; remove flooring, ceiling, and wall components to expose void spaces without compromising structural integrity; apply water for maximum effectiveness; expose and extinguish hidden fires in walls, ceilings, and subfloor spaces; recognize and preserve obvious signs of area of origin and arson; and evaluate for complete extinguishment.

3-3.13 Conserve property as a member of a team, given salvage tools and equipment and an assignment, so that the building and its contents are protected from further damage.

(a) *Prerequisite Knowledge:* The purpose of property conservation and its value to the public, methods used to protect property, types of and uses for salvage covers, operations at properties protected with automatic sprinklers, how to stop the flow of water from an automatic sprinkler head, identification of the main control valve on an automatic sprinkler system, and forcible entry issues related to salvage.

(b) *Prerequisite Skills:* The ability to cluster furniture; deploy covering materials; roll and fold salvage covers for reuse; construct water chutes and catch-alls; remove water; cover building openings, including doors, windows, floor openings, and roof openings; separate, remove, and relocate charred material to a safe location while protecting the area of origin for cause determination; stop the flow of water from a sprinkler with sprinkler wedges or stoppers; and operate a main control valve on an automatic sprinkler system.

3-3.14* Connect a fire department pumper to a water supply as a member of a team, given supply or intake hose, hose tools, and a fire hydrant or static water source, so that connections are tight and water flow is unobstructed.

(a) *Prerequisite Knowledge:* Loading and off-loading procedures for mobile water supply apparatus; fire hydrant operation; and suitable static water supply sources, procedures, and protocol for connecting to various water sources.

(b) *Prerequisite Skills:* The ability to hand lay a supply hose, connect and place hard suction hose for drafting operations, deploy portable water tanks as well as the equipment necessary to transfer water between and draft from them, make hydrant-to-pumper hose connections for forward and reverse lays, connect supply hose to a hydrant, and fully open and close the hydrant.

3-3.15* Extinguish incipient Class A, Class B, and Class C fires, given a selection of portable fire extinguishers, so that the correct extinguisher is chosen, the fire is completely extinguished, and proper extinguisher-handling techniques are followed.

(a) *Prerequisite Knowledge:* The classifications of fire; the types of, rating systems for, and risks associated with each class of fire; and the operating methods of, and limitations of portable extinguishers.

(b) *Prerequisite Skills:* The ability to operate portable fire extinguishers, approach fire with portable fire extinguishers, select an appropriate extinguisher based on the size and type of fire, and safety carry portable fire extinguishers.

3-3.16 Illuminate the emergency scene, given fire service electrical equipment and an assignment, so that designated areas are illuminated and all equipment is operated within the manufacturer's listed safety precautions.

(a) *Prerequisite Knowledge:* Safety principles and practices, power supply capacity and limitations, and light deployment methods.

(b) *Prerequisite Skills:* The ability to operate department power supply and lighting equipment, deploy cords and connectors, reset ground-fault interrupter (GFI) devices, and locate lights for best effect.

3-3.17 Turn off building utilities, given tools and an assignment, so that the assignment is safely completed.

(a) *Prerequisite Knowledge:* Properties, principals, and safety concerns for electricity, gas, and water systems; utility disconnect methods and associated dangers; and use of required safety equipment.

(b) *Prerequisite Skills:* The ability to identify utility control devices, operate control valves or switches, and assess for related hazards.

3-3.18* Combat a ground cover fire operating as a member of a team, given protective clothing, SCBA if needed, hose lines, extinguishers or hand tools, and an assignment, so that threats to property are reported, threats to personal safety are recognized, retreat is quickly accomplished when warranted, and the assignment is completed.

(a) *Prerequisite Knowledge:* Types of ground cover fires, parts of ground cover fires, methods to contain or suppress, and safety principles and practices.

(b) *Prerequisite Skills:* The ability to determine exposure threats based on fire spread potential, protect exposures, construct a fire line or extinguish with hand tools, maintain integrity of established fire lines, and suppress ground cover fires using water.

3-4 Rescue Operations. This duty involves no requirements for Fire Fighter I.

3-5 Prevention, Preparedness, and Maintenance. This duty involves performing activities that reduce the loss of life and property due to fire through hazard identification, inspection, education and response readiness, according to the following job performance requirements.

3-5.1 Perform a fire safety survey in a private dwelling, given survey forms and procedures, so that fire and life-safety hazards are identified, recommendations for their correction are made to the occupant, and unresolved issues are referred to the proper authority.

(a) *Prerequisite Knowledge:* Organizational policies and procedures, common causes of fire and their prevention, the importance of a fire safety survey and public fire education programs to the fire department public relations and the community, and referral procedures.

(b) *Prerequisite Skills:* The ability to complete forms, recognize hazards, match findings to preapproved recommendations, and effectively communicate findings to occupants or referrals.

3-5.2* Present fire safety information to station visitors or small groups, given prepared materials, so that all information is presented, the information is accurate, and questions are answered or referred.

(a) *Prerequisite Knowledge:* Parts of informational materials and how to use them, basic presentation skills, and departmental standard operating procedures for giving fire station tours.

(b) *Prerequisite Skills:* The ability to document presentations and to use prepared materials.

3-5.3 Clean and check ladders, ventilation equipment, self-contained breathing apparatus (SCBA), ropes, salvage equipment, and hand tools, given cleaning tools, cleaning supplies, and an assignment, so that equipment is clean and maintained according to manufacture's or departmental guidelines, maintenance is recorded, and equipment is placed in a ready state or reported otherwise.

(a) *Prerequisite Knowledge:* Types of cleaning methods for various tools and equipment, correct use of cleaning solvents, and manufacturer's or departmental guidelines for cleaning equipment and tools.

(b) *Prerequisite Skills:* The ability to select correct tools for various parts and pieces of equipment, follow guidelines, and complete recording and reporting procedures.

3-5.4 Clean, inspect, and return fire hose to service, given washing equipment, water, detergent, tools, and replacement gaskets, so that damage is noted and corrected, the hose is clean, and the equipment is placed in a ready state for service.

(a) *Prerequisite Knowledge:* Departmental procedures for noting a defective hose and removing it from service, cleaning methods, and hose rolls and loads.

(b) *Prerequisite Skills:* The ability to clean different types of hose; operate hose washing and drying equipment; mark defective hose; and replace coupling gaskets, roll hose, and reload hose.

CHAPTER 4 FIREFIGHTER II

4-1 General

4-1.1 For certification at Level II, the Fire Fighter I shall meet the job performance requirements defined in Sections 4-2 through 4-5 of this standard and the requirements defined in Chapter 3, Competencies for the First Responder at the Operational Level, of NFPA 472, *Standard for Professional Competence of Responders to Hazardous Materials Incidents.*

4-1.1.1 General Knowledge Requirements. Responsibilities of the Fire Fighter II in assuming and transferring command within an incident management system, performing assigned duties in conformance with applicable NFPA and other safety regulations and authority having jurisdiction procedures, and the role of a Fire Fighter II within the organization.

4-1.1.2 General Skill Requirements. The ability to determine the need for command, organize and coordinate an incident management system until command is transferred, and function within an assigned role in the incident management system.

4-2 Fire Department Communications. This duty involves performing activities related to initiating and reporting responses, according to the following job performance requirements.

4-2.1 Complete a basic incident report, given the report forms, guidelines, and information, so that all pertinent information is recorded, the information is accurate, and the report is complete.

(a) *Prerequisite Knowledge:* Content requirements for basic incident reports, the purpose and usefulness of accurate reports, consequences of inaccurate reports, how to obtain necessary information, and required coding procedures.

(b) *Prerequisite Skills:* The ability to determine necessary codes, proof reports, and operate fire department computers or other equipment necessary to complete reports.

4-2.2* Communicate the need for team assistance, given fire department communications equipment, standard operating procedures (SOPs), and a team, so that the supervisor is consistently informed of team needs, departmental SOPs are followed, and the assignment is accomplished safely.

(a) *Prerequisite Knowledge:* Standard operating procedures for alarm assignments and fire department radio communication procedures.

(b) *Prerequisite Skills:* The ability to operate fire department communications equipment.

4-3 Fireground Operations. This duty involves performing activities necessary to insure life safety, fire control, and property conservation, according to the following job performance requirements.

4-3.1* Extinguish an ignitable liquid fire, operating as a member of a team, given an assignment, an attack line, personal protective equipment, a foam proportioning device, a nozzle, foam concentrates, and a water supply, so that the proper type of foam concentrate is selected for the given fuel and conditions, a properly proportioned foam stream is applied to the surface of the fuel to create and maintain a foam blanket, fire is extinguished, reignition is prevented, team protection is maintained with a foam stream, and the hazard is faced until retreat to safe haven is reached.

(a) *Prerequisite Knowledge:* Methods by which foam prevents or controls a hazard; principles by which foam is generated; causes for poor foam generation and corrective measures; difference between hydrocarbon and polar solvent fuels and the concentrates that work on each; the characteristics, uses, and limitations of fire-fighting foams; the advantages and disadvantages of using fog nozzles versus foam nozzles for foam application; foam stream application techniques; hazards associated with foam usage; and methods to reduce or avoid hazards.

(b) *Prerequisite Skills:* The ability to prepare a foam concentrate supply for use, assemble foam stream components, master various foam application techniques, and approach and retreat from spills as part of a coordinated team.

4-3.2* Coordinate an interior attack line team's accomplishment of an assignment in a structure fire, given attack lines, personnel, personal protective equipment, and tools, so that crew integrity is established; attack techniques are selected for the given level of the fire (for example, attic, grade level, upper levels, or basement); attack techniques are communicated to the attack teams; constant team coordination is maintained; fire growth and development is continuously evaluated; search, rescue and ventilation requirements are communicated or managed; hazards are reported to the attack teams; and incident command is appraised of changing conditions.

(a) *Prerequisite Knowledge:* Selection of the proper nozzle and hose for fire attack given different fire situations; selection of adapters and appliances to be used for specific fire ground situations; dangerous building conditions created by fire and fire suppression activities; indicators of building collapse; the effects of fire and fire suppression activities on wood, masonry (brick, block, stone), cast iron, steel, reinforced concrete, gypsum wall board, glass, and plaster on lath; search and rescue and ventilation procedures; indicators of structural instability; suppression approaches and practices for various types of structural fires; and the association between specific tools and special forcible entry needs.

(b) *Prerequisite Skills:* The ability to assemble a team, choose attack techniques for various levels of a fire (e.g., attic, grade level, upper levels, or basement), evaluate and forecast a fire's growth and development, select proper tools for forcible entry, incorporate search and rescue procedures and ventilation procedures in the completion of the attack team efforts, and determine developing hazardous building or fire conditions.

4-3.3* Control a flammable gas cylinder fire operating as a member of a team, given an assignment, a cylinder outside of a structure, an attack line, personal protective equipment, and tools, so that crew integrity is maintained, contents are identified, safe havens are identified prior to advancing, open valves are closed, flames are not extinguished unless the leaking gas is eliminated, the cylinder is cooled, cylinder integrity is evaluated, hazardous conditions are recognized and acted upon, and the cylinder is faced during approach and retreat.

(a) *Prerequisite Knowledge:* Characteristics of pressurized flammable gases, elements of a gas cylinder, effects of heat and pressure on closed cylinders, boiling liquid expanding vapor explosion (BLEVE) signs and effects, methods for identifying contents, how to identify safe havens before approaching flammable gas cylinder fires, water stream usage and demands for pressurized cylinder fires, what to do if the fire is prematurely extinguished, valve types and their operation, alternative actions related to various hazards and when to retreat.

(b) *Prerequisite Skills:* The ability to execute effective advances and retreats, apply various techniques for water application, assess cylinder integrity and changing cylinder conditions, operate control valves, choose effective procedures when conditions change.

4-3.4* Protect evidence of fire cause and origin, given a flashlight and overhaul tools, so that the evidence is properly noted and protected from further disturbance until investigators can arrive on the scene.

(a) *Prerequisite Knowledge:* Methods to assess origin and cause; types of evidence; means to protect various types of evidence; the role and relationship of Fire Fighter IIs, criminal investigators, and insurance investigators in fire investigations; and the effects and problems associated with removing property or evidence from the scene.

(b) *Prerequisite Skills:* The ability to locate the fire's origin area, recognize possible causes, and protect the evidence.

4-4 Rescue Operations. This duty involves performing activities related to accessing and disentangling victims from motor vehicle accidents and helping special rescue teams, according to the following job performance requirements.

4-4.1* Extricate a victim entrapped in a motor vehicle as part of a team, given stabilization and extrication tools, so that the vehicle is stabilized, the victim can be disentangled without undue further injury, and hazards are managed.

(a) *Prerequisite Knowledge:* The fire department's role at a vehicle accident, points of strength and weakness in auto body construction, dangers associated with vehicle components and systems, the uses and limitations of hand and power extrication equipment, and safety procedures when using various types of extrication equipment.

(b) *Prerequisite Skills:* The ability to operate hand and power tools used for forcible entry and rescue in a safe and efficient manner; use cribbing and shoring material; and choose and apply appropriate techniques for moving or removing vehicle roofs, doors, windshields, windows, steering wheels or columns, and the dashboard.

4-4.2* Assist rescue operation teams, given standard operating procedures, necessary rescue equipment, and an assignment, so that procedures are followed, rescue items are quickly recognized and retrieved, and the assignment is completed.

(a) *Prerequisite Knowledge:* The fire fighter's role at a special rescue operation, the hazards associated with special rescue operations, types and uses for rescue tools, and rescue practices and goals.

(b) *Prerequisite Skills:* The ability to identify and retrieve various types of rescue tools, establish public barriers, and assist rescue teams operating as a member of the team when assigned.

4-5 Prevention, Preparedness, and Maintenance. This duty involves performing activities related to reducing the loss of life and property due to fire through hazard identification, inspection, and response readiness, according to the following job performance requirements.

4-5.1* Prepare a preincident survey, given forms, necessary tools, and an assignment, so that all required occupancy information is recorded, items of concern are noted, and accurate sketches or diagrams are prepared.

(a) *Prerequisite Knowledge:* The sources of water supply for fire protection; the fundamentals of fire suppression and detection systems; common symbols used in diagramming construction features, utilities, hazards, and fire protection systems; departmental requirements for a preincident survey and form completion; and the importance of accurate diagrams.

(b) *Prerequisite Skills:* The ability to identify the components of fire suppression and detection systems; sketch the site, buildings, and special features; detect hazards and special considerations to include in the preincident sketch; and complete all related departmental forms.

4-5.2 Maintain power plants, power tools, and lighting equipment, given appropriate tools and manufacturers' instructions, so that equipment is clean, maintained according to manufacturer and departmental guidelines, maintenance is recorded, and equipment is placed in a ready state or reported otherwise.

(a) *Prerequisite Knowledge:* Types of cleaning methods, correct use of cleaning solvents, manufacturer and departmental guidelines for maintaining equipment and its documentation, and problem-reporting practices.

(b) *Prerequisite Skills:* The ability to select correct tools; follow guidelines; complete recording and reporting procedures; and operate power plants, power tools, and lighting equipment.

4-5.3 Perform an annual service test on fire hose, given a pump, a marking device, pressure gauges, a timer, record sheets, and related equipment, so that procedures are followed, the condition of the hose is evaluated, any damaged hose is removed from service, and the results are recorded.

(a)* *Prerequisite Knowledge:* Procedures for safely conducting hose service testing, indicators that dictate any hose be removed from service, and recording procedures for hose test results.

(b) *Prerequisite Skills:* The ability to operate hose testing equipment and nozzles and to record results.

4-5.4* Test the operability of and flow from a fire hydrant, given a Pitot tube, pressure gauge, and other necessary tools, so that the readiness of the hydrant is assured and the flow of water from the hydrant can be calculated and recorded.

(a) *Prerequisite Knowledge:* How water flow is reduced by hydrant obstructions; direction of hydrant outlets to suitability of use; the effect of mechanical damage, rust, corrosion, failure to open the hydrant fully, and susceptibility to freezing; and the meaning of the terms *static, residual,* and *flow pressure.*

(b) *Prerequisite Skills:* The ability to operate a pressurized hydrant, use a Pitot tube and pressure gauges, detect damage, and record results of test.

APPENDIX A EXPLANATORY MATERIAL

A-3-2.1 The Fire Fighter I should be able to receive and accurately process information received at the station. Fire Fighters used as telecommunicators (dispatchers) should meet the requirements of NFPA 1061, *Standard for Professional Qualifications for Public Safety Telecommunicator,* for qualification standards and job performance requirements.

A-3-3.1 The Fire Fighter I should already be wearing full protective clothing prior to the beginning of this SCBA-donning procedure. In addition to fully donning and activating the SCBA, the Fire Fighter I should also replace any personal protective clothing (i.e., gloves, protective hood, helmet, etc.) displaced during the donning procedure and activate the PASS device within the specified 1-min time limit.

A-3-3.2 Other personal protective equipment might include hearing protection in cabs that have a noise level in excess of 90 dBa, eye protection for fire fighters riding in jump seats that are not fully enclosed, and SCBAs for those departments that require fire fighters to don SCBAs while en route to the emergency.

A-3-3.3 The Fire Fighter I should be able to force entry through wood, glass, and metal doors that open in and out; overhead doors; and windows common to the community or service area.

A-3-3.4 When training exercises are intended to simulate emergency conditions, smoke-generating devices that do not create a hazard are required. Several accidents have occurred when smoke bombs or other smoke-generating devices that produce a toxic atmosphere have been used for training exercises. All exercises should be conducted in accordance with the requirements of NFPA 1404, *Standard for a Fire Department Self-Contained Breathing Apparatus Program.*

A-3-3.5 The fire fighter should be able to accomplish this task with each type and length of ground ladder carried by the department.

A-3-3.6 Passenger vehicles include automobiles, light trucks, and vans.

A-3-3.7 The Fire Fighter I should be able to extinguish fires in stacked or piled materials such as hay bales, pallets, lumber, piles of mulch, sawdust, other bulk Class A materials, or small unattached structures that are attacked from the exterior. The tactics for extinguishing each of these types of fires are similar enough to be included in one JPR.

Live fire evolutions should be conducted in accordance with the requirements of NFPA 1403, *Standard on Live Fire Training Evolutions.* It is further recommended that prior to involvement in live fire evolutions, the fire fighter demonstrate the use of SCBA in smoke and elevated temperature conditions.

In areas where environmental or other concerns restrict the use of Class A fuels for training evolutions, properly installed and monitored gas-fueled fire simulators might be substituted.

A-3-3.9 The Fire Fighter I should be proficient in the various attack approaches for room and contents fires at three different levels (at grade, above grade, and below grade). Maintenance of body posture in the standard refers to staying low during initial attack, protecting oneself from falling objects, and otherwise using common sense given the state of the fire's growth or suppression.

Live fire evolutions should be conducted in accordance with the requirements of NFPA 1403, *Standard on Live Fire Training Evolutions*. It is further recommended that prior to involvement in live fire evolutions, the fire fighter demonstrate the use of SCBA in smoke and elevated temperature conditions.

In areas where environmental or other concerns restrict the use of Class A fuels for training evolutions, properly installed and monitored gas-fueled fire simulators might be substituted.

A-3-3.14 Static water sources can include portable water tanks, ponds, creeks, and so forth.

A-3-3.15 The Fire Fighter I should be able to extinguish incipient Class A fires such as wastebaskets, small piles of pallets, wood, or hay; Class B fires of approximately 9 ft^2 (.84 m^2); and Class C fires where the electrical equipment is energized.

A-3-3.18 Protective clothing is not personal protective clothing as used throughout the rest of this document. Some jurisdictions provide fire fighters with different clothing for ground cover fires than is worn for structural fires. This clothing can be substituted for structural protective clothing in order to meet the intent of this job performance requirement.

A-3-5.2 The Fire Fighter I should be able to present basic information on how to (a) stop, drop, and roll when one's clothes are on fire; (b) crawl low in smoke; (c) perform escape planning; (d) alert others of an emergency; (e) call the fire department; and (f) properly place, test, and maintain residential smoke detectors. The Fire Fighter I is not expected to be an accomplished speaker or instructor.

A-4-2.2 The Fire Fighter II could be assigned to accomplish or coordinate tasks away from direct supervision. Many of these tasks could result in the need for additional or replacement personnel due to the ever-changing conditions on the scene of an emergency. The Fire Fighter II is expected to identify these needs and effectively communicate this information within an incident management system. Use of radio communication equipment necessitates that these communications be accurate and efficient.

A-4-3.1 The Fire Fighter II should be able to accomplish this task with each type of foam concentrate used by the jurisdiction. This could include the use of both Class A and B foam concentrates on appropriate fires. When using Class B foams to attack flammable or combustible liquid fires, the Fire Fighter II should extinguish a fire at least 100 ft^2 (9 m^2). The Fire Fighter II is not expected to calculate application rates and densities. The intent of this JPR can be met in training through the use of training foam concentrates or gas-fired training props.

A-4-3.2 The Fire Fighter II should be able to coordinate the actions of the interior attack line team at common residential fires and small business fires in the fire department's district. Complex or large interior fire management should be left to the officers; however, this job performance requirement will facilitate the development of the Fire Fighter II towards effectively handling specific assignments within large fires.

Jurisdictions that use Fire Fighter IIs as acting company officers should comply with the requirements of NFPA 1021, *Standard for Fire Officer Professional Qualifications.*

A-4-3.3 Controlling flammable gas cylinder fires can be a very dangerous operation. The Fire Fighter II should act as a team member, under the direct supervision of an officer, during these operations.

A-4-3.4 The Fire Fighter II should be able to recognize important evidence as to a fire's cause and maintain the evidence so that further testing can be done without contamination or chain-of-custody problems. Evidence should be left in place (when possible, otherwise chain-of-custody must be established), not altered by improper handling, walking, and so forth, and not destroyed. Possible means to protect evidence is to avoid touching, protect with salvage covers during overhaul, or rope off the area where the evidence lies. The Fire Fighter II is not intended to be highly proficient at origin and cause determination.

Jurisdictions that use Fire Fighter IIs to determine origin and cause should comply with the requirements of NFPA 1021, *Standard for Fire Officer Professional Qualifications.*

A-4-4.1 In the context of this standard, the term "extricate" refers to those activities required to allow emergency medical personnel access to the victim, stabilization of the vehicle, the displacement or removal of vehicle components obstructing victim removal, and the protection of the victim and response personnel from hazards associated with motor vehicle accidents and the use of hand and power tools on a motor vehicle.

As persons performing extrication can be different from those performing medical functions, this standard does not address medical care of the victim. An awareness of the needs and responsibilities of emergency medical functions is recommended to allow for efficient coordination between the "extrication" team and the "medical" team.

A-4-4.2 The Fire Fighter II is not expected to be proficient in special rescue skills. The Fire Fighter II should be able to help special rescue teams in their efforts to safely manage structural collapses, trench collapses, cave and tunnel emergencies, water and ice emergencies, elevator and escalator emergencies, energized electrical line emergencies, and industrial accidents.

A-4-5.1 The Fire Fighter II should be able to compile information related to potential emergency incidents within their community for use by officers in the development of preincident plans. Jurisdictions that use Fire Fighter IIs to develop preincident plans should comply with the requirements of NFPA 1021, *Standard for Fire Officer Professional Qualifications.*

A-4-5.3(a) Procedures for conducting hose testing can be found in Chapter 5, Service Testing, of NFPA 1962, *Standard for the Care, Use, and Service Testing of Fire Hose Including Couplings and Nozzles.*

A-4-5.4 All fire fighters should be able to flow test a hydrant. While not all fire departments have hydrants in their jurisdiction, departments without hydrants in their jurisdiction can effectively train and evaluate a Fire Fighter II's flow testing skills by using hose streams.

Index

Indexed by Kari J. Bero

F

T

U

V

CORRELLATION CHART
NFPA 1001

1997 EDITION VS. 2002 EDITION

The IFSTA *Essentials of Fire Fighting* (4th edition) was originally written to address the 1997 edition of NFPA 1001, Standard for Fire Fighter Professional Qualifications. The NFPA released a new edition of this standard in 2002. With the exception of one new job performance requirement (JPR) in the Fire Fighter I level (5.3.3), there were no major substantive changes to this standard. However, because of a new style format that the NFPA has adopted for their standards, all of the JPRs were renumbered. The following charts will allow you to locate individual JPRs between the two editions of the standard.

Fire Fighter I

Objective Number in the 1997 Edition of . NFPA 1001...	...Is now found in this objective number in the 2002 Edition of NFPA 1001.
Chapter 3	**Chapter 5**
3-1	5.1
3-1.1	5.1.1
3-1.1.1	5.1.1.1
3-1.1.2	5.1.1.2
3-2	5.2
3-2.1	5.2.1
3-2.1(a)	5.2.1(A)
3-2.1(b)	5.2.1(B)
3-2.2	5.2.2
3-2.2 (a)	5.2.2(A)
3-2.2 (b)	5.2.2(B)
3-2.3	5.2.3
3-2.3 (a)	5.2.3(A)
3-2.3 (b)	5.2.3(B)
3-3	5.3
3-3.1	5.3.1
3-3.1 (a)	5.3.1(A)
3-3.1 (b)	5.3.1(B)
3-3.2	5.3.2
3-3.2 (a)	5.3.2(A)
3-3.2 (b)	5.3.2(B)
	5.3.3 (New)
	5.3.3(A) (New)
	5.3.3 (B) (New)
3-3.3	5.3.4
3-3.3 (a)	5.3.4(A)
3-3.3 (b)	5.3.4(B)
3-3.4	5.3.5
3-3.4 (a)	5.3.5(A)
3-3.4 (b)	5.3.5(B)
3-3.5	5.3.6

Objective Number in the 1997 Edition.... 2002 Edition.	Is now found in this objective number in the
Chapter 3	**Chapter 5**
3-3.5 (a)	5.3.6(A)
3-3.5 (b)	5.3.6(B)
3-3.6	5.3.7
3-3.6 (a)	5.3.7(A)
3-3.6 (b)	5.3.7(B)
3-3.7	5.3.8
3-3.7 (a)	5.3.8(A)
3-3.7 (b)	5.3.8(B)
3-3.8	5.3.9
3-3.8 (a)	5.3.9(A)
3-3.8 (b)	5.3.9(B)
3-3.9	5.3.10
3-3.9 (a)	5.3.10(A)
3-3.9 (b)	5.3.10(B)
3-3.10	5.3.11
3-3.10 (a)	5.3.11(A)
3-3.10 (b)	5.3.11(B)
3-3.11	5.3.12
3-3.11 (a)	5.3.12(A)
3-3.11 (b)	5.3.12(B)
3-3.12	5.3.13
3-3.12 (a)	5.3.13(A)
3-3.12 (b)	5.3.13(B)
3-3.13	5.3.14
3-3.13 (a)	5.3.14(A)
3-3.13 (b)	5.3.14(B)
3-3.14	5.3.15
3-3.14 (a)	5.3.15(A)
3-3.14 (b)	5.3.15(B)
3-3.15	5.3.16
3-3.15 (a)	5.3.16(A)
3-3.15 (b)	5.3.16(B)
3-3.16	5.3.17
3-3.16 (a)	5.3.17(A)
3-3.16 (b)	5.3.17(B)
3-3.17	5.3.18
3-3.17 (a)	5.3.18(A)
3-3.17 (b)	5.3.18(B)
3-3.18	5.3.19
3-3.18 (a)	5.3.19(A)
3-3.18 (b)	5.3.19(B)
3-4	5.4
3-5	5.5
3-5.1	5.5.1
3-5.1 (a)	5.5.1(A)

Objective Number in the 1997 Edition....	Is now found in this objective number in the 2002 Edition.
Chapter 3	**Chapter 5**
3-5.1 (b)	5.5.1(B)
3-5.2	5.5.2
3-5.2 (a)	5.5.2(A)
3-5.2 (b)	5.5.2(B)
3-5.3	5.5.3
3-5.3 (a)	5.5.3(A)
3-5.3 (b)	5.5.3(B)
3-5.4	5.5.4
3-5.4 (a)	5.5.4(A)
3-5.4 (b)	5.5.4(B)

Fire Fighter II

Objective Number in the 1997 Edition of NFPA 1001....	...Is now found in this objective number in the 2002 Edition of NFPA 1001.
Chapter 4	**Chapter 6**
4-1	6.1
4-1.1	6.1.1
4-1.1.1	6.1.1.1
4-1.1.2	6.1.1.2
4-2	6.2
4-2.1	6.2.1
4-2.1(a)	6.2.1(A)
4-2.1(b)	6.2.1(B)
4-2.2	6.2.2
4-2.2 (a)	6.2.2(A)
4-2.2 (b)	6.2.2(B)
4-3	6.3
4-3.1	6.3.1
4-3.1 (a)	6.3.1(A)
4-3.1 (b)	6.3.1(B)
4-3.2	6.3.2
4-3.2 (a)	6.3.2(A)
4-3.2 (b)	6.3.2(B)
4-3.3	6.3.3
4-3.3 (a)	6.3.3(A)
4-3.3 (b)	6.3.3(B)
4.3.4	6.3.4
4-3.4 (a)	6.3.4(A)
4-3.4 (b)	6.3.4(B)
4-4	6.4
4-4.1	6.4.1
4-4.1(a)	6.4.1(A)
4-4.1(b)	6.4.1(B)
4-4.2	6.4.2

Objective Number in the 1997 Edition of NFPA 1001....	...Is now found in this objective number in the 2002 Edition of NFPA 1001.
Chapter 4	**Chapter 6**
4-4.2(a)	6.4.2(A)
4-4.2(b)	6.4.2(B)
4-5	6.5
4-5.1	6.5.1
4-5.1(a)	6.5.1(A)
4-5.1(b)	6.5.1(B)
4-5.2	6.5.2
4-5.2(a)	6.5.2(A)
4-5.2(b)	6.5.2(B)
4-5.3	6.5.3
4-5.3(a)	6.5.3(A)
4-5.3(b)	6.5.3(B)
4-5.4	6.5.4
4-5.4(a)	6.5.4(A)
4-5.4(b)	6.5.4(B)

ESSENTIALS of Fire Fighting
4th EDITION

COMMENT SHEET

DATE ____________________ NAME ______________________________________

ADDRESS ____

ORGANIZATION REPRESENTED ____________________________________

CHAPTER TITLE ______________________________ NUMBER __________

SECTION/PARAGRAPH/FIGURE ____________________ PAGE __________

1. Proposal (include proposed wording, or identification of wording to be deleted), OR PROPOSED FIGURE:

2. Statement of Problem and Substantiation for Proposal:

RETURN TO: IFSTA Editor
Fire Protection Publications
Oklahoma State University
930 N. Willis
Stillwater, OK 74078-8045

SIGNATURE ______________________________

Use this sheet to make any suggestions, recommendations, or comments. We need your input to make the manuals as up to date as possible. Your help is appreciated. Use additional pages if necessary.

Your Training Connection...

The International Fire Service Training Association

We have a free catalog describing hundreds of fire and emergency service training materials available from a convenient single source: the International Fire Service Training Association (IFSTA).

Choose from products including IFSTA manuals, IFSTA study guides, IFSTA curriculum packages, Fire Protection Publications manuals, books from other publishers, software, videos, and NFPA standards.

Contact us by phone, fax, U.S. mail, e-mail, Internet web page, or personal visit.

Phone
1-800-654-4055

Fax
405-744-8204

U.S. mail
IFSTA, Fire Protection Publications
Oklahoma State University
930 N. Willis
Stillwater, OK 74078-8045

E-mail
editors@osufpp.org

Internet web page
www.ifsta.org

Personal visit
Call if you need directions!